ORGANIC REACTION MECHANISMS · 1984

ORGANIC REACTION MECHANISMS · 1984

An annual survey covering the literature dated December 1983 through November 1984

Edited by

A. C. KNIPE and W. E. WATTS,
University of Ulster,
Northern Ireland

An Interscience® Publication

JOHN WILEY & SONS
Chichester · New York · Brisbane · Toronto · Singapore

Library of Congress Catalog Card Number 66–23143

British Library Cataloguing in Publication Data:

Organic reaction mechanisms : an annual survey
covering the literature dated December 1983
through November 1984.
—1984
1. Chemistry, Physical organic—Periodicals
2. Chemical reactions—Periodicals
I. Knipe, A. C. II. Watts, W. E.
547.1′394′05 QD476

ISBN 0 471 90797 9

Phototypeset by Macmillan India Ltd.
Printed and bound in Great Britain by the Bath Press, Bath, Avon

Contributors

A. ALBERTI	Istituto dei composti del carbonio, Contenenti eteroatomi e loro applicazioni, Consiglio Nationale delle Ricerche. Bologna, Italy
D. C. BILLINGTON	Merck Sharp & Dohme Research Laboratories, Neuroscience Research Centre, Harlow, Essex
C. CHATGILIALOGLU	Istituto dei composti del carbonio, Contenenti eteroatomi e loro applicazioni, Consiglio Nationale delle Ricerche, Bologna, Italy
D. J. COWLEY	Department of Chemistry, University of Ulster
R. A. COX	Department of Chemistry, University of Toronto, Canada
M. R. CRAMPTON	Department of Chemistry, Durham University
G. W. J. FLEET	Dyson Perrins Laboratory, Oxford University
R. B. MOODIE	Department of Chemistry, University of Exeter
C. J. MOODY	Department of Chemistry, Imperial College of Science and Technology, London
R. A. MORE O'FERRALL	Department of Chemistry, University College, Dublin, Ireland
A. W. MURRAY	Department of Chemistry, University of Dundee
M. I. PAGE	Department of Chemical Sciences, Huddersfield Polytechnic
R. M. PATON	Department of Chemistry, University of Edinburgh
J. SHORTER	Department of Chemistry, University of Hull
W. J. SPILLANE	Chemistry Department, University College, Galway, Ireland
C. I. F. WATT	Department of Chemistry, University of Manchester

Preface

The present volume, the twentieth in the series, surveys research on organic reaction mechanisms described in the literature dated December 1983 to November 1984. In order to limit the size of the volume, we must necessarily exclude or restrict overlap with other publications which review specialist areas (e.g. photochemical reactions, biosynthesis, electrochemistry, organometallic chemistry, surface chemistry and heterogeneous catalysis). In order to minimize duplication, while ensuring a comprehensive coverage, the editors conduct a survey of all relevant literature and allocate publications to appropriate chapters. While a particular reference may be allocated to more than one chapter, we do assume that readers will be aware of the alternative chapters to which a border-line topic of interest may have been preferentially assigned.

We welcome one new contributor, Dr. R. A. More O'Ferrall who is well known for his use of energy contour diagrams as an aid to interpretation of transition state character. His expertise is particularly appropriate to the elimination chapter which was previously written by Professor A. F. Hegarty, of the same department and to whom we extend our thanks for his annual contribution since we assumed editorship in 1977.

On behalf of the contributors we are pleased to acknowledge the many very favourable reviews of this now well established series which have appeared in professional journals. A common feature of such reviews has been an appreciation that the volume provides a comprehensive, yet readable, insight into the recent literature without recourse to other information retrieval techniques which may be relatively unsuitable for mechanistic topics. This was of course the aim of Brian Capon, John Perkins and Charles Rees when they launched *Organic Reaction Mechanisms* twenty years ago. We have attempted to maintain the standards set by the original editors and must acknowledge the sustained efforts of our team of experienced contributors, the publication and production staff of John Wiley and Sons, and Dr. N. Cully who compiled the subject index.

A.C.K.
W.E.W.

Contents

Organic Reaction Mechanisms 1984
Edited by A. C. Knipe and W. E. Watts

CHAPTER 1

Reactions of Aldehydes and Ketones and their Derivatives

M. I. PAGE

Department of Chemical & Physical Sciences, Huddersfield Polytechnic

Formation and Reactions of Acetals and Ketals

The length of the C—O bond in ethers and esters increases with increasing electron withdrawal in the substituent attached to oxygen and as the carbon changes through methyl, primary, secondary, and tertiary. There is an inverse linear correlation between the bond length C—OR and the pK_a of ROH. Although increasing the polarity of bonds is traditionally associated with shorter and stronger bonds the evidence does not support this for carbon–heteroatom bonds in general. It is therefore suggested that longer C—O bonds are associated with an increase in the contribution from the ionic valence bond tautomer $\overset{+}{C}\ \overset{-}{O}R$.[1] The trends in the lengths of the bonds at the acetal centre of axial aryl tetrahydropyranyl acetals (**1**) noted earlier have been confirmed. The C—OAr bond length increases whilst the ring C—O bond length decreases with decreasing pK_a of ArOH. These effects on bond lengths practically disappear on going to the α-glucopyranosides, presumably because of the four oxygen substituents. For equatorial tetrahydropyranyl acetals, the more electron-withdrawing OR group is associated with a longer exocyclic C—OR bond and a shorter endocyclic bond. Stereoelectronically this cannot be attributable to oxygen lone-pair donation and is thought to arise from σ C—O bond donation.[2]

The rate constants for the spontaneous hydrolysis of aryl acetals is a linear function of the pK_a of ArOH and therefore also of the C—O bond length. This implies that there is a linear region in the first half of the reaction coordinate for cleavage of the C—O bond.[3]

Kinetic solvent isotope effects for the acid-catalysed hydrolysis of benzaldehyde acetals show little variance, $k_{D^+}/k_{H^+} = 3.2 \pm 0.2$, over a 10^3 change in reactivity. Together with the invariance of the Brønsted α value for general acid catalysis this is taken to indicate rate-limiting diffusional separation of the aggregate (**2**). Weakly basic leaving alcohols, ROH, and stable oxo-carbocations cause a change in the rate-limiting step to formation of (**2**). These arguments would not be valid if the hydronium ion and general-acid-catalysed reactions occurred by different mechanisms.[4]

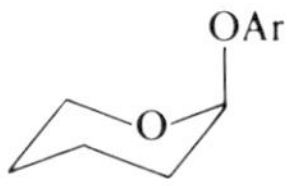

OAr

O

(**1**)

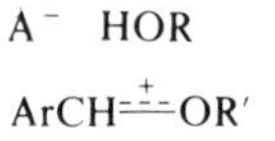

A⁻ HOR

ArCH$\overset{+}{=\!=}$OR′

(**2**)

The secondary kinetic deuterium isotope effect for general-acid-catalysed hydrolysis of benzaldehyde ethyl phenyl acetal increases from 1.06 to 1.26 as the pK_a of the catalyst increases. This is inconsistent with rate-limiting diffusion apart of the ion-pair formed rapidly and reversibly.[5] As previously shown, the Brønsted α value decreases from 0.7 to 0.5 as substituents in the phenoxy leaving group become more electron-withdrawing. Also as described earlier the Brønsted coefficient is thought not to be a measure of the degree of proton transfer but is an indication of the ease of cleavage of the C—OAr bond.[5]

Although the rate of the acid-catalysed hydrolysis of benzaldehyde di-*tert*-butyl acetal is 10^2 faster than the corresponding diethyl acetal, the Brønsted α values for general acid catalysis are the same. It seems that the rate acceleration associated with relief of steric strain does not affect the degree of concertedness of the proton transfer and the C—O bond-breaking processes. The electronic effect appears to be the dominant feature controlling concertedness.[6]

Stereoelectronic effects in acetal hydrolysis have been reviewed.[7]

The solvent dependence of secondary deuterium isotope effects on the hydronium-ion-catalysed hydrolysis of acetaldehyde diethyl acetal and ethyl vinyl ether suggests that the ethoxyethyl cation intermediate (**3**) becomes sufficiently unstable in aqueous dioxan for the mechanism to become concerted ((**4**) and (**5**)).[8]

The rate of the pH-independent hydrolysis of 2-(4-nitrophenoxy)tetrahydropyran in water–aprotic solvent mixtures may be correlated with a variety of parameters. The solvent effects are dominated by changes in entropies of activation. Dispersion effects are not dominant and the solvent exerts its effect mainly by hydrogen bonding in the dipolar transition state.[9]

Sulphonaphthoxyacetic acids act as general acid catalysts for the hydrolysis of 2-(*p*-nitrophenoxy)tetrahydropyran (**6**) and are 20-fold more effective than the reactivity predicted from their pK_a. This could be attributable to the sulphonate

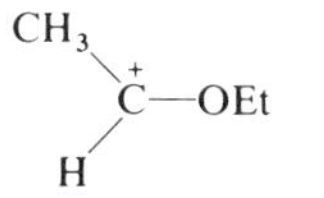

(3) 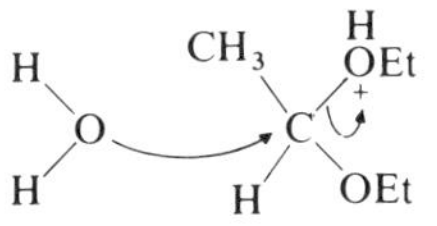 (4)

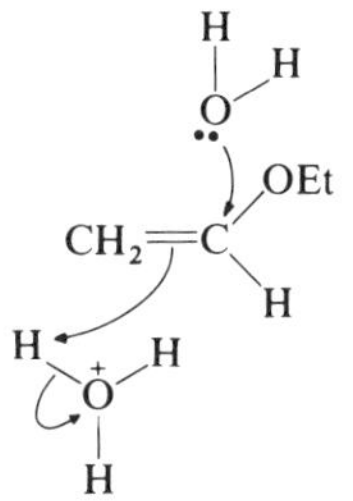

(5) 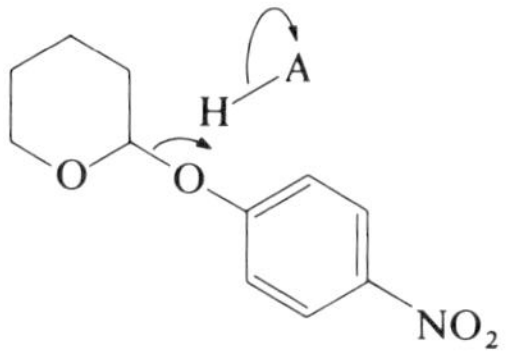 (6)

anion stabilizing the incipient carbonium ion or to a hydrophobic interaction between the catalyst and substrate.[10]

The general-acid-catalysed hydrolysis of 2-(2,2,2-trifluoroethoxy)tetrahydropyran shows a Brønsted α value of 0.7, similar to that for benzaldehyde *O*-ethyl *O*-2,2,2-trifluoroethyl acetal hydrolysis. There is no general acid catalysis observed in the hydrolysis of the tetrahydropyran ethyl acetal and it is suggested that hydrolysis occurs by the classical *A*1 mechanism.[11]

The effect of substituents on the equilibrium constant for dimethyl acetal formation from substituted acetophenones has been determined in methanol, water, and dodecane. The Hammett ρ^n values in these solvents are 1.73, 0.99, and 1.81, respectively, which is interpreted by the inductive effect on the stability of the ketone and specific inhibition of acetal solvation.[12]

The ^{13}C-NMR spectrum of *N*-acetyl-L-phenylalanil in the presence of α-chymotrypsin shows a signal attributable to the formation of a hemiacetal. At pH > 7 two signals for the hemiacetal appear which are pH-dependent but the reason for this is not known.[13]

The acid-catalysed hydrolysis of 2-methoxy-2-phenyltetrahydrofuran at pH 6 involves rate-limiting formation of the oxo-carbocation with exocyclic cleavage of the C—OMe bond. Below pH 5 the rate-limiting step changes to breakdown of the hemiketal intermediate (**7**). Hydroxide ion catalysis makes hemiketal decomposition faster at higher pH whereas acid catalysis for this step is slower than formation of the oxo-carbocation.[14]

Monomeric 3-mercaptopropanol has been characterized in the gas phase and is not in equilibrium with the thio-hemiacetal (**8**).[15]

The chemistry of cyclopropanone hemiacetals has been reviewed.[16]

Acid-catalysed dehydration of the 4-methylene-1,3-dioxolane (**9**) or the keto vinyl ether (**10**) gives the benzofuran (**11**) presumably *via* the same intermediate oxo-carbocations.[17]

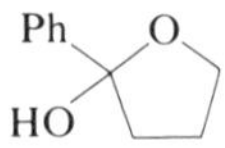

Ph O HO

(7)

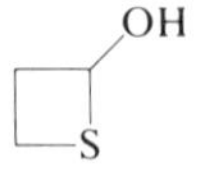

OH S

(8)

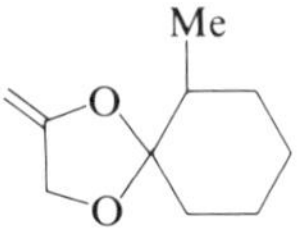

Me O O

(9)

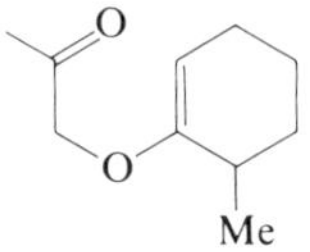

O O Me

(10)

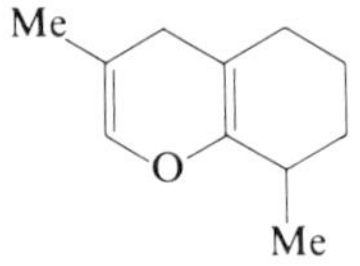

Me O Me

(11)

+ CH_2OR O

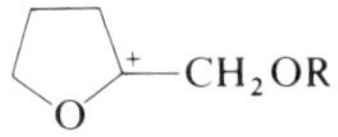

(12)

Tetrahydrofurfural acetals give the expected carbocations in fluorosulphonic acid which then rearrange to **(12)**.[18]

Lactam acetals **(13)** undergo exchange of the methoxyl groups through the imminium ion **(14)** or the corresponding enamine.[19]

The 1,3-dioxolane **(15)** ring-opens upon treatment with alane $RAlMe_2$. Nucleophilic attack on the ketal methylene allows one of the ketal oxygens to open the epoxide aided by the Lewis acid catalyst.[20]

The kinetic evidence for the mercury(II)-promoted hydrolysis of 2,2-diphenyl-1,3-thiolane is compatible with rate-limiting intermolecular attack of water on the Hg^{2+}–*S,S*-acetal adduct **(16)**.[21] The kinetics differ completely from those previously reported for the *O,S*-acetal.[22]

Hydrolysis of 2-methylene-1,3-dithiolane proceeds by rate-limiting carbon protonation to give the carbocation as an intermediate above pH 3 and at zero buffer concentration. With increasing buffer concentration hydration of the carbocation becomes progressively rate-limiting. Below pH 2 breakdown of the orthoester **(17)** is the slowest step.[23]

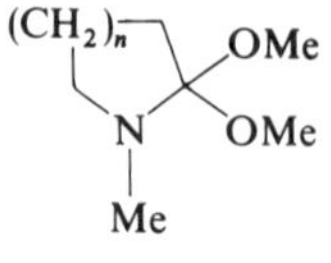

$(CH_2)_n$ OMe N OMe Me

(13)

$(CH_2)_n$ + N OMe Me

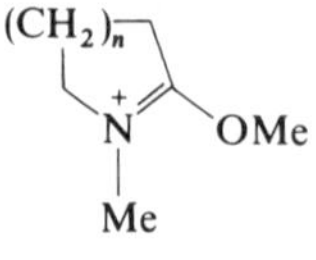

(14)

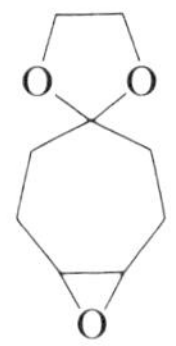

(15)

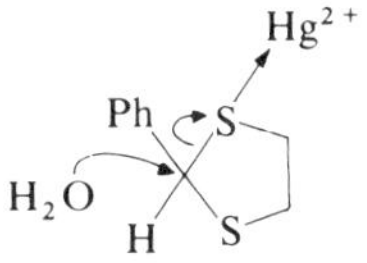

(16)

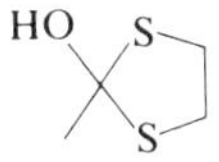

(17)

α,β-Unsaturated ketones react selectively in the presence of a saturated ketone with ethylene glycol if 2,4,6-collidinium *p*-toluenesulphonate is used as a catalyst. Steric shielding of the proton in the salt is suggested to be responsible for this chemoselectivity.[24]

Hemiacetal formation from substituted aromatic aldehydes and methanol in methanol–dioxan mixtures has been observed.[25]

Chemoselective reductive cleavage of ketals and acetals is also observed using monochloroborane etherate as the reducing agent.[26]

The kinetics and mechanisms of radical reactions of cyclic acetals have been reviewed.[27]

Reactions and Formation of Glycosides

Non-enzymic Reactions

Examination of 111 carbohydrate derivatives from the Cambridge Crystallographic Data Base shows that all α- and β- glycosides exist with the *exo*-oxygen in a conformation with lone pairs antiperiplanar to the C(1)–ring-oxygen bond. It is sometimes forgotten that the *exo* anomeric effect exists and in terms of double-bond–no-bond resonance (**18**) would be expected to lead to a shorter C(1)–*exo*-oxygen bond and a longer C(1)–ring-oxygen bond. This is indeed found in the β-glycosides although ring C(1)—0 bonds are near the standard value. In the axial α-glycosides there is relatively little difference between ring and *exo* C(1)—0 bonds, although both are shorter than the standard but longer than that found in the β-anomers. Of course, when interpreting these observations it should be remembered that any CX_n grouping has a reduced C—X bond length because of the electronegativity of X.[28]

The ratio of rates of departure of pyridines from α- and β-glycosyl and -xylopyranosyl pyridinium salts (**19**) and (**20**) are about 80 and 10, respectively. In glycosyl transfer there is thus no evidence to support the too readily accepted notion that antiperiplanar lone pairs of electrons on oxygen favour departure of the leaving group. An often neglected point in the use of stereoelectronic effects to rationalize experimental observations is the presumably variable interaction between orbitals which is dependent upon the degree of bond breaking in the transition state. Since the transition state for glycosyl transfer strongly resembles the intermediate oxo-carbocation the antiperiplanar lone-pair hypothesis is of limited importance. It is suggested that the antiperiplanar lone-pair hypothesis is a special case of the principle of least nuclear motion.[29]

The acid-catalysed hydrolysis of 9-(β-D-ribofuranosyl)purine proceeds by rate-limiting departure of the mono- and di-protonated purine and formation of the glycosyl oxo-carbocation at high acid concentration. In less acidic solution opening of the imidazole ring occurs.[30] The alkaline hydrolysis of 9-(1-alkoxyethyl)purines involves rate-limiting attack of hydroxide ion on C(8) of the purine.[31]

The rate of the acid-catalysed hydrolysis of 3-β-C-ribofuranosylwye (**21**), the most probable structure for wyosine, is *ca.* 10^6 faster than common nucleosides which is attributable to the 3-methylguanosine residue. A reduction in steric repulsion between the 4-methyl and 3-ribofuranosyl groups is thought to be responsible for the rapid hydrolysis. Under alkaline conditions the glycosidic bond and the base moiety of (**21**) are cleaved competitively.[32]

The energy of activation for the acid-catalysed hydrolysis of sucrose does not change with temperature over the range 0–60°. This is in contrast to previous results

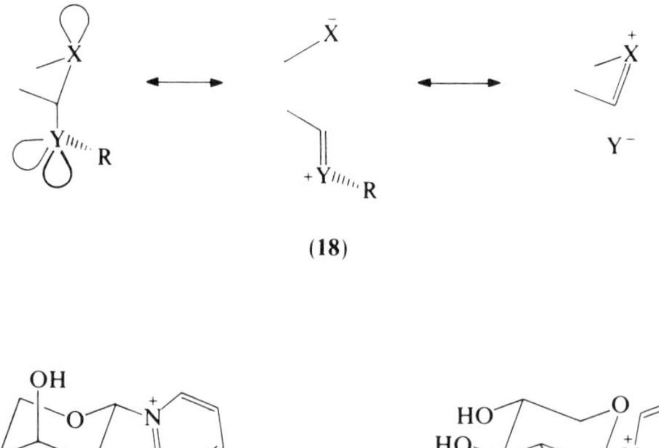

(**18**)

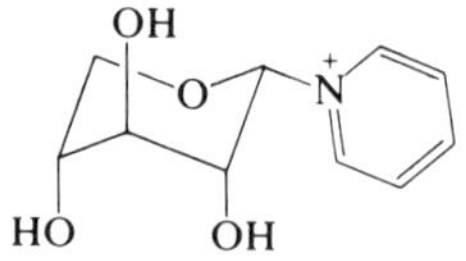

(**19**)

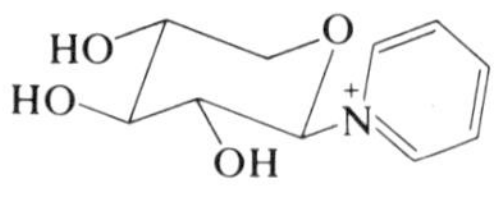

(**20**)

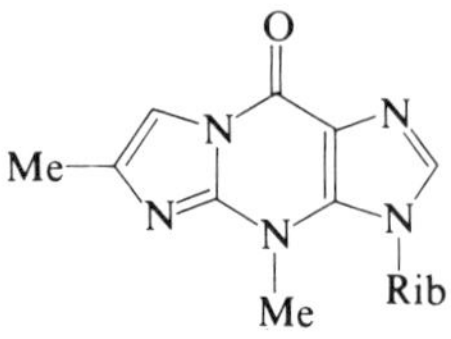

(**21**)

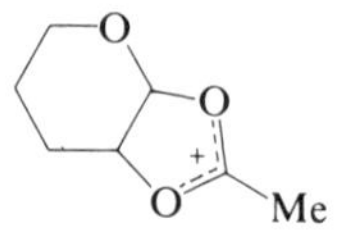

(**22**)

which often involved a comparison of observations made at different concentrations of sucrose and acid.[33]

NMR evidence suggests that 1,4-anhydro-6-azido-2,3-di-*O*-benzoyl-6-deoxy-β-D-galactopyranose from 1-*O*-acetyl-2,3-di-*O*-benzoyl-4,6-bis-*O*-(methylsulphonyl)-α-D-glycopyranose by treatment with sodium azide occurs by neighbouring-group participation of the C(1) oxy-anion.[34]

A kinetic study of the conversion of carbohydrate 1,2-orthoesters to 1,2-*trans*-glycosides suggests that the isomerization occurs by an intramolecular mechanism *via* the acyloxonium ion (**22**).[35]

The kinetics of the base-catalysed rearrangement of glyoxal and 3-deoxy-D-*erythro*-hexos-2-ulose have been compared in different solvents.[36]

In the final phase of the formose reaction, sugars are formed by the reaction of glycolaldehyde, glyceraldehyde, and dihydroxyacetone. A quantitative analysis of intermediates suggest that metal-ion-catalysed aldol reactions occur.[37]

The Knoevenagel reaction of aldehydo sugars with active methylene compounds has been investigated.[38]

The acid-catalysed hydrolysis of *O*-(2-hydroxypropyl)cellulose forms 1,2-*O*-(1-2-*O*-(1-methyl-1,2-ethanedicyl-α-D-glucose acetals. The mechanism of this transformation has been discussed.[39]

The influence of amine basicity on the thermal degradation of *N*-substituted 1-amino-1-deoxyfructoses has been examined.[40]

The stereochemistry of the addition of nucleophiles to nitroalkene derivatives of pyranosides has been investigated.[41]

Enzymic Reactions

Better leaving groups decrease the kinetic isotope effect (k_H/k_D) for isotopic substitution at the anomeric carbon for the acid-catalysed hydrolysis of aryl-β-D-glycopyranosides, as expected for an A1 mechanism. However, the α-deuterium effect for the β-D-glucosidase- and β-D-xylosidase-catalysed hydrolysis of substrates increases with better leaving aglycon groups, which is interpreted in terms of a mechanism with S_N2 characteristics.[42]

Glucokinase catalyses the phosphorylation of glucose with positive cooperativity. The solvent isotope effect at low concentrations of glucose is inverse, $k(D_2O)/k(H_2O) = 3.5$, but at high concentrations it is normal, $k(H_2O)/k(D_2O) = 1.3$. These observations are consistent with two forms of the enzyme with either an increse in the affinity in D_2O of glucose for the enzyme form with the lower affinity in H_2O or a decrease in their rate of interconversion in D_2O.[43]

There seems to be increasing evidence of single enzymes showing bifunctional activity. A recent example is an enzyme with separate kinase and phosphatase catalytic sites which is important in carbohydrate metabolism.[44]

The activity of α-amylase immobilized by coupling to cellulose or carboxymethylcellulose depends on the pH of the coupling reaction and is interpreted in terms of regulation of electrostatic interactions.[45]

A new class of *exo*-glucanases which give malto-oligosaccharides having the α-configuration have been classified as *exo*-α-amylases.[46]

Reactions and Formation of Nitrogen Bases

Schiff Bases and Related Species

The chemistry of Schiff bases has been reviewed.[47]

It has been claimed that the pH-independent hydrolysis of the imines of 2-hydroxybenzaldehyde and of 2-hydroxy-1-naphthaldehyde occur by water attack on the neutral imine (the diploar ion). The generally accepted pathway of hydroxide ion attack on the protonated imine is excluded on the basis of relative rates between the two imines although this arises largely from a difference in pK_a of the two imines. Similarly the phenoxide ion is only 1.5-fold more reactive than the undissociated phenol in the rate of imine formation and yet this has been interpreted as evidence for intramolecular general base catalysis by phenoxide ion in imine formation.[48]

The pH-independent hydrolysis of the imine (**23**) is thought to proceed by the addition of water to the aminoquinone tautomer.[49]

The presence of enolimine tautomers formed during the hydrolysis of Schiff bases of β-diketones retards the rate of reaction. An intermediate containing adjacent positive charges is thought to be responsible for a rate acceleration in other cases by enhancing the rate of carbinolamine formation.[50]

The hydrolysis of bis(2-pyridylamino)phenylmethane involves initial transformation to the Schiff base 2-(benzylideneamino)pyridine.[51]

Intramolecular general base catalysis has been suggested for carbinolamine formation during the hydrolysis of Schiff bases formed from pyridoxal phosphate (**24**). However, no experimental evidence has been presented to support this claim. Hydrolysis of the imine occurs by attack of hydroxide ion and water on the iminium ion at alkaline pH. Compared with other aldehydes pyridoxal phosphate gives imines with a more basic nitrogen and a more favourable equilibrium constant for their formation.[52]

The reactions of hydroxide ion with ferrocenyliminium ions (**25**) are faster in 1 : 1 water : acetonitrile than in water and proceed by rate-limiting addition to the iminium ion. The ferrocenyl group is strongly electron-donating but the difference in reactivity between aryl and ferrocenyl substrates is not large. Water is *ca.* 10^8-fold less reactive towards (**25**) than is hydroxide ion.[53]

The hydrolysis of the Schiff base derived from 2-methoxyethylamine and α-hydroxyisobutyrophenone is catalysed by borate ion and shows a saturation phenomenon with respect to increasing concentration of borate. These specific effects must involve the hydroxyl group as they are not observed in other systems and are attributed to the intermediate formation of a borate–substrate complex (**26**), although hydrolysis is suggested to occur by intramolecular transfer of boron-coordinated hydroxide ion (**27**).[54]

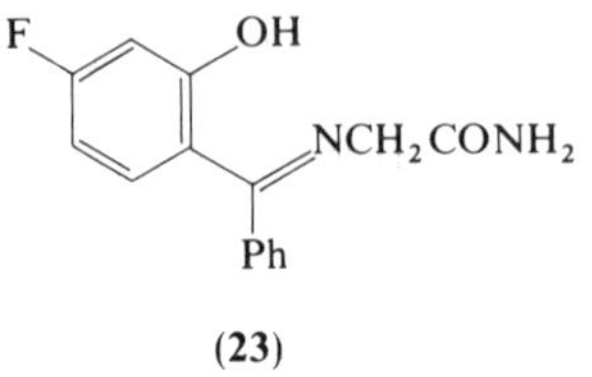

F, OH, NCH_2CONH_2, Ph

(**23**)

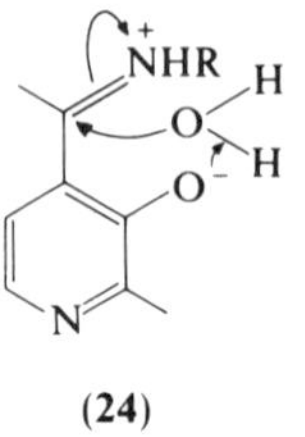

$\overset{+}{N}HR$, H, O, H, O^-, N

(**24**)

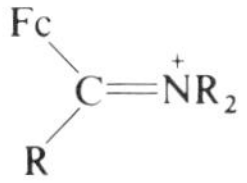

Fc, R, C=N⁺R₂

(25)

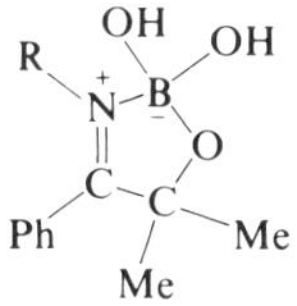

R, OH, OH, N, B, O, C, C, Ph, Me, Me

(26)

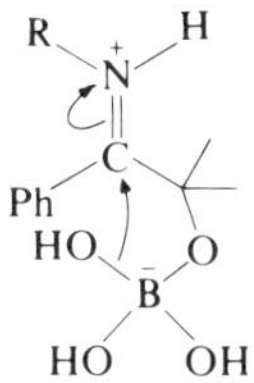

R, H, N, C, Ph, HO, O, B, HO, OH

(27)

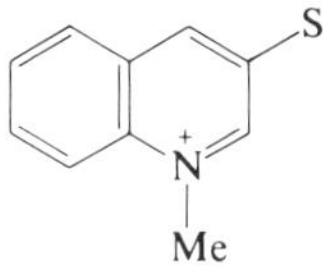

S, N, Me

(28)

The activation volumes for hydroxide ion attack on Schiff bases coordinated by iron(II) complexes in binary aqueous solvent mixtures vary with the nature of the complex but are similar in water. Solution of hydroxide and the complex are presumably important in determining reactivity as a function of pressure.[55]

The nucleophilic addition to tetrahydropyridinium salts has been reviewed with a particular emphasis on stereoelectronic effects.[56]

Hydroxide ion attack at C(2) of the substituted quinolinium cation (**28**) is faster than addition to C(4), the thermodynamically favoured site. Equilibrium constants for these pseudo-base formations have been determined.[57]

The pyridoxamine model compound (**29**) with an intramolecular general base catalyst catalyses the conversion of the ketimine to the aldimine intermediate. Although the rate enhancement is small the product amino acid shows an enantiomeric excess of up to 96%. This has been attributed to intramolecular general-acid-catalysed protonation of the ketimine (**30**).[58]

It has been suggested that the amine exchange reaction of the Schiff base (**31**) in ethanol cyclohexane is facilitated by the alcohol hydrogen bonding to the phenolic group.[59]

The tetrahedral intermediates formed by nucleophilic addition of amines to benzaldehyde in dimethyl sulphoxide have been observed directly by ^{1}H-NMR spectroscopy.[60]

Various linear free energy relationships and their variation with solvent have been reported for the formation of Schiff bases.[61]

The formation of the Schiff base from pyridoxal 5′-phosphate and *n*-hexylamine is thought to be intramolecularly catalysed.[62]

The partitioning of the imine intermediate formed from ammonia and benzaldehyde to benzonitrile and benzoic acid has been described.[63]

Based on substituent effects, it has been suggested that Schiff base formation from (**32**) and amines proceeds by rate-limiting general-acid-catalysed addition of the amine to the enol of (**32**).[64]

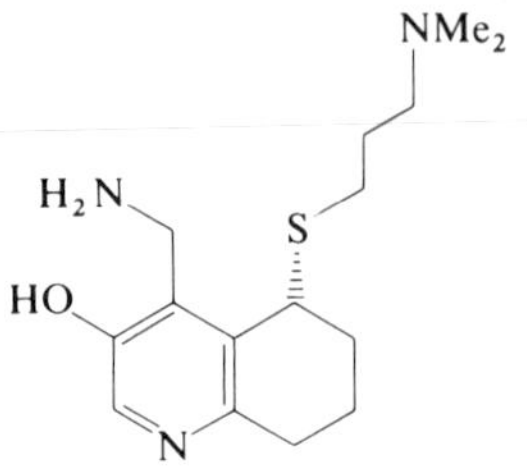

(29)

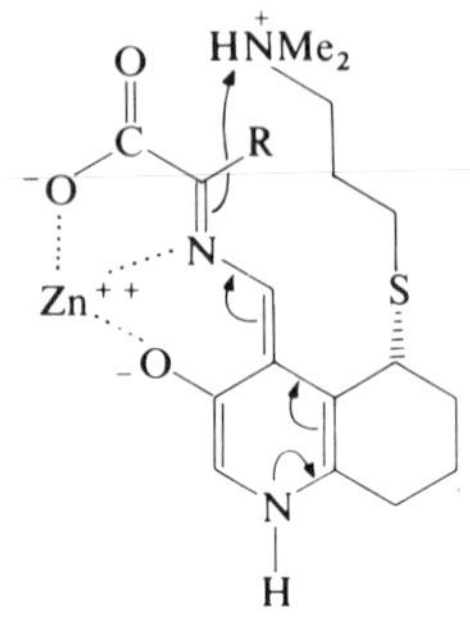

(30)

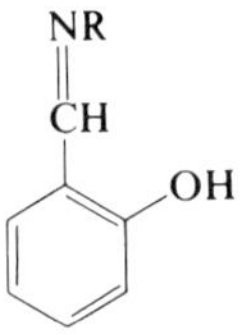

(31)

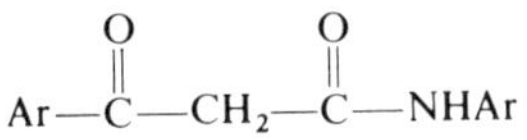

(32)

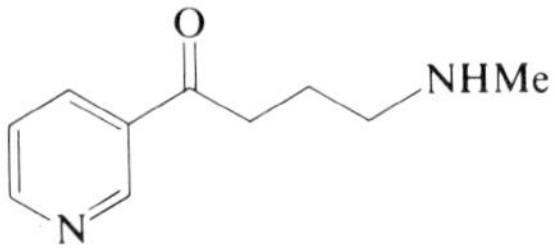

(33)

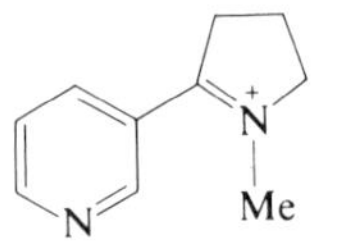

(34)

The nicotine metabolite **(33)** is in equilibrium ($K \simeq 1$) with the iminium ion **(34)** at neutral pH. There is no evidence for intermediate formation of the carbinolamine.[65] The corresponding primary aminoketone exists predominantly (> 99%) as the corresponding cyclic imine.[66]

The Wadsworth–Emmons reaction of phosphoramidates to give C=N systems has been reviewed.[67]

Solvent and substituent effects on the isomerization of imines have been reported.[68]

In vertebrate vision, 11-*cis*-retinal bound to opsin *via* a protonated Schiff base (rhodopsin) is photochemically isomerized to its *all-trans*-congener which is then hydrolysed to *all-trans*-retinal and -opsin. In order for rhodopsin regeneration to occur, *all-trans*-retinal must be thermally isomerized to its 11-*cis*-congener. Contrary to an earlier report, unprotonated retinal Schiff bases are isomerized very slowly at room temperature whereas isomerization is acid-catalysed. As expected, Schiff bases formed from secondary amines readily undergo isomerization by a mechanism involving nucleophilic catalysis.[69, 70]

Cram selectivity is enhanced in the reaction of imines with alkyl-9-BBN. The low Cram selectivity in the reaction of allylic organometallic compounds with chiral aldehydes having no ability to be chelated is presumably determined by steric factors at the chiral centre. Imines may casue the α-chiral centre to take up the axial position and selectivity depends on both the original steric factor of the chiral centre and the steric influence of ligand L (**35**).[71]

DL-γ-Carboxyglutamic acid reacts quantitatively with formaldehyde to give 4,4-dicarboxyproline in a pH-independent reaction. It has been suggested that cyclization occurs by an intramolecular Mannich-type mechanism in which a carbanion attacks the initially formed iminium ion (**36**).[72]

Recent advances in the chemistry of enamines have been reviewed.[73]

Enamines with β-hydrogens react with trifluoroacetonitrile to give 2,4-bis(trifluoromethyl)pyrimidines *via*, it is thought, the intermediate formation of four-membered ring amidines.[74]

Mono-protonated primary–tertiary diamines catalyse deprotonation of ketones by forming iminium ions which are transformed into enamines by intramolecular general base catalysis using the tertiary amino group. The pro-S-deuteron is selectively removed 12–20 times more rapidly than the pro-R-deuteron in deuterated methoxyacetone using (**37**) as a catalyst. This is the result of a steric effect of the methoxy substituent.[75]

Changing the α-carbon from primary to secondary has little effect upon the pK_a for α-deprotonation of aldimines but increases the pK_a for ketimines; this has been attributed to steric effects. The *syn*-configuration of the anions (**38**) is estimated to be at least 4 kcal mol^{-1} more stable than the *anti*-isomer, although the latter is preferred in endocyclic imines.[76]

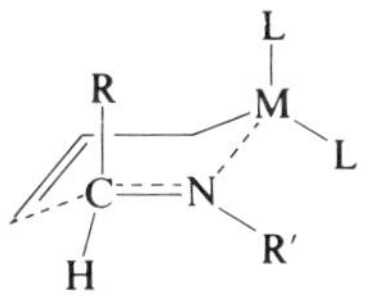

R, L, M, L, C, N, R′, H

(**35**)

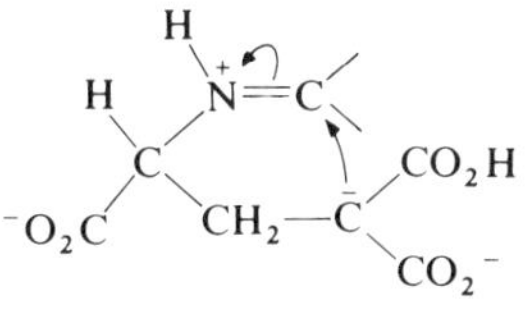

H, H, N, C, C, CO_2H, ^-O_2C, CH_2, C, CO_2^-

(**36**)

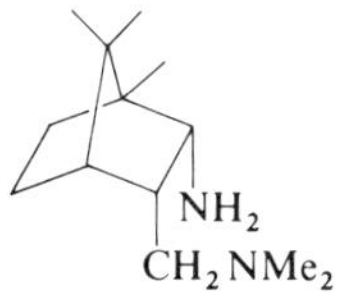

NH_2, CH_2NMe_2

(**37**)

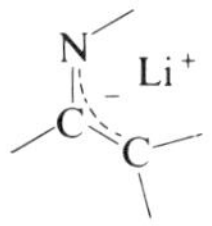

N, Li^+, C, C

(**38**)

The rate of the reaction of *N*,2,6-trichlorobenzoquinone imine with sodium thiosulphate in water–ethylene glycol mixtures increases with increasing dielectric constant of the medium and has been interpreted in terms of a highly solvated transition state.[77]

Hydrazones, Oximes, and Related Compounds

The acid-catalysed hydrolysis of benzylidenesalicylohydrazides shows a ρ^+ of -0.9 and is thought to involve rate-limiting attack of water on the protonated substrate (**39**).[78]

The bimolecular reaction of methoxide ion with (z)-hydrazonoyl chlorides gives stereospecifically the (z)-methylhydrazonate (**40**). However, with poorer leaving groups than chloride a mixtures of stereoisomers is formed. If it is assumed that the addition and elimination steps are stereoelectronically controlled then the tetrahedral intermediate (**41**) should be formed which should collapse rapidly to (**40**).[79]

2-Aminoalkylhydrazones, but not the 3-derivatives, exist in both the acyclic and the cyclic perhydrotriazine form. The equilibrium depends on the nature of the substituents.[80] The acylhydrazones of benzoylacetone and benzoylacetaldehyde exist as pyrazolinol (**42**) and enamine tautomers.[81]

The reaction of α-*N*-methylaminopropionamide with formaldehyde in aqueous solution gives imidazolidinones by rate-limiting attack of the amido group on formaldehyde with intramolecular general base catalysis (**43**).[82]

1-Phenylazo-2-naphthols exist in equilibrium with the corresponding hydrazone-ketone.[83]

The equilibrium constant between some amino-oximes (**44**; X = NOH) and amino-aldehydes (**44**; X = 0) and their corresponding cyclic tautomers (**45**) in aprotic polar solvents have been determined.[84]

The rate of E–Z isomerization of *O*-acylaldoximes in glacial acetic acid is competitive with dehydration to give nitriles.[85]

The acid-catalysed reaction of the isooxazole (**46**) proceeds through rate-limiting enolization of ring-opened oxime or may give the amidonitrile as product depending mainly on the substituent R^2.[86]

Under Beckmann conditions α-azidosteroidal oximes give mono- and di-cyano derivatives, presumably by initial C—C bond cleavage (**47**).[87]

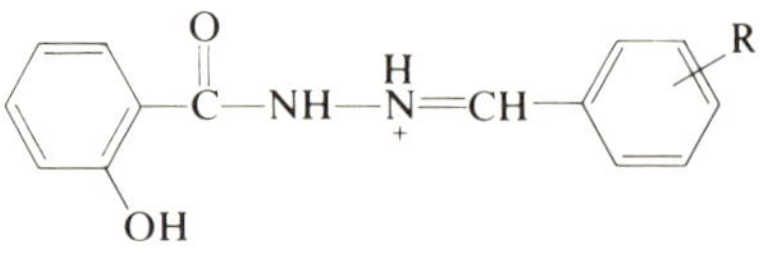

(39)

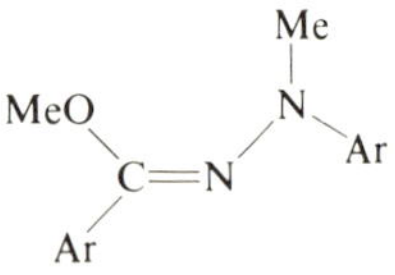

(40)

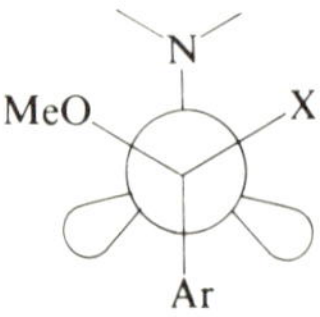

(41)

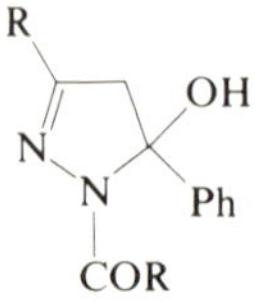

(42)

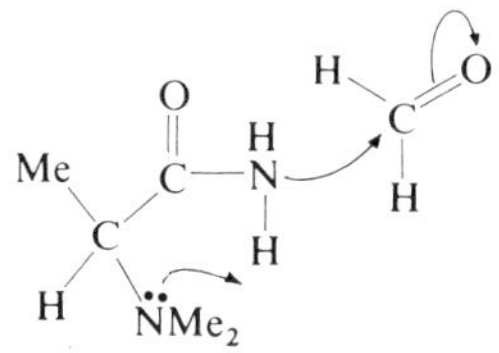

(43) 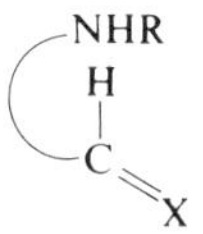 (44) 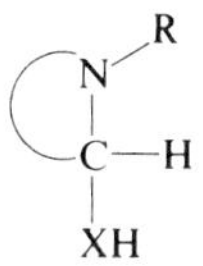 (45)

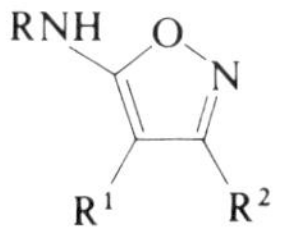

(46) 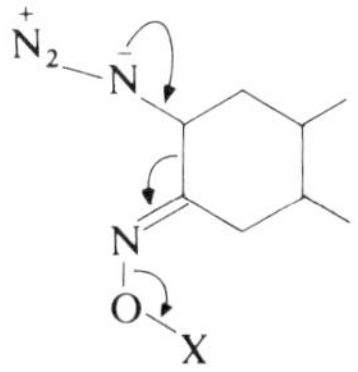 (47) 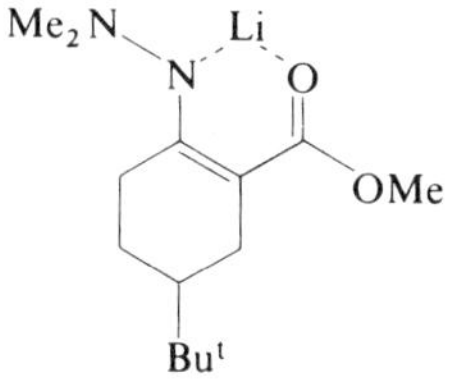 (48)

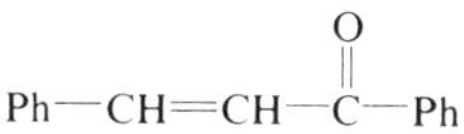

(49) 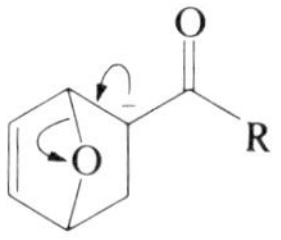 (50)

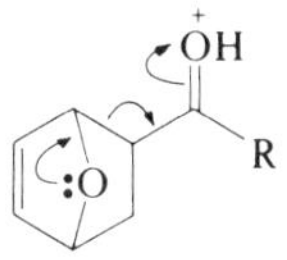

(51) 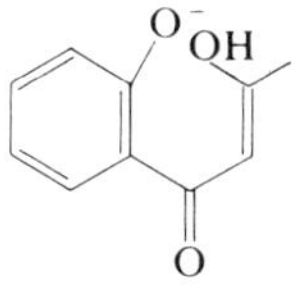 (52)

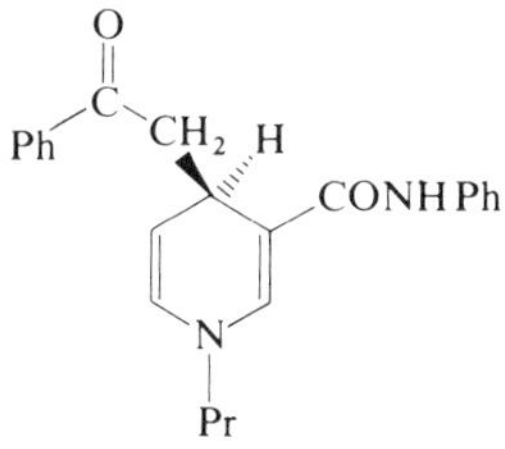

(53) 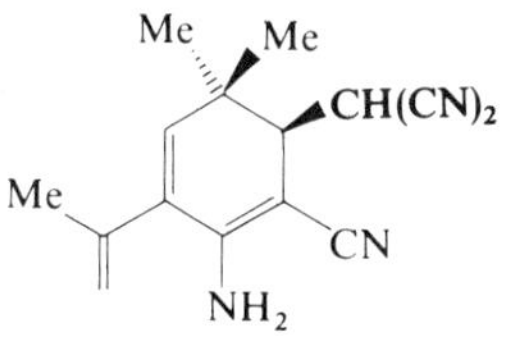 (54)

High axial selectivity is observed in the alkylation of the dimethylhydrazones of 2-cyano-4-*tert*-butylcyclohexanone but not with the 2-alkoxycarbonyl-substituted derivatives. It has been suggested that selectivity may be related to coordination face selectivity. A carbomethoxy group in the 2-position may force the dimethylamino group *anti* by chelation (**48**).[88]

The lithium salts of *tert*-butyl and trityl hydrazones react with Michael acceptors by both ionic and thermal pathways.[89]

Aldol and Related Reactions

The kinetics of hydration and retro-aldol reactions of chalcone (**49**) to benzaldehyde and acetophenone have been determined in aqueous solutions of sodium hydroxide. The equilibrium constant for aldol adduct formation is 4.3 M^{-1} and that for dehydration 25.[90]

The rates of the thermal retro-aldol reactions of β-hydroxy-esters correlate with calculated strain energies. The release of steric strain has previously been thought to be unimportant in these reactions.[91]

The reverse Michael cleavage of 7-oxabicyclo[2.2.1]-heptanes and -heptenes is a formally unlikely 5-*endo–trig* process (**50**). Similarly retro-aldol reactions occur in acid solution (**51**). The ease of these reactions has been attributed to favourable geometric alignments of the bonding and anti-bonding orbitals of the bridging oxygen and its neighbouring carbons rather than to the release of strain energy.[92]

The degradation of β-dicarbonyl phenolates (**52**) is inhibited by hydroxide ion because of ionization of the enol.[93]

Pyridinium salts can be used as an enolate transferring agent. For example, the base-catalysed addition of acetophenone to an NAD analogue gives the intermediate (**53**) which can then be reacted with carbonyl acceptors to give aldol products. No hydride transfer from (**53**) is observed.[94]

The effect of solvent and cation on the ratio of *C*- to *O*-alkylation of the enolate of ethyl acetoacetate in the presence of picric acid has been investigated.[95]

Condensation of 2,5,5-trimethylhexa-2,3-dien-6-al with malonitrile gives the unexpected product (**54**). The initially formed aldol product is thought to undergo a thermal cyclization to give an imino-2,5-cyclohexadienone which then reacts with more malonitrile.[96]

A kinetic analysis of the reaction of formaldehyde and acetaldehyde to give pentaerythritol has allowed for a non-steady concentration of the carbanion intermediate.[97]

The development of highly stereoselective aldol and related C—C bond-forming reactions continues to be a dominant theme. Most methods utilize enolate monoanions of a ketone, an ester, or their equivalent. Selective aldol condensations occur with quaternary ammonium enethiolates generated from (z)-*N*,*N*-dimethyl-*S*-trimethylsilylketene *S*,*N*-acetals (**55**) with benzaldehyde in the presence of a Lewis acid. There is very little hard evidence to elucidate the mechanisms of these reactions.[98]

It appears to be very difficult to achieve asymmetric carbon–carbon bond-forming reactions controlled by the non-covalent interaction of chiral ligands to substrates. The magnesium salt of allyl tolyl sulphone undergoes asymmetric addition to acetone under the influence of a chiral ligand derived from L-proline which is the first example of ligand controlled chiral induction.[99]

z-Lithium enolates, formed from *n*-alkyl *tert*-butyl ketones and lithium diisopropylamide, react with benzaldehyde to give *syn*-aldols. The enolization step is slow with sterically hindered ketones. In non-polar solvents *syn–anti* equilibration occurs which can be rationalized by extensive negative charge delocalization in the transition state compared with that in the reactants or products.[100]

Crotyl- and α-methylallyl-dichlorotin derivatives can react with aldehydes to give four isomeric adducts: *threo*, *trans*, *erythro*, and *cis*. Their rearrangements, probably through transition states with bicyclononane structures such as (**56**), give stereospecifically *trans*- or *cis*-chlorotetrahydropyrans. A *cis*-stereoconvergent synthesis is obtained using 1-buten-3-yl-*n*-butyldichlorotin.[101]

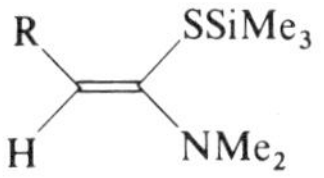

R, H, $SSiMe_3$, NMe_2

(**55**)

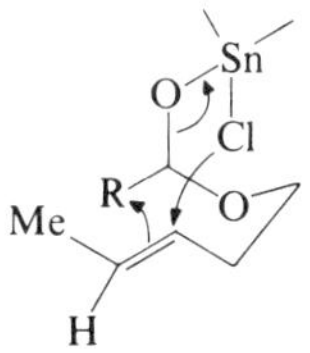

Sn, O, Cl, R, Me, O, H

(**56**)

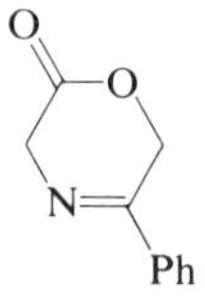

O, O, N, Ph

(**57**)

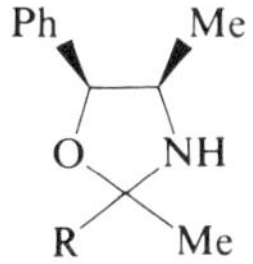

Ph, Me, O, NH, R, Me

(**58**)

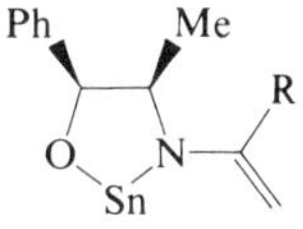

Ph, Me, R, O, N, Sn

(**59**)

The stereochemical course of stannous-triflate-mediated aldol-type reactions is dramatically altered by the addition of tetramethylethylenediamine as a ligand.[102]

The Erlenmeyer–Ploechl reaction of hippuric acid and benzaldehyde gives a benzylideneoxazolone from the intermediate aldol product (**57**) and excess aldehyde.[103]

It is still a problem to achieve high asymmetric induction in aldol reactions of methyl ketones and acetic acid derivatives. A chiral oxazolidine (**58**), prepared from a chiral 1,2-amino-alcohol and a methyl ketone, treated with lithium diisopropylamide generates a lithium azoenolate which, upon addition of stannous chloride, gives the chiral cyclic azoenolate (**59**). Subsequent reaction with aldehydes gives aldol products of 58–86% enantiomeric excess.[104]

Examples of titanium-tetrachloride-promoted aldol reactions of silyl enol ethers have been reviewed.[105]

Ethylene chloroboronate generates enolboronates from carbonyl compounds which show very high *erythro*-diastereoselectivity with both aliphatic and aromatic aldehydes. The transition state is assumed to be the usual pericyclic boat-like configuration.[106]

The aldol reaction of the ethyl mercaptoacetate dianion with carbonyl compounds gives adducts which may be treated with ethyl chloroformate in the presence of trivalent phosphorus to give E-isomers of α,β-unsaturated esters. The mechanism is thought to involve the intermediate formation of a thiirane.[107]

There have been many other reports and reviews of the stereoselective aldol reactions.[108]

Other Addition Reactions[109]

Aldehyde analogues of normal alcohol substrates induce adenosinetriphosphatase activities of various kinases. It is thought that the hydrated aldehydes are the activators of these reactions.[110] This observation provides a basis for measuring the rates of hydration and dehydration of aldehydes by ^{18}O-transfer from the aldehyde or its hydrate to phosphate. For a series of heavily hydrated aldehydes there is little variation in the dehydration rate constants over a 10^3 change in equilibrium constants for hydration. This presumably reflects the greater effect of substituents α to the aldehyde on the electrophilicity of the carbonyl carbon of the free aldehyde than on the pK of the *gem*-diol of the hydrated form.[111]

The reversible hydration of 1,3-dichloroacetone in the presence of aerosol-OT reversed micelles in hexane shows different kinetic behaviour from that in aqueous dioxan. Proton inventories indicate the participation of a second water molecule acting as a general base catalyst.[112]

The rate constant for the uncatalysed hydration of trifluoroacetophenone is $3.2\ s^{-1}$. The reaction is not subject to acid catalysis and although base-catalysed the Brønsted β value is very small. The reaction of hydroxylamine shows a change in rate-limiting step with increasing concentration of catalysing acid. Several rate and equilibrium constants for the addition of nucleophiles to trifluoroacetophenone have been reported.[113]

Calculation of the deprotonating factor (the degree of activation brought about by proton removal) and the proton-activating factor (the degree of activation brought about by protonation of the substrate) for the hydration of aldehydes confirms earlier interpretations that this occurs by general acid catalysis (**60**) rather than the kinetically equivalent general base mechanism.[114]

Molecular orbital calculations suggest that the hydration of ketene to form the ketone hydrate is favoured over formation of acetic acid. This is in contrast to protonation which favours proton transfer to the β-carbon.[115]

The transition-state structure for the gas-phase addition of water to formaldehyde has been characterized by quantum-mechanical calculations.[116]

The participation of solvent water in the hydration of formaldehyde, ketenimine, and formamidines and in keto–enol tautomerism has been reviewed.[117]

1-Arylmethyl substituents favour the open-chain tautomer if steric effects occur in the ring-closed form of l-hydroxyphthalan (**61**) and related structures.[118]

The first report of the efficiency of intramolecular hydride transfer indicates that the effective molarity is high, as expected on the basis of entropy changes between inter- and intra-molecular systems. Intramolecular hydride transfer in the alkoxide (**62**) shows an E.M. of 6×10^6 M compared with an analogous intermolecular process.[119] In the crystal structures of related polycyclic hydroxyketones the hydrogen to be transferred is 2.3–2.5 Å from the carbonyl carbon and shows an angle $O{=}\hat{C}{-}{-}{-}{-}H$ of 96–99°. The carbonyls show small pyramidalizations of 0.01–0.02 Å.[119]

The diastereo-controlled reduction of 2-amino- or 2-hydroxy-ketones to give optically active *threo*-1,2-amino-alcohols or 1,2-diols, respectively, may be performed using hydrosilanes in the presence of a catalytic amount of tetrabutylammonium fluoride in hexamethylphosphoric acid. In remarkable contrast *erythro*-selective reduction may be achieved under acidic conditions of trifluoroacetic acid. The *threo*-selectivity has been attributed to the bulky ($Bu^n_4N^+$) (R_3SiHF^-) species which attacks the ketone carbonyl carbon according to the Felkin transition-state model. The acid-catalysed pathway may be rationalized by Cram's proton-bridged cyclic model.[120]

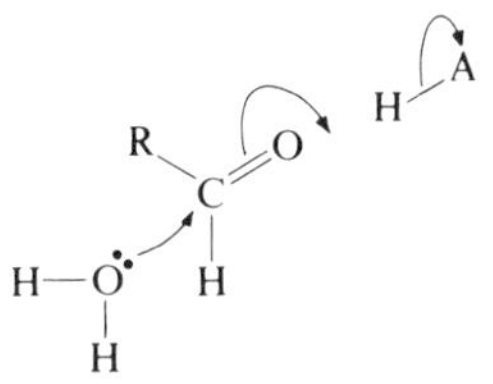

(60)

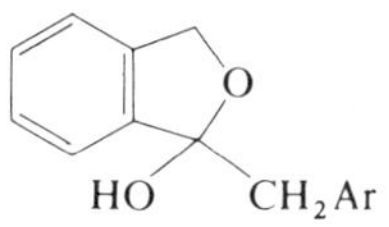

(61)

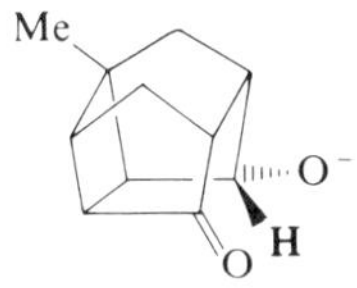

(62)

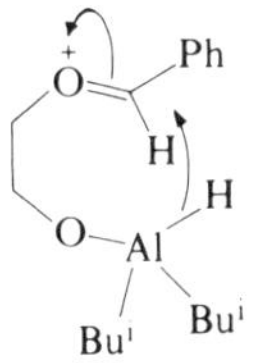

(63)

Lithium aluminium hydride modified by chiral amines reduces prochiral cyclic ketones to give optically active cyclic alcohols in high enantiomeric excess.[121]

Benzylidene acetals of 1,2- and 1,3-glycols are easily cleaved by diisobutylaluminium hydride in toluene to give the corresponding monobenzyl ethers. In most cases cleavage occurs selectively at the least hindered site. Although this has rationalized by intramolecular hydride transfer **(63)** an intermolecular mechanism is not excluded.[122]

The hydroxide-ion-catalysed Cannizzaro reaction of 4-nitrosobenzaldehyde in aqueous dioxane gives equimolar quantities of 4,4′-diformylazoxybenzene and 4,4′-dicarboxyazoxybenzene by rate-limiting hydride ion transfer from the normal initially formed adduct.[123]

Zinc ion accelerates the reduction of 2-pyridinecarbaldehyde by dihydroquinoline derivatives in aqueous solution. Kinetic complexities in catalysis disappear when the reaction is studied under deoxygenated dark conditions.[124]

The optical yield in the asymmetric reduction of ethyl benzoylformate with a chiral NADH model compound varies with the extent of reaction. This has been attributed to the oxidized NADH model forming a magnesium ion complex with unreacted NADH model compound.[125]

A previously proposed correlation between the stereospecificity of nicotinamide coenzyme-dependent dehydrogenases that reduce carbonyls and the value of the equilibrium constant for the catalysed reactions has been justifiably criticized.[126]

The pH dependence of the reaction catalysed by lactate dehydrogenase, where pyruvate adds covalently to NAD to give a NAD–pyruvate adduct, suggests similar binding modes for enol pyruvate, ketopyruvate, and lactate. It has been suggested that, although the enzyme provides a general catalyst, general acid catalysis occurs externally by buffers.[127]

Steric inhibition of resonance has been invoked in interpretation of the dissociation constants of cyanohydrins of substituted acetophenones.[128]

The reaction of aromatic ketones with diethylphosphorocyanidate in the presence of lithium cyanide gives cyanophosphates which yield α,β-unsaturated nitriles on treatment with boron trifluoride etherate.[129]

Surprisingly, hydroxynitrile lyase, the enzyme from bitter almond which catalyses the formation and breakdown of cyanohydrins, uses flavin adenine dinucleotide as a coenzyme. Flavoproteins usually catalyse true redox processes. 3-Oxo-3-phenylpropyne and 3-oxo-3-phenylpropene are active-site-directed inhibitors of the flavoprotein hydroxynitrile lyase. In cyanohydrin formation the enzyme is thought to form a thio-hemiacetal with the aldehyde substrate using a cysteine thiol group. Cyanide binding to N(5) of the isoalloxazine moiety of the coenzyme and carbonyl carbon binding to N(4a) gives a species (**64**) which is thought to rearrange to cyanohydrin. The chemical logic of this mechanism is not obvious.[130]

Pyrylium cations (**65**) coexist in water with the enedione pseudo-base (**66**) and its anion and a small amount of its enol.[131] With amines the cations give pyridinium ions through the intermediate formation of the enamine-ketone (**67**), but are less reactive than the less nucleophilic aromatic amines. This has been attributed to a larger proportion of the imine tautomer with aromatic amines.[132] The synthetic utility of these reactions and the optimum conditions required have been discussed.[133, 134] Water-soluble pyrylium salts react with the lysine residues of gelatin and α-chymotrypsin, but although the pseudo-base (**66**) reacts with free lysine there is no reaction with the proteins.[135]

The reaction of secondary amines with 2,4,6-triphenylpyrylium ions to give ring-opened divinylogous amides is base-catalysed, whereas that with primary amines is not. It has been suggested that for the latter formation of the intermediate (**68**) is rate-limiting, whereas for secondary amines proton transfer from the analogous intermediate becomes the rate-limiting step.[136]

Pentane-2,4-dionate salt reacts with dimethyldichlorosilane to give a pyrylium salt, while the trifluoromethyl analogue gives the corresponding neutral pyranylidene complex. The proximity of the highly electronegative trifluoromethyl group reduces the basicity of the ring oxygen which probably hinders the formation of the pyrylium salt.[137] The use of pyrylium salts for transforming amino groups into other functionalities has been reviewed.[138]

The Cram Rule has been reviewed.[139]

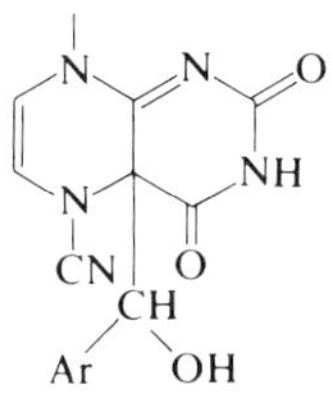

(64) 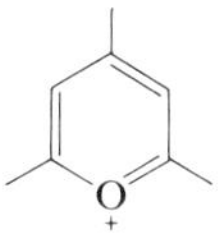 (65) 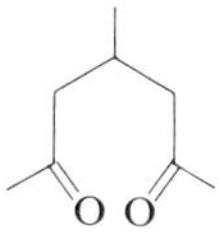 (66) 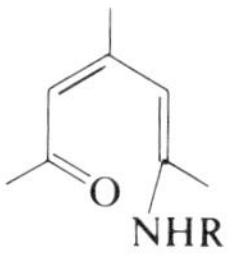 (67) 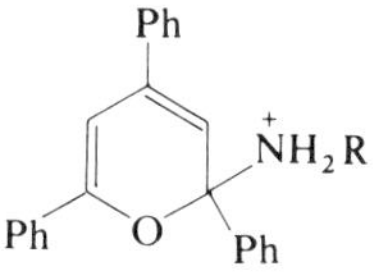 (68)

There is a decrease in the gas-phase proton affinities (*PA*) on going from cyclooctanone to cyclobutanone. There is an abrupt decrease in *PA* of 3.6 kcal mol^{-1} between cyclohexanone and cyclopentanone and 5.4 kcal mol^{-1} between the latter and cyclobutanone. This has been attributed to changes in strain energy and repulsion between the α-substituents of the carbonyl group.[140]

The influence of solvent on the substituent effect on the stretching frequency of the carbonyl group in 4-substituted camphor has been reported.[141]

A simple model has been proposed for describing the trajectory along the reaction coordinate of a nucleophile attacking a carbonyl group. Hard nucleophiles (those with low-lying HOMOs) approach carbonyl carbon at a smaller angle than corresponding soft nucleophiles (those with high-lying HOMOs).[142]

Ab initio calculations suggest that hydride ion transfer between methoxide ion and formaldehyde occurs through a centro-symmetrical transition state.[143]

According to theoretical calculations nucleophilic addition of Me^- is more exothermic than that of H^- to coordinated carbon monoxide.[144]

The self-condensaton of formaldehyde catalysed by thiazolium salts gives selectively dihydroxyacetone (triose). As no glycolaldehyde is formed it has been suggested that the carbanion (**69**) formed by the addition of two formaldehyde molecules to the thiazolium salt reacts very rapidly with another molecule of formaldehyde to give the three-carbon fragment which can give triose.[145]

Thiamine diphosphate is the essential coenzyme for pyruvate decarboxylase. Attack of the thiazolium ring on a conjugated β-keto-acid followed by loss of carbon dioxide gives an enamine intermediate (**70**) which has been identified spectroscopically. *C*-Protonation of the enamine therefore occurs in a separate step.[146]

The nature of the hetero-atom attached to the nucleophilic carbon of ylides is important in determining the course of their reaction with carbonyl groups. It has been suggested, on the basis of theoretical calculations that a cyclic intermediate is formed with all ylides and that differences arise from the relative strengths of P—C and S—C bonds.[147]

The Hammett relationship has been used to estimate the equilibrium between the betaine and the oxaphosphetane in the Wittig reaction of substituted benzaldehydes.[148]

Wittig-reaction intermediates generated from stabilized triphenylphosphonium ylides and aromatic aldehydes are primarily oxaphosphetanes formed reversibly. This has now been shown for the reaction of non-stabilized ylides with aliphatic

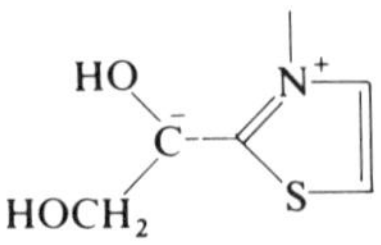

(69) 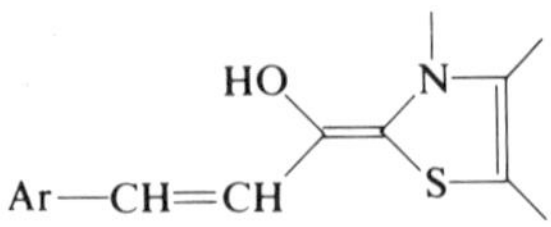 (70)

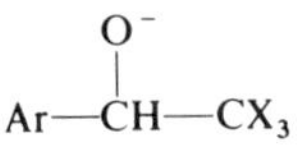

(71) 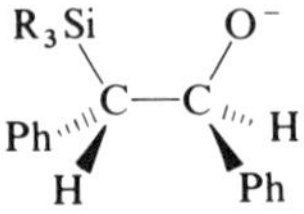 (72)

aldehydes. Reversibility may be used to account for the stereoselectivity observed in some Wittig reactions.[149]

The Wittig reaction of benzaldehyde with phosphonium ylides shows that carboxylate and oxido functionalities near the ylide centre give anomalously high stereoselective formation of E-alkenes.[150]

β-Amino-alcohols can be prepared from the reaction of aromatic non-enolizable ketones with dimethyloxosulphonium methylide to give an oxiran followed by treatment with an appropriate amine.[151]

The stereoselective control of the addition of *C*-nucleophiles such as Grignard reagents to carbonyl compounds or enolates to chiral α- or β-alkoxy-aldehydes or -ketones has been reviewed.[152]

The base-catalysed decompositions of 1-aryl-2,2,2-trihalogenoethanols in water proceed through the alkoxide (**71**). The Hammett ρ value for the decomposition of (**71**) is -0.6 and -0.8 for the trichloro and tribromo derivatives, respectively. The tribromomethyl anion is a better leaving group than the trichloro species which is presumably attributable to a steric effect.[153]

Regiospecific retro-Claisen-type cleavage of β-keto-acetals derived from non-enolizable β-diketones has been rationalized on the basis of intermediate formation of a dialkoxy-carbenium ion.[154,155]

α-Silyl carbanions react with aldehydes to form alkenes. This Peterson reaction represents an alternative to the Wittig reaction for the synthesis of alkenes. The reaction of $Ph\bar{C}HSiMe_3$ with benzaldehyde gives a constant ratio of *cis*:*trans*-stilbenes with different alkoxides and media. This is interpreted as indicative of an irreversible attack of the carbanion on the aldehyde to give an intermediate (**72**) which undergoes stereospecific *syn*-elimination to give the stilbene.[156]

The diastereoselectivity of the cyclocondensation of a siloxydiene with α-alkoxy-aldehydes is improved by using europium(III) complexes as Lewis acid catalysts.[157]

α-Selenoalkyllithiums bearing two alkyl groups on the carbanionic centre may be used to ring enlarge cyclic ketones in the presence of thallium ethoxide in chloroform. It has been suggested that the latter generates dichlorocarbene which facilitates rearrangement of the intermediate β-hydroxyselenide.[158]

The rôle of selenonyl groups in the reaction of vinyl selenones with enolates has been described.[159]

The stereochemistry of the condensation product formed from the reaction of aldehydes with α-sulphenyl-, α-sulphinyl-, and α-sulphonyl-carbonyl compounds with aldehydes is controlled by the steric requirements of the two functional groups in a sulphur-stabilized carbanion intermediate.[160]

The kinetics of 4-aminobenzylalcohol formation from foraldehyde and substituted aromatic amines in acid solution have been reported.[161]

erythro-Selectivity can be achieved in the reaction of 2-alkylpyridines with benzaldehyde in the presence of dialkylboryl triflate and triethylamine.[162]

The gas-phase addition of the trifluoromethyl anion to ketones and other carbonyl compounds is analogous to that in the condensed phase.[163]

The Hammett ρ^- value for the reaction of trichloromethyl anions with substituted benzaldehydes in dimethyl sulphoxide is 1.2 whereas that for the more selėctive tribromomethyl anion is 1.4. The reaction rate is therefore presumably not encounter-controlled.[164]

The annelation of benzocyclobutenedione (**73**) monoketals with vinyllithium reagents is thought to proceed by addition to form a benzocyclobutenol followed by ring-opening and cyclization.[165]

The kinetics of the reaction of ketones with alkyllithiums suggest that the product is formed by two competing pathways. These involve either an initially formed complex between the two reactants which undergoes a rearrangement or a reaction between uncomplexed ketone with the more reactive alkyllithium monomer.[166]

Acetaldehyde and propionaldehyde, but not trichloroacetaldehyde, are trimerized readily in sulphur dioxide by the initial formation of a charge-transfer complex.[167]

Alkylation of carbanions generated from ketimine *N,S*-acetals react at either the keto or the aldehyde carbon depending on the alkyl halide used.[168]

α,β-Unsaturated acetals and ketals react with organolithium reagents by addition, substitution, and proton abstraction pathways.[169]

Treatment of cyclic ketones with tris(methylthio)methyllithium and tetrakis(acetonitrile)copper(I) tetrafluoroborate gives ring-expanded 1,2-keto-thioketals. The mechanism is thought to involve the intermediate formation of an epoxide followed by stereospecific migration of the carbon–carbon bond.[170]

The toxicity and degenerative chemistry of malondialdehyde may be a result of its ability to covalently bond and cross-link biological macromolecules. The reaction of malondialdehyde with nucleic acid bases gives novel cyclopropane derivatives (**74**).[171]

The ethynylation of 4-substituted acetophenones in liquid ammonia containing potassium hydroxide shows a Hammett ρ value of 0.93 and is thought to involve several pathways involving different stoichiometries of the catalyst.[172]

Electron-withdrawing substituents increase the rate of the base-catalysed Darzens-like condensation of $RC_6H_4COCH_2Cl$ with phenanthrenequinone.[173]

The Stobbe condensation between *ortho*-substituted aromatic aldehydes gives the (E)-monobenzylidenesuccinates stereoselectively.[174]

The Abramov reaction, the addition of dialkyl phosphites to aldehydes or ketones,

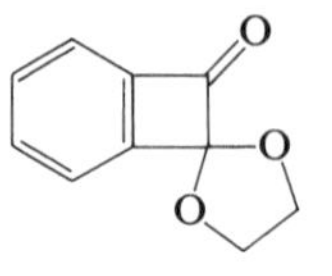

(73) 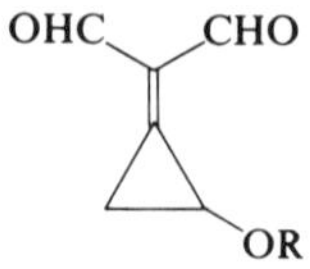 (74)

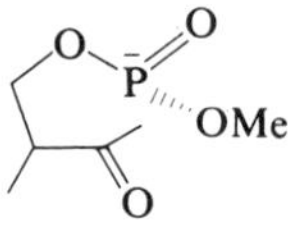

(75) 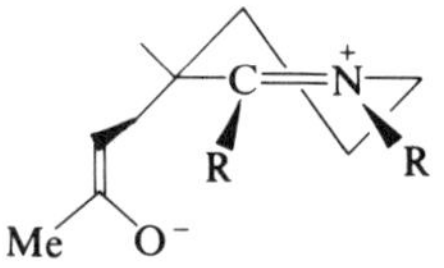 (76)

is a particularly useful way to form P—C bonds. The retro-Abramov pathway is thought to occur in the methoxide-ion-catalysed equilibration of diastereomeric 3,4-dimethylequilibration of diastereomeric 3,4-dimethyl-2-methoxy-2-oxo-1,2-oxaphospholan-3-ols by P—C cleavage to give the carbonyl dialkyl phosphite anion (**75**).[175]

The acid-catalysed conversion of furans and hydrogen sulphide to thiophenes may proceed through the reaction of the diketone as an intermediate. The rate-limiting step is thought to be the addition of H_2S to the protonated ketone.[176]

The annelation of Δ^2-tetrahydrophyridines with methyl vinyl ketones invariably gives hydroquinolones. Deviations from this route suggest that the Michael addition step occurs in the axial fashion on the same face of the molecule as the lone electron pair on nitrogen to give (**76**). Conformational inversion must then occur before intramolecular enolate anion attack on the iminium ion can take place.[177]

The Hammett ρ value for the thermal ene reactions of diethyl oxomalonate with 1-arylcyclopentenes changes from -1.2 to -3.9 on going from the uncatalysed to the tin(IV)-chloride-catalysed reaction. Structural selectivity can be dramatically reversed by catalysts as the influence of electronic factors is amplified by Lewis acids and steric approach control becomes less important.[178]

The dimethylaluminium-chloride-catalysed ene reactions of aliphatic aldehydes with geometrical isomers of alkenes gives complex mixtures of *erythro*- and *threo*-adducts.[179]

There has been a report of a high-pressure ene reaction.[180]

The addition of the methyl radical to trifluoroacetone in the gas phase occurs with a second-order rate constant of 1.8×10^4 $\text{M}^{-1}\,\text{s}^{-1}$ at 430 K which is about 10-fold slower than the reaction with hexafluoroacetone.[181]

The absolute rate constants for the reactions of tri-*n*-butylgermyl and tri-*n*-butylstannyl radicals with carbonyl compounds have been reported.[182]

Enolization and Related Reactions

The microwave spectra of *syn*-vinyl alcohol shows that the carbon oxygen bond is *ca.* 0.06 Å shorter than that in saturated alcohols. Double-bond character of the

C—O bond is also reflected in a high C—O stretching force constant. The interesting bond angles in this simple enol are shown in (77).[183]

The solvent dependence of the conformation of the hydroxyl group in thermodynamically stable, sterically hindered enols has been determined by NMR spectroscopy. In non-polar non-hydrogen-bond-accepting solvents the enols exist predominantly in the *syn*-planar arrangement. In hydrogen-bond-accepting solvents the *anti*-clinal conformer also exists.[184] There are some differences between these results and the conclusions previously reached for simple aliphatic enols.[185]

The crystal structure of a vinyl alcohol has been reported.[186]

There have been more reports of stable enols[187] and tautomerism of phenols in the quinoid form.[188]

Based on no unknown quantities, the pK_a of acetone ionizing as a carbon acid in aqueous solution at 25° is 19.16 and the pK_E for enolization is 8.22.[189]

$H_2C{=}C(OH)H$ (C=C–H angle 129°, C–O angle 105°)

(77)

$ArOCH_2CH_2{-}C(=O){-}CH_3$

(78)

Acetophenone enol has been generated by flash photolysis and its rate of ketonization in aqueous acid solution has been measured. The keto–enol equilibrium constant pK_E = 7.90 and the ionization constant for acetophenone as a carbon acid pK_a = 18.2. Protonation of the β-carbon of the enolate anion by the hydronium ion shows a rate constant of 4×10^{10} $\text{M}^{-1}\,\text{s}^{-1}$ which is so great as to suggest proton transfer down solvent bridges.[190]

The equilibrium constants for ketone–enol tautomerism in aliphatic ketones, cycloalkanones, and substituted acetophenones have been summarised together with the associated enthalpy and entropy changes.[191]

Brønsted plots for proton abstraction from carbon acids by oxy-anions, but not thiol anions, exhibit pronounced curvature such that highly basic oxy-anions are non-selective (β = 0.2) whereas weakly basic oxy-anions are very selective (β = 0.8). That this curvature is attributable to solvation rather than a change in transition-state structure is supported by the very small changes in k_H/k_D for proton abstraction from (**78**) by a variety of oxy-anions. An inverse correlation is found between the magnitude of the isotope effect for proton abstraction by X^- and the energy change upon conversion of HX to DX.[192]

Tertiary amines are more effective than primary and secondary amines which are more effective than oxy-anions toward proton abstraction from nitroethane. The Brønsted β values are, respectively, 0.45, 0.60, and 0.71. The rate ratios, for a given basicity, are much smaller than those previously observed in the enolization of oxaloacetate which supports the suggestion of a nucleophilic mechanism for the latter reaction.[193]

Primary and secondary kinetic isotope effects are small (k_H/k_D = 1.13) for proton exchange in malondialdehyde. The rate of exchange increases with the proportion of dimethyl sulphoxide in binary mixtures with water and occurs rapidly at pH values well above the pK_a of the dialdehyde.[194]

The enolization of ammonium ketones is facilitated by the positively charged nitrogen due to electrostatic stabilization with a negatively charged base catalyst and to intramolecular stabilization of the enolate anion (**79**). There is no evidence for intramolecular general acid catalysis although intramolecular general-base-catalysed enolization is suggested for (**80**).[195]

It has been previously proposed that magnesium and maganese(II) ions catalyse the enolization of methyl 2-oxo-1-phosphonate (**81**) by coordination to the carbonyl group, although the thermodynamically favoured binding site is the phosphonate group. However, it has now been shown that the Brønsted β values for the general-base-catalysed enolization of (**81**) and the magnesium and manganese complexes are 0.62, 0.69, and 0.82, respectively. This trend is in the opposite direction to that expected from the reactivity–selectivity principle. It has been suggested that catalysis occurs *via* a coordinated water molecule which forms a bridge between the base and the phosphonate-coordinated substrate.[196]

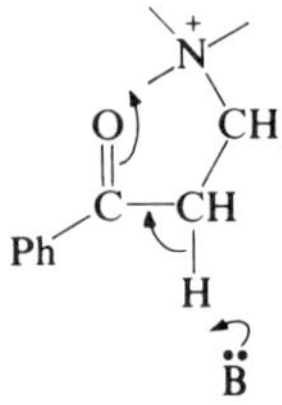

(**79**)

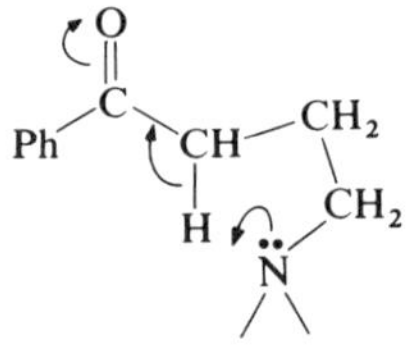

(**80**)

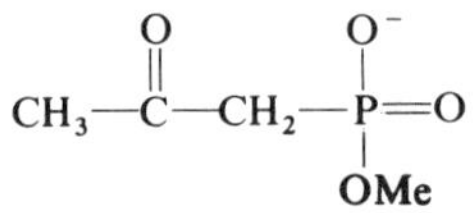

(**81**)

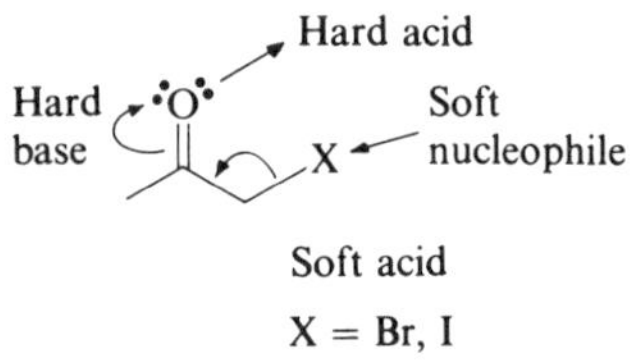

(**82**)

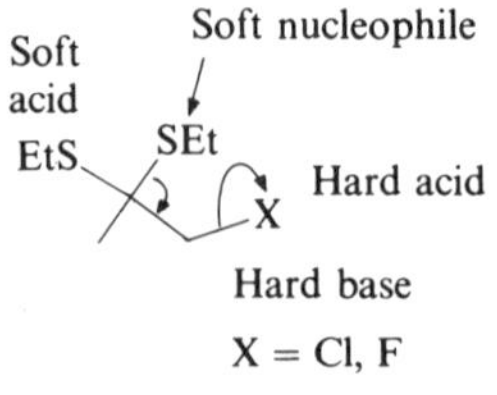

(**83**)

The rates of metal-ion-catalysed enolization and hydration of oxalacetate conform to the Marcus equation for group transfer. The influence of Mg(II) is essentially accounted for by the thermodynamics of the interactions with the strong chelating centres formed during enolization and hydration, whereas the intrinsic activation barriers are metal-ion-independent.[197]

In alkaline solution there is a term in the rate law for the chlorination of ketones which is first order in ketone and hypochlorite but zero order in hydroxide ion. If it is assumed that this term represents the reaction of enolates with hypochlorous acid at a diffusion controlled rate the pK_a of ketones in aqueous solution may be determined; for example, that of acetone is calculated to be 19.1. Hypobromite anion reacts with acetone enolate *ca.* 10^3-fold faster than does hypochlorite anion.[198]

The chemistry of α-halogenated ketones and imines has been reviewed.[199]

As the carbonyl oxygen is a hard centre in α-bromoketones whereas the bromine is soft, a suitable combination of hard acid and soft nucleophile can be used for reductive debromination of these systems (**82**). This method does not work for dechlorination or defluorination but conversion to the dithioacetal enables ready dehalogenation of α-chloro- and α-fluoro-ketones (**83**) with ethanethiol and aluminium chloride.[200]

The rate of the acid-catalysed bromination of 2,4,6-trimethylacetophenone depends upon bromine concentration. The reaction of bromine with the sterically hindered enol is thought to be rate-limiting.[201]

The rate-limiting step in the chlorination of ketones by 1-chlorobenzotriazole is thought to be enolization of the ketone.[202]

The chlorination of acetophenone by chloramine-T is catalysed by cationic micelles.[203]

Aldehydes and ketones react with *tert*-butyl bromide/dimethyl sulphoxide to give α-brominated products. The brominating species is thought to be bromodimethylsulphonium bromide ($Me_2S^+Br\ Br^-$).[204]

The haloform reaction has been reviewed.[205]

Based on model studies, the B_{12}-dependent diol dehydratase conversion of 1,2-diols to aldehydes is thought not to involve Co participation in the rearrangement step.[206]

Hydrolysis and Other Reactions of Enol Ethers and Related Compounds

It has been previously reported[207] that the nine-membered ring vinyl ether (**84**) is 500 times more reactive towards acid-catalysed hydrolysis than its six-membered ring analogue and, unusually, proceeds by reversible carbon protonation. The nine-membered ring vinyl ether shows a normal isotope effect of $k_{H^+}/k_{D^+} = 1.94$ and is consistent with rate-limiting carbon protonation. A comparison of rate constants shows that the unusual reactivity of (**84**) is not due to the nine-membered ring.[208]

Anionic micelles enhance the rate of the acid-catalysed hydrolysis of vinyl ethers but the rate constant passes through a maximum with increasing surfactant concentration. The pseudo-phase ion-exchange model can be used to quantitatively account for the observations. The substrate–micelle binding constant increases with increasing hydrophobicity of the vinyl ether.[209]

β-Monomethyl and β,β-dimethyl substitution in ketene dithioacetals decreases their rate of acid-catalysed hydrolysis by factors of 26 and 7×10^5, respectively. This large reduction is similar to that observed for ketene acetals and explanations based on the stability of the carbocation intermediate (**85**) are not convincing.[210] Cyclic ketene thioacetals (**86**) undergo hydrolysis at similar rates to those for analogous acyclic derivatives. Unlike ketene acetals, protonation of the thio derivatives is reversible and breakdown of the hemi-thioacetal can be rate-limiting at low pH.[211]

The reaction of silyl enol ethers with aldehydes under high pressure gives an adduct (**87**) which may be hydrolysed to the corresponding aldol. The stereoselectivity at low pressure is in agreement with a chair transition state whereas the more compact boat arrangement is favoured at high pressure.[212]

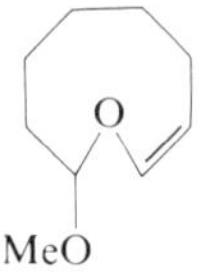

O

MeO

(**84**)

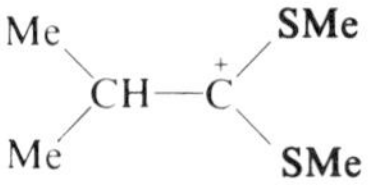

$Me_2CH{-}\overset{+}{C}(SMe)_2$

(**85**)

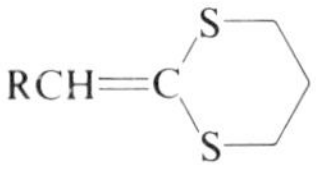

$RCH{=}C$ (S, S ring)

(**86**)

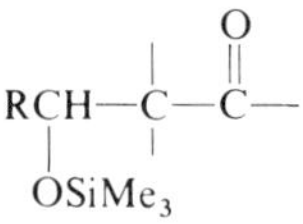

$RCH(OSiMe_3){-}C{-}C(=O){-}$

(**87**)

$CH_2{=}C(OPO_3^{2-})(CO_2^-)$

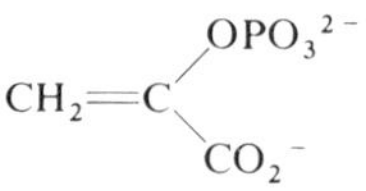

(**88**)

The rate of hydrolysis of pyruvic acid enol phosphate (88) in iodine solutions is first order in each substrate and increases with increasing pH and in the presence of silver(I) ions.[213]

The production of phosphoenolpyruvate from oxalacetate and guanosine 5′-triphosphate is catalysed by a carboxykinase but a stereochemical study indicates that the reaction does not involve a covalent phosphoryl–enzyme intermediate.[214]

Other Reactions

The oxidation of maltose and lactose by Tl(III) in acidic media is thought to proceed through rate-limiting formation of an acylium–thallium complex.[215]

The Hammett ρ value for the alcoholysis of *N,N′*-diarylsulphurdiimides is 0.62 in the presence of copper (II) chloride.[216]

The mechanism of the hydroformylation reaction catalysed by cobalt carbonyls has been reviewed.[217]

The reaction of trimesitylvanadium with ketones to give alkenes, alcohols, or radicals has been studied.[218]

References

1 Allen, F. H., and Kirby, A. J., *J. Am. Chem. Soc.*, **106,** 6197 (1984).
2 Briggs, A. J., Glenn, R., Jones, P. G., Kirby, A. J., and Ramaswamy, P., *J. Am. Chem. Soc.*, **106,** 6200 (1984).
3 Jones, P. G., and Kirby, A. J., *J. Am. Chem. Soc.*, **106,** 6207 (1984).
4 Jensen, J. L., and Yamaguchi, K. S., *J. Org. Chem.*, **49,** 2613 (1984).
5 Lamaty, G., Menut, C., and Nguyen, M., *Recl. Trav. Chim. Pays-Bas,* **103,** 75 (1984); see *Org. Reaction Mech.*, **1979,** 2; see *Org. Reaction Mech.*, **1982,** 1; Lamaty, G., and Menut, C., *Recl. Trav. Chim. Pays-Bas,* **103,** 54 (1984).
6 Jensen, J. L., Martinez, A. B., and Shimazu, C. L., *J. Org. Chem.*, **48,** 4175 (1983).
7 Kirby, A. J., *Acc. Chem. Res.*, **17,** 305 (1984).
8 Kresge, A. J., and Weeks, D. P., *J. Am. Chem. Soc.*, **106,** 7140 (1984).
9 Haak, J. R., and Engberts, J. B. F. N., *J. Org. Chem.*, **49,** 2387 (1984).
10 O'Leary, S., *Can. J. Chem.*, **62,** 1320 (1984).
11 Jensen, J. L., and Wuhrman, W. B., *J. Org. Chem.*, **48,** 4686 (1983).
12 Toullec, J., El-Alaoui, M., and Kleffert, P., *J. Org. Chem.*, **48,** 4808 (1983).
13 Shah, D. O., Lai, K., and Gorenstein, D. G., *J. Am. Chem. Soc.*, **106,** 4272 (1984).
14 McClelland, R. A., and Seaman, N. E., *Can. J. Chem.*, **62,** 1608 (1984).
15 Carlsen, L., Egsgaard, H., Jorgensen, F. S., and Nicolaisen, F. M., *J. Chem. Soc., Perkin Trans. 2,* **1984,** 609.
16 Salaun, J., *Chem. Rev.*, **83,** 619 (1983).
17 Meier, L., Runsink, J., and Scharf, H.-D., *Liebigs Ann. Chem.*, **1984,** 1298.
18 Karakhanov, R. A., Skurko, M. R., Ramazanov, O. M., Kantor, E. A., Bartok, M., and Bucsi, I., *Acta Phys. Chem.*, **29,** 181 (1983); *Chem. Abs.*, **101,** 22826 (1984).
19 Singh, J., Sardana, V., and Anand, N., *Indian J. Chem.*, **22B,** 1141 (1983); *Chem. Abs.*, **101,** 6490 (1984).
20 Weinhardt, K. K., *Tetrahedron Lett.*, **25,** 1761 (1984).
21 Penn, D., and Satchell, D. P. N., *J. Chem. Soc., Perkin Trans. 2,* **1984,** 933.
22 See *Org. Reaction Mech.*, **1980.**
23 Okuyama, T., *J. Am. Chem. Soc.*, **106,** 7134 (1984).
24 Nitz, T. J., and Paquette, L. A., *Tetrahedron Lett.*, **25,** 3047 (1984).
25 Karpinets, A. P., Bezuglyi, V. D., Pivnenko, N. S., and Lizenko, N. V., *Zh. Obshch. Khim.*, **54,** 1889 (1984); *Chem. Abs.*, **101,** 190808 (1984).
26 Borders, R. J., and Bryson, T. A., *Chem. Lett.*, **1984,** 9.
27 Rakhmankulov, D. L., Zorin, V. V., Pastushenko, E. V., and Zlotskii, S. S., *Heterocycles,* **22,** 817 (1984); *Chem. Abs.*, **101,** 22621 (1984).
28 Fuchs, B., Schleifer, L., and Tartakovsky, E., *Nouv. J. Chim.*, **8,** 275 (1984).
29 Hosie, L., Marshall, P. J., and Sinnott, M. L., *J. Chem. Soc., Perkin Trans. 2,* **1984,** 1121.
30 Lonnberg, H., and Heikkinen, E., *Acta Chem. Scand.*, **38B,** 673 (1984).
31 Lonnberg, H., Lukkari, J., and Lehikoinen, P., *Acta Chem. Scand.*, **38B,** 573 (1984).
32 Itaya, T., and Harada, T., *J. Chem. Soc., Chem. Commun.*, **1984,** 858.
33 Buchanan, S., Kubler, D. G., Meigs, C., Owens, M., and Tallman, A., *Int. J. Chem. Kinet.*, **15,** 1229 (1983).
34 Dessinges, A., Castillon, S., Okesker, A., Thang, T. T., and Lukacs, G., *J. Am. Chem. Soc.*, **106,** 450 (1984).
35 Banoub, J. H., Michon, F., Rice, J., and Rateb, L., *Carbohydr. Res.*, **123,** 109 (1983).
36 Vuorinen, T., *Carbohydr. Res.*, **127,** 327 (1984).
37 Harsch, G., Bauer, H., and Voelter, W., *Liebigs Ann. Chem.*, **1984,** 623.
38 Herrera, F. J. L., Fernandez, M. V., and Segura, R. G., *Carbohydr. Res.*, **127,** 217 (1984).
39 Lee, D.-S., and Perlin, A. S., *Carbohydr. Res.*, **126,** 101 (1984).
40 Birch, E. J., Lelievre, J., and Richards, E. L., *Carbohydr. Res.*, **128,** 351 (1984).
41 Sakakabara, T., Tachimori, Y., and Sudoh, R., *Carbohydr. Res.*, **131,** 197 (1984).
42 Doorslaer, E. Van, Opstal, O. Van, Kersters-Hilderson, H., and De Bruyne, C. K., *Biorg. Chem.*, **12,** 158 (1984).

[43] Pollard-Knight, D., and Cornish-Bowden, A., *Eur. J. Biochem.*, **141,** 157 (1984).
[44] Pilkis, S. J., Regen, D. M., Stewart, H. B., Pilkis, J., Pate, T. M., and El-Maghrabi, M. R., *J. Biol. Chem.*, **259,** 949 (1983).
[45] Coruslu, E., and Pekin, B., *Carbohydr. Res.*, **127,** 297 (1984).
[46] Nakakuki, T., Azuma, K., and Kainuma, K., *Carbohydr. Res.*, **128,** 297 (1984).
[47] Dhar, D. N., and Taploo, C. L., *J. Sci. Ind. Res.*, **41,** 501 (1982); *Chem. Abs.*, **98,** 33895 (1983).
[48] Singh, R. M. B., and Main, L., *Aust. J. Chem.*, **36,** 2327 (1983).
[49] Maupas, B., Fleury, M. B., and Mompon, B., *Analusis*, **12,** 72 (1984); *Chem. Abs.*, **101,** 38054 (1984).
[50] Saeed, A. A. H., Watton, M. H., and Sultan, A. W. A., *Thermochim. Acta*, **67,** 17 (1983); *Chem. Abs.*, **100,** 5459 (1984).
[51] Dash, A. C., Patra, M., Dash, B., and Mahapatra, P. K., *Indian J. Chem.*, **22A,** 944 (1983); *Chem. Abs.*, **100,** 208771 (1984).
[52] Gout, E., Zador, M., and Béguin, C. G., *Nouv. J. Chim.*, **8,** 243 (1984).
[53] Bunton, C. A., Davoudzadeh, F., Jagdale, M. H., and Watts, W. E., *J. Chem. Soc., Perkin Trans. 2*, **1984,** 395.
[54] Matsuda, H., Nagamatsu, H., Okuyama, T., and Fueno, T., *Bull. Chem. Soc. Jpn.*, **57,** 500 (1984).
[55] Burgess, J., and Hubbard, C. D., *J. Am. Chem. Soc.*, **106,** 1717 (1984).
[56] Stevens, R. V., *Acc. Chem. Res.*, **17,** 289 (1984).
[57] Bunting, J. W., and Fitzgerald, N. P., *Can. J. Chem.*, **62,** 1301 (1984).
[58] Zimmerman, S. C., and Breslow, R., *J. Am. Chem. Soc.*, **106,** 1490 (1984).
[59] Nagy, P., *Acta Chim. Hung.*, **112,** 461 (1983); *Chem. Abs.*, **99,** 194071 (1983).
[60] Forlani, L., Marianucci, E., and Todesco, P. E., *J. Chem. Res. Synop.*, **1984,** 126.
[61] Prasad, K. S., and Chowdary, M. C., *J. Indian Inst. Sci.*, **63,** 217 (1981); *Chem. Abs.*, **98,** 34154 (1983).
[62] Llor, J., Sanchez-Ruiz, J. M., Rodriguez-Pulido, J. M., and Cortijo, M., *An. Quim.*, **80A,** 27 (1984); *Chem. Abs.*, **101,** 109964 (1984).
[63] Shik, G. L., Khodzhaev, O. M., Chernikov, V. V., and Shakhtakhtinskii, T. N., *Neftekhimiya*, **23,** 536 (1983); *Chem. Abs.*, **99,** 175065 (1983).
[64] Andreichikov, Yu. S., Kozlov, A. P., and Kurdina, L. N., *Zh. Org. Khim.*, **19,** 1709 (1983); *Chem. Abs.*, **99,** 211880 (1983).
[65] Brandange, S., Lindblom. L., Pilotti, A., and Rodriguez, B., *Acta Chem. Scand.*, **37B,** 617 (1983).
[66] Brandange, S., and Rodriguez, B., *Acta Chem. Scand.*, **37B,** 643 (1983).
[67] Stec, W. J., *Acc. Chem. Res.*, **16,** 411 (1983).
[68] Alcaide, B., Lago, A., Perez-Ossorio, R., and Plumet, J., *An. Quim.*, **78C,** 220 (1982); *Chem. Abs.*, **97,** 161935 (1982).
[69] Lukton, D., and Rando, R. R., *J. Am. Chem. Soc.*, **106,** 258 (1984).
[70] Lukton, D., and Rando, R. R., *J. Am. Chem. Soc.*, **106,** 4525 (1984).
[71] Yamamoto, Y., Komatsu, T., and Maruyama, K., *J. Am. Chem. Soc.*, **106,** 5031 (1984).
[72] Capasso, R., Randazzo, G., and Pecci, L., *Can. J. Chem.*, **61,** 2657 (1983).
[73] Hickmott, P. W., *Tetrahedron*, **40,** 2989 (1984).
[74] Burger, K., Wassmuth, U., Hein, F., and Rottegger, S., *Liebigs Ann. Chem.*, **1984,** 991.
[75] Hine, J., and Sinha, A., *J. Org. Chem.*, **49,** 2186 (1984).
[76] Fraser, R. R., Bresse, M., Chuaqui-Offermanns, N., Houk, K. N., and Rondan, N. G., *Can. J. Chem.*, **61,** 2729 (1983).
[77] Harfoush, A. A., Zaghloul, A., and Abdel-Halim, F. M., *Indian J. Chem.*, **23A,** 102 (1984); *Chem. Abs.*, **101,** 71937 (1984).
[78] Temerk, Y. M., Kamal, M. M., and Ahmed, M. E., *J. Chem. Soc., Perkin Trans. 2*, **1984,** 337.
[79] Rowe, J. E., and Hegarty, A. F., *J. Org. Chem.*, **49,** 3083 (1984).
[80] Lobanor, P. S., Solod, O. V., and Potekhin, A. A., *Zh. Org. Khim.*, **19,** 2310 (1983); *Chem. Abs.*, **100,** 102531 (1984).
[81] Yakimovich, S. I., Nikolaev, V. N., and Kutsenko, E. Yu., *Zh. Org. Khim.*, **19,** 2333 (1983); *Chem. Abs.*, **100,** 102532 (1984).
[82] Pascal, R., Lasperas, M., Taillades, J., Commeyras, A., and Perez-Rubalcaba, A., *Bull. Soc. Chim. Fr. II*, **1984,** 329.
[83] Miyahara, M., *Eisei Shikensho Hokoku*, **1982,** 135; *Chem. Abs.*, **100,** 5716 (1984).
[84] Kessler, H., Oschkinat, H., Zimmermann, G., Möhrle, H., Biegholdt, M., Arz, W., and Förster, H., *Chem. Ber.*, **117,** 702 (1984).
[85] Mollin, J., and Holakovska, A., *Chem. Zvesti*, **37,** 633 (1983); *Chem. Abs.*, **100,** 102525 (1984).
[86] Ahumada, A. A., Manzo, R. H., and Martinez de Bertorello, M., *An. Asoc. Quim. Argent.*, **71,** 139 (1983); *Chem. Abs.*, **100,** 33898 (1984).

[87] Takahashi, T. T., Nomura,, K., and Satoh, J. Y., *J. Chem. Soc., Chem. Commun.*, **1984,** 1441.
[88] Collum, D. B., Kahne, D., Gut, S. A., DePue, R. T., Mohamadi, F., Wanat, R. A., Clardy, J., and Van Duyne, G., *J. Am. Chem. Soc.*, **106,** 4865 (1984).
[89] Baldwin, J. E., Adlington, R. M., Bottaro, J. C., Jain, A. U., Kolhe, J. N., Perry, M. W. D., and Newington, I. M., *J. Chem. Soc., Chem. Commun.*, **1984,** 1095.
[90] Guthrie, J. P., Cossar, J., Cullimore, P. A., Kamkar, N. M., and Taylor, K. F., *Can. J. Chem.*, **61,** 2621 (1983).
[91] Houminer, Y., Kao, J., and Seeman, J. I., *J. Chem. Soc., Chem. Commun.*, **1984,** 1608.
[92] Keay, B. A., Rajapaksa, D., and Rodrigo, R., *Can. J. Chem.*, **62,** 1093 (1984).
[93] Zsuga, M., Szabo, V., and Korodi, F., *Stud. Org. Chem. (Amsterdam)*, **11,** 25 (1981); *Chem. Abs.*, **97,** 162042 (1982).
[94] Mavhrauw, S. H., and Kellogg, R. M., *J. Am. Chem. Soc.*, **105,** 7792 (1983).
[95] Serpone, N., Ignacz, T., Sayer, B. G., and McGlinchey, M. J., *Inorg. Chim. Acta*, **89,** 139 (1984).
[96] Lankin, D. C., Griffin, G. W., Bernal, I., Korp, J., Watkins, S. F., DeLord, T. J., and Bhacca, N. S., *Tetrahedron*, **40,** 2829 (1984).
[97] Koudelka, L., *Chem. Zvesti*, **37,** 369 (1983); *Chem. Abs.*, **99,** 175153 (1983).
[98] Goasdone, C., Goasdone, N., and Gaudemar, M., *J. Organomet. Chem.*, **263,** 273 (1984).
[99] Akiyama, T., Shimizu, M., and Mukaiyama, T., *Chem. Lett.*, **1984,** 611.
[100] Heathcock, C. H., and Lampe, J., *J. Org. Chem.*, **48,** 4330 (1983).
[101] Gambaro, A., Boaretto, A., Marton, D., and Tagliavini, G., *J. Organomet. Chem.*, **260,** 255 (1984).
[102] Mukaiyama, T., and Iwasawa, N., *Chem. Lett.*, **1984,** 753.
[103] Kavrakova, I., Koleva, V., Simova, E., and Kurter, B., *Dokl. Bolg. Akad. Nauk*, **37,** 601 (1984); *Chem. Abs.*, **101,** 190777 (1984).
[104] Narasaka, K., Miwa, T., Hayashi, H., and Ohta, M., *Chem. Lett.*, **1984,** 1399.
[105] Mukaiyama, T., *Isr. J. Chem.*, **24,** 162 (1984).
[106] Gennari, C., Colombo, L., and Poli, G., *Tetrahedron Lett.*, **25,** 2279 (1984); Gennari, C., Cardani, S., Colombo, L., and Scolastico, C., *Tetrahedron Lett.*, **25,** 2279 (1984).
[107] Matsui, S., *Bull. Chem. Soc. Jpn.*, **57,** 426 (1984).
[108] Mukaiyama, T., *Org. React.*, **28,** 203 (1982). Heathcock, C. H., *Asymmetric Synth.*, **3,** 111 (1984); *Chem. Abs.*, **101,** 37789 (1984). Heathcock, C. H., *Stud. Org. Chem. (Amsterdam)*, **5B,** 177 (1984); *Chem. Abs.*, **101,** 54081 (1984). Evans, D. A., *Asymmetric Synth.*, **3,** 1 (1984); *Chem. Abs.*, **101,** 71793 (1984). Liebeskind, L. S., and Welker, M. E., *Tetrahedron Lett.*, **25,** 4341 (1984). Reetz, M. T., Kesseler, K., and Jung, A., *Tetrahedron Lett.*, **25,** 729 (1984). Kauffmann, T., Konig, R., and Wensing, M., *Tetrahedron Lett.*, **25,** 637 (1984). Kauffmann, T., Konig, R., Kriegesmann, R., and Wensing, M., *Tetrahedron Lett.*, **25,** 641 (1984). Posner, G. H., and Hulce, M., *Tetrahedron Lett.*, **25,** 379 (1984). Posner, G. H., Kogan, T. P., and Hulce, M., *Tetrahedron Lett.*, **25,** 379 (1984). Labadie, S. S., and Stille, J. K., *Tetrahedron*, **40,** 2309 (1984). Meyers, A. I., and Yamamoto, Y., *Tetrahedron*, **40,** 2309 (1984). Yamamoto, Y., Yatagai, H., Ishihara, Y., Maeda, N., and Maruyama, K., *Tetrahedron*, **40,** 2239 (1984). Hofmann, R. W., and Kemper, B., *Tetrahedron*, **40,** 2219 (1984). Agami, C., and Sevestre, H., *J. Chem. Soc., Chem. Commun.*, **1984,** 1385. Annunziata, R., Cinquini, M., Cozzi, F., and Restelli, A., *J. Chem. Soc., Chem. Commun.*, **1984,** 1253. Sato, F., Takeda, Y., Uchiyama, H., and Kobayashi, Y., *J. Chem. Soc., Chem. Commun.*, **1984,** 1132. Sato, F., Kusakabe, M., and Kobayashi, Y., *J. Chem. Soc., Chem. Commun.*, **1984,** 1130. Reetz, M. T., and Kesseler, K., *J. Chem. Soc., Chem. Commun.*, **1984,** 1079. Jephcote, V. J., Pratt, A. J., and Thomas, E. J., *J. Chem. Soc., Chem. Commun.*, **1984,** 800. Kienzle, F. von, Fellman, J.-Y., and Stadlwieser, J., *Helv. Chim. Acta*, **67,** 789 (1984). Zhao, C., and Wen, Z., *Huaxue Tongbao*, **1984,** 8; *Chem. Abs.*, **101,** 54077 (1984). Banfi, L., *Quad. Tec. Sint. Spec. Org.*, **1983,** 1; *Chem. Abs.*, **101,** 22624 (1984). Prokopenkova, L. V., Garshina, S. I., and Perelygin, V. M., *Izv. Vyssh, Uchebn. Zaved., Khim. Khim. Teknol.*, **27,** 405 (1984); *Chem. Abs.*, **101,** 72215 (1984). Suzuki, H., Takahashi, M., and Morooka, Y., *Kenkyu Hokoku-Asahi Garasu Kogyo Gijutsu Shoreikai*, **41,** 169 (1983); *Chem. Abs.*, **99,** 194143 (1983).
[109] Kharchenko, V. G. (Ed.), *Nucleophilic Reactions of Carbonyl Compounds* (1982); *Chem. Abs.*, **101,** 6559 (1984). Solladie, G., *Asymmetric Synth.*, **2,** 157 (1983); *Chem. Abs.*, **101,** 37796 (1984). Tomioka, K., and Koga, K., *Asymmetric Synth.* **2,** 201 (1983); *Chem. Abs.*, **101,** 37797 (1984).
[110] Rendina, A. R., and Cleland, W. W., *Biochemistry*, **23,** 5157 (1984).
[111] Rendina, A. R., Hermes, J. D., and Clelland, W. W., *Biochemistry*, **23,** 5148 (1984).
[112] El Seoud, O. A., Vieira, R. C., and Farah, J. P. S., *Solution Behav. Surfactants: Theor. Appl. Aspects*, **2,** 867 (1982); *Chem. Abs.*, **98,** 33988 (1983).
[113] Ritchie, C. D., *J. Am. Chem. Soc.*, **106,** 7187 (1984).
[114] Stewart, R., *Can. J. Chem.*, **62,** 907 (1984).

[115] Nguyen, M. T., and Hegarty, A. F., *J. Am. Chem. Soc.*, **106,** 1552 (1984).
[116] Spangler, D., Williams, I. H., and Maggiora, G. M., *J. Comput. Chem.*, **4,** 524 (1983); *Chem. Abs.*, **100,** 138247 (1984).
[117] Tachibana, A., and Yamabe, T., *Kagaku (Kyoto)*, **39,** 59 (1984); *Chem. Abs.*, **100,** 208655 (1984).
[118] Smith, J. G., and Dibble, P. W., *Tetrahedron*, **40,** 1667 (1984).
[119] Davis, A. M., Page, M. I., Mason, S. C., and Watt, I., *J. Chem. Soc., Chem. Commun.*, 1671 (1984); Cernik, R. V., Craze, G.-A., Mills, O. S., Watt, I., and Whittleton, S. N., *J. Chem. Soc., Perkin Trans. 2*, **1984,** 685.
[120] Fujita, M., and Hiyama, T., *J. Am. Chem. Soc.*, **106,** 4629 (1984).
[121] Kawasaki, M., Suzuki, Y., and Tereshima, S., *Chem. Lett.*, **1984,** 239.
[122] Takano, S., Ariyama, M., Sato, S., and Ogasawara, K., *Chem. Lett.*, **1983,** 1593.
[123] Riad, Y., Asaad, A. N., Gohar, G. A. S., and Abdallah, A. A., *Z. Naturforsch.*, **39a,** 893 (1984); *Chem. Abs.*, **101,** 190917 (1984).
[124] Tabushi, I., Kuroda, Y., and Mizutani, T., *J. Am. Chem. Soc.*, **106,** 3377 (1984).
[125] Baba, N., Amano, M., Oda, J., and Inouye, Y., *J. Am. Chem. Soc.*, **106,** 1481 (1984).
[126] Oppenheimer, N. J., *J. Am. Chem. Soc.*, **106,** 3032 (1984).
[127] Burgner, J. W., and Ray, W. J., *Biochemistry*, **23,** 3626, 3636 (1984).
[128] Baliah, V., and Sundari, V., *Indian J. Chem.*, **22B,** 804 (1983); *Chem. Abs.*, **100,** 102660 (1984).
[129] Harusawa, S., Yoneda, R., Kurihara, T., Hamada, Y., and Shiori, T., *Tetrahedron Lett.*, **25,** 427 (1984).
[130] Jaenicke, L., and Preun, J., *Eur. J. Biochem.*, **138,** 319 (1984).
[131] Katritzky, A. R., De Rosa, M., and Grzeskowiak, N. E., *J. Chem. Soc., Perkin Trans. 2*, **1984,** 841.
[132] Katritzky, A. R., Mokrosz, J. L., and De Rosa, M., *J. Chem. Soc., Perkin Trans. 2*, **1984,** 849.
[133] Katritzky, A. R., Yang, Y.-K., Gabrielsen, B., and Marquet, J., *J. Chem. Soc., Perkin Trans. 2*, **1984,** 857.
[134] Katritzky, A. R., and Leahy, D. E., *J. Chem. Soc., Perkin Trans. 2*, **1984,** 867.
[135] Katritzky, A. R., Mokrosz, J. L., and Lopez-Rodriguez, M. L., *J. Chem. Soc., Perkin Trans. 2*, **1984,** 875.
[136] Doddi, G., Illuminati, G., Mecozzi, M., and Nunziante, P., *J. Org. Chem.*, **48,** 5268 (1983).
[137] Kurts, A. L., Davydov, D. V., and Bundel, Yu. G., *Vestn. Mosk. Univ., Khim.*, **24,** 385 (1983); *Chem. Abs.*, **99,** 174995 (1983).
[138] Katritzky, A. R., and Marson, C. M., *Angew. Chem. Int. Ed.*, **23,** 420 (1984).
[139] Eliel, E. L., *Asymmetric Synth.*, **2,** 125 (1983); *Chem. Abs.*, **101,** 109812 (1984). Mulzer, J., *Nachr. Chem., Tech. Lab.*, **32,** 16 (1984); *Chem. Abs.*, **100,** 120128 (1984).
[140] Bouchoux, G., and Houriet, R., *Tetrahedron Lett.*, **25,** 5755 (1984).
[141] Laurence, C., Berthelot, M., Lucon, M., Helbert, M., Morris, D. G., and Gal, J.-F., *J. Chem. Soc., Perkin Trans. 2*, **1984,** 705.
[142] Liotta, C. L., Burgess, E. M., and Eberhardt, W. H., *J. Am. Chem. Soc.*, **106,** 4849 (1984).
[143] Sheldon, J. C., Bowie, J. H., and Hayes, R. N., *Nouv. J. Chim.*, **8,** 79 (1984).
[144] Dedieu, A., and Nakumura, S., *Nouv. J. Chim.*, **8,** 317 (1984).
[145] Matsumoto, T., Yamamoto, H., and Inoue, S., *J. Am. Chem. Soc.*, **106,** 4829 (1984).
[146] Kuo, D. U., and Jordan, F., *J. Biol. Chem.*, **258,** 13415 (1983).
[147] Volatron, F., and Eisenstein, O., *J. Am. Chem. Soc.*, **106,** 6117 (1984).
[148] Piskala, A., Rehan, A. H., and Schlosser, M., *Collect. Czech. Chem. Commun.*, **49,** 3539 (1984).
[149] Reitz, A. B., and Maryanoff, B. E., *J. Chem. Soc., Chem. Commun.*, **1984,** 1548.
[150] Maryanoff, B. E., Duhl-Emswiler, B. A., and Reitz, A. B., *Phosphorus Sulfur*, **18,** 187 (1983); *Chem. Abs.*, **100,** 191021 (1984).
[151] Crooks, P. A., and Sommerville, R., *Chem. Ind. (London)*, **1984,** 350.
[152] Arjona, O., Gonzalez Raurich, M., Perez-Ossorio, R., Perez-Rubalcaba, A., and Quiroga, M. L., *An. Quim.*, **79C,** 184 (1983); *Chem. Abs.*, **101,** 190750 (1984). Reetz, M. T., *Angew. Chem. Int. Ed.*, **23,** 556 (1984).
[153] Lins, H. S., Nome, F., Rezende, M. C., and de Sousa, I., *J. Chem. Soc., Perkin Trans. 2*, **1984,** 1521.
[154] Duthaler, R. O., and Maienfisch, P., *Helv. Chim. Acta.*, **67,** 832 (1984).
[155] Duthaler, R. O., and Maienfisch, P., *Helv. Chim. Acta.*, **67,** 845 (1984).
[156] Bassindale, A. R., Ellis, R. J., and Taylor, P. G., *Tetrahedron Lett.*, **25,** 2705 (1984).
[157] Midland, M. M., and Graham, R. S., *J. Am. Chem. Soc.*, **106,** 4294 (1984).
[158] Laboureur, J. L., and Krief, A., *Tetrahedron Lett.*, **25,** 2713 (1984).
[159] Ando, R., Sugawara, T., and Kuwajima, I., *J. Chem. Soc., Chem. Commun.*, **1983,** 1514.
[160] Tanikaga, R., Tamura, T., Nozaki, Y., and Kaji, A., *J. Chem. Soc., Chem. Commun.*, **1984,** 87.
[161] Nayar, M. R. G., and Francis, J. D., *Indian J. Chem.*, **22B,** 776 (1983); *Chem. Abs.*, **100,** 102455 (1984).

[162] Hamana, H., and Sugasawa, T., *Chem. Lett.*, **1984,** 1591.
[163] McDonald, R. N., and Chowdhury, A. K., *J. Am. Chem. Soc.*, **105,** 7267 (1983).
[164] Atkins, P. J., Gold, V., and Wassef, W. N., *J. Chem. Soc., Perkin Trans. 2,* **1984,** 1247.
[165] Spangler, L. A., and Swenton, J. S., *J. Org. Chem.*, **49,** 1800 (1984).
[166] Al-Aseer, M. A., and Smith, S. G., *J. Org. Chem.*, **49,** 2608 (1984).
[167] Krasnov. V. L., Ozherel'eva, N. K., Trub, E. P., Tsvetkov, V. G., and Bodrikov, I. V., *Zh. Obshch. Khim.*, **53,** 2367 (1983); *Chem. Abs.*, **100,** 33909 (1984).
[168] Bohme, H., and Braun, G., *Arch. Pharm. (Weinheim, Ger.),* **317,** 408 (1984).
[169] Mioskowski, C., Manna, S., and Falck, J. R., *Tetrahedron Lett.*, **25,** 519 (1984).
[170] Knapp, S., Trope, A. F., Theodore, M. S., Hirata, N., and Barchi, J. J., *J. Org. Chem.*, **49,** 608 (1984).
[171] Nair, V., Turner, G. A., and Offerman, R. J., *J. Am. Chem. Soc.*, **106,** 3370 (1984).
[172] Esikova, I. A., Kuz'michev, A. I., and Yufit, S. S., *Kinet. Katal.*, **25,** 50 (1984); *Chem. Abs.*, **101,** 6288 (1984).
[173] Awad, W. I., Taha, A. A., Wasef, W. N., and Shindy, S. M., *Egypt. J. Chem.*, **26,** 65 (1983); *Chem. Abs.*, **101,** 22745 (1984).
[174] Anjaneyulu, A. S. R., Raghu, P., Rao, K. V. R., Sastry, C. V. M., Umasundari, P., and Satyanarayana, P., *Curr. Sci.*, **53,** 239 (1984); *Chem. Abs.*, **101,** 71878 (1984).
[175] Wróblewski, A. E., and Konieczko, W. T., *Monatsh. Chem.*, **115,** 785 (1984).
[176] Voronin, S. P., Gubina, T. I., Markushina, I. A., and Kharchenko, V. G., *Nukleofil'nye Reakts. Karbonil'nykh Soedin,* **1982,** 7; *Chem. Abs.*, **101,** 71968 (1984). Martem'yanova, N. I., and Karasev, E. A., *Nukleofil'nye Reakts Karbonil'nykh Soedin,* **1982,** 80; *Chem. Abs.*, **101,** 71970 (1984).
[177] Stevens, R. V., and Hrib, N., *J. Chem. Soc., Chem. Commun.*, **1984,** 1422; Stevens, R. V., and Pruitt, J. R., *J. Chem. Soc., Chem. Commun.*, **1984,** 1425.
[178] Salomon, M. F., Pardo, S. N., and Salomon, R. G., *J. Org. Chem.*, **49,** 2446 (1984).
[179] Cartaya-Marin, C. P., Jackson, A. C., and Snider, B. B., *J. Org. Chem.*, **49,** 2443 (1984).
[180] Jurczak, J., Kozluk, T., Pikul, S., and Salanski, P., *J. Chem. Soc., Chem. Commun.*, **1984,** 1447.
[181] Kerr, J. A., and Wright, J. P., *Int. J. Chem. Kinet.*, **16,** 1321, 1327 (1984).
[182] Ingold, K. U., Lusztyk, J., and Scaiano, J. C., *J. Am. Chem. Soc.*, **106,** 343 (1984).
[183] Rodler, M., and Bauder, A., *J. Am. Chem. Soc.*, **106,** 4025 (1984); Rodler, M., Blom, C. E., and Bauder, A., *J. Am. Chem. Soc.*, **106,** 4029 (1984).
[184] Biali, S. E., and Rappoport, Z., *J. Am. Chem. Soc.*, **106,** 5641 (1984).
[185] Siddhanta, A. K., and Capon, B., *J. Org. Chem.*, **49,** 80 (1984).
[186] Simonen, T., and Kivekas, R., *Acta Chem. Scand.*, **38B,** 679 (1984).
[187] Biali, S. E., and Rappoport, Z., *J. Am. Chem. Soc.*, **106,** 477 (1984).
[188] Andreeva, I. M., Babeshko, O. M., Bondarenko, E. M., Medyantseva, E. A., and Minkin, V. I., *Zh. Org. Khim.*, **19,** 2146 (1983); *Chem. Abs.*, **100,** 138352 (1984).
[189] Chiang, Y., Kresge, A. J., Tang, Y. S., and Wirz, J., *J. Am. Chem. Soc.*, **106,** 460 (1984).
[190] Chiang, Y., Kresge, A. J., and Wirz, J., *J. Am. Chem. Soc.*, **106,** 6392 (1984).
[191] Toullec, J., *Tetrahedron Lett.*, **25,** 4401 (1984).
[192] Hupe, D. J., and Pohl. E. R., *J. Am. Chem. Soc.*, **106,** 5634 (1984).
[193] Bruice, P. Y., *J. Am. Chem. Soc.*, **106,** 5959 (1984); see *Org. Reaction Mech.*, **1983,** 24.
[194] Chuaqui-Offermanns, N., and Chuaqui, C. A., *Chem. Phys. Lipids,* **33,** 215 (1983); *Chem. Abs.*, **100,** 138524 (1984).
[195] Cox, B. G., De Maria, P., and Fini, A., *J. Chem. Soc., Perkin Trans. 2,* **1984,** 1647.
[196] Kluger, R., Wong, M. K., and Dodds, A. K., *J. Am. Chem. Soc.*, **106,** 1113 (1984).
[197] Leussing, D. L., and Emly, M., *J. Am. Chem. Soc.*, **106,** 443 (1984).
[198] Guthrie, J. P., Cossar, J., and Klym, A., *J. Am. Chem. Soc.*, **106,** 1351 (1984).
[199] De Kimpe, N., and Verhe, R., *The Chemistry of Halides, Pseudo-Halides and Azides* (Eds. S. Patai and Z. Rappoport), vol. 1, Wiley, Chichester, 1983, pp. 549, 813.
[200] Fuji, K., Node, M., Kawabata, T., and Fujimoto, M., *Chem. Lett.*, **1984,** 1153.
[201] Pinkus, A. G., and Gopalan, R., *J. Am. Chem. Soc.*, **106,** 2630 (1984).
[202] Jayaraman, P. S., Sundaram, S., and Venkatasubramanian, N., *Proc. Indian Acad. Sci., Chem. Sci.*, **91,** 421 (1982); *Chem. Abs.*, **98,** 88529 (1983).
[203] Raghavan, P. S., and Srinivasan, V. S., *Proc. Indian Acad. Sci., Chem. Sci.*, **93,** 849 (1984); *Chem. Abs.*, **101,** 151147 (1984).
[204] Armani, E., Dossena, A., Marchelli, R., and Casnati, G., *Tetrahedron,* **40,** 2035 (1984).
[205] Brossi, M., and Kaenel, H. R., *SLZ, Schweiz. Lab.-Z,* **40,** 277, 281 (1983); *Chem. Abs.*, **99,** 193957 (1983).
[206] Finke, R. G., and Schiraldi, D. A., *J. Am. Chem. Soc.*, **105,** 7605 (1983).
[207] See *Org. Reaction Mech.*, **1980.**
[208] Burt, R. A., Chiang, Y., Kresge, A. J., and Szilagyi, S., *Can. J. Chem.*, **62,** 74 (1984).

[209] Velázquez, M. M., Garcia-Mateos, I., Herraez, M. A., and Rodriguez, L. J., *Int. J. Chem. Kinet.*, **16,** 269 (1984).
[210] Okuyama, T., Kawao, S., and Fueno, T., *J. Org. Chem.*, **49,** 85 (1984).
[211] Okuyama, T., Kawao, S., Fujiwara, W., and Fueno, T., *J. Org. Chem.*, **49,** 89 (1984).
[212] Yamamoto, Y., Maruyama, K., and Matsumoto, K., *J. Am. Chem. Soc.*, **105,** 6963 (1983).
[213] Kanivets, N. P., Volkova, N. V., and Yasnikov, A. A., *Dopov. Akad. Nauk Ukr. RSR, Ser. B: Geol. Khim. Biol. Nauki,* **1983,** 30; *Chem. Abs.*, **99,** 211863 (1983).
[214] Sheu, K.-F., Ho, H.-T., Nolan, L. D., Markovitz, P., Richard, J. P., Utter, M. F., and Frey, P. A., *Biochemistry,* **23,** 1779 (1984).
[215] Shukla, S. N., and Kesarwani, R. N., *Carbohydrate Res.*, **133,** 319 (1984).
[216] Carpanelli, C., and Gaiani, G., *Gazz. Ital.* **113,** 503 (1983); *Chem. Abs.*, **100,** 138325 (1984).
[217] Marko, L., *Fundam. Res. Homogeneous Catal.*, **4,** 1 (1984); *Chem. Abs.*, **100,** 138221 (1984).
[218] Kreisel, G., and Seidel, W., *J. Organomet. Chem.*, **260,** 301 (1984).

Organic Reaction Mechanisms 1984
Edited by A. C. Knipe and W. E. Watts

CHAPTER 2

Reactions of Acids and their Derivatives

W. J. Spillane

Chemistry Department, University College, Galway, Ireland

Carboxylic Acids

Tetrahedral Intermediates

The direct observation and study of hemiorthoester intermediates of type (**1**) has been possible by the use of low-temperature NMR by the Capon group and UV spectroscopy by the McClelland group. A few years ago the former group reviewed their work[1] and now the latter in a timely and well-written account have reviewed their kinetic work.[2] This present account[2] brings together data on the formation and decomposition of intermediates (**1**) from orthoesters of types (**2**) and (**3**). Equilibrium constants for the formation of tetrahedral intermediates of types (**1**), (**4**), and (**5**) are also given.

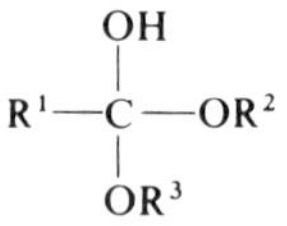

(**1**) 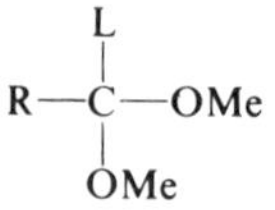 (**2**) 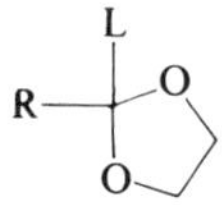 (**3**)

R = H, Me, Ph; L = OMe, OAc, *etc.*

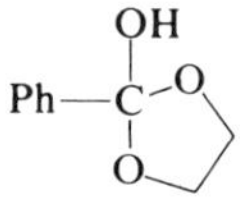

(**4**) 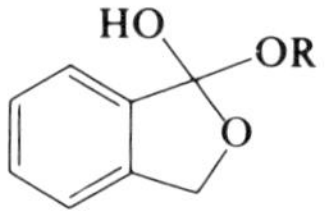 (**5**)

R = H, Et

In new work the first example where equilibration between an orthoester or acetal precursor and the oxocarbocation involved in the first stage of orthoester or acetal hydrolysis[3] has been studied is reported.[4] With most systems the probability that the oxocarbocation once formed will undergo the reverse reaction back to starting material is low. However, the 3-aryl-2,4,10-trioxaadamantane system (**6**) appears to be ideally set up for a reversible ring-opening since the departing hydroxy group is forced to remain close to the charged centre in the carbocation (**7**). Equilibrium constants, $pK_{(7)}$ for the system (**7**)$\rightleftharpoons$(**6**), have been measured in 3–48% H_2SO_4 using an appropriate (H_T) acidity function and the overlapping method.

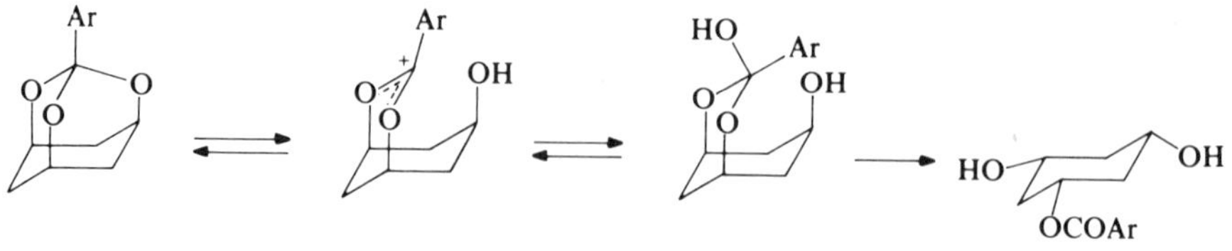

(**6**) (**7**) (**8**) (**9**)

In an accompanying paper[5] the same workers have accomplished a detailed kinetic analysis of the hydrolysis of compounds (**6**). The pH–rate profiles obtained have been interpreted in terms of a change in rate-determining step between high and low pH. At high pH hydration of the cation (**7**) is the slow step but at low pH the decomposition of the hemiorthoester (**8**) to product (**9**) is rate-limiting.

Detailed kinetic studies of the breakdown of hemiorthoesters (**10**), (**11**), and (**12**) have been reported.[6,7] These are the tetrahedral intermediates involved in the hydrolysis of the dialkoxyalkyl acetates (**13**) and (**14**) and the ketene acetals (**15**).

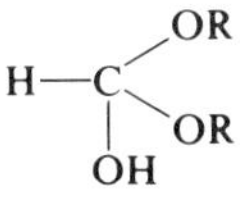

(**10**) R = Me, Et

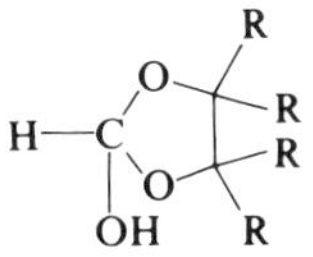

(**11**) R = H, Me

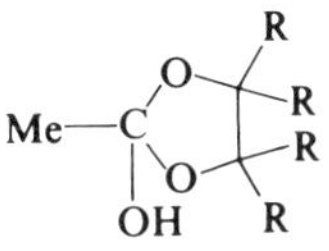

(**12**) R = H, Me

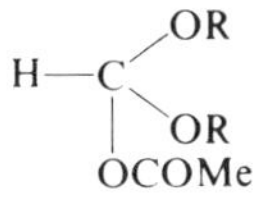

(**13**) R = Me, Et

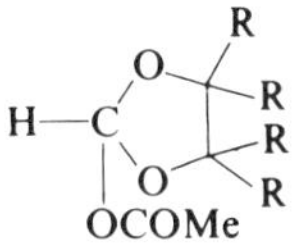

(**14**) R = H, Me

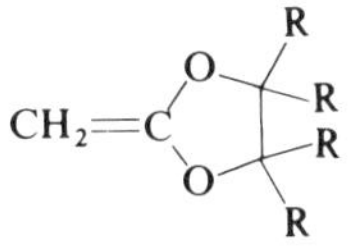

(**15**) R = H, Me

Mechanistic changes accompanying pH changes, as observed with (**6**), have also been found[8] in the hydrolysis of the cyclic orthoester 2-methyl-2-methoxy-1,3-dioxolane (**16**), but not in the hydrolysis of the acyclic orthoesters, trimethyl orthoacetate (**17**) and tris(2-methoxyethyl) orthoacetate (**18**). The same group[9] has measured heats of

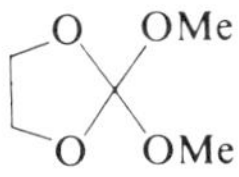

(**16**)

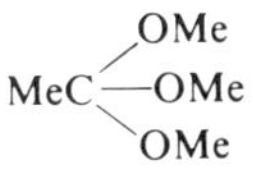

(**17**)

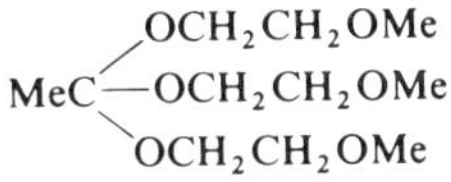

(**18**)

ionization and solution of the dioxolanes (**19**) in 97% H_2SO_4. Under these conditions compounds (**19**) form the 1,3-dioxolenium ions (**20**) which would in turn be the precursors of the appropriate hemiorthoesters.

The rate-controlling step in the poly(ethylenimines)-catalysed hydrolysis of 4-nitro-3-carboxytrifluoroacetanilide (**21**) is the formation of a tetrahedral intermediate.[10] Both formation and breakdown of the tetrahedral intermediates in the basic hydrolysis of the methyl, ethyl, and phenyl *N*-aryl-*N*-methylcarbamates (**22**) are proposed as slow steps (depending on the particular carbamate).[11] The acid hydrolysis of ketene dithioacetals of type $R^1R^2CHC(SR^3)_3$ involving protonated tetrahedral intermediates of type (**23**) has been studied.[12] The rates of hydrolysis of a series of 1,1-diethoxy-3,4-dihydrobenzo-2-pyrans (**24**) and of a related series of 2,2-diethoxy-3,4-dihydrobenzo-1-pyrans have been determined in water:dioxan (2:1 v/v).[13,14] The hemiorthoester produced from the 2,2-diethoxy series is of type (**25**).

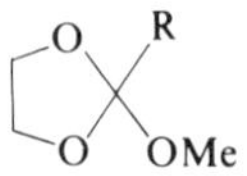

(19)

R = H, Me, Ph

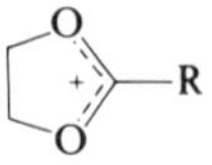

(20)

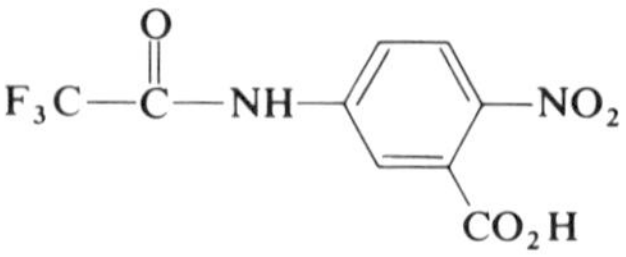

$F_3C-C(=O)-NH-C_6H_3(CO_2H)-NO_2$

(21)

$ArN(Me)CO_2R$

(22)

R = Me, Et, Ph

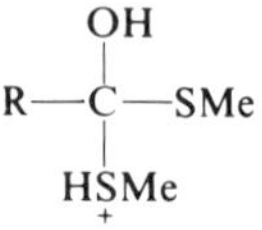

(23)

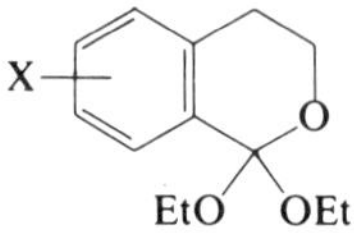

(24)

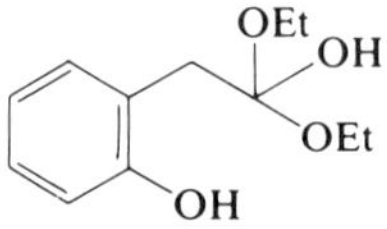

(25)

Anionic tetrahedral intermediates have been studied in a variety of systems. Several requirements which need to be first fulfilled, if tetrahedral intermediates are to be observed, have been given in a recent paper.[15] A hemiorthoester anion (**26**) in equilibrium with its tetrahedral intermediate (**27**) and generated by cyclization in base of pinacol mono-*p*-nitrobenzoate (**28**) has been observed from UV and rate studies.[15]

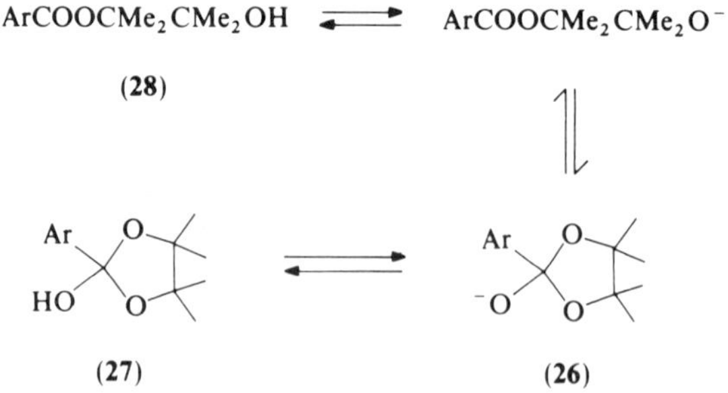

$$ArCOOCMe_2CMe_2OH \rightleftharpoons ArCOOCMe_2CMe_2O^-$$

(28)

(27) (26)

Ar = p-$O_2NC_6H_4$

Base hydrolysis of *N*-acetylpyrrole involves the anionic tetrahedral intermediate (**29**).[16] The free energies of activation for the formation of (**29**) and its breakdown (to substrate and to product) have been determined. The hydrolysis of 2-methoxy- and 2-ethoxy-2-phenyltetrahydrofuran (**30**) does not involve an anionic tetrahedral intermediate but again changes in the rate-determining step are recorded with

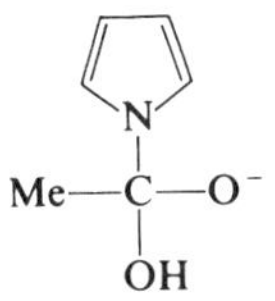

Me—C(N-pyrrolyl)(OH)—O⁻

(29)

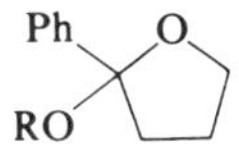

Ph, RO-substituted tetrahydrofuran

(30)

R = Me, Et

R—C(O⁻)(HO)—OCH$_3$

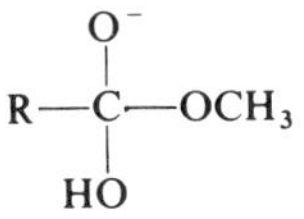

(31)

R = F$_3$C, Me

changes in pH.[17] The life-times of anionic tetrahedral intermediates (**31**) in the hydrolysis of methyl trifluoroacetate and methyl acetate catalysed by a zinc hydroxide complex (see p. 60) are related to the catalytic efficiency of the catalyst.[18] Carbonyl addition adducts for the reaction of F_3C^- with ketones and esters in the gas phase have been detected.[19]

Intermolecular Catalysis and Reactions

Reactions in Hydroxylic Solvents

(*a*) *General.* A review on the mechanisms of polar reactions including substitution, elimination, ester hydrolysis, and micellar reactions has appeared.[20] Using combined crystallographic and kinetic data for eleven different systems Kirby's group have used increasing electron demand at oxygen as a probe to test the relationship between bond-length and reactivity.[21] In nearly all cases the lengths of the C—O or P—O bonds in the systems studied increase significantly with increasing electron demand and there is a linear correlation between bond lengths and the free energies of activation for cleavage—the longer the bond the lower the energy required to cleave it. These relationships are seen as genuine structure–reactivity relationships and undoubtedly as further crystallographic and kinetic data become available there will most likely be many additional useful correlations of the type reported.

The problem of dealing with multiple substituents in systems such as X, Y—ArZ or X—Ar—Z—Ar—Y has been approached using an interactive free energy relationship (IFER)[22] of type:

$$\log k_{xy}/k_{HH} = \rho_1^H \sigma_x + \rho_2^H \sigma_y + q\sigma_x\sigma_y$$

This equation has been tested for different types of substituent effects on various types of reaction and equilibria involving 35 data sets of multiple substituent effects. The interaction constant, q, varies from $+1$ to -8 and reflects in part a change in the transition state induced by the substituent. High q values are found for systems having high ρ values and the converse is also true. The method presented is a significant improvement in dealing with multiple substituent effects in rates and equilibria.

The two-point method for the evaluation of the Brønsted α has been shown to be of generally limited value and a re-examination of the effect of solvent on the α-effect has been made.[23] Using a combination of dynamic ^{1}H- and ^{2}H-NMR spectroscopy an "NMR proton inventory" for the proton exchange and H-bonding reaction between acetic acid and methanol in THF-d_8 has been carried out.[24]

(*b*) *Esters*. A large number of papers have appeared on the alkaline hydrolysis of aryl esters. Basic hydrolysis of phthalic (**32**), isophthalic, and terephthalic acid methyl esters has been studied conductiometrically.[25] The activation parameters support a $B_{AC}2$ mechanism. The basicity of the leaving phenolate ion decides the mechanism (*E*1*cB* or $B_{AC}2$) in the hydrolysis of aryl 4-hydroxybenzoates (**33**).[26] The rate of alkaline hydrolysis of methyl salicylate in DMSO decreases with the DMSO concentration as the aqueous–DMSO mixture increases. Activation data indicates that the transition state is desolvated relative to the reactants.[27] Alkaline hydrolysis of ethyl 7-nitro-7-bromo-9-oxyfluorene-1-carboxylates and the ethyl fluorene-1-carboxylates have been studied in aqueous dioxane.[28] A rate enhancement due to the oxo group has been observed. A structure–reactivity study of the basic hydrolysis of halo derivatives of aryl alkyl esters has been made.[29] Saponification rates of methyl biphenyl-4-carboxylates in 85% methanol–water (w/w) have been measured.[30] Activation parameters and Hammett ρ values have been obtained. Hydrogen exchange occurs with methoxide ion in methanol and tritium-labelled methyl benzoates, $XC_6H_4CO_2CH_2T$, dimethyl carbonate, and menthyl methyl carbonate.[31] Measurements of $\Delta H^{\ne}$ and $\Delta S^{\ne}$ for the exchange indicates that the transition states are dominated by bond formation and fission characterized by $\Delta H^{\ne} = 10\text{–}20\ \text{kcal mol}^{-1}$ and $\Delta S^{\ne} = -15$ to -30 e.u.

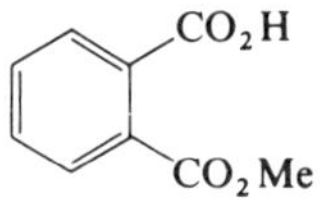

CO_2H / CO_2Me (**32**)

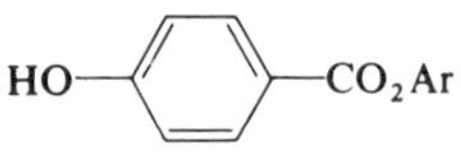

HO—C_6H_4—CO_2Ar (**33**)

$RCCl_2CO_2Ar$ (**34**)

R = H, Ar = *p*-$MeOC_6H_4$
R = Me, Ar = *p*-$MeOC_6H_4$

An S_N1 mechanism is proposed for the solvolysis of *tert*-butyl hydrogen phthalate in various aqueous organic solvent mixtures.[32] This conclusion is based on a Grunwald–Winstein *m* of *ca.* 0.8 and a solvent number of *ca.* 5.3. Rates of hydrolysis and methanolysis of *N,N'*-carbonylbis(imidazole), -(benzimidazole), -(1,2,4-triazole), and -(benzotriazole) have been reported together with similar rates for their thiocarbonyl analogues.[33] The CO compounds are more reactive than the CS compounds and the 1,2,4-triazoles were the most reactive in each series. Further work[34] has been reported on the water-catalysed hydrolysis of *p*-methoxyphenyl dichloracetate (**34**) and of 2,2-dichloropropionate (**34**)[35] in water-rich 2-*n*-butoxyethanol. The large positive heat capacities of activation obtained contrast sharply with the negative $\Delta C_p^{\ne}$ values found for S_N solvolysis reactions in highly aqueous mixed solvents. The heat-capacity data has been rationalized in terms of a temperature-dependent partitioning of the acyl-activated esters between a water-rich and a cosolvent-rich micro-phase. The effects of pressure (1 → 900 bars) on the hydrolysis of 2,4-dinitro 4-hydroxybenzoate in acetone–water give a positive volume of activation ($+12\ cm^3\ mol^{-1}$) and this clearly indicates that its hydrolytic decomposition does not involve a $B_{AC}2$ path (as with most esters) but rather a dissociation one involving the oxoketene (**35**).[36]

Several papers have appeared on the nucleophilic hydrolysis of aryl esters by heterocycles and polymeric materials. 2-Methyl-, 2-amino-, 2,4-dimethyl-, 2-amino-4-methyl-, 4-methoxy-2-methyl-, and 4-amino-2-methyl-pyridine have proved effective catalysts for the hydrolysis of *p*-nitrophenyl and 2,4-dinitrophenyl acetates.[37] Strong nucleophilic attack by HO^-, H_2O, benzotriazole, and deprotonated benzotriazole is indicated in hydrolysis of compounds (**36**) and of *m*-$NO_2C_6H_4OAc$.[38] Poly(4(5)-vinylimidazole) was a better catalyst than imidazole for the hydrolysis of the series of esters (**37**).[39] The catalytic effects of poly(acrylic acid-*N*-acryloylhistamine) or a terpolymer of acrylic acid, *N*-acryloylhistamine, and *N*-acryloylhexylamine and the hydrolysis of *p*-nitrophenyl acetate and *N*-carbobenzoxy-L-phenylalanine *p*-nitrophenyl ester have been examined at differing temperatures and pressures.[40]

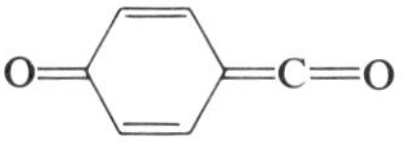

O=C₆H₄=C=O (**35**)

p-$O_2NC_6H_4OR$ (**36**) R = Ac, PhCO, *p*-$O_2NC_6H_4CO$

p-$NO_2C_6H_4O_2C(CH_2)_nMe$ (**37**) $n = 0, 4, 10$

The synthesis of various aryl esters with retention of configuration at the alcoholic carbon atom has been achieved in the presence of diethyl azodicarboxylate (**38**) and 3-methylbenzothiazole-2-selone (**39**);[41] (**38**) and (**39**) form a betaine adduct which functions as an activating agent for carboxylic acids and/or alcohols. The kinetics of acetylation of *p*-phenetidine with acetic acid have been reported.[42] The hydrolysis of aspirin in a wide variety of media has been studied.[43]

In an interesting recent paper, the rates of alkaline hydrolysis of a series of bi- and tri-cyclic bridgehead carboxylic acid esters are found to vary inversely with the size of the bridge.[44] Base-catalysed ester hydrolysis is suggested as a means of determining the effective size of alkyl substituents. The basic hydrolysis and trans-esterification reactions of *n*-alkyl esters of both acetic and (2,4-dichlorophenoxy)acetic acids have been studied in ethanol–water.[45] Not surprisingly the betaine ester, (carbethoxymethyl)trimethylammonium chloride (**40**), shows similar behaviour to acetylchloline[46] in alkaline hydrolysis in ethanol–water and dimethyl sulphoxide–water in that extensive desolvation of the two transition states does not take place.[47] This conclusion is based on the determination of the enthalpies of solvent transfer. The alkaline hydrolysis of vinyl acetate in the presence of carbazoles has been studied.[48] Semi-theoretical INDO calculations of the activation energy in the basic hydrolysis of esters in solution have been reported.[49]

Several papers have appeared on the acid-catalysed hydrolysis of aliphatic esters.

Acid hydrolysis of diethyl malonate in dioxane–water,[50] of methyl and ethyl acetates in the presence of urea and substituted ureas,[51] of multiply halogenated propanoate methyl esters,[52] and of 5,6-anhydro-1,2,*O*-isopropylidene-*β*-L-idofuranose (**41**)[53]

$EtO_2C—N{=}N—CO_2Et$

(**38**)

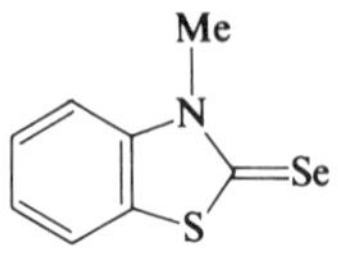

Me N Se S

(**39**)

$Me_3\overset{+}{N}CH_2CO_2Et$

(**40**)

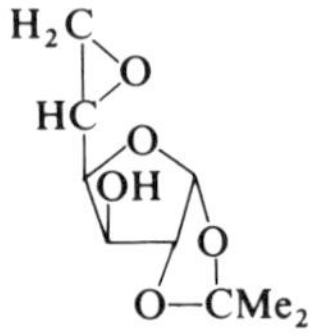

H_2C O HC O OH O O—CMe_2

(**41**)

have been reported. The kinetics of epimerization and hydration of *exo*- and *endo*-5-acetyl-2-norbornenes and the hydration of 3-acetylnortricyclane indicate that the epimerization occurs via enolic intermediates and the hydration by an $A\text{–}S_E2$ mechanism.[54] The acetolysis of enol esters of type (**42**) shows general acid catalysis and follows an $A_{SE}2$ mechanism.[55] Proton inventories are downward curved and $k(H_2O)/k(D_2O)$ are 2.46 and 2.60 respectively for the hydrolysis of the carbonates (**43**) and (**44**).[56] The transition states involve three and two water molecules, respectively. For the hydrolysis of (**44**), one water molecule acts as a general base, removing a proton from the nucleophilic water molecule.

$CH_2{=}C(OCOZ)Ph$

(**42**)

O_2N O O =O

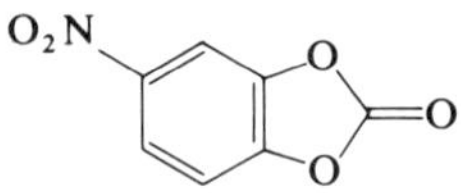

(**43**)

$p\text{-}O_2NC_6H_4OC(O)OC_6H_4NO_2\text{-}p$

(**44**)

Z = H, Me, CH_2Cl, $CHCl_2$, CCl_3

Solvolytic studies on aliphatic esters activated by halo substituents have been carried out by many groups. The kinetics of hydrolysis of methyl chloroformate,[57] of alcoholysis of alkyl chloroformates,[58] of the ethanolysis of the haloformates RCO_2Cl (with R = C_{1-4} *n*-alkyl, Me_2CH, Ph, $PhCH_2$, *etc.*), $EtCO_2F$, $PhCO_2F$, and $EtS(O)CCl$,[59] and of multiply halogenated propanoate esters[52] such as $F_3CCH(F)CO_2Me$ and $F_2(Cl)CC(Cl)_2CO_2Me$ have been reported.

Earlier work[46] by a Finnish group on the hydrolysis of methyl trifluoroacetate and chloromethyl dichloroacetate has been extended[60] and in new work[61] the acid-catalysed hydrolysis of chloromethyl chloroacetate in aqueous dioxane containing perchlorate has been studied.

A number of papers have appeared on the kinetics of esterification. Thus, the rates of esterification of isobutyric acid by 1-pentanol,[62] acetic acid by benzyl alcohol,[62] of RCO_2H (R = Me, Et, Me_2CH, $ClCH_2$, and Cl_3C) with EtOH over hydrated ferric or cupric sulphates,[63] of RCO_2H (R = C_{1-3} alkyl, halomethyl) with EtOH (leading to a LFER for these esterifications),[64] of RCO_2H (R = $ClCH_2CHOH$, $NCCH_2$, Me_3C, *etc.*) with allyl alcohol giving a LFER with Taft E_s values,[65] of lower aliphatic carboxylic acids with *n*-propyl alcohol giving again a LFER with Taft E_s values,[66] and of lauric acid with ethanolamine giving initially, in a rapid equilibrium step, compound (**45**) and then sequential conversion into (**46**) and (**47**),[67] have been studied. The kinetics and mechanism of formation of dimethyl oxalate by oxidative carbonylation of methanol using a Pd(II)–LiCl–*p*-benzoquinone system have been explored.[68] The trans-esterification kinetics of *n*-butyl chloracetate by *sec*-butyl orthotitanate have been determined in eleven solvents and in some binary mixtures of these solvents.[69] The effects of solvents have been correlated using Kamlet/Taft and Koppel/Palm parameters.

$HOCH_2CH_2\overset{+}{N}H_3\,{}^{-}O_2C(CH_2)_{10}Me$ (**45**)

$Me(CH_2)_{10}CONHCH_2CH_2OH$ (**46**)

$Me(CH_2)_{10}CONHCH_2CH_2OCO(CH_2)_{10}Me$ (**47**)

The acid-catalysed amination of ethyl acetoacetate with aniline in MeOH to give (**48**) has been studied.[70] A three-parameter (inductive, steric, and solvent) Taft-type equation has been found for the reaction of phosgene with aliphatic alcohols.[71] An addition–elimination mechanism is proposed. The stereoselective hydrolysis of amino-acid esters by modified poly(ethylenimine)s with optically active groups has been examined.[72,73] A plausible intermediate (**49**) in the Meerwein orthoester reaction has been prepared but ruled out as an intermediate in the conversion of (**50**) into (**51**).[74]

A highly stereoselective four-step preparation of the *trans*-5,6-disubstituted valerolactone (**52**) from (**53**) has been reported.[75] The stereochemistry of hydrogen

PhNH(Me)C=C(H)CO_2Et (**48**)

OEt, O (**49**)

OEt, + BF_4^-, O (**50**)

MeO, OMe, O (**51**)

R′, R, Me, Me, HO_2C, O, O, R, R′ (**52**)

Me, Me, R, R′, O, O, O, R, R′, O (**53**)

abstraction in vitamin-K-dependent carboxylations has been elucidated by using stereospecifically labelled 4-deuterated L-glutamate containing peptides as substrate.[76] A rationale has been provided for the appreciable mono-oxygen donation potentials of the biochemically important 4a-hydroperoxyflavins.[77] Now a kinetic pK_a for a flavin 4a-pseudobase (**54**) has been obtained.[78]

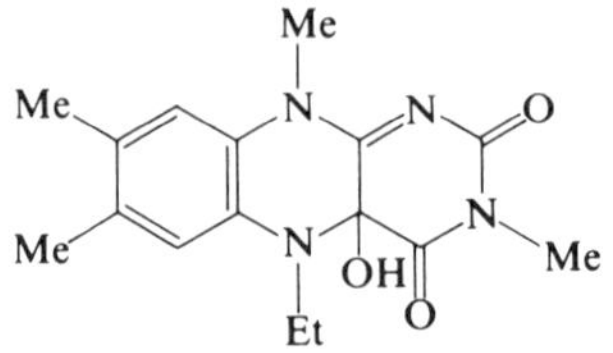

(**54**)

(*c*) *Acids, Anhydrides, and Anilides.* The dissociation constants of a series of 3-substituted bicyclo[2,2,2]octane-1-carboxylic acids and of some benzoic acids have been measured in DMSO, MeOH, *n*-BuOH, and $HOCH_2CH_2OH$.[79] The data cannot be reconciled with predictions from the electrostatic theory. The effects of about one hundred different solvents on the pK_a values of a large number of aliphatic and aromatic carboxylic acids (including dicarboxylic acids), phenols, and thiophenols have been examined and the relationship:

$$pK(AHS_2) - pK(AHS_1) = \text{constant}$$

which describes the effect of solvent changes on p*K*, was found to hold in certain limited cases.[80] The epimerization of benzylpenicilloic acid in alkaline media proceeds *via* the imine (rather than the enamine) tautomer of penamaldic acid.[81] The lactonization of (E)-β-*tert*-butylcinnamic acid (**55**) to 3,4-dimethyl-3-phenyl-γ-valerolactone (**56**) may involve a bridged, non-classical carbocation (**57**).[82] This result is supported by deuteration and ^{13}C-NMR experiments.

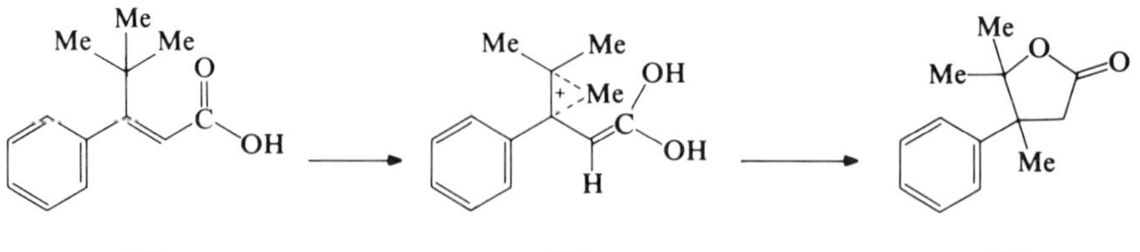

(**55**) (**57**) (**56**)

The calculated preferred conformations compare well with existing experimental data when the *gem*-dimethyl effect is quantitatively assessed for the succinic acid (**58**)$\rightleftharpoons$anhydride (**59**) equilibrium using the 2-methyl, racemic 2,3-dimethyl, tetramethyl, and racemic 2,3-di-*tert*-butyl derivatives.[83]

The hydrolysis of acetic anhydride catalysed by 2-methylpyridine has been studied and the catalysis is considered to be nucleophilic rather than general base.[37]

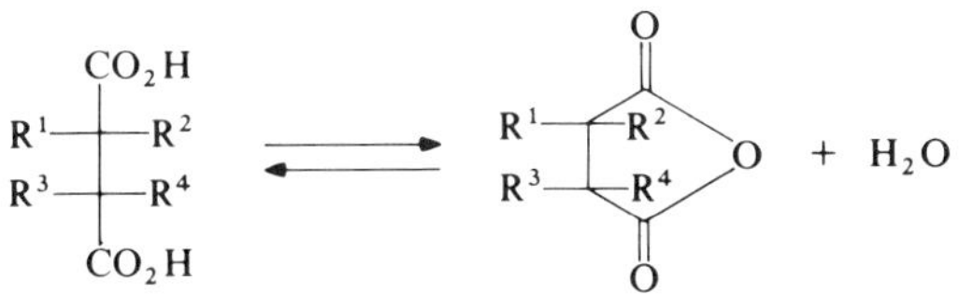

(58) **(59)**

$R^1 = R^2 = R^3 = R^4 = H$; $R^1 = Me$, $R^2 = R^3 = R^4 = H$;
$R^1 = R^4 = Me$, $R^2 = R^3 = H$; $R^1 = R^2 = R^3 = R^4 = Me$;
$R^1 = R^4 = Bu^t$, $R^2 = R^3 = H$

The hydrolysis of the same anhydride in a flow system has also been reported.[84] The aminolysis of maleic anhydride by a series of primary straight-chain amines to give maleamic acids has been studied at pH 4 and 15° and 25°.[85] The slow step for amines, where conjugate acids have $pK_a < 7.7$, is proton transfer within the zwitterionic tetrahedral intermediate, *i.e.* **(60)** → **(61)**; the reactions of more basic amines involves rate-determining formation of **(60)**.

(60) **(61)**

The negative Hammett ρ value (-2.54) for the reaction of a series of *meta*- and *para*-substituted phenols with *N*-chloroacetanilide has been interpreted in favour of a transition state for the reaction in which there is considerable chlorine–oxygen bonding and the majority of the charge has been transferred from oxygen to nitrogen.[86] Several interpretations are shown to be possible for the observed isotope effects in the basic hydrolysis of *p*-$NO_2C_6H_4NHCOCH_3$ and *p*-$NO_2C_6H_4NHCOCD_3$[87] at 30°, with $[HO^-]$ varying from 0.002 to 2.31 M. Three different types of tetrahedral intermediate have been invoked, *viz.*, a uninegative adduct **(62)**, a dinegative adduct **(63)**, and a third which arises from overall leaving-group expulsion from **(63)**. A Brønsted α of 0 (substituted phenols as general acids) suggests that the general acid is a spectator at spontaneous expulsion of the leaving group, producing catalysis by fast subsequent trapping of MeNAr in the rate-limiting (at high acid concentrations) additions of MeO^- to *m*-$NO_2C_6H_4N(Me)COCF_3$ in methanol at 25°.[88] The Jencks clock shows that the minimum life-time of the tetrahedral intermediate **(64)** is 1–10 ns.

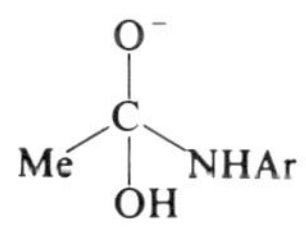

(62) 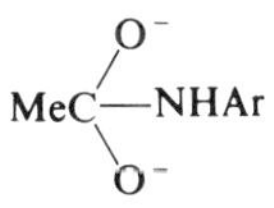 **(63)** 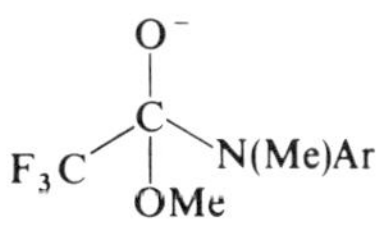 **(64)**

(*d*) *Acid Halides.* The rate constants for the reaction of a variety of nucleophiles with acetyl chloride in water containing 2.5% (v/v) dioxane have been determined.[89] For primary amines, $\beta_{nuc} = 0.25$ and attack of the nucleophile is the slow step. Pyridines with $pK_a > 5$ behave similarly but less basic pyridines react more slowly. The rate constants do not fit equations using Ritchie's N_+ values. Several papers of mechanistic interest have involved work on benzoyl chloride. Further work by Bentley's group[90] indicates from an examination of solvent effects on reactivity that the hydrolysis/alcoholysis of benzoyl chloride involves a direct displacement (S_N2 character), in competition with a carbonyl addition reaction which becomes dominant in less aqueous media. These conclusions are arrived at from careful analysis of solvent effects using an equation which takes account of solvent ionizing power (Grunwald–Winstein Y values) and solvent nucleophilicity (N):

$$\log (k/k_0)_{RX} = mY + lN$$

The same group[91] has, using HPLC, been able to carry out product studies in the same solutions that were used to follow kinetically (conductimetrically) the competing methanolysis/aminolysis of benzoyl chloride in methanol in the presence *o*-, *m*-, and *p*-nitroanilines. The amines are acting as nucleophiles not as base catalysts. A nucleophilic rôle for tertiary amines in reaction with benzoyl chloride in water–dioxane is also indicated from $\Delta H^{\neq}$ and $\Delta S^{\neq}$ measurements.[92] A LFER is used to calculate rate constants for the reaction of the benzoyl chlorides RC_6H_4COCl (R = 4-MeO, 4-Me, 3-Cl, 3-NO_2) with ring-substituted primary anilines.[93] A Russian group have studied the kinetics of the reaction of butyl halides, RCOX, with *n*-butanol in acetic acid catalyst[94] and of butyl chloride with *n*-butanol in diphenylphosphinic acid.[95] The acids act as bifunctional catalysts in each case. The solvolyses of nine different acid chlorides in aqueous acetone have been examined.[96] Isomers solvolyse at different rates and give the original acids. No evidence of tautomerism or isomerism between the different isomeric pairs of acid chlorides could be detected.

(*e*) *Ureas, Carbamates, and Derivatives.* Boiling-temperature (ebulliometric) measurements have been used to study the kinetics of the decomposition of urea in neutral, acidic, and basic media.[97] The shapes of the ebulliometric curves are different for the different media and these profiles are correlated with the species present. Ebulliometric measurements are seen as being useful for kinetic/mechanistic studies. The kinetics of the alkaline hydrolysis of *tert*-butylurea have been measured at various temperatures.[98] The rates of hydrolysis of *N*,*N'*-bis(methoxymethyl)urea in aqueous 0.053M NaOAc have been measured at varying pH.[99] The methanolysis rate constants and the dissociation constants have been measured for a series of benzoyl derivatives of phenylureas and phenylthioureas (**65**). The dissociation constants of the thio derivatives are higher by one order of magnitude and the rate constants are higher by two orders of magnitude.[100] Changes in the slow step account for the fact that the Hammett ρ for the methanolysis of the ureas (*ca.* 1.50) is about twice that for the thioureas (*ca.* 0.70).

The kinetics of alkaline hydrolysis at 80° of the spiro-compound (**66**) and of 5,5-dimethylbarbituric acid have been determined.[101] Ring strain causes a five-fold increase in the hydrolysis rate of (**66**; $n = 3$) compared to those of the other compounds. The alkaline hydrolysis of 5,5-diethylbarbituric acid is an example of an irreversible consecutive reaction of the type: A → B → C, where B and C are diethylmalonuric acid (**67**) and ammonia respectively.[102] The pH-rate profiles

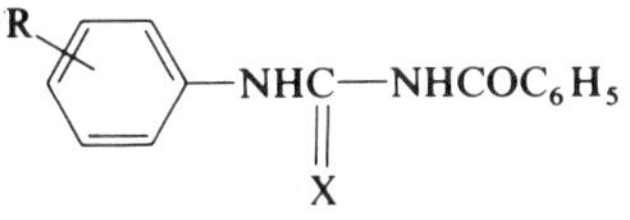

(**65**) X = O, S

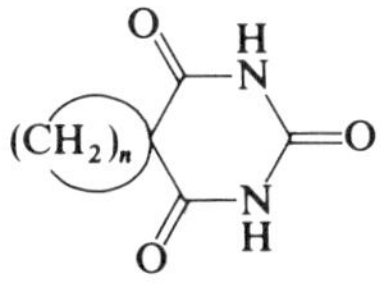

(**66**) n = 3–5

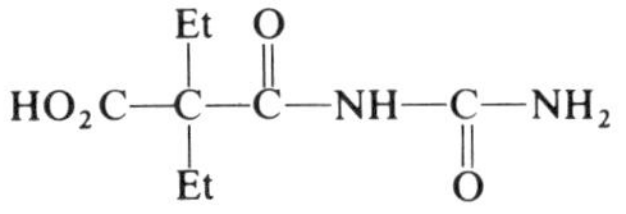

(**67**)

obtained at several temperatures reveal distinct phases and these are attributed to changes in the slow step with changes in [HO^-]. A changeover in mechanism from *ElcB* (aryl) to $B_{AC}2$ (alkyl) for the hydrolysis of aryl and alkyl 4-phenylallophanates (**68**) to phenylurea and phenol or alcohol is indicated on the basis of studies of acid–base catalysis, deuterium solvent isotope effects, and thermodynamic data.[103] Hammett, Brønsted (non-linear), and Yukawa–Tsuno plots have been carried out and the parameters ρ (using σ^-), α and r are in agreement with the proposed mechanisms.

The carboxylate group depresses the rate of alkaline hydrolysis of the dihydrouracil (**69**) by a neighbouring-group effect[104] (see section on Intramolecular Catalysis). Photolysis of 1,3-dimethyluracil (**70**) in MeOH results in the formation of the enamine 2-methoxycarbonyl-*N*-methyl-3-methylaminopropenamide (**71**), *via* a novel ring-opening reaction.[105] The rate profiles for the alkaline hydrolysis of some dihydrouracils of type (**72**; R^2 = H, Me, Ph) and dihydroorotic acids (**72**;

$C_6H_5NH—CO—NHCO_2R$

(**68**) R = aryl, alkyl

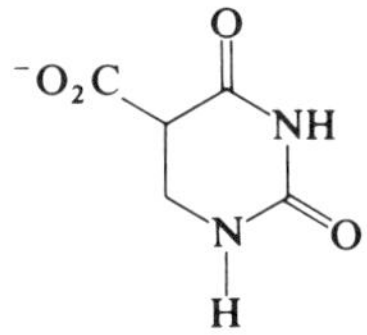

(**69**)

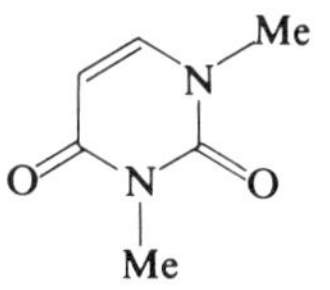

(**70**)

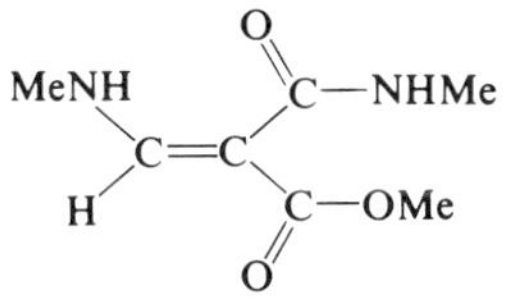

(**71**)

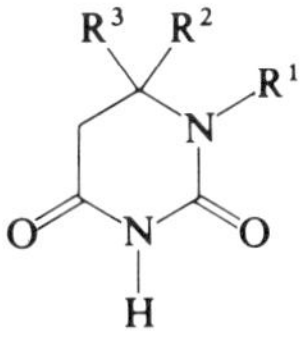

(**72**) R^2 = H, Me, Ph, CO_2H

R = CO_2H) have been determined in order to assess the effect of allylic strain on the ring-opening of the 1,6-dihydrouracils.[106]

Seven papers of mechanistic interest in carbamate chemistry have appeared during the period covered by the review. The base hydrolysis of methyl, ethyl, and phenyl *N*-aryl-*N*-methylcarbamates (**22**) and the involvement of tetrahedral intermediates has been referred to above.[11] The same group has reported on the basic hydrolysis of alkyl and aryl *N*-(4-nitrophenyl)carbamates (**73**).[107] Both $B_{AC}2$ and *E*1*cB* mechanisms have been recognized (depending on carbamate) from the hydrolysis. For compounds that react by the $B_{AC}2$ mechanism, the tetrahedral intermediate (**74**) partitions in favour of C—OR bond-breaking rather than C—N bond-breaking. Stopped-flow techniques have been used for the study of carbamate formation in the reaction of CO_2 with diethanolamine.[108] A recent study[109] of the hydrolysis of benzhydryl *N*-arylthiocarbamates and *S*-benzhydryl *N*-arylthiocarbamates (**75**) has established an *E*1*cB* mechanism for the hydrolysis of the former (substituent effects, phenyl isocyanate trapped). The reaction kinetics of the base-catalysed reaction of formaldehyde with cinnamyl carbamate in aqueous dioxane at 30° indicate that an S_E1 mechanism is being followed.[110] The decomposition kinetics of *N*-hydroxymethylcinnamyl carbamate in base have also been followed. Water has a marked influence on the selenium-catalysed conversion of nitrobenzene into methyl *N*-phenylcarbamate.[111] A dithiocarbamic acid intermediate (**76**), leading to the dithiocarbamate anion, forms in the reaction of carbon disulphide with morpholine, benzylamine, and pyrrolidine in ethanol at 25°.[112] Reversibility is observed with the first two amines, but with pyrrolidine the formation of (**76**) is rate-determining.

(*f*) *Amides and β-Lactams.* The mechanism of cyclization of *ortho*-substituted aromatic amides involves the anionic form of the amide group according to MINDO/3 calculations.[113] α-Chloroamides (**77**) are isolable intermediates (high yields) in the basic hydrolysis of α-chloronitriles to give ketones.[114] Two papers have appeared on the alkaline hydrolysis of carboxamides. Rates are determined polarographically for the hydrolysis of the pyridinecarboxamides, nicotinamide, and picolinamide.[115] The second paper[116] describes the base hydrolysis of the indole-1-carboxamides (**78**) and of 5*H*-dibenz[*b*,*f*]azepine-5-carboxamide in water at 60°. The rate constants for tetrahedral intermediate formation are strongly increased by *N*-substitution.

The tautomerism of pyran-2,4-diones bearing 5-acetyl and 5-benzoyl substituents have been studied by IR, UV, ^1H- and ^{13}C-NMR spectroscopy (see Scheme 1).[117] The hydrolysis at 100° in aqueous H_2SO_4 of five 1-aryl-2-phenylimidazolines (**79**)[118] have been reported. The effects of substituents X in the N(1)-phenyl ring can be correlated in the Hammett equation. Electron-withdrawing substituents speed the rates (the reverse is the case for alkaline hydrolysis). A mechanism involving two molecules of water in the transition state is proposed for the hydrolysis of 1-benzoylbenzotriazole in the pH range 2-7 where the hydrolysis is independent of pH. A Hammett ρ value of 1.7 (phenyl ring of benzoyl) and a Brønsted α value of -1.7 have been obtained. Above pH 7, the hydrolysis is base-catalysed.[119] The effect of

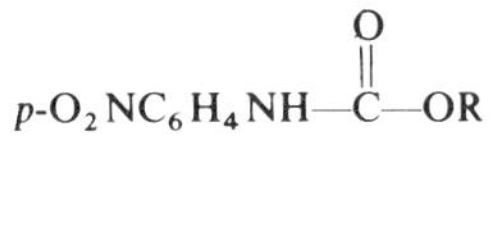

p-$O_2NC_6H_4NH$—C(=O)—OR

(73)

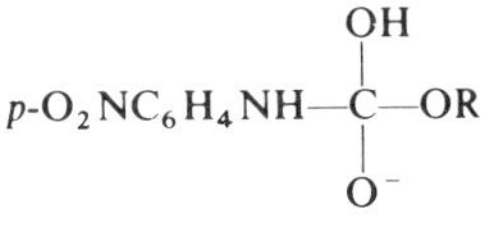

p-$O_2NC_6H_4NH$—C(OH)(O^-)—OR

(74)

ArNH—C(=Y)—X—$CH(C_6H_5)Ar'$

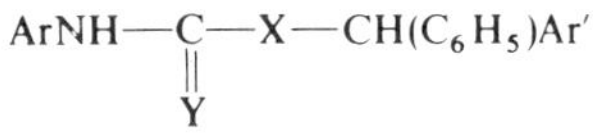

(75)

Y = S, X = O; Y = X = S

$R^1R^2NCS_2H$

(76)

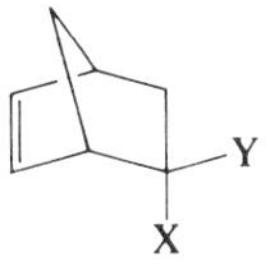

(77)

Y = Cl, X = $CONH_2$
Y = $CONH_2$, X = Cl

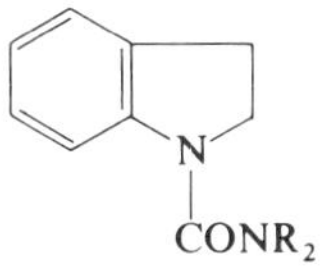

(78)

R = H, Me

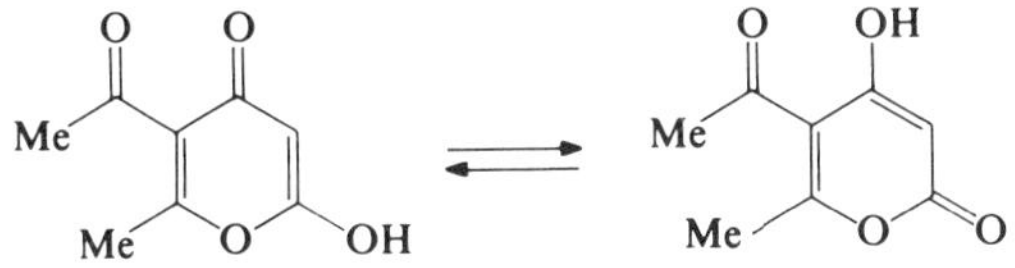

SCHEME 1

substrate hydrophobicity has been probed in the water and HO^--ion-catalysed hydrolysis of the 1-acyl-1,2,4-triazoles (**80**) in water-rich Bu^tOH–H_2O mixtures.[120] The excess acidity method has been applied to hydrolysis data for nine acyl(benzoyl)hydrazines (**81**) in aqueous H_2SO_4.[121] In dilute acid, the transfer of a (second) proton to oxygen in the mono-protonated substrate is the rate-determining step followed by a rapid water attack (*A*2). In concentrated acid, nitrogen protonation is rate-determining and this is concerted most likely with C—N bond-breaking to give (in some cases) acylium ions. Rates of hydrolysis and activation parameters are available for the hydrolysis of 9-(1-acetoxyethyl)carbazole and some ring-substituted derivatives.[122] First- and second-order dependence on $[HO^-]$ have been found in the hydrolysis of indomethacin (**82**) and some derivatives.[123]

An important account of the mechanisms of reactions of β-lactam antibiotics has appeared.[124] The same group has examined the mechanism of this ring-opening in cephalosporins.[125] The non-linear dependence of the rate of aminolysis of some cephalosporins upon $[HO^-]$ is interpreted in favour of the formation of a tetrahedral intermediate (**83**). At high $[HO^-]$ the slow step is the formation of (**83**); however, at low $[HO^-]$ it is the diffusion-controlled encounter of the intermediate and HO^-. Rate constants for the formation and breakdown of the intermediate are reported. Cleavage of the β-lactam ring with acid has also been studied for penicillanate 1,1-dioxides (**84**).[126] The degradation (also involving β-lactam ring-

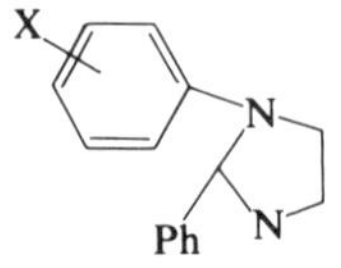

(79)

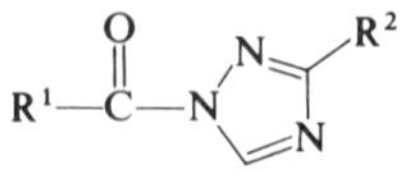

(80)

R^1 = Me, R^2 = H
R^1 = Me, R^2 = Ph
R^1 = R^2 = Ph

(81)

R = alkyl, Ph, $ClCH_2$, H

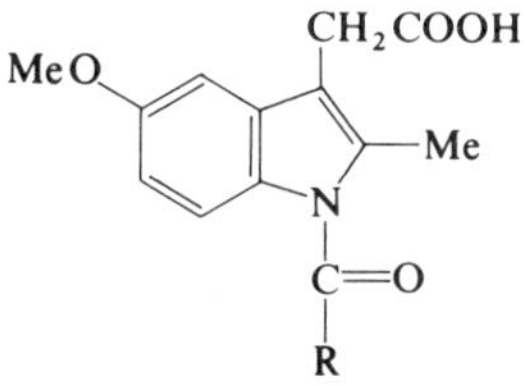

(82)

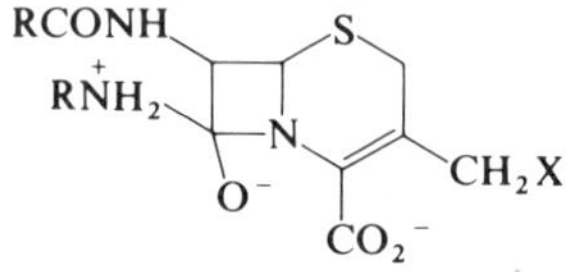

(83)

opening) of the third-generation cephalosporin, sodium cefotaxime (**85**), has been studied at varying pH.[127]

Ultrasound can promote β-lactam formation in excellent yields in the Reformatsky-type reaction of ethyl bromoacetate, Zn, and a Schiff base.[128] Ester enolates and *N*-trimethylsilylimines react to give *N*-protio-β-lactams.[129]

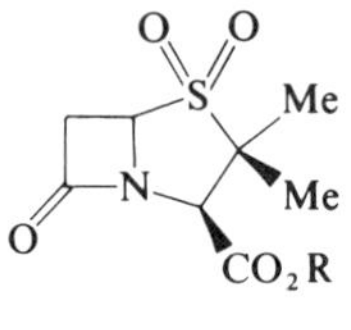

(84)

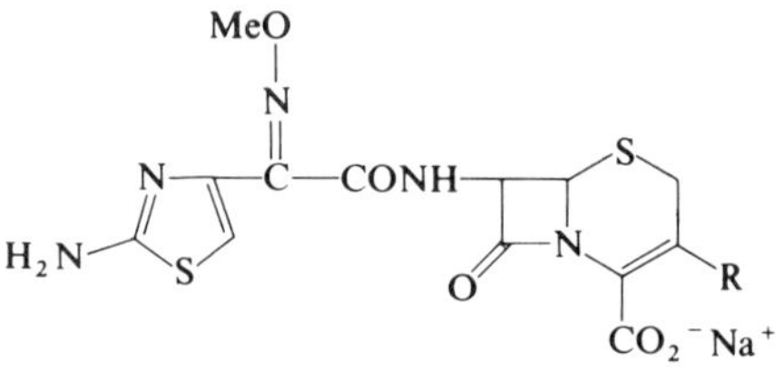

(85)

(*g*) *Other Nitrogen Centres.* Ultrasound increases the hydrolysis rate of various nitriles in basic solution.[130] Kinetic data have been obtained for the three-step reaction of *o*-phthalodinitrile (**86**) with aniline in phenol.[131] Chlorination of alanine in dilute aqueous media gives both acetaldehyde and acetonitrile and subsequent reactions then occur using *N*-chloroamines which are also formed.[132] *N,O*-Diarylhydroxylamines (ArNHOAr) do not undergo solvolysis *via* the reversible formation of ion-pairs.[133] This conclusion, based on ^{18}O-labelling studies, is in conflict with previous conclusions based on indirect kinetic evidence. The mechanism of hydrolysis of phenylhydroxamic acids (PhCONHOH) in basic media is more complex than that arising in acidic media; k_{obs} *vs.* [HO^-] plots show a maximum and this is attributed to the existence of one or more pre-equilibria steps prior to the slow step.[134] These pre-equilibria complicate the situation and it is unclear whether the mono-anion (in one of its tautomeric forms) $PhCONHO^-$ or the dianion (**87**) are the critical intermediates *en route* to the tetrahedral intermediates. Enhanced stabilization of the transition state accounts for the fact that the anion of α,α,α-trifluoroacetophenone oxime is much more reactive towards *p*-nitrophenyl acetate than (the similarly basic) *p*-chlorophenoxide anion on addition of DMSO to aqueous solutions.[135]

The hydrolysis of 2-ethoxy-*N*-vinylpyrrolidiniminium tetrafluoroborate (**88**) to give acetaldehyde and (**89**) and the alcoholysis of (**88**) giving (**90**) can be explained by the initial formation of an "enamine-like" intermediate.[136] Kinetic and thermodynamic control of C(2) (**91**) and C(4) pseudo-base formation from 3-X-1-methylquinolinium ions has been studied for various substituents X;[137] (**91**) predominates at equilibrium for X = H, Br, while for X = $CONH_2$, CO_2Me, CN, and NO_2 the C(4) pseudo-bases are the thermodynamically preferred species. The C(2) pseudo-bases are the kinetically controlled products. Equilibrium constants

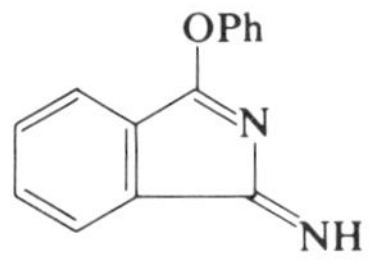

(**86**) 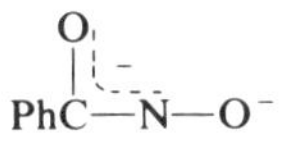 (**87**) 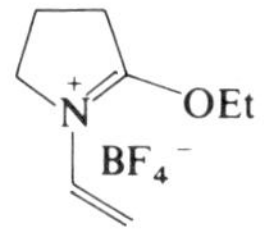 (**88**)

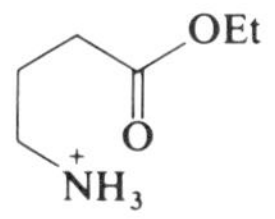

(**89**) 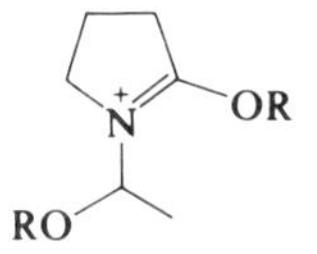 (**90**) (**91**)

R = Me, Et

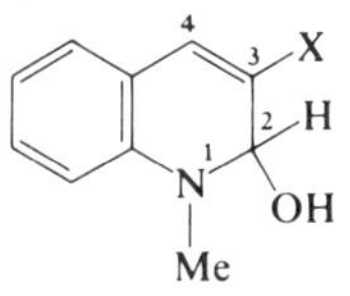

have been obtained for the formation of both pseudo-bases. The logarithms of the rates of lactamization of the zwitterionic benzylstrychnine (**92**) in aqueous solution at 25° show a maximum at about 5–10% H_2SO_4. The maximum is explained by taking account of the extent of protonation of the NH group.[138]

The kinetics and mechanism of the reaction of phenyl isocyanate and alcohols in benzene has been the subject of a further study.[139] Denitrosation of *N*-methyl-*N*-nitrosoaniline and *N*-nitrosodiphenylamine in acid solutions is catalysed by nucleophiles such as thiocyanate, halide, and thiourea but, at high nucleophile concentrations, the catalytic effect disappears.[140] A reaction mechanism involving protonation of the nitroso compound followed by nucleophilic attack (either stage being rate-limiting depending on the reactivity and concentration of the nucleophile) is proposed. The nitrosation of 3-mercaptopropanoic acid, mercaptosuccinic acid, and cysteine in aqueous acid solution at 25° have been studied by stopped-flow

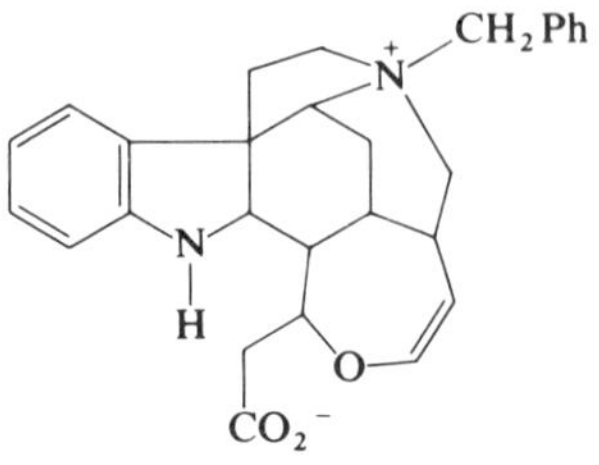

(92)

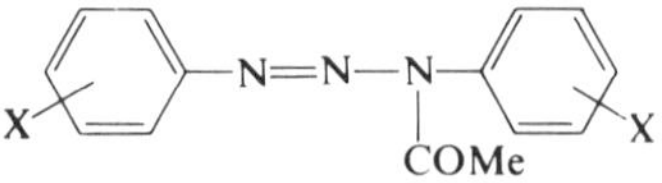

(93)

spectrophotometry.[141] These three compounds are significantly more reactive (giving thionitrites RSNO) than conventional nitrite traps and they appear to offer good potential in this area.

The kinetics of solvolysis of the 3-acetyl-1,3-diphenyltriazenes (**93**) in 20% aqueous ethanol have been determined.[142] The methoxide-ion-catalysed isomerization of 4-acyloxy-1,3,4-oxadiazines to *N*-aminooxazolidonylhydrazones has been studied.[143] The effect of EtOH in EtOH–benzene mixtures on the rate of amine exchange of hydroxynaphthylideneanilines with benzylamine to give the Schiff base (**94**) has been examined.[144] *N*-Cyanourea ($NH_2CONHCN$) is an intermediate in the hydrolysis of dicyanamide (NCNHCN).[145] The hydrolysis of both of these materials and of *N*,*N*′-dimethylcyanoynanidine $(MeNH)_2C{=}NCN$ is specific-acid-catalysed. Both *N*-cyanourea and dicyanamide appear to react *via* their carbodimide forms. The reaction of anilines with dicyanamide in aqueous solution has also been examined.[146] This reaction is subject to general acid catalysis with Brønsted exponents as follows: $\alpha = -0.54$ with sulphanilic acid as the aniline; for aniline variation, $\beta_{nuc} = 0.63$ (acetic acid) and 0.66 (oxonium ion). The mechanism is seen as involving an attack by aniline on the anionic dicyanamide assisted by concerted proton transfer from the general acid to the substrate N.

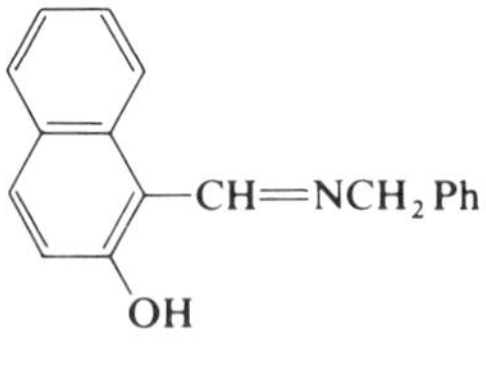

(94)

Reactions in Aprotic Solvents

(*a*) *Ester formation and reactions.* The kinetics of formation of isoamyl butyrate and amyl acetate by direct esterification have been studied.[147] Data on the esterification of benzyl chloride and benzenesulphonyl chloride by substituted phenols have been re-interpreted by a Russian group.[148] A review in Russian on the mechanism of action and catalytic activity of tertiary amines in alkyl-transfer reactions has appeared.[149] Trans-esterification of carboxylic acid esters by titanium alcoholates is the subject of two papers from the same Russian group.[150,151] In the first paper,[150] the kinetics of trans-esterification of three butyl benzoates *m*-$XC_6H_4CO_2Bu^n$ (X = H, Cl, NO_2) in heptane by *sec*-butyl orthotitanate $Ti(OBu^s)_4$ have been measured. In the second paper,[151] the trans-esterification kinetics of RCO_2Bu^n (R = Me, Bu^n, Cl_2CH, Cl_3C, $PhOCH_2$, $ClCH_2CH_2$, Ph_2CH, *etc.*) by the titanium compound have been determined in heptane. Both sets of kinetics are analysed in terms of R group effects and temperature.

The preparation and study of the reactivity of *O*-[alkoxy(or phenoxy)-alkoxy]alkyl *O,O*-*tert*-butyl carbonates $R(OCH_2CHR')_2OCO_2OCMe_3$ (R = H, Ph, $PhCH_2$; R′ = H, Me) have been reported.[152] The kinetics of the benzoylperoxide-initiated chlorination of pentaerythritol tetrapropionate and tetradecanoate, neopentyl glycol dipropionate, and 1,1,1-trimethylolpropane tripropionate have been examined.[153] The kinetics of the cyclization of (2-pyrimidinylthio)phenylacetic acids with acetic anhydride in acetic acid/Et_3N to give dihydrothiazolopyrimidines (**95**) followed by elimination of acetic acid to give thiazolopyrimidines has been studied.[154] Alkylation of methyl esters of 4-methyl 5-oxo-5-phenylpentanoates (**96**) gives *cis*- and *trans*-tetrahydro-5,6-dimethyl-6-phenyl[2*H*]pyran-2-ones (**97**) and (**98**), respectively (Scheme 2).[155]

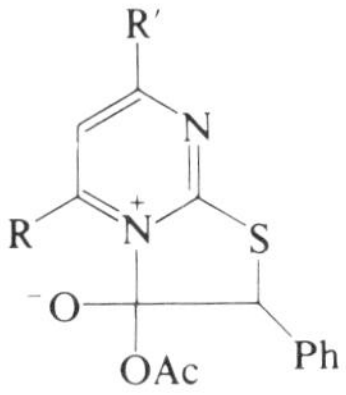

(**95**)

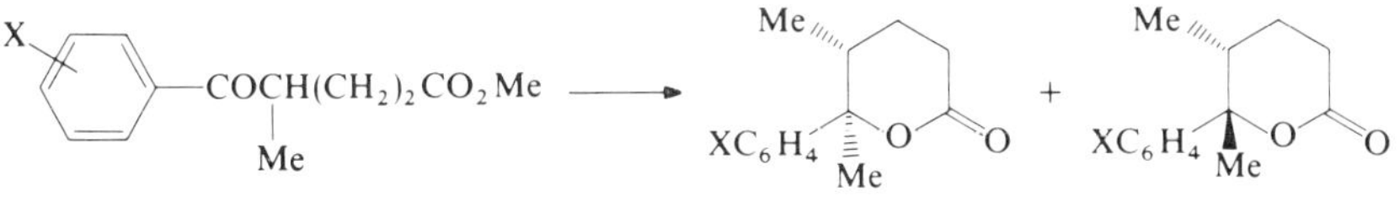

(**96**) (**97**) (**98**)

SCHEME 2

The kinetics of the reaction of alkyl chloroformates with aliphatic/aromatic amines to give acylated amines have been probed and a Taft–Pavelich equation (σ^* and E_S) derived.[156] The reactions proceed by an addition–elimination mechanism. A new synthesis of optically active *β*-hydroxy-carboxylic acids has been developed based on the diastereoselective addition of the chiral nitrile oxide (**99**) to 1,2-disubstituted alkenes.[157] Kinetics of the trans-esterification in inert organic solvents of phenyl acylates $AcOC_6H_4R$ (R = H, NO_2, halo, alkyl, *etc.*) and the aromatic amides $RC_6H_4CONH_2$ (R = H, RO, halo, NO_2) have been probed.[158]

Treatment of the ethyl *o*-phenylbenzoate with potassium superoxide in benzene in 18-crown-6 at room temperature for 24 h gives 88 % of acid but no acyl peroxide is detected, although the peroxycarboxylate anion is involved.[159] The kinetics and mechanism of methanolysis in benzene of methyl perchlorate ($MeOClO_3$) have been investigated.[160] The mechanism is essentially S_N2 in character and an alternative mechanism is given for the methanolysis of the previously studied triphenylmethyl chloride. Symmetrical and unsymmetrical stilbenes (ArCH=CHAr′) are found on pyrolysis of aryl cinnamates (**100**).[161] The lactonization of 2-aroylpropionic acids (**101**) to give the butenolides (**102**) has been studied and the ease of lactonization depends on the substituents in the aryl group; electron-releasing substituents speed

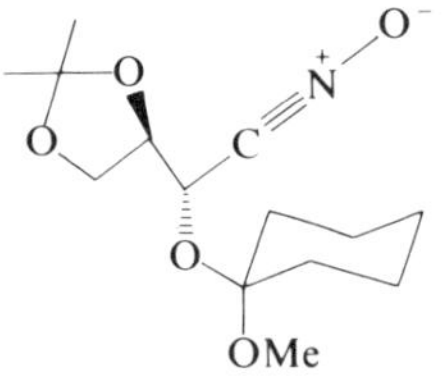

(**99**)

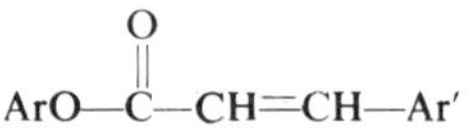

ArO—C(=O)—CH=CH—Ar′

(**100**)

the reaction.[162] The best reagent for the reaction is acetic anhydride/H_2SO_4, with acetyl chloride next, and acetic anhydride alone being next. Azo compounds are unexpectedly formed in the basic decomposition of ethyl *N*-(*o*-azidoaryl)carbamates (**103**).[163] The interesting hemiacetal anion (**104**) is an intermediate in these reactions.

$ArCOCH_2CH_2CO_2H$

(**101**)

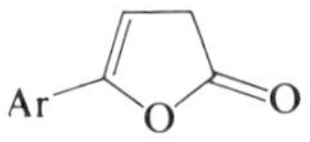

(**102**)

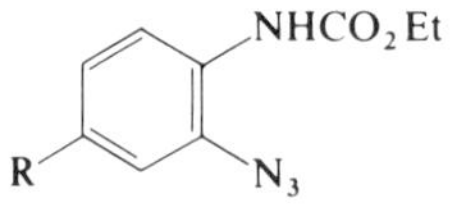

(**103**)

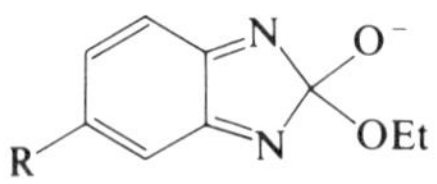

(**104**)

Vapour-phase pyrolysis of tertiary alkyl acetates ($AcOCMe_2Bu^t$ and $AcOCMePr^i_2$) to give alkenes and acetic acid has provided an opportunity to assess steric (E_S values used) and polar effects in these *β*-elimination reactions.[164] The thermal decomposition of acetic acid in the vapour phase in argon has been studied in the temperature range 1300–1950 K in a single-pulse shock tube.[165] Lithium borohydride/triethylborane reduced RCO_2Et ($R = BuCH_2CH_2, Me_3C$, Ph, *etc.*) readily.[166] The reaction involves complexes of RCO_2Et with BEt_3 and $LiBHEt_3$. The kinetics of aminolysis of 1-thio-*β*-D-glucopyranosyl esters of N-protected alanines (**105**) in dichloromethane at 26° have been reported.[167] Vinylogous Dieckmann condensations involving extended enol-*endo* [with (**106**)] and enol-*exo* [with (**107**)] have been described recently.[168] An *ab initio* Hartree–Fock calculation on the benzoyl peroxide/$C_6H_5NMe_2$ reaction has been reported.[169] Fluorescence spectra have been used for the direct observation of the pyridinol form (**108**) in tautomeric 2-[1*H*]pyridones.[170]

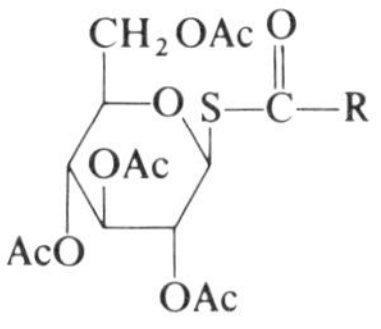

(**105**) 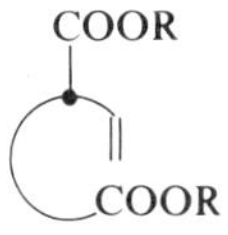 (**106**)

RC(=O) = Ac-L-Ala, Ac-D-Ala, *etc.*

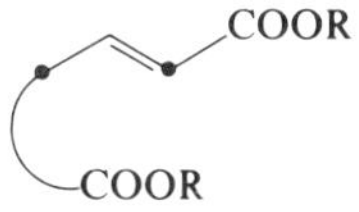

(**107**) 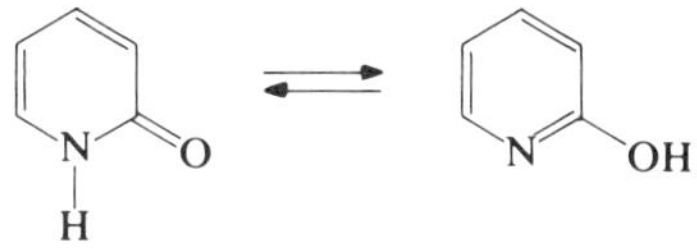 (**108**)

(*b*) *Other reactions.* Rate constants for the acylation of aromatic amines by phenyl chloroformate (PhOCOCl) in the presence of pyridine have been determined.[171a] A nucleophilic mechanism of catalysis involving an acylammonium intermediate is proposed. The same Russian group has obtained kinetic data for the acylation of aniline by $(4\text{-}XC_6H_2CO)_2O$ (X = Ph, NO_2) and Ac_2O in the presence of tertiary bases and again acylammonium intermediates are proposed.[171b] A correlation between the basicity of the amine and its reactivity in acylation (using phthalic anhydride) has been reported.[172] The acylation in Ac_2O pyridine of both N (first) and O (second) centres occurs with certain sympathomimetic amines and the reaction is catalysed strongly by 4-(dimethylamino)pyridine.[173] The kinetics of *N*-formylation of piperidine, 3-methylpiperidine, and 1,2,3,4-tetrahydroisoquinoline by chloral and chloral hydrate in toluene show that the reactions are first-order in both reactants.[174]

Kinetic data for the transamination of *N*,*N*′-diphenylurea by diisobutylamine to give $PhNHCON(CH_2CH_2Me_2)_2$ and aniline indicate that two mechanisms operate: (1) direct reaction and (2) dissociation to PhNCO and $PhNH_2$ followed by addition of diisobutylamine to the isocyanate.[175] Activation enthalpy and entropy data for both processes have been calculated. A kinetic and thermodynamic study of the reaction of 2,4,6-triphenylthiopyrylium ion with *n*-butylamine and cyclohexylamine has been reported. The reaction involves a two-step process in which the formation of protonated thiopyrans is the rate-limiting step.[176]

Several papers involving reactions of acid halides in aprotic media have appeared. The reactions of $(MeO)_2NCMe_2CO_2Me$ with $SOCl_2$, $BF_3 \cdot OEt_2$, Me_3SiCl, HCl, and 3-$ClC_6H_4CO_2OH$ have been investigated.[177] Rate constants and activation energies for the acylation of aniline in Me_2NCOMe by XC_6H_4COCl (X = 4-MeO, 4-Me, 4-Cl, 3-CO_2Me, 3-Me, H) have been reported.[178] A dual-parameter (σ, temperature) equation correlates the log k values. The aminolysis rates of benzimidoyl chlorides (**109**) by substituted aromatic amines correlate well with the Hammett equation using $\Sigma\sigma$, yielding $\rho = -4.1$ with (**109**; R = SO_2Ph) and -2.2 with (**109**; R = Me, PhCO).[179] A bimolecular nucleophilic mechanism is proposed. The same Russian group has probed the *N*-methylimidazole-catalysed aminolysis of benzoyl halides. The imidazolium ion (**110**) is an intermediate.[180] The kinetics of the reaction of *N*-methyl-*m*-nitrophenylcarbamoyl chloride (**111**) with aniline in the presence of pyridine have been reported.[181] The rates of acylation of aromatic amines by phenyl chloroformate have been studied.[170] The reactions of acid chlorides RCOCl (R = Me_3C, $C_4H_9CH_2$, c-C_3H_5) with tri-*n*-butyltin hydride yield RCHO and *n*-Bu_3SnCl as initial products. Other products are formed by subsequent reactions of the aldehyde, *e.g.* $Bu_3{}^nSnOCH_2R$, $RC(O)OCH_2R$, RC(O)OCHClR.[182] Radicals are not involved in the reaction. The reactions of 1-naphthoyl chlorides (with 4- and 6-substituents) and of 2-naphthoyl chlorides (with 3- and 6-substituents) with aniline correlate well with Hammett ρ values of 1.38 and 1.55 for the 6- and 3-substituted 2-naphthoyl chlorides, respectively.[183]

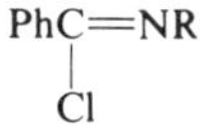

PhC(Cl)=NR (**109**) R = Me, PhCO, SO_2Ph

PhCON⁺ imidazolium NMe (**110**)

PhN(Me)COCl (**111**)

Intramolecular Catalysis and Neighbouring-group participation

Intramolecular general acid catalysis has been observed in the hydrolysis of 5-[184] and 4-[185] nitro(trifluoroacetylamino)benzoic acids (**112**) in the pH range 0–3. Catalysis is by the unionized carboxyl group, whose pK_a is about 2.5. Significant buffer catalysis is observed at intermediate pH. Neighbouring carboxylate ion depresses the rate of alkaline hydrolysis of the dihydrouracil (**69**).[104] This rate decrease is attributed to the electrostatic effect exerted by this group in an enforced axial conformation in the doubly negative transition state. An interesting case has been reported[186] of a neighbouring hydroxy group (first) and a neighbouring carboxylate anion (second) aiding a reaction. 4,4,4,-Trichloro-3-hydroxybutanoic acid (**113**) in aqueous base forms the carboxylate anion and then, with the assistance of the hydroxy group, expels one of the chlorine atoms to give the epoxide (**114**); (**114**) is then ring-opened by backside attack of the carboxylate anion leading to the oxetanone (**115**) which is hydrolysed to malic acid (**116**). This sequence of reaction provides an example of double anchimeric assistance. Intramolecular catalysis by

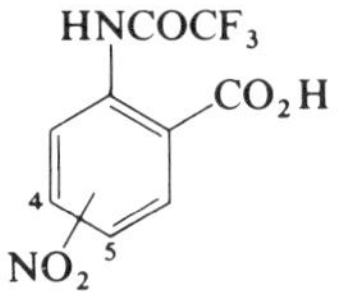

(112) 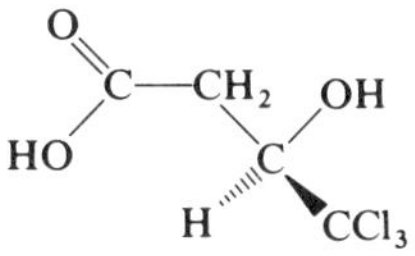 **(113)** 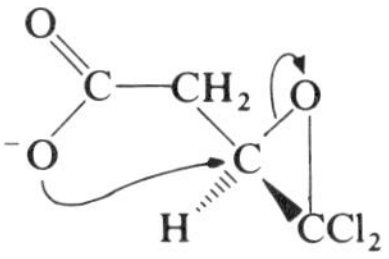 **(114)**

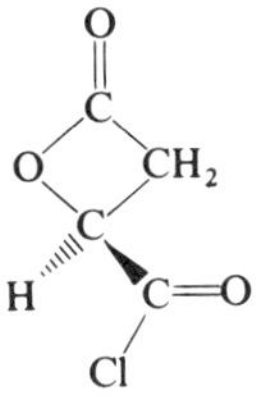

(115) 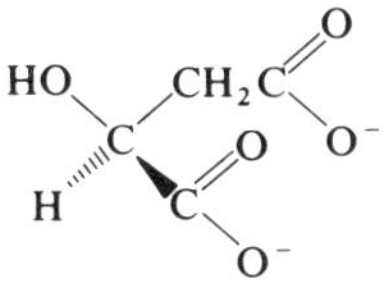 **(116)**

terminal succinate carboxyl groups, which have been added to certain drugs to increase their solubility in water, has been studied recently using methylprednisolone 21-succinate and acetate (**117**) as substrates;[187] [**117**; R = $(CH_2)_2CO_2Na$] undergoes both hydrolysis and 21 $\rightleftharpoons$ 17 acyl transfer, both processes being illustrated in Scheme 3; the product of hydrolysis is methylprednisolone (**118**) and the product of acyl transfer is the "17-ester" (**119**). Intramolecular catalysis by the terminal carboxylate group is seen in both reactions for [**117**; R = $(CH_2)_2CO_2Na$]. While acyl transfer can only be catalysed *via* the carboxyl group acting as a general acid or general base, hydrolysis may occur by nucleophilic or general acid–base catalysis. Using aniline buffers (to trap succinic anhydride as the anilide), about 15–20% of the overall hydrolysis is shown to occur by the nucleophilic mechanism. Intramolecular catalysis favours acyl migration over hydrolysis. Thus, the hydrolysis

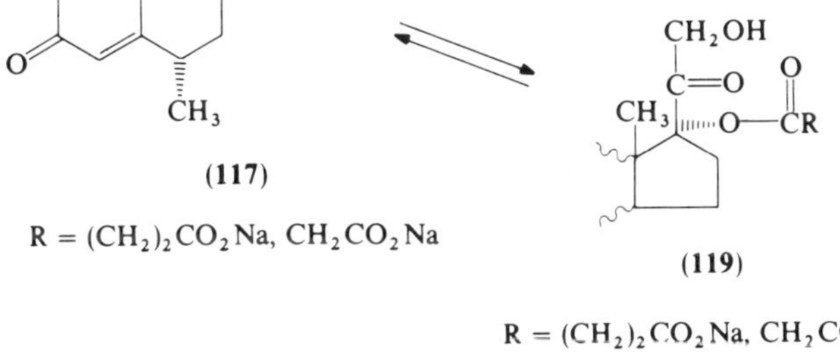

(117)

R = $(CH_2)_2CO_2Na$, CH_2CO_2Na

(118)

(119)

R = $(CH_2)_2CO_2Na$, CH_2CO_2Na

SCHEME 3

of [**117**; R = $(CH_2)_2CO_2Na$] is faster in basic solutions, while acyl migration is the dominant reaction in the catalysed region of the pH–rate profile between pH 7.4 and 3.6.

An amide group is capable of intramolecular participation in the formation of tricyclic 1,3-dioxolan-2-ylium cation (**120**), which can be considered as a model for an electrophilic intermediate which might form from a peptidyl-t-RNA during protein biosynthesis;[188] (**120**) reacts with water and methanol to give (**121**) and (**122**), respectively, and with dimethylamine the amide acetal (**123**) is formed. Intramolecular nucleophilic attack by ureide anion on carboxylate anion has been

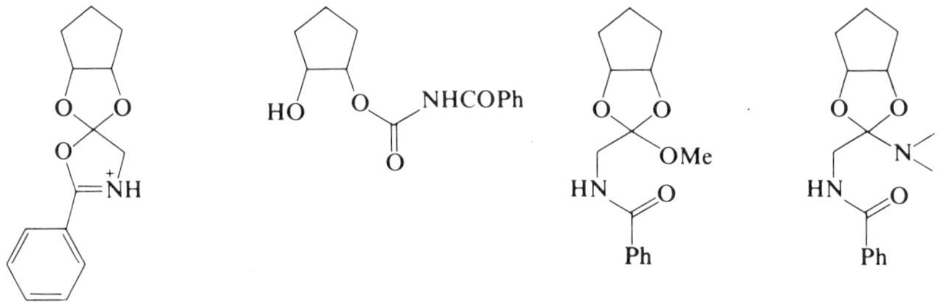

(**120**) (**121**) (**122**) (**123**)

observed[189] in the cyclization of 2,2,3,5-tetramethylhydantoic acid (**124**) to the corresponding hydantoin (**125**).

Keto participation is the predominant pathway in the hydrolysis of the ester (**126**; $n = 2$).[190] This is evident *inter alia* by the high negative entropy of activation for the basic hydrolysis of (**126**; $n = 2$) compared to (**126**; $n = 3, 4$). In (**126**; $n = 4$) direct attack on the carbomethoxy group by base is the only pathway of hydrolysis involved.

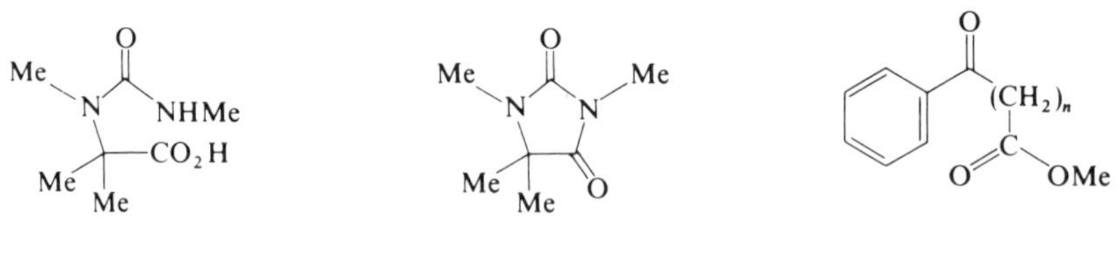

(**124**) (**125**) (**126**)

$n = 2$–4

Intramolecular catalysis of *o*-hydroxy groups is well known and further examples in the *n*-butylaminolysis of catechol monoacetate and methyl *o*-hydroxyphenylacetate have been given.[191] A *cis*-2-hydroxy group in cyclopentyl benzoylglycinate (a simple model for peptidyl-t-RNA) catalyses the *n*-butylaminolysis of that compound by a factor of 300 in acetonitrile;[192] *trans*-2-hydroxy and *cis*- or *trans*-2-methoxy groups are less efficient. The amide group of the benzoylglycinates does not assist the reaction. Intramolecular nucleophilic attack by neighbouring hydroxyl is responsible for the preparation in good yields of substituted tetrahydrofurans.[193] The reactions are NBS oxidations in $CHCl_3$ or CCl_4 at room temperature of 1,3-dioxolanes and -dioxanes. The crucial transition state (**127**) in the oxidation involves hydroxy attack at the highly stabilized carbocation intermediate.

The topic of proton transfer from intramolecularly hydrogen-bonded acids has been reviewed.[194] An intramolecular proton transfer produces a reversal in acylation of 1,3-dimethyl-6-aminouracil (**128**).[195]

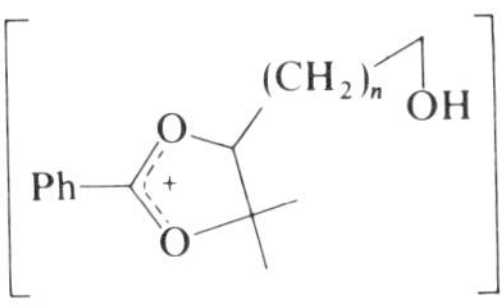

(**127**) 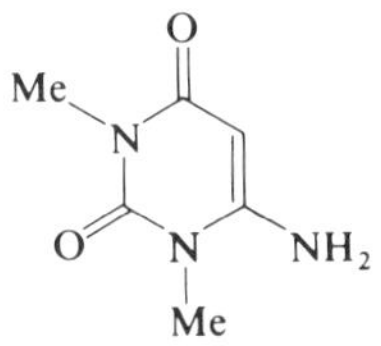 (**128**)

Intramolecular migration (participation?) of the carbonyl π-orbital of the capronium chloride $Me_3\overset{+}{N}(CH_2)_3CO_2MeCl^-$ to the atom adjacent to the $Me_3\overset{+}{N}$ group *via* a five-membered cyclic transition state is proposed for the formation of γ-butyrolactone (and Me_4NCl) in the pyrolysis of the capronium compound.[196]

Association-prefaced Catalysis

The reactions of *N*-acyl-pyrrole, -indole, and -carbazole with hydroxyl ion are inhibited by anionic micelles of sodium lauryl sulphate and accelerated by cationic micelles of CTABr.[197] Rate–surfactant profiles are analysed in terms of the distribution of HO^- and heterocycle between aqueous and micellar pseudo-phase. The kinetics of the hydrolysis of α-aminobenzylpenicillin in acid in the presence of sodium lauryl sulphate micelles obeys the pseudo-phase model and the mechanism involves a protonated oxazolone (**129**) as intermediate.[198]

The kinetics of the basic hydrolysis of carbamates (**22**)[11] and (**73**)[107] in the presence of CTABr have been studied. Rate inhibitions and enhancements and enantioselectivities ranging from 1.0 to 3.3 are observed in the cleavages of the *p*-nitrophenyl derivatives of carbamate (**130**), 1-methylheptyl carbonate (**131**), and α-methoxyphenyl acetate (**132**) in the presence of homomicelles of *N*-hexadecyl-*N*-methylephedrinium bromide (**133**) and comicelles composed of N^α-myristoyl-L-histidine (**134**) and hexadecyltrimethylammonium bromide or (**133**).[199] The micellar (CTABr and sodium 1-dodecanesulphonate) hydrolysis of octano-, *N*-methyloctano-, and a series of ring-substituted benzohydroxamic acids RC(=O)NHOH under acidic and basic conditions has been studied.[200] Normal reaction-rate orders are observed except for the alkaline hydrolysis of octanohydroxamic acid above the cmc when novel pseudo-zero-order kinetics are obtained.

The hydrolysis of *p*-nitrophenyl acetate and hexanoate at pH 8.7 is catalysed by

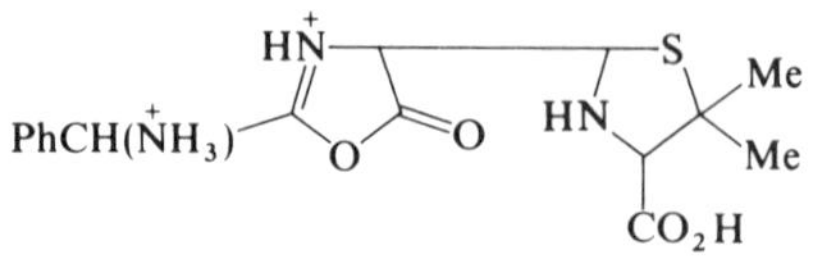

PhCH($\overset{+}{N}H_3$)-oxazolone / thiazolidine structure (129)

(129)

PhCH(OH)CH(Me)N(Me)CO_2PNP

(130)

PNP = *p*-nitrophenyl

C_6H_{13}CH(Me)OCO_2PNP

(131)

PhCH(OMe)CO_2PNP

(132)

PhCH(OH)CH(Me)$\overset{+}{N}$(Me)$_2C_{16}H_{33}$ Br^-

(133)

ImCH_2CH(CO_2H)NHCO$C_{13}H_{27}$

(134)

Im = imidazolyl

CTA micelles and $HS(CH_2)_nCO_2^-$ ($n = 2, 5, 11$) or the imidazoles (**135**) as anions.[201] The results obtained in this study indicate that functionalized counter-ion surfactants may provide simple but effective micellar catalytic systems whose kinetic features compare well with those of functional comicelles. Compound (**112**; 4-nitro) undergoes rapid alkaline hydrolysis in the presence of CTABr and a change of mechanism (compared to non-surfactant conditions) occurs.[185] The alkaline hydrolysis of aspirin (acetylsalicylic acid) in the presence of CTABr has been explained in terms of the pseudo-phase model.[202]

The effects of metal ions Co(II), Zn(II), and Ni(II) on the kinetics of hydrolysis of the *p*-nitrophenyl ester (**136**), the hydroxamic acid (**137**), and the imidazole amide (**138**) in the presence of CTABr and other cationic surfactants have been

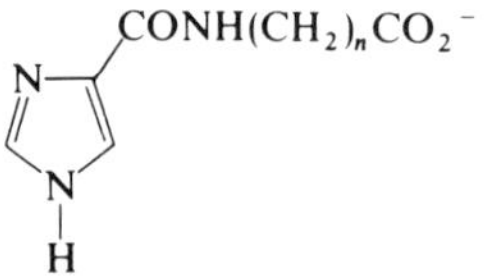

Imidazole-$CONH(CH_2)_nCO_2^-$

(135)

p-$O_2NC_6H_4O_2C$CH(NHCO$(CH_2)_n$H)CH_2Ph

(136)

$n = 1, 5, 9, 11, 15$

Me$(CH_2)_7$CH[$(CH_2)_5$Me]CONHOH

(137)

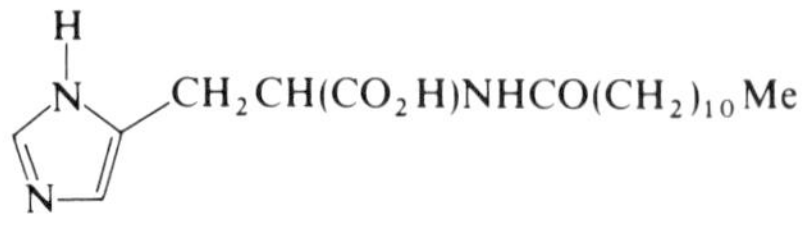

Imidazolyl-CH_2CH(CO_2H)NHCO$(CH_2)_{10}$Me

(138)

X-substituted isochroman with EtO, OEt

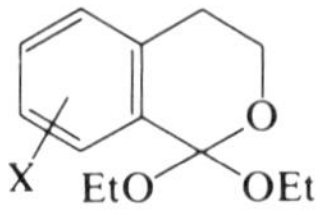

(139)

explored.[203] The rates of hydrolysis of 6- or 7-substituted (MeO and NO_2 groups) 1,1-diethoxy-3,4-dihydrobenzo-2-pyrans (**139**) in the presence of sodium dodecyl sulphate (0.024 M) are 76 times faster than in water and the Hammett ρ values for the hydrolysis of these orthoesters are both about -1.90.[204]

The pseudo-phase model of micellar catalysis fails for the hydrolysis of the phosphate ester p-$O_2NC_6H_4OP(O)(OET)OLi$ in aqueous CTABr at high concentrations of sodium hydroxide.[205] A mechanism similar to that operating in phase-transfer catalysis is more suitable. Further studies are reported on the catalytic cleavage of *p*-nitrophenyldiphenyl phosphate, *p*-nitrophenyl acetate, and *p*-nitrophenyl hexanoate by *o*-iodosobenzoate, *o*-iodoxybenzoate, and 5-(*n*-octyloxy)-2-iodosobenzoate (**140**) in aqueous CTACl at pH 8.[206] The system (**140**)/CTACl is the best catalyst and the phosphate is the best substrate. Dephosphorylation of *p*-nitrophenyldiphenyl phosphate by HO_2^- or *m*-chloroperoxybenzoate is markedly speeded by CTA micelles.[207] The pseudo-phase model again fails at high HO^- concentrations for the aqueous dechlorination of DDT, DDD, and $(p\text{-}ClC_6H_4)_2CHCH_2Cl$ at 25° in the presence of CTABr micelles.[208]

In a study of various *p*-nitrophenyl esters as substrates, zwitterionic long-chain fluorocarbon hydroxamates, *e.g.* (**141**), are shown to be more efficient micellar nucleophiles than the simpler hydrocarbon hydroxamates.[209]

Polymer–surfactant binding effects on the rates of the pH-independent hydrolysis of 1-benzoyl-1,2,4-triazole have been investigated.[210] Atactic poly(*N*-vinylpyrrolidone) and sodium dodecyl sulphate binding has been studied by conductivity measurements. An additional pseudo-phase of mixed micelles of polymer and surfactant is proposed. The same group[211] has shown that this polymer or poly(acrylic acid) do not retard the water-catalysed hydrolysis of compounds (**34**), but neutral atactic and syndiotactic poly(methacrylic acid) do. Poly(ethylenimines) increase the rate of hydroxide attack by 6–180 times in the hydrolysis of (**21**).[10] This rate enhancement is largely a reflection of the increased local concentration of HO^- in the vicinity of the polymeric cation sites.

The basic hydrolysis of *p*-nitrophenyl laurate in the presence of surfactant and polymerized surfactant vesicles, prepared from $[n\text{-}C_{15}H_{31}CO_2(CH_2)_2]\text{-}\overset{+}{N}(Me)(CH_2C_6H_4CH{=}CH_2)Cl^-$ has been explored.[212] Aminolysis of the substrate by ethylenediamine has also been studied. Previously reported[213] endovesicular reactions in the cleavage of *p*-nitrophenyl diphenylphosphate by HO^- or F^- in vesicles of dihexadecyldimethylammonium bromide may be in error[214] and other examples[214] of complex kinetics in vesicular reactions may be artifactual.

Further work has been reported on the catalysis (by various linear and cyclic peptides containing histidine and other amino-acid residues) of hydrolysis of several *p*-nitrophenyl esters.[215] A thermodynamic study involving measurement of the isokinetic temperature (β) has been made for the deacylation of *p*-nitrophenyl *N*-dodecanoyl-D(L)-phenylalaninates catalysed by L-histidine derivatives in artificial membrane systems.[216] Bilayer surfactants, didodecyldimethylammonium bromide and di(tetradecyl)dimethylammonium bromide, with and without cholesterol and CTABr, have been used. The β values obtained have been related to changes in the hydrophobic micro-environments. Further similar work by the same group[217]

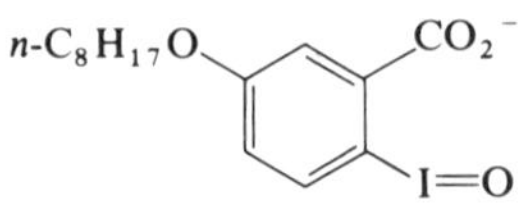

$n\text{-}C_8H_{17}O$ CO_2^- I=O

(140)

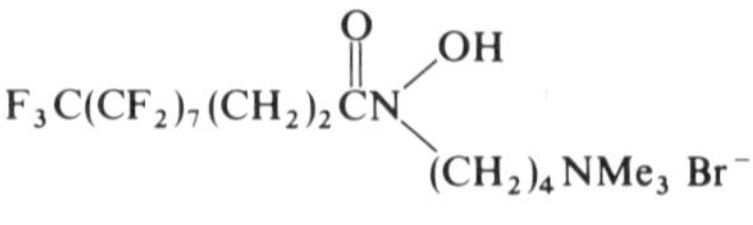

$F_3C(CF_2)_7(CH_2)_2C(=O)N(OH)(CH_2)_4NMe_3\ Br^-$

(141)

R^1 O N R^2

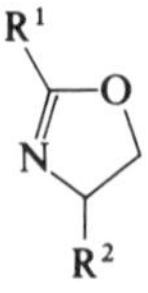

(142)

involving stereoselective deacylations of *p*-nitrophenyl *N*-acyl-D(or L)-phenylalaninates D(or L)-S_n ($n = 2$, 12, 16) and *N*-benzyloxycarbonyl-D(L)-phenylalaninates D(L)-ZS has been reported.

A number of papers have appeared involving catalysis by cyclodextrins. Fastrez[218] has continued studies[219] on the hydrolysis of oxazolones of type (**142**), which form inclusion complexes with the α- and β-cyclodextrins used. The reaction is enantioselective. α-Cyclodextrin reduces the rate of bromination of anisole and *p*-methylanisole in aqueous KBr solutions (0.05–0.1 M).[220] The retardation is principally attributed to the formation of a cyclodextrin–tribromide-ion complex. The intermediacy of a ternary complex, cyclodextrin–anisole–Br_2 cannot be eliminated. The hydrolysis of *m*- and *p*-nitrophenyl acetates by 6-deoxy-6-alkylthio-β-cyclodextrins and the corresponding sulphoxides has been reported.[221] The same group[222] has hydrolysed α-naphthyl acetates with cyclohexylamylose with the observation of "α-selectivity"; *i.e.* the α-isomers hydrolyse more rapidly than the β-. However, cyclooctaamylose shows "β-selectivity" and the selectivity of cycloheptaamylose falls between the other two amyloses. Another Japanese group has reported on the effect of pressure (up to 2 kbar) at 25° and pH 6.9 in 0.05 M Tris buffer on the hydrolysis of *p*-nitrotrifluoroacetanilide catalysed by α- and β-cyclodextrins.[223]

Metal-Ion Catalysis

$Co(NH_3)_5(H_2O)(ClO_4)_3$ catalyses the hydrolysis of 1-benzoylbenzotriazole and the ratio of rate constants for the cobalt-catalysed reaction compared to the hydroxyl ion-catalysed reaction is $(1.2\text{–}5.7) \times 10^3$.[119] The tin compound $Bu_2Sn(OBz)_2$ catalyses the aminolysis by $PhCH_2NH_2$, $(PhCH_2)_2NH$, and $(Me_2CH)_2NH$ in C_6H_6 of 2,4-$R(O_2N)C_6H_3OAc$ (R = H, O_2N). $Et_2Pb(OBz)_2$ catalyses the aminolysis of $PhCH_2O_2CNHCH_2CO_2C_6H_4NO_2$-*p* by $H_2NCH_2CO_2CMe_3$.[224] Cyclic transition states are proposed for both reactions.

The synthesis of the D-arabino-(**143**) and D-ribo-(**144**)ortholactones and their treatment with activated zinc in boiling aqueous alcohols to give the unsaturated esters (**145**) and (**146**), respectively, has been described.[225]

The life-times of the anionic tetrahedral intermediates (**31**) in the hydrolysis of methyl acetate and methyl trifluoroacetate are related to the catalytic efficiencies of the zinc hydroxide catalyst (**147**).[18] $Fe_2(SO_4)_3 \cdot 7H_2O$, $CuSO_4 \cdot 5H_2O$, $CoSO_4 \cdot 7H_2O$, and $NiSO_4 \cdot 7H_2O$ have been used as catalysts for the esterification of various RCO_2H (R = simple alkyl, $BrCH_2$, Cl_2CH, Cl_3C, *etc.*).[226] Two papers[227,228] describe the use of metal catalysts for the decomposition of

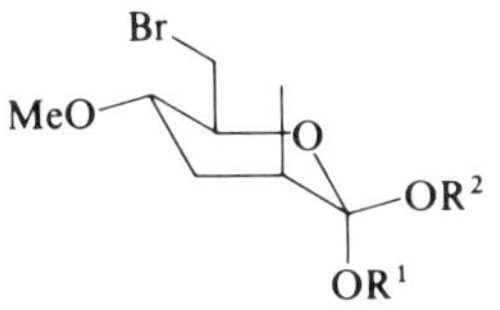

Br
MeO
O
OR²
OR¹

(143)

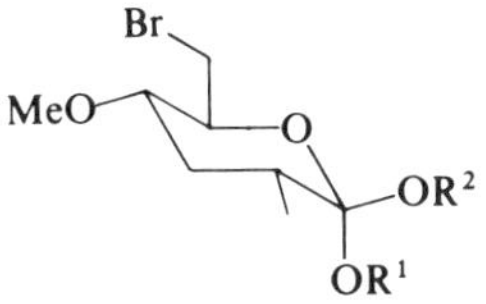

Br
MeO
O
OR²
OR¹

(144)

R^2 = Et; R^1 = Et, Me
R^2 = Me; R^1 = Et, Me

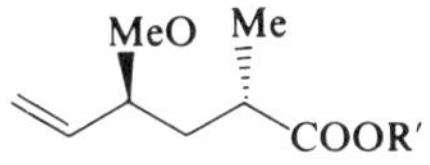

MeO Me
COOR′

(145)

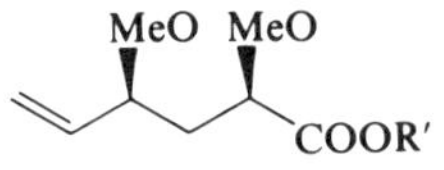

MeO MeO
COOR′

(146)

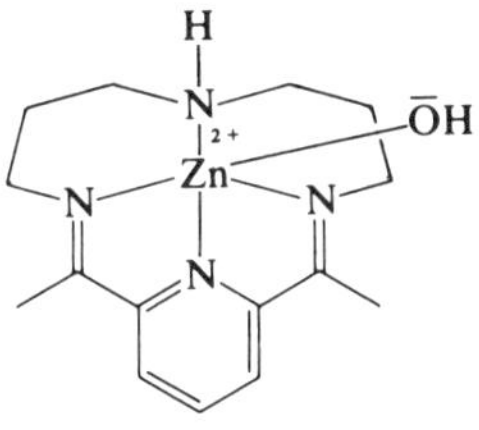

H
N
Zn
2+
OH
N
N
N

(147)

phosphorus esters. Nucleophilic displacement of ethoxide ion on *p*-nitrophenyl diphenylphosphinate, $Ph_2P(O)OC_6H_4NO_2$-*p*, is catalysed strongly by Li^+ and K^+ and to a lesser extent by mixtures of Li^+ and K^+ with [2.1.1]- or [2.2.2]-cryptand or dicylohexano-18-crown-6.[227] The second paper[228] deals with the basic hydrolysis of coordinated 4-nitrophenyl phosphate esters in *cis*-$[Co(en)_2(OH)O_3POC_6H_4NO_2]$ (**148**) over the pH range 7–14. The release of *p*-nitrophenol is monitored at 400 nm and ^{31}P-NMR spectroscopy is also employed to probe the reaction. The hydrolysis is 10^5-fold faster than that of the uncoordinated ester under the same conditions.

Metallic catalysis of a variety of synthetic procedures has also been reported. Thus, internal macrocyclic esterification of ω-hydroxycarboxylic acids,

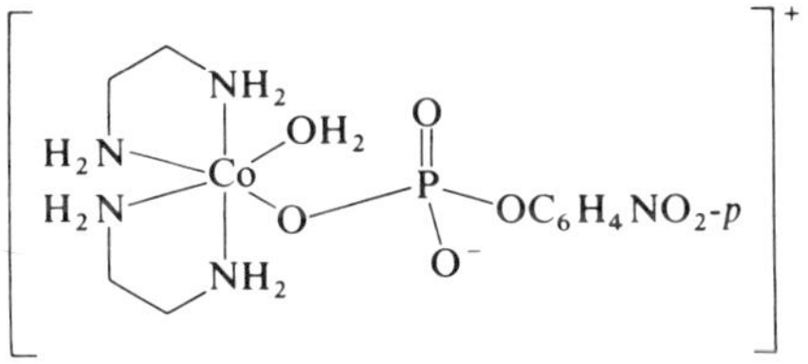

NH_2
OH_2
O
H_2N
Co
H_2N
O
P
$OC_6H_4NO_2$-*p*
O^-
NH_2
+

(148)

$HO(CH_2)_nCO_2H$ (n = 7, 8, 10, 11, 14, 15) has been achieved using a tin-mediated "template-driven" esterification process.[229] The application of this method to the synthesis of the macrolide antibiotics is discussed. The kinetics of the alkoxycarbonylation of RCH_2Cl (R = Bu^n, $HOCH_2$, Ph, Ac, MeO_2C, CN) with CO in MeOH containing $Co_2(CO)_8$ to give the methyl esters RCH_2CO_2Me follow a LFER with σ^*.[230] Double carbonylation of benzoylformyl complexes of Pd(II) and Pt(II) with CO in the presence of nucleophiles γ^- (γ^- = HO^-, R'_2N^-) (Scheme 4) has been

$$RX + 2CO + \gamma^- \xrightarrow{\text{cat.}} RCOCOY + X^-$$

R = hydrocarbyl, X = halide

SCHEME 4

probed.[231] Oxidative addition of the organic halide to the metal is followed by CO insertion into the metal–carbon bond. Thus, a metal–acyl species forms as an intermediate and nucleophilic attack on the coordinated CO molecule of RCO–Pd–CO gives RCO–Pd–$CONR'_2$ which then undergoes reductive elimination to give $RCOCONR'_2$. Finally, the regiospecific rhodium-catalysed ring-expansion of aziridines to β-lactams (Scheme 5) has been achieved in quantitative yield in C_6H_6 at

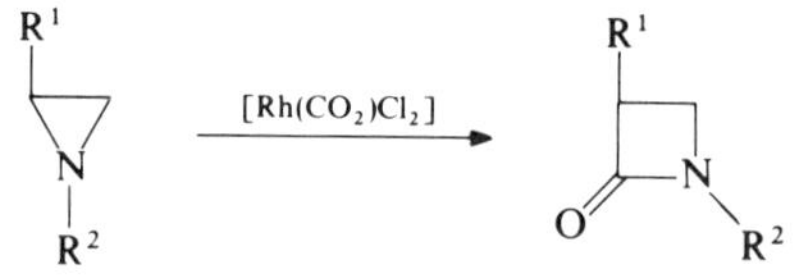

[Rh(CO$_2$)Cl$_2$]

SCHEME 5

90° and 20 atm with the catalyst shown and also 1,5-cyclooctadienerhodium(I) chloride dimer.[232] The mechanism of this first example of direct ring-expansion and carbonylation of an aziridine probably involves the sequence shown in Scheme 6.

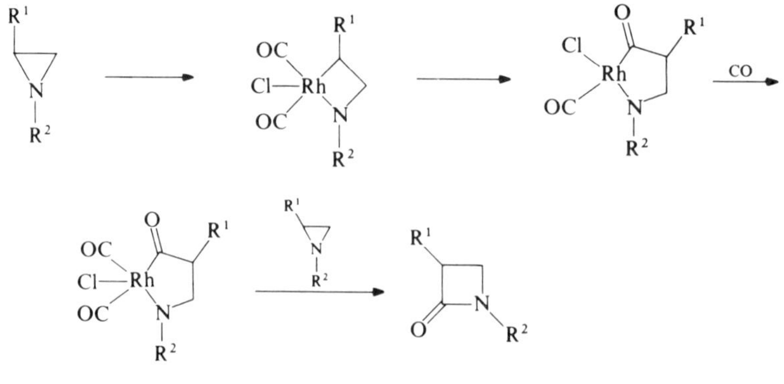

SCHEME 6

Decarboxylation

Decarboxylations which have been studied include that of methyl(ethyl)malonic acid (**149**) in resorcinol or pyrocatechol and of $RCH(CO_2Et)_2$ (R = Me, Et).[233] The isokinetic temperatures for the reactions have been determined. The decarboxylation of phthalide-3-carboxylic acid (3-oxo-1,3-dihydroisobenzofuran-1-carboxylic acid) (**150**) has been examined in DMSO, neat and in the presence of aromatic aldehydes,[234] which are efficiently trapped to produce mixtures of 3-arylidene phthalides (**151**) and 3-(arylhydroxmethyl) phthalides (**152**). Kinetic and other evidence suggests that a tight cyclic transition state is involved in the reaction with aldehydes. Imidazo[1,2-*a*]pyridines play an important rôle as intermediates in

$MeC(Et)(CO_2H)_2$

(**149**)

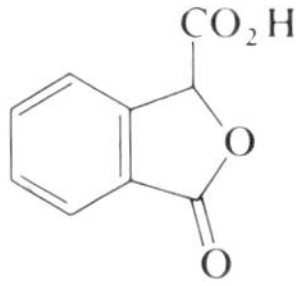

(**150**)

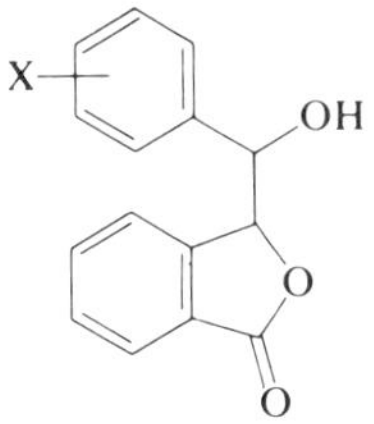

(**151**)

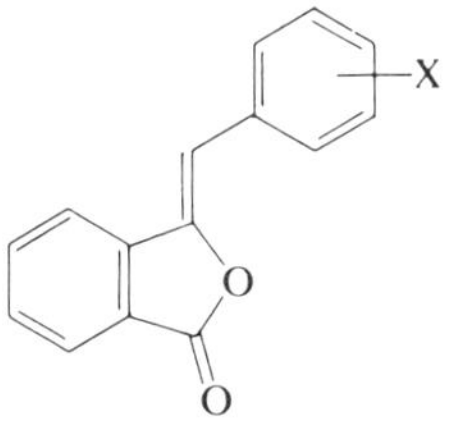

(**152**)

the decarboxylative dimerization of maleic anhydride to dimethylmaleic anhydride (**153**); the key step involves a Michael addition.[235] The decarboxylation of trichloroacetic acid to CO_2 and trichloromethane has been the subject of a kinetic study performed in solution of 1,3,5-trinitrobenzene.[236] The slow step is the unusually rapid unimolecular breakdown of the trichloroacetate ion. In two accompanying papers,[237] Grigg's group has shown how decarboxylative transamination leads to new syntheses of various heterocycles and they also show that the presently accepted mechanism for decarboxylative transamination is incorrect. They propose[237a] a mechanism in which the imine undergoes decarboxylation *via* the zwitterionic form (**154**) generating a 1,4-dipole (**155**). This mechanism is demonstrated by trapping intermediates (**155**) with a range of dipolarophiles.

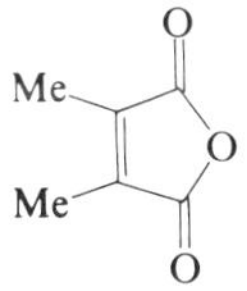

(**153**)

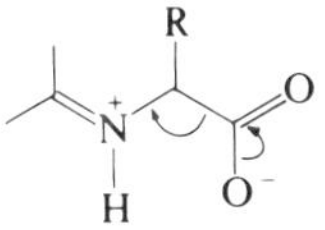

(**154**)

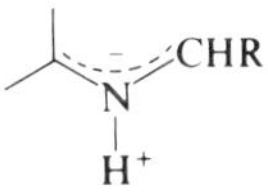

(**155**)

Enzymic Catalysis

General

Several useful reviews have appeared during the period under review here. In a review titled "Binding forces, equilibria, and rates: new models for enzymic catalysis," Rebek has discussed the shift in an enzyme's conformational equilibrium by activators and inhibitors (*allosteric* effects).[238] Scott and coworkers provide an excellent review of their own and other work on the use of high-resolution ^{13}C-NMR spectroscopy to study enzyme–substrate and enzyme–inhibitor complexes.[239] The review also deals to some extent with other vital techniques in the observation of these complexes, *i.e.* X-ray and cryoenzymology. Proton transfer from intramolecularly hydrogen-bonded acids, a process of cruicial importance in enzyme action, has been reviewed.[194]

A restriction in the use of enzymes in organic synthesis, namely that they are incompatible with the systems used in organic reactions, appears to have been overcome at least in certain well-defined instances. Zaks and Klibanov[240] have shown that the well-characterized enzyme porcine pancreatic lipase, when added as a powder to the reaction of tributyrin (glyceryl tributyrate) with various alcohols, catalyses the trans-esterification reaction to give the appropriate mono-ester of butyric acid and dibutyrin. The enzyme acquires some new properties in this almost anhydrous medium. One important property is increased substrate specificity. This work has recently been extended to the preparation of optically active alcohols and their propionic esters in large amounts using hog liver carboxyl esterase as a stereoselective catalyst and methyl propionate in the matrix ester.[241]

$$\text{Tributyrin} + HOCHR^1R^2 \longrightarrow CH_3CH_2CH_2\overset{\displaystyle O}{\overset{\|}{C}}OCHR^1R^2 + \text{Dibutyrin}$$

$$R^1 = (CH_2)_n\ (n = 2, 5, 8\ldots),\ R^2 = CH_3;$$
$$R^1 = CH_3,\ R^2 = (CH_2)_5CH_3,\ \textit{etc.}$$

Serine Proteinases

Further steps have been made in the synthesis of hosts that mimic serine transacylases.[242] The hosts are 20-membered macrocycles made up by attaching to one another aryloxy, cyclic urea, pyridyl, biphenyl ethylene, methylene, and oxygen units.

The pH dependence of pre-steady-state and steady-state kinetics for the porcine pancreatic β-kallikrein-B-catalysed hydrolysis of *N*-α-carbobenzoxy-L-arginine *p*-nitrophenyl ester have been studied between pH 2.4 and 8 and the results have been interpreted in favour of a three-step mechanism involving an acyl-enzyme intermediate.[243] The acylation transition state of the acetylcholinesterase-catalysed hydrolysis of *o*-nitroacetanilide has been examined with the aid of pH– and pD–rate profiles and solvent isotope effects.[244] In further studies of trypsin-catalysed hydrolysis at low temperatures, Scott and coworkers have detected by ^{13}C-NMR and cryoenzymology an acyl-trypsin intermediate in the hydrolysis of the highly specific substrate, *N*-α-benzyloxycarbonyl-L-lysine *p*-nitrophenyl ester.[245] The kinetics of the reaction are in agreement with those determined by UV. The effects of the cryoenzymological conditions on rate constants K_m and K_s have been measured.[246] Other low-temperature work (0 to $-30°$) has probed the reactions of trypsin with the *p*-nitroanilides of *N*-α-carbobenzoxy-L-lysine and -L-arginine and *N*-α-benzoyl-L-arginine.[247] A "burst" phase in the reactions has been interpreted in favour of a temperature/solvent-induced isomerization of trypsin to a less catalytically efficient from rather than a tetrahedral intermediate.

Several papers have appeared on chymotrypsins. Interest in enzyme-activated irreversible inhibitors ("suicide" inhibitors) continues and the haloenol lactone systems, 5-halomethylene tetrahydrofuran-2-ones and 6-halomethylene tetrahydropyran-2-ones,[248] and the benzoxazinones (**156**), 2-ethoxy- and 2-(trifluoromethyl)-4*H*-3,1-benzoxazin-4-ones,[249] are effective "suicide" inhibitors of chymotrypsin. Several characterization methods including EPR spectroscopy have been used to study the deactivation of α-chymotrypsin–CNBr–Sepharose 4B conjugates in aliphatic alcohols.[250] ^{13}C-NMR has revealed two interconverting forms of the hemiacetal complex formed by α-chymotrypsin and *N*-acetyl-L-phenylalaninol.[251] Rate constants for the acylation of α-chymotrypsin by substituted *N*-benzoylimidazoles of type (**157**) have been determined by proflavin displacement from the active site.[252] *N*-(4-Fluorophenyl)-*N*-phenylcarbamoyl chloride, *p*-$FC_6H_4(C_6H_5)$NCOCl, reacts with α-chymotrypsin to give a covalently modified protein which has no catalytic activity.[253] ^{1}H- and ^{19}F-NMR have been used to probe the active site of the enzyme and the fluorophenyl ring is found preferentially there. Active-site mapping of the serine proteases, Human leukocyte elastase, thepsin G, porcine pancreatic elastase, rat mast cell proteases I and II, bovine chymotrypsin A-α, and *Staphyloccoccus aureus* protease V-8 has been carried out with a series of tripeptide thiobenzyl ester substrates of general type Boc-Ala-Ala-AA-SBzl, where AA represents one of 13 amino-acids.[254]

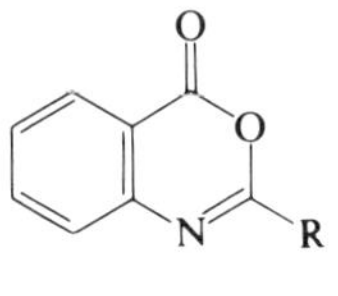

(**156**)

R = EtO, CF_3

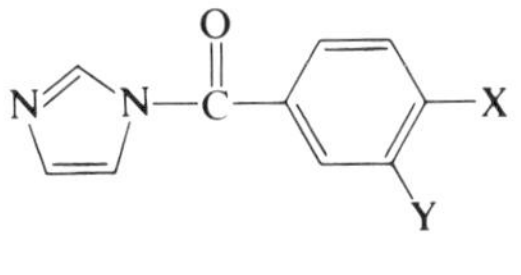

(**157**)

X = NMe_2, OMe, H, Me, Cl, CN, Y = H
X = H, Y = NO_2

Deacylation is the rate-limiting step for the Human leukocyte elastase-catalysed hydrolysis of specific peptide esters, amides, and *p*-nitroanilides.[255] The inhibition of the same enzyme and of porcine pancreatic elastase, human leukocyte Cathepsin G, and chymotrypsin by 3-chloroisocoumarin and 3,3-dichloropthalide has been reported.[256] The catalytic activity of HCHO–alkylamine–salicylic acid polymers (structurally similar to serine hydrolase) during the hydrolysis of p-$NO_2C_6H_4OAc$ has been examined.[257]

Thiol Proteinases

Treatment of ethanolamine ammonia-lyase, which is an adenosylcobalamin-dependent enzyme that catalyses the rearrangement of ethanolamine and other vicinal amino-alcohols to oxo compounds and ammonia, with the sulphydryl group blocking reagent, methyl methanethiosulphonate, produces a compound with reduced catalytic activity.[258]

Metallo-proteinases

Carboxypeptidase A is a zinc metalloprotease and is competitively inhibited by DL-2-benzyl-3-formylpropanoic acid which exists as an equilibrium mixture of free aldehyde, hydrated aldehyde, and lactone hemiacetal (**158**) (Scheme 7).[259] The carboxypeptidase-A-catalysed ester hydrolysis has been examined by multi-channel resonance Raman (RR) spectroscopy under cryoenzymological conditions.[260] Two papers have appeared from the same group[261, 262] on the zinc-ion-catalysed hydrolysis of esters and the implications of this work for the actions of metalloenzymes, *e.g.* carbonic anhydrase and carboxypeptidase A, are discussed. Further carbonic anhydrase models which facilitate the $HCO_3^- \rightleftharpoons CO_2$ conversion have been prepared by another group.[263] They are tris(4,5-di-*n*-propyl-2-imidazolyl)-phosphinezinc(II) and bis(4,5-di-isopropyl-2-imidazolyl)-2-imidazolyphosphine-zinc(II).

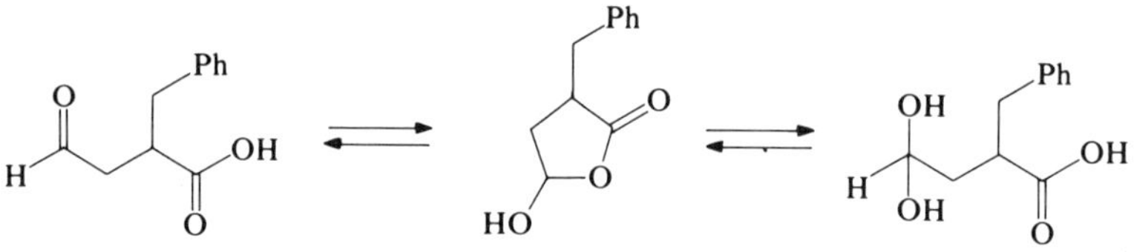

(**158**)

SCHEME 7

Deiters and Holmes[264] have continued their work[265] on the enzyme–substrate system Ca^{2+}-staphylococcal nuclease–pdTp, where pdT_p represents deoxythymidine diphosphate. The 2-oxoglutarate binding site of prolyl 4-hydroxylase was studied by assaying the inhibitory potential of 24 aliphatic and aromatic compounds.[266]

Other Enzymes

Four papers have appeared on β-lactamase work. The covalent enzyme–substrate complex found during β-lactamase-catalysed hydrolysis of sodium 3-dansylamidomethyl-7-β-(thienyl-2′)-acetamido-ceph-3-em-4-oate has been studied.[267] The same group has recently examined the β-lactamase-catalysed hydrolysis of pyridine-2-azo-4′-(N',N'-dimethylaniline)cephalosporin (**159**) and of cephaloridine (**160**).[268] *Bacillus cereus* β-lactamase-catalysed hydrolysis of benzylpenicillin, furylacryloylpenicillin, cephaloridine, cephalothin, nitrocefin, and carbeniccillin (**161**) have been looked at;[269, 270] pH profiles and solvent kinetic isotope effects have been used to probe the mechanisms involved.

Kinetic, magnetic, primary, and secondary kinetic deuterium isotope effects, and secondary equilibrium isotope effects have been examined for the *cis*, *cis*-muconate cycloisomerase-catalysed interconversion of *cis*, *cis*-muconate and (+)-muconolactone (Scheme 8).[271, 272] The results indicate that a carbanion intermediate is involved in which proton removal appears to be the slow step.

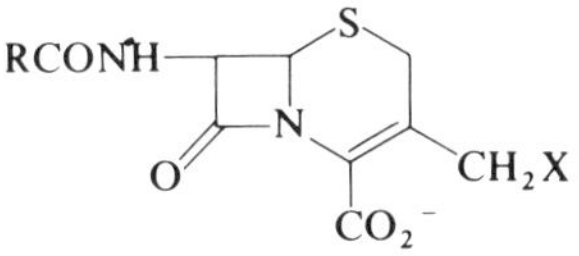

(**159**) R = 2′-thienylacetyl, X = pyridinium–N=N–$C_6H_4NMe_2$-*p*

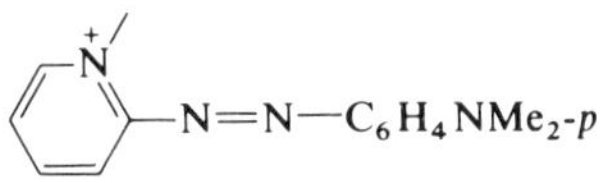

(**160**) R = 2′-thienylacetyl, X = –N^+ (pyridinium)

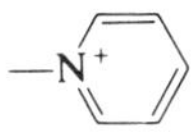

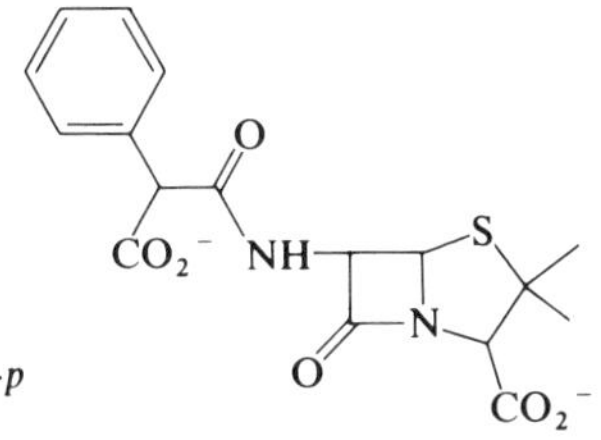

(**161**)

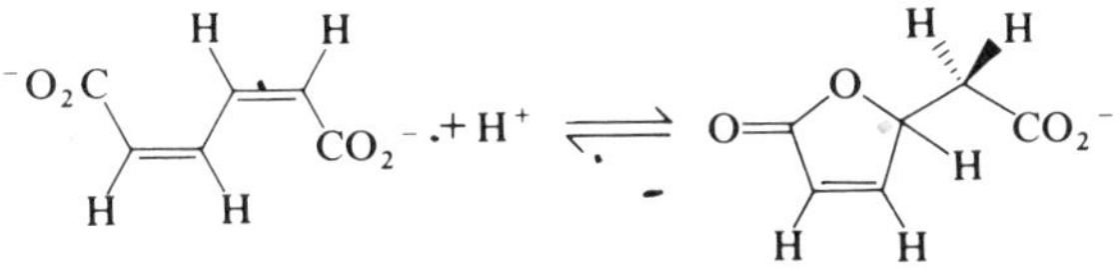

SCHEME 8

Sulphydryl groups are released by nucleophilic attack of amine on the thio-ester bond of α_2-macroglobulin in a bimolecular reaction in which the access of the amines is sterically hindered.[273] The reaction of bovine opsin with 9-*cis*-retinoyl fluoride (**162**) inactivates the enzyme. *All-trans*-retinoyl fluoride is four-fold less potent as an inactivator of bovine opsin.[274] The dipeptidyl peptidase-IV-catalysed hydrolysis of aminoacyl *p*-nitroanilides has been studied recently.[275] Stereospecific pig liver esterase-catalysed hydrolysis of monocyclic *meso*-1,2-diesters provides a good route to enantiomers of chiral lactones.[276] In an extension of this work the same group have prepared both enantiomers of 3-substituted valerolactones (**163**) of 100% e.e. starting from 3-mono-substituted glutaric acid diesters and using the same enzyme.[277]

Nanosecond laser flash photolysis has produced and identified the vitamin K semiquinone radical from vitamin K dihydroquinone.[278] Its formation and decay in the presence of vitamin-K-dependent carboxylase (epoxidase) has been studied. By using samples of optically active 2-tritioacetoacetate, the stereochemical course (retention or inversion) of the decarboxylation of acetoacetate catalysed by acetoacetate decarboxylase (AAD) has been examined.[279]

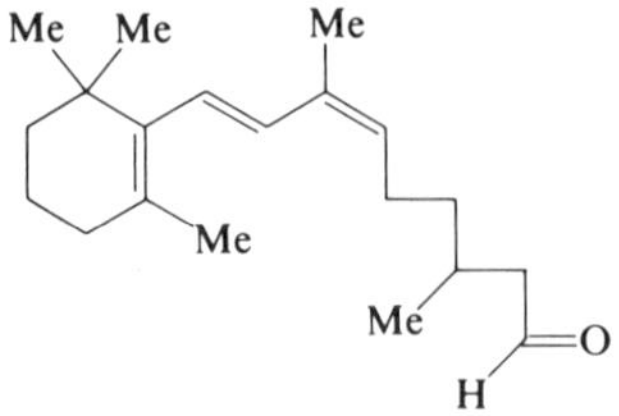

(162) 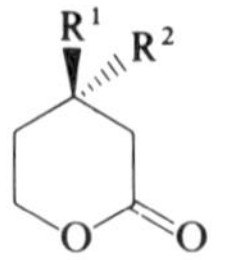 (163)

α-Amino-acid decarboxylases have been mimicked by the reaction of *N*-substituted and α,α-disubstituted amino-acids with carbonyl compounds to produce 1,3-dipolar species.[237b] The cyclization of acid (**124**) to the hydantoin (**125**) provides a model for the carboxylation of the cofactor biotin by hydrogen carbonate.[189] The dioxolon cation (**120**) can be considered a model for the electrophilic intermediate which might form from a peptidyl t-RNA during protein biosynthesis.[188] *Cis*-2-Hydroxycyclopentyl benzoylglycinate is a simple model for peptidyl t-RNA itself (see p. 56). Finally, valine methylamide $Me_2CHCH(NH_2)CONHMe$ has been proposed as a model for the *N*-terminal valine residues of hemoglobin.[280]

Non-carboxylic Acids

Phosphorus-containing Acids

Non-enzymic Reactions

Phenyl [(*R*)-^{16}O,^{17}O,^{18}O]phosphate, 2,4-dinitrophenyl[(*R*)-^{16}O,^{17}O,^{18}O]-phosphate (**164**), and [(*R*)-^{16}O,^{17}O,^{18}O]phosphocreatine have been synthesized and in each case the stereochemical course of methanolysis has been examined.[281] Complete inversion of configuration at phosphorus is observed for each substrate and it is clear that metaphosphate ion, PO_3^-, if it is involved as a true intermediate, does not leave the solvent cage in which it is generated. A mechanistic distinction between a "pre-associative concerted" path (S_N2-like) and a "pre-associative step-wise" path (*via* a short lived PO_3^- intermediate) cannot be made. The same group[282] has studied the migration of the phospho group of 2-phosphopropane-1,2-diol to the 1-position on heating in aqueous acid. Two mechanistic pathways have been discerned. The "direct route" proceeds with retention of configuration at phosphorus and involves a pseudo-rotating penta-coordinated intermediate and in the "phospho-diester route" a cyclic 1,2-phospho-diester is formed. Stereoelectronic effects have been probed in the hydrolysis of methyl ethylene phosphate (**165**).[283] The alkaline hydrolysis of S_P-methyl methylphenylphosphinate (**166**) in 60% $H_2{}^{18}O$ occurs with > 90% inversion of configuration at phosphorus.[284]

Kirby's method[21] involving study of increasing electron demand at oxygen as a probe to examine the relationship between bond-length and reactivity (see p. 37) has been applied to phosphate monoester dianions, $ROPO_3^{2-}$ and phosphate triesters (**167**).

Participation by neighbouring carboxamide occurs in the hydrolysis of the nitrogen-substituted carboxamides of phosphoenolpyruvate (**168**).[285] The reactive cyclic intermediate proposed is of type (**169**), oxygen rather than nitrogen participation being favoured. This work shows that amides can become phosphorylated during processes that involve interaction of peptides and nucleotides or during phosphate-transfer processes.

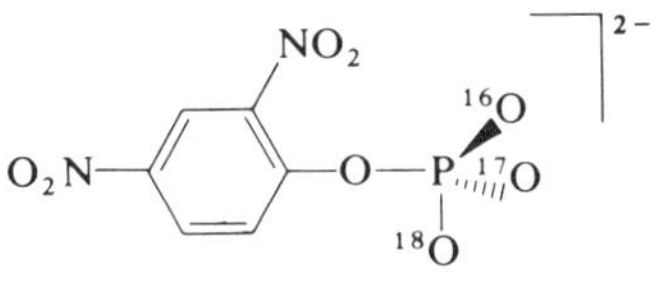

(**164**) 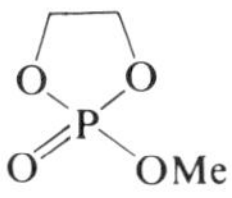 (**165**) 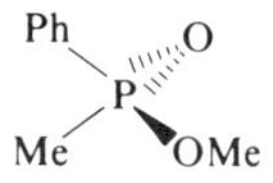 (**166**)

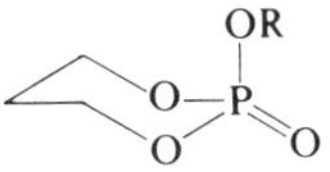

(**167**) 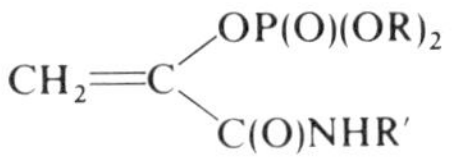 (**168**) (**169**)

R = Et, R′ = Pr^n
R = Et, R′ = Ph

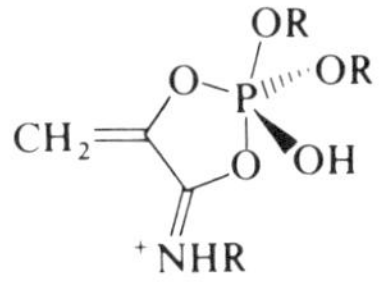

Nucleophilic assistance to nucleophilic substitution at phosphorus, specifically by fluoride ion, in the activation of alcoholysis of the P–X (X = Cl, F) bond in *cis*- and *trans*-2-halogeno-5-(chloromethyl)-5-methyl-2-oxo-1,3,2-dioxaphosphorinanes and related compounds has been reported by Corriu and coworkers.[286] Ketophosphonium and vinyloxyphosphonium bromides $RO(R^1)(R^2)\ {}^+P\ CH_2COPh\ Br^-$ have been isolated as intermediates in the reactions of α-bromoacetophenone with trineopentyl phosphite, dineopentyl phenylphosphonite, and neopentyl diphenylphosphinate.[287]

Enzymic Reactions

Direct spectroscopic observation (at 440 nm) of a Brewer's yeast pyruvate decarboxylase-bound enamine produced from the conjugated α-keto-acid (E)-4-(4-chlorophenyl)-2-oxo-3-butenoic acid and the enzyme has been reported. The band was attributed to a thiamine-diphosphate-bound intermediate.[288] Cyclodiphosphates have been postulated as intermediates in the reactions of cyanogen bromide with adenosine 5′-*O*-(1-thiodiphosphate) and 5′-*O*-(2-thiotriphosphate) which produce in neutral to alkaline aqueous solutions ADP and ATP, respectively, in good yields.[289] A stable serine ester complex showing a single peak at −4.7 ppm in the ^{31}P-NMR spectrum has been formed in the reaction of α-chymotrypsin with the axial triester 2-(2,4-dinitrophenoxy)-2-exo-*trans*-5-6-tetramethylene-1,3,2-dioxaphosphorinane (**170**).[290] Carbamoylphosphate synthetase from *E. coli* catalyses the bicarbonate-dependent β,γ-bridge to β-nonbridge positional oxygen exchange in β,γ-^{18}O-bridge ATP. This has been shown by ^{31}P-NMR spectroscopy.[291]

The stereochemical course of phospho transfer catalysed by adenylosuccinate synthetase[292] and by human prostatic acid phosphatase[293] has been probed using ^{31}P-NMR. The stereochemical course of a thiophosphoryl group transfer catalysed by mitochondrial phosphoenolpyruvate carboxykinase has also been charted[294] as has that of the restrction endonuclease EcoRl hydrolysis of the R_p diastereomer of d(pGGsAATTCC), an analogue of d(pGGAATTCC) containing a chiral phosphorothioate group at the cleavage site.[295]

At least two intermediates have been formed in detectable amounts in the D-ribulose-1,5-biphosphate carboxylase/oxygenase reaction,[296] when a [^{32}P]-orthophosphate substrate is used. The step-wise mechanism for phosphoenolpyruvate carboxylasc action has been supported by recent studies of the bicarbonate-dependent dephosphorylation of phosphoenol-α-ketobutyrate.[297]

A recent paper in which (**148**) is hydrolysed in base has implications for the mechanism of action of the Zn(II)-containing enzyme *E. coli* alkaline phosphatase (see p. 61).[228] The α-hydroxy-carboxylic acids, L- and D-lactates, DL-α-hydroxyvalerate, L- and D-glycerates, DL-nitrolactate, and DL-β-chlorolactate are all phosphorylated on the α-hydroxy group in a pyruvate kinase-catalysed reaction to give the corresponding phospho-esters. All of these acids are new substrates for this ATP-dependent phosphorylation.[298] *N*-Hydroxycarbamate, $HONHCO_2^-$, is the substrate for phosphorylation of hydroxylamine by the same enzyme at high pH.[299]

Elegant deuterium/tritium isotope studies have indicated that the addition and elimination steps in the 5-enolpyruvyl shikimate-3-phosphate (ESP) synthetase reaction proceed with opposite stereochemistry.[300] ESP is the immediate precursor of chorismate (**171**), the last common intermediate in the biosynthesis of aromatic substances through the shikimate pathway in fungi, bacteria, and higher plants. [9-^{2}H,^{3}H]-(**171**) has been synthesized[301a] and found to undergo a [3,3]-sigmatropic Claisen rearrangement to prephenate (**172**), catalysed by chorismate mutase, *via* a chair transition state.[301b]

Three papers[302–305] from the same group consider various aspects of ribonuclease A catalysis.The first report[302] considers the hydrolysis by ribonuclease A of cytidine cyclic 2′,3′-phosphate (cCMP) (**173**) at varying pH, ionic strength, and in D_2O/H_2O, while the second paper[303] considers the non-enzymatic hydrolysis. Finally, the temperature dependence of the enzyme-catalysed reaction has been studied.[304]

Deletions of large portions of the carboxyl-terminal end of the adenylate cyclase ATP pyrophosphate lyase (cyclizing) from *Salmonella typhimurium* does not significantly affect the enzymatic activity of the enzyme.[305]

There is no inhibiting effect by mono-and di-cations on the rates of phosphate transfer by metaphosphate mechanisms in the hydrolysis of phosphoroguanidines and this is in agreement with a transition state that has predominently S_N1(P)

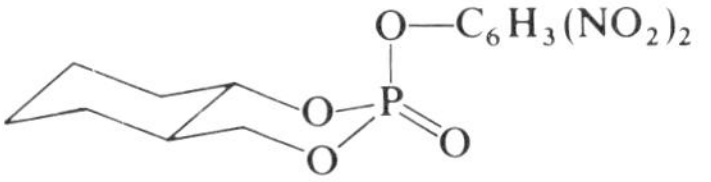

$O{-}C_6H_3(NO_2)_2$

(170) 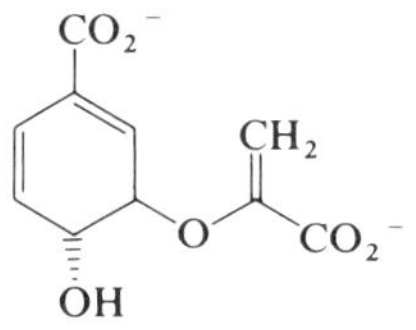 **(171)**

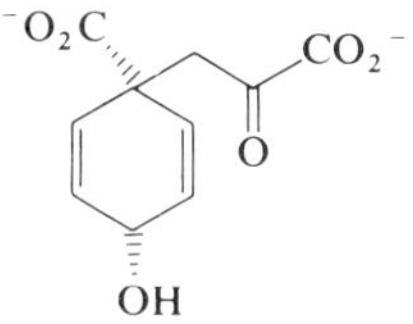

(172) 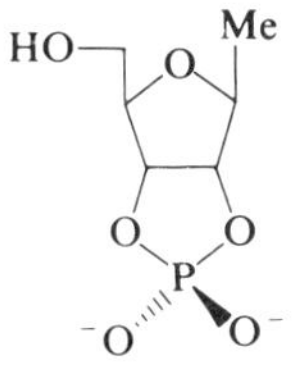 **(173)**

character. If Mg^{2+} coordinates to the phosphate that is being transferred in a kinase-catalysed reaction, there should therefore be no rate inhibition.[306]

In the phosphotransferase system, histidine residues act as nucleophilic catalysts transferring phosphoryl groups between enzymes. Second-order rate constants have now been determined for the reaction of N^2-acetylhistidine methylamide and N^2-acetylhistidine with a series of alkylating agents.[307]

Sulphur-containing Acids

The reaction of ethene- and *trans*-1-propene-sulphonyl chlorides with various tertiary amines proceeds principally *via* a vinylogous substitution reaction to form the cationic sulphene (**174**) which subsequently reacts with water to form either the betaine (**175**) (addition and deprotonation) or the alkenesulphonate anion (**176**) (vinylogous substitution and deprotonation).[308] The kinetics and mechanism of hydrolysis of benzenesulphonyl halides in aqueous acid[309] and the mechanism of the reaction of thionyl chloride with $RC_6H_4CO_2H$[310] have been studied. Thionyl chloride reacts with *N*-benzoyl-*o*-azidoanilines (**177**) to give chlorobenzimidazoles (**178**; R = Cl) *via* 2-chloroisobenzimidazoles (**179**).[311] 2-Phenylbenzimidazole (**178**;

$R_3\overset{+}{N}CHRCH{=}SO_2$ **(174)**

$R_3\overset{+}{N}CHRCH_2SO_3^-$ **(175)**

$CHR{=}CHSO_3^-$ **(176)**

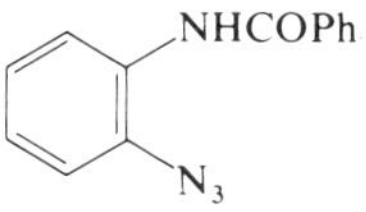

(177) 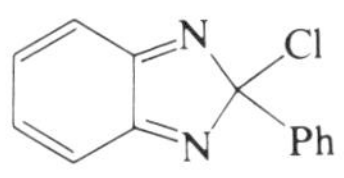 **(179)** 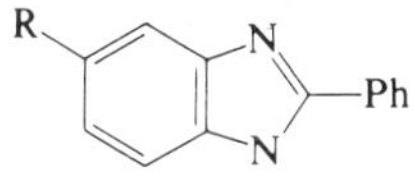 **(178)**

R = H, Cl

R = H) is also formed. The reactions of $PhSO_2X$ (X = Cl, Br, SO_3Ph) with 4-dimethylaminopyridine and its 1-oxide have been studied.[312] The 1-oxides are more reactive due to the greater thermodynamic stability of the onium product (**180**; Q = 0) compared to (**180**; Q = a bond). The novel peroxides (**181**) are obtained by reaction of sulphamoyl chloride with the appropriate hydroperoxides in pyridine.[313] The hydrolysis and ammonolysis of (**181**) generates the hydroperoxide and sulphamic acid or sulphamide.

The kinetics and mechanism of solvolysis in concentrated and fuming H_2SO_4 of $4\text{-}XC_6H_4SO_3Me$ (X = H, Me,Cl) have been reported.[314] The rate-determining step is the decomposition of an ester $H_2S_2O_7$ complex. Nucleophilic substitution of $XCF_2OCF_2CF_2SO_3CF_2CF_2OCF_2X$ (X = ICF_2, $ClCF_2$, HCF_2, Cl_2CF) gives $XCF_2OCF_2CF_2SO_2R$ and XCF_2OCF_2COR (R = nucleophile) *via* S–O bond cleavage.[315] The same Chinese group has more recently prepared the perfluorophenyl-3-oxaperfluoroalkanesulphonates (**182**) and found that the reactivity of these towards nucleophiles parallels the pK_a values of the corresponding acids.[316] The slow step in the acid-catalysed hydrolysis of anilinomethanesulphonates $R'NHCH(R)SO_3^-$ is the formation of the S_N2 complex (**183**).[317] The saponification of *threo*-(**184**) with NaOH or NaOMe gives *threo*-

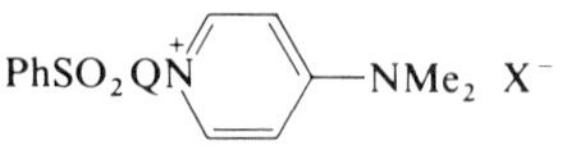

$PhSO_2Q\overset{+}{N}C_5H_4\text{—}NMe_2\ X^-$

(**180**)

Q = O or an S—N bond

$H_2NSO_2OOCH_2R$

(**181**)

R = Et, Pr^n

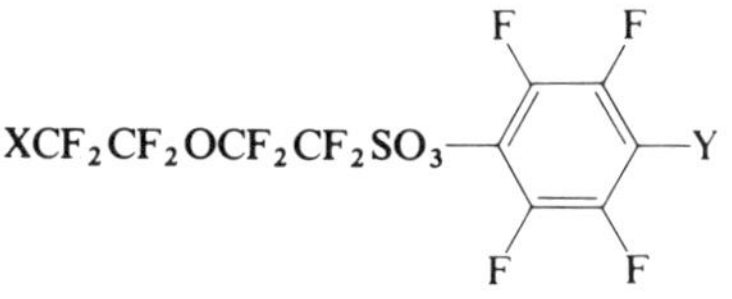

$XCF_2CF_2OCF_2CF_2SO_3\text{—}C_6F_4\text{—}Y$

(**182**)

X = I, Cl; Y = F, Cl

$\overset{\delta+}{H_2O}\cdots C(R)(H)(SO_3^-)\cdots \overset{\delta+}{NHR'}$

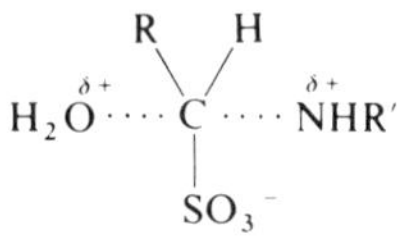

(**183**)

$HO_2C(CH_2)_7CH(OH)CH(OH)(CH_2)_7CO_2H$ with retention of configuration;[318] (**184**) is unreactive to Cl^- or Br^- but it can be acylated with retention of configuration. The kinetics of the reaction of 1-tosyl-3-methylimidazolium chloride (**185**) with various amines give a Brønsted β value of 0.48 and large negative entropies.[319] A small degree of bonding between the amine and the sulphur atom in the transition state is considered likely.

The alkaline hydrolysis of *S*-aryl esters of thiocarboxylic acids $RCOSC_6H_4Br$-4 (R = Me, Et, Pr^n, Me_2CH, Bu^n, Me_2CHCH_2) gives a correlation between log *k* and E_S. A $B_{AC}2$ mechanism has been proposed.[320] A Hammett–Taft relationship has been used to correlate rate constants for the hydrolysis of RC_6H_4SAc (R = 4-NO_2, 4-Br, 3-MeO, 4-F, *etc.*).[321] The alkaline hydrolysis of thian-*r*-4-ol acetates in aqueous dioxane has been reported.[322] *p*-Toluenesulphonic acid is esterified by $CH_2=C(Me)CO_2CH_2CH_2OH$.[323]

pK_a values have been measured by potentiometric and UV methods for various types of sulphamides[324, 325] and arylsulphonamides (**186**).[326]

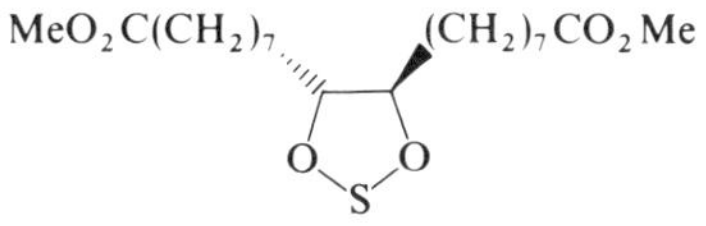

$MeO_2C(CH_2)_7$ … $(CH_2)_7CO_2Me$ (cyclic sulphite) (**184**)

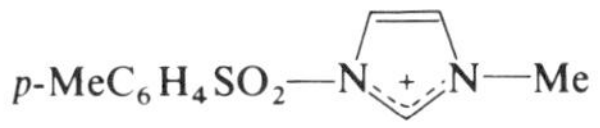

p-$MeC_6H_4SO_2$—N⁺—Me (imidazolium) (**185**)

$XC_6H_4SO_2NH_2$ (**186**)

The involvement of a bicyclic[3,2,1]-intermediate (**187**) in the reaction of diethylaminosulphur trifluoride with γ-keto-acids to give γ-fluorobutyrolactones (**188**) has been postulated (Scheme 9).[327]

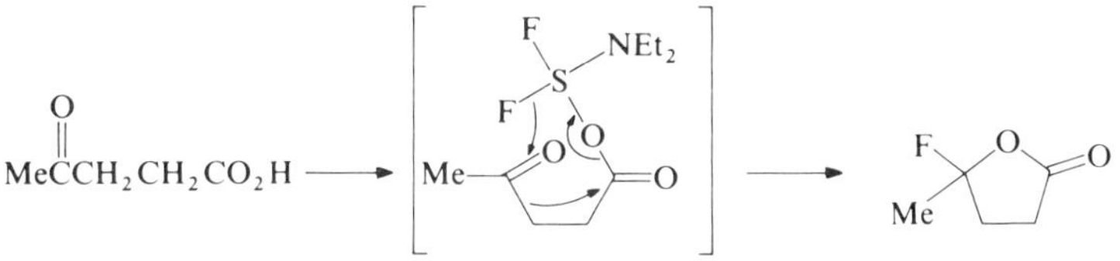

$MeCOCH_2CH_2CO_2H$ → [intermediate with F, NEt_2, S, F, O, Me, =O] (**187**) → F, Me lactone (**188**)

SCHEME 9

The kinetics of the reaction of benzoxazoline-2-thione with XC_6H_4OCN (X = 3-Me, 4-MeO, 3-NO_2, *etc.*) to give the addition products (**189**) and 2-thiocyanatobenzoxazole support a mechanism involving nucleophilic attack by the sulphur atom on the cyanate followed by rearrangement *via* a four-centre transition state.[328] Kinetics and isotope studies indicate that the rate-limiting step in the alkaline hydrolysis of 2-amino-4-oxothiazolines (**190**) in water is the attack by HO^- at the carbonyl carbon.[329]

A review of progress in the chemistry of sulphinic esters RSO_2R', their synthesis, and reactivity has appeared in Japanese.[330]

On ionization *N*-acylglycine ethyl dithio-esters form enethiolate anions of type (**191**).[331] The papain-catalysed hydrolysis of methyl thion-esters give *N*-acylglycine papain dithio-esters—no evidence has been obtained for the formation of ions of type (**191**).

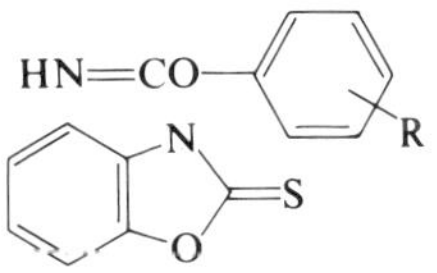

HN=CO—C_6H_4R, benzoxazole =S (**189**)

R, O, S, N, NH_2 (**190**)

R = H, Me, Et

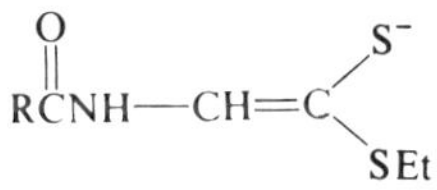

$RC(=O)NH—CH=C(S^-)SEt$ (**191**)

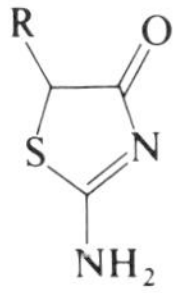

The kinetics of the aminolysis of the *N,N'*-thiocarbonylbisimidazole (**192**) by $XC_6H_6NH_2$ (X = 2-NO_2, 3-NO_2, H, 4-NO_2) have been determined and an addition–cleavage mechanism proposed.[332]

The rates of decomposition in base of xanthates $ROCS_2H$ to alcohol and carbon disulphide have been correlated in a Taft-type equation.[333] The rate-limiting step is the hydrolysis of the xanthate ion. *S-t*-Butyl thioates $RCOSBu^t$ have been proposed as convenient protecting groups for carboxylic acids because their deprotection under neutral conditions can be achieved by electro-oxidation using Br^- salts as electrolytes in aqueous acetonitrile.[334] The hydrolysis of 2-(*p*-nitrophenoxy)tetrahydropyran (**193**) has been measured in the presence of the sulphonaphthoxyacetic acids (**194**) which act as catalysts;[335] (**193**) is a model substrate for the phenyl glycosides and compounds (**194**) have been chosen as catalysts because they bear a structural resemblance to the direct dyes which bind to cellulose, they are water-soluble, and they have a carboxylic acid group which will be protonated at a higher pH than the sulphonate anions, thus enabling the molecules to act as general acid catalysts. Reaction of (**195**) with diphenylketene gives the 1,3-

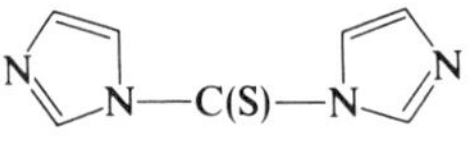

N—C(S)—N

(**192**)

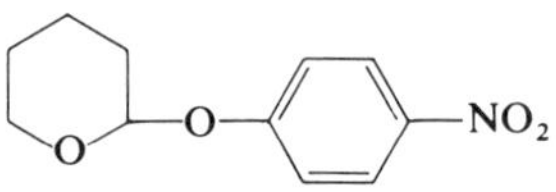

O—NO_2

(**193**)

Y, O, COOH, NaO_3S, X

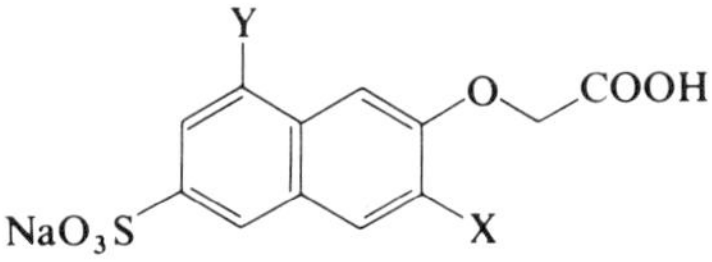

(**194**)

X = SO_3Na, Y = H
X = H, Y = SO_3Na

disubstituted 5-diphenylmethylidene-Δ^3-pyrrolin-2-ones (**196**) *via* a [2+2]-cycloaddition reaction.[336] The same type of cycloaddition occurs with compounds (**197**). In further work on transfer of sulphur entities,[337] Williams and coworkers[338] describe a linear Brønsted plot for the transfer of a sulphonyl group between a series of oxy-anions with basicities both smaller and larger than the donor 4-nitrophenolate group. This evidence is consistent with a concerted, one-transition-state

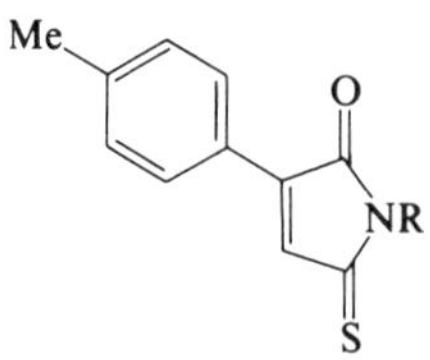

Me, O, NR, S

(**195**)

R = Ph, *c*-C_6H_{11}

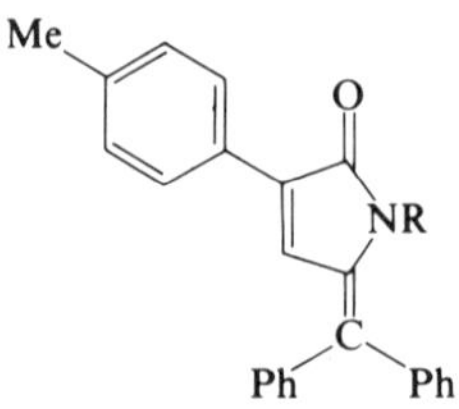

Me, O, NR, C, Ph, Ph

(**196**)

R = Ph, *c*-C_6H_{11}

RNH, S, N, S, Me

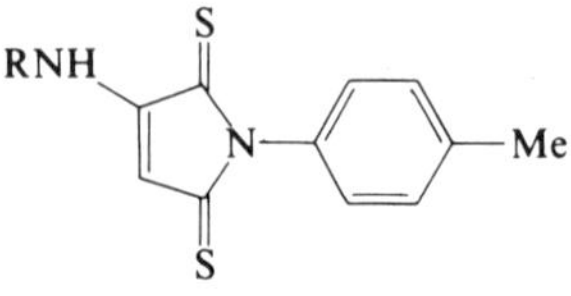

(**197**)

R = *p*-MeC_6H_4, *c*-C_6H_{11}

reaction, represented by (**198**). A step-wise mechanism (Scheme 10) would lead to a non-linear Brønsted relationship due to two distinct transition states.

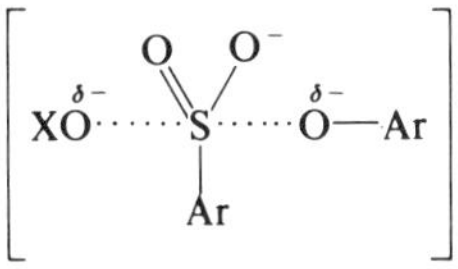

(**198**)

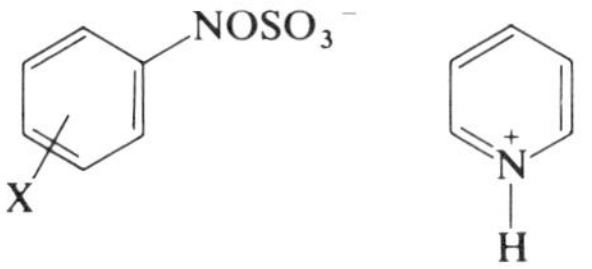

(**199**)

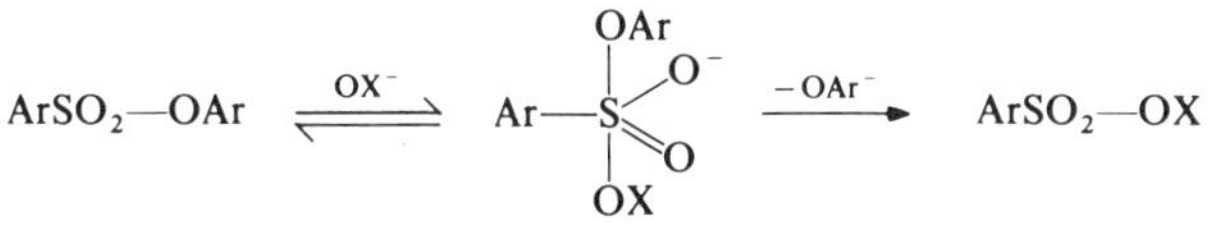

$$ArSO_2{-}OAr \underset{}{\overset{OX^-}{\rightleftharpoons}} Ar{-}S(OAr)(O^-)(=O)(OX) \xrightarrow{-OAr^-} ArSO_2{-}OX$$

SCHEME 10

Ring-substituted *N*-sulphonoxyacetanilides (**199**) are models for carcinogenic metabolites of polynuclear aromatic amides. The solvolysis of these compounds in aqueous solvents occurs by N–O cleavage to give intimate and solvent-separated ion-pairs.[339] The intimate ion-pairs cannot be trapped by nucleophiles or reducing agents and give rise to the rearranged materials, *o*-sulphonoxyacetanilides (**200**), but the solvent-separated ion-pairs can be trapped to give ring-substituted compounds and reduction products. In ethanol these esters solvolyse exclusively *via* S–O bond cleavage with apparent production of SO_3.

Intramolecular carboxyl-group catalysis of a series of 1-mono- and 1,1-di-substituted carboxy-*N*-methyl-*N*-phenylmethanesulphonamides (**201**) has been probed and rate constants and thermodynamic parameters determined.[340] A four-membered cyclic transition state (**202**) is involved and this is particularly favoured by alkyl and *gem*-dialkyl substituents at C(1) (see Scheme 11). The mechanism of hydrolysis of the sulphonamides involves intramolecular nucleophilic catalysis. Intramolecular nucleophilic catalysis also occurs in the acid-catalysed hydrolysis of the *N*,*N*-dialkylbenzenesulphonamide systems (**203**).[341] The neighbouring hydro-

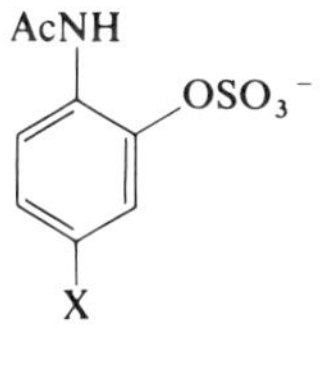

(**200**)

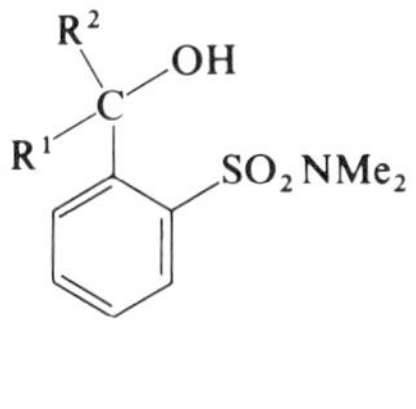

(**203**)

(**204**)

$R^1 = R^2 = H, Me$

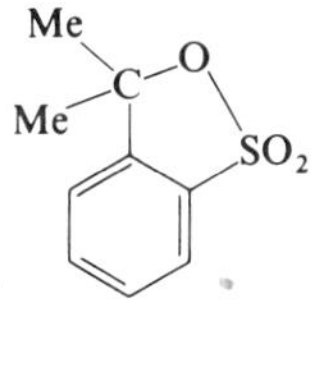

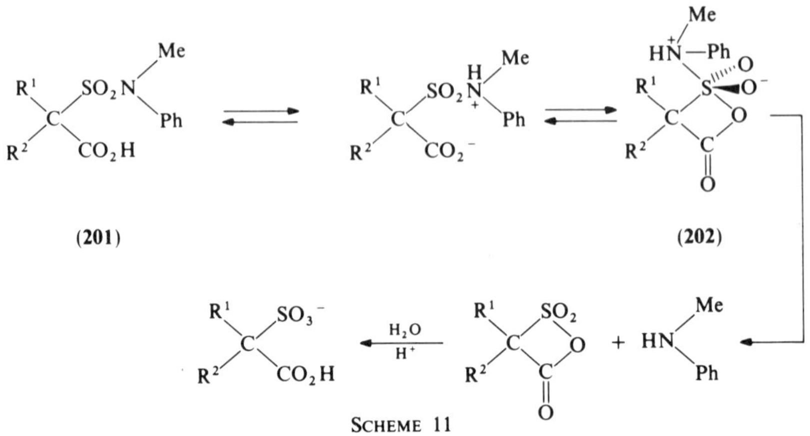

(**201**) (**202**)

SCHEME 11

xyl group participates in the hydrolysis of these compounds. With (**203**; $R^1 = R^2 = H$) the dimethylsulphonamide moiety is stable almost indefinitely in 1M HCl, whereas the corresponding compound with *o*-Me_2COH, *i.e.* (**203**; $R^1 = R^2 = Me$) cyclizes to the sultone (**204**) very rapidly.

A number of papers have appeared on thiourea and thiocarbamate systems. There has been a report[342] of the facile conversion of 1,3-disubstituted thioureas into the corresponding ureas in DMSO at 20° using superoxide radical anion *via* peroxysulphur intermediates (such as -sulphenate, -sulphinate, or -sulphonate) (**205**), in generally good yields.

The alkaline hydrolysis of a series of *S*-arylthiocarbamate esters (**206**) has been studied and the mechanism is Elc*B*.[343] The hydrolysis of benzhydryl and *S*-benzhydryl *N*-arylthiocarbamates has been dealt with earlier (see p. 46).[109] The aminolysis by piperidine of (**207**) occurs *via* a six-membered cyclic H-bonded complex.[344] The same Russian group[345] has reported on the aminolysis of compounds (**207**; R = C_6Cl_5, 2-benzothiazolyl, 2-benzoxazolyl, *etc.*) by piperidine. The acid-catalysed hydrolysis of ketene dithioacetates has been dealt with elsewhere (see p. 35).[12]

Two papers from Kice's group deal with the hydrolysis of organoselenium compounds.[346, 347] The hydrolysis of *o*-nitro- and *o*-benzoyl-benzeneselenic anhydrides (**208**) to the corresponding selenic acids ArSeOH has been studied kinetically in 60% aqueous dioxane over a range of pH in various buffers. The hydrolysis of (**208**; Ar = *o*-$NO_2C_6H_4$) exhibits general acid catalysis with a non-linear Brønsted plot. This behaviour is ascribed to a pre-association mechanism. The hydrolysis of (**208**; Ar = *o*-$PhC(O)C_6H_4$) is much faster and shows specific hydronium ion catalysis.[346] The second paper deals with the reaction of thiols with compounds (**208**) and their corresponding acids, *i.e.* ArSeOH. Acid catalysis is not important in these reactions in which (**208**) are involved but it is of importance in the reactions involving the selenic acids.[347] The reaction of *o*-$PhCOC_6H_4SeOH$ with Bu^nSH involves acid-catalysed reversible formation of the hemiketal (**209**) by acid- and buffer-catalysed addition of the thiol to the carbonyl group of the acid, followed by specific hydronium-ion-catalysed intramolecular break-up of (**209**).

$RNH-C(SO_nO^-)=NR$

(205)

$n = 1$–3

RNHCOSAr

(206)

$RSCONH(CH_2)_6NHCOSR$

(207)

R = Ph, amyl

ArSeOSeAr

(208)

Ar = o-$O_2NC_6H_4$, o-$PhC(O)C_6H_4$

(209)

$O_3NCH_2CH(NO_3)(CH_2)_2NO_3$

(210)

Other Acids

The thermal decomposition of the nitrate esters of type (**210**), $NO_3(CH_2)_2NO_3$, and $NO_3CH_2CH(NO_3)CH_2NO_3$ has been studied and mechanisms proposed.[348] The kinetics of hydrolysis in 0.5–60% H_2SO_4 of 2,3- (and 1,4-)butanediol-1,2-propanediol dinitrates and of glycerol trinitrate have been reported; log k plots are linear with H_0 and the structure of the substrates have little effect on the rates of hydrolysis.[349] The basic hydrolysis of 1,2-glyceryl and 1,3-glyceryl dinitrate[350] and of trinitroglycerin[351] has been reported. Second-order kinetics are observed at 25° and at 25°, 18°, and 10° in the case of trinitroglycerin esters. The 1,2-dinitrate isomerizes to the 1,3-dinitrate in the basic solution and this then hydrolyses principally to glycidyl nitrate, but some nitrites and nitrates are also formed. Trinitroglycerin gives mainly inorganic nitrite and nitrate. Other products include calcium oxalate [$Ca(OH)_2$ is used as the base], nitrate esters, calcium formate, and some polymeric material.

References

1 See *Org. Reaction Mech.*, **1981,** 23.
2 McClelland, R. A., and Santry, L. J., *Acc. Chem. Res.*, **16,** 394 (1983).
3 See *Org. Reaction Mech.*, **1983,** 34.
4 McClelland, R. A., and Lam, P. W. K., *Can. J. Chem.*, **62,** 1068 (1984).
5 McClelland, R. A., and Lam, P. W. K., *Can. J. Chem.*, **62,** 1074 (1984).
6 Capon, B., and Sanchez, M. de N. de M., *Tetrahedron*, **39,** 4143 (1983).
7 Capon, B., and Sanchez M. de N. de M., *An. Conf. Fis. -Quim. Org. 1st.*, **1982,** 199; *Chem. Abs.*, **99,** 211844 (1983).
8 Chiang, Y., Kresge, A. J., Lahti, M. O., and Weeks, D. P., *J. Am. Chem. Soc.*, **105,** 6852 (1983).
9 Burt, R. A., Chambers, C. A., Chiang, Y., Hillock, C. S., Kresge, A. J., and Larsen, J. W., *J. Org. Chem.*, **49,** 2622 (1983).
10 Suh, J., and Klotz, I. M., *J. Am. Chem. Soc.*, **106,** 2373 (1984).
11 Broxton, T. J., *Aust. J. Chem.*, **36,** 2203 (1983).
12 Okuyama, T., Kawao, S., and Fueno, T., *J. Org. Chem.*, **49,** 85 (1984).
13 Lamaty, G., Moreau, C., and Mouloungui, Z., *Can. J. Chem.*, **61,** 2643 (1983).
14 Lamaty, G., Lorente, P., and Moreau, C., *Can. J. Chem.*, **61,** 2651 (1983).
15 McClelland, R. A., Seaman, N. E., and Cramm, D., *J. Am. Chem. Soc.*, **106,** 4511 (1984).
16 Cipiciani, A., Savelli, G., and Bunton, C. A., *J. Heterocycl. Chem*, **21,** 975 (1984).
17 McClelland, R. A., and Seaman, N. E., *Can. J. Chem.*, **62,** 1608 (1984).
18 Chin, J., and Zou, X., *J. Am. Chem. Soc.*, **106,** 3687 (1984).

[19] McDonald, R. N., and Chowdhury, A. K., *J. Am. Chem. Soc.*, **105,** 7267 (1983).
[20] Morris, D. G., *Annu. Rep. Prog. Chem., Sect. B,* **79,** 51 (1983); *Chem. Abs.*, **100,** 173903 (1984).
[21] Jones, P. G., and Kirby, A. J., *J. Am. Chem. Soc.*, **106,** 6207 (1984).
[22] Dubois, J. -E., Ruasse, M. -F., and Argile, A., *J. Am. Chem. Soc.*, **106,** 4840 (1984).
[23] Hoz, S., and Buncel, E., *Tetrahedron Lett.*, **25** 3411 (1984).
[24] Gerritzen, D., and Limbach, H. -H., *J. Am. Chem. Soc.*, **106,** 869 (1984).
[25] Komadel, P., and Holba, V., *Chem. Zvesti,* **38,** 151 (1984); *Chem. Abs.*, **101,** 90030 (1984).
[26] Cevasco, G., Guanti, G., Thea, S., and Williams, A., *J. Chem. Soc., Chem. Commun.*, **1984,** 783.
[27] Singh, L., Gupta, A. K., Singh, R. T., Verma, D. K., and Jha, R. C., *React. Kinet. Cat. Lett.*, **24,** 161 (1984); *Chem. Abs.*, **101,** 54172 (1984).
[28] Gershom, H. R., Raval, D. A., *J. Inst. Chem. (India)*, **56,** 33 (1984); *Chem. Abs.*, **101,** 109973 (1984).
[29] Babaev, L. G., Efendiev, Z. B., and Gaidbekova, P. G., *Fiz-Khim. Metody Analiza i Kontrolya Pr-va, Makhachkala,* **1982,** 137; *Chem. Abs.*, **100,** 208786 (1984).
[30] Ananthakrishnanadar, P., and Kannan, N., *J. Chem. Soc., Perkin Trans. 2,* **1984,** 35.
[31] Hoops, S. C., van Oosterdiep, J., and Schowen, R. L., *J. Org. Chem.*, **49,** 435 (1984).
[32] Saramma, K., and Kurian, J., *Indian J. Chem.*, **23A,** 522 (1984); *Chem. Abs.*, **101,** 151135 (1984).
[33] Purygin, P. P., Laletina, Z. P., Sinitsina, N. K., and Shibaev, V. N., *Izv. Akad. Nauk SSSR, Ser. Khim.*, **1983,** 1676; *Chem., Abs.*, **99,** 174993 (1983).
[34] See *Org. Reaction Mech.*, **1983,** 38.
[35] Holterman, H. A. J., and Engberts, J. B. F. N., *J. Org. Chem.*, **48,** 4025 (1983).
[36] Isaacs, N. S., and Najem, T., *J. Chem. Commun.*, **1984,** 1361.
[37] Deady, L. W., and Finlayson, W. L., *Aust. J. Chem.*, **36,** 1951 (1983).
[38] Yesodha, G., and Thiagarajan, V., *Indian J. Chem.*, **23B,** 146 (1984); *Chem. Abs.*, **101,** 151107 (1984).
[39] Baeza, J., Freer, J., Gomez, E., and Winkler, A., *An. Conf. Fis.-Quim. Org., 1st,* **1982,** 170; *Chem. Abs.*, **99,** 211843 (1983).
[40] Kitano, H., Sun, Z., and Ise, N., *Macromolecules,* **16,** 1823 (1983); *Chem. Abs.*, **100,** 5461 (1984).
[41] Mitsunobu, O., Takemasa, A., and Endo, R., *Chem. Lett.*, **1984,** 855.
[42] Popov, V. V., Ulanov, V. A., and Laz'yan, Yu. I., *Khim.-Farm., Zh.*, **17,** 848 (1983); *Chem. Abs.*, **99,** 194076 (1983).
[43] Bakar, S. K., and Niazi, S., *J. Pharm. Sci.*, **72,** 1024, (1983).
[44] Perlmutter, H. D., Kristol, D. S., and Tomkins, R. P. T., *J. Am. Chem. Soc.*, **106**, 340 (1984).
[45] Schmidt, J., and Mitzner, R., *Wiss. Z. Paedagog. Hochsch. "Karl Liebknecht" Potsdam,* **28,** 105 (1984); *Chem. Abs.*, **101,** 90046 (1984).
[46] See *Org. Reaction Mech.*, **1983,** 40.
[47] Haberfield, P., and Fortier, D., *J. Org. Chem.*, **48,** 4554 (1983).
[48] Sutyagin, V. M., Lopatinskii, V. P., and Tugovskaya, S. A., *Vopr. Kinet. i Kataliza, Ivanovo,* **1982,** 107; *Chem. Abs.*, **100,** 50783 (1984).
[49] Rakshit, S. C., and Datta, S., *J. Indian Chem. Soc.*, **60,** 456 (1983); *Chem. Abs.*, **99,** 211874 (1983).
[50] Assad, A. N., and Khalil, F. Y., *Z. Naturforsch.*, **39a,** 95 (1984); *Chem. Abs.*,**100,** 173985 (1984).
[51] Dasgupta, P. K., Bhattacharya, P. K., and Moulik, S. P., *Indian J. Chem.*, **23A,** 192 (1984); *Chem. Abs.*, **101,** 71936 (1984).
[52] Paleta, O., Kvičala, J., and Dědek, V., *Collect. Czech. Chem. Commun.*, **48,** 2805 (1983).
[53] Baggett, N., and Samra, A. K., *Carbohydr. Res.*, **127,** 149 (1984).
[54] Lajunen, M., and Hintsanen, A., *Acta Chem. Scand,* **37A,** 545 (1983); *Chem. Abs.*, **100,** 138301 (1984).
[55] Monthéard, J. -P., Camps, M., Chatzopoulos, M., and Benzaid, A., *Bull. Soc. Chim. Fr. II,* **1984,** 109.
[56] Gopalakrishnan, G., and Hogg, J. L., *J. Org. Chem.*, **49,** 3161 (1984).
[57] Berman, M. Yu., Kukalenko, S. S., Mogilyanskii, A. I., Burmakin, N. M., and Kulagin, N. I., *Pestitsidy i ikh Primenenie,* **1983,** 30; *Chem. Abs.*, **99,** 194125 (1983).
[58] Orlov, S. I., Chimishkyan, A. L., Elinevskii, A. V., and Grabarnik, M. S., *Zh. Org. Khim.*, **20,** 464 (1984); *Chem. Abs.*, **101,** 22716 (1984).
[59] Orlov, S. I., Chimishkyan, A. L., and Grabarnik, M. S., *Zh. Org. Khim.*, **19,** 2271 (1983); *Chem. Abs.*, **100,** 67577 (1984).
[60] Kanerva, L. T., *Acta Chem. Scand.*, **37B,** 755 (1983).
[61] Kanerva, L. T., Euranto, E. K., and Cleve, N. J., *Acta Chem. Scand.*, **38B,** 529 (1984).
[62] Petro, M., Mravec, D., and Ilavsky, J., *Chem. Zvesti,* **37,** 461 (1983); *Chem. Abs.*, **100,** 33868 (1984).
[63] Huang, H., Liu, F., Xu, Z., and Zhong, X., *Huaxue Xuebao,* **41,** 1081 (1983); *Chem. Abs.*, **100,** 155998 (1984).

[64] Huang, H., Xu, Z., Liu, F., and Wang, M., *Huaxue Xuebao*, **41**, 1087 (1983); *Chem. Abs.*, **100**, 155999 (1984).
[65] D' yachkov, A. I., Likhterov, V. R., and Etlis, V., *Zh. Org. Khim.*, **20**, 913 (1984); *Chem. Abs.*, **101**, 129899 (1984).
[66] Vojtko, J., *Chem. Zvesti*, **37**, 467 (1983); *Chem. Abs.*, **100**, 50771 (1984).
[67] Kabanov, V. P., Kuz'mina, V. A., Belyaev, A. A., Khar'kov, S. N., and Chegolva, A. S., *Zh. Prikl. Khim (Leningrad)*, **56**, 2109 (1983); *Chem. Abs.*, **100**, 33873 (1984).
[68] Zhir-Lebed, L. N., and Temkin, O. N., *Kinet. Katal.*, **25**, 316 (1984); *Chem. Abs.*, **101**, 72209 (1984).
[69] Uri, A., and Tuulmets, A., *Org. React. (Tartu)*, **20**, 578 (1983).
[70] Furuya, Y., Hayashida, S., Tomiyama, M., Ohtsuka, N., and Ueda, H., *Yakugaku Zasshi*, **103**, 867 (1983); *Chem. Abs.*, **100**, 85162 (1984).
[71] Orlov, S. I., Chimishkyan, A. L., and Grabarnik, M. S., *Zh. Org. Khim.*, **19**, 2266 (1983); *Chem. Abs.*, **100**, 67576 (1984).
[72] Kimura, Y., Nango, M., Ihara, Y., and Kuroki, *Chem. Lett.*, **1984**, 429.
[73] Kimura, Y., Kanda, S., Nango, M., Ihara, Y., Koga, J., and Kuroki, N., *Chem. Lett.*, **1984**, 433.
[74] Mir-Mohamad-Sadeghy, B., and Rickborn, B., *J. Org. Chem.*, **49**, 1477 (1984).
[75] Hoye, T. R., Peck, D. R., and Swanson, T. A., *J. Am. Chem. Soc.*, **106**, 2738 (1984).
[76] Decottignies Le Maréchal, P., Ducrocq, C., Marguet, A., and Azerad, R., *C. R. Hebd. Seances Acad. Sci., Ser. 2*, **298**, 343 (1984).
[77] See *Org. Reaction Mech.*, **1983**, 47.
[78] Venkataram, U. V., and Bruice, T. C., *J. Chem. Soc., Chem. Commun.*, **1984**, 899.
[79] Kalfus, K., Friedl, Z., and Exner, O., *Collect. Czech. Chem. Commun.*, **49**, 179 (1984).
[80] Nummert, V., *Org. React. (Tartu)*, **20**, 279 (1983).
[81] Ghebre-Sellassie, I., Hem, S. L., and Knevel, A. M., *J. Pharm. Sci.*, **73**, 125 (1984).
[82] Jalander, L., *Tetrahedron Lett.*, **25**, 457 (1984).
[83] Ivanov, P. M., and Pojarlieff, I. G., *J. Chem. Soc., Perkin Trans. 2*, **1984**, 245.
[84] Zhao, T., Li, D., Guo, W., Xiao, M., Wen, M., Jiang, K., Liu, R., and Yao, C., *Huaxue Shijie*, **24**, 271 (1983); *Chem. Abs.*, **100**, 191037 (1984).
[85] Kluger, R., and Hunt, J. C., *J. Am. Chem. Soc.*, **106**, 5667 (1984).
[86] Dietze, P. E., and Underwood, G. R., *J. Org. Chem.*, **49**, 2492 (1984).
[87] Stein, R. L., Fujihara, H., Quinn, D. M., Fischer, G., Küllertz, G., Barth, A., and Schowen, R. L., *J. Am. Chem. Soc.*, **106**, 1457 (1984).
[88] Venkatasubban, K. S., and Schowen, R. L., *J. Org. Chem.*, **49**, 653 (1984).
[89] Palling, D. J., and Jencks, W. P., *J. Am. Chem. Soc.*, **106**, 4869 (1984).
[90] Bentley, T. W., Carter, G. E., and Harris, H. C., *J. Chem. Soc., Chem. Commun.*, **1984**, 387.
[91] Bentley, T. W., and Freeman, A. E., *J. Chem. Soc., Perkin Trans. 2*, **1984**, 1115.
[92] Shebanova, O. K., and Krotova, G. G., *Izv. Vyssh. Uchebn. Zaved. Khim Khim. Tekhnol.*, **27**, 279 (1984); *Chem. Abs.*, **101**, 6308 (1984).
[93] Bobko, L. A., Kuritsyn, L. V., and Nesterova, L. N., *Izv. Vyssh. Uchebn. Zaved. Khim Khim. Tekhnol.*, **26**, 1295 (1983); *Chem. Abs.*, **100**, 120218 (1984).
[94] Semenyuk, G. V., Zhiltsov, N. P., and Litvinenko, L. M., *Org. React. (Tartu)*, **20**, 103 (1983).
[95] Semenyuk, G. V., and Razumova, N. G., *Zh. Org. Khim.*, **20**, 468 (1984); *Chem. Abs.*, **101**, 54177 (1984).
[96] Bhatt, M. V., Somayaji, V., and Rao, M. V., *Proc. Indian Acad. Sci. Chem. Sci.*, **93**, 593 (1984); *Chem. Abs.*, **101**, 151165 (1984).
[97] De Azevedo, F. G., and De Oliveira, W. A., *Int. J. Chem. Kinet.*, **16**, 793 (1984).
[98] Kalistratova, T. A., Antipanova, V. E., Chermyanina, E. I., and Sklyar, S. Ya., *Zh. Prikl. Khim.*, **57**, 190 (1984); *Chem. Abs.*, **100**, 156000 (1984).
[99] Toivonen, H., and Björkquist, B., *Acta Chem. Scand.*, **38B**, 563 (1984).
[100] Kaválek, J., El Bahaie, S., and Štěrba, V., *Collect. Czech. Chem. Commun.*, **49**, 2103 (1984).
[101] Mokrosz, J., and Bojarski, J., *Pol. J. Chem.*, **56**, 491 (1982); *Chem. Abs.*, **100**, 191036 (1984).
[102] Khan, M. N., and Khan, A. A., *Int. J. Chem. Kinet.*, **16**, 1301 (1984).
[103] Al Sabbagh, M. M., Calmon, M., and Calmon, J. -P., *J. Chem. Soc., Perkin Trans. 2*, **1984**, 1233.
[104] Blagoeva, I., Kozhdikov, A., and Podzharliev, I., *Dokl. Bolg. Akad. Nauk.*, **37**, 333 (1984); *Chem. Abs.*, **101**, 151115 (1984).
[105] Arys, M., Christensen, T. B., and Eriksen, J., *Tetrahedron Lett.*, **25**, 1521 (1984).
[106] Koedjikov, A. H., Blagoeva, I. B., Pojarlieff, I. G., and Stankevic, E. J., *J. Chem. Soc., Perkin Trans. 2*, **1984**, 1077.
[107] Broxton, T. J., *Aust. J. Chem.*, **37**, 47 (1984).
[108] Barth, D., Tondre, C., and Delpuech, J. -J., *Int. J. Chem. Kinet.*, **15**, 1147 (1983).

[109] Mindl, J., Štěrba, V., Kadeřábek, Jr. V., and Klicnar, J., *Collect. Czech. Chem. Commun.*, **49,** 1577 (1984).
[110] Takeuchi, S., and Noguchi, Y., *Toyama Daigaku Kyoikugakubu, Kiyo, B,* **1984,** 11; *Chem. Abs.*, **101,** 109991 (1984).
[111] Besenyei, G., Viski, P., Simandi, L. I., and Nagy, F., *React. Kinet. Catal. Lett.*, **24,** 293 (1984); *Chem. Abs.*, **101,** 22711 (1984).
[112] Castro, E. A., Peña, S. A., Santos, J. G., and Vega, J. C., *J. Org. Chem.*, **49,** 863 (1984).
[113] Arkhipova, I. A., Zhubanov, B. A., Gabdrakipov, V. Z., and Shalabaeva, I. D., *Dokl. Akad. Nauk SSSR,* **270,** 1396 (1983); *Chem. Abs.*, **100,** 5498 (1984).
[114] Shiner, C. S., Fisher, A. M., and Yacoby, F., *Tetrahedron Lett.*, **24,** 5687 (1983).
[115] Del Valle, C., Crovetto, G., and Thomas, J., *Ars. Pharm.*, **24,** 177 (1983); *Chem. Abs.*, **101,** 71912 (1984).
[116] Ebert, C., Lovrecich, M., and Rubessa, F., *J. Heterocycl. Chem.*, **21,** 271 (1984).
[117] Tan, S. -F., Ang, K. -P., Jayachandran, H., and Sammes, M. P., *J. Chem. Soc., Perkin Trans. 2,* **1984,** 1317.
[118] Fernández, B. M., Reverdito, A. M., Perillo, I. A., and Lamdan, S., *J. Heterocycl. Chem.*, **20,** 1585 (1983).
[119] Yeshoda, G., Alwar, S. B. S., and Thiagarajan, V., *Indian J. Chem.*, **23A,** 564 (1984); *Chem. Abs.*, **101,** 151133 (1984).
[120] Karzijn, W., and Engberts, J. B. F. N., *Recl. Trav. Chim. Pays-Bas,* **102,** 513 (1983).
[121] Cox, R. A., and Yates, K., *Can. J. Chem.*, **62,** 1613 (1984).
[122] Lopatinskii, V. P., Sutyagin, V. M., Tuzovskaya, S. A., *Zh. Org. Khim.*, **19,** 2429 (1983); *Chem. Abs.*, **100,** 102469 (1984).
[123] Cipiciani, A., Ebert, C., Linda, P., Rubessa, F., and Savelli, G., *J. Pharm. Sci.*, **72,** 1075 (1983).
[124] Page, M. I., *Acc. Chem. Res.*, **17,** 144 (1984).
[125] Page, M. I., and Proctor, P., *J. Am. Chem. Soc.*, **106,** 3820 (1984).
[126] Crackett, P. H., and Stoodley, R. J., *Tetrahedron Lett.*, **25,** 1295 (1984).
[127] Fabre, H., Eddine, N. H., and Berge, G., *J. Pharm. Sci.*, **73,** 611 (1984).
[128] Bose, A. K., Gupta, K., and Manhas, M. S., *J. Chem. Soc., Chem. Commun.*, **1984,** 86.
[129] Ha, D. -C., Hart, D. J., and Yang, T. -K., *J. Am. Chem. Soc.*, **106,** 4819 (1984).
[130] Elguero, J., Goya, P., Lissavetzky, J., and Valdeomillos, A. M., *C. R. Hebd. Seances Acad. Sci., Ser. C.*, **298,** 877 (1984).
[131] (a) Siling, S. A., Ponomarev, I. I., Kuznetsov, V. V., Lokshin, B. V., Korshak, V. V., and Vinogradova, S. V., *Izv. Akad. Nauk. SSSR, Ser. Khim.*, **1983,** 1755; *Chem. Abs.*, **100,** 5506 (1984); (b) Siling, S. A., Ponomarev, I. I., Kuznetsov, V. V., Lokshin, B. V., Vinogradova, S. V., and Korshak, V. V., *Deposited Doc.*, **1983,** VINITI 2858-83; *Chem. Abs.*, **101,** 72219 (1984).
[132] Le Cloirec, C., Poncin, J., and Martin, G., *C. R. Hebd. Seances Acad. Sci., Ser. 2,* **298** 559 (1984).
[133] Scott, C. M., Underwood, G. R., and Kirsch, R. B., *Tetrahedron Lett.*, **25,** 499 (1984).
[134] Bhuva, V. S., and Buglass, A. J., *Tetrahedron Lett.*, **25,** 777 (1984).
[135] Laloi-Diard, M., Verchere, J. -F., Gosselin, P., and Terrier, F., *Tetrahedron Lett.*, **25,** 1267 (1984).
[136] Smith, M. B., and Shroff, H. N., *J. Org. Chem.*, **49,** 2900 (1984).
[137] Bunting, J. W., and Fitzgerald, N. P., *Can. J. Chem.*, **62,** 1301 (1984).
[138] Edward, J. T., Wong, S. C., and McClelland, R. A., *Can. J. Chem.*, **62,** 144 (1984).
[139] Sivakamasundari, S., and Ganesan, R., *J. Org. Chem.*, **49,** 720 (1984).
[140] Al-Kaabi, S. S., Hallett, G., Meyer, T. A., and Williams, D. L. H., *J. Chem. Soc., Perkin. Trans. 2,* **1984,** 1803.
[141] Dix, L. R., and Williams, D. L. H., *J. Chem. Soc., Perkin Trans. 2,* **1984,** 109.
[142] Pytela, O., Pilny, M., and Večeřa, M., *Collect. Czech. Chem. Commun.*, **49,** 1173 (1984).
[143] Mackay, D., and McIntyre, D. D., *Can. J. Chem.*, **62,** 355 (1984).
[144] Guylas, I., and Nagy, P., *Magy, Kem. Lapja,* **39,** 234 (1984); *Chem. Abs.*, **101,** 151116 (1984).
[145] Hill, S. V., Williams, A., and Longridge, J. L., *J. Chem. Soc., Perkin Trans. 2,* **1984,** 1009.
[146] Hill, S. V., Longridge, J. L., and Williams, A., *J. Org. Chem.*, **49,** 1819 (1984).
[147] Vojtko, J., Cihova, M., and Ilavsky, J., *Petrochemia,* **23,** 81 (1983); *Chem. Abs.*, **100,** 67559 (1984).
[148] Savelova, V. A., Oleinik, N. M., Vizgert, R. V., and Maksimenko, N. N., *Ukr. Khim. Zh. (Russ. Ed.),* **50,** 288 (1984); *Chem. Abs.*, **101,** 54189 (1984).
[149] Dadali, V. V., *Mezhmol. Vzaimodeistviy a Reakts Sposobn. Org. Soedin,* **1983,** 33; *Chem. Abs.*, **101,** 89916 (1984).
[150] Uri, A., Tuulmets, A., and Palm, V., *Org. React. (Tartu),* **20,** 122 (1983).
[151] Uri, A., *Org. React. (Tartu),* **20,** 444 (1983); *Chem. Abs.*, **101,** 54168 (1984).

[152] Fomin, V. A., Etlis, V. S., Aleksandrova, Z. I., and Ivanova, N. N., *Zh. Org. Khim.*, **20,** 282 (1984); *Chem. Abs.*, **100,** 208948 (1984).
[153] Pozdeeva, N. N., and Denisov, E. T., *Izv. Akad. Nauk SSSR, Ser. Khim.*, **1983,** 2029; *Chem. Abs.*, **99,** 211885 (1983).
[154] Stetsyuk, G. A., Dyadyusha, G. G., Tolmachev, A. I., Fedotov, K. V., and Romanov, N. N., *Ukr. Khim. Zh.*, **49,** 1092 (1983); *Chem. Abs.*, **100,** 5524 (1984).
[155] Di Maio, G., Vecchi, E., Zeuli, E., and Delfini, M., *Tetrahedron,* **40,** 749 (1984).
[156] Lapin, A. S., and Grabarnik, M. S., *Zh. Org. Khim.*, **19,** 2190 (1983); *Chem. Abs.*, **100,** 22215 (1984).
[157] Kozikowski, A. P., Kitagawa, Y., and Springer, J. P., *J. Chem. Soc., Chem. Commun.*, **1983,** 1460.
[158] Kametani, T., Kigasawa, K., Shimizu, H., Hayashida, S., and Saitou, S., *Yakugaku Zasshi,* **104,** 11 (1984); *Chem. Abs.*, **101,** 6274 (1984).
[159] Forrester, A. R., and Purushotham, V., *J. Chem. Soc., Chem. Commun.*, **1984,** 1505.
[160] Kevill, D. N., and Posselt, H. S., *J. Chem. Soc., Perkin Trans. 2,* **1984,** 909.
[161] Chamchaang, W., Chantarasiri, N., Chaona, S., Thebtaranonth, C., and Thebtaranonth, Y. *Tetrahedron,* **40,** 1727 (1984).
[162] Tsolomitis, A., and Sandris, C., *J. Heterocycl. Chem.*, **20,** 1545 (1983).
[163] Smalley, R. K., and Stocker, A. W., *Tetrahedron Lett.*, **25,** 1389 (1984).
[164] Louw, R., Vermeeren, H. P. W., and Vogelzang, M. W., *J. Chem. Soc., Perkin Trans. 2,* **1983,** 1875.
[165] Mackie, J. C., and Doolan, K. R., *Int. J. Chem. Kinet.*, **16,** 525 (1984).
[166] Yoon, N. M., Park, H. M., Cho, B. T., and Oh, I. H., *Bull. Korean Chem. Soc.*, **4,** 287 (1983); *Chem. Abs.*, **100,** 208837 (1984).
[167] Horvat, Š., Tomíc, S., and Jeričević, Ž., *Tetrahedron,* **40,** 1047 (1984).
[168] Kodpinid, M., and Thebtaranonth, Y., *Tetrahedron Lett.*, **25,** 2509 (1984).
[169] Sakai, S., Imoto, M., and Oiwa, M., *Kobunshi Ronbunshu,* **41,** 45 (1984); *Chem. Abs.*, **100,** 191079 (1984).
[170] Kuzuya, M., Noguchi, A., and Okuda, T., *J. Chem. Soc., Chem. Commun.*, **1984,** 435.
[171] (a) Savchenko, A. S., Bilobrova, A. I., and Kostin, A. I., *Deposited Doc.*, **1982,** SPSTL 334 Khp-D82; *Chem. Abs.*, **100,** 208790 (1984); (b) Savchenko, A. S., Bilobrova, A. I., and Kostin, A. I., *Deposited Doc.*, **1982,** SPSTL 335 Khp-D82; *Chem. Abs.*, **101,** 22707 (1984).
[172] Sugg, E., and Mason, J. G., *Nasa* [*Contract Rep.*] CR 1983; Chem. Abs., **99,** 211891 (1983).
[173] Sysmalainen, M., and Halmekoski, J., *Acta Pharm. Fenn.*, **92,** 155 (1983); *Chem. Abs.*, **100,** 138307 (1984).
[174] Maat, L., and Peereboom, M., *Bull. Soc. Chim. Belg.*, **92,** 877 (1983).
[175] Gulyaev, N. D., Leonova, T. V., and Chimishkyan, A. L., *Zh. Org. Khim.*, **19,** 2194 (1983); *Chem. Abs.*, **100,** 50795 (1984).
[176] Doddi, G., and Ercolani, G., *J. Org. Chem.*, **49,** 1806 (1984).
[177] Rubchenko, V. F., Shtamburg, V. G., Pleshkova, A. P., Nasibov, Sh. S., Chervin, I. I., and Kostyanovskii, R., *Izv. Akad. Nauk SSSR, Ser. Khim.*, **1983,** 1578; *Chem. Abs.*, **99,** 194097 (1983).
[178] Kuzmin, N. I., and Zhizdyuk, B. I., *Zh. Org. Khim.*, **20,** 134 (1984); *Chem. Abs.*, **100,** 173992 (1984).
[179] Litvinenko, L. M., Mikhailov, V. A., Drizhd, L. P., Savelova, V. A., and Kryuchkova, E. N., *Zh. Org. Khim.*, **20,** 1253 (1984); *Chem. Abs.*, **101,** 190788 (1984).
[180] Dadali, V. A., Zubareva, T. M., Litvinenko, L. M., and Simanenko, Yu. S., *Zh. Org. Khim.*, **20,** 1246 (1984); *Chem. Abs.*, **101,** 190787 (1984).
[181] Savchenko, A. S., Bilobrova, A. I., and Kostin, A. I., *Deposited Doc.*, **1982,** SPSTL 336 Khp-D82; *Chem. Abs.*, **101,** 22706 (1984).
[182] Lusztyk, J., Lusztyk, E., Maillard, B., and Ingold, K. U., *J. Am. Chem. Soc.*, **106,** 2923 (1984).
[183] Ananthakrishnanadar, P., and Varghesedharumaraj, G., *Indian, J. Chem.*, **22B,** 506 (1983); *Chem. Abs.*, **99,** 194075 (1983).
[184] Broxton, T. J., *Aust. J. Chem.*, **36,** 1885 (1983).
[185] Broxton, T. J., *Aust. J. Chem.*, **37,** 977 (1984).
[186] Wynberg, H., and Staring, E. G. J., *J. Chem. Soc., Chem. Commun.*, **1984,** 1181.
[187] Anderson, B. D., Conradi, R. A., and Lambert, W. J., *J. Pharm. Sci.*, **73,** 604 (1984).
[188] Mestdagh, H., and Pancrazi, A., *Tetrahedron,* **40,** 3399 (1984).
[189] Blagoeva, I. B., Pojarlieff, I. G., and Kirby, A. J., *J. Chem. Soc., Perkin Trans. 2,* **1984,** 745.
[190] Bhatt, M. V., Ravindranathan, M., Somayaji, V., and Venkoba Rao, G., *J. Org. Chem.*, **49,** 3170 (1984).
[191] Snell, R. L., Chandler, C. D., Leach, J. T., and Lossin, R., *J. Org. Chem.*, **48,** 5106 (1983).
[192] Julia, M., and Mestdagh, H., *Tetrahedron,* **40,** 327 (1984).
[193] Williams, D. R., Harigaya, Y., Moore, J. L., and D'sa, A., *J. Am. Chem. Soc.*, **106,** 2641 (1984).
[194] Hibbert, F., *Acc. Chem. Res.*, **17,** 115 (1984).

[195] Bernier, J. -L., Henichart, J. -P., and Warin, V., *J. Heterocycl. Chem.*, **21**, 1129 (1984).
[196] Ohya, K., Kitaoka, H., Yotsui, Y., and Sano, M., *Org. Mass. Spectrom.*, **18**, 139 (1983); *Chem. Abs.*, **99**, 175076 (1983).
[197] Cipiciani, A., Linda, P., Savelli, G., and Bunton, C. A., *J. Phys. Chem.*, **87**, 5262 (1983).
[198] Ortega, F., Vera, S., Rodenas, E., and Cachaza, J. M., *An. Quim. Ser. A*, **80**, 82 (1984); *Chem. Abs.*, **101**, 109965 (1984).
[199] Fornasier, R., and Tonellato, U., *J. Chem. Soc., Perkin Trans. 2*, **1984**, 1313.
[200] Berndt, D. C., Utrapiromsuk, N., and Conran, D. E., *J. Org. Chem.*, **49**, 106 (1984).
[201] Gobbo, M., Fornasier, R., and Tonellato, U., *Surfactants Solution, Proc. Int. Symp., 4th 1982*, **2**, 1169 (1984); *Chem. Abs.*, **101**, 109960 (1984).
[202] Vera, S., Rodenas, E., Ortega, F., and Otero, C., *J. Chim. Phys. Phys.-Chim. Biol.*, **80**, 543 (1983); *Chem. Abs.*, **100**, 102463 (1984).
[203] Ueoka, R., Matsumoto, Y., Sugihara, E., Miyamoto, T., and Nagahama, S., *Kenkyu Hokoku., Kumamoto Kogyo Daigaku*, **9**, 67 (1984); *Chem. Abs.*, **101**, 54181 (1984).
[204] Boyer, B., Lamaty, G., Moreau, C., and Mouloungui, Z., *Tetrahedron Lett.*, **25**, 4499 (1984).
[205] Ionescu, L. G., Rubio, D. A. R., Henriques, A., and Terezinha, M., *An. Conf. Fis.-Quim. Org. Ist.*, **1982**, 217; *Chem. Abs.*, **99**, 174971 (1983).
[206] Moss, R. A., Alwis, K. W., and Shin, J. -S., *J. Am. Chem. Soc.*, **106**, 2651 (1984).
[207] Bunton, C. A., Mhala, M. M., Moffatt, J. R., Monarres, D., and Savelli, G., *J. Org. Chem.*, **49**, 426 (1984).
[208] Ionescu, L. G., and Nome, F., *Surfactants Solution, Proc. Int. Symp., 4th 1982*, **2**, 1107 (1984); *Chem. Abs.*, **101**, 129891 (1984).
[209] Kunitake, T., Ihara, H., and Hashiguchi, Y., *J. Am. Chem. Soc.*, **106**, 1156 (1984).
[210] Fadnavis, N., and Engberts, J. B. F. N., *J. Am. Chem. Soc.*, **106**, 2636 (1984).
[211] Jager, J., and Engberts, J. B. F. N., *J. Am. Chem. Soc.*, **106**, 3331 (1984).
[212] Ishiwatari, T., and Fendler, J. H., *J. Am. Chem. Soc.*, **106**, 1908 (1984).
[213] Moss, R. A., and Bizzigotti, G. O., *J. Am. Chem. Soc.*, **103**, 6512 (1981); Moss, R. A. and Schreck, R. P., *ibid.*, **105**, 6767 (1983).
[214] Moss, R. A., Swarup, S., Hendrickson, T. F., and Hui, Y., *Tetrahedron Lett.*, **25**, 4079 (1984).
[215] Kodaka, M., *Bull. Chem. Soc. Jpn.*, **56**, 3857 (1983).
[216] Ueoka, R., Matsumoto, Y., Nagamatsu, T., and Hirohata, S., *Tetrahedron Lett.*, **25**, 1363 (1984).
[217] Matsumoto, Y., and Ueoka, R., *Bull. Chem. Soc. Jpn.*, **56**, 3370 (1983).
[218] Daffe, V., and Fastrez, J., *Pol. J. Chem.*, **56**, 327 (1982); *Chem. Abs.*, **99**, 211811 (1983).
[219] See *Org. Reaction Mech.*, **1983**, 57.
[220] Tee, O. S., and Bennett, J. M., *Can. J. Chem.*, **62**, 1585 (1984).
[221] Fujita, K., Ejima, S., Ueda, T., Imoto, T., and Schulten, H. -R., *Tetrahedron Lett.*, **25**, 3711 (1984).
[222] Fujita, K., Ejima, S., and Imoto, T., *Tetrahedron Lett.*, **25**, 3587 (1984).
[223] Makimoto, S., Suzuki, K., and Taniguchi, Y., *Bull. Chem. Soc. Jpn.*, **57**, 175 (1984).
[224] Litvinenko, L. M., Oleinik, N. M., and Garkusha-Bozhko, I. P., *Zh. Org. Khim.*, **19**, 2353 (1983); *Chem. Abs.*, **100**, 102467 (1984).
[225] Bernet, B., and Vasella, A., *Helv. Chim. Acta.*, **67**, 1328 (1984).
[226] Xu, Z., Wang, M., and Huang, H., *Jilin Daxue Ziran Kexue Xuebao*, **1983**, 70; *Chem. Abs.*, **99**, 211870 (1983).
[227] Buncel, E., Dunn, E. J., Bannard, R. A. B., and Purdon, J. G., *J. Chem. Soc., Chem. Commun.*, **1984**, 162.
[228] Jones, D. R., Lindoy, L. F., and Sargeson, A. M., *J. Am. Chem. Soc.*, **105**, 7327 (1983).
[229] Steliou, K., and Pouport, M. -A., *J. Am. Chem. Soc.*, **105**, 7130 (1983).
[230] Suzuki, T., Matsuki, T., Kudo, K., and Sugita, N., *Nippon Kagaku Kaishi*, **1983**, 1482; *Chem. Abs.*, **100**, 33956 (1984).
[231] Chen, J. -T., and Sen, A., *J. Am. Chem. Soc.*, **106**, 1506 (1984).
[232] Alper, H., Urso, F., and Smith, D. J. H., *J. Am. Chem. Soc.*, **105**, 6737 (1983).
[233] (a) Haleem, M. A., Hakeem, M. A., Jarallah, M., and Basit, M. A., *Muslim Sci.*, **12**, 3 (1983); *Chem. Abs.*, **100**, 50810 (1984); (b) Haleem, M. A., Hakeem, M. A., Jarallah, M., and Basit, M. A., *Pak. J. Sci. Ind. Res.*, **26**, 1 (1983); *Chem. Abs.*, **100**, 102601 (1984).
[234] Prager, R. H., and Schiesser, C. H., *Tetrahedron*, **40**, 1517 (1984).
[235] Baumann, M. E., Bosshard, H., Breitenstein, W., Rihs, G., and Winkler, T., *Helv. Chim. Acta*, **67**, 1897 (1984).
[236] Atkins, P. J., Gold, V., and Marsh, R., *J. Chem. Soc., Perkin Trans. 2*, **1984**, 1239.
[237] (a) Grigg, R., and Thianpatanagul, S., *J. Chem. Soc., Chem. Commun.*, **1984**, 180; (b) Grigg, R., Aly, M. F., Sridharan, V., and Thianpatanagul, S., *J. Chem. Soc., Chem. Commun.*, **1984**, 182.

[238] Rebek, J., *Acc. Chem. Res.*, **17,** 258 (1984).
[239] Mackenzie, N. E., Malthouse, J. P. G., and Scott, A. I., *Science,* **225,** 883 (1984).
[240] Zaks, A., and Klibanov, A. M., *Science,* **224,** 1249 (1984).
[241] Cambou, B., and Klibanov, A. M., *J. Am. Chem. Soc.*, **106,** 2687 (1984).
[242] Cram, D. J., Katz, H. E., and Dicker, I. B., *J. Am. Chem. Soc.*, **106,** 4987 (1984).
[243] Ascenzi, P., Amiconi, G., Bolognesi, M., Guarneri, M., Menegatti, E., and Antonini, E., *Biochem. Biophys. Acta,* **785,** 75 (1984).
[244] Quinn, D. M., and Swanson, M. L., *J. Am. Chem. Soc.*, **106,** 1883 (1984).
[245] Mackenzie, N. E., Malthouse, J. P. G., and Scott, A. I., *Biochem. J.*, **219,** 437 (1984).
[246] Malthouse, J. P. G., and Scott, A. I., *Biochem. J.*, **215,** 555 (1984).
[247] Compton, P. D., and Fink, A. L., *Biochemistry,* **23,** 2989 (1984).
[248] Daniels, S. B., Cooney, E., Sofia, M. J., Chakravarty, P. K., and Katzenellenbogen, J. A., *J. Biol. Chem.*, **258,** 15046 (1983).
[249] Hedstrom, L., Moorman, A. R., Dobbs, J., and Abeles, R. H., *Biochemistry,* **23,** 1753 (1984).
[250] Clark, D. S., and Bailey, J. E., *Biochim. Biophys. Acta,* **788,** 181 (1984).
[251] Shah, D. O., Lai, K., and Gorenstein, D. G., *J. Am. Chem. Soc.*, **106,** 4272 (1984).
[252] Kogan, R. L., and Fife, T. H., *Biochemistry,* **23,** 2983 (1984).
[253] Cairi, M., and Gerig, J. T., *J. Am. Chem. Soc.*, **106,** 3640 (1984).
[254] Harper, J. W., Cook, R. R., Roberts, C. J., McLaughlin, B. J., and Powers, J. C., *Biochemistry,* **23,** 2995 (1984).
[255] Stein, R. L., Viscarello, B. R., and Wildonger, R. A., *J. Am. Chem. Soc.*, **106,** 795 (1984).
[256] Harper, J. W., Hemmi, K., and Powers, J. C., *J. Am. Chem. Soc.*, **105,** 6518 (1983).
[257] Ho, P., and Huang, W., *Proc., IUPAC Macromol. Symp., 28th 1982,* 86; *Chem. Abs.*, **100,** 84959 (1984).
[258] Kopczynski, M. G., and Babior, B. M., *J. Biol. Chem.*, **259,** 7652 (1983).
[259] Galardy, R. E., and Kortylewicz, Z. P., *Biochemistry,* **23,** 2083 (1984).
[260] Hoffman, S. J., Chu, S. S. -T., Lee, H. -H., Kaiser, E. T., and Carey, P. R., *J. Am. Chem. Soc.*, **105,** 6971 (1983).
[261] Suh, J., and Han, H., *Bioorg. Chem.*, **12,** 177 (1984).
[262] Suh, J., Cheong, M., and Han, H., *Bioorg. Chem.* **12,** 188 (1984).
[263] Slebocka-Tilk, H., Cocho, J. L., Frakman, Z., and Brown, R. S., *J. Am. Chem. Soc.*, **106,** 2421 (1984).
[264] See *Org. Reaction Mech.*, **1983,** 62.
[265] Deiters, J. A., and Holmes, R. R., *J. Am. Chem. Soc.*, **106,** 3307 (1984).
[266] Majamaa, K., Hanauske-Abel, H. M., Günzler, V., and Kivirikko, K. I., *Eur. J. Biochem.*, **138,** 239 (1984).
[267] Anderson, E. G., and Pratt, R. F., *J. Biol. Chem.*, **258,** 13120 (1983).
[268] Faraci, W. S., and Pratt, R. F., *J. Am. Chem. Soc.*, **106,** 1489 (1984).
[269] Hardy, L. W., and Kirsch, J. F., *Biochemistry,* **23,** 1282 (1984).
[270] Hardy, L. W., Nishida, C. H., and Kirsch, J. F., *Biochemistry,* **23,** 1288 (1984).
[271] Ngai, K. -L., Ornston, N., and Kallen, R. G., *Biochemistry,* **22,** 5223 (1983).
[272] Ngai, K. -L., and Kallen, R. G., *Biochemistry,* **22,** 5231 (1983).
[273] Larsson, L. -J., and Björk, I., *Biochemistry,* **23,** 2802 (1984).
[274] Wong, C. G., and Rando, R. R., *Biochemistry,* **23,** 20 (1984).
[275] Heins, J., Neubert, K., Barth, A., Canizaro, P. C., and Běhal, F. J., *Biochim. Biophys. Acta,* **785,** 30 (1984).
[276] Sabbioni, G., Shea, M. L., and Jones, J. B., *J. Chem. Soc., Chem. Commun.*, **1984,** 236.
[277] Francis, C. J., and Jones, J. B., *J. Chem. Soc., Chem. Commun.*, **1984,** 579.
[278] Canfield, L. M., and Ramelow, U., *Arch. Biochem. Biophys.*, **230,** 389 (1984).
[279] Rozzell, J. D., and Benner, S. A., *J. Am. Chem. Soc.*, **106,** 4937 (1984).
[280] Poirier, V., and Calleman, C. J., *Acta Chem. Scand.*, **37B,** 817 (1983).
[281] Buchwald, S. L., Friedman, J. M., and Knowles, J. R., *J. Am. Chem. Soc.*, **106,** 4911 (1984).
[282] Buchwald, S. L., Pliura, D. H., and Knowles, J. R., *J. Am. Chem. Soc.*, **106,** 4916 (1984).
[283] Taira, K., Fanni, T., and Gorenstein, D. G., *J. Am. Chem. Soc.*, **106,** 1521 (1984).
[284] Trippett, S., and White, C. L., *J. Chem. Soc., Chem. Commun.*, **1984,** 251.
[285] Kluger, R., Chow, J. F., and Croke, J. J., *J. Am. Chem. Soc.*, **106,** 4017 (1984).
[286] Corriu, R. J. P., Dutheil, J. -P., and Lanneau, G. F., *J. Am. Chem. Soc.*, **106,** 1060 (1984).
[287] Petneházy, I., Szakál, G., Töke, L., Hudson, H. R., and Powroznyk, L., *Tetrahedron,* **39,** 4229 (1983).
[288] Kuo, D. J., and Jordan, F., *J. Biol. Chem.*, **258,** 13415 (1983).

289 Iyengar, R., Ho, H. -T., Sammons, R. D., and Frey, P. A., *J. Am.Chem. Soc.*, **106,** 6038, (1984).
290 Shah, D. O., Kallick, D., Rowell, R., Chen, R., and Gorenstein, D. G., *J. Am. Chem. Soc.*, **105,** 6942 (1983).
291 Reynolds, M. A., Oppenheimer, N. J., and Kenyon, G. L., *J. Am. Chem. Soc.*, **105,** 6663 (1983).
292 Webb, M. R., Reed, G. H., Cooper, B. F., and Rudolph, F. B., *J. Biol. Chem.*, **259,** 3044 (1983).
293 Buchwald, S. L., Saini, M. S., Knowles, J. R., and Van Etten, R. L., *J. Biol. Chem.*, **259,** 2208 (1983).
294 Shen, K. -F., Ho, H. -T., Nolan, L. D., Markovitz, P., Richard, J. P., Utter, M. F., and Frey, P. A., *Biochemistry*, **23,** 1779 (1984).
295 Connolly, B. A., Eckstein, F., and Pingoud, A., *J. Biol. Chem.*, **259,** 10760 (1983).
296 Jaworowski, A., Hartman, F. C., and Rose, I. A., *J. Biol. Chem.*, **259,** 6783 (1983).
297 Fujita, N., Izui, K., Nishino, T., and Katsuki, H., *Biochemistry*, **23,** 1774 (1984).
298 Ash, D. E., Goodhart, P. J., and Reed, G. H., *Arch. Biochem. Biophys.*, **228,** 31 (1984).
299 Weiss, P. M., Hermes, J. D., Dougherty, T. M., and Cleland, W. W., *Biochemistry*, **23,** 4346 (1984).
300 Lee, J. J., Asano, Y., Shieh, T. -L., Spreafico, F., Lee, K., and Floss, H. G., *J. Am. Chem. Soc.*, **106,** 3367 (1984).
301 (a) Hoare, J. H., and Berchtold, G. A., *J. Am. Chem. Soc.*, **106,** 2700 (1984); (b) Sogo, S. G., Widlanski, T. S., Hoare, J. H., Grimshaw, C. E., Berchtold, G. A., and Knowles, J. R., *J. Am. Chem. Soc.*, **106,** 2701 (1984).
302 Eftink, M. R., and Biltonen, R. L., *Biochemistry*, **22,** 5123 (1983).
303 Eftink, M. R., and Biltonen, R. L., *Biochemistry*, **22,** 5134 (1983).
304 Eftink, M. R., and Biltonen, R. L., *Biochemistry*, **22,** 5140 (1983).
305 Leib, T. K., and Gerlt, J. A., *J. Biol. Chem.*, **258,** 12982 (1983).
306 Prigodich, R. V., and Haake, P., *J. Org. Chem.*, **49,** 2090 (1984).
307 Calleman, C. J., and Poirier, V., *Acta Chem. Scand.*, **37B,** 809 (1983).
308 King, J. F., Hillhouse, J. H., and Skonieczny, S., *Can. J. Chem.*, **62,** 1977 (1984).
309 Ivanov, S. N., and Gnedin, B. G., *Vopr. Kinet. i. Kataliza, Ivanovo*, **1982,** 132; *Chem. Abs.*, **100,** 50784 (1984).
310 Vulakh, E. L., Nemleva, S. A., Ivanova, V. M., Kaminskaya, E. G., and Gitis, S. S., *Zh. Org. Khim.*, **19,** 1898 (1983); *Chem. Abs.*, **100,** 22210 (1984).
311 Smalley, R. K., and Stocker, A. W., *Chem. Ind. (London)*, **1984,** 222.
312 Savelova, V. A., Belousova, I. A., Litvinenko, L. M., and Yakovets, A. A., *Dokl. Akad. Nauk SSSR*, **274,** 1393 (1984); *Chem. Abs.*, **101,** 72211 (1984).
313 Blaschette, A., and Safari, H., *Monatsh. Chem.*, **115,** 875 (1984).
314 Gnedin, B. G., Karmanova, T. V., Chumakoya, M. V., and Ivanov, S. N., *Zh. Org. Khim.*, **20,** 557 (1984); *Chem. Abs.*, **101,** 22717 (1984).
315 Chen, Q., and Zhu, S., *Huaxue Xuebao*, **41,** 1044 (1983); *Chem. Abs.*, **100,** 102491 (1984).
316 Chen. Q., and Zhu, S., *Huaxue Xuebao*, **41,** 1153 (1983); *Chem. Abs.*, **100,** 173989 (1984).
317 Senapeschi, A. N., De Groote, R. A. M. C., and Neumann, M. C., *Tetrahedron Lett.*, **25,** 2313 (1984).
318 Sans, V., and Seoane, E., *An. Quim. Ser. C*, **79,** 52 (1983); *Chem. Abs.*, **100,** 50735 (1984).
319 Monjoint, P., and Ruasse, M. -F., *Tetrahedron Lett.*, **25,** 3183 (1984).
320 Bozhenov, B. N., and Baranskii, V. A., *Zh. Org. Khim.*, **20,** 1594 (1984); *Chem. Abs.*, **101,** 190797 (1984).
321 Bazhenov, B. N., Baranskii, V. A., and Aliev, I. A., *Zh. Org. Khim.*, **19,** 2628 (1983); *Chem. Abs.*, **100,** 102503 (1984).
322 Meera, A., Nanjappan, P., Ramalingam, K., and Berlin, K. D., *Indian J. Chem.*, **22B,** 718 (1983); *Chem. Abs.*, **100,** 50765 (1984).
323 Charelishvili, B. I., Vladimirov, L. V., and Tiger, R. P., *Zh. Obshch. Khim.*, **54,** 688 (1984); *Chem. Abs.*, **101,** 37907 (1984).
324 McDermott, S. D., Burke, P. O., and Spillane, W. J., *J. Chem. Soc., Perkin Trans. 2*, **1984,** 499.
325 Burke, P. O., McDermott, S. D., Hannigan, T. J., and Spillane, W. J., *J. Chem. Soc., Perkin Trans. 2*, **1984,** 1851.
326 Ludwig, M., Pytela, O., Kalfus, K., and Večeřa, M., *Collect. Czech. Chem. Commun.*, **49,** 1182 (1984).
327 Patrick, T. B., and Poon, Y. -F., *Tetrahedron Lett.*, **25,** 1019 (1984).
328 Glatt, H. H., Bacaloglu, R., Csunderlik, C., Munteau, D., and Martin, D., *J. Prakt. Chem.*, **326,** 129 (1984); *Chem. Abs.*, **101,** 6281 (1984).
329 Fedoseev, V. M., Mandrugin, A. A., and Semenenko, M. N., *Khim. Geterotsikl. Soedin.*, **1984,** 44; *Chem. Abs.*, **100,** 102520 (1984).
330 Hiroi, K., Sato, S., and Kitayama, T., *Annu. Rep. Tohoku Coll. Pharm.*, **1983,** 1; *Chem. Abs.*, **101,** 89884 (1984).

[331] Lee, H., Carey, P. R., and Storer, A. C., *Can. J. Chem.*, **62,** 763 (1984).
[332] Purygin, P. P., and Pron, A. N., *Izy Vyssh. Uchebn. Kaved. Khim. Khim. Tekhnol.*, **27,** 736 (1984); *Chem. Abs.*, **101,** 151131 (1984).
[333] Harris, P. J., *S. Afr. J. Chem.*, **37,** 91 (1984); *Chem. Abs.*, **101,** 190804 (1984).
[334] Kimura, M., Matsubara, S., and Sawaki, Y., *J. Chem. Soc., Chem. Commun.*, **1984,** 1619.
[335] O'Leary, S., *Can. J. Chem.*, **62,** 1320 (1984).
[336] Augustin, M., and Köhler, M., *Z. Chem.*, **23,** 402 (1983).
[337] See *Org. Reaction Mech.*, **1983,** 71.
[338] D' Rozario, P., Smyth, R. L., and Williams, A., *J. Am. Chem. Soc.*, **106,** 5027 (1984).
[339] Novak, M., Pelecanou, M., Roy, A. K., Andronico, A. F., Plourde, F. M., Olefirowicz, T. M., and Curtin, T. J., *J. Am. Chem. Soc.*, **106,** 5623 (1984).
[340] Jager, J., Graafland, T., Schenk, H., Kirby, A. J., and Engberts, J. B. F. N., *J. Am. Chem. Soc.*, **106,** 139 (1984).
[341] Wagenaar, A., Kirby, A. J., and Engberts, J. B. F. N., *J. Org. Chem.*, **49,** 3445 (1984).
[342] Kim, Y. H., Yon, G. H., and Kim, H. J., *Chem. Lett.*, **1984,** 309.
[343] Bourne, N., Williams, A., Douglas, K. T., and Penkava, T. R., *J. Chem. Soc., Perkin Trans. 2*, **1984,** 1827.
[344] Salnikova, G. A., and Nenasheva, T. N., *Deposited Doc.*, **1982** VINITI 781 Khp-D82; *Chem. Abs.*, **101,** 6229 (1984).
[345] Nenasheva, T. N., and Salnikova, G. A., *Deposited Doc.*, **1982,** VINITI 783 Khp-D82; *Chem. Abs.*, **101,** 22700 (1984).
[346] Kice, J. L., McAfee, F., and Slebocka-Tilk, H., *J. Org. Chem.*, **49,** 3100 (1984).
[347] Kice, J. L., McAfee, F., and Slebocka-Tilk, H., *J. Org. Chem.*, **49,** 3106 (1984).
[348] Amer, A. A., *Propellants, Explos., Pyrotech.*, **8,** 149 (1983); *Chem. Abs.*, **100,** 50918 (1984).
[349] Lur'e, B. A., Stepanova, N. A., and Fedonina, L. M., *Deposited Doc.*, **1982,** VINITI 6052; *Chem. Abs.*, **100,** 102507 (1984).
[350] Capellos, C., Fisco, W. J., Ribando, C., Hogan, V. D., Campisi, J., Murphy, F. X., Castorina, T. C., and Rosenblatt, D. H., *Int. J. Chem. Kinet.*, **16,** 1009 (1984).
[351] Capellos, C., Fisco, W. J., Ribando, C., Hogan, V. D., Campisi, J., Murphy, F. X., Castorina, T. C., and Rosenblatt, D. H., *Int. J. Chem. Kinet.*, **16,** 1027 (1984).

Organic Reaction Mechanisms 1984
Edited by A. C. Knipe and W. E. Watts

CHAPTER 3

Radical Reactions: Part 1

A. ALBERTI and C. CHATGILIALOGLU

Istituto dei Composti del Carbonio Contenenti Eteroatomi e loro Applicazioni, C.N.R., 40064 Ozzano Emilia, Italy

Introduction

A monograph has been published on neutral reactive molecules in organic chemistry.[1] Radical reactions have been the subject of several reviews: attention has been focused on stereoselective reactions of radicals,[2] the use of homolytic reactions leading to formation of carbon–carbon bonds,[3] the radical addition of acetals to multiple carbon–carbon bonds,[4] and the free-radical transformation of cyclic acetals.[5] Reviews have also been published on the radical intermediates formed in the photoreactions of carbonyl compounds with a variety of organometallic derivatives,[6] on stable pyridinyl radicals,[7] and on the reactions of organic sulphur(II) compounds.[8] A survey of the radical reactions of peroxyl phoxyl radicals as studied by kinetic ESR has been reported.[9]

The reactivity of oxy-radicals towards biological substrates has been examined, and the two hypotheses of "oxygen fixation" and "active oxygen" have been discussed;[10] the rôle of the oxy-radical as well as of other organic and inorganic radicals in the radiation chemistry of cytochrome C has been illustrated, and it is stressed that the reactions involved are more complicated than simple reduction and oxidation processes.[11]

A very useful compilation of rate constants for reactions, in liquids, of radicals centred at nitrogen, sulphur, phosphorus and other hetero-atoms has been published.[12]

Review articles have also appeared describing the use of spectroscopic techniques in the study of radical reactions; thus the use of modulation spectroscopy[13] and of Laser Flash Photolysis have been illustrated.[14]

Structure, Stereochemistry, and Stability

Carbon-centred Radicals

Several theoretical studies on free radicals have been published. A computational method has been described which allows, *via* the additive group scheme, the determination of the heats of formation of hydrocarbon radicals[15,16] and also of radicals containing oxygen, nitrogen, and sulphur atoms[17] with deviations from experimental data of less than 0.5–1.0 kcal mol^{-1}. Heats of formation have also been determined for methyl (35.17),[18] ethyl (28.5),[18,19] aminomethyl (31.1), and *N,N*-dimethylaminomethyl (29.7 kcal mol^{-1}) radicals;[20] for the last two species the stabilization energy (*SE*) relative to methyl has been reported to be 16.4 and 17.6 kcal mol^{-1}, respectively,[20] and the effect of electron correlation on *SE* has been studied by MINDO/3-RHF and *ab initio* calculations.[21]

Resonance energies of 3.7 kcal mol^{-1} and 7.8 kcal mol^{-1} have been determined for tricyclopropylmethyl[22] and cyanodicyclopropylmethyl[23] radicals, by making use of the relationship between strain enthalpy and free energy of activation; the activation parameters have also been reported for the dissociation of hexacyclopropylethane ($\Delta G^{\ddagger}_{300} = 42.4$ kcal mol^{-1}, $\Delta H^{\ddagger} = 53.0$ kcal mol^{-1} and $\Delta S^{\ddagger} = 18.5$ e.u.)[22] and 1,2-dicyanotetracyclopropylethyl radical ($\Delta G^{\ddagger}_{300} = 40.2$ kcal mol^{-1}, $\Delta H^{\ddagger} = 48.8$ kcal mol^{-1} and $\Delta S^{\ddagger} = 15.0$ e.u.).[23] It has been pointed out that bond dissociation energies are not to be directly equated to radical stability, it being possible to define the net stabilization energy $SE^{\circ}(\mathrm{R}\cdot, \mathrm{RX})$ (Table 1) only relative to the component R of closed-shell RX. Indeed, methyl is more destabilized and *n*-propyl, isopropyl, and *tert*-butyl are more stabilized relative to ethyl than predicted by the corresponding Bond Dissociation Enthalpies.[24]

Allyl resonance energy has been calculated by the MINDO/3-HF method for cycloalkenyl and cycloalk-1-enyl radicals and a mean value of 13.1 ± 2.9 kcal mol^{-1} has been derived;[25] an *ab initio* study on cyclopropenyl radical indicates that the equilibrium structure should be an ethylenic form that can exist in three equivalent conformations which interconvert *via* a non-planar allylic transition state (ΔE

TABLE 1. The relative net stabilization energies $S^\circ(R\cdot, RX)$ of alkyl radicals ($R\cdot$) in the gas phase ($kcal\ mol^{-1}$)[24]

·R	X = H	X = Me	X = R
·Me	8.85	7.65	5.50
·Et	0	0	0
·Pr^n	−0.78	−0.48	−0.48
·Pr^i	−4.55	−3.35	−2.15
·Bu^t	−9.09	−7.65	−5.98

~ 3–4 kcal mol^{-1}).[26] Both ESR results and semi-empirical (MINDO/3 and UMNDO) calculations indicate that most of the radical character in (**1**) is concentrated in the centre of the molecule and that stabilization energies for polyenyl radicals increase with increasing number of carbon atoms and can be related to the ESR hfs constants of the protons of the terminal methylene group through the relation:[27] SE (kcal mol^{-1}) = $(89.7 \pm 0.9) - (65 \pm 2.4)\ \log[a(H_1)/G]$.

(**1**)

The importance of allylic resonance has been predicted to decrease with increasing difference between the electronegativities of the central and terminal atoms;[28] it has also been pointed out that electron repulsion will make allylic resonance more important in diradicals than in the corresponding radicals containing one less electron.[28] Calculations have also been carried out on the resonance energies of aromatic hydrocarbon radicals[29] and have permitted a quantitative analysis of the interaction energies of substituents in benzenoid systems.[30] Stabilization energies of 18.89 and 22.72 kcal mol^{-1} have been derived from kinetic studies for the cyclohexadienyl radical $\cdot C_6H_5OH$ and its naphthalene analogue $\cdot C_{10}H_8OH$, respectively.[31]

Several photoelectron spectroscopic studies have allowed the determination of adiabatic and vertical ionization potentials for a variety of radicals, including $Br\dot{C}H_2$, $Br\dot{C}D_2$, $Br_2\dot{C}H$, $I\dot{C}H_2$,[32] ethyl,[33] 1-propyl, 1-butyl, isobutyl, 2-butyl, neopentyl,[34] 1-methylallyl, 2-methylallyl, allylcarbinyl, and cyclobutyl;[35] IP of allyl (8.18 eV) and perfluoroallyl (8.44 eV) have also been determined[36] in a mass spectrometric study of the pyrolysis of 1,5-hexadiene and its perfluorinated analogue. The electron affinities of $Ph\dot{N}H$ (39.3) and $Ph\dot{C}H_2$ (19.9 kcal mol^{-1}) have also been determined.[37]

In other theoretical studies, the attention has been focused on the free energy of solvation of the simple methyl radical in methane and methylene chloride,[38] on the properties of the as yet undetected thioformyl radical $H\dot{C}(S)$,[39] on the spin density distribution and IP of benzyl, anilino, and phenoxyl radicals,[40] on the prediction of isotropic and anisotropic hfs constants for small radicals, including the isoelectronic

H_2CN, $H_2\overset{+}{C}O$, and H_2BO,[41] and on the frequencies and intensities of absorptions in the IR spectrum of *tert*-butyl radical.[42] It has also been shown that the dissociation energy of the CH bond in RCH_2—H correlates linearly with the rotational barrier (V_0) about the R—C bond in $RCH_2\cdot$ radicals:[43]

$$\Delta H^\circ(RCH_2\text{—}H)/\text{kcal mol}^{-1} = (97.7 \pm 0.7) - (0.75 \pm 0.04)\ V_0$$

through this relation the dissociation energy of the C—H bond in H—CH_2SR (R = Me, Bu^t) has been determined as 93 kcal mol^{-1} (*i.e. ca.* 2 kcal mol^{-1} less than for the corresponding H—CH_2OR species).[43]

As usual a significant body of information on radical species has been derived through the use of spectroscopic techniques, such as IR, UV, ESR, ENDOR, and CIDNP. The behaviour of muonium radicals in pure liquid benzene, pure liquid styrene, and in their mixtures has been investigated, and it has been shown that the radicals observed in the mixtures are the same detected in the pure liquids, thus indicating that no slow intermolecular exchange (10^{-9}–10^{-5} s) occurs.[44] A number of new α-aminoalkyl radicals, *viz.* $\cdot CH_2NMe_2$, $\cdot CH_2N(Et)Me$, $\cdot CH_2NEt_2$, $Me\dot{C}HNEt_2$, $Et\dot{C}HNPr_2$, and $Pr\dot{C}HNBu_2$, have been observed in solution by laser flash photolysis ESR:[45] previous reports of the extremely short life-time of α-aminoalkyl radicals have been confirmed.[45] ENDOR measurements at 6 K allowed the determination, with high accuracy, of the geometry at the radical centre in $\cdot CH_2OH$: the radical presents a very small deviation from planarity (2.3°) with the unpaired electron residing mainly on the central carbon ($\rho_C \geqslant 0.83$ and $\rho_O \leqslant 0.17$).[46] The solid-state ESR spectra of 2-bromotetramethylethyl radical indicate that no bridging occurs between the C(1) and C(2) carbons *via* the bromine atom, and the temperature dependence of the spectral pattern suggests the existence of an equilibrium, where the fluxional process is attributed to a rocking motion of the

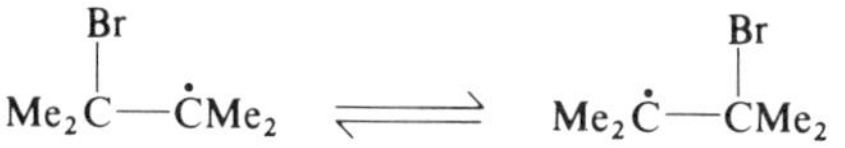

$$Me_2C(Br)\text{—}\dot{C}Me_2 \rightleftharpoons Me_2\dot{C}\text{—}C(Br)Me_2$$

organic unit rather than to a direct migration of bromine.[47] Perfluoroethyl radicals have been generated through discharge sampling experiments and their IR spectra have been recorded in solid argon.[48]

ESR results have indicated that the most favoured conformation for radicals $(R_3MCH_2)_2\dot{C}X$ (**2**; R_3M = SiR_3, GeR_3, SnR_3 and X = H, Me, $OSiMe_3$), is the one in which the organometallic groups eclipse the singly occupied *p*-orbital on the central carbon atom, and a study of the spectral line-shape dependence on temperature has afforded the activation barriers to the flip-flop motion (**2a**) ⇌ (**2b**).[49] Thiomalonylmethyl radical, $\dot{C}H_2CH(COOEt)COXEt$ (X = S), has

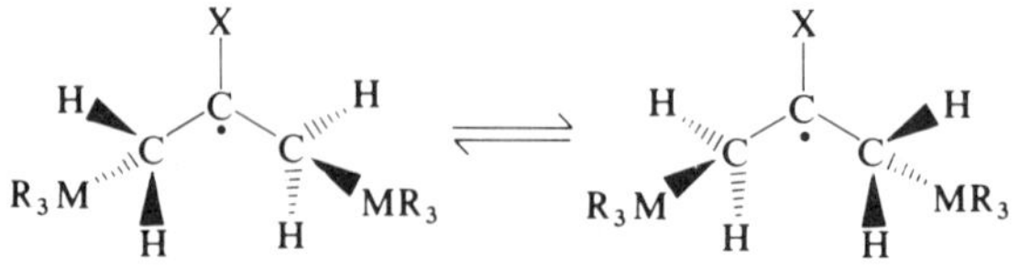

(**2a**) ⇌ (**2b**)

been shown to adopt a conformation similar to that preferred by malonylmethyl (X = O), although in the former evidence for incipient bridging of the β-(ethylthio)carbonyl group to the radical centre has been obtained;[50] the preferred conformation of 2-cumylethyl radical at 153 K is the one where the phenyl ring is located *gauche* to the radical centre as indicated in structure **3**. This finding has been considered to provide support to the suggestion that there is a weak attractive interaction between the singly occupied orbital and the π-cloud of the aromatic ring.[51]

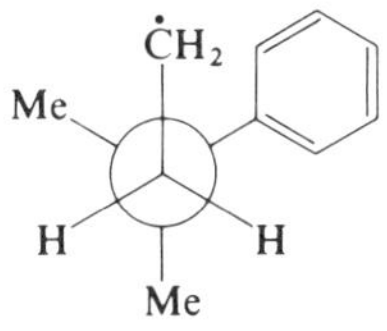

(**3**)

The gradual sharpening of the broad absorption spectrum of benzyl radical obtained after excitation of benzyl chloride with an ArF excimer laser has been attributed to collisional relaxation between hot benzyl radicals;[52] the excited state doublet–doublet and fluorescence spectra of diphenylmethyl have been reported and the photochemistry of this species has been discussed also in relation to that of the triphenylmethyl radical.[53] Absorption maxima at 365 and 380 nm have been reported for 1-naphthylmethyl and 2-naphthylmethyl radicals, respectively:[54] these species are thought to form *via* homolytic cleavage of the carbon–halogen bond of a (σ, σ^*) dissociative triplet state arising from intersystem crossing following photo-excitation of the S_2 (π, π^*) state of 1- and 2-halomethylnaphthalenes.[54] Conformational isomers have been detected for a number of furoyl, tenoyl, and pyrroyl radicals and their structural assignement has been obtained from the magnitude of the hfs constants of the ring protons.[55]

On the basis of ESR results collected for a series of five symmetrically deuterated triphenylmethyl radicals it has been suggested that a thorough analysis of the interactions between the electronic states and the vibrational modes of the molecule as a whole, rather than just fragments of the molecule, is required for an adequate understanding of the observed deuterium isotope effect.[56] Some pentaarylmethyl radicals have been prepared: in particular, 9-(triphenylmethyl)fluorenyl can be isolated as a brown solid upon reaction of 9,9-dichlorofluorene with triphenyl-methylsodium; ESR spectral parameters have been reported for these species.[57] New members have been added to the family of perchlorinated di- or tri-phenylmethyl-like stable free radicals, *viz.* perchloro-2-phenyldiphenylmethyl (**4**) and perchloro-9-phenylfluorenyl (**5**) and some related species, and have been characterized by IR, UV–visible, and ESR spectroscopy;[58, 59] fluorescence in (**4**) has been attributed to the formation of an internal π-complex between two aromatic rings.[58] ESR, ENDOR, and TRIPLE spectroscopic studies have been performed on various ^{13}C-labelled and deuterium-substituted phenalenyls.[60]

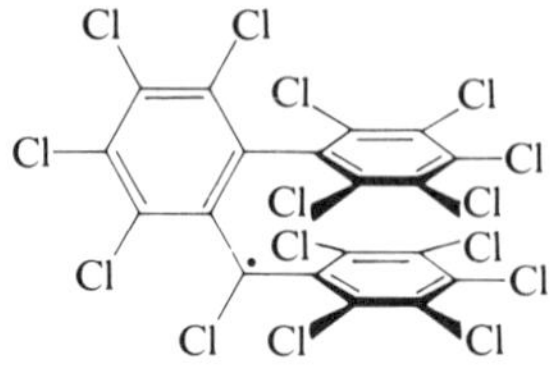

(4)

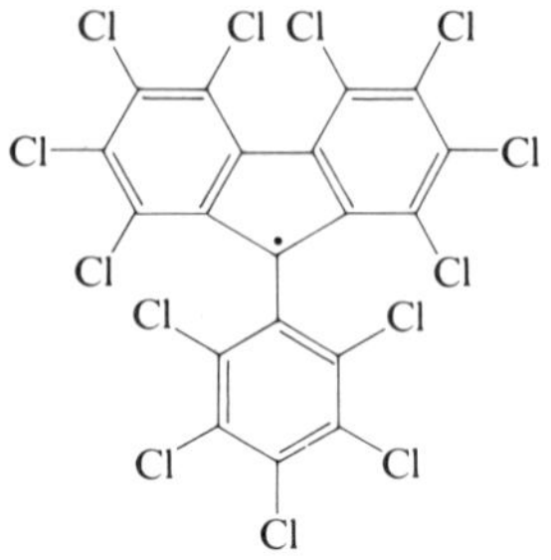

(5)

The π-electronic ground state of phenylethynyl radical, $Ph\dot{C}{=}C$:, has been confirmed in a recent ESR investigation in the solid state;[61] the ESR spectral parameters for $Bu^tS\dot{C}H{-}CH{=}CH_2$ and $Bu^tS\dot{C}H{-}C{\equiv}CH$ have been reported; comparison with data for the analogous alkoxy-substituted radicals indicate a better electron acceptor ability of the alkylthio group; this is confirmed by the values of the barriers to rotation found for $Bu^tS{-}\dot{C}H_2$ (6.29 kcal mol^{-1}), $AdS{-}\dot{C}H_2$ (> 7 kcal mol^{-1}), $Bu^tO{-}\dot{C}H_2$ (4.45 kcal mol^{-1}), and $AdO{-}\dot{C}H_2$ (6.07 kcal mol^{-1}).[62] The barrier to rotation in 1-cyano-1-methoxyallyl radical is *ca.* 6.0 kcal mol^{-1}, *i.e.* some 3 kcal mol^{-1} less that expected on the basis of additivity of the substituent effects, and the difference has been attributed to a slight captodative stabilization of the transition state.[63] Only the *all-trans* conformation could be observed in an ESR investigation of heptatrienyl radical;[64] conversely, both the *cisoid*(**6**)- and *transoid*(**7**)-isomers of the radical from propanoic anhydride could be observed;[65] the existence of a similar isomerism was also evident in the radical from butanoic anhydride, *viz.* $PrC(O)OC(O)\dot{C}HEt$; MINDO/3 calculations suggest that (**7**) is the marginally more stable (1.6 kcal mol^{-1}) isomer.[65]

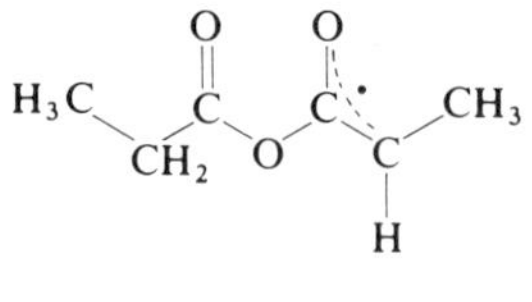

(6)

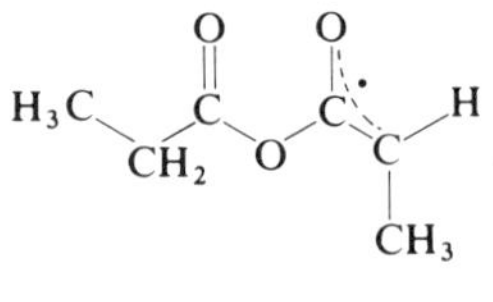

(7)

Under Hunsdiecker conditions (−)-(R)-2,2-diphenyl-1-(trimethylsilyl)cyclopropane carboxylic acid (**8**) is converted into the bromide (**9**) with complete loss of configuration.[66] The failure of the α-silylated radical to retain configuration has

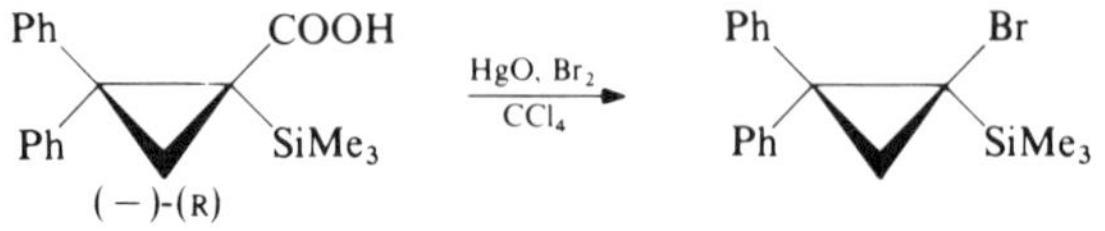

(−)-(R)

(8)

(9)

been attributed (*a*) to stabilization of the radical through d_π–p_π overlap and (*b*) to a decrease of the *s*-character of the singly occupied orbital which would facilitate racemization; this behaviour is in contrast with that of the corresponding α-silylated carbanion which is found to be configurationally stable under Haller–Bauer conditions **(10)** → **(11)**.[66] Monosubstitution of pentaphenylcyclopentadienyl radical

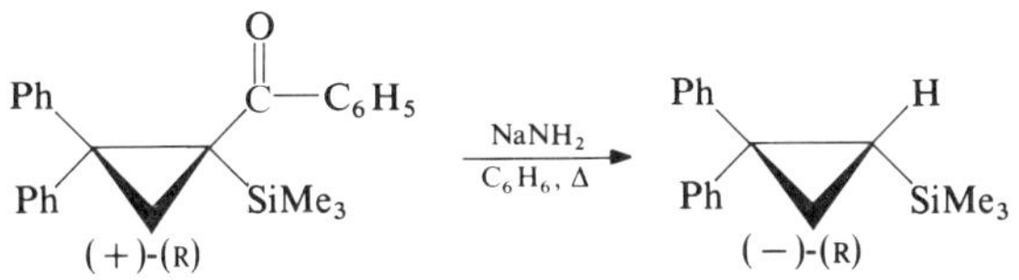

(+)-(R) **(10)** → $NaNH_2$, C_6H_6, Δ → (−)-(R) **(11)**

lifts the degeneracy of the HOMO, as indicated by the results of ^{13}C-, ^{19}F-, and ^{1}H-ENDOR measurements.[67] UV absorptions and molar extinction coefficients have been determined for 1-hydro-4-acetylpyridinyl radical in a simultaneous time-resolved ESR-optical absorption spectroscopic study and a mechanism has been proposed for the photoreduction of 4-acetylpyridine in isopropanol.[68] Several *N*-methyl- and *N*-(triphenylgermyl)-2-cyanopyridinyls have been examined in a combined polarographic and ESR investigation in which the following scale of stability could be derived:[69] **(12d)** ≃ **(12f)** > **(12c)** ≫ **(12e)** > **(12b)** > **(12a)**. Under reducing conditions 2-cyano-substituted pyridinyls appear to evolve by loss of cyanide anions.[69] The results of a CIDEP investigation indicate that quinoxaline and quinazoline react *via* their polarized triplet states, phtalazine *via* its unpolarized singlet state, and cinnoline *via* both its singlet and triplet states.[70]

It has been suggested that the transient intermediate detected in the photocyclization of 5-(2-chlorophenyl)-1,3-diphenylpyrazole should be assigned the cyclohexadienilic structure **(13)**.[71]

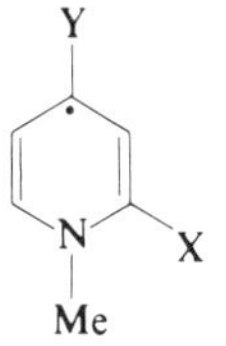

(12)

	X	Y
a:	CN	H
b:	CN	Me
c:	CN	Bu^t
d:	CN	CN
e:	CO_2Et	H
f:	COMe	H

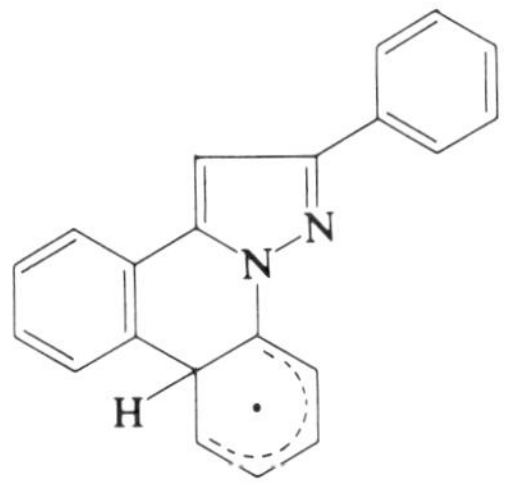

(13)

Miscellaneous Radicals

A number of phosphineboryl radicals ($R_3\overset{+}{P}$–$\dot{\bar{B}}H_2$; R = MeO, CF_3CH_2O, Me_2N, Et, Bu^n, Bu^t) have been investigated by ESR and found to be planar at boron with the unpaired electron mainly contained in a B $2p_\pi$-orbital ($a(^{11}B) = 13.4$–20.5 G).[72]

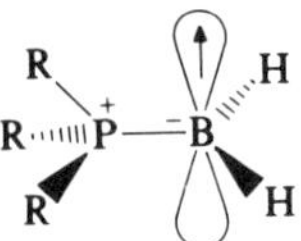

A series of organometallic iminyls, MCH=N· (M = Ag, Au, Cu) has been characterized by ESR spectroscopy at 77 K, and it has been suggested that these species exist as *cis*- and *trans*-isomers.[73]

An investigation on the ^{17}O-labelled alkylamidyl radical $Bu^t\dot{N}C(O)Me$ has ascertained that most of the spin density is concentrated on the nitrogen atom ($\rho^\pi_N \simeq 0.9$, $\rho^\pi_O \simeq 0.13$, $\rho^\pi_C \simeq 0.04$);[74] values predicted by the STO3G method for a π-allyl state are in very good agreement.[74] *N*-Alkyl-*N*-(trialkylsilyl)aminyls have been shown to have a π-electronic ground state and it appears that the $+I/-M$ effect of the trialkylsilyl group is not sufficient to induce a switch from a π- to a σ-SOMO.[75] The hfs constants measured in some *N*-alkylthio-*N*-alkylaminyls ($a_N = 11.6$–12.3 G and $a(^{33}S) = 5.7$–6.7 G) indicate that the unpaired electron mainly resides on the nitrogen and sulphur atoms;[76] rather low enthalpies of dissociation have been measured for the ArS(R)N—N(R)SAr dimers, which has been attributed to the destabilizing action of the sulphur atoms.[76] Delocalization on the whole molecule was found by ESR spectroscopy in the new persistent radicals 1,2,3-dithiazolyls;[77] an ENDOR study has allowed the assignment of hfs constants in 1,4-dihydro-1,3-diphenyl-1,2,4-benzotriazinyl radical;[78] ESR data indicate that the unpaired electron is equally distributed over three nitrogen atoms in 3,5-diphenyl-1,2,4,6-thiatriazinyl radical.[79]

Consistent with ESR data, results from a time-resolved resonance Raman investigation indicate that the structure of *p*-aminophenoxyl radical is more like that of a semiquinone than—that of a phenoxyl radical;[80] also, the high second-order rate constant ($2k_t = 2 \times 10^9$ M^{-1} s^{-1}) observed for the decay of *p*-$H_2NC_6H_4O$· indicates that the NH_2 group modifies significantly the redox properties of the radical.[80] ^{17}O-labelled radical (**14**), the first example of a new class of peroxyl radicals, has been characterized by ESR spectroscopy as an intermediate in the oxidation of 2,2,6,6-tetramethylpiperidinyl.[81]

MNDO calculations predict, in agreement with spectroscopic results, that methylthiyl and ethylthiyl radicals have a π-structure, and that the disulphur radicals

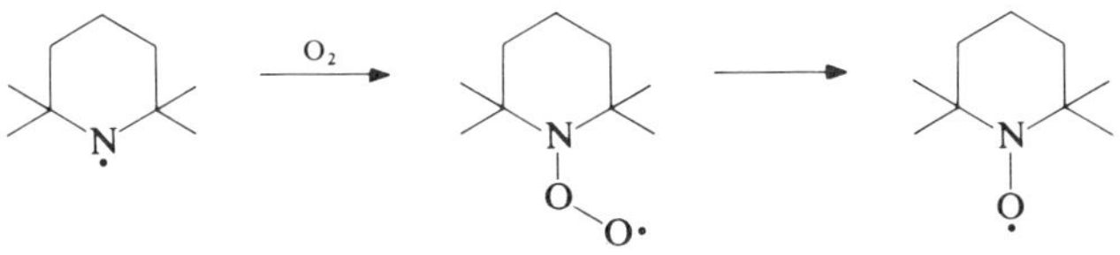

(14)

previously reported to derive from MeS· and EtS· are better interpreted as sulphuranyls $MeS\dot{S}Me_2$ and $EtS\dot{S}Et_2$.[82] Absorption spectra have been reported for a variety of $R_2\dot{S}Br$ radicals;[83] the absorption at 390 nm detected for sulphuranyl radical (**15**) has been assigned to a structure which possesses a $\sigma^2\sigma^{*1}$ three-electron bond between sulphur and oxygen.[84]

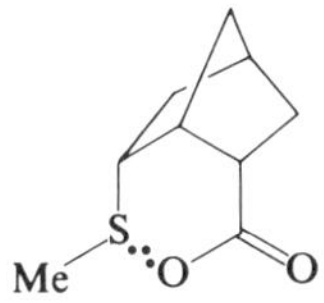

(15)

The heat of formation of trimethylsilyl radical has been calculated as -11.72 ± 2.4 kcal mol^{-1}, this value being considerably lower than that reported previously ($+2.39$ kcal mol^{-1});[85] this species has been reconfirmed through *ab initio* calculations to possess a pyramidal structure, and barriers to inversion through a planar transition state have been shown to vary drastically upon substitution, *viz.* 5 kcal mol^{-1} ($\cdot SiH_3$), 13.3 kcal mol^{-1} ($\cdot SiMe_3$), and 68 kcal mol^{-1} ($\cdot Si(CF_3)_3$).[86] The bond dissociation energies for Mes_3Si–$SiMes_3$ and Mes_3Ge–$GeMes_3$ (Mes = mesityl) have been measured as *ca.* 19.0 kcal mol^{-1} and 20.7 kcal mol^{-1}, respectively;[87] similar values have been reported for other digermanes.[87]

ESR results and extended Hückel MO calculations on new acyltetracarbonyl iron radicals, $RC(O)\dot{F}e(CO)_4$, provide evidence for a trigonal–bipyramidal structure.[88]

Rearrangements

The equilibrium configurations, transition state, and reaction path have been studied by using *ab initio* calculations for the isomerization reaction $CH_3O\cdot \rightarrow \cdot CH_2OH$;[89] the 1,2-hydrogen shift in $CH_3S\cdot$ and $CH_3CH_2S\cdot$ radicals has been predicted by MINDO calculations to be exothermic by 0.14 and 6.31 kcal mol^{-1}, respectively.[90] MINDO/3 calculations indicate that activation energies of 1,2-hydrogen migration in cyclohexadienyl radicals vary with substituents.[91] A series of ω-formylalkyl radicals, $H(O)C(CH_2)_nCH_2\cdot$ ($n = 3$–5), have been generated in vapour and liquid phases;[92] in the former medium intramolecular aldehyde H- (or D-) atom transfer is a facile process proceeding through six- or seven-membered transition states, while in the latter intramolecular aldehyde H-atom transfer occurs to an extent of 74% *via* a six-membered transition state but of only 29% or 38% *via* seven- or eight-membered transition states, respectively.[92] Evidence for the isomerization process $ArSC(Cl)_2\dot{C}Cl_2 \rightarrow ArS\dot{C}(Cl)CCl_3$ has been given.[93] Calculated energies for 1,2-radical rearrangement have been obtained

using the MINDO/3 method for a variety of X substituents:[91] for electron-withdrawing groups, such as NO_2 or CHO, migration has been predicted to be a relatively easy process due to stabilization of the transition state through delocalization of the unpaired electron onto the migrating group, while for donating (methyl) substituents the barrier to migration should be higher than for hydrogen. Some rearrangements of free radicals and alkyl cations in solution have been reviewed.[94]

The rearrangements $RCMe_2\dot{C}H_2 \rightarrow RCH_2\dot{C}Me_2$ (R = Ph, $Me_3CC{\equiv}C$, $Me_3CC{=}O$, $N{\equiv}C$) have been studied over a wide temperature range by product analyses using a competing reaction;[95] the Arrhenius parameters, which are summarized in Table 2, have been discussed in relation to the results of thermochemical kinetic calculations and to earlier work on the 1,2-migration of unsaturated groups in radicals. Arrhenius parameters have also been provided for the neophyl rearrangement (see Table 2) and for the neophyl-like rearrangement of (**16**) to (**17**) relative to hydrogen abstraction from Bu_3SnH.[97]

TABLE 2. Arrhenius parameters and rate constants for some $RCMe_2\dot{C}H_2 \rightarrow RCH_2\dot{C}Me_2$ rearrangements[a]

R	log A (s^{-1})	E_a (kcal mol^{-1})	k at 25° (s^{-1})	Ref.
Ph	11.55 ± 0.28	11.82 ± 0.40	762	97
$Me_3CC{\equiv}C$	11.34 ± 0.70	12.78 ± 1.08	93	95
$H_2C{=}CH$	12.5 ± 0.6	6.6 ± 0.4	4.3×10^7	96
$Me_3CC{=}O$	10.94 ± 0.49	7.77 ± 0.75	1.7×10^5	95
$N{\equiv}C$	12.05[b]	16.4	0.9	95

[a]Values preferred over those already available from kinetic ESR measurements.
[b]Assigned value.

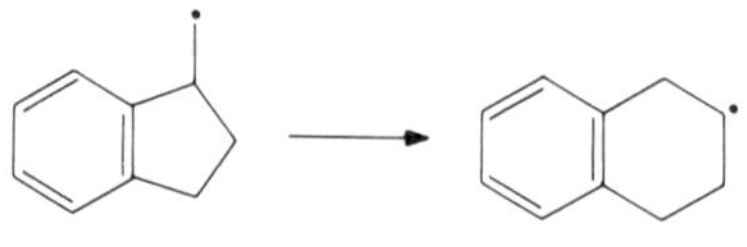

(**16**) (**17**)

ESR and product studies indicate that homolytic 1,2-rearrangement of ethoxycarbonyl and (ethylthio)carbonyl groups are relatively slow processes.[98] Conclusive evidence that the rearrangement of (**18**) to (**19**) involves a charge-separated

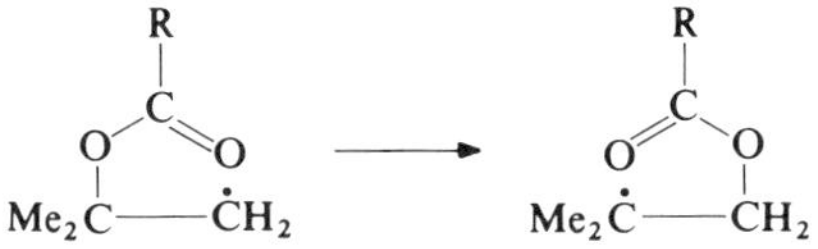

(18) (19)

transition state rather than a cyclic 1,3-dioxan-2-yl radical intermediate has been derived from ESR kinetic data,[99] and the conclusions drawn from an *ab initio* study for 1,2-acyloxy migration in the prototype system, β-(formyloxy)ethyl radical, are in agreement with the experimental findings.[100] From the results of an investigation on the isomerization **(20)** $\rightleftharpoons$ **(22)** it has been inferred that no intermediate is formed when the unpaired electron is located at the α-carbon atom **(23)**; actually the reaction seems to proceed *via* the intermediate **(21)** in which four electrons are involved in the bonding shown by dotted lines and an electron is in a π^*-orbital of oxygen.[101]

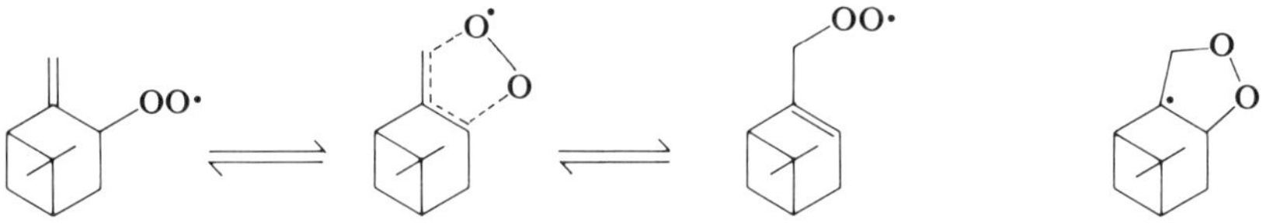

(20) (21) (22) (23)

By addition of muonium to appropriate substrates, Arrhenius parameters have been obtained for the cyclization of **(24)** to **(25)** using the Muonium Spin Rotation technique.[102]

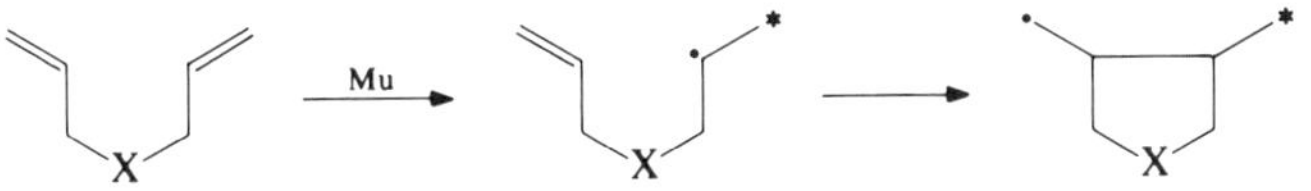

(24) (25)

X = O, CMe_2; the asterisk denotes the position of muonium

The rate constant for ring-closure of **(26)** to **(27)** has been found to be *ca.*10^7 s^{-1} at 338 K which is a higher value than that found for the 1,5-cyclization of 5-hexenyl radical; such rate enhancement makes this system an attractive radical probe.[103]

(26) (27)

Several well-established free-radical probes have been used to understand the nature of the reactions involving 2-lithio-1,3-dithiane.[104]

Cyclizations of 1-methyl-5-hexenyl anion and radical to give 1,2-dimethylcyclopentanes can be distinguished since they occur with different *cis/trans* ratios and since products of anion cyclization are accompanied by those of 1,4-proton shifts; such results should prove useful in avoiding discrepancies in the elucidation of reaction mechanisms.[105]

Relative rates of five- and six-membered ring-formation from 2-allylbenzyl (**28**) and 2-(2-vinylbenzyl)ethyl (**29**) radicals have been reported; closure of (**29**) to the five-membered ring occurs with a 7.3 ± 0.6 kcal mol^{-1} barrier and is preferred over closure to the six-membered ring by 2.1 ± 1 kcal mol^{-1}.[97]

$\dot{C}H_2$ $\dot{C}H_2$

(**28**) (**29**)

Cyclization *via* intramolecular addition of radicals (**30**) generated by reaction of a suitable bromide with Bu_3SnH provides a good route for the preparation of tetrahydroindanes (**31**).[106]

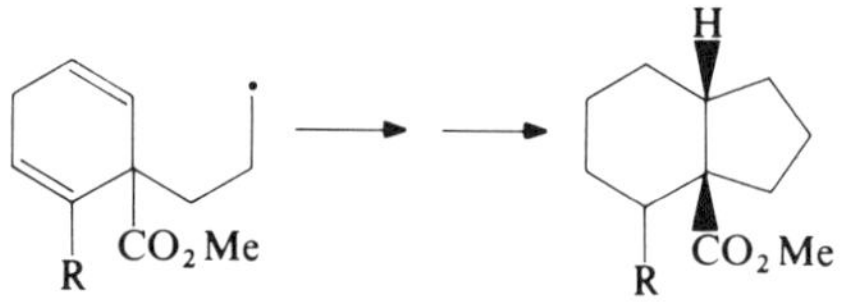

H

R CO_2Me R CO_2Me

(**30**) (**31**)

The addition of an acetyl radical to caryophyllene leads to four stereoisomeric ketones through the cyclization outlined in Scheme 1; as the acidic catalysis induces a strongly different skeletal rearrangement, caryophyllene may serve as a general indicator to test the intervention of a radical species in a chemical reaction.[107]

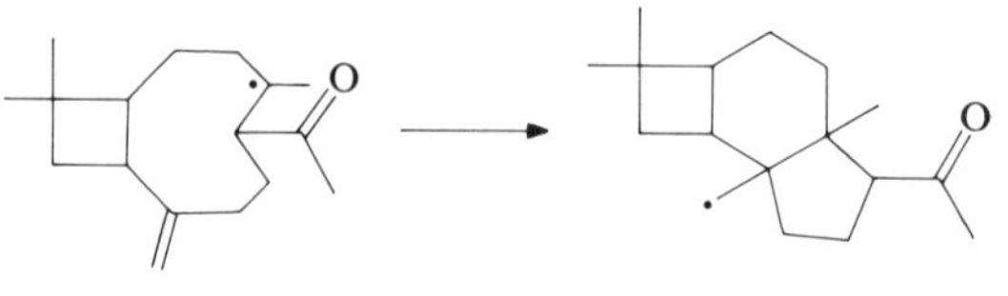

O O

SCHEME 1

The radical intermediates generated by deoxygenation of alcohols (Barton reaction) undergo intramolecular ring-closure when suitably located triple bonds (C≡C or C≡N) are present; the reaction provides a new synthesis of bicyclic ketones from either monocyclic ketones or cyclic olefins.[108] Several ω-allenyl radicals, $Me_2C{=}C{=}CH(CH_2)_nCH_2\cdot$ ($n = 2$–5), have been generated by reaction of the corresponding halides with Bu_3SnH: the observed hydrocarbon products indicate that (*i*) when $n = 2$, cyclization occurs only by attack to central *sp*-carbon of the allene group and is very fast, (*ii*) when $n = 3$, cyclization becomes slower and occurs to both the *sp*-carbon and to sp^2-carbon closer to the radical centre, (*iii*) when $n = 4$, cyclization becomes even slower and occurs only at the closer sp^2-carbon, and (*iv*) when $n = 5$, no cyclization is observed.[109] Olefins can be converted to γ-lactones (**33**) under neutral conditions by thermal reaction with iodostannyl esters (**32**) in the presence of AIBN; two alternative mechanisms have been proposed for the process involving either cyclization to an acetal with loss of $SnBu_3$ (contrary to Baldwin's rules) or a S_Hi reaction (Scheme 2).[110]

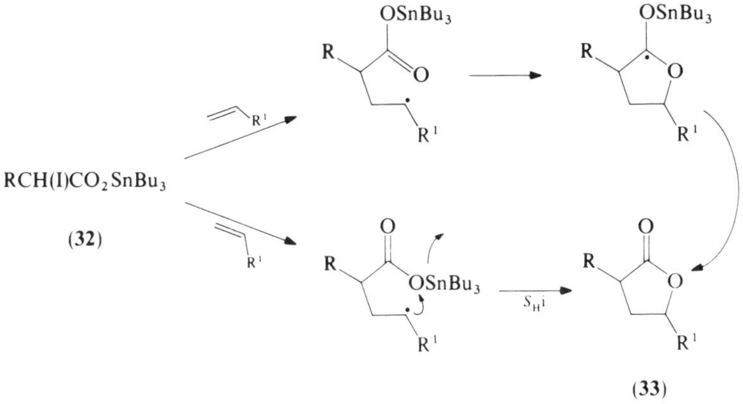

$RCH(I)CO_2SnBu_3$ (**32**) → (**33**)

SCHEME 2

Oxidation of the sodium salt of tropone-tosylhydrazone with Ag_2CrO_4 leads to 2-tosyl-2*H*-indazole; the reaction is thought to proceed through the cyclization of the hydrazyl radical intermediate, *viz.*:[111]

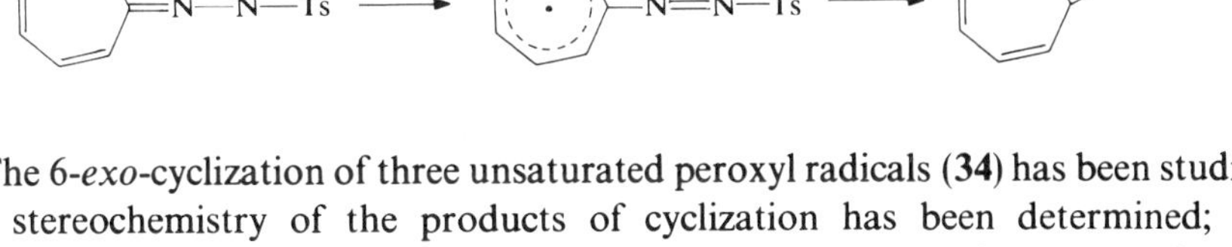

The 6-*exo*-cyclization of three unsaturated peroxyl radicals (**34**) has been studied; the stereochemistry of the products of cyclization has been determined; the preferred stereoselectivity of cyclization may be understood by assuming an equatorially substituted chair-like transition state.[112] The silylradicals produced by hydrogen abstraction from butenylsilane, an allyldisilane, a pentenylsilane, a butenyldisilane, and a (butenyloxy)silane all cyclize in an *endo*-fashion at 418 K with yields varying from 13 to 24%.[113] *N*-Allyl-2-aminothiyl radicals (**35**) cyclize to either thiomorpholine or thiazolidine derivatives; while the former compounds are

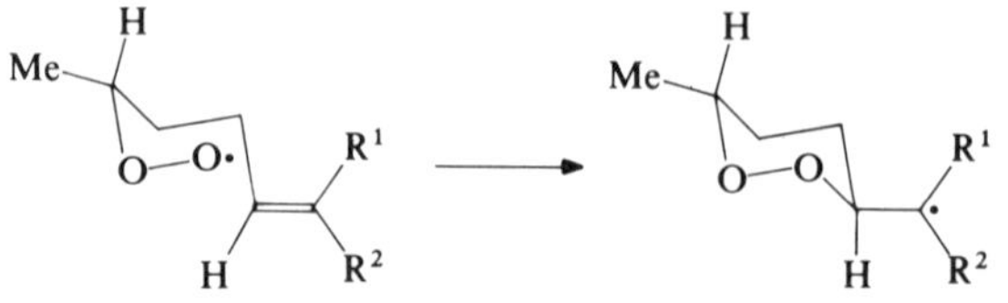

(**34**)

$R^1 = R^2 = H$
$R^1 = Et, R^2 = H$
$R^1 = H, R^2 = Et$

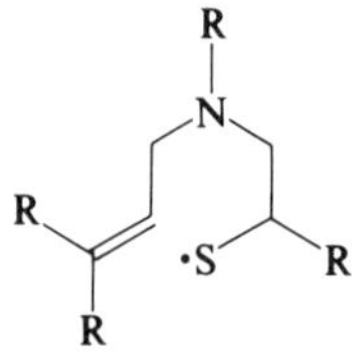

(**35**)

obtained by direct cyclization, the latter derive from a more complex mechanism.[114]

Thermolysis (400°) of homoadamantane affords 1-methyl- and 2-methyladamantane and is accelerated by the radical initiator AIBN; it has been suggested that isomerization is initiated by hydrogen abstraction from either position 3 or 4 followed by β-scission and intramolecular addition as outlined in Scheme 3;[115,116] further evidence for this mechanism was obtained by ^{13}C and ^{2}H isotopic substitution.

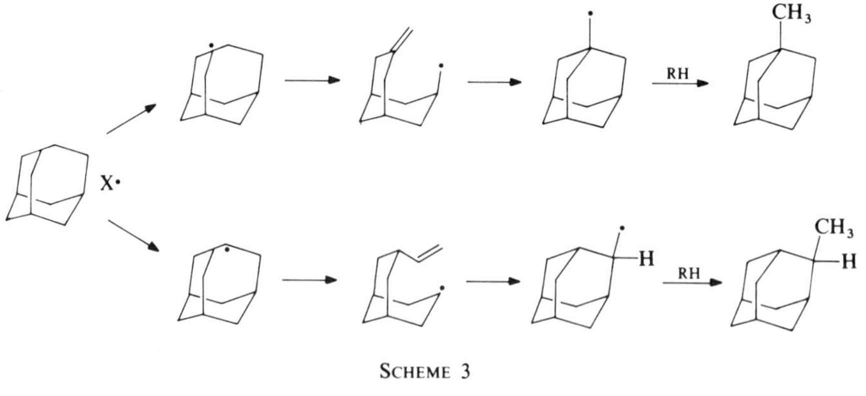

SCHEME 3

A new facile synthesis of tropone (**36**) has been reported which involves the reduction of dihalogenomethylcyclohexadienones with tributyltin hydride; the mechanism of this interesting new reaction has been proposed as indicated in Scheme 4.[117]

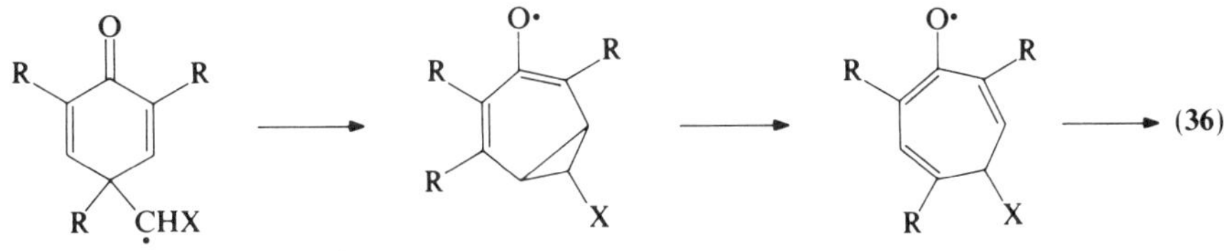

→ (36)

SCHEME 4

It has been shown that the thermal rearrangement of azulene to naphthalene which obeys first-order kinetics does not demand a unimolecular mechanism, but may be compatible with a radical-chain mechanism initiated by the coupling of two azulene molecules.[118]

An ESR spectroscopic investigation of (37) and (38) in the temperature range from −85 to +100° has shown that (37) rearranges irreversibly to (38) above 0°; the behaviour of these radicals differs from that of the corresponding carbanions which, at room temperature, isomerize to benzene.[119]

(37) (38)

The Arrhenius equations $\log(k_1/\mathrm{s}^{-1}) = (11.0 \pm 0.6) - (5.3 \pm 0.7)/\theta$ and $\log(k_2/\mathrm{s}^{-1}) = (14.5 \pm 0.4) - (10.4 \pm 0.6)/\theta$, where $\theta = 2.3\,RT$ kcal mol^{-1}, have been obtained for the ring-opening of (39) and (40), respectively, using the Muon Spin Rotation technique; comparison of these data with those of unsubstituted cyclopropylcarbinyl systems reveals that one methyl group at the radical centre causes only small changes while a second methyl group increases both the frequency factor and the activation energy considerably; a reasonable explanation for this finding does not as yet exist.[120]

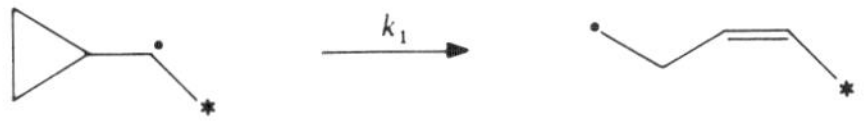

(39)

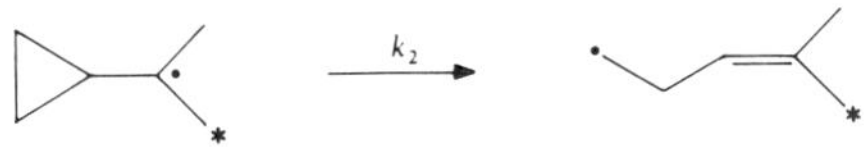

(40)

The asterisk denotes the position of muonium

The results presented on the reactions of cyclopropylcarbinyl halides with (trimethylstannyl)alkalis, *i.e.* Me_3SnLi, Me_3SnNa, Me_3SnK, are consistent with the conclusion that formation of rearranged products in reaction of cyclopropylcarbinyl halides should not be taken as *prima facie* evidence for free-radical intermediates.[121] The rearrangement (**41**) → (**42**) has been postulated in order to explain the product derived from the reaction of photogenerated acetonyl radicals with 1,7-dimethyltricyclo[4.1.1.0]heptane.[122] Ring-opening of *trans*-2-alkylcyclopropylmethyl radicals (**43**) leads mainly to the secondary alkyl radicals (**44**) when R and/or CXY carries electronegative substituents, whereas, in the absence of electronegative substituents, ring-opening occurs in favour of the primary alkyl radical (**45**); this regioselectivity has been interpreted in terms of the frontier orbital interactions which are involved.[123]

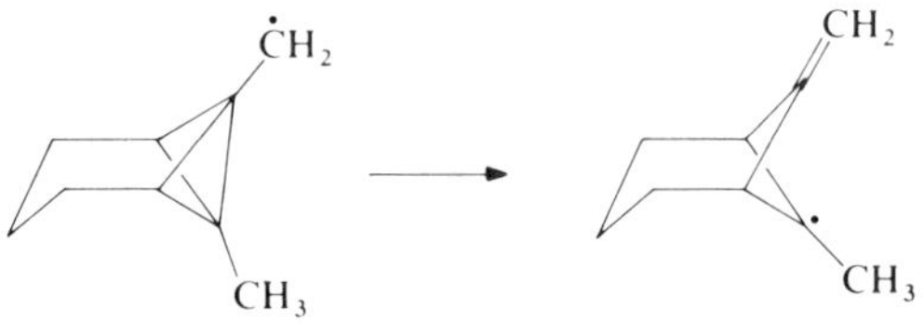

(**41**) (**42**)

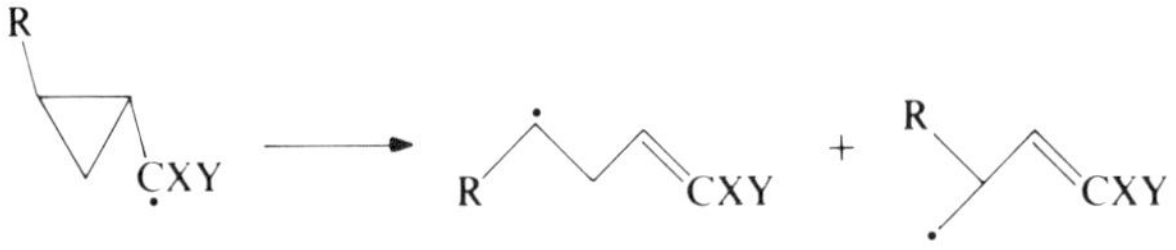

(**43**) (**44**) (**45**)

Free-radical addition of $PhSeSO_2Ar$ to β-pinene gives solely the product (yield 95%) derived from the ring-opening of the four-membered ring, *viz.*:[124]

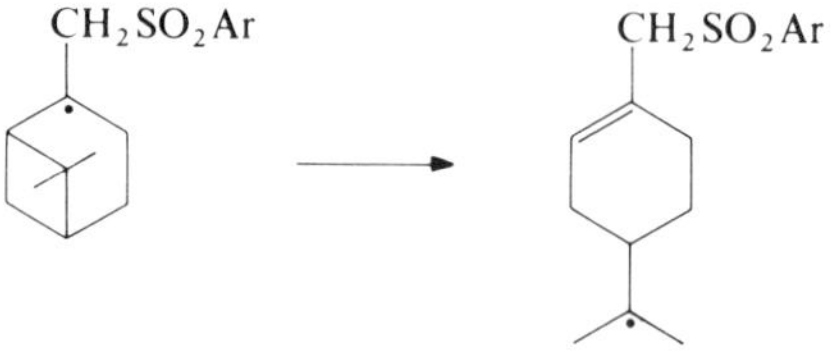

Cis- and *trans*-9-decalinoxyl radicals, generated from the corresponding hypobromide, give products which imply that the ring-opening (**46**) → (**47**) occurs at 0° whereas at temperature higher than 60° the rearrangement (**46**) → (**48**) predominates; these results indicate (*i*) that 9,10-bond fission of radicals (**46**) is rapid but reversible, (*ii*) that 1,9-bond fission of (**46**) is relatively slow and is essentially irreversible, and (*iii*) that ring-closure of the radical (**47**) favours formation of *cis*-(**46**).[125]

Products detected in the photoreactions of *B*-homocholest-5-en-7a-ol hypoiodite,[126] 7-cyano-2,3-benzobicyclo[4.2.0]octa-2,4-dien-6-yl hypoiodite,[127] and other fused-ring derivatives[128] have been rationalized according to the initial formation of alkoxyl radicals followed by β-scission or intramolecular addition to

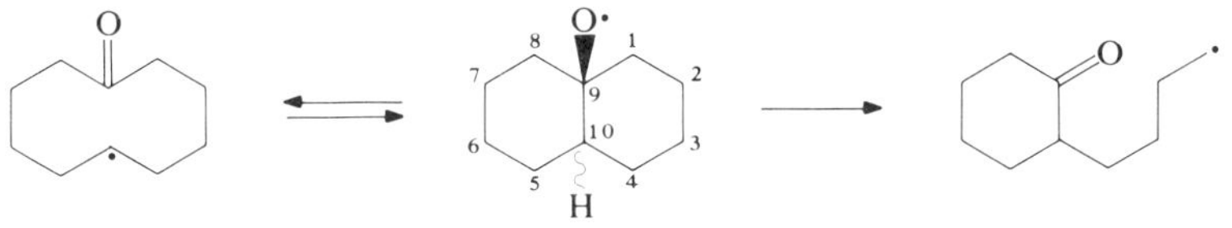

(47) (46) (48)

suitably located double bonds; the iodosobenzene diacetate–iodine system is found to be an excellent reagent for intramolecular functionalization reactions of hydroxy compounds *via* alkoxyl radicals.[129]

An ESR study has shown that *cis*- and *trans*-2-methylaziridineboryl radicals (**49**) fragment to give both the secondary (**50**) and primary (**51**) radicals;[130] similarly to what has been found for the isoelectronic 2-methylcyclopropylmethyl radical, the *cis*-isomer mainly gives the secondary radical [(**50**):(**51**) $\simeq$ 95:5] while the *trans*-isomer leads preferentially to the primary radical [(**50**):(**51**) $\simeq$ 35:65]; the regioselective ring-openings have been interpreted on the basis of frontier molecular orbital theory.

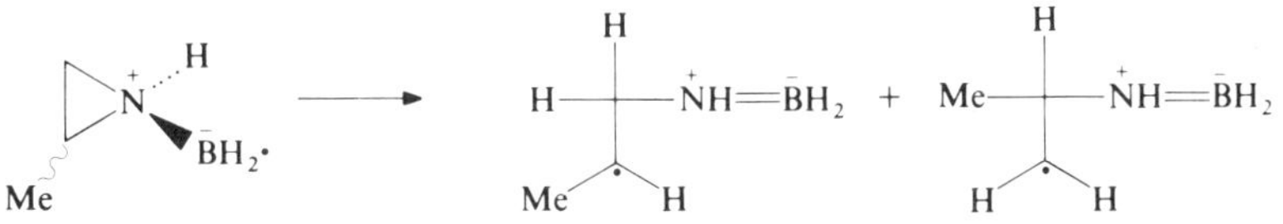

(49) (50) (51)

Fragmentations

The unimolecular decomposition of model alkyl radicals to alkenes has been theoretically investigated using classical trajectories;[131] chemically activated $\cdot C_2H_4{}^{18}F$ radicals, generated by addition of thermal ^{18}F to ethylene, at low pressure, have been found to decompose both by H-atom loss and by elimination of HF molecules.[132] The mechanism of olefin formation from alkyl radicals with leaving groups in the β-position in the presence of Fe^{2+}–EDTA, Fe^{2+}, or Cu^{+} in acidic solution has been proposed to proceed through a reductive elimination (*i.e.* simultaneous electron transfer and elimination of the leaving group).[133]

In tetraethoxysilane the rate constant for the decarbonylation of the 2-methylpropanoyl radical, *viz.*:

$$(CH_3)_2CH\dot{C}O \xrightarrow{k} (CH_3)_2\dot{C}H + CO$$

can be represented by $\log (k/s^{-1}) = (14.0 \pm 0.5) - (13.0 \pm 0.5)/\theta$, where $\theta = 2.3\,RT$ kcal mol^{-1};[134] the activation energy for such decarbonylation is intermediate between that of propanoyl (14.6 kcal mol^{-1}) and 2,2-dimethylpropanoyl (9.3 kcal mol^{-1}), demonstrating the influence of the degree of stabilization of the resulting alkyl radical.

Dediazoniation rates have been determined for a number of alkyldiazenyl radicals R—N=N·:[135] some typical values are

$$k = 1.1 \times 10^6 \mathrm{s}^{-1} \quad (\mathrm{R = Ph}), \quad k = 3.7 \times 10^7 \mathrm{s}^{-1} \quad (\mathrm{R} = \text{cyclopropyl}),$$
$$k = 9.0 \times 10^7 \mathrm{s}^{-1} \quad (\mathrm{R = Me}), \quad k = 1.9 \times 10^9 \mathrm{s}^{-1}, \quad (\mathrm{R = Pr^i}),$$
$$k = 1.5 \times 10^{11} \mathrm{s}^{-1} \quad (\mathrm{R = Bu^t}), \quad k = 6.2 \times 10^{12} \mathrm{s}^{-1} \quad (\mathrm{R} = \omega\text{-butenyl}).^{135}$$

The effect of temperature and pressure on the homolytic dediazoniation of four substituted benzenediazonium ions (*m*-Cl, *m*-Br, *p*-CN, *p*-NO_2) has been studied in acidic methanol; the competition between homolytic and heterolytic mechanisms has also been discussed.[136]

Rate constants for the reaction $(CH_3)_2CHO\cdot \rightarrow CH_3CHO + CH_3\cdot$ have been determined for different initial pressures (20–760 Torr), in the range 333–434 K, and it has been found that they increase rapidly with increasing pressure up to 200 Torr then level off;[137] trifluoro-*t*-butoxyl radicals, generated in the gas phase by two different routes, decompose exclusively by loss of the CF_3 group over the temperature range 361–600 K,

viz.:[138,139] $$CF_3C(O)CH_3 + \cdot CH_3 \not\leftarrow CF_3C(\dot{O})(CH_3)_2 \rightarrow CH_3C(O)CH_3 + \cdot CF_3.$$

The rate constant for the *β*-scission of cumyloxyl radicals, *viz.*: $PhC(CH_3)_2O\cdot \rightarrow PhC(O)CH_3 + CH_3\cdot$, has been found to be $k = (1.0 \pm 0.2) \times 10^7\ \mathrm{s}^{-1}$ at 298 K in aqueous solution; this is an order of magnitude larger than in organic solvents.[140] $BrCF_2CF_2O\cdot$ decomposes *via* cleavage of the C—C bond with a rate constant $k > 10^6\ \mathrm{s}^{-1}$.[141] α-Peroxyl radicals obtained by radiochemical oxidation of alcohols can decompose via two different paths involving either a five-membered cyclic transition state (**52**) leading to carbonyl derivates or a six-membered cyclic transition state (**53**) leading to unsaturated compounds; the effect of pH and ratio of reactants has also been discussed.[142]

$$R\dot{C}HOH + O_2 \longrightarrow [\text{R—CH(O—H)(O—}\dot{O})] \longrightarrow RCHO + HO_2\cdot$$

(**52**)

$$CH_3(CH_2)_2\dot{C}HOH + O_2 \longrightarrow [CH_3CH(H\cdots\dot{O}\text{—O—})CH_2CHOH] \longrightarrow CH_3CH{=}CH_2 + HCO_2H + HO\cdot$$

(**53**)

The gaseous reaction of vinyl radicals with molecular oxygen has been studied between 297 and 602 K; the magnitude of the rate constant, its temperature and pressure dependence, and the identity of the products of this reaction indicate that it proceeds as follows:[143,144]

$$\cdot C_2H_3 + O_2 \longrightarrow \underset{H}{\overset{H}{>}}C{=}C\overset{O-\dot{O}}{\underset{H}{<}} \longrightarrow \left[\underset{H}{\overset{H}{>}}C{\cdots}C\overset{O\cdots O}{\underset{H}{<}} \right]^{\ddagger} \longrightarrow H_2CO + H\dot{C}O$$

Kinetic data on the decomposition of $PhCH(CH_3)OOH$ indicate that the formation of PhCHO proceeds *via* the following reaction:[145]

$$PhCH(CH_3)OO^{\cdot} \rightarrow PhCHO + CH_3O^{\cdot}$$

Combination and Disproportionation Reactions

Products of fast radical–radical reactions have been studied and the rate constants determined using far-infrared laser magnetic resonance spectroscopy in isothermal systems.[146] The rate constant for the disproportionation reaction of formyl radicals, *viz.*: $HCO^{\cdot} + HCO^{\cdot} \rightarrow H_2CO + CO$, has been measured as $k = (2.02 \pm 0.51) \times 10^{10}\,\text{M}^{-1}\,\text{s}^{-1}$ and has been found to be independent of temperature in the range 298–475 K.[147] Cross-disproportionation/combination ratios for $^{\cdot}CFH_2$, $^{\cdot}CF_2H$, and $^{\cdot}CF_3$ with $CH_3CH_2^{\cdot}$ radicals have been determined to be $\Delta = 0.032 \pm 0.012$, $\Delta = 0.066 \pm 0.020$, and $\Delta = 0.098 \pm 0.020$, respectively, over the temperature range 25–75°; following these results it has been concluded that the increases in Δ are due primarily to the changes in disproportionation rate constants and that polar effects are important for the transfer of a β-hydrogen from the ethyl to the attacking fluoroalkyl radicals.[148] For the pathways yielding CFH plus CH_3CH_3 and CF_2 plus CH_3CH_3 cross-disproportionation/combination ratios have been determinated to be $\Delta = 0.020 \pm 0.005$ and $\Delta = 0.265 \pm 0.038$, respectively, at 25°.[148] Rates of reaction between $MeOCO^{\cdot}$, $Me^{\cdot}$, and $MeO^{\cdot}$ radicals formed in flash photolysis of dimethyl oxalate in gas phase have been reported.[149] Rate constants for the self-termination of $(CH_3)_2\dot{C}H$ radicals, generated by photolysis of $(CH_3)_2CHCOCH(CH_3)_2$ in various liquids, have been determined by kinetic ESR spectroscopy and found to be diffusion-controlled; the ratio of disproportionation to combination is nearly independent of temperature and solvent viscosity.[150] The disproportionation/combination ratio for *tert*-butyl radicals, generated by photo- or thermo-decomposition of *trans*-2,2′-azoisobutane in the gas phase, has been determined from product analysis.[151] Rate constants and activation parameters have been determined for the recombination of linear C_nH_{2n+1} (n = 6–10, 12) radicals and cyclohexyl radicals in the corresponding alkane; the k values for the linear radical decreased with increasing η and were larger than those for cyclohexyl.[152]

Disproportionation of cycloalkyl radicals in the liquid phase results in the formation of cycloalkenes and cycloalkanes; cycloalkyls C_5–C_8 only lead to *cis*-cycloalkenes, cycloalkyls C_9 and C_{10} lead to *cis*- and *trans*-cycloalkenes, and cycloalkyls larger than C_{11} only lead to *trans*-cycloalkenes.[153] The decay of 2-cyclohexanonyl radical in adamantane matrix has been monitored by kinetic ESR spectroscopy and it has been found to be biexponential, viz.:

$$I(t) = I_f^\circ e^{-k_f t} + I_s^\circ e^{-k_s t}$$

where $I(t)$ is the intensity of the ESR signal as a function of time, thus indicating that the decay proceeds by two different channels with fast (k_f) and slow (k_s) rate constants; the former is thought to be a radical–radical recombination while the latter is hydrogen abstraction from the host; the activation parameters have been determined for both processes in both $C_{10}H_{16}$ and $C_{10}D_{16}$ matrices.[154] The escape efficiency, which represents the fraction of radicals escaping cage-recombination, for the radicals produced by decomposition of

$$C_6H_{11}(CN)[(CH_2)_nCOOH]C{-}N{=}N{-}C[(CH_2)_mCOOH](CN)C_6H_{11}$$

with $n = m = 4$, $n = m = 10$, $n = 4$ and $m = 10$ has been found to decrease (ten-fold) drastically when passing from chlorobenzene to multi-lamellar vescicles (liposomes) of dipalmitoylphosphatidyl choline.[155] The rate constant and activation parameters for the second-order self-termination reaction of the *N*-hydropyridinyl radicals have been measured by kinetic ESR spectroscopy.[156] This technique has also been used to obtain rate constants for various simultaneous reactions in systems containing more than one kind of transient radicals; thus, it has been applied to the reaction of *tert*-butyl and isopropylol radicals, *viz.*:[157]

$$(CH_3)_3CC(O)C(CH_3)_3 \xrightarrow{h\nu,\, I_1} 2\,(CH_3)_3C^\cdot + CO$$

$$(CH_3)_2CO + (CH_3)_2CHOH \xrightarrow{h\nu,\, I_2} 2(CH_3)_2\dot{C}OH$$

$$2\,(CH_3)_3C^\cdot \xrightarrow{k_t^1} \text{products}$$

$$2\,(CH_3)_2\dot{C}OH \xrightarrow{k_t^2} \text{products}$$

$$(CH_3)_3C^\cdot + (CH_3)_2\dot{C}OH \xrightarrow{k_t^x} \text{products}$$

$$(CH_3)_3C^\cdot + (CH_3)_2CHOH \xrightarrow{k} (CH_3)_3CH + (CH_3)_2\dot{C}OH$$

The extent of diffusion control of the termination constants has also been discussed.[157] Rate constants for the self- and cross-terminations of $(CH_3)_2\dot{C}OH$ and its anion, $(CH_3)_2\dot{C}O^-$, in aqueous solution have been determined; whereas the self-termination of the neutral radical [$(0.97 \pm 0.03) \times 10^9\ \text{M}^{-1}\text{s}^{-1}$] occurs close to the diffusion-controlled limit, the cross- [$(0.53 \pm 0.03) \times 10^9\ \text{M}^{-1}\text{s}^{-1}$] and self-terminations [$(0.23 \pm 0.02) \times 10^9\ \text{M}^{-1}\text{s}^{-1}$] involving the anion are slower and reflect effects of charge repulsion and steric constraints by solvation.[158] Expressions for the recombination probability of uncharged radicals in zero magnetic field have been derived taking into account the singlet–triplet mixing induced by isotopic hyperfine interactions;[159] experimental data on the magnetic ^{13}C isotope enrichment of dibenzyl ketone under photolysis have been discussed. Relative rates for radical disproportionation and recombination in the liquid phase at 150° have been determined for a series of reactions involving resonance-stabilized hydroaromatic radicals.[160] Results of laser flash photolytic studies have indicated that ketyl radicals of acetophenone dimerize with diffusion-controlled rates $[2k_t = (3.2 \pm 0.5) \times 10^9 \text{ M}^{-1}\text{s}^{-1}]$ to give pinacols and isopinacols;[161] the kinetics for recombination of $Ph_2\dot{C}OH$ radicals and their disproportionation with stable aminoxyls ($=N\text{–}\dot{O}$) in individual solvents and binary mixtures of different viscosities have been investigated.[162] Asymmetric allyl radicals, generated *via* H-atom abstraction by $Bu^t\dot{O}$ are trapped by 1,1,3,3-tetramethylisoindolin-2-yloxyl; combination mainly occurs at the more substituted terminus of the π-system, this being attributed to the electrophilic character of isoindolinoxyl.[163] Rate constants for recombination of $Ph_2N^\cdot$ radicals with themselves by N—N and C—N bond formation were 0.75 $\times 10^7$ and $1.05 \times 10^7 \text{ M}^{-1}\text{s}^{-1}$, respectively at 293 K.[161,164] The kinetics of disproportionation of several sterically hindered phenoxyl radicals has been studied and the activation energies have been derived.[165]

The termination rate constants have been determined for $CH_3CH_2CH(CH_3)O_2^\cdot$ ($k = 4.5 \times 10^6 \text{ M}^{-1}\text{s}^{-1}$), $CH_3CH_2CD(CH_3)O_2^\cdot$ ($k = 1.0 \times 10^6 \text{ M}^{-1}\text{s}^{-1}$), and $c\text{-}C_6H_{11}O_2^\cdot$ ($k = 2.0 \times 10^6 \text{ M}^{-1}\text{s}^{-1}$): these values are 2–4 orders of magnitude larger than those for tertiary alkylperoxyls, and *ca.* 10 times smaller than those of primary peroxyls with the same number of atoms.[166] Second-order rate constants have been determined for the reaction of $Ph_2N^\cdot$ with $Bu^tOO^\cdot$ at 283–303 K, the activation energy being less than 3.6 kcal mol^{-1}.[164] The rate constants for the reactions of the cyclohexylsulphonyl radical with cyclohexylperoxyl and tridecylperoxyl radicals are 1.5×10^8 and $3.0 \times 10^8 \text{ M}^{-1}\text{s}^{-1}$, respectively at 293 K.[167] Oxidation of vitamin E (α-tocopherol) at 50° in acetonitrile and hexane by alkylperoxyl radicals gives ~ 50% yield of (**54**) derived from combination of $RO_2^\cdot$ and vitamin E radical; this product rapidly hydrolyses to tocophenylquinone (**55**).[168]

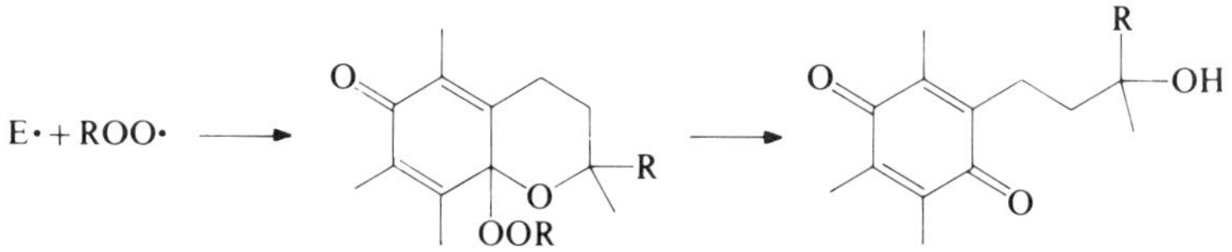

$$E^\cdot + ROO^\cdot \longrightarrow (\mathbf{54}) \longrightarrow (\mathbf{55})$$

(**54**) (**55**)

Atom Abstraction Reactions

Hydrogen Atom Abstraction

A new theoretical model has been developed to calculate rate constants for reactions in which hydrogen transfer occurs by tunnelling, *i.e.* by a non-classical mechanism; instead of formulating the hydrogen transfer in terms of barrier penetration (conventional tunnelling formalism) it has been described as a radiationless transition between potential energy surfaces.[169,170] Correspondingly the input parameters are not the height and width of the barrier through which the hydrogen atom tunnels but are vibrational parameters describing the initial and final states of the transfer process together with a coupling. This approach has been applied to a number of hydrogen-transfer reactions, reported in the literature, that exhibit a non-classical behaviour, $AH(D) + B \rightarrow A + H(D)B$, where A and B are carbon, nitrogen, or oxygen atoms attached to other atoms or groups.[171] The kinetics of the reaction $H^{\bullet} + C_2H_6 \rightarrow H_2 + C_2H_5^{\bullet}$ has been studied in the temperature region of 1000 K;[172] hot H-atoms generated by photolysis of HI with UV light have been shown to abstract D and Br from CD_3Br and the energy dependence of both processes has been examined.[173] The reactions $CH_4 + R^{\bullet} \rightarrow {}^{\bullet}CH_3 + HR$, where R = H, CH_3, NH_2, OH, F, have been studied using *ab initio* SCF and CI methods and geometric, energetic, and kinetic aspects have been discussed.[174]

Hydrogen Atom Abstraction by Carbon-centred Radicals

The reactions of methylidene radicals, $:\dot{C}H$, with a number of saturated hydrocarbons have been studied; in particular, Arrhenius expressions have been determined for the reactions with methane, ethane, and *n*-butane.[175] Computational studies of the reaction of methane with $^{\bullet}CH_3$ have been performed using the MINDO/3 approach.[176,177] The temperature and site dependence of the rate constants of hydrogen and deuterium abstraction by methyl radicals in CH_3OH and CD_3OD glasses have been investigated;[178] deuterium isotope effects in the reaction of photolytically generated $Me^{\bullet}$ with MeOH at 77–110 K have been interpreted in terms of two processes, *i.e.* $Me^{\bullet}$ conversion to a reactive state and H-atom transfer *via* tunnelling.[179] Arrhenius parameters have been obtained for both hydrogen abstraction and addition reactions of methyl radicals with 1-hexene.[180] The hydrogen abstraction from $(CH_3)_4Si$ by $^{\bullet}CH_3$ radicals has been investigated relative to the recombination of methyl radicals;[181] a combination of the obtained values with existing data gives the following equation:

$$\log(k/\text{M}^{-1}\text{s}^{-1}) = (8.82 \pm 0.07) - (10.56 \pm 0.07)/\theta$$

where $\theta = 2.3\,RT$ kcal mol^{-1}. It has been found that $CH_3^{\bullet}$, $C_2H_5^{\bullet}$, and $CF_3^{\bullet}$ radicals abstract hydrogen from heptane 3.3, 3.2, and 2.2 times faster than from isooctane, respectively.[182] An LFER of the modified Swain–Lipton parameters for the H-atom abstraction kinetics of RC_6H_4COMe (R = H, *p*-NO_2, *p*-OMe, *p*-Cl) with polyacrylonitrile radicals shows that the transitions state is non-polar.[183] The initial rates of formation of the products for the reaction between isopropyl radicals and molecular hydrogen have been determined in the range 473–559 K and a mechanism has been proposed;[184] the Arrhenius parameters for the metathesis reaction

$$Bu^{t\bullet} + HI\ (\text{or DI}) \rightarrow Bu^tH\ (\text{or } Bu^tD) + I^{\bullet}$$

have been obtained.[185]

The Arrhenius parameters of hydrogen abstraction from H_2S by $^{\bullet}CH_3$ and $^{\bullet}CF_3$ have been found to be $\log(k/\text{M}^{-1}\text{s}^{-1}) = (8.00 \pm 0.10) - (2.09 \pm 0.02)/\theta$ and

$\log(k/\text{M}^{-1}\text{s}^{-1}) = (8.87 \pm 0.10) - (4.19 \pm 0.17)/\theta$, respectively;[186,187] these results show that $\cdot CF_3$ is less reactive than $\cdot CH_3$ in attacking H_2S, thus indicating that polar effects play a significant role in the hydrogen-transfer reactions of $\cdot CF_3$ radicals. Arrhenius parameters have been determined for $R\cdot + HBr \rightarrow RH + Br\cdot$ where R $= CF_3$, C_2F_5, i-C_3F_7 in the vapour phase.[188] The gas-phase reactions of $\cdot CF_3$ and $\cdot C_2F_5$ with HCN have been examined both quantitatively and qualitatively over a wide conversion range using CF_3I or C_2F_5I as thermal and photolytic sources of radicals;[189,190] the possible mechanistic implications have been discussed. Arrhenius parameters for the reaction $\cdot CF_3 + CX_3CN \rightarrow CF_3X + \cdot CX_2CN$, where X = H or D, have been measured relative to $\cdot CF_3$ recombination; a high primary kinetic isotope effect and the non-curved Arrhenius plot found confirm the absence of an addition reaction.[191] The rates of hydrogen abstraction by $PhCH_2\cdot$ from 23 different hydrogen donors have been determined relative to the rate of deuterium abstraction from Ph_3SiD; values are in the range from 0.12 for mesithylene (slower rate) to 37 for 9,10-dihydroanthracene (faster rate); the ln (k_H/k_D) is linearly related to the resonance stabilization energy of arylmethyl donors and evidence that steric, entropic, and stereoelectronic factors can play an important rôle has been given.[192] The reaction $R\cdot + HSnBu_3\ (DSnBu_3) \rightarrow RH(RD) + \cdot SnBu_3$ has been studied for eight R$\cdot$ radicals, namely oct-1-yl, 2,2-dimethylpent-1-yl, 3-methylpent-1-yl, phenyl, benzyl, *p*-fluorobenzyl, *p*-methylbenzyl, and *p*-(trifluoromethyl)benzyl: in all cases no evidence of tunnelling could be gained;[193] it has also been shown that the transition states of the H-atom transfer become increasingly unsymmetrical in the order benzyl < secondary < primary < phenyl raidcals, and the important rôle played by polar contributions in determining free-radical reactivity patterns has been established.[193]

For several hetaryl radicals, generated photochemically from their corresponding halides, rate constant ratios for H-atom abstraction from either methoxide ion or methanol and addition to thiophenoxide ion have been obtained; for example, 3-pyridyl radical addition to thiolate ion is 2.2 and 102 times faster than hydrogen abstraction from methoxide ion and methanol, respectively.[194] The acetonylation of esters, acids, and nitriles in solution by the free-radical decomposition of *O-tert*-butyl and *O*-isopropenyl peroxycarbonate gives fairly good yields although the acetonylation of functional substrates often gives mixtures of isomers.[195]

Hydrogen Atom Abstraction by Oxygen-centred Radicals

The reactions between $O(^3P)$ and butanols have been studied in the temperature range 300–600 K; the obtained activation energies were found to be two to four times smaller than those for abstraction from alkanes probably due to the inductive effect of the hydroxy group.[196] Arrhenius parameters have been reported for the gas-phase reactions between $\cdot OH$ radicals with ethane and a series of methanes and ethanes containing chlorine and/or fluorine;[197,198] some of these data have been compared with data calculated using semi-empirical transition-state theory.[198] Relative rate constants for the reaction of $\cdot OH$ radicals with a series of branched alkanes have been determined at 297 K, using methyl nitrite photolysis in air as source of $\cdot OH$ radicals.[199] The rate constant for the reaction $HO\cdot + CH_2O \rightarrow \dot{C}HO$

$+H_2O$ has been determined at room temperature using either the FTIR spectroscopic method[200] or laser magnetic resonance.[201] The kinetics of the reaction X· $+CH_3CN \rightarrow {}^{\cdot}CH_2CN + XH$ (X = OH, Cl) has been investigated in the gas phase by the flash photolysis resonance fluorescence technique, and the obtained kinetic data have been discussed.[202]

Rate constants have been determined at 296 K for the gas-phase reactions of $^{\cdot}NO_3$ radicals with formaldehyde $(1.97 \pm 0.16) \times 10^5\ \mathrm{M^{-1}s^{-1}}$, methanol $(1.34 \pm 0.28) \times 10^5\ \mathrm{M^{-1}s^{-1}}$,[203] a series of alkanes,[204] aromatic hydrocarbons, and hydroxy- and methoxy-substituted aromatics,[205] all these reactions almost certainly proceed *via* H-atom abstraction. Rate constants for the reaction of carbonate radical with aniline and some *para*-substituted anilines have been determined by the flash photolysis technique; the rate constants at pH = 8.5 have a Hammett LFER with $\rho^+ = 1$.[206]

The gas-phase H-atom abstraction reactions by thermally generated butoxyl radicals with a variety of substrates have been studied, and the activation parameters determined and compared to those calculated theoretically from spectroscopic data; some values found for $\log A(\mathrm{M^{-1}s^{-1}})$ and E_a ($\mathrm{kcal\,mol^{-1}}$) are 9.25 and 7.74 (2,2-dimethylpropane), 9.59 and 5.98 (*cis*-2-butene), and 9.97 and 5.71 (2-methylpropane).[207] Relative Arrhenius parameters for decomposition of Bu^{t}O· radicals and H-atom abstraction reactions with isobutane and cyclohexane in the gas phase have been obtained.[208] *tert*-Butoxyl and *tert*-butyl radicals show a negative and a positive ρ, respectively, in the hydrogen abstraction reaction with a number of substituted toluenes.[209] The hydrogens at C(2) in spiro[2,*n*]alkanes (**56**) are abstracted about five times more rapidly than "normal" secondary hydrogens by *tert*-butoxyl radicals, because of favourable overlap of the *p*-orbital at C(2) with the HOMO of the adjacent cyclopropyl ring.[210] In hydrogen abstraction from fatty acids by Bu^{t}O· radicals, abstraction from the carbon α to the carboxylic group predominates in acids with up to six carbon atoms, while in higher terms abstraction also occurs from the next to last $(\omega - 1)$-carbon atom of the chain;[211] it would appear that in these higher terms abstraction from the α- and $(\omega - 1)$-methylenic group is twice as fast as from other methylenic groups in the molecule.[211] The reaction of Bu^{t}O· radicals with eight different olefins have been studied by trapping the reaction products with 1,1,3,3-tetramethylisoindolin-2-yloxyl; the relative rates of addition to

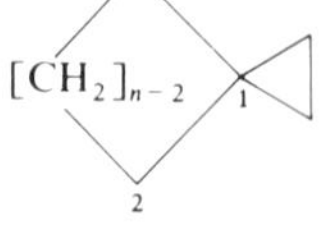

$[CH_2]_{n-2}$ 1 2

(56)

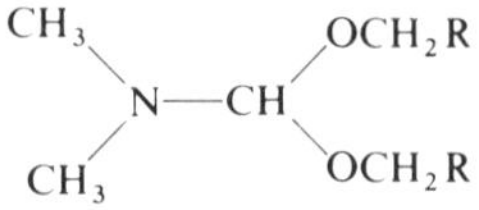

$(CH_3)_2N{-}CH(OCH_2R)_2$

(57)

the double bond *vs.* hydrogen abstraction have been measured and discussed.[212] The reaction between Bu^{t}O· and methyl methacrylate has been studied in 11 different solvents; it has been found that more polar solvents increase the ratio of abstraction to addition and also favour β-fragmentation of *tert*-butoxyl over the two above processes.[213] Absolute rate constants for the reaction of Bu^{t}O· radicals with tertiary amines, leading to the formation of α-aminoalkyl radicals, have been shown to be solvent-dependent; *e.g.* rate constants for Me_3N at 299 K are 1.2×10^7 and 1.1×10^8 $M^{-1}s^{-1}$ in wet acetonitrile and benzene–peroxide, respectively.[214] A kinetic study of the Bu^{t}OO Bu^t-initiated homolysis of dimethylformamide acetals (**57**; R = H, Et, Bu, pentyl) and cyclic acetals have been described;[215,216] the difference in E_a for cleavage of the C—H bonds attached to three heteroatoms and one heteroatom were 5 kcal mol^{-1} for (**57**; R = H) and 8kcal mol^{-1} for (**57**; R = Bu).[215] The reactivity of *tert*-butoxyl radicals in aqueous solution has been studied using the pulse radiolysis technique.[217] Absolute rate constants for the reaction Bu^{t}O· + RCHO → Bu^{t}OH + R$\dot{C}$O have been measured in solution by using laser flash photolysis techniques;[218] at 297 K, the rate constants are 8.9, 6.8, 1.0, and 0.05 × 10^7 $M^{-1}s^{-1}$ for EtCHO, PhCHO, Me_2NCHO, and EtOCHO, respectively; this decrease in the rate constants has been attributed to the importance of the polar contribution to the transition state.[218] Absolute rate constants for H-atom abstraction from MeOH, EtOH, and Pr^{i}OH by $PhC(CH_3)_2O$· radicals have been determined as $(8.5 \pm 1.0) \times 10^5$, $(3.8 \pm 0.5) \times 10^6$, and $(9.9 \pm 1.1) \times 10^6$ $M^{-1}s^{-1}$, respectively;[219] the trend is similar to that observed for similar abstraction reactions by Bu^{t}O· radicals in organic solvents, but values are 3–5 times higher reflecting a higher reactivity of cumyloxyl radicals. The rate constant for H-atom abstraction from methane by CF_3O· radicals is $\geqslant 1.5 \times 10^5$ $M^{-1}s^{-1}$ at room temperature.[220]

Rate constants for the reactions of CH_3OO· with CH_3OH and CH_2O have been obtained at 600 K using molecular modulation spectroscopy;[221] hydrogen abstraction by Bu^{t}OO· radicals from a variety of alkanes, aralkanes, alkenes, and alkyl halides have also been reviewed.[222] Absolute rate constants and activation parameters have been determined for the reaction of $(CF_3)_2NO$· with a variety of organic substrates; the pre-exponential factors were found to be much smaller than the values generally considered "normal" for radical/H-atom abstractions and radical/C═C double-bond additions, *viz.* $10^{8.5 \pm 0.5}$ $M^{-1}s^{-1}$; it has been suggested that they are probably due to geometric constraints on the transition states rather than to quantum-mechanical tunnelling.[223]

Hydrogen Atom Abstraction by Halogen Atoms

The abstraction of hydrogen/deuterium from gaseous CH_3CH_2Cl, CH_3CHDCl, and CH_3CD_2Cl by photochemically generated ground-state chlorine atoms has been investigated over the temperature range of 8–94° using methane as a competitor; the temperature dependence of the relative rate constants and of the primary and secondary kinetic isotope effects have been obtained and the reaction mechanism has been discussed.[224,225] The reaction of bromine atoms with isobutane,[226] diethyl ether,[227] and acetaldehyde[228] have been studied in the temperature range 298–333 K using the VLPR technique; rate constant and activation parameters have been obtained.

Halogen Atom Abstraction

The reaction Cl· + RI → ICl + R· where R = CF_3, C_2F_5, *i*-C_3F_7) have been studied competitively in the gas phase over the range 27–231°.[229] Arrhenius parameters for halogen atom abstraction from haloalkanes by methyl radical have been determined relative to those for methyl radical combination.[230] It has been shown that photolysis of di-*tert*-butyl peroxide is not quenched by CCl_4; methyl and trichloromethyl radicals, observed by ESR, are almost certainly formed *via* the following reactions:[231]

$$(CH_3)_3CO\cdot \rightarrow (CH_3)_2CO + CH_3\cdot$$
$$CH_3\cdot + CCl_4 \rightarrow CH_3Cl + CCl_3\cdot$$

The Arrhenius equation $\log(k/\text{M}^{-1}\,\text{s}^{-1}) = (8.14 \pm 0.42) - (5.52 \pm 0.63)/\theta$, where $\theta = 2.3\,RT\,\text{kcal mol}^{-1}$, has been obtained for the reaction:

$$PhC(Me)_2\dot{C}H_2 + CCl_4 \rightarrow PhC(Me)_2CH_2Cl + Cl_3C\cdot$$

using the neophyl rearrangement as a "clock" reaction,[232] while for the corresponding reaction of cyclopropyl radicals, *viz.*:

$$\triangleright\!\cdot + CCl_4 \longrightarrow \triangleright\!\!-\!Cl + \cdot CCl_3$$

the equation $\log(k/\text{M}^{-1}\,\text{s}^{-1}) = (8.7 \pm 0.4) - (3.50 \pm 0.6)/\theta$ was derived following a detailed kinetic investigation;[233] such information can be utilized as "standard" in future competitive kinetic studies of cyclopropyl radical reactivity. The gas-phase reactions of $CF_3\cdot$, $C_2F_5\cdot$, and $C_3F_7\cdot$ radicals with CCl_4 have been studied using the photolysis of the corresponding perfluoroalkyl iodide as the free-radical source;[234] the Arrhenius parameters, based on the value of $2.3 \times 10^{10}\ \text{M}^{-1}\,\text{s}^{-1}$ for the self-combination rate constant of all radicals, are almost identical, *viz.* $\log A = 9.8\ \text{M}^{-1}\,\text{s}^{-1}$ and $E_a = 11.6 \pm 0.4\ \text{kcal mol}^{-1}$. The kinetics of gas-phase reactions of $CF_3CF_2CF_2\cdot$ and $(CF_3)_2CF\cdot$ with ClCN have been studied in the temperature range 70–330°; the main product is formed *via* Cl-atom abstraction, although at lower temperature the addition reaction also occurs.[235,236] Phosphine–boryl radicals, $X_3\overset{+}{P}\text{–}\dot{\bar{B}}H_2$, abstract bromine atoms from alkyl bromides but do not abstract chlorine from alkyl chlorides, with the only exception of $PhCH_2Cl$, and are less reactive than the corresponding aminoboryl radicals, $X_3\overset{+}{N}\text{–}\dot{\bar{B}}H_2$.[237] $Bu_3Sn\cdot$ radicals abstract chlorine and bromine atoms from *trans*-3-*X*,5-butylcyclohexene (**58**) 10.1 and 6.0 times faster, respectively, than from the *cis*-epimers (**59**); a similar although less marked effect was detected for the *trans*- and *cis*-saturated derivatives, (**60**) and (**61**).[238] This rate enhancement has been attributed to stereoelectronic effects in the cyclohexene derivatives, *i.e.* the cleavage is easier for the bond which is closer to the plane of the π-orbital of the double bond, while the

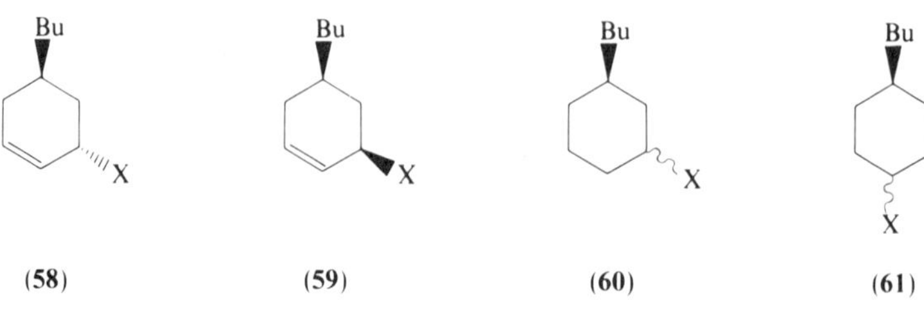

(**58**) (**59**) (**60**) (**61**)

most likely explanation for the behaviour of the saturated halides is the difference in free energy of activation between the corresponding transition complexes.[238]

From the results of a mechanistic study which has been carried out on the formation of α-diketones by reaction of acyl chlorides with diiodosamarium, *viz.*:

$$2\,RCOCl + 2SmI_2 \rightarrow RCOCOR + 2SmI_2Cl$$

it would appear that the first-formed acyl radicals are reduced to acyl anions by SmI_2 before undergoing decarbonylation.[239] Convincing evidence has been reported showing that the reaction of organotin hydrides with acid chlorides is not a radical-chain process, and that radicals are not involved as intermediates.[240]

Halogenation

Alkyl chlorides can be fluorinated *via* aereosol direct fluorination; tertiary alkyl chlorides lead to fluorinated primary alkyl chlorides through a 1,2-chlorine shift, while secondary alkyl chlorides lead to a mixture of fluorinated primary and secondary alkyl chlorides.[241] The radical chlorination of alkanes with chlorinating agents as Cl_2, *N*-chlorosuccinimide and *tert*-butyl hypochlorite in the presence of trialkylboranes has been reported, although further studies are needed to ascertain the rôle of the borane and the actual mechanism of the reaction.[242] Chlorination of cycloalkanes by means of Cl_2 in CS_2 or of iododichlorobenzene and some of its derivatives in CCl_4, CS_2, or benzene has been studied and the observed regio- and stereo-selectivities discussed for the different chlorinating agents, solvents, and substrates.[243] The mechanism and kinetics of photochlorination of $ClCHF_2$ have been investigated.[244] The variations in product ratios with temperature and in solvent–reactant ratio have been determined in the chlorination of chloroalkanes with Cl· and its complex with *o*-dichlorobenzene.[245] In the side-chain halogenation of benzyl radical and benzyl chlorides, substitutive chlorination is favoured over additive chlorination by low Cl_2 concentrations and higher temperatures, and it has been proposed that a π-complex is first formed, which then rearranges to a σ-complex.[246]

The chlorination of 1-hexene at 148–295 K occurred by both radical and molecular mechanisms; the radical path became more important either by increasing the concentration of 1-hexene or by decreasing the polarity of the medium;[247] *in situ* generated Mn(III) chloride species have been used in chlorination of alkenes; the reaction has proved to be particularly effective in chlorinating non-conjugated systems.[248] Kinetic studies have been carried out on the chlorination of 2-propenyl compounds,[249] polyhydric esters,[250] and some chloro derivatives of 2-picoline;[251] the reaction mechanisms have been discussed.

The controversy about radical-chain bromination of alkyl bromides has been surveyed; from the available data it is concluded that the β-bromo substituent facilitates hydrogen abstraction and constrains the configuration at the vicinal site.[252] Bromination of 2-bromobutane has been carried out in sealed tubes in the vapour phase and the results have been compared with those of bromination in the liquid phase; in the former case significant amounts of polybrominated compounds are obtained.[253] The relative rate constants for bromination of *p*-RC_6H_4X

(R = 1-adamantyl; X = CH_3, CH_2Br, $CHBr_2$) have been obtained relative to C_6H_5X and discussed in relation to LFER.[254] Further evidence has been presented for the existence of the three distinctive hydrogen abstractors in systems containing Br_2 and NBS, *i.e.* Br·, S_π·, and S_σ· (S· = succinimidyl).[255,256]

Addition Reactions

Carbon-centred Radicals

Kinetic data have been obtained and a mechanism has been proposed for formation of propylene *via* addition of Me· radicals to ethylene;[257] Me· has been found to add to the terminal carbon of allenes with high regioselectivity and the following kinetic expression has been derived for the process:[258]

$$\log(k/\text{M}^{-1}\,\text{s}^{-1}) = 7.76 - (12.42/\theta)$$

where $\theta = 2.3RT\,\text{kcal mol}^{-1}$; it has also been shown that the behaviour of methyl, and more generally, of carbon-centred radicals differs from that of such hetero-radicals as Br·, RS·, $ArSO_2$·, and R_3Sn· which preferentially add to the central carbon atom.[259] The Arrhenius parameters for the gas-phase addition of methyl radicals to trifluoroacetone[260] and hexafluoroacetone[261] to give the corresponding alkoxyl radicals can be represented by $\log(k/\text{M}^{-1}\,\text{s}^{-1}) = (7.7 \pm 0.2) - (6.7 \pm 0.2)/\theta$ and $\log(k/\text{M}^{-1}\,\text{s}^{-1}) = (8.0 \pm 0.3) - (5.1 \pm 0.7)/\theta$, respectively, where $\theta = 2.3\,RT\,\text{kcal mol}^{-1}$; these results have been discussed in relation to existing data for radical addition to carbonyl groups and the addition of methyl radicals to olefins. The build-up of $CFCl_2O_2$· concentration has been used to monitor the rate constant for the reaction of ·$CFCl_2$ with molecular oxygen;[262] $BrCCl_3$ adds across the double bond of 1-decene in the absence of radical initiators to give $CH_3(CH_2)_7CHBrCH_2CCl_3$ in a 50% yield;[263] the spontaneous formation of ·CCl_3 (one-electron transfer by the solvent, THF) has been established by means of the spin-trapping technique.[263] In the addition of CCl_4 across ethylenic double bonds in the presence of ruthenium or chromium complexes the formation of radical pairs [{$RuCl_3(PPh_3)$·} {·CCl_3}] and [{CrCl(THF) $(CO)_3$·} {CCl_3·}] has been postulated.[264,265] Difluorodiiodomethane has been proved to be an excellent reagent for free-radical chain addition to alkenes (Scheme 5).[266]

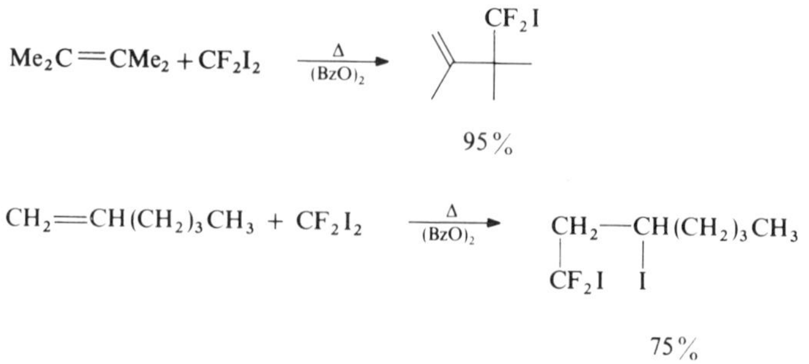

$$Me_2C{=}CMe_2 + CF_2I_2 \xrightarrow[(BzO)_2]{\Delta} CH_2{=}C(Me){-}CMe_2{-}CF_2I \quad 95\%$$

$$CH_2{=}CH(CH_2)_3CH_3 + CF_2I_2 \xrightarrow[(BzO)_2]{\Delta} CF_2I{-}CH_2{-}CHI(CH_2)_3CH_3 \quad 75\%$$

SCHEME 5

The laser photolysis resonance absorption technique has been used to measure absolute rate constants for the reaction of formyl and deuteroformyl radicals with NO and O_2 at 295 K;[267] deuterium substitution was found to increase the rate constants by 25% and 10% in reactions with NO and O_2, respectively.[267] Rates of addition of $\cdot CH_2CH_3$,[268] $\cdot CH_3$, $\cdot CH_2CF_3$, $\cdot CH(CF_3)_2$, and $\cdot CF_2CH_3$[269] to ethylene and its fluorinated derivatives have been determined relative to H-atom abstraction from the solvent (heptane or isooctane). The rate of addition of $Bu^t\cdot$ radical to some unsaturated substrates has been found to decrease in the order maleic anhydride > fluoranyl > benzoquinone > methyl acrylate > butyl acrylate > styrene > vinyl acetate > vinyl butyl ether.[270] Captodative effects induce an increase beyond expectation of the rate constant of addition of $(CH_3)_2\dot{C}(CN)$ radicals to some substituted ethylenes and styrenes.[271] The activation energies have been determined for the reactions of undecyl radical with 1-acetoxycyclohexene (6.50), 3-acetoxycyclohexene (8.20), 2-cyclohexen-1-one (4.59), 2,6,6-trimethyl-2-cyclohexen-1-one (5.50), and 2-cyclohexen-1-ol (3.9 kcal mol^{-1}).[272]

The electronic nature of free radicals has been studied by the CNDO/2 method with respect to their addition reactions to substituted styrenes, and it has been shown that, in agreement with experimental data, the calculations predict correctly the change of sign of the Hammett ρ constant on going from electronegative to electropositive radicals.[273] The radical-chain decarboxylation of the carboxylic acid esters derived from *N*-hydroxypyridine-2-thione and similar substrates has been elegantly used for synthetic purposes;[274-278] see for example Scheme 6.[276]

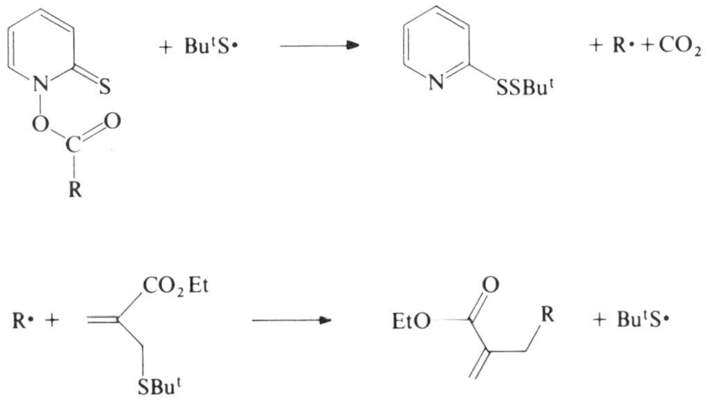

$$\text{(pyridine-2-thione } N\text{-O-C(=O)R)} + Bu^tS\cdot \longrightarrow \text{(2-pyridyl)}SSBu^t + R\cdot + CO_2$$

$$R\cdot + CH_2{=}C(CO_2Et)CH_2SBu^t \longrightarrow EtO\text{-}C(=O)C(=CH_2)CH_2R + Bu^tS\cdot$$

Scheme 6

Addition of alkyl radicals to unsaturated stannanes has proved to be a valuable tool for synthetic purposes; thus, addition of $R\cdot$ to β-stannyl acrylates, affords β-unsaturated esters in a regiospecific manner,[279] while addition to triphenylprop-2-ynyl stannane provides a general route to terminal allenes (Scheme 7).[280]

The formation of *trans*-isomers predominates in the solvomercuriation and reductive C—C bond formation reaction of cyclic alkenes (cyclopentene, dihydrofuran, cyclohexene, dihydropyran) with alkenes HXC═CYZ (Scheme 8), the stereoselectivity increasing with decreasing reactivity of alkenes and being greater

$R\cdot + HC{\equiv}CCH_2SnPh_3 \longrightarrow RCH_2C{\equiv}CH + \cdot SnPh_3$

$RX + \cdot SnPh_3 \longrightarrow R\cdot + XSnPh_3$

SCHEME 7

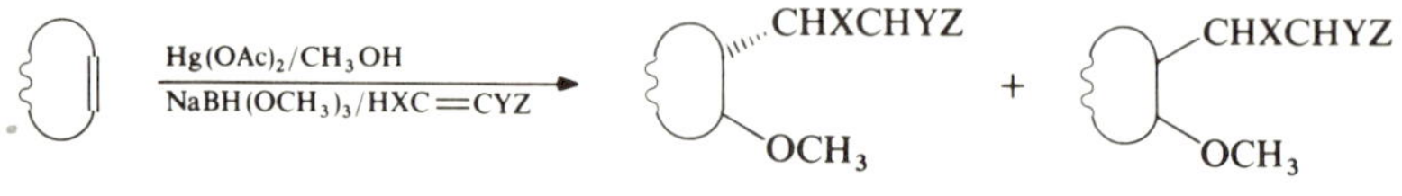

Hg(OAc)$_2$/CH$_3$OH ; NaBH(OCH$_3$)$_3$/HXC=CYZ → CHXCHYZ / OCH$_3$ (*trans*) + CHXCHYZ / OCH$_3$ (*cis*)

SCHEME 8

for five-membered than for six-membered substrates.[281] The rate constants for addition of 5-hexenyl radicals to methyl acrylate have been determined in several media: a *ca.* seven-fold increase has been observed when going from aqueous THF (0.79×10^5 M^{-1} s^{-1}) to a 1:1 mixture of acetonitrile and acetic acid (5.0×10^5 M^{-1} s^{-1});[282] with high concentration of $CH_2{=}CHCO_2Me$ the rate of addition was greater than that of cyclization of the 5-hexenyl radical.[282]

A theoretical study has indicated that in the addition of vinyl and cyclopropyl radicals to ethylene a reactant-like transition state (σ-configuration) is involved, and that the formation of a π-complex, as previously suggested for the addition of hydrocarbon σ-radicals, can be discarded.[283] Rate constants for addition of cyclopropyl radicals to olefins have been measured;[284] in particular, rates of addition to styrene, β-methylstyrene, and cyclohexene at 298 K are $(1.5 \pm 0.3) \times 10^7$, $(2.0 \pm 0.5) \times 10^6$, and $(1.3 \pm 0.3) \times 10^6$ M^{-1} s^{-1}, respectively, thus making the reactivity of the cyclopropyl radical intermediate between that of phenyl and acyclic alkyl radicals.[284] Cyclohexyl radicals add predominantly to the unsubstituted carbon of mono-substituted maleic anhydrides and show a contrathermodynamic stereoselectivity ranging between 2.3 and 19.[285] Rate constants of 2.3×10^6 and 4.5×10^6 M^{-1} s^{-1} have been determined for the addition of phenyl radicals to $PhCH_2C(S)OPh$ and $Ph_2CHC(S)OPh$, respectively, at 45°.[286]

Alkylation of β-styrenyl or β,β-diphenylvinyl derivatives (**62**) and (**63**) *via* a free-radical chain reaction with alkyl mercury halides has been reported:[287]

$$PhCY{=}CHX + RHgCl \rightarrow PhCY{=}CHR$$

(**62**) Y=H

(**63**) Y=Ph

$X = Bu_3Sn, HgCl, HgBr, I, PhSO_2, PhSO, PhS$

The process involves regioselective addition of R· to (**62**) or (**63**) followed by β-elimination of X· radicals which regenerate the species R· by attack of RHgX.[287] The rates for the reaction of substituted benzenes, *p*-benzoquinone, duroquinone,[288] and 2,3-dimethyl-1,3-butadiene[289] with cyclohexadienyl radicals from addition of muonium to substituted benzenes have been determined.

Atoms

Addition of atomic hydrogen to ethene to give ethyl radical has been studied using quasi-classical trajectories and agreement has been found with the rate constants predicted through the activated complex theory with a tunnelling correction;[290] addition to acetylene leads to ethylene, 1,3-butadiene, and benzyl radical through a mechanism involving chain-reaction of vinyl and butadienyl radicals.[291] Photolysis of a number of methyl- and dimethyl-quinolines at 20 K led to the detection of fluorescence spectra of species derived by loss of a H-atom from one of the methyl group and its addition to another molecule.[292] The kinetics for the addition reaction of atomic chlorine to ethylene[293] and vinyl bromide[294] has been studied and discussed.

Oxygen- and Sulphur-centred Radicals

Several studies have been carried out on the reaction of hydroxyl radicals with alkenes; gas-phase studies included addition to ethene, propene,[295,296] a number of alkyl-substituted olefins,[296,297] some cycloalkenes,[298] and a series of conjugated dienes and trienes;[299] techniques have also been developed which allow prediction of rate constants for addition of ·OH radicals to alkenes and cycloalkenes.[297,298] In liquid-phase studies, the ·OH adducts of a number of olefins have been observed by ESR and their reactivity compared with that of $SO_4^{\cdot-}$ and $Cl_2^{\cdot-}$;[300] it has been found that hydroxyl radicals predominantly add to the carbon–carbon double bond of the side-chain of *p*-hydroxycinnamic acid to form benzyl-type radical adducts.[301]

The photohydration of acetylene in the presence of polyoxymetallate(VI) anions as photocatalysts has been reported,[302] and rate constants have been obtained for the addition of hydroxyl to acethylene, propyne, and 1-butyne.[303]

tert-Butoxyl radicals have been reacted with captodative substituted olefins and the yields of addition of $Bu^tO\cdot$ *vs.* addition of Me· (from decomposition of butoxyl radicals) have been determined;[304] the higher amounts of butoxyl adducts obtained when starting from enamines (X = CN, Y = NR_2) rather than from olefin ethers (X = CN, Y = OR) or olefin thioethers (X = CN, Y = SR) indicate the important role played by electronic and polar effects on radical addition.[304]

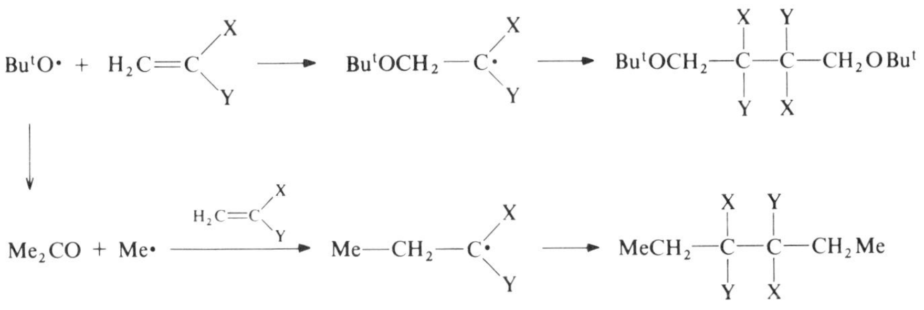

$$Bu^tO\cdot + H_2C{=}C(X)(Y) \longrightarrow Bu^tOCH_2{-}\dot{C}(X)(Y) \longrightarrow Bu^tOCH_2{-}C(X)(Y){-}C(Y)(X){-}CH_2OBu^t$$

$$\downarrow$$

$$Me_2CO + Me\cdot \xrightarrow{H_2C=C(X)Y} Me{-}CH_2{-}\dot{C}(X)(Y) \longrightarrow MeCH_2{-}C(X)(Y){-}C(Y)(X){-}CH_2Me$$

SCHEME 9

The gas-phase reaction of *tert*-butylperoxyl radicals with five different alkenes has been investigated at 393 K, and the rate constants determined as 3.36 (ethene), 2.60 (2-methylpropene), 3.60 (2-methylbut-1-ene), 0.38 (2-methylbut-2-ene), and 148 $M^{-1} s^{-1}$ (2,3-dimethyl-but-2-ene).[305]

Methylthiyl and *tert*-butylthiyl radicals add to aromatic aldoximes, but the paramagnetic species detected by ESR are those arising from rearrangement of the initially formed adducts.[306] α-Thioalkyl-α-silyloxy-substituted radicals

$$\text{Ar—CH=N}\sim\text{OH} \xrightarrow{\text{RS}\cdot} \text{Ar—}\underset{\text{SR}}{\text{CH}}\text{—}\dot{\text{N}}\text{—OH} \longrightarrow \text{Ar—}\underset{\text{SR}}{\text{CH}}\text{—}\underset{\text{H}}{\text{N}}\text{—}\dot{\text{O}}$$

$R\dot{C}(OSiR_3)SR'$ were unexpectedly detected when adding alkylthiyl (MeS·, Bu^nS·, Bu^tS·) and arylthiyl radicals to acyl- or aroyl-silanes ($XC(O)SiR_3$ (where X = Me, Bu^t, Ph) and although three different reaction pathways have been discussed to account for the observed species, an unambiguous choice was not possible on the basis of the experimental data.[307] The reactivity of phenylthiyl radicals towards unsaturated substrates has been further investigated; addition of p-$NH_2C_6H_4S\cdot$ to styrene has been discussed,[308] rate constants for addition of p-$NMe_2C_6H_4S\cdot$ to α-methylstyrene have been shown to be strongly dependent on solvent (a 35-fold increase on going from DMSO to cyclohexene),[309] and the absolute rate constants for the reversible addition of p-$XC_6H_4S\cdot$(X = H, Br, Cl, Me, OMe) to vinylpyridines have been found to be 3–5 times lower than for styrene.[310] Rate constants have also been determined for addition of these radicals to vinyl sulphides, (2.6–48) $\times 10^6$ $M^{-1} s^{-1}$, vinyl ethers, (0.8 to 1.0) $\times 10^5$ $M^{-1} s^{-1}$, vinylsulphoxide, 1.4 $\times 10^4$ $M^{-1} s^{-1}$, and vinyl sulphone, 1.6 $\times 10^4$ $M^{-1} s^{-1}$. It was suggested that the reactivity of each vinyl monomer is determined by the thermodynamic stability of the adduct radicals p-$XC_6H_4SCH_2\dot{C}HY$;[311] also, the reactivity of p-chlorophenylthiyl radicals towards cycloalkenes appears to be controlled by the resonance stabilization of the transition state, its polar nature, and the release of strain energy;[312] absolute rate constants ($M^{-1} s^{-1}$) vary in the sequence: conjugate dienes (10^6–10^8) > bicycloalkenes (10^5–10^7) > cyclomonoalkenes (10^2–10^5).

The free-radical addition of alkaneselenosulphonates and bis(p-tolysulphonyl)selenide to alkenes has been investigated.[313] A free-radical mechanism has been proposed for the ruthenium-catalysed reaction of arenesulphonyl chlorides with *cis*- and *trans*-β-methylstyrenes in the presence of Et_3N which accounts for the detection of only (E)-2-arylsulphonyl-1-phenylpropene.[314]

Miscellaneous Radicals

The gas-phase reaction of NO_2 with a series of conjugated and non-conjugated alkenes proceeds *via* initial addition of NO_2 to a double bond: rate constants were determined for 2,3-dimethyl-2-butene, 1,3-cyclohexadiene, and α-phellanthrene,[315] and the rate for addition of $\cdot NO_3$ to propene was also measured.[316] *N*-Halo-*N*-alkylamides (—C(O)—N(X)—R) add to cyclohexene and 1-octene much more slowly than *N*-halo-*N*-acylamides (—C(O)—N(X)—C(O)R);[317] this finding has been attributed to the more electrophilic nature of —C(O)—$\dot{N}$—C(O)— with respect to —C(O)—$\dot{N}$—R and to the better halogen donating ability of —C(O)—N(X)—C(O)— with respect to —C(O)—N(X)—R; the mechanism (Scheme 10) is thought to be similar (radical chain) to that already reported for the addition to other alkenes.[317]

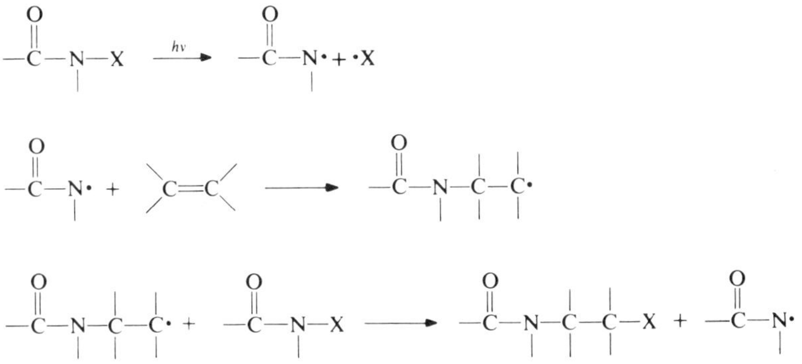

$$-\overset{O}{\overset{\|}{C}}-\underset{|}{N}-X \xrightarrow{h\nu} -\overset{O}{\overset{\|}{C}}-\underset{|}{N}{\bullet} + {\bullet}X$$

$$-\overset{O}{\overset{\|}{C}}-\underset{|}{N}{\bullet} + \rangle C{=}C\langle \longrightarrow -\overset{O}{\overset{\|}{C}}-\underset{|}{N}-\underset{|}{\overset{|}{C}}-\underset{|}{\overset{|}{C}}{\bullet}$$

$$-\overset{O}{\overset{\|}{C}}-\underset{|}{N}-\underset{|}{\overset{|}{C}}-\underset{|}{\overset{|}{C}}{\bullet} + -\overset{O}{\overset{\|}{C}}-\underset{|}{N}-X \longrightarrow -\overset{O}{\overset{\|}{C}}-\underset{|}{N}-\underset{|}{\overset{|}{C}}-\underset{|}{\overset{|}{C}}-X + -\overset{O}{\overset{\|}{C}}-\underset{|}{N}{\bullet}$$

SCHEME 10

Kinetically stabilized phosphinyl radicals $\cdot P[CH(SiMe_3)_2]_2$ add readily to double and triple bonds: in particular, the reaction with 2,3-dimethyl-1,3-butadiene leads to a 1,4-diphosphino-2-butene derivative which has been identified on the basis of ^{31}P-, ^{1}H-, and ^{13}C-NMR data.[318]

A free-radical mechanism has been proposed (Scheme 11) for the reduction of aldehydes and ketones to the corresponding alcohols by triphenyltin formate.[319]

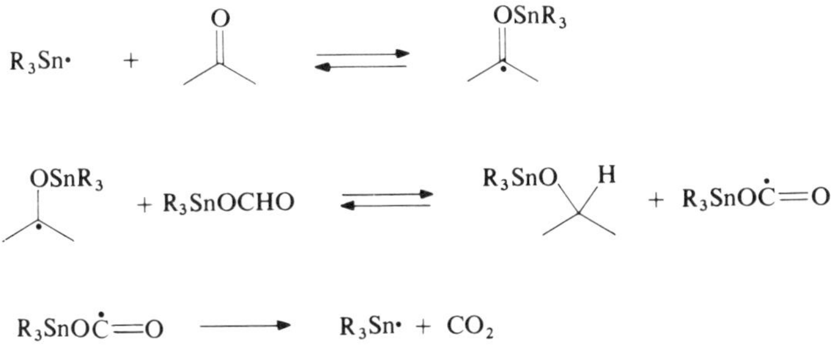

$$R_3Sn\cdot + \overset{O}{\overset{\|}{C}}\langle \rightleftharpoons \overset{OSnR_3}{\overset{|}{\dot{C}}}\langle$$

$$\overset{OSnR_3}{\overset{|}{\dot{C}}}\langle + R_3SnOCHO \rightleftharpoons R_3SnO\!\!\underset{}{\overset{}{\rangle}}\!C\!\langle H + R_3SnO\dot{C}{=}O$$

$$R_3SnO\dot{C}{=}O \longrightarrow R_3Sn\cdot + CO_2$$

SCHEME 11

The addition of Group IVB organometallic radicals to chromanone and thiochromanone has been studied by ESR spectroscopy: the SiR_3, GeR_3, and SnR_3 adducts have been characterized and compared to the related species observed with chromone and thiochromone.[320] The absolute rate constants for the reactions of $Bu_3{}^nGe\cdot$ and $Bu_3{}^nSn\cdot$ radicals with a large number of carbonyl compounds, unsaturated derivatives, and organic halides have been measured through laser flash photolysis experiments;[321] in most cases germanium- and tin-centred radicals showed a similar reactivity, both species being less reactive than the corresponding silicon-centred radicals.[322] UMNDO calculations have been used to elucidate the radical-chain mechanism of hydrostannylation: the key step of the process (addition of $Me_3Sn\cdot$ to ethylene) was predicted to be exothermic by 16.2 kcal mol^{-1}.[322]

Homolytic Aromatic Substitution

Cyclohexadienyl radicals are formed at rates close to the diffusion-controlled limit, *viz.* $k \sim 7 \times 10^9\,\text{M}^{-1}\,\text{s}^{-1}$, by addition of muonium to the aromatic rings of substituted benzenes;[323, 324] a kinetic isotope effect of about 3 was found with respect to hydrogen, which reflects the mean thermal velocity ratio of Mu/H;[323] the rate constants for the corresponding hydrogen or deuterium addition to C_6H_6 and C_6D_6 have also been measured.[325]

The rate constants for the addition of $\cdot CH_3$, $\cdot CF_2CH_3$, and $\cdot CF_3$ to hexafluorobenzene, perfluorotoluene, $C_6F_5O(CF_2)_4OC_6F_5$, *p*-$(CF_3)_2C_6H_4$, and benzene have been determined relative to hydrogen abstraction from the solvent (isooctane).[326]

Three different sources of methyl radicals have been tested in the methylation of protonated pyrimidines, *i.e.* reaction with the systems $Bu^tOOH–Fe^{2+}$, $MeCO_2H–S_2O_8{}^{2-}–Ag^+$, and $MeS(O)Me–H_2O_2–Fe^{2+}$: the last method was found to be the most efficient;[327] the selectivity in the observed mono- and polymethylated derivatives has been explained in terms of the low nucleophilicity of methyl radical and of the electron deficiency of the protonated hetero-base.[327] Homolytic alkylation of protonated hetero-aromatic bases can also be readily achieved through reaction of the substrates with aroyl peroxides and alkyl iodides, the key steps of the process (Scheme 12) being iodine abstraction by aryl radicals and decomposition of the peroxide induced by the pyridinyl radical intermediate;[328]

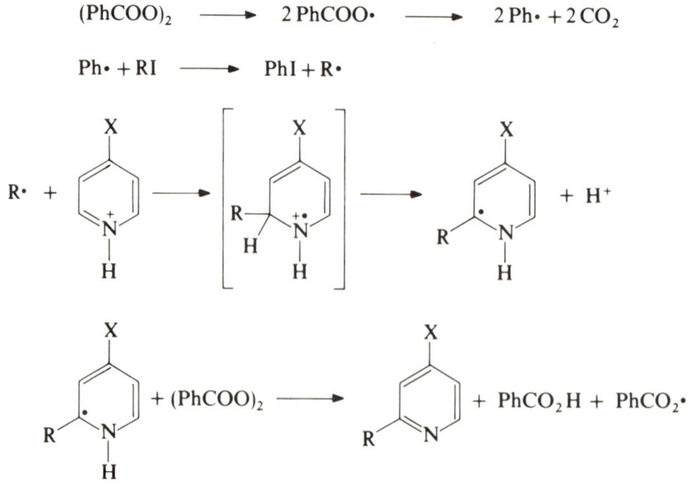

$$(PhCOO)_2 \longrightarrow 2\,PhCOO\cdot \longrightarrow 2\,Ph\cdot + 2\,CO_2$$

$$Ph\cdot + RI \longrightarrow PhI + R\cdot$$

R· + [4-X-pyridinium (N–H)] ⟶ [R,H-adduct radical cation] ⟶ [2-R-4-X-pyridinyl radical (N–H)] + H⁺

[2-R-4-X-pyridinyl radical (N–H)] + (PhCOO)₂ ⟶ [2-R-4-X-pyridine] + PhCO₂H + PhCO₂·

SCHEME 12

yields are in the range 70% to 100% with primary or secondary alkyl iodides but fall to much smaller values for tertiary halides. This method has also been used to obtain selective substitution at the 2-position of protonated lepidine by a variety of nucleophilic free radicals.[329] Photolysis of chloroacetonitrile in aromatic solvents leads to cyanomethylated aromatic compounds, but experimental evidence seem to exclude a homolytic aromatic substitution reaction, and the mechanism outlined in Scheme 13 has been proposed instead.[330]

$$ArH^{*1} + ClCH_2CN \longrightarrow [ArH \ldots\ldots ClCH_2CN]^* \longrightarrow ArH^{+\bullet} + {}^{\bullet-}ClCH_2CN$$

$${}^{\bullet-}ClCH_2CN \longrightarrow {}^{\bullet}CH_2CN + Cl^-$$

$$ArH^{+\bullet} + {}^{\bullet}CH_2CN \longrightarrow H\overset{+}{A}rCH_2CN \xrightarrow{-H^+} ArCH_2CN$$

SCHEME 13

Acyl radicals attack lumazines at position 7 to give acylated products, the presence of a substituent in this position directing substitution to the adjacent position 6;[331] aromatic acetoxylation is promoted by Mn (III) or Ce(III) in the reaction of aromatic compounds with acetone (Scheme 14).[332] The k_H/k_D (3.8) and Hammett ρ (-2.4 ± 0.3) values indicate that the rate-determining step is the proton loss of acetone.

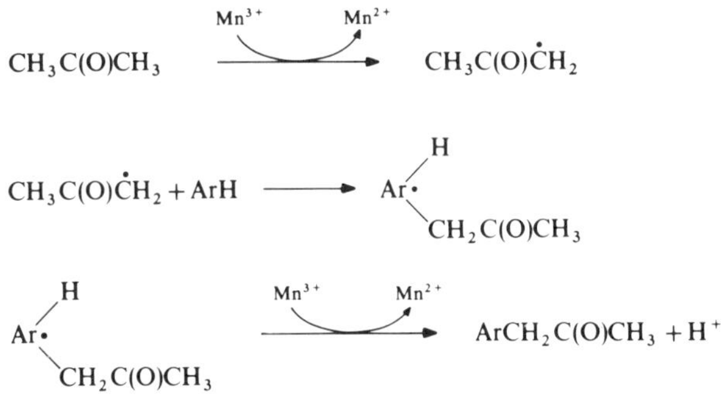

$$CH_3C(O)CH_3 \xrightarrow{Mn^{3+} \rightarrow Mn^{2+}} CH_3C(O)\dot{C}H_2$$

$$CH_3C(O)\dot{C}H_2 + ArH \longrightarrow Ar^{\bullet}(H)(CH_2C(O)CH_3)$$

$$Ar^{\bullet}(H)(CH_2C(O)CH_3) \xrightarrow{Mn^{3+} \rightarrow Mn^{2+}} ArCH_2C(O)CH_3 + H^+$$

SCHEME 14

4-R-1,2,4-Triazoline-3,5-diones homolytically substitute the ring of electron-rich heterocycles such as *N*-methylpyrrole or *N*-methylindole, and it has been proposed that the process involves urazolyl radicals and the formation of different CT complexes.[333]

Nitration of compounds (**64**) with nitric acid/sulphuric acid mixtures leads to the *ipso*-substituted dinitro derivatives (**65**), and kinetic studies indicate a radical mechanism involving formation and dissociation of nitroquinalidine compounds.[334]

Absolute rate constants for the gas-phase addition of ${}^{\bullet}OH$ radical to benzene have been determined;[335–337] it has been shown that in the range 298–330 K the adduct

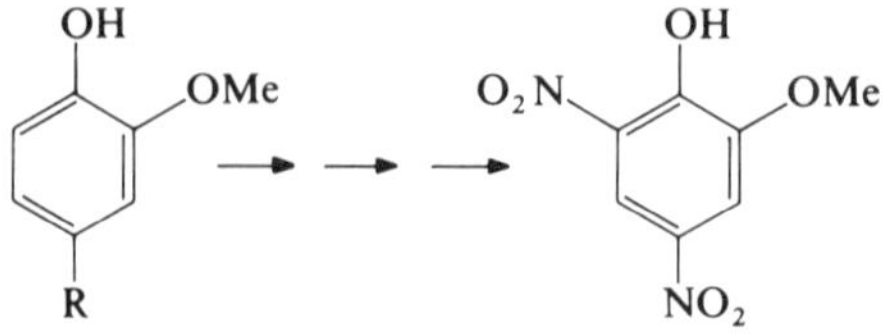

(64) → → → (65)

R = CHO, CH_2O, CO_2H

$^{\bullet}C_6H_6OH$ is formed, in the range 330–420 K the reaction becomes reversible, and at > 420 K hydrogen abstraction from benezene occurs leading to phenyl radicals.[335] Rate constants have also been measured for the reaction of ·OH radicals with a variety of aromatic compounds; a LFER between the rate constants and the ionization potenzial of the aromatic compounds has been obtained.[336,337] Relative rate constants have been determined for the radical hydroxylation of a number of aromatic substrates including benzenes, pyridines, and furans, and appropriate Hammett correlations have been derived.[338] Arrhenius expressions have been derived for the reaction of ·OH radicals with furan and thiophene; the negative activation energies observed suggest that both reactions proceed *via* addition routes.[339]

In neutral aqueous solutions tyrosine undergoes addition to the aromatic ring (*ortho: meta: para* = 50:35:10) almost exclusively (> 90%), while hydrogen abstraction from the side-chain is relatively unimportant;[340] hydroxybenzoic acids (*p* ⩾ *o* > *m*) are obtained along with small amounts of phenol and chlorophenols in the photoreaction of benzoic acid with sodium hypochlorite in aqueous alkali (pH ⩾ 12);[341] in this process the product distribution is wavelength-dependent owing to the fact that photodecomposition of ClO^- leads to different starting species with

$$ClO^- \xrightarrow{h\nu} \begin{cases} Cl^- + O(^3P) \\ Cl\cdot + O^{\bar{\cdot}} \\ Cl^- + O(^1D) \end{cases}$$

quantum yields which vary with wavelength. It has been suggested that the detected phenols (chlorophenols) are formed either by *ipso*-attack by $O(^3P)$ on the carboxylic acid or by electron transfer from the carboxylate anion to ·OH (or Cl·) radicals followed by decarboxylation.[341]

Examples of intramolecular homolytic aromatic substitution have been reported: thus, reaction **(66)** → **(67)** seems to be the key step for a new synthetic route to

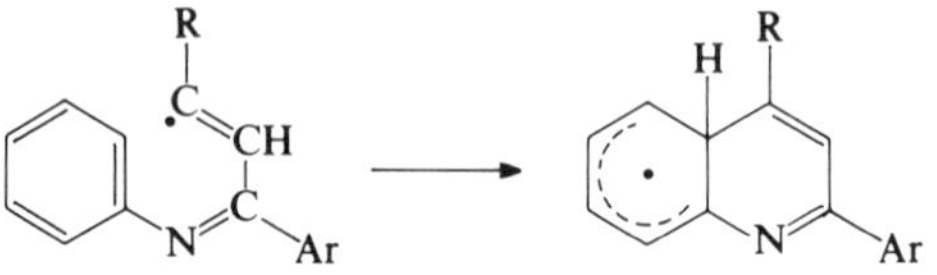

(66) ⟶ (67)

quinolines,[342] and the formation of perchloro-9-phenylfluorenyl occurs upon heating of perchloro-2-phenyldiphenylmethyl radical.[343] Thermolysis of 1,2,5-triazoheptadiene leads to substantial amounts (*ca.* 25%) of cyclization products which are conceivably formed through the two simultaneous mechanism depicted in Scheme 15. From the relative amounts of products (**68**) and (**71**) the spirodienyl component of the overall cyclization was estimated as 57% (X = Me, Y = H) and 72% (X = H, Y = Me);[344] substitution of a 2,6-dimethylphenyl group for the tosyl group leads to the formation of 7-methylindole (55–62%);[345] the equilibration of radicals (**69**) and (**70**) *via* a spirodienyl intermediate has also been established by using [1-^{15}N]- and [4-^{15}N]-labelled starting compounds.[346] Transformation of *o*-bis(phenylsulphonyl)benzene derivatives into dibenzothiophene dioxide, *via* an intramolecular homolytic aromatic substitution, is an efficient electrocatalytic process in the presence of a base.[347]

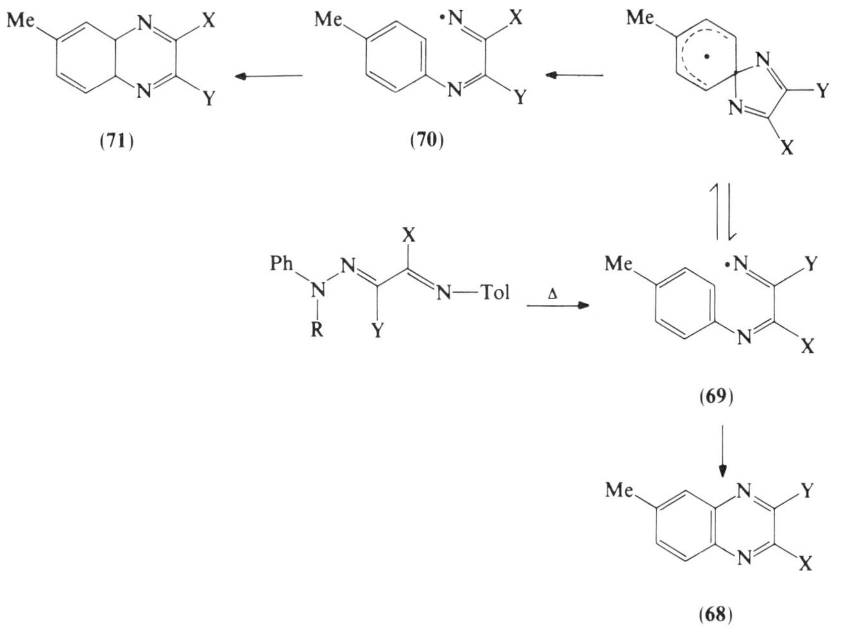

(71) (70) (69) (68)

SCHEME 15

S_Hi and S_H2 Reactions

The formation of oxiranes by the intramolecular substitution reaction at oxygen of *β*-*tert*-butylperoxyalkyl radicals has been studied and the absolute rates of ring-closure have been determined by the relative ratios of oxiranes to peroxides (Scheme 16). Typical values for k_C/s^{-1} are 6.2×10^6 ($R^1 = R^3$ = Et, R^2 = H), 1.96×10^6 ($R^1 = R^3$ = Me, R^2 = H), 2.0×10^6 (2-*tert*-butylperoxycyclopentyl), and 1.12×10^6 (2-*tert*-butylperoxycyclohexyl).[348] Although the decomposition of *tert*-

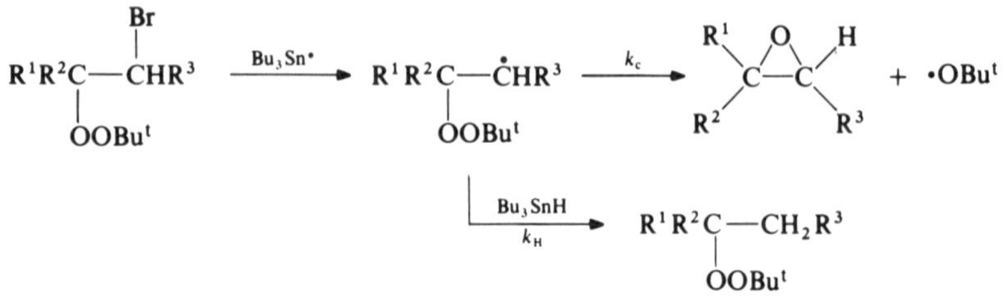

$$R^1R^2C(OOBu^t)—CHR^3Br \xrightarrow{Bu_3Sn\cdot} R^1R^2C(OOBu^t)—\dot{C}HR^3 \xrightarrow{k_c} \text{epoxide } (R^1, R^2, H, R^3) + \cdot OBu^t$$

$$R^1R^2C(OOBu^t)—\dot{C}HR^3 \xrightarrow[k_H]{Bu_3SnH} R^1R^2C(OOBu^t)—CH_2R^3$$

SCHEME 16

butylperpent-4-enoate to 5-(2-tetra-hydrofuryl)-4-pentanolide in THF in the presence of radical initiators may be thought to proceed *via* addition followed by intramolecular displacement, the mechanism has not been established unambiguously and the possibility of a concerted process cannot be ruled out.[349] Biomolecular homolytic displacement of transition-metal complexes from carbon has been reviewed; its potentiality in synthetic organic chemistry is apparent.[350] Pent-4-enylcobaloximes react with Cl_3CSO_2Cl, in inert solvents under irradiation with tungstens lamps, to give good yields of 2-(β,β,β-trichloroethyl)sulpholanes; the same products are formed in the thermal reactions in the presence of an excess of sulphur dioxide.[351] The process is thought to occur by initial attack of $\cdot CCl_3$ at the terminal olefinic carbon of the pent-4-enyl moiety; in the presence of SO_2 a sulphonyl radical is then formed that can undergo an S_Hi reaction on the α-carbon of the organic ligand (**72** represents the transition state), thereby displacing cobaloxime (II).[351]

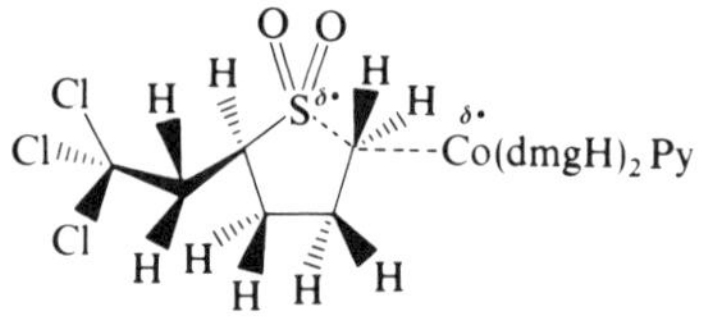

(72)

The gas-phase reaction of H· (or D·) and thietane leading to H_2S has been shown to involve the intermediacy of the allyl radical.[352] The rate constant for the following S_H2 displacement reactions:

$$H\cdot + CF_3SSCF_3 \rightarrow CF_3SH + \cdot SCF_3$$
$$H\cdot + Me_3SiSiMe_3 \rightarrow Me_3SiH + \cdot SiMe_3$$

have been determined;[353,354] the former reaction is characterized by a high frequency factor ($3 \times 10^{11}\,M^{-1}s^{-1}$) which can be partially accounted for by assuming free rotation about the S—CF_3 and the S—S bonds in the transition state.[353] Rate constants have also been reported for the reaction of atomic hydrogen and of atomic oxygen with ethylene: the 79% yield of atomic hydrogen in the latter process indicates that formation of C_2H_3O is a major product channel.[355] The fractional extents of the displacement route in the reaction of F· with ethylene, vinyl chloride, and vinyl bromide have been determined to be (0.65 ± 0.06) for C_2H_4 and (0.72 ± 0.14) for C_2H_3Cl (or C_2H_3Br);[356] the rate of displacement of atomic bromine by reaction of Cl· on C_2H_3Br is $(1.43 \pm 0.19) \times 10^7\ \text{M}^{-1}\text{s}^{-1}$ at 298 K, this reaction providing a very good mean for titrating atomic chlorine down to extremely low concentrations (5×10^{10} atoms cm^{-3}).[357] An FTIR study of photolysis of the system Cl_2/CH_3HgCH_3 has been reported and the results are consistent with the predominant occurrence of the following displacement reaction:

$$Cl\cdot + CH_3HgCH_3 \rightarrow CH_3HgCl + \cdot CH_3$$

the rate constant being $k = (1.66 \pm 0.18) \times 10^{11}\ \text{M}^{-1}\text{s}^{-1}$.[358]

Quantitative CIDNP evidence has been reported that isobutyl radical reacts with ethylmagnesium bromide by displacing the ethyl radical with a rate constant $k > 10^5\ \text{M}^{-1}\text{s}^{-1}$.[359] In the displacement reaction PhSeX + Ph· → PhSePh + X·, relative rate constants have been determined as 1.00 (X = p-$NO_2C_6H_4S$), 1.28 (X = SPh), 3.34 (X = SePh), and 8.34 (X = $SO_2C_6H_4Me$-p).[360]

The Arrhenius expression for the gas-phase reaction of ·OH with tetrahydrothiophene, $\log(k/\text{M}^{-1}\text{s}^{-1}) = (9.83 \pm 0.75) - (-0.33 \pm 0.19)/\theta$, where $\theta = 2.3RT$ kcal mol^{-1}, if compared with the analogous expression for the reaction with THF, suggests an important role of attack by hydroxyl on the sulphur atom.[361]

The characterization of alkoxythiocarbonyl radicals ROĊ(S) in the reaction of $\cdot SnR_3$ with O-alkyl-S-methyl dithiocarbonate suggests that the deoxygenation of alcohols *via* the Barton reaction proceeds according to Scheme 17, rather than by addition of stannyl radicals to the thiocarbonyl group.[362]

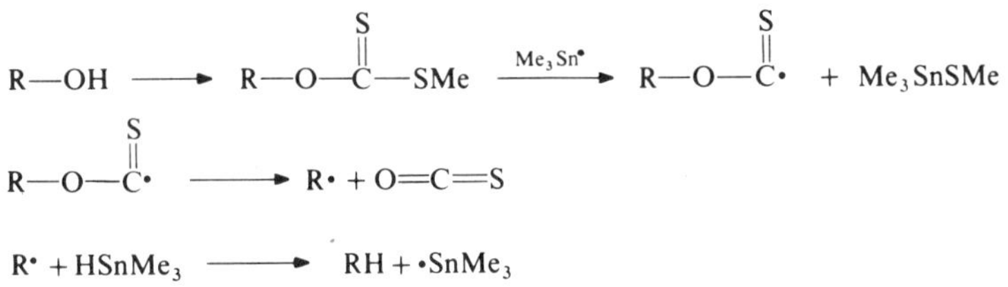

$$R{-}OH \longrightarrow R{-}O{-}C({=}S){-}SMe \xrightarrow{Me_3Sn\cdot} R{-}O{-}\dot{C}({=}S) + Me_3SnSMe$$

$$R{-}O{-}\dot{C}({=}S) \longrightarrow R\cdot + O{=}C{=}S$$

$$R\cdot + HSnMe_3 \longrightarrow RH + \cdot SnMe_3$$

Scheme 17

Nitroxides and Spin-trapping

Ab initio SCF/CI calculations have been used to rationalize vibrational effects on the ESR spectrum of the simplest aminoxyl, $H_2NO\cdot$;[363] the structures of adducts of hydroxyl and several *ortho*-, *meta*-, or *para*-substituted aryl radicals with a number of nitrones have been investigated by using GLC-mass spectrometry;[364] when the resulting aminoxyls were too unstable for their mass spectrum to be detected, they were trimethylsilylated by reaction with N,O-bis(trimethylsilyl)acetamide. From the values of the β-hydrogen hfs in radicals ArCH(X)N(O·)H it has been inferred that whereas sulphur-containing (X = SR) aminoxyls adopt a conformation where the C—S bond eclipses the SOMO, the Group IVB adducts (X = SiR_3, GeR_3) are the average of a variety of torsional conformers.[365] The β-bromoalkylaminoxyl

$Me_3CN(O)$=$CHCH(Br)N(\dot{O})CMe_3$ exhibiting large $^{79}Br/^{81}Br$ hfs constants has been characterized and it has been shown that its ESR parameters fit well with those of the corresponding chlorine or fluorine derivatives.[366] The use of out-of-phase ESR spectroscopy has allowed the identification of species not readily distinguishable;[367] analysis of the super-hyperfine coupling constants of γ- and δ-nuclei in some piperidine derivatives indicated that tempol (**73**) and tempamine (**74**) are stable in chair conformations, while tempo (**75**) and tempone (**76**) rapidly interconvert between two identical chair and twist conformations, respectively.

(**73**) X = OH
(**74**) X = NH_2
(**75**) X = H
(**76**) X = O

The results of a kinetic pulse radiolytic study suggest that the reactions between α-(alkylthio)alkyl radicals and tetranitromethane proceeds not by a simple electron-transfer process but *via* a two-step mechanism where addition of the α-(alkylthio)alkyl radical to the tetranitromethane occurs to give an adduct (**77**) which

$$R^1\dot{C}HSR^2 + C(NO_2)_4 \longrightarrow (R^1)(R^2S)CH{-}ON(O\cdot){-}C(NO_2)_3 \longrightarrow (R^1CH{=}SR^2)^+ + NO_2 + \bar{C}(NO_2)_3$$

(**77**)

then dissociates into the reaction products.[368] The activation energies for the unimolecular decay of (**77**) have been determined for $R^1 = H$, $R^2 = Me$ and $R^1 = H$, $R^2 = Et$ to be 13.87 and 13.61 kcal mol^{-1}, respectively. Alkyldiethoxyl phosphonyl nitroxides (**78**) have been observed by ESR upon photoreaction of tetraethyl pyrophosphite and the appropriate nitroalkane ($MeNO_2$, $EtNO_2$, Pr^nNO_2, Pr^iNO_2, Bu^tNO_2) in the presence of di-*tert*-butyl peroxide, and it has been proposed that they are formed *via* the following reactions:[369]

$$R{-}NO_2 + (EtO)_2POP(OEt)_2 \longrightarrow R{-}NO + (EtO)_2P(O)OP(OEt)_2$$

$$Bu^tO\cdot + (EtO)_2POP(OEt)_2 \longrightarrow (EtO)_2\dot{P}(O) + Bu^tOP(OEt)_2$$

$$R{-}NO + \cdot P(O)(OEt)_2 \longrightarrow R{-}N(O\cdot){-}P(O)(OEt)_2$$

(**78**)

A combination of ESR, ENDOR, and TRIPLE results has shown that the photorearrangement of 4-aryl-2,2,5,5-tetramethyl-3-imidazoline-3-oxide (**79**) to the bicyclic nitroxide (**80**) does not involve the nitroxide function;[370] nitration of (**79**) with metal nitrates or nitric acid in sulphuric acid solutions leads to the introduction of a NO_2 group in the *ortho* (60 %)- or *para* (30 %)-position of the aromatic ring, thus indicating the *ortho–para* orientating activity of the nitrone function.[371]

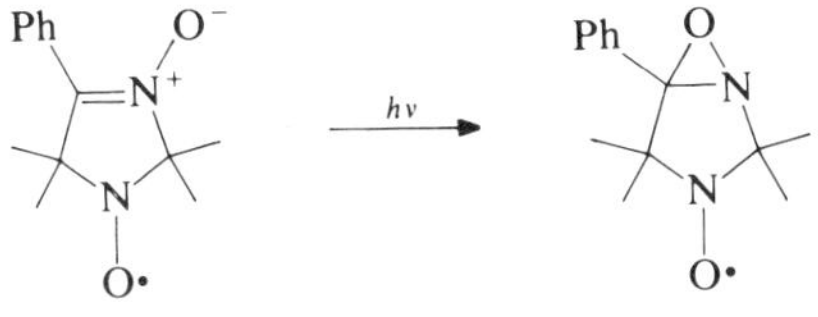

(**79**) (**80**)

The addition of benzoyloxyl radicals to vinyl acetate has been studied by ESR in the presence of radical scavengers (2-methyl-2-nitrosopropane, MNP, and 2,4,6-tri-*tert*-butyl-nitrosobenzene, TNB): with MNP only the adduct from tail addition was trapped but with TNB a small amount of the adduct from head addition was also trapped;[372] although there is no guarantee that both radicals are trapped with equal efficiency, the ratio of head-to-tail addition appears to be 3:10.

The spin-trapping technique has been used in a combined HPLC–ESR study of γ-irradiation of aqueous solutions of *cis*-4-chloro-L-proline and *cis*-4-hydroxy-L-proline: among the various spin adducts formed $Bu^tN(O^\cdot)CH(COO^-)CH_2CHClCH_2NH_3^+$, $Bu^tN(O^\cdot)\overline{CHCH_2NH_2^+CH(COO^-)CH_2}$, $Bu^tN(O^\cdot)CH(COO^-)CH_2CH(OH)CH_2NH_3^+$ and $Bu^tN(O^\cdot)\overline{CHCH_2NH_2^+CH(COO^-)CH_2CH}OH$ could be positively identified.[373] A similar investigation of L-glutamine, L-asparagine, sodium L-glutamate, sodium L-aspartate, L-serine, and L-theonine indicated that the trapped radicals result either from reaction of the amino-acid with the hydrated electron or by attack of $HO^\cdot$ radicals.[374] γ-Irradiation of polycrystalline α-D-glucose in the presence of MNP allowed the ESR characterization of five long-lived nitroxides.[375] Spin-trapping experiments indicate that thermolysis of *N*-pivaloxyacetanilides (**81**) occurs through a mechanism involving homolytic cleavage of the N—O bond.[376]

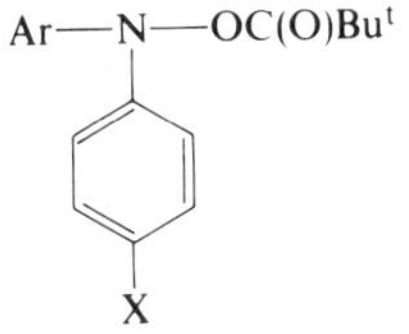

(**81**)

X = Me, H, NO_2

The spin-trapping of cyanatyl radical (·NCO) by nitrosodurene leads to the same ESR spectrum obtained by trapping azide radicals (·N_3); on this basis the observed aminoxyl has been assigned to structure (**82**)[377] and the mechanism shown in Scheme 18 has been proposed.

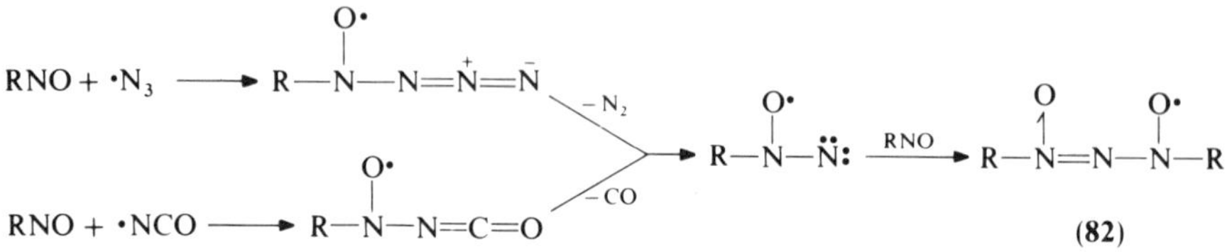

$$\mathrm{RNO} + {}^{\bullet}\mathrm{N_3} \longrightarrow \mathrm{R{-}N(O^{\bullet}){-}N{=}\overset{+}{N}{=}\overset{-}{N}} \xrightarrow{-\mathrm{N_2}} \mathrm{R{-}N(O^{\bullet}){-}\ddot{N}\!:} \xrightarrow{\mathrm{RNO}} \mathrm{R{-}N(\rightarrow O){=}N{-}N(O^{\bullet}){-}R}\ (\mathbf{82})$$

$$\mathrm{RNO} + {}^{\bullet}\mathrm{NCO} \longrightarrow \mathrm{R{-}N(O^{\bullet}){-}N{=}C{=}O} \xrightarrow{-\mathrm{CO}} \mathrm{R{-}N(O^{\bullet}){-}\ddot{N}\!:}$$

SCHEME 18

Sonolysis (55 kHz) of organometallic tin halides in benzene solutions containing ND leads to trapping of R· radicals resulting from cleavage of R—SnR_2X and of cyclohexadienyl radicals XR_2Sn—C_6H_6· formed *via* addition of XR_2Sn· to solvent molecules.[378] A novel radical scavenger, pentamethoxynitrosobenzene, has recently been isolated as a dimer which readily dissociates in solution into the active monomer;[379] the small line-width that usually characterizes the spectra of its adducts makes this compound a spin-trapping agent useful for structural ESR studies. Glow discharge of polyethylene powder doped with PBN (10 %) leads to the trapping of alkyl radicals;[380] spin-trapping experiments have been carried out in SDS micelles using several nitrones of different solubility.[381] The rate constants have been determined for the scavenging of phenylthiyl radicals by cyclic (DMPO) and acyclic (PBN) nitrones: it has been found that trapping by the cyclic spin trap is faster than for the acyclic one, and that electron-withdrawing X substituents (*p*—XC_6H_4S·) increase the rate constants of the process for both DMPO and PBN;[382] a remarkable slowing effect is also exerted by polar solvents such as methanol, due to hydrogen bonding to the nitrone function (Table 3). Alkylthiyl radicals CH_3CH_2S·, $HOCH_2CH_2S$·, $(CH_3)_3CS$·, and $(CH_3)_2CHS$· have been trapped by decomposing alkyl thionitrites RSNO in the presence of either DMPO or MNP.[383] The detection of [DMPO–OH]· aminoxyl upon irradiation (visible light) of solutions of

TABLE 3. Rate constants for the trapping of XPhS· radicals by DMPO and PBN ($k \times 10^{-6}/\mathrm{M}^{-1}\,\mathrm{s}^{-1}$)

X	DMPO benezene	DMPO methanol	PBN benzene
Br	750	72	17
Cl	700	60	12
H	520	54	11
Bu^t	260	26	5.6
Me	300	19	5.0
MeO	110	5.4	1.1

daunomicine or adriamicine has been attributed to decomposition of the first-formed $[DMPO\text{–}O_2^{-}]^{\cdot}$ adduct rather than to the presence of free hydroxyl radicals.[384]

The absolute rate constants have been determined by a competition method for the trapping of ·OH radicals from 2- and 6-mercaptopurines.[385]

A new arsenoaminoxyl spin adduct has been obtained by trapping $^{\cdot}AsO_2$ with 2-methyl-2-nitrosopropane, MNP.[386]

References

1 Wentrup, C., *Reactive Molecules. The Neutral Reactive Intermediates in Organic Chemistry*, John Wiley and Sons, New York, 1984.

2 Henning, H. G., Pragst, F., and Fuhrmann, J., *Z. Chem.*, **24,** 1 (1984); *Chem. Abs.*, **100,** 138234 (1984).

3 Hart, D. J., *Science*, **223,** 883 (1984); *Chem. Abs.*, **100,** 138178 (1984).

4 Rakhmankulov, D. L., Zorin, V. V., Sapiev, O. G., Zlotskii, S. S., and Bartok, M., *Acta Phys. Chem.*, **29,** 165 (1983); *Chem. Abs.*, **101,** 89903 (1984).

5 Rakhmankulov, D. L., Zorin, V. V., Pastushenko, E. V., and Zlotskii, S. S., *Heterocycles*, **22,** 817 (1984); *Chem. Abs.*, **101,** 22621 (1984).

6 Creber, K. A. M., Chen, K. S., and Wan, J. K. S., *Reviews of Chemical Intermediates*, **5,** 37 (1984).

7 Kosower, E. M., *Top. Curr. Chem.*, **112,** 117 (1983); *Chem. Abs.*, **99,** 193994 (1983).

8 Freidlina, R. Kh., Kandror, I. I., and Kopylova, B. V., *Izv. Akad. Nauk SSSR, Ser. Khim.*, **1983,** 1799; *Chem. Abs.*, **99,** 174917 (1983).

9 Howard, J. A., *Reviews of Chemical Intermediates*, **5,** 1 (1984).

10 Greenstock, C. L., *Isr. J. Chem.*, **24,** 1 (1984).

11 Koppenol, W. H., and Butler, J., *Isr. J. Chem.*, **24,** 11 (1984).

12 Ingold, K. U., and Roberts, B. P., *Landolt–Börnstein, Group II*, Vol. 13, Springer Verlag, Berlin, 1983; *Chem. Abs.*, **101,** 38083 (1984).

13 Burkey, T. J., and Griller, D., *Reviews of Chemical Intermediates*, **5,** 21 (1984).

14 Scaiano, J. C., *Contrib. Cient. Tecnol*, **1983,** 9; *Chem. Abs.*, **101,** 72062 (1984).

15 Orlov, Yu. D., Lebedev, Yu. A., Pavlinov, L. I., Khrapkovskii, G. M., and Marchenko, G. N. *Dokl. Akad. Nauk SSSR*, **271,** 1433 (1983); *Chem. Abs.*, **100,** 51017 (1984).

16 Orlov, Yu. D., and Lebedev, Yu. A., *Izv. Akad. Nauk SSSR, Ser. Khim.*, **1984,** 1074; Chem. Abs., **101,** 71870 (1984).

17 Orlov, Yu. D., and Lebedev, Yu. A., *Izv. Akad. Nauk SSSR, Ser. Khim.*, **1984,** 1335; *Chem. Abs.*, **101,** 109898 (1984).

18 Pacey, P. D., and Wimalasena, J. H., *J. Phys. Chem.*, **88,** 5657 (1984).

19 Cao, J.-R., and Back, M. H., *Int. J. Chem. Kinet.*, **16,** 961 (1984).

20 Grela, M. A., and Colussi, A. J., *J. Phys. Chem.*, **88,** 5995 (1984).

21 Lee, I., Lee, B. S., and Song, C. H., *Taehan Hwahakhoe Chi.*, **27,** 320 (1983); *Chem. Abs.*, **100,** 120139 (1984).

22 Bernlöhr, W., Beckhaus, H.-D., Peters, K., von Schnering, H.-G., and Rüchardt, C., *Chem. Ber.*, **117,** 1013 (1984).

23 Bernlöhr, W., Beckhaus, H.-D., and Rüchardt, C., *Chem. Ber.*, **117,** 1026 (1984).

24 Nicholas, A. M., De P., and Arnold, D. R., *Can. J. Chem.*, **62,** 1850 (1984).

25 Ziegler, U., and Ondruschka, B., *J. Prakt. Chem.*, **326,** 139 (1984); *Chem. Abs.*, **100,** 208683 (1984).

26 Chipman, D. M., and Miller, K. E., *J. Am. Chem. Soc.*, **106,** 6236 (1984).

27 Green, I. G., and Walton, J. C., *J. Chem. Soc., Perkin Trans. 2*, **1984,** 1253.

28 Feller, D., Davidson, E. R., and Borden, W. T., *J. Am. Chem. Soc.*, **106,** 2513 (1984).

29 Yang, P., *Kexue Tongbao*, **28,** 1711 (1983); *Chem. Abs.*, **101,** 6215 (1984).

30 Nicholas, A. M., De P., and Arnold, D. R., *Can. J. Chem.*, **62,** 1860 (1984).

31 Lorenz, K., and Zellner, R., *Ber. Bunsen-Ges. Phys. Chem.*, **87,** 629 (1983); *Chem. Abs.*, **99,** 194093 (1983).

32 Andrews, L., Dyke, J. M., Jonathan, N., Keddar, N., and Morris, A., *J. Phys. Chem.*, **88,** 1950 (1984).

33 Dyke, J. M., Ellis, A. R., Keddar, N., and Morris, A., *J. Phys. Chem.*, **88,** 2565 (1984).

34 Schultz, J. C., Houle, F. A., and Beauchamp, J. L., *J. Am. Chem. Soc.*, **106,** 3917 (1984).

[35] Schultz, J. C., Houle, F. A., and Beauchamp, J. L., *J. Am. Chem. Soc.*, **106,** 7336 (1984).
[36] Kagramanov, N. D., Ujszaszy, K., Tamas, J., Mal'tsev, A. K., and Nefedov,O. M., *Izv. Akad. Nauk SSSR, Ser. Khim.*, **1983,** 1683; *Chem. Abs.*, **99,** 175091 (1983).
[37] Drzaic, P. S., and Brauman, J. I., *J. Phys. Chem.*, **88,** 5285 (1984).
[38] Stratt, R. M., and Desjardins, S. G., *J. Am. Chem. Soc.*, **106,** 256 (1984).
[39] Goddard, J. D., *Chem. Phys. Lett.*, **102,** 224 (1983); *Chem. Abs.*, **100,** 33824 (1984).
[40] Mestechkin, M. M., Poltavets, V. N., Degtyarev, L. S., and Pokhodenko, V. D., *Teor. Eksp. Khim.*, **20,** 81 (1984); *Chem. Abs.*, **100,** 208689 (1984).
[41] Feller, D., and Davidson, E. R., *J. Chem. Phys.*, **80,** 1006 (1984); *Chem. Abs.*, **100,** 173921 (1984).
[42] Schrader, B., Pacansky, J., and Pfeiffer, U., *J. Phys. Chem.*, **88,** 4069 (1984).
[43] Nonhebel, D. C., and Walton, J. C., *J. Chem. Soc., Chem. Commun.*, **1984,** 731.
[44] Stadlbauer, J. M., Ito, Y., Miyake, Y., and Walker, D. C., *Hyperfine Interact.*, **19,** 821 (1984); *Chem. Abs.*, **101,** 38081 (1984).
[45] McLauchlan, K. A., and Ritchie, A. J. D., *J. Chem. Soc., Perkin Trans. 2,* **1984,** 275.
[46] Bernhard, W. A., Hornig, T. L., and Mercer, K. R., *J. Phys. Chem.*, **88,** 1317 (1984).
[47] Maj, S. P., Symons, M. C. R., and Trousson, M. R., *J. Chem. Soc., Chem. Commun.*, **1984,** 561.
[48] Jacox, M. E., *J. Phys. Chem.*, **88,** 445 (1984).
[49] Kira, M., Akiyama, M., and Sakurai, H., *J. Organomet. Chem.*, **271,** 23 (1984).
[50] Aeberhard, U., Keese, R., Stamm, E., Vögeli, U.-C., Law, W., and Kochi, J. K., *Helv. Chim. Acta*, **66,** 2740 (1983).
[51] Ingold, K. U., Nonhebel, D. C., and Wildman, T. A., *J. Phys. Chem.*, **88,** 1675 (1984).
[52] Ikeda, N., Nakashima, N., and Yoshihara, K., *J. Phys. Chem.*, **88,** 5803 (1984).
[53] Bromberg, A., Schmidt, K. H., and Meisel, D., *J. Am. Chem. Soc.*, **106,** 3056 (1984).
[54] Hilinski, E. F., Huppert, D., Kelley, D. F., Milton, S. V., and Rentzepis, P. M., *J. Am. Chem. Soc.*, **106,** 1951 (1984).
[55] Chatgilialoglu, C., Lunazzi, L., Macciantelli, D., and Placucci, G., *J. Am. Chem. Soc.*, **106,** 5252 (1984).
[56] Trapp, C., and Kulkarui, S. V., *J. Phys. Chem.*, **88,** 2703 (1984).
[57] Smith, W. B., and Harris, M. C., *J. Org. Chem.*, **48,** 4957 (1983).
[58] Ballester, M., Castañer, J., Riera, J., and Pujadas, J., *J. Org. Chem.*, **49,** 2884 (1984).
[59] Ballester, M., Castañer, J., Riera, J., Pujadas, J., Armef, O., Onrubia, C., and Rio, J. A., *J. Org. Chem.*, **49,** 770 (1984).
[60] Hass, C. H., Kirste, B., Kurreck, H., and Schlömp, G., *J. Am. Chem. Soc.*, **105,** 7375 (1983).
[61] Kasai, P. H., and McBay, H. C., *J. Phys. Chem.*, **88,** 5932 (1984).
[62] Griller, D., Nonhebel, D. C., and Walton, J. C., *J. Chem. Soc., Perkin Trans. 2,* **1984,** 1817.
[63] Korth, H.-G., Lommes, P., and Sustmann, R., *J. Am. Chem. Soc.*, **106,** 663 (1984).
[64] Green, I. G., and Walton, J. C., *J. Chem. Soc., Perkin Trans. 2,* **1984,** 1253.
[65] Bascetta, E., Gunstone, F. D., and Walton, J. C., *J. Chem. Soc., Perkin Trans. 2,* **1984,** 401.
[66] Paquette, L. A., Uchida, T., and Gallucci, J. C., *J. Am. Chem. Soc.*, **106,** 335 (1984).
[67] Kieslich, W., and Kurreck, H., *J. Am. Chem. Soc.*, **106,** 4328 (1984).
[68] Leuschner, R., and Dohrmann, J. K., *Ber. Bunsen Ges. Phys. Chem.*, **88,** 50 (1984); *Chem. Abs.*, **100,** 191131 (1984).
[69] Greci, L. Alberti, A., Carelli, I., Trazza, A., and Casini, A., *J. Chem. Soc., Perkin Trans. 2,* **1984,** 2013.
[70] Basu, S., McLauchlan, K. A., and Ritchie, A. J. D., *Chem. Phys.*, **79,** 95 (1983); *Chem. Abs.*, **100,** 5691 (1984).
[71] Grimshaw, J., and Trocha-Grimshaw, J., *J. Chem. Soc., Chem. Commun.*, **1984,** 744.
[72] Baban, J. A., and Roberts, B. P., *J. Chem. Soc., Perkin Trans. 2,* **1984,** 1717.
[73] Howard, J. A., Sutcliffe, R., and Mile, B., *J. Phys. Chem.*, **88,** 171 (1984).
[74] Forrester, A. R., Irikawa, H., and Soutar, G., *Tetrahedron Lett.*, **25,** 5445 (1984).
[75] Brand, J. C., Cook, M. D., and Roberts, B. P., *J. Chem. Soc., Perkin Trans. 2,* **1984,** 1187.
[76] Miura, Y., and Kinoshita, M., *J. Org. Chem.*, **49,** 2724 (1984).
[77] Mayer, R., Domschke, G., Bleisch, S., Fabian, J., Bartl, A., and Staško, A., *Collect. Czech. Chem. Commun.*, **49,** 684 (1984).
[78] Kadirov, M. K., Il'yasov, A. V., Vafina, A. A., Buzykin, B. I., Gazetdinova, N. G., and Kitaev, Yu. P., *Izv. Akad. Nauk SSSR, Ser. Khim.*, **1984,** 698; *Chem. Abs.*, **101,** 72157 (1984).
[79] Cordes, A. W., Hayes, P. J., Josephy, P. D., Koenig, H., Oakley, R. T., and Pennington, W. T., *J. Chem. Soc., Chem. Commun.*, **1984,** 1021.
[80] Tripathi, G. N. R., and Schuler, R. H., *J. Phys. Chem.*, **88,** 1706 (1984).
[81] Faucitano, A., Buttafava, A., Martinotti, F., and Bortolus, P., *J. Phys. Chem.*, **88,** 1187 (1984).

[82] Glidewell, C., *J. Chem. Soc., Perkin Trans. 2,* **1984,** 407.
[83] Göbl, M., Bonifačić, M., and Asmus, K.-D., *J. Am. Chem. Soc.,* **106,** 5984 (1984).
[84] Glass, R. S., Hojjatie, M., Wilson, G. S., Mahling, S., Göbl, M., and Asmus, K.-D., *J. Am. Chem. Soc.,* **106,** 5382 (1984).
[85] Szepes, L., and Baer, T., *J. Am. Chem. Soc.,* **106,** 273 (1984).
[86] Cartledge, F. K., and Piccione, R. V., *Organometallics,* **3,** 299 (1984).
[87] Neumann, W. P., Schultz, K.-D., and Vieler, R., *J. Organomet. Chem.,* **264,** 179 (1984).
[88] Krusic, P. J., Cote, W. J. and Grand, A., *J. Am. Chem. Soc.,* **106,** 4642 (1984).
[89] Colwell, S. M., *Mol. Phys.,* **51,** 1217 (1984); *Chem. Abs.,* **101,** 110040 (1984).
[90] Glidewell, C., *J. Chem. Soc., Perkin Trans. 2,* **1984,** 407.
[91] Kikuchi, O., and Shimomura, S., *Bull. Chem. Soc. Jpn.,* **57,** 2995 (1984).
[92] Druliner, J. D., Kitson, F. G., Rudat, M. A., and Tolman, C. A., *J. Org. Chem.,* **48,** 4951 (1984).
[93] Martynov, A. V., Mirskova, A. N., and Voronkov, M. G., *Zh. Org. Khim.,* **19,** 1869 (1983); *Chem. Abs.,* **100,** 33876 (1984).
[94] Reutov, O. A., *Usp. Khim.,* **53,** 462 (1984); *Chem. Abs.,* **101,** 71818 (1984).
[95] Lindsay, D. A., Lusztyk, J., and Ingold, K. U., *J. Am. Chem. Soc.,* **106,** 7087 (1984).
[96] Chatgilialoglu, C., Ingold, K. U., Tse-Sheepy, I., and Warkentin, I., *Can. J. Chem.,* **61,** 1077 (1983).
[97] Franz, J. A., Barrows, R. D., and Camaioni, D. M., *J. Am. Chem. Soc.,* **106,** 3964 (1984).
[98] Aeberhard, U., Keese, R., Stamm, E., Vögeli, U.-C., Lau, W., and Kochi, J. K., *Helv. Chim. Acta,* **66,** 2740 (1983).
[99] Barclay, L. R. C., Lusztyk, J., and Ingold, K. U., *J. Am. Chem. Soc.,* **106,** 1793 (1984).
[100] Saebo, S., Beckwith, A. L. J., and Radom, L., *J. Am. Chem. Soc.,* **106,** 5119 (1984).
[101] Brill, W. F., *J. Chem. Soc., Perkin Trans. 2,* **1984,** 621.
[102] Burkhard, P., Roduner, E., Hochmann, J., and Fischer, H., *J. Phys. Chem.,* **88,** 773 (1984).
[103] Ashby, E. C., and Pham, T. N., *Tetrahedron Lett.,* **25,** 4333 (1984).
[104] Chung, S. K., and Dunn, L. B., *J. Org. Chem.,* **49,** 935 (1984).
[105] Garst, J. F., and Hines, J. B., *J. Am. Chem. Soc.,* **106,** 6443 (1984).
[106] Beckwith, A. L. J., O'Shea, D. M., and Roberts, D. H., *J. Chem. Soc., Chem. Commun.,* **1983,** 1445.
[107] van der Linde, L. M., and van der Weerdt, A. J. A., *Tetrahedron Lett.,* **25,** 1201 (1984).
[108] Clive, D. L. J., Beaulieu, P. L., and Set, L., *J. Org. Chem.,* **49,** 1313 (1984).
[109] Apparu, M., and Crandall, J. K., *J. Org. Chem.,* **49,** 2125 (1984).
[110] Kraus, G. A., and Landgrebe, K., *Tetrahedron Lett.,* **25,** 3939 (1984).
[111] Saito, K., Toda, T., and Mukai, T., *Bull. Chem. Soc. Jpn.,* **57,** 1567 (1984).
[112] Porter, N. A., and Zuraw, P. J., *J. Org. Chem.,* **49,** 1345 (1984).
[113] Barton, T. J., and Revis, A., *J. Am. Chem. Soc.,* **106,** 3802 (1984).
[114] Crozet, M. P., Kaafarani, M., and Surzur, J.-M., *Bull. Soc. Chim. Fr. II,* **1984,** 390.
[115] Zeller, K.-P., Müller, R., and Alder, R. W., *J. Chem. Soc., Perkin Trans. 2,* **1984,** 1711.
[116] Zeller, K.-P., Müller, R., and Alder, R. W., *J. Chem. Soc., Chem. Commun.,* **1984,** 330.
[117] Barbier, M., Barton, D. H. R., Devys, M., and Satish Topgi, R., *J. Chem. Soc., Chem. Commun.,* **1984,** 743.
[118] Scott, L. T., *J. Org. Chem.,* **49,** 3021 (1984).
[119] Dern, H.-J., Lange, F., and Sustmann, R., *Chem. Ber.,* **116,** 3316 (1983).
[120] Burkhard, P., Roduner, E., Hochmann, J., and Fischer, H., *J. Phys. Chem.,* **88,** 773 (1984).
[121] Alnajjar, M. S., Smith, G. F., and Kuivila, H. G., *J. Org. Chem.,* **49,** 1271 (1984).
[122] Gassman, P. G., and Smith, J. L., *Tetrahedron Lett.,* **25,** 3051 (1984).
[123] Ratier, M., Pereyre, M., Davies, A. G., and Sutcliffe, R., *J. Chem. Soc., Perkin Trans. 2,* **1984,** 1907.
[124] Kang, Y.-H., and Kice, J. L., *J. Org. Chem.,* **49,** 1507 (1984).
[125] Beckwith, A. L. J., Kazlauskas, R., and Syner-Lyons, M. R., *J. Org. Chem.,* **48,** 4718 (1983).
[126] Suginome, H., Ohtsuka, T., Orito, K., Jaime, C., and Ōsawa, E., *J. Chem. Soc., Perkin Trans. 2,* **1984,** 575.
[127] Suginome, H., Liu, C. F., and Tokuda, M., *J. Chem. Soc., Chem. Commun.,* **1984,** 334.
[128] Suginome, H., Liu, C. F., and Furusaki, A., *Chem. Lett.,* **1984,** 911.
[129] Concepciòn, J. I., Francisco, C. G., Hernàndez, R., Salazar, J. A., and Suàrez, E., *Tetrahedron Lett.,* **25,** 1953 (1984).
[130] Baban, J. A., and Roberts, B. P., *J. Chem. Soc., Chem. Commun.,* **1984,** 850.
[131] Hase, W. L., Duchovic, R. J., Swamy, K. N., and Wolf, R. J., *J. Chem. Phys.,* **80,** 714 (1984); *Chem. Abs.,* **100,** 174093 (1984).
[132] Xue, Z. L., Tomiyoshi, K., Mathis, C. A., Knickelbein, M. B., and Root, J. W., *Chem. Phys. Lett.,* **103,** 73 (1983); *Chem. Abs.,* **100,** 173974 (1984).
[133] Eberhardt, M. K., *J. Org. Chem.,* **49,** 3720 (1984).

[134] Lipscher, J., and Fischer, H., *J. Phys. Chem.*, **88,** 2555 (1984).
[135] Engel, P. S., and Gerth, D. B., *J. Am. Chem. Soc.*, **105,** 6849 (1983).
[136] Kuokkanen, T., *Finn. Chem. Lett.*, **1984,** 38.
[137] Al Akeel, N. Y., and Waddington, D. J., *J. Chem. Soc., Perkin Trans. 2,* **1984,** 1575.
[138] Kerr, J. A., and Wright, J. P., *Int. J. Chem. Kinet.*, **16,** 1321 (1984).
[139] Kerr, J. A., and Wright, J. P., *Int. J. Chem. Kinet.*, **16,** 1327 (1984).
[140] Neta, P., Dizdaroglu, M., and Simic, M. G., *Isr. J. Chem.*, **24,** 25 (1984).
[141] Vedeneev, V. I., and Chernysheva, A. V., *Khim. Fiz.*, **1983,** 1521; *Chem. Abs.*, **100,** 50891 (1984).
[142] Pertyaev, E. P., Maslovskaya, L. A., and Shadyro, O. I., *Zh. Org. Khim.*, **19,** 2031 (1983); *Chem. Abs.*, **100,** 22244 (1984).
[143] Slagle, I. R., Park, J.-Y., Heaven, M. C., and Gutman, D., *J. Am. Chem. Soc.*, **106,** 4356 (1984).
[144] Park, J. Y., Heaven, M. C., and Gutman, D., *Chem. Phys. Lett.*, **104,** 469 (1984); *Chem. Abs.*, **100,** 191163 (1984).
[145] Vidoczy, T. Gal, D., Tavadjan, L. A., and Nalbandyan, A. B., *Magy. Kem. Foly.*, **89,** 427 (1983); *Chem. Abs.*, **100,** 50857 (1984).
[146] Temps, F., *Report* **1983,** MPIS-4/1983; *Chem. Abs.*, **101,** 151149 (1984).
[147] Veyret, B., Roussel, P., and Lesclaux, R., *Chem. Phys. Lett.*, **103,** 389 (1984); *Chem. Abs.*, **100,** 173979 (1984).
[148] Pritchard, G. O., Nilsson, W. B., Marchant, P. E., Case, L. C., Parmer, J. F., and Youngs, R. F., *Int. J. Chem. Kinet.*, **16,** 69 (1984).
[149] Hassimen, E., Riepponen, P., Blomqvist, K., Kalliorinne, K., and Koskikallio, J., *Bull. Soc. Chim. Belg.*, **92,** 847 (1983); *Chem. Abs.*, **100,** 138422 (1984).
[150] Lipscher, J., and Fischer, H., *J. Phys. Chem.*, **88,** 2555 (1984).
[151] Osborne, D. A., and Waddington, D. J., *J. Chem. Soc., Perkin Trans. 2,* **1984,** 1861.
[152] Nikolaev, A. I., Enikeeva, L. R., and Safiullin, R. L., *Khim. Fiz.*, **3,** 711 (1984); *Chem. Abs.*, **101,** 72218 (1984).
[153] Wojnarovits, L., *J. Chem. Soc., Perkin Trans. 2,* **1984,** 1449.
[154] Tegowski, A. T., and Pratt, D. W., *J. Am. Chem. Soc.*, **106,** 64 (1984).
[155] Porter, A. D., Petter, R. C., and Brittain, W. J., *J. Am. Chem. Soc.*, **106,** 813 (1984).
[156] Leuschner, R., Holger, K., and Dohrmann, J. K., *Ber. Bunsen Ges. Phys. Chem.*, **88,** 462 (1984); *Chem. Abs.*, **101,** 72168 (1984).
[157] Münger, K., and Fischer, H., *Int. J. Chem. Kinet.*, **16,** 1213 (1984).
[158] Wu, L.-M., and Fischer, H., *Int. J. Chem. Kinet.*, **16,** 1111 (1984).
[159] Salikhov, K. M., *Chem. Phys.*, **82,** 145 (1983); *Chem. Abs.*, **100,** 120381 (1984).
[160] Manka, M. J., and Stein, S. E., *J. Phys. Chem.*, **88,** 5914 (1984).
[161] Khudyakov, I. V., and Koroli, L. L., *Chem. Phys. Lett.*, **103,** 383 (1984); *Chem. Abs.*, **100,** 191050 (1984).
[162] Koroli, L. L., Kuzmin, V. A., and Khudyakov, I. V., *Int. J. Chem. Kinet.*, **16,** 379 (1984).
[163] Cuthbertson, M. J., Rizzardo, E., and Solomon, D. H., *Aust. J. Chem.*, **36,** 1957 (1983).
[164] Varlamov, V. T., Safiullin, R. L., and Denisov, E. T., *Khim. Fiz.*, **1983,** 408; *Chem. Abs.*, **101,** 130144 (1984).
[165] Yarkov, S. P., Roginskii, V. A., and Zaikov, G. E., *Khim. Fiz.*, **1983,** 1410; *Chem. Abs.*, **100,** 85177 (1984).
[166] Howard, J. A., *Isr. J. Chem.*, **24,** 33 (1984).
[167] Nikolaev, A. I., Safiullin, R. L., and Komissarov, V. D., *Khim. Fiz.*, **3,** 257 (1984); *Chem. Abs.*, **100,** 209063 (1984).
[168] Winterle, J., Dulin, D., and Mill, T., *J. Org. Chem.*, **49,** 491 (1984).
[169] Siebrand, W., Wildman, T. A., and Zgierski, M. Z., *Chem. Phys. Lett.*, **98,** 108 (1983).
[170] Siebrand, W., Wildman, T. A., and Zgierski, M. Z., *J. Am. Chem. Soc.*, **106,** 4083 (1984).
[171] Siebrand, W., Wildman, T. A., and Zgierski, M. Z., *J. Am. Chem. Soc.*, **106,** 4089 (1984).
[172] Cao, J. R., and Back, M. H., *Can. J. Chem.*, **62,** 86 (1984).
[173] Oldershaw, G. A., and Smith, A., *Int. J. Chem. Kinet.*, **16,** 213 (1984).
[174] Sana, M., Leroy, G., and Villaveces, J. L., *Theor. Chim. Acta*, **65,** 109 (1984); *Chem. Abs.*, **101,** 54212 (1984).
[175] Berman, M. R., and Lin, M. C., *Chem. Phys.*, **82,** 435 (1983); *Chem. Abs.*, **100,** 85005 (1984).
[176] Gilliom, R. D., *J. Comput. Chem.*, **5,** 237 (1984); *Chem. Abs.*, **101,** 54211 (1984).
[177] Canadell, E., Olivella, S., and Poblet, J. M., *J. Phys. Chem.*, **88,** 3545 (1984).
[178] Doba, T., Ingold, K. U., Siebrand, W., and Wildman, T. A., *J. Phys. Chem.*, **88,** 3165 (1984).
[179] Zaitsev, S. A., Vyazovkin, V. L., Pantilova, E. V., Bal'shakov, B. N., and Tolkachev, V. A., *Khim. Fiz.*, **1983,** 517; *Chem. Abs.*, **101,** 90341 (1984).

[180] Dobe, S., Reti, F., and Berces, T., *React. Kinet. Catal. Lett.*, **24,** 413 (1984); *Chem. Abs.*, **101,** 22713 (1984).
[181] Arican, H., and Arthur, N. L., *Aust. J. Chem.*, **36,** 2185 (1983).
[182] Zhuravlev, M. V., Serov, S. I., Sass, V. P., and Sokolov, S. V., *Zh. Org. Khim.*, **19,** 2027 (1983); *Chem. Abs.*, **100,** 22213 (1984).
[183] Patnaik, L. N., Ront, S. P., Samal, N. C., Senapati, M., Mishra, R., and Rout, M. K., *Colloid. Polym. Sci.*, **262,** 119 (1984); *Chem. Abs.*, **100,** 208772 (1984).
[184] Szirovicza, L., *Acta Phys. Chem.*, **29,** 35 (1983); *Chem. Abs.*, **100,** 174090 (1984).
[185] Rossi, M. J., and Golden, D. M., *Int. J. Chem. Kinet.*, **15,** 1283 (1983).
[186] Arican, H., and Arthur, N. L., *Aust. J. Chem.*, **36,** 2195 (1983).
[187] Arican, H., and Arthur, N. L., *Int. J. Chem. Kinet.*, **16,** 335 (1984).
[188] Weeks, I., and Whittle, E., *Int. J. Chem. Kinet.*, **15,** 1329 (1983).
[189] Lane, S. I., Oexler, E. V., and Staricco, E. H., *Int. J. Chem. Kinet.*, **16,** 1357 (1984).
[190] Pasteris, L., Oexler, E. V., and Staricco, E. H., *An. Asoc. Quim. Argent.*, **71,** 255 (1983); *Chem. Abs.*, **100,** 33882 (1984).
[191] Pasteris, L., Oexler, E. V., and Staricco, E. H., *Ber. Bunsen Ges. Phys. Chem.*, **88,** 568 (1984); *Chem. Abs.*, **101,** 71959 (1984).
[192] Bockrath, B., Bittner, E., and McGrew, J., *J. Am. Chem. Soc.*, **106,** 135 (1984).
[193] Strong, H. L., Brownawell, M. L., and San Filippo, J., Jr., *J. Am. Chem. Soc.*, **105,** 6526 (1983).
[194] Zoltewicz, J. A., and Locko, G. A., *J. Org. Chem.*, **48,** 4214 (1983).
[195] Lalande, R., Filliatre, C., Villenave, J.-J., and Jaouhari, R., *Helv. Chim. Acta*, **67,** 149 (1984).
[196] Roscoe, J. M., *Can. J. Chem.*, **61,** 2716 (1983).
[197] Bjarnov, E., Munk, J., Nielsen, O. J., Pagsberg, P., and Sillesen, A., *INIS Atomindex*, **14,** Abstr. no.778399 (1983); *Chem. Abs.*, **100,** 67575 (1984).
[198] Jeong, K.-M., Hsu, K.-J., Jeffries, J. B., and Kaufman, *J. Phys. Chem.*, **88,** 1222 (1984).
[199] Atkinson, R., Carter, W. P. L., Aschmann, S. M., Winer, A. M., and Pitts, J. N., *Int. J. Chem. Kinet.*, **16,** 469 (1984).
[200] Niki, H., Maker, P. D., Savage, C. M., and Breitenbach, L. P., *J. Phys. Chem.*, **88,** 5342 (1984).
[201] Temps, F., and Wagner, H. G., *Ber. Bunsenges. Phys. Chem.*, **88,** 415 (1984); *Chem. Abs.*, **101,** 37911 (1984).
[202] Kurylo, M. J., and Knable, G. L., *J. Phys. Chem.*, **88,** 3305 (1984).
[203] Atkinson, R., Plum, C. N., Carter, W. P. L., Winer, A. M., and Pitts, J. N., *J. Phys. Chem.*, **88,** 1210 (1984).
[204] Atkinson, R., Plum, C. N., Carter, W. P. L., Winer, A. M., and Pitts, J. N., *J. Phys. Chem.*, **88,** 2361 (1984).
[205] Atkinson, R., Carter, W. P. L., Plum, C. N., Winer, A. M., and Pitts, J. N., *Int. J. Chem. Kinet.*, **16,** 887 (1984).
[206] Elango, T. P., Ramakrishnan, V., Vancheesan, S., and Kuriacose, J. C., *Proc. Indian Acad. Sci., Chem. Sci.*, **93,** 47 (1984); *Chem. Abs.*, **100,** 208855 (1984).
[207] Sway, M. I., and Waddington, D. J., *J. Chem. Soc., Perkin Trans. 2*, **1984,** 63.
[208] Song, S. A., and Choo, K. Y., *Bull. Korean Chem. Soc.*, **5,** 16 (1984); *Chem. Abs.*, **101,** 54470 (1984).
[209] Cheun, Y. G., Hwang, M. S., and Lee, I., *Taehan Hwahakhoe Chi*, **27,** 391 (1983); *Chem. Abs.*, **100,** 85186 (1984).
[210] Roberts, C., and Walton, J. C., *J. Chem. Soc., Chem. Commun.*, **1984,** 1109.
[211] Bascetta, E., Gunstone, F. D., and Walton, J. C., *J. Chem. Soc., Perkin Trans. 2*, **1984,** 401.
[212] Cuthbertson, M. J., Rizzardo, E., and Solomon, D. H., *Aust. J. Chem.*, **36,** 1957 (1983).
[213] Grant, R. D., Griffiths, P. G., Moad, G., Rizzardo, E., and Solomon, D. H., *Aust. J. Chem.*, **36,** 2447 (1983).
[214] Kim-Thuan, N., and Scaiano, J. C., *Int. J. Chem. Kinet.*, **16,** 371 (1984).
[215] Kurbanov, D., Pastushenko, E. V., Zlotskii, S. S., and Rakhmankulov, D. L., *Zh. Org. Khim.*, **20,** 936 (1984); *Chem. Abs.*, **101,** 110142 (1984).
[216] Pastushenko, E. V., Kurbanov, D., Zlotskii, S. S., and Rakhmankulov, D. L., *Khim. Geterotsikl. Soedin.*, **1983,** 1614; *Chem. Abs.*, **100,** 85021 (1984).
[217] Bors, W., Tait, D., Erben-Russ, M., and Saran, M. C., *Oxygen Radicals Chem. Biol., Proc., Int. Conf., 3rd*, **1983,** 49; *Chem. Abs.*, **101,** 38013 (1984).
[218] Chatgilialoglu, C., Lunazzi, L., Macciantelli, D., and Placucci, G., *J. Am. Chem. Soc.*, **106,** 5252 (1984).
[219] Neta, P., Dizdaroglu, M., and Simic, M. G., *Isr. J. Chem.*, **24,** 25 (1984).
[220] Shoikhet, A. A., and Teitel'boim, M. A., *Deposited Doc.*, **1982,** VINITI 4800; *Chem. Abs.*, **99,** 194265 (1983).

[221] Anastasi, C., and Hancock, D., *J. Chem. Soc., Faraday Trans. 1*, **80**, 935 (1984).
[222] Howard, J. A., *Isr. J. Chem.*, **24**, 33 (1984).
[223] Doba, T., and Ingold, K. U., *J. Am. Chem. Soc.*, **106**, 3958 (1984).
[224] Niedzielski, J., Tschuikow-Roux, E., and Yano, T., *Int. J. Chem. Kinet.*, **16**, 621 (1984).
[225] Niedzielski, J., Yano, T., and Tschuikow-Roux, E., *Can. J. Chem.*, **62**, 899 (1984).
[226] Islam, T. S. A., and Benson, S. W., *Int. J. Chem. Kinet.*, **16**, 995 (1984).
[227] Kondo, O., and Benson, S. W., *Int. J. Chem. Kinet.*, **16**, 949 (1984).
[228] Islam, T. S. A., Marshall, R. M., and Benson, S. W., *Int. J., Chem. Kinet.*, **16**, 1161 (1984).
[229] Ahonkhai, S. I., and Whittle, E., *Int. J. Chem. Kinet.*, **16**, 543 (1984).
[230] Sidebottom, H., and Treacy, J., *Int. J. Chem. Kinet.*, **16**, 579 (1984).
[231] Davis, S. A., Gilbert, B. C., Griller, D., and Nazran, A. S., *J. Org. Chem.*, **49**, 3415 (1984).
[232] Lindsay, D. A., Lusztyk, J., and Ingold, K. U., *J. Am. Chem. Soc.*, **106**, 7087 (1984).
[233] Johnston, L. J., Scaiano, J. C., and Ingold, K. U., *J. Am. Chem. Soc.*, **106**, 4877 (1984).
[234] de Vöhringer, C. M., and Staricco, E. H., *Int. J. Chem. Kinet.*, **16**, 1351 (1984).
[235] de Vöhringer, C. M., and Staricco, E. H., *J. Chem. Soc., Faraday Trans. 1*, **80**, 2631 (1984).
[236] de Vöhringer, C. M., and Staricco, E. H., *An. Asoc. Quim. Argent.*, **71**, 539 (1983); *Chem. Abs.*, **101**, 72061 (1984).
[237] Baban, J. A., and Roberts, B. P., *J. Chem. Soc., Perkin Trans. 2*, **1984**, 1717.
[238] Beckwith, A. L. J., and Westwood, S. W., *Aust. J. Chem.*, **36**, 2123 (1983).
[239] Souppe, J., Namy, J.-L., and Kagan, H. B., *Tetrahedron Lett.*, **25**, 2869 (1984).
[240] Lusztyk, J., Lusztyk, E., Maillard, B., and Ingold, K. U., *J. Am. Chem. Soc.*, **106**, 2923 (1984).
[241] Adcock, J. L., and Evans, W. D., *J. Org. Chem.*, **49**, 2719 (1984).
[242] Hoshi, M., Masuda, Y., and Arase, A., *Chem. Lett.*, **1984**, 195.
[243] Schneider, H.-J., and Philippi, K., *Chem. Ber.*, **117**, 3056 (1984).
[244] Chelibanov, V. P., Komarov, V. S., Evseeva, I. Yu., and Kozlinev, M. Z., *Deposited Doc.*, **1983**, 2470; *Chem. Abs.*, **101**, 71947 (1984).
[245] Aver'yanov, V. A., Zarytovskii, V. M., and Shvets, V. F., *Zh. Org. Khim.*, **19**, 2112 (1983); *Chem. Abs.*, **100**, 50794 (1984).
[246] Aver'yanov, V. A., Kirichenko, S. E., and Shvets, V. F., *Zh. Org. Khim.*, **19**, 2120 (1983); *Chem. Abs.*, **100**, 22214 (1984).
[247] Sergeev, G. B., Smirnov, V. V., and Porodenko, E. V., *Vestn. Mosk. Univ., Ser. 2: Khim.*, **25**, 170 (1984); *Chem. Abs.*, **101**, 37914 (1984).
[248] Donnelly, K. D., Fristad, W. E., Gellerman, B. J., Peterson, J. R., and Selle, B. J., *Tetrahedron Lett.*, **25**, 607 (1984).
[249] Shinoda, K., and Yasuda, K., *Toyama Kogyo Koto Senmon Gakko Kiyo*, **18**, 1, (1984); *Chem. Abs.*, **101**, 151128 (1984).
[250] Pozdeeva, N. N., and Denisov, E. T., *Izv. Akad. Nauk SSSR, Ser. Khim.*, **1983**, 2029; *Chem. Abs.*, **99**, 211885 (1984).
[251] Emel'yanov, V. I., Stul, B. Yu., and Korolev, B. N., *Kinet. Katal.*, **25**, 546 (1984); *Chem. Abs.*, **101**, 151126 (1984).
[252] Skell, P. S., and Traynham, J. G., *Acc. Chem. Res.*, **17**, 160 (1984).
[253] Tanner, D. D., Ruo, T. C. S., Kosugi, Y., and Potter, A., *Can. J. Chem.*, **62**, 2310 (1984).
[254] Rakhimov, A. I., Ozerov, A. A., and Litinskii, A. O., *Zh. Org. Khim.*, **19**, 2630 (1983); *Chem. Abs.*, **100**, 138318 (1984).
[255] Skell, P. S., and Seshadri, S., *J. Org. Chem.*, **49**, 1650 (1984).
[256] Skell, P. S., *J. Am. Chem. Soc.*, **106**, 1838 (1984).
[257] Scherzer, K., Claus, P., and Dabbagh, M., *J. Prakt. Chem.*, **325**, 680 (1983); *Chem. Abs.*, **100**, 5489 (1984).
[258] Nikisha, L. V., Moshkina, R. I., Polyak, S. S., and Vedeneev, V. I., *Khim. Fiz.*, **3**, 569 (1984); *Chem. Abs.*, **101**, 54240 (1984).
[259] Pasto, D. J., *Tetrahedron*, **40**, 2811 (1984).
[260] Kerr, J. A., and Wright, J. P., *Int. J. Chem. Kinet.*, **16**, 1327 (1984).
[261] Batt, L., and Mowat, S. I., *Int. J. Chem. Kinet.*, **16**, 603 (1984).
[262] Caralp, F., and Lesclaux, R., *Chem. Phys. Lett.*, **102**, 54 (1983); *Chem. Abs.*, **100**, 50889 (1984).
[263] Heintz, M., Leny, G., and Nedelec, J. Y., *Tetrahedron Lett.*, **25**, 5767 (1984).
[264] Bland, W. J., Davis, R., and Durrant, J. L. A., *J. Organomet. Chem.*, **260**, C75 (1984).
[265] Bland, W. J., Davis, R., and Durrant, J. L. A., *J. Organomet. Chem.*, **267**, C45 (1984).
[266] Elsheimer, S., Dolbier, W. R., Murla, M., Seppelt, K., and Paprott, G., *J. Org. Chem.*, **49**, 205 (1984).

[267] Langford, A. O., and Moore, C. B., *J. Chem. Phys.*, **80,** 4211 (1984); *Chem. Abs.*, **101,** 22737 (1984).
[268] Zhuravlev, M. V., Serov, S. I., Sass, V. P., and Sokolov, S. V., *Zh. Org. Khim.*, **19,** 2027 (1983); *Chem. Abs.*, **100,** 22213 (1984).
[269] Zhuravlev, M. V., Sass, V. P., and Sokolov, S. V., *Zh. Org. Khim.*, **19,** 2022 (1983); *Chem. Abs.*, **100,** 33889 (1984).
[270] Mun, G. A., Golubev, V. B., Skorikova, E. E., and Zubov, V. P., *Vestn. Mosk. Univ., Ser. 2: Khim.*, **24,** 280 (1983); *Chem. Abs.*, **99,** 194111 (1983).
[271] Lahousse, F., Merényi, R., Desmurs, J. R., Allaime, H., Borghese, A., and Viehe, H. G., *Tetrahedron Lett.*, **25,** 3823 (1984).
[272] Budeiko, N. L., Gudimenko, Yu. I., Agabekov, V. E., Kosmacheva, T. G., and Mitskevich, N. I., *Dokl. Akad. Nauk SSSR*, **28,** 347 (1984); *Chem. Abs.*, **101,** 6309 (1984).
[273] Ponec, R., Hàjek, M., and Màlek, J., *Collect. Czech. Chem. Commun.*, **48,** 2469 (1983).
[274] Barton, D. H. R., Hervé, Y., Potier, P., and Thierry, J., *J. Chem. Soc., Chem. Commun.*, **1984,** 1298.
[275] Barton, D. H. R., Crich, D., and Motherwell, W. B., *J. Chem. Soc., Chem. Commun.*, **1984,** 242.
[276] Barton, D. H. R., and Crich, D., *Tetrahedron Lett.*, **25,** 2787 (1984).
[277] Barton, D. H. R., Crich, D., and Kretzschmar, G., *Tetrahedron Lett.*, **25,** 1055 (1984).
[278] Barton, D. H. R., and Kretzschmar, G., *Tetrahedron Lett.*, **24,** 5889 (1983).
[279] Baldwin, J. E., Kelly, D. R., and Ziegler, C. B., *J. Chem. Soc., Chem. Commun.*, **1984,** 133.
[280] Baldwin, J. E., Adlington, R. M., and Basak, A., *J. Chem. Soc., Chem. Commun.*, **1984,** 1284.
[281] Giese, B., Heuck, K., Lenhardt, H., and Lüning, U., *Chem. Ber.*, **117,** 2132 (1984).
[282] Giese, B., and Kretzschmar, G., *Chem. Ber.*, **117,** 3160 (1984).
[283] Olivella, S., Canadell, E., and Poblet, J. M., *J. Org. Chem.*, **48,** 4696 (1983).
[284] Johnston, L. J., Scaiano, J. C., and Ingold, K. U., *J. Am. Chem. Soc.*, **106,** 4877 (1984).
[285] Giese, B., and Kretzschmar, G., *Chem. Ber.*, **117,** 3175 (1984).
[286] Gasanov, R. G., Petrova, R. G., Churkina, T. D., and Freidlina, R. Kh., *Dokl. Akad. Nauk SSSR*, **274,** 353 (1984); *Chem. Abs.*, **100,** 208774 (1984).
[287] Russell, G. A., Tashtoush, H., and Ngoviwatchai, P., *J. Am. Chem. Soc.*, **106,** 4622 (1984).
[288] Roduner, E., Brinkman, G. A., and Louvrier, P. W. F., *Hyperfine Interact.*, **19,** 797 (1984); *Chem. Abs.*, **101,** 72201 (1984).
[289] Roduner, E., and Muenger, K., *Hyperfine Interact.*, **19,** 793 (1984); *Chem. Abs.*, **101,** 38080 (1984).
[290] Swamy, K. N., and Hase, W. L., *J. Phys. Chem.*, **87,** 4715 (1983).
[291] Callear, B., and Smith, G. B., *Chem. Phys. Lett.*, **105,** 119 (1984); *Chem. Abs.*, **100,** 19116 (1984).
[292] Mamedov, Kh. I., and Niyazova, E. N., *Khim. Vys. Energ.*, **17,** 452 (1983); *Chem. Abs.*, **100,** 5636 (1984).
[293] Iyer, R. S., Rogers, P. J., and Rowland, F. S., *J. Phys. Chem.*, **87,** 3799 (1983).
[294] Iyer, R. S., and Rowland, F. S., *Chem. Phys. Lett.*, **103,** 213 (1983); *Chem. Abs.*, **100,** 138286 (1984).
[295] Zellner, R., and Lorenz, K., *J. Phys. Chem.*, **88,** 984 (1984).
[296] Atkinson, R., and Aschmann, S. M., *Int. J. Chem. Kinet.*, **16,** 1175 (1984).
[297] Ohta, T., *Int. J. Chem. Kinet.*, **16,** 879 (1984).
[298] Atkinson, R., Aschmann, S. M., and Carter, W. P. L., *Int. J. Chem. Kinet.*, **15,** 1161 (1983).
[299] Atkinson, R., Aschmann, S. M., and Carter, W. P. L., *Int. J. Chem. Kinet.*, **16,** 967 (1984).
[300] Davies, M. J., and Gilbert, B. C., *J. Chem. Soc., Perkin Trans. 2*, **1984,** 1809.
[301] Bobrowski, K., *J. Chem. Soc., Faraday Trans. 1*, **80,** 1377 (1984).
[302] Yamase, T., and Kurozumi, T., *Inorg. Chim. Acta.*, **83,** L25 (1984).
[303] Atkinson, R., and Aschmann, S. M., *Int. J. Chem. Kinet.*, **16,** 259 (1984).
[304] Mignani, S., Janousek, Z., Merenyi, R., Viehe, H. G., Riga, J., and Verbist, J., *Tetrahedron Lett.*, **25,** 1571 (1984).
[305] Morgan, M. E., Osborne, D. A., and Waddington, D. J., *J. Chem. Soc., Perkin Trans. 2*, **1984,** 1869.
[306] Grossi, L., Lunazzi, L., and Placucci, G., *J. Chem. Soc., Perkin Trans. 2*, **1983,** 1831.
[307] Alberti, A., Degl'Innocenti, A., Grossi, L., and Lunazzi, L., *J. Org. Chem.*, **49,** 4615 (1984).
[308] Svoboda, P., Pytela, O., and Večeřa, M., *Collect. Czech. Chem. Commun.*, **48,** 3287 (1983).
[309] Ito, O., and Matsuda, M., *J. Phys. Chem.*, **88,** 1002 (1984).
[310] Ito, O., and Matsuda, M., *Int. J. Chem. Kinet.*, **16,** 909 (1984).
[311] Ito, O., Furuya, S., and Matsuda, M., *J. Chem. Soc., Perkin Trans. 2*, **1984,** 139.
[312] Ito, O., and Matsuda, M., *J. Org. Chem.*, **49,** 17 (1984).
[313] Kang, Y.-H., and Kice, J. L., *J. Org. Chem.*, **49,** 1507 (1984).
[314] Kamigata, N., Narushima, T., Sawada, H., and Kobayashi, M., *Bull. Chem. Soc. Jpn.*, **57,** 1421 (1984).
[315] Atkinson, R., Aschmann, S. M., Winer, A. M., and Pitts, J. N., *Int. J. Chem. Kinet.*, **16,** 697 (1984).

[316] Atkinson, R., Plum, C. N., Carter, W. P. L., Winer, A. M., and Pitts, J. N., *J. Phys. Chem.*, **88**, 1210 (1984).
[317] Lessard, J., Couture, Y., Mondon, M., and Touchard, D., *Can. J. Chem.*, **62**, 105 (1984).
[318] Cowley, A. H., Norman, N. C., Stewart, C. A., and Whittlesey B. R., *Tetrahedron Lett.*, **25**, 3519 (1984).
[319] Wuest, J. D., and Zacharie, B., *J. Org. Chem.*, **49**, 163 (1984).
[320] Adeleke, B. B., Weir, D., Depew, M. C., and Wan, J. K. S., *Can. J. Chem.*, **62**, 117 (1984).
[321] Ingold, K. U., Lusztyk, J., and Scaiano, J. C., *J. Am. Chem. Soc.*, **106**, 343 (1984).
[322] Dewar, M. J. S., Grady, G. L., Kuhn, D. R., and Merz, K. M., *J. Am. Chem. Soc.*, **106**, 6773 (1984).
[323] Stadlbauer, J. M., Ng, B. W., Ganti, R., and Walter, D. C., *J. Am. Chem. Soc.*, **106**, 3151 (1984).
[324] Roduner, E., Brinkman, G. A., and Lauvrier, P. W. F., *Hyperfine Interact.*, **19**, 803 (1984); *Chem. Abs.*, **101**, 72202 (1984).
[325] Nicovich, J. M., and Ravishankara, A. R., *J. Phys. Chem.*, **88**, 2534 (1984).
[326] Komendantov, A. M., Rondarev, D. S., Sass, V. P., and Sokolov, S. V., *Zh. Org. Khim.*, **19**, 1920 (1983); *Chem. Abs.*, **100**, 33878 (1984).
[327] Giordano, C., Minisci, F., Tortelli, V., and Vismara, E., *J. Chem. Soc., Perkin Trans. 2*, **1984**, 293.
[328] Castaldi, G., Minisci, F., and Vismara, E., *Tetrahedron Lett.*, **25**, 3897 (1984).
[329] Minisci, F., Giordano, C., Vismara, E., Levi, S., and Tortelli, V., *J. Am. Chem. Soc.*, **106**, 7146 (1984).
[330] Kurz, M. E., Lapin, S. C., Mariam, K., Hagen, T. J., and Qian, X. Q., *J. Org. Chem.*, **49**, 2728 (1984).
[331] Baur, R., Kleiner, E., and Pfleiderer, W., *Liebigs Ann. Chem.*, **1984**, 1798.
[332] Kurz, M. E., Baru, V., and Nguyen, P.-N., *J. Org. Chem.*, **49**, 1603 (1984).
[333] Hall, J. H., Kaler, L., and Herring, R., *J. Org. Chem.*, **49**, 2579 (1984).
[334] Bazanova, G. V., and Stotskii, A. A., *Zh. Org. Khim.*, **19**, 2124 (1983); *Chem. Abs.*, **100**, 33890 (1984).
[335] Lorenz, K., and Zellner, R., *Ber. Bunsen Ges. Phys. Chem.*, **87**, 629 (1983); *Chem. Abs.*, **99**, 194093 (1983).
[336] Wahner, A., and Zetzsch, C., *J. Phys. Chem.*, **87**, 4945 (1983).
[337] Rinke, M., and Zetzsch, C., *Ber. Bunsen Ges. Phys. Chem.*, **88**, 55 (1984); *Chem. Abs.*, **100**, 191048 (1984).
[338] Vysotskaya, N. A., Shevchuk, L. G., Gavrilova, S. P., Badovskaya, L. A., and Kul'nevich, V. G., *Ukr. Khim. Zh.* (*Russ. Ed.*), **49**, 865 (1983); *Chem. Abs.*, **99**, 194092 (1983).
[339] Wine, P. H., and Thompson, R. J., *Int. J. Chem. Kinet.*, **16**, 867 (1984).
[340] Solar, S., Solar, W., and Getoff, N., *J. Phys. Chem.*, **88**, 2091 (1984).
[341] Ogata, Y., and Tomizawa, K., *J. Chem. Soc., Perkin Trans. 2*, **1984**, 985.
[342] Leardini, R., Pedulli, G. F., Tundo, A., and Zanardi, G., *J. Chem. Soc., Chem. Commun.*, **1984**, 1320.
[343] Ballester, M., Castañer, J., Riera, J., and Pujadas, J., *J. Org. Chem.*, **49**, 2884 (1984).
[344] McNab, H., *J. Chem. Soc., Perkin Trans. 1*, **1984**, 371.
[345] McNab, H., *J. Chem. Soc., Perkin Trans. 1*, **1984**, 377.
[346] McNab, H., and Smith, G. S., *J. Chem. Soc., Perkin Trans. 1*, **1984**, 381.
[347] Nori, M., Garbarino, G., Dell'Erba, C., and Petrillo, G., *J. Chem. Soc., Chem. Commun.*, **1984**, 1205.
[348] Bloodworth, A. J., Courtneidge, J. L., and Davies, A. G., *J. Chem. Soc., Perkin Trans. 2.*, **1984**, 523.
[349] Kharrat, A., Gardrat, C., and Maillard, B., *Can. J. Chem.*, **62**, 2385 (1984).
[350] Johnson, M. D., *Acc. Chem. Res.*, **16**, 343 (1983).
[351] Ashcroft, M. R., Bougeard, P., Bury, A., Cooksey, C. J., Johnson, M. D., Hungerford, J. M., and Lampman, G. M., *J. Org. Chem.*, **49**, 1751 (1984).
[352] Yamada, M., Kamo, T., Nishino, J., and Amano, A., *Nippon Kagaku Kaishi*, **1983**, 1475; *Chem. Abs.*, **100**, 138335 (1984).
[353] Ekwenchi, M. M., Safarik, I., and Strausz, O. P., *Int. J. Chem. Kinet.*, **16**, 741 (1984).
[354] Ellul, R., Potzinger, P., and Reimann, B., *J. Phys. Chem.*, **88**, 2793 (1984).
[355] Sridharan, U. C., and Kaufman, F., *Chem. Phys. Lett.*, **102**, 45 (1983); *Chem. Abs.*, **100**, 50888 (1984).
[356] Slagle, I. R., and Gutman, D., *Energy Res. Abstr.*, **8**, Abstr. no. 39159 (1983); *Chem. Abs.*, **100**, 50776 (1984).
[357] Park, J. Y., Slagle, I. R., and Gutman, D., *Energy Res. Abstr.*, **8**, Abs. no. 39158 (1983); *Chem. Abs.*, **100**, 50777 (1984).
[358] Niki, H., Maker, P. S., Savage, C. M., and Breitenbach, L. P., *J. Phys. Chem.*, **87**, 3722 (1983).

359 Lehr, G. F., and Lawler, R. G., *J. Am. Chem. Soc.*, **106,** 4048 (1984).
360 Yoshida, M., Cho, T., and Kobayashi, M., *Chem. Lett.*, **1984,** 1109.
361 Wine, P. H., and Thompson, R. J., *Int. J. Chem. Kinet.*, **16,** 867 (1984).
362 Barker, P. J., and Beckwith, A. L. J., *J. Chem. Soc., Chem. Commun.*, **1984,** 683.
363 Pauzat, F., Gritli, H., Ellinger, Y., and Subra, R., *J. Phys. Chem.*, **88,** 4581 (1984).
364 Abe, K., Suezawa, H., Hirota, M., and Ishii, T., *J. Chem. Soc., Perkin Trans. 2,* **1984,** 29.
365 Grossi, L., Lunazzi, L., and Placucci, G., *J. Chem. Soc., Perkin Trans. 2,* **1983,** 1831.
366 Janzen, E. G., Rehorek, D., and Stronks, H. J., *J. Magn. Res.*, **56,** 174 (1984).
367 Mossoba, M. M., Makino, K., Riesz, P., and Perkins, R. C., *J. Phys. Chem.*, **88,** 4717 (1984).
368 Göbl, M., and Asmus, K.-D., *J. Chem. Soc., Perkin Trans. 2,* **1984,** 691.
369 Alberti, A., Hudson, A., and Pedulli, G. F., *Tetrahedron,* **40,** 4955 (1984).
370 Sagdeev, R. Z., Gogolev, A. Z., Grigor'ev, I. A., Shchukin, G. I., Volodarsky, L. B., Möhl, W., and Möbius, K., *Chem. Phys. Lett.*, **105,** 223 (1984).
371 Volodarsky, L. B., Grigor'ev, I. A., Grogr'eva, L. N., and Kirilyuk, I. A., *Tetrahedron Lett.*, **25,** 5809 (1984).
372 Lane, J., and Tabner, B. J., *J. Chem. Soc., Perkin Trans. 2,* **1984,** 1823.
373 Moriya, F., Makino, K., Iguchi, N., Suzuki, N., Rokushika, S., and Hatano, H., *J. Phys. Chem.*, **88,** 2373 (1984).
374 Iguchi, N., Moriya, F., Makino, K., Rokushika, S., and Hatano, H., *Can. J. Chem.*, **62,** 1722 (1984).
375 Thléry, C. J., Agnel, J. P. L., Fréjaville, C. M., and Raffl, J. J., *J. Phys. Chem.*, **87,** 4485 (1983).
376 Novak, M., and Brodeur, B. A., *J. Org. Chem.*, **49,** 1142 (1984).
377 Rehorek, D., and Janzen, E. G., *Can. J. Chem.*, **62,** 1598 (1984).
378 Rehorek, D., and Janzen, E. G., *J. Organomet. Chem.*, **268,** 135 (1984).
379 Villa, F., Boyer, M., Gronchi, G., Duccini, Y., Santero, O., and Tordo, P., *Tetrahedron Lett.*, **25,** 2215 (1984).
380 Poulin-Dandurand, S., Wertheimer, M. R., and Yelon, A., *Polymer,* **24,** 1581 (1983).
381 Janzen, E. G., and Coulter, G. A., *J. Am. Chem. Soc.*, **106,** 1962 (1984).
382 Ito, O., and Matsuda, M., *Bull. Chem. Soc. Jpn.*, **57,** 1745 (1984).
383 Josephy, P. D., Rehorek, D., and Janzen, E. G., *Tetrahedron Lett.*, **25,** 1685 (1984).
384 Carmichael, A. J., Mossoba, M. M., and Riesz, P., *FEBS Lett.*, **164,** 403 (1983).
385 Czauderna, M., and Wagner-Czauderna, E., *J. Radioanal. Nucl. Chem.*, **81,** 283 (1984); *Chem. Abs.*, **100,** 191278 (1984).
386 Rehorek, D., and Janzen, E. G., *Polyhedron,* **3,** 631 (1984); *Chem. Abs.*, **101,** 130034 (1984).

Organic Reaction Mechanisms 1984
Edited by A. C. Knipe and W. E. Watts

CHAPTER 4

Radical Reactions: Part 2

D. J. Cowley

Department of Chemistry, University of Ulster (Coleraine)

Homolytic Oxidation and Reduction

Photoassisted nitro-oxylation of substituted toluenes by ceric ammonium nitrate in MeCN gives good yields of benzyl nitrates without complicating aromatic nitration products.[1] The relative reactivity of Me and Pr^i groups in the oxidation of *p*-cumene by ceric ammonium nitrate in acetic acid at 80° has been investigated.[2] A radical cation mechanism for the oxidation of 1,2-diphenylethane and of 2,3-dimethyl-2,3-diphenylbutane is plausible for ceric ammonium nitrate but not for cobalt(III) acetate as oxidant.[3]

The balance between side-chain and nuclear substitution in the oxidation of *p*-methoxytoluene by tris(phenanthroline)iron(III) in MeCN depends crucially on the nature of any pyridine bases present, *via* the competition between base-catalysed proton removal from *p*-$MeOC_6H_4CH_3^{+\cdot}$, favoured by, *e.g.*, 2,6-lutidine, and nucleophilic addition of base to the radical cation, favoured by, *e.g.*, pyridine.[4] The kinetics of the oxidation of *para*-substituted phenols with tris (phenanthroline)iron (III) and with $IrCl_6^{2-}$ in aqueous acid media have been used to estimate standard redox potentials for the corresponding radical cations.[5]

The transition state in the oxidation of 2- and 3-methoxy-*N*-methylacridans (DH_2) by π-electron acceptors resembles the acridinium ion product D^+ while with tris(bipyridyl)cobalt(III) it resembles the radical cation $DH^{+\cdot}$.[6] Oxidations of mesidine by $Fe(CN)_6{}^{3-}$ *etc.*, resulting in some unusual migration products,[7] of propan-2-ol and cyclobutanol by sodium ruthenate,[8] and of tertiary amines by lead tetraacetate[9] have been reported.

Developments in organic synthesis by electrolysis have been reviewed.[10] Anodic oxidation of trialkylboranes R_3B complexed with nucleophiles (*e.g.* pyridine, SCN^-, OH^-, THF) can provide better yields of dimers R—R than Kolbé processes but coupling of $R^1\cdot$ and $R^2\cdot$ radicals is statistical and thereby unselective for unsymmetrical products.[11] A correlation of standard oxidation potentials for CF_3COOH solution and gas-phase vertical ionization potentials of alkylbenzenes and their π-complexes with $Cr(CO)_3$ holds good.[12] The dication radicals formed initially from oxidation of protonated methyl-substituted anthracenes and naphthalenes in FSO_3H are unstable to proton loss and further oxidation.[13] The dication radicals on anodic oxidation of 1,*n*-bis(3,4-dimethoxyphenyl)alkanes, where $n = 6$, 8, 10, 11, and 12, undergo intermolecular cyclization to yield medium- and large-ring compounds. When $n = 11$ and 12, some intramolecular cyclisation also occurs.[14] Anodic oxidation of aldehyde enol ethers or enamines in methanol provides a useful first step in the synthesis of octaalkylporphyrins;[15] a later step involves addition of CN^- to a porphyrin radical cation. Regeneration of carbonyl compounds by carbon–sulphur bond-cleavage following one-electron oxidation of dithioacetals and 1,3-dithianes (*e.g.* Scheme 1) occurs as a convenient deprotection method.[16]

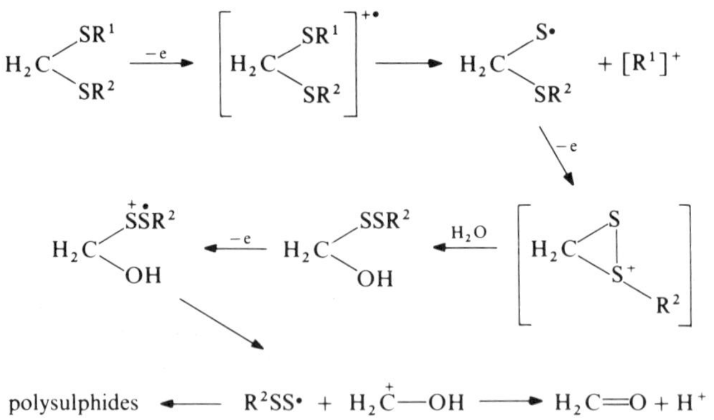

$H_2C(SR^1)(SR^2)$ $\xrightarrow{-e}$ $[H_2C(SR^1)(SR^2)]^{+\cdot}$ $\longrightarrow$ $H_2C(S\cdot)(SR^2)$ + $[R^1]^+$

$H_2C(S\cdot)(SR^2)$ $\xrightarrow{-e}$ $[H_2C(S)(S^+R^2)]$ (cyclic) $\xrightarrow{H_2O}$ $H_2C(SSR^2)(OH)$ $\xrightarrow{-e}$ $H_2C(\overset{+\cdot}{S}SR^2)(OH)$

polysulphides $\longleftarrow$ $R^2SS\cdot + H_2\overset{+}{C}—OH$ $\longrightarrow$ $H_2C{=}O + H^+$

SCHEME 1

Comparison of hydroxyl radical with ferryl ion, FeO^{2+}, allows a value of 1.77 V to deduced for the redox couple, $\cdot OH_{aq}/OH_{aq}{}^-$, [17] much higher than earlier estimates of 1.40 V. Radical(**1**), formed by addition of $\cdot$OH radical to 3-phenylpropanol, undergoes oxidation of the side-chain terminus to yield (**2**) or loss of a benzylic hydrogen to give (**3**), depending on the pH. The ESR signal of (**4**) is replaced by that of (**5**) below pH 3 which in turn is replaced by that of (**6**) below pH 1.8.[18]

$-(CH_2)_3OH$ (1) $-CH_2CH_2\dot{C}HOH$ (2) $-\dot{C}HCH_2CH_2OH$ (3)

HO, H $-(CH_2)_nCH_2CO_2H$ (4) $-(CH_2)_nCH_2\cdot$ (5) $-\dot{C}H(CH_2)_nCO_2H$ (6)

$n = 1, 2$

Oxidation of anisole by H_2O_2/Fe^{2+} or with $S_2O_8{}^{2-}/Fe^{2+}$ yields phenol *via* the anisole radical cation generated by two different reaction paths.[19] Mono- and di-substituted alkenes yield *trans*-vicinal-diacetates on oxidation with $S_2O_8{}^{2-}/Fe^{2+}$ in acetic acid solution[20] by the sequence

$$SO_4{}^{\overline{\cdot}} + RCH{=}CH_2 \rightarrow RCHCH_2OSO_3{}^{\overline{\cdot}} \rightarrow [RCH{=}CH_2]^{+\cdot} + SO_4{}^{2-}$$

$$[RCH{=}CH_2]^{+\cdot} + HOAc \rightarrow R\dot{C}HCH_2OAc + H^+$$

$$R\dot{C}HCH_2OAc + SO_4{}^{\overline{\cdot}} \rightarrow RCH(OSO_3{}^-)CH_2OAc \rightarrow etc.$$

In MeCN the oxidation of alkyl-substituted toluenes by $S_2O_8{}^{2-}/Cu^{2+}$ to benzaldehydes occurs, while in AcOH benzyl acetates are obtained.[21,22] Oxidation of 1-arylethanols by $S_2O_8{}^{2-}/Cu^{2+}$ occurs mainly to ketones *via* loss of a benzylic proton in the aromatic radical cation, even when Ar = p-$O_2NC_6H_4$; lower ease of oxidation to the radical cation is compensated by more rapid proton loss from the radical cation. Carbon–carbon bond-cleavage is favoured in the oxidation of 2-phenylethanol, however, since the stable benzylic radicals result.[23] The first-order reaction of phenylacetic acid with persulphate has an activation energy of 25 kcal mol^{-1}.[24] 18-Crown-6 ether greatly accelerates the decomposition of persulphate due, in part, to enhanced reactivity of a potassium-complexed crown ether radical towards $S_2O_8{}^{2-}$.[25] Simple oxidation of aliphatic sulphides by $SO_4{}^{\overline{\cdot}}$ is found but oxidation of sulphoxides is more complex, *e.g.*:

$$Me_2S{=}O + SO_4{}^{\overline{\cdot}} \rightarrow Me_2S(O)^{+\cdot}$$

$$Me_2S(O)^{+\cdot} + H_2O \rightarrow Me_2\dot{S}(O)OH + H^+$$

$$Me_2\dot{S}(O)OH \rightarrow MeSO_2H + Me\cdot$$

$$MeSO_2H + Me\cdot \rightarrow MeH + MeSO_2\cdot$$

Persulphate ion acts as an oxidant towards unsubstituted (Me·, $k = 3.3 \times 10^4\ \text{M}^{-1}\,\text{s}^{-1}$) and oxygen-substituted (·CH_2OH, $k = 1.3 \times 10^5\ \text{M}^{-1}\,\text{s}^{-1}$) alkyl radicals.[26]

Oxidation of benzyl radicals, generated from attack of $SO_4{}^{\overline{\cdot}}$ on toluene, by nitrobenzene occurs *via* an alkoxynitroxide adduct radical.[27] A formally non-oxidative fluorine/acetoxy exchange in 1- and 2-fluoronaphthalenes is induced by electron-transfer chain catalysis by an $S_{\mathrm{ON}}2$ mechanism involving one-electron oxidants such as $S_2O_8{}^{2-}$ in the presence of acetic acid.[28] Oxidation during bromination of triphenylamines[29] and oxidative dimerization of *o*-alkylphenols with iodine in alkaline methanol[30] have been examined. Molecular iodine or bromine oxidation of heptafulvenes leads to dimerization *via* radical cation intermediates.[31] Hydrogen-atom transfer from tin formates to aldehydes is assisted by butan-1-ol[32] but reduction is slower with even reactive ketones.[33] Reductive elimination of nitro groups from 9,9′-dinitro-9,9′-bifluorenyl by Sn(II)Cl_2 in DMF at 20° gives 9,9′-bifluorenylidene by an $E_{\mathrm{RC}}1$ mechanism.[34] Dimerization following one-election reduction at a mercury electrode occurs irreversibly and reversibly for 2,4,6-trimethyl- and 2,4,6-tri-*tert*-butyl-pyrylium salts, respectively.[35]

Radical Ions

Radical Cation Structure and Properties

ESR studies of isotopically substituted ethene radical cation at 77 K indicate a twist of $45 \pm 5°$ about the central bond.[36] This is also indicated from the ESR spectra of $Me_2C{=}CHMe$ and $MeCH{=}CH_2$ radical cations,[37] and from the free rotation of the methyl group at 77 K in the latter species.[38] Calculated barriers to rotation about the C=C bond in ethene radical cations range from 35 kJ mol^{-1} for 1,1-dimethoxy substitution to 230 kJ mol^{-1} for 1,2-dimethoxy substitution.[39]

Cobalt-60 γ-irradiation of substrates in $CFCl_3$ and related matrices at 77 K has been exploited for structural studies on the corrsponding radical cations of haloalkanes,[40] benzyl chloride and bromide,[41] 2-cyano-2-methylpropane,[42] methyl- and silyl-benzenes,[43] substituted pyridines,[44,45] acyclic and cyclic ethers and acetals,[46] acetaldehyde and aliphatic ketones,[47–49] methyl and ethyl formates and acetates,[50,51] tertiary aliphatic amines and 1,1-diamines,[52] dialkyl sulphides and disulphides,[53] and trimethyl phosphate.[54] Attention has been focussed on the interaction of a solvent chlorine atom with the radical cation "hole",[40,42,47–49,51,54] the π- or σ-character of the singly occupied MO in the radical cation,[42,44–46,48,51,53] and on large ESR hyperfine splittings arising from conformational constraints.[41,43,46,48,50,51] Of mechanistic interest are the rearrangements of $[(RO)_2CH{-}CH{=}CH_2]^{+\cdot}$ to give $(RO)_2\overset{+}{C}{-}\dot{C}HMe$[46] and of $CH_3OP(OCH_3)_2O\cdot$ to give $\cdot CH_2OP(OCH_3)_2OH$,[54] and the observation at 77 K of the unimolecular breakdown products of *tert*-butyl and isopropyl acetate radical cations, $[H_2C{=}CMe_2]^{+\cdot}$ and $[H_2C{=}CHMe]^{+\cdot}$, respectively.[51]

Both theoretical and experimental work on novel isomeric forms of small radical cations, in the gas phase mainly, continues apace. The structures and chemistry of radical-ion dipole complexes $Y_2CXH^{+\cdot}$ have been reviewed.[55] ESR evidence,[56] supported by calculations,[57] indicates a ring-opened symmetrical C_{2v} structure for ethylene oxide radical cation. Both $^2B_1$ and $^2A_1$ states of oxirane radical cation are predicted to have very small barriers to ring-opening.[58]

The $[CH_3COH]^{+\cdot}$ species is predicted to be a stable isomer but is also a key intermediate in the isomerization of vinyl alcohol radical cation to $[CH_3CHO]^{+\cdot}$.[59] The dissociation energies of the complexes $[CH_2{=}CHOH]^{+\cdot}CH_3OH$ and $[CH_2{=}CHOH]^{+\cdot}$, H_2O in the gas phase are 21 and 13 kcal mol^{-1}, respectively.[60] The radical cations $[CH_3CHNH_3]^{+\cdot}$ and $[CH_2CH_2NH_3]^{+\cdot}$ appear to exist as distinct stable species in the gas phase[61] while *ab initio* calcultions indicate that the isomers of $C_2NH_5^{+\cdot}$ are similar in relative energies to those of $C_2H_4O^{+\cdot}$.[62] A *m*-xylene radical cation bromine atom pair, pertinent to aromatic bromination, has been detected in the gas phase.[63] Addition of NO_2 to radical cations of methylbenzenes yield σ-bonded intermediates. Such does not occur with tetrafluorobenzenes, furan, or pyridine and reaction with naphthalene radical cation is surprisingly slow.[64] Carbon-atom scrambling is almost complete in metastable thiophene radical cations; $C_2H_2S^{+\cdot}$ has a thioketene structure.[65]

Methyl-group rotation appears to be unhindered even at $-80°$ in solution in the radical cations $Bu^tMe\overset{+}{C}{-}\dot{C}MeBu^t$ and $Bu_2{}^t\dot{C}{-}\overset{+}{C}Me_2$.[66] Irradiation of σ-bonded aluminium halide–cyclobutadiene complexes (method A) leads to more persistent cyclobutadiene radical cations than photolysis of alkynes with Lewis acids (method B). Thus $Et_4C_4{}^{+\cdot}$ generated at $-55°$ had half-lives of 14 minutes (method A) and 15 seconds (method B). However, tetraneopentyl- and tetra-1-adamantyl-cyclobutadiene radical cations could not be generated by method A.[67] ESR hyperfine splittings (^{13}C) for $Et_4C_4{}^{+\cdot}$ indicate a π-type radical with no out-of-plane or in-plane distortions; *cis*- and *trans*-isomers of $R_2Bu_2{}^tC_4{}^{+\cdot}$ (R = Me or Et) have been identified but with more bulky R groups the alkyl substituents are pushed out of the ring plane.[68] One-electron oxidation of tetra-(*tert*-butyl)tetrahedrane leads to the corresponding cyclobutadiene radical cation, the intermediacy of the *tert*-butyl(tri-*tert*-butylcyclopropenyl)carbene radical cation being feasible on the basis of MNDO calculations.[69] Hexamethylbutadiene radical cation can be generated by a novel molecular rearrangement of $Bu^tC{\equiv}CBu^t$ induced by $AlCl_3$ in CH_2Cl_2.[70]

Pentamethylcyclopentadiene radical cation is generated efficiently in CF_3CO_2H at $-10°$ by photolytic homolysis of pentamethylcyclopentadiene followed by protonation.[71] Biphenyl radical cation, formed from either benzene or biphenyl, appears to be planar.[72] Semi-empirical calculations on the ion radicals from [3.3.3] propellanes anticipate the presence of a σ-(one-electron)-bond linking the central atoms of the radical cation.[73]

Striking changes in the nodal properties of the SOMO of (**7**) systems are caused by changes in the bridge perturbation, *viz.* A = $>CH_2$, $>C{=}O$, or bond.[74] NMR, ESR, ENDOR, and TRIPLE spectroscopy of various chroman radical cations has supplied detailed information on the signs of the hyperfine couplings and on the

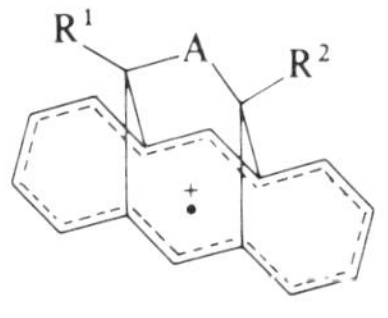

(7) 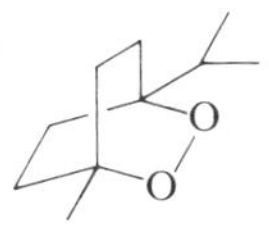 (8) 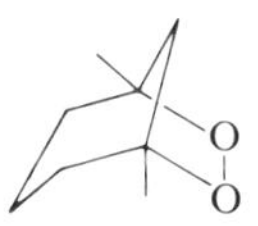 (9)

rates and energetics of electron exchange between the radicals and their substrates.[75]

Peroxides (**8**) and (**9**) are harder to oxidize electrochemically than the corresponding hydrazines, and thus undergo no rehybridization at oxygen of lone-pair orbitals, but produce long-lived radical cations.[76] The optical absorption spectrum of the radical cation of (**10**) indicates stabilization by N . . . N bonding interactions.[77] The pK_a of (**11**), produced by pulse-radiolysis reduction of 2,3-diazabicyclo[2.2.1]hept-2-ene and thus avoiding complications of redox equilibria, is 5.3 as compared to 7.93 for the parent hydrazine. Since the pK_a values of $[Me_2NNH_2]^{+\cdot}$ and $[Me_2NNH_3]^{+\cdot}$ are 7.9 and 7.2, respectively, it is clear that no correlation of parent and radical cation pK_a values can be made.[78] Reduction of 1-alkylpyrazinium iodides in acidic aqueous media yields persistent 4-hydro-1-alkylpyrazinium radical cations (**12**) which have a slight asymmetry of unpaired spin density.[79]

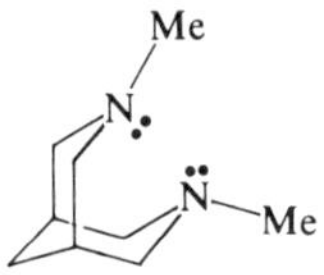

(**10**) 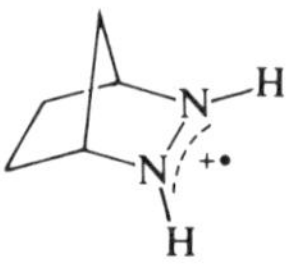 (**11**) 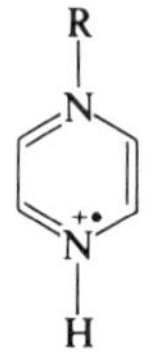 (**12**)

Radical Cation Reactions

Calculations indicate that the formation of 1,2-dioxetane radical cation from $C_2H_4^{+\cdot}$ and O_2 is exothermic[80] by -225.4 kJ mol^{-1} and proceeds *via* a peroxy cation intermediate[81] with an activation enthalpy of 57.7 kJ mol^{-1}. In the gas phase the ion dipole complex $\cdot CH_2—\overset{+}{X}—CH_3$ (X = Cl or Br) is reactive towards nucleophiles and electrophiles. Thus attack by MeCN and NO leads to displacement of CH_3^+ and $CH_2^{+\cdot}$, respectively.[82] In a low-temperature matrix, methyl-substituted propane radical cations photodissociate to olefinic radical cations while methyl-substituted butene radical cations deprotonate at a methyl group to yield primary neutral alkyl radicals.[83] An activation energy of 30 kcal mol^{-1} for the interconversion of the toluene and cycloheptatriene radical cations in the gas phase indicates that, in solution, proton loss will be pre-emptive of any such interconversion.[84] Electronic spectra and photorearrangements of cyclic diene and acyclic triene radical cations of C_8H_8, $C_{10}H_{10}$, and C_8H_{12} hydrocarbons at 20 K have been studied.[85] Activation energies for isomerization *via* ring-opening of gas-phase 1-methyl- and 3-methyl-cyclobutene radical cations are much lower than those of the parent hydrocarbons but substituent effects run parallel.[86] Cycloaddition of ketene radical cation to ethene yields an adduct having a symmetrical cyclobutanone radical cation structure.[87] The barrier to retro-Diels–Alder reaction of 4-vinylcyclohexene radical cation is 1140 kJ mol^{-1}. The calculated reaction path involves sequential rupture of the C(3)—C(4) and C(5)—C(6) bonds.[88] Photolysis of protonated cyclic unsaturated hydrocarbons in CF_3CO_2H solution is a convenient route to radical cations.[71,72,89]

In acidic 70% $MeCN-H_2O$ solution the cleavage reaction

$$[PhMe_2CCMe_2Ph]^{+\cdot} \rightarrow PhCMe_2\cdot + PhCMe_2{}^{+}$$

is faster than the normal benzylic proton loss in related 1,2-diarylethane radical cations. In the latter, the diaryl alcohol products undergo further (radical) oxidation to aldehydes/ketones.[90]

Fluorination of anthracene derivatives with I_2/AgF in MeCN solution occurs by *meso*-substitution of the radical cations in a charge-controlled manner. Phenanthrene, perylene, and pyrene failed to react.[91] Reaction pathways of the radical cations of anthracene and its nitrogen/sulphur analogues have been reviewed.[92] Oxidation of poly(methylthio)- and poly(isopropylthio)-benzenes does not lead in all cases to the corresponding radical cations since, when R = Me, the fragmentation

$$ArSR^{+\cdot} \rightarrow Ar\cdot + RS^{+}$$

occurs. For R = Pr^i, the easiest fragmentation path should be

$$ArSR^{+\cdot} \rightarrow ArS^{+} + R\cdot$$

but this was not observed to occur.[93] The 9-arylfluoren-9-yl cations, $ArFl^+$, generated from the alcohols in CF_3CO_2H solution, are reduced to ArFlH by polymethylated aromatic hydrocarbons, $Ar'CH_3$, with the possible intermediacy of the radical pair ArFl·, $Ar'CH_3{}^{+\cdot}$.[94] A novel photo-induced ring-enlargement reaction of the radical cation of (**13**) occurs in CF_3CO_2H or $HClO_4$ media to give (**14**).[95]

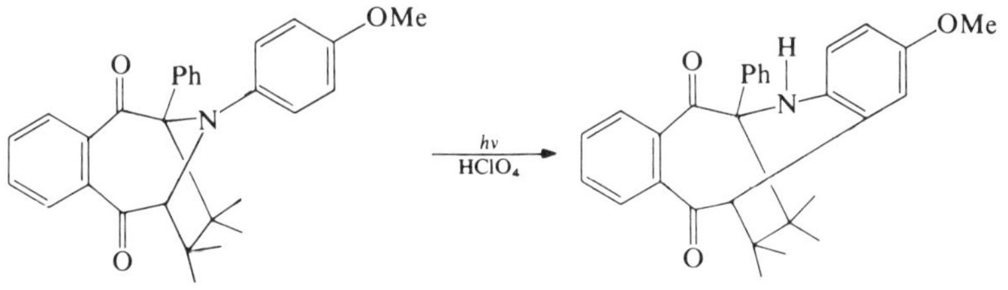

(**13**) —$h\nu$, $HClO_4$→ (**14**)

Captodative stabilization of (**15**) radical cation is found to inhibit dimerization even at room temperature but under kinetic control heterocumulene radical cations like $(\mathbf{16})^{+\cdot}$ are 50% dimerized at −80°.[96]

Photoinduced electron transfer to quinones has been used to generate radical cations for mechanistic and structural studies with extensive use of the information provided by the CIDNP method.[97–102] In tri- and tetra-alkylcyclopropane radical cations the most highly substituted ring bond cleaves or is weakened as judged by CIDNP effects consequent on ion-pair recombination.[97] The radical cation (**17**) is trapped by O_2 more efficiently than cyclic trimethylenemethane triplet biradicals.[98]

(15)

Ar = p-$Me_2NC_6H_4$

(16)

(17)

(18)

(19)

a: $R^1 = R^2 = H$
b: $R^1 = H$, $R^2 = Me$
c: $R^1 = R^2 = Me$

(20)

(21)

Rearrangement of (**18a**), following one-electron oxidation, unexpectedly requires the intermediacy of (**19**) or (**20**) which then undergoes a 1,3-hydrogen shift to give (**21**) and thence norcar-2-ene product. This reaction is suppressed in (**18b**) and (**18c**).[99] Two different radical cation species of hexamethylbenzene, (**22**) and (**23**), are implicated from CIDNP studies and theoretical calculations, (**22**) being 8 kcal mol^{-1} lower in energy than (**23**) and 70 kcal mol^{-1} above $C_6H_6{}^{+}$.[100]

(22) (23)

One-electron oxidation of (**24**), in which the transannular bond is orthogonal to both π-bonds but the 1–6 bond is doubly allylic, yields radical action (**25**) rapidly on

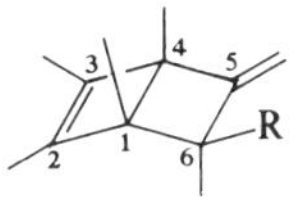

(24) 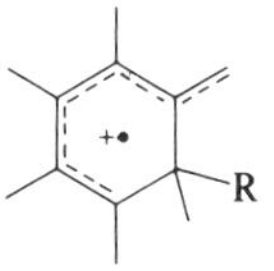 (25) 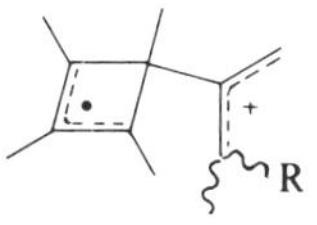 (26)

the CIDNP time-scale when R = Cl, but the doubly allylic radical cation (**26**) when R = OMe.[101]

The cage radical cations (**27**) have shallow energy minima and undergo fast ring-opening to (**28**); hence cyclization of diolefin radical cations does not occur.[102]

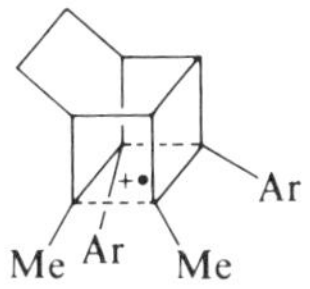

(27) 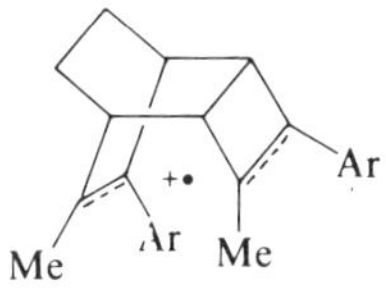 (28)

Ar = Ph, p-MeC_6H_4, p-$MeOC_6H_4$

The path energetics of prototype olefin/olefin radical cation cycloaddition have been evaluated. The long-band complex (**29**) is not readily dissociated (E_a = *ca.* 19.8 kcal mol^{-1}), requires negligible activation for its formation from (**30**), and has a low (*ca.* 7 kcal mol^{-1}) activation barrier to forming the long-band cyclobutane (**31**); (**31**) may form (**32**) if a positive hole-stabilizing substituent is present.[103]

(30) (29) (31) (32)

The "collision-activated decomposition" (CAD) spectrum of the adduct of styrene and styrene radical cation in the gas phase approximates closely to the CAD spectra of *cis*- and *trans*-diphenylcyclobutane radical cations.[104] Photoinduced electron-transfer dimerization of phenyl vinyl ether (PVE) involves equilibration of a PVE, $PVE^{+\cdot}$ adduct with the related *cis*-cyclobutane radical cation; the rate constant for ring-opening is estimated at 1.8×10^6 s^{-1}.[105] Chemoselective cyclobutane ring-formation in preference to Diels–Alder addition is found in the (photo)electron-transfer-catalysed reaction of 1,1-dicyclopentenyl with electron-rich olefins or styrenes.[106]

Hydroxylamine-*O*-sulphonic acid with ferrous ion yields $H_3N^{+\cdot}$ radical cations; attack of $H_3N^{+\cdot}$ on amides generates nucleophilic carbamoyl and α-*N*-amidoalkyl radicals, which selectively add to protonated heteroaromatic bases.[107] α-Deprotonation occurs competitively with intramolecular (1,4)hydrogen-atom abstraction in the $(c\text{-}C_6H_{11})_2NH^{+\cdot}$ radical cation, being more important at lower acidities.[108] 3,4-Dimethoxybenzyl radicals act as chain carriers in the Knabe reaction of 1,2-dihydro-2-methylpapaverine.[109] The oxidative dimerization of a range of substituted diphenylamines has received theoretical study.[110] The kinetics and mechanism of the decomposition of water by triarylamine radical cations,[111] of the reaction of thiourea with $p\text{-}Et_2NC_6H_4Cl^{+\cdot}$ radical cation,[112] and of the formation and decay of hydroquinone radical cations in concentrated sulphuric acid[113] have been investigated. Thianthrene radical cation associates increasingly with FCH_2CO_2H, F_2CHCO_2H, and F_3CCO_2H, in that order.[114] Oxidation of 1,2-dithiete derivatives occurs readily in $AlCl_3/CH_2Cl_2$ mixtures; the resulting radical cations have both spin and positive charge essentially localized on the disulphide bridge of the ring system.[115]

Radical Anion Structure and Properties

The generation, optical, photoelectron, and ESR spectroscopy, and structural relaxations of radical ions have been reviewed,[116] as has the ESR/ENDOR spectroscopy of phane radical ions.[117] Gas-phase electron-capture studies yield standard heats of formation for $(CF_3)C^{\cdot -}$ and $(CF_3)_2\dot{C}H$ of -282.2 and -280.0 kcal mol^{-1}, respectively, and an electron affinity of 50.1 kcal mol^{-1} for $(CF_3)_2\dot{C}H$.[118] The electronic ground states of the radical anions of C_6F_6, C_6HF_5, and 1,2,4,5-tetrafluorobenzene are pseudo-π in character; the C—F bonds are displaced out of the ring plane due to σ-electron-state participation.[119] Electron addition to *p*-nitrobenzyl bromide and chloride at 77 K yields radical anions more stable than those of 2-halo-2-nitropropane.[120] The mechanism of formation of nitrobenzene radical anion in ethanol supersaturated with NaOH has been elucidated.[121] The IR spectra of 38 aromatic nitro radical anions have been surveyed.[122] The generation and ESR characteristics of pentalene radical anion,[123] the aromatic octafluorocyclooctatetraene radical anion,[124] and the [16]annulene radical anion (from [8]annulene radical anion)[125] have been reported.

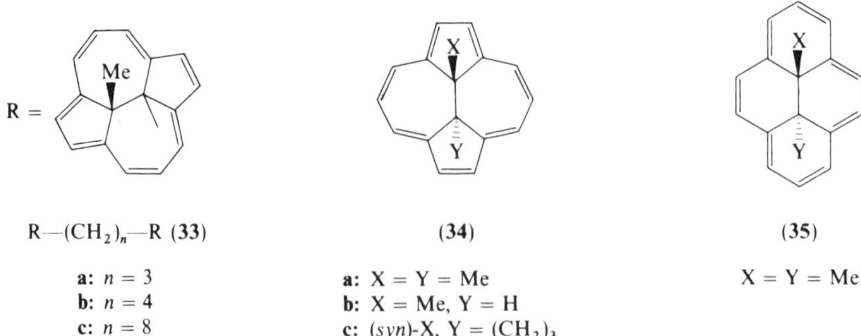

The radical anion of (**33a**) has unpaired spin distributed over both annulene moieties whereas in (**33c**) radical anion spin is localized in one π-system. (**33b**) radical anion is intermediate in character but the formation of tight ion-pairs in MTHF leads to localization.[126]

The anion radicals of (**34a, b, and c**) are akin to those of (**35**),[127] the change in the relative arrangement of the saturated bridge at 90° to, and within, the 14-membered π-electron perimeter leaving the sequence of SOMOs and LUMOs unaltered. The ease of formation and spin density distributions in 1,2-bis(phenalen-1-ylidene)-ethane and -ethene radical anions and cations, and related systems, have been linked to the correlation of the frontier orbitals with the non-bonding π-orbitals of two phenalenyl systems.[128]

A new spin-transfer mechanism to account for paramagnetic solvent NMR shifts in solutions of aromatic hydrocarbon and ketone radical anions has been proposed.[129] The ESR spectrum of cycloheptatrienyl radical dianion in hexane containing Bu^nLi and potassium *tert*-amyloxide is characteristic of K^+ ion association.[130] Good correlations exist between solvation enthalpies, crystal lattice energies, and total π-electron energies of polyacene radical anions.[131] The thermodynamics of the association of $PhCHO^{\overset{\cdot}{-}}$ with alcohols in DMF,[132] and of $PhCOMe^{\overset{\cdot}{-}}$ with water,[133] have been elucidated.

Radical Anion Reactions

The kinetics of recombination of (*inter alia*) radical ions in the liquid phase have been reviewed.[134] Disproportionation of $Ph_2CO^{\overset{\cdot}{-}}$ radical anions is activation-energy-controlled, but electron transfer to nitroxide radicals is nearly diffusion-controlled.[135] Reaction of *cis*- or *trans*-1,2-diphenylcyclopropane with Na/K in THF yields after protonation 1,3-diphenylpropane and 1,3-diphenylpropene and involves the ring-opened radical anion but no intramolecular 1,2-hydrogen migration.[136] Electrochemical reduction of azibenzil, $PhC(O)C(N_2)Ph$, to the radical anion is followed by a slow loss of N_2 to form the very reactive carbene radical anion, $Ph\dot{C}(O^-)CPh$.[137] Radical anion (**36**), formed on sodium reduction of the parent tetraene, disproportionates to give (**37**) and possibly the acepentalenediide (**38**).[139] Radical anions prepared by potassium reduction in DME of Ph_3SiE (E = chloro, cyclopropyl, vinyl, 1-propynyl, *etc.*) are more labile than those from methylphenylsilanes. Coupling and C–Si cleavage reactions occur at sites of highest unpaired spin density predictable from ESR data.[139]

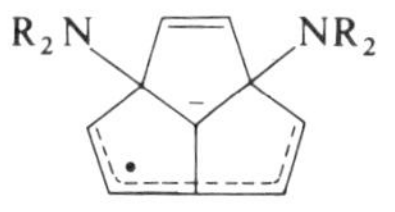

R_2N NR_2 — (**36**)

R_2N NR_2 NR_2 — (**37**)

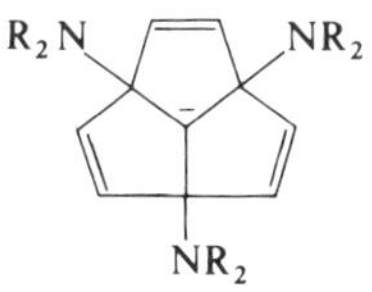

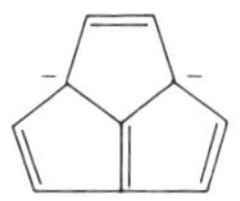

(**38**)

Halogen-cleavage reactions in radical anions have received continued attention. The relationship between reduction potentials and aromatic radical anion cleavage rates has been surveyed.[140] For fragmentation to occur in a π^*-aromatic radical anion, the unpaired electron has to be transferred to a σ^*-orbital by an orbital crossing induced by stretching the nucleofugal-group–aryl carbon bond.[141] In condensed phases, the state of $CH_3Cl^{\overset{\cdot}{-}}$ is purely dissociative.[142] The fluorocyclooctatetraene radical anion and dianion are unusually stable towards F^- elimination;[143] this is attributed to the non-bonding nature of the LUMO and to the substantial resonance energy of the planar anions. The radical anion of *p*-$IC_6H_4NO_2$ is stabilized at a mercury cathode but decomposes to iodine atom and $PhNO_2$ with the consumption of two electrons *per* $PhNO_2$ formed.[144] α-Substitution in nitrobenzyl halide radical anions generally increases halide fragmentation rates. However, Bu^t-substitution decreases the rate by enforcing rotation of the carbon–halogen bond towards the aromatic ring plane, thus reducing the π–σ overlap on which the electron-transfer rate depends.[145] Reactions of $ArCH_2^{\cdot}$ radicals, following fragmentation of the radical anions $ArCH_2X^{\cdot -}$, have been studied where Ar = 5-nitro-2-furfuryl and X = Br, SO_2Ph, SO_2CHCl_2, *p*-$CH_3C_6H_4SO_2$, and ONO_2.[146] A series of 2-nitrofuran radical anions has been analysed for spin distribution and transmission of substituent effects, by means of INDO calculations.[147]

Reduction of 9-chloro-9-[α-(9-fluorenylidene)benzyl]fluorene and 9-chloro-9-mesitylfluorene, RCl, in MeCN is catalysed by ferrocene and reversible organic redox couples P/Q,[148] reduction occurring up to 2 V ahead of the direct electrochemical route, thus:

$$P + e \rightleftharpoons Q; \quad RCl \rightleftharpoons R^+ + Cl^- \text{ (rate-determining step)}$$

$$R^+ + Q \rightarrow R\cdot + P \text{ (fast, irreversible step)}$$

Assessment of aprotic solvent effects on activation parameters of halide ion cleavage of 9-cyano-10-haloanthracene radical anions leads to the conclusion that unimolecular cleavage of the chloro radical anion is a minor pathway; instead, a higher order of reaction leads to 9,9′-bianthryl-10,10′-dicarbonitrile.[149]

Loss of $PhSO_2^-$ from the radical anion of *o*-bis(phenylsulphonyl)benzene is faster than protonation in Me_2SO,[150] the resulting aryl radical undergoing intramolecular arylation or hydrogen abstraction from PhSH or solvent.

The reactivity of substituted anthraquinone radical anions towards O_2 has been examined.[151] The general reactions and activation of $O_2^{\cdot -}$ by transformation in the presence of H^+ or electrophiles have been surveyed.[152] With regard to hydrogen abstraction, $O_2^{\cdot -}$ is reckoned to be as unreactive as iodine atoms.[153] Oxidation of tetralin to tetralone by KO_2 (in *n*-hexane containing 18-crown-6 at −10°) only occurs in the presence of "effectors" such as $COCl_2$. The key oxidant in this case is the acyl peroxyl radical ·OOC(=O)OO· formed by reaction of $COCl_2$ and $O_2^{\cdot -}$.[154] Oxidation of either 1,2- or 1,3-dihydroxynaphthalenes with KO_2 in aprotic media takes place heterogeneously and yields 2-hydroxy-1,4-naphthoquinone.[155] The same product is obtained in reasonable yield by similar oxidation of α- and β-tetralones.[156] Cleavage of the disulphide bond in (DNP)SSR compounds by $O_2^{\cdot -}$ in aprotic media (DNP = 2,4-dinitrophenyl and R = alkyl or aryl) is an S_N2-like process with some one-electron transfer character; the Hammett ρ value is +1.07. In aqueous media, hydroxyl radicals are the active cleaving agents.[157] Facile conversion of 1,3-disubstituted thioureas to the corresponding ureas occurs by reaction with $O_2^{\cdot -}$ in DMSO.[158]

Electron-transfer Reactions

Despite severe steric restrictions, spin exchange in the radical cation of **(39)** is high and must occur by a charge-exchange effect since spin exchange in the corresponding diradical of **(39)** is absent.[159] The electron-transfer reaction between polymer **(40)** and $HOCH_2\dot{C}HNH_2$ has been monitored by ESR.[160] Rate constants for electron transfer from $Ar_2\dot{C}OH$ radicals to paraquat dication correlate well with the electron-releasing nature of the aryl ring substituents.[161] Rates of electron transfer from α-aminoalkyl radicals to paraquat dication in wet MeCN and in CD_3OD are found to be near diffusion-controlled.[162]

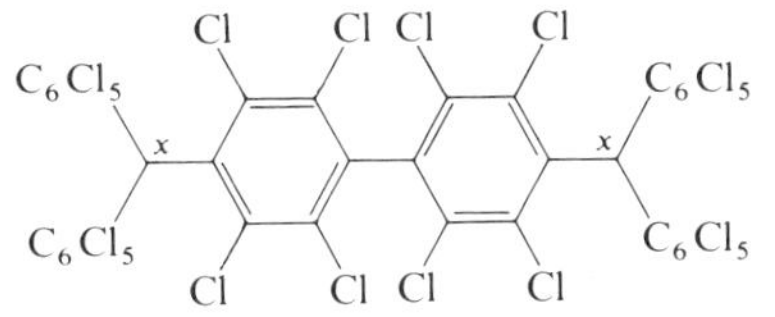

(39)

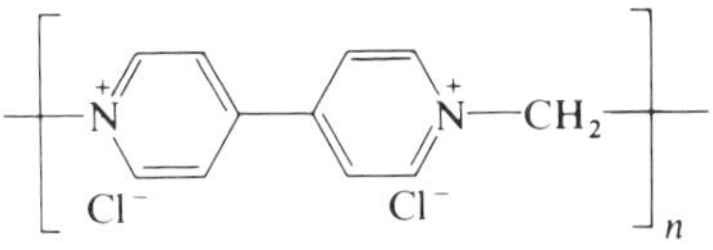

i: $x = \frac{1}{2}\bullet, \frac{1}{2}+$ (radical cation)
ii: $x = \bullet$ (diradical)

(40)

Rates of intramolecular long-distance (1 nm) electron transfer in radical anions $A—Sp—B^{\overline{\cdot}}$, where Sp is a 5α-androstane rigid spacer and A and B are aromatic hydrocarbon systems, have revealed for the first time the existence of the "inverted region" of the log *k* *vs.* reaction free-energy correlation for electron-transfer reactions in solution.[163] Pulse radiolytic study of the intermolecular reaction, $D^{\overline{\cdot}} + A \rightarrow D + A^{\overline{\cdot}}$, indicates that the maximum rate for optimally exothermic reactions between species in contact at 296 K is $10^{13.3}\ s^{-1}$, and thus also the conclusion that fluid-phase studies of intermolecular electron transfer can never provide evidence of the existence of the "inverted region" of Marcus theory.[164] Intramolecular electron-transfer rates between naphthalene systems separated by 1,3- or 5-spiro-bonded cyclobutane rings have been deduced from ESR and ENDOR studies of the radical anions.[165] Solvent effects on the self-exchange of PhCN and $PhCN^{\overline{\cdot}}$ were well reproduced by calculations.[166] The rate of electron transfer from biphenyl radical anion to pyrene is reduced in the presence of $Bu_4N^+.PF_6^-$ in THF.[167] Solvent effects on the rates of electron transfer involving substituted *p*-phenylenediamine, quinone diimines, and semiquinone diimines in aprotic media accorded well with Marcus theory predictions. Deviations from the expected energy surface were accounted for by a temperature-dependent reaction distance and a preceding association of counter-ions with the radical cations.[168] Comparison of rates for attack of $Cl_2^{\overline{\cdot}}$ and of ·OH on substituted pyrimidines suggests different modes of reaction.[169] A reduction potential of 240 mV (*vs.* NHE) for 5-aminophthalazine-1,4-dione has been deduced from a study of the equilibration of luminol radical and O_2.[170]

$Ar_3N^{+\cdot}$ radicals are useful for the mild and selective removal of dithioketal protecting groups.[171] Verdazyl radicals complex with Hg(II)Cl_2 before undergoing electron transfer with a second verdazyl radical.[172] Reaction rates with tropylium tetrafluoroborate increase with increasing electron-donor ability of verdazyl radicals and of tetracyanoquinodimethane and *p*-chloranil radical anions.[173] The onium systems Ph_2I^+, 4-$CH_3C_6H_4I^+C_3F_7$, $Me_2S^+CF_3$, $PhS^+(CF_3)Me$, *etc.*, are much weaker electron acceptors than the corresponding diazonium systems towards reduction by verdazyl radicals or semiquinone radical anions.[174]

Contrary to many previous misidentifications of persistent radicals in such systems, a direct simultaneous observation of a radical anion–radical cation pair arising from a SET reaction of (**41**) and TCNE or TCNQ has been made.[175]

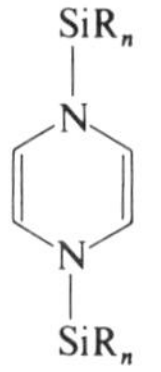

SiR$_n$ / N / N / SiR$_n$

(**41**)

R$_n$ = Me$_3$, Et$_3$, Pr$^i_3$, Me$_2$Bu^t

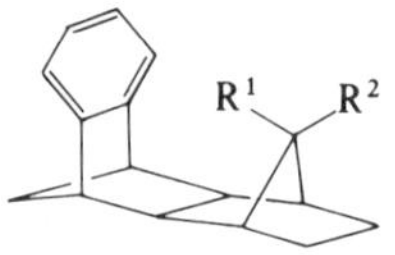

(**42**)

a: R^1 = H, R^2 = H
b: R^1 = OH, R^2 = H
c: R^1 = OMe, R^2 = H

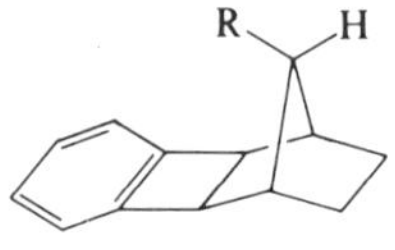

(**43**)

a: R = H
b: R = OH

(**44**)

a: R^1 = OH, R^2 = H
b: R^1 = H, R^2 = OH

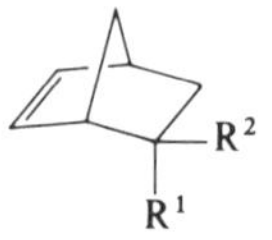

(**45**)

a: R^1 = OH, R^2 = H
b: R^1 = H, R^2 = OH

Hydrolysis of both *N*-vinylphenothiazines and 9-vinylcarbazole in the presence of organic electron acceptors involves a SET step.[176] Chelation of 1,5-dihydroflavin with divalent metal ions, especially Ni^{2+} and Co^{2+}, greatly enhances the rate of electron transfer to *p*-chloronitrobenzene in EtOH, probably by a flattening of the flavin ring system.[177] A low deuterium k.i. e. of 2.33 is found for the reduction of *m*-nitro-α,α,α-trifluoroacetophenone by 1-(3-phenylpropyl)- and 1-(2-phenylthioethyl)-1,4-dihydronicotinamide in the presence of magnesium ions. This is not the case for the 1-(4-phenylthiobutyl) derivative and illustrates the proximity effect of a phenyl group on the electron-transfer process.[178]

The spatially proximate OH groups in the *syn*-alcohols (**42b**) and (**43b**) cause a marked enhancement of the rate of Birch reduction, while in (**44a**) and (**45a**) a moderate acceleration occurs. A second-order kinetics contribution is attributed to intramolecular protonation of the radical anion intermediate, and is supported by *ab initio* calculations.[179]

The diene (**46**) undergoes Birch reduction 2000 times more rapidly than does norbornene and 20,000 times faster than does (**47**) via π^*-orbital interactions. Smaller rate enhancements are found for (**48**) where the interactions occur through four bonds.[180]

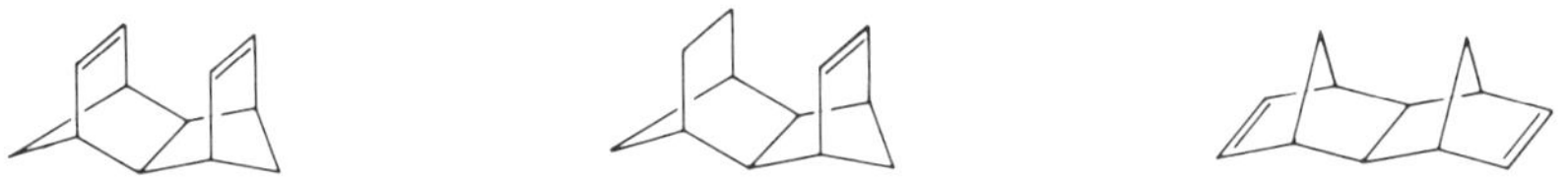

(**46**) (**47**) (**48**)

Alkali metal reduction of oxiranes such as (**49**) involves deoxygenation to olefin in the case of lithium, but with sodium the key intermediate (**50**) leads especially to dimeric alcohol products (**51**) when R = Ar.[181] The action of sodium metal on carboxylic esters in the presence of HMPA provides a simple and efficient method to transform alcohols to alkanes. The intermediate ester radical anion is stabilized by HMPA and Bu^{t}OH.[182]

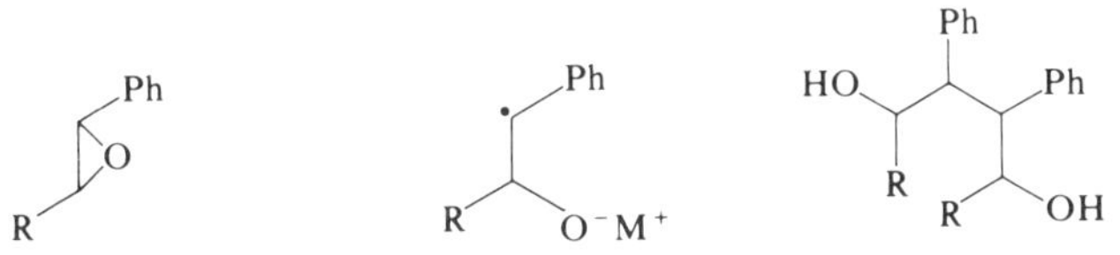

(**49**) (**50**) (**51**)

A SET mechanism prevails over an S_N2 mechanism in the reaction of *N*-benzoylaziridines with sodium metal or naphthalene radical anion, as in Scheme 2.[183] The significance of SET reactions in organic chemistry has been reviewed.[184,185] Other reviews on electron-transfer reactions have surveyed the activation and propagation of chain-reactions,[186] the donor abilities of anionic and carbanionic nucleophiles[187] (concluding that electron transfer from alkoxide ions to nitroaromatics, ketones, halides, *etc*., is not feasible), transition states involving Grignard reagents,[188] dehydroaromatization of heterocyclic compounds,[189] nitro compounds,[190] and (*inter alia*) the copper-assisted nucleophilic substitution of aryl halogen.[191]

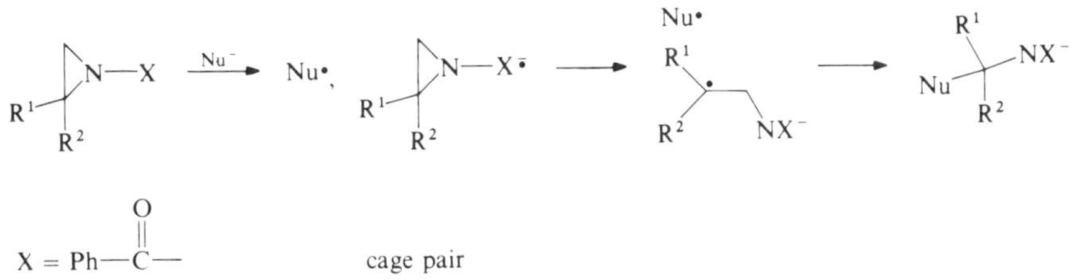

X = Ph—C(=O)—

cage pair

SCHEME 2

Preparation of α-nitrosulphides (*e.g.* $Me_2C(SR)NO_2$; R = 2-pyridyl, 1,3-benzothiazol-2-yl, 1-methylimidazol-2-yl, *etc.*) occurs only slowly by $S_{RN}1$ reaction of RS^- and 2-substituted 2-nitropropanes; the oxidative addition route thus:[192]

$$Me_2CNO_2^- + Fe(CN)_6^{3-} \rightarrow Me_2CNO_2^{\cdot} + Fe(CN)_6^{4-}$$

$$Me_2CNO_2^{\cdot} + RS^- \rightarrow Me_2C(SR)NO_2^{\cdot-}$$

$$Me_2C(SR)NO_2^{\cdot-} + Fe(CN)_6^{3-} \rightarrow Me_2C(SR)NO_2 + Fe(CN)_6^{4-}$$

is preferable.

α-Nitronitriles can be transformed efficiently to β-nitronitriles by an electron-transfer chain-reaction involving nitro-paraffin anions.[193] No monosubstitution products are detectable in the $S_{RN}1$ chain-reaction of 7,7-dibromobicyclo-[4.1.0]heptane with thiophenylate or pinacolone enolate anions.[194] The electrochemically initiated $S_{RN}1$ reaction of PhBr with $Bu^n_4N^+$ ^-SPH in DMSO yields 67% Ph_2S, the addition of PhCN suppressing the competitive cleavage of the $Ph_2S^{\cdot-}$ radical anion intermediate.[195] The regiochemistry of the $S_{RN}1$ photoarylation of 2- and 3-phenylindenyl anions with PhBr in DMSO has been examined.[196]

Evidence for a SET-mediated metal–halogen interchange in the reaction of Bu^tLi with α,ω-dibromo- and -bromoiodo-alkanes has been obtained. The life-times of the alkyl radical,halide ion-pairs have been probed and the departure rate of the halide ion has been shown to affect the product partitioning since the metal–halogen interchange shares a common mechanism with Wurtz coupling.[197] CIDNP effects are observable in the reaction of vinylmagnesium bromide with Ph_3CCl and Ph_2CHCl but not with $PhCH_2Cl$.[198] *m*-Alkylphenol products are formed by SET processes in the reaction of *halogenated* orthoquinol acetates with RMgBr or R_2Mg (R = Pr^i).[199] Alkylation/desulphonylation of $PhC{\equiv}CSO_2Ph$ in THF at 0° with the SET probe, 5-hexenylmagnesium chloride, has confirmed a SET mechanism which also appears to hold for vinylic sulphones with allylic, benzylic, or *tert*-alkyllithium reagents.[200] Application of Marcus electron-transfer theory, using redox potentials, indicates a high SET rate constant of 2×10^5 M^{-1} s^{-1} for reaction of 5-hexenylmagnesium bromide with 2-phenyl-3-(phenylimino)-3*H*-indole, in accord with experimental estimates.[201]

A comprehensive investigation of the occurrence of electron transfer in the reduction of alkyl halides by $LiAlH_4$ and by AlH_3 has been made utilizing the cyclization of 5-hexenyl radicals as a clock, the trapping of radicals by dicyclohexylphosphine, the direct ESR determination of trityl radicals from reduction of trityl bromide, and stereochemical studies with $LiAlD_4$.[202]

Persistent binuclear radical anions result from SET reactions of AlH_3 with pyrazine, quinoxaline, phenazine, and 4,4′-bipyridyl, of the form (heterocycle) $(AlH_3)_2^{\cdot-}$ but 2,2′-bipyridyl yields the radical anion complexed with AlH_2^+.[203] Pyrazine with lithium triethylhydridoborate gives the persistent radical anion (pyrazine) $(BEt_3)_2^{\cdot-}$, boron h.f.s. being detected in the ESR spectrum for the first time.[204]

Stereochemical evidence exists for a SET mechanism in the reduction of methylcyclohexanones, camphor, and menthone by alkoxyaluminium dichlorides, $ROAlCl_2$, rather than a polar cyclic mechanism.[205]

The reagent *N*-lithio-*N*-butyl-5-methyl-1-hex-4-enamine (**52**), typical of lithium diamides, has been evaluated as a diagnostic probe of electron transfers *via* the cyclization of radical (**53**) to give (**54**) and other products.[206] Similarily *N*-lithio-*N*-propylcyclobutylamine has been tested, the ring-opening of the *N*-propylcyclobutylamine radical being the crucial process.[207] A new synthesis of chiral nitroxides involves SET between lithium amides and molecular oxygen.[208]

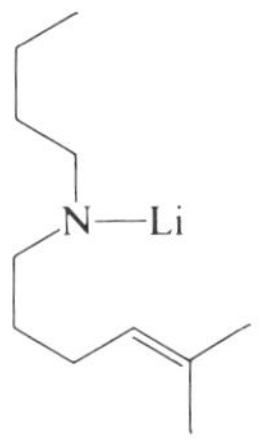

N—Li N• N

(52) 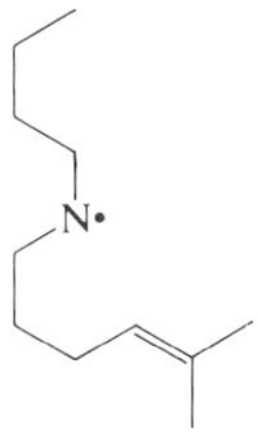 (53) 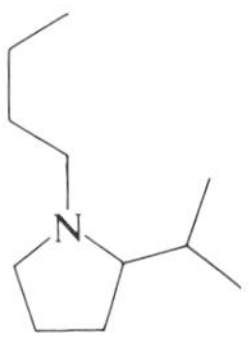 (54)

Reaction of the lithium enolate of propiophenone with primary alkyl iodide SET probes has been shown to involve SET; the corresponding bromides and tosylates are unreactive.[209] The slow SET reaction of $Pr^iO^-Li^+$ with 2,2-dimethyl-1-iodo-5-hexene in HMPA gives cyclic hydrocarbon products even in the presence of dicyclohexylphosphine radical trap; the latter actually enhances the rate *via* an $S_{RN}1$ chain sequence; $Bu^tO^-Li^+$ acted in similar fashion but no substitution products (Bu^tO for I) were found.[210] Both intra- and inter-molecular electron transfer is suspected in the indirect cyclization of the alkoxyallene (**55**) induced by Bu^tO^- in DMSO.[211]

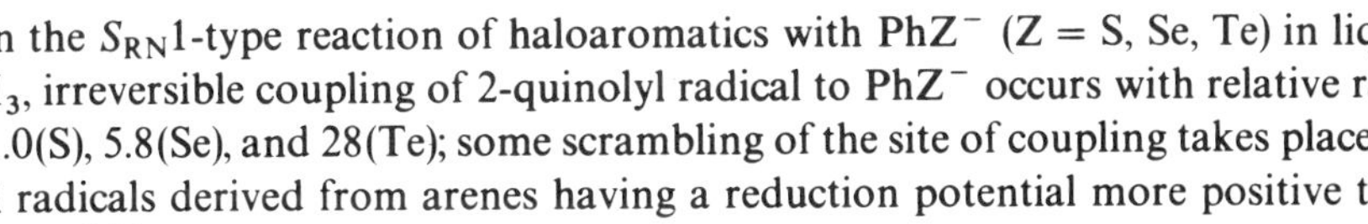

R^1 OH R^2 OMe $\xrightarrow[\text{DMSO}]{Bu^tO^-}$ R^1 OH R^2 OMe $\xrightarrow[\text{abstraction}]{H}$ R^1 O• R^2 OMe $\longrightarrow$ R^1 O R^2 OMe

(55) DMSO•

In the $S_{RN}1$-type reaction of haloaromatics with PhZ^- (Z = S, Se, Te) in liquid NH_3, irreversible coupling of 2-quinolyl radical to PhZ^- occurs with relative rates of 1.0(S), 5.8(Se), and 28(Te); some scrambling of the site of coupling takes place for aryl radicals derived from arenes having a reduction potential more positive than −2.6 V.[212]

Photoinduced Electron Transfers

This expanding area has been reviewed in several aspects, including photoinduced electron-transfer catalysis,[213] organic heterogeneous photocatalysis,[214] and electron-transfer-induced photofragmentation.[215] Theoretical aspects discussed include endothermic electron transfers and the use of More O'Ferrall diagrams,[216] solvent reorganization energy effects in the Marcus inverted region,[217] and rate constants of highly exothermic electron transfers in solution for aromatic systems.[218]

A radical ion-pair is formed predominantly on irradiation of cyclohex-3-enone with 1,4-diazabicyclo[2.2.2]octane.[219] Increasing solvent viscosity by two orders of magnitude caused only small increase and decrease, respectively, in the rates of ion-pair recombination and dissociation for the pyrene-*N*,*N*-dimethylaniline system.[220] Amine dimer cation radical formation has been detected when *trans*-stilbene is quenched by short-chain α,ω-alkanediamines; *i.e.* the initial exciplex yields a "triplex".[221] Inter- and intra-molecular photo-charge-transfer to radical cations of alkyl esters of anthracene-2-carboxylic acid has been studied using ns laser photolysis techniques.[222]

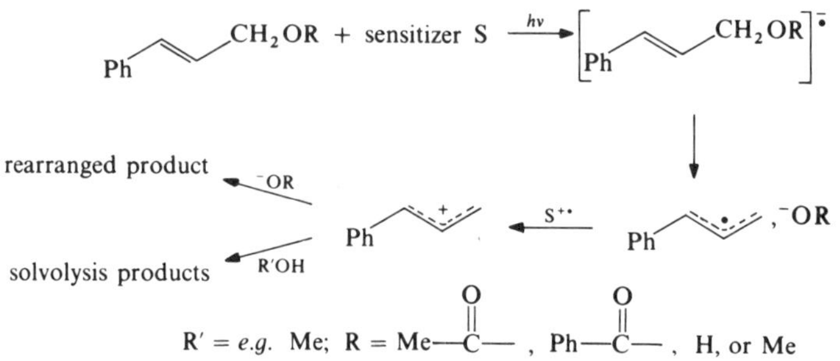

Ph–CH=CH–CH_2OR + sensitizer S $\xrightarrow{h\nu}$ [Ph–CH=CH–CH_2OR]$^{\bar{\bullet}}$, $S^{+\bullet}$

rearranged product ← $^-$OR — Ph(allyl cation) ← $S^{+\bullet}$ — Ph(allyl radical), $^-$OR

solvolysis products ← R′OH

R′ = *e.g.* Me; R = Me—C(=O)— , Ph—C(=O)— , H, or Me

SCHEME 3

Cinnamyl alcohol derivatives undergo rearrangement and solvolysis as shown in Scheme 3.[223] Electron-donor naphthalene derivatives sensitize the photosolvolysis of oxiranes.[224] Irradiation of *cis*- and *trans*-2,3,-dicarbomethoxy-2,3-bis(4-methoxyphenyl)oxirane with 1,4-dicyanonaphthalene (DCN) in MeCN leads to the slow ($t_{1/2}$ = *ca.* 350 ns) build-up of the *cis*-oxirane radical cation (**56**) either by a slow ring-opening of (**57**) or by isomerization of an initial open-ring strained configuration.[225] Ylide (**58**) is formed rapidly by reverse electron transfer in the ion-pair produced on photolysis of *cis*- and *trans*-stilbene oxide with DCN.[226]

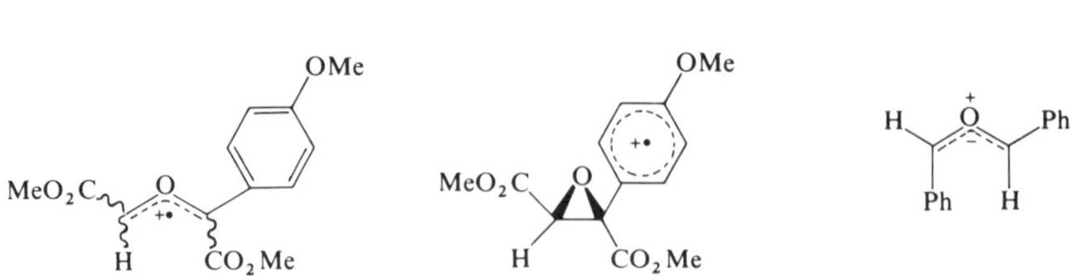

(**56**) (**57**) (**58**)

Studies of the photoreaction of biacetyl and 1,3-dioxole derivatives have indicated that the photoaddition of carbonyl compounds to electron-rich olefins should include ionic dissociation of the exciplex in the mechanism.[227] While 1,4-

dicyanobenzene and 2,3-dimethylbut-2-ene undergo photoaddition, 1,3-dicyanobenzene is unreactive despite a favourable free-energy change. The difference in behaviour is attributed to the nature of the 1,3-dicyanobenzene radical anion.[228] An inversion of reactivity between the thermal and the photochemical charge-transfer reaction of biacetyl and 1,1-diethoxyethene has been noted.[229]

Initial electron transfer on photolysis of *N*-methyl-1,8-naphthalimide and α-methylstyrene, followed by addition of methanol to the α-methylstyrene radical cation, yields a radical pair which couples to give (**59**) exclusively. By contrast, *N*-methylphthalimide gives a product coupled at its imide carbonyl group.[230]

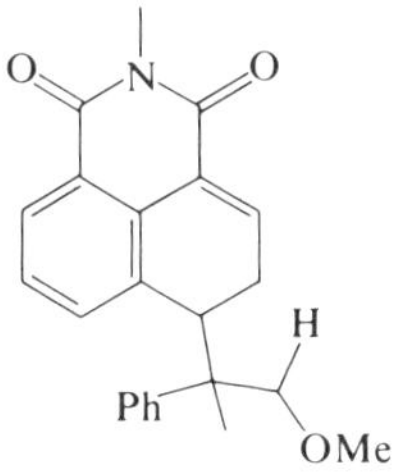

O
N
O
H
Ph
OMe

(**59**)

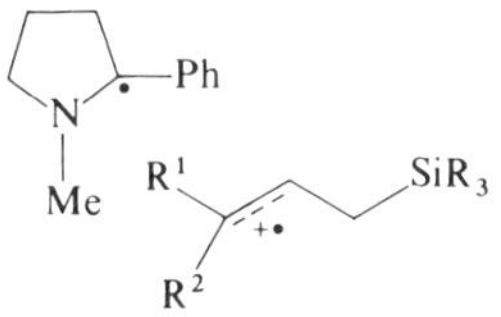

Ph
N
Me
R¹
SiR₃
R²

(**60**)

R = Me, Et, *etc.*
R^1, R^2 = H, Me, *etc.*

Electron-transfer-initiated photoaddition of 1-methyl-2-phenyl-1-pyrrolinium perchlorate to allylsilanes proceeds *via* the radical pair (**60**), which couples at the least-substituted allylic radical centre with desilylation.[231]

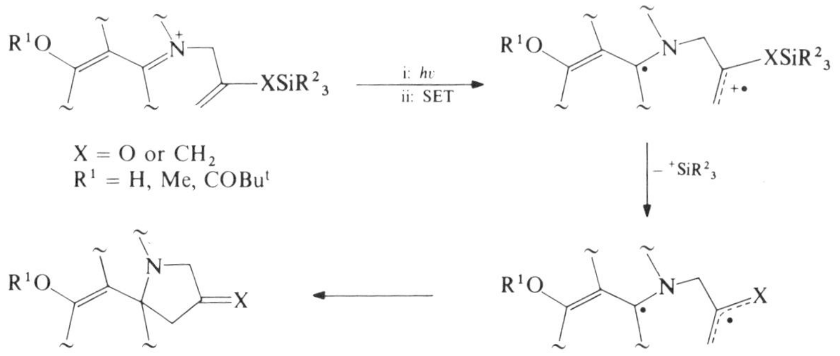

R¹O
N
XSiR²₃
i: hν
ii: SET
X = O or CH₂
R¹ = H, Me, COBuᵗ
− ⁺SiR²₃
X

SCHEME 4

β-Enaminone-derived allyliminium salts, especially when trialkylsilyl-substituted, undergo a photospirocyclization (Scheme 4),[232] useful in the synthesis of the Harringtonine alkaloids.[233] Carbon–carbon double-bond reduction of aryl-substituted enones, methyl cinnamate derivatives, and cinnamonitriles by 1-benzyl-1,4-dihydronicotinamide (BNAH), photosensitized by $Ru(bpy)_3{}^{2+}$, involves as key intermediates the BNA· radical and a half-reduced species formed by a series of electron and proton transfers.[234]

Photo-oxidations

Hydroxyl-radical-initiated oxidation of acrylonitriles in the gas phase with NO has been studied by FTIR spectroscopy.[235] Significant yields of C_4 and C_5 dicarbonyl aldehydes and furans are found on photo-oxidation of toluene and of *o*-xylene in a CH_3ONO/NO/air system.[236]

Bond-cleavage of electron-rich 1,2-diaryl-substituted ethanes occurs in the radical cation obtained by photo charge transfer to 9,10-dicyanonaphthalene sensitizer, the neutral radical fragment reacting with O_2 to lead to aryl ketone products. The cation fragment is reduced by sensitizer radical anion to a neutral radical.[237] Ozonides are formed nearly quantitatively on photo-oxidation of electron-rich stilbene oxides in the presence of 9,10-dicyanoanthracene (DCA).[238] The rate of ozonide formation from *e.g.* stilbene oxide in O_2-saturated MeCN containing DCA increases dramatically if biphenyl is also present. The ring-opened stilbene oxide radical cation receives an electron from $O_2^{\overline{\cdot}}$, forming a carbonyl ylide which then reacts with singlet O_2 giving *cis*-ozonide product exclusively.[239]

Photo-oxygenation of 1,1-di(*p*-anisyl)ethene by DCA/O_2 in MeCN gives very high yields of 3,3,6,6-tetra(*p*-anisyl)-1,2-dioxane.[240] The reaction involves a radical-cation chain and is found for other aryl-substituted ethenes.[241] Benzoquinone intercepts $O_2^{\overline{\cdot}}$ in the DCA-sensitized photo-oxidation of *trans*-stilbene.[242] The nucleophilic character of $O_2^{\overline{\cdot}}$ in addition to olefin radical cations and to ester carbonyl groups has been examined theoretically.[243] In polar solvent, photo-oxygenation of methylnaphthalenes sensitized by DCA involves both $^1O_2$ and $O_2^{\overline{\cdot}}$.[244] A caution has been sounded concerning the assumption that DCA-sensitized oxygenations involve $O_2^{\overline{\cdot}}$ since many excimers/exciplexes can cross to the triplet manifold and from thence generate singlet oxygen even in polar media.[245] Phenanthrenequinone gives diphenic acid and its peroxy acid on photo-oxidation whilst acenaphthenequinone forms mainly naphthalene-1,8-dicarboxylic anhydride.[246]

Photolysis

The rotating sector technique for intermittent illumination has been improved to a time resolution of 20 μs, and applied to the determination of the rate constants for self- and cross-termination reactions of $Bu^{t\cdot}$ and $Me_2\dot{C}OH$ radicals.[247]

The photobehaviour of alkyl halides in solution has been reviewed.[248] Photolysis of vinyl halides (notably iodides) provides a convenient route to highly strained vinyl cations in particular. Electron transfer occurs in the radical pair produced by carbon–halogen bond homolysis.[249] The radical anion formed by charge transfer from Pr^iOH to photoexcited 2-fluorocyclohexanone undergoes heterolytic cleavage of the carbon–fluorine bond faster than protonation by the solvent radical cation.[250] Pulsed-laser-photolysis studies have been made to clarify the photodissociation mechanisms of benzyl halides[251] and of naphthylmethyl halides.[252,253] Electron transfer from excited veratrole or naphthalene rings in compounds such as (**61**) occurs to both *syn* and *anti* C–Cl bonds (*anti* being preferred). Separation of Cl^- from the zwitterionic biradical (**62**) is accompanied by a mainly *syn* "Wagner–Meerwein" rearrangement, *i.e.* concerted migration with retention of configuration.[254]

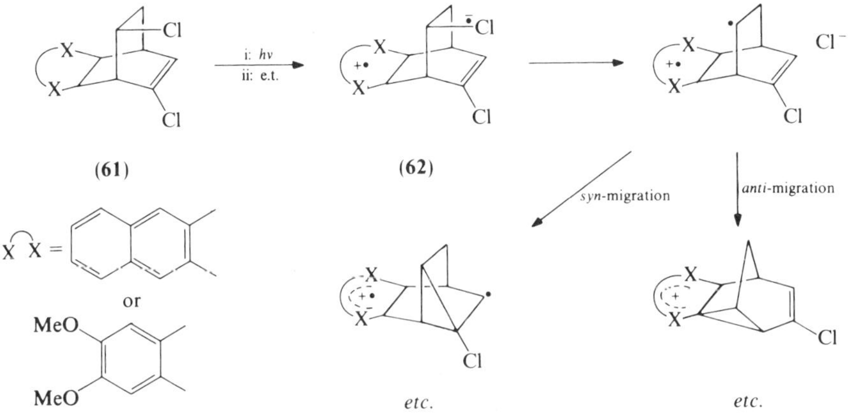

(**61**) (**62**)

Picosecond laser studies of the photo-dissociation of the triarylmethanes Ph_3CR (R = OH, OMe, Cl, Br, SMe) and $(p\text{-}Me_2NC_6H_4)_2CPhR$ (R = H, OH, OMe) reveal that the amount of cleavage is more dependent on the electron affinity of R· than on the R—C bond energy. Thus, even in non-polar media, homolytic cleavage may occur *via* initial heterolytic cleavage followed by reverse charge transfer.[255]

Flash photolysis of benzoin provides a convenient source of diphenylketyl radicals (monitored by their absorption at 540 nm).[256] The photoinduced polymerization by these radicals of diaryliodonium and triarylsulphonium salts has been investigated quantitatively.[257] A Norrish Type II reaction, having a high Arrhenius factor of $5 \times 10^{12}\ s^{-1}$, dominates the triplet decay of excited alkyl esters of benzoylformic acids which possess a γ-hydrogen.[258] Photolysis of the oxime esters $Ph_2C{=}N{-}O{=}C({-}O)Ar$ in benzene or pyridine provides a simple synthesis of biaryls, and is especially useful for generating bipyridyls.[259]

CIDNP studies continue to provide detailed evidence on in-cage and out-of-cage reactions in radical pairs. The limits of validity of the qualitative sign rules of Kaptein can be decided by a simple graphical approach.[260] CIDNP methods have been applied to the photodeuteration of benzaldehydes,[261] the photoinduced dissociation and rearrangement of angular dibenzacridines,[262] and photoreactions of acridines and quinoxalines in proton donor solvents.[263] Surface-exposed cysteine residues in peptides and proteins, labelled with a 4-hydroxyphenyl probe, give strong CIDNP emission from the phenol C(3) and C(5) protons when they interact with photoexcited dye.[264]

The photoreduction of chromone and chromanone with phenol or Et_3N has been investigated using ESR and CIDEP effects.[265] The spin polarization of α-

aminoalkyl radicals, generated by photolysis of tertiary amines and benzene-1,2:4,5-tetracarboxylic dianhydride, facilitate their observation by ESR/CIDEP.[266]

Photoreduction by $NaBH_4$ in ethanol of cata- and peri-condensed polynuclear aromatic compounds, *via* their exciplexes with $PhNEt_2$, provides an alternative route to the Birch reduction method for compounds of low solubility in liquid ammonia.[267] In benzene solution, photolysis of 9-cyanophenanthrene with Et_2NH leads exclusively to $Et_2N\cdot$ radicals by N—H abstraction, favoured by hydrogen bonding in a relatively non-polar exciplex. Increase in solvent polarity leads to production of α-aminoalkyl radicals also.[268] The latter route is favoured by increasing α-alkylation in the aliphatic amine donor.[269] The major product of the photoreduction of cyclohexanone or acetone by $Bu_3{}^nSnH$ is the tri-*n*-butylstannyl ether of the corresponding alcohol, formed by a chain-reaction with $_3{}^nSn\cdot$ radicals as carriers.[270]

Photolysis of the 1,3′-bicyclopropenyl system (**63**) leads *via* a vinyl carbene to the spiro-biradical (**64**) and thence to (**65**).[271]

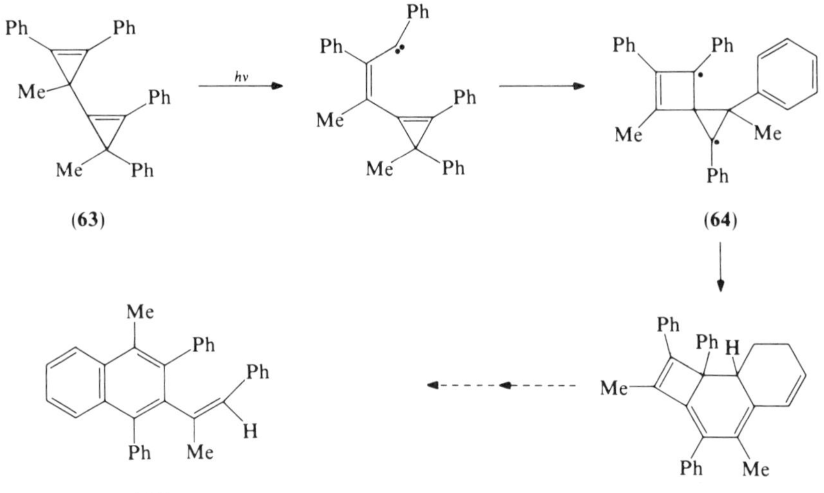

(**63**)

(**64**)

(**65**)

The 2,5-dioxo-1-piperazinyl radical acts as a chain carrier in the photoaddition of 1,4-dibromo-2,5-piperazinedione to 1-alkenes, 1 : 1 and 1 : 2 adducts being obtained.[272] Photoaddition in a 1,6-fashion of the triplet excited chromophore in (**66**) gives the chair conformation of (**67**) and thence *via* the boat conformation (**67**) to (**68**).[273]

Acetone-sensitized irradiation of 2-allyl-3-oxo-2,3-dihydro-l**H**-pyrrole-2-carboxylates leads to the formation of 2-allyl-3-hydroxy-1**H**-pyrrole-2-carboxylates.[274] A long-range hydrogen migration in the adduct of bicyclohept-1-en-1-yl and 2,5-dimethyl-*p*-benzoquinone in the solid state and in benzene solution induced by visible light has been reported.[275] Diverse radical-chain reaction pathways are found in the photoaddition of aliphatic ketones to bicyclo[1.1.0]butanes.[276]

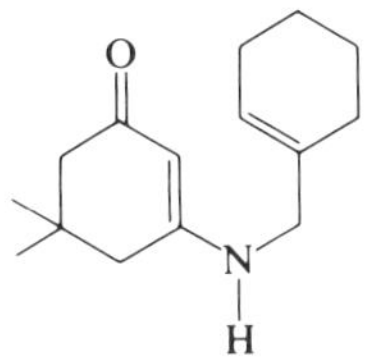

(66) 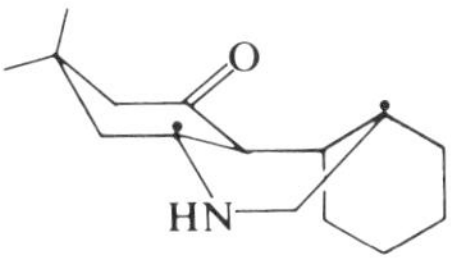 (67) 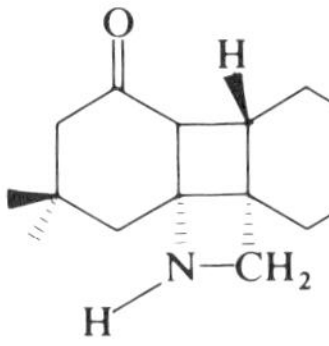 (68)

Radiolysis

The radiation chemistry of halocarbons[277] and the rôle of oxy radicals in the radiobiological oxygen effect[278] have been reviewed. Cyclopentyl radicals are produced selectively on radiolysis of neopentane–cyclopentane mixtures at 77 K but neopentyl radicals form also at 4 K.[279]

Radiolysis of ethene in oxygenated aqueous solution generates 2-hydroxyethylperoxyl radicals which undergo a rate-determining second-order decay ($2k = 2 \times 10^8\ \mathrm{M}^{-1}\,\mathrm{s}^{-1}$) to a tetroxide. This decomposes by three major routes.[280] The catechol antioxidant, nordihydroguaiaretic acid, traps radiolytically produced alkoxy radicals competitive with the rapid intramolecular β-fragmentation.[281] The rate constant of the decomposition[282]

$$\mathrm{PhCMe_2O\cdot \rightarrow PhC({=}O)Me + Me\cdot}$$

in water ($1 \times 10^7\ \mathrm{s}^{-1}$) is higher than that found in organic solvents ($1 \times 10^6\ \mathrm{s}^{-1}$). The dominant species in the radiolysis of dihydrouracil is **(69)**. At lower pH and/or higher dose rates **(69)** decays bimolecularly to give barbituric acid, but at higher pH base-catalysed $O_2^{\cdot -}$ elimination gives isouracil and thus uracil.[283] The radical anions of 5-bromo- and 5-iodo-uracil in aqueous solution have half-lives of 7.0 and 1.7 ns, respectively, much shorter than that (4.9 μs) of the 5-chlorouracil radical anion. Uracilyl radical forms an adduct with 5-bromouracil ($k = 2.66 \times 10^8\ \mathrm{M}^{-1}\,\mathrm{s}^{-1}$) having a convenient absorption at 365 nm; the reaction serves as a reference for hydrogen-abstraction rate determinations.[284] Radical cations of 5-halouracil and 5-halocytosine and their deprotonated forms have been examined in irradiated 12 M LiCl glass at low temperatures.[285]

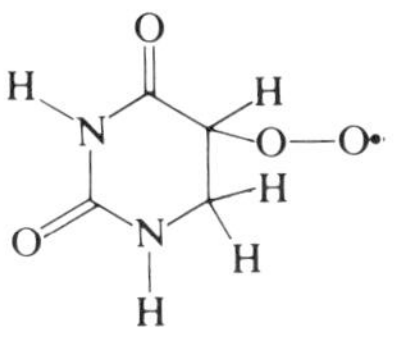

(69) 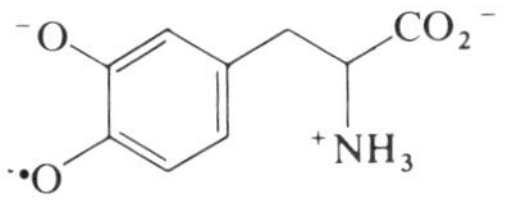 (70)

The kinetics of disproportionation of dopasemiquinone **(70)** to melanin precursors have been measured as a function of pH.[286] One-electron oxidation of CCl_4

solutions of ZnTPP, TMPD, and phenols[287] during pulse radiolysis has been monitored, as has the reduction of a promising leukemia cytotoxic drug, 9-(2-methoxy-4-methylsulphonylaminoanilino)acridinium salt.[288] A chain isomerization of 2-(2-furyl)-3-(5-nitro-2-furyl)acrylamide occurs on radiolysis, propagated by electron transfers between the isomers and their radical anions.[289]

Reduction of $Me_2S{=}O$ by radiolytically produced hydrogen atoms generates dimethyl sulphide radical cations whose reactions can be studied undisturbed by complexation with Me_2S.[290] A considerable red-shift in the optical absorption spectra of the radical cations $R_2SSR_2^{+\cdot}$ (R = H, Me, Et, Pr^n, Pr^i, Bu^n) is attributed to the inductive effects of the substituents on the $1\sigma^*$-orbital of the three-electron bond.[291] The redox reactions of flavins in the presence of 2-mercaptoethanol are sensitive to pH; EtS· radicals oxidize dihydroflavin at pH 6.3, but when complexed with EtS^- strongly reduce flavin[292] and lumiflavin.[293] Radiolytically prepared dialkoxy(thiophosphoryl)thio radicals are very short-lived but relatively unreactive towards O_2 or aliphatic alkenes separately. In the presence of both, peroxyl radicals are rapidly formed *via* rapid reversible addition to the double bond followed by O_2 attack.[294]

Pyrolysis

Evidence against high rates for dissociative recombination reactions of methyl radicals has been obtained from the pyrolysis of ethane in high-temperature shock-tube experiments.[295] Pyrolysis studies of isobutene yield a value of 0.8 ± 2 kcal mol^{-1} for the heat of formation of 2-methyl-2-pentyl radical.[296] Thermolysis of tetracyclopropylsuccinonitrile in mesitylene at 260° indicates a resonance energy of 7.8 kcal mol^{-1} for the $R_2\dot{C}CN$ radical.[297] Correlations of activation free energies and ground-state strain energies/strain changes on dissociation have been made for the thermolysis of symmetrical tetraalkyldiarylethanes, and a resonance energy of 8.4 kcal mol^{-1} deduced for α,α-dialkylbenzyl radicals.[298] Similar correlations with force-field-calculated strain-energy changes are found for the thermolysis of 1,1′-diphenyl-1,1′-bicycloalkyls (C_4–C_8 ring sizes).[299]

Radical yields of *sym*-proportionation at 450° and 200 bar pressure from cyclohexane with 1-octene agree with those calculated on the basis of the equilibrium thermodynamics.[300] Thermolysis of (**71**) obeys first-order kinetics and gives (**72**) *via* diradical (**73**), which is electronically more stabilized than (**74**) by 25.8 kJ mol^{-1}.[301] Model studies of coal liquefaction have concerned the decompositions of 1,3-diphenylpropane, dibenzyl ether, and phenethyl phenyl ether.[302] Addition of 1- and 2-tetralyl radicals to 1,2-dihydronaphthalene is followed by intramolecular cyclization and hydrogen abstraction.[303] Flash vacuum pyrolysis of phenyl-substituted benzobicyclo[3.1.0]hex-2-enes cause cleavage gener-

(**71**) (**72**) (**73**) (**74**)

ally at the C(1)–C(5) "internal" bond, followed by 1,2-shift of H or Ph in the resulting biradical.[304]

The homogeneous chain-reaction of CF_3OF and $CF_3CF{=}CF_2$ becomes explosive in the presence of large amounts of O_2.[305] The ratio of silacyclohexene to silacyclopentene products in the thermolysis of 3-silabicyclo[3.1.0]heptanes depends strongly on the nature of the alkyl substituents at the C(1) and C(5) bridgehead positions.[306] Deuteration studies on the radical-induced extrusion reaction of 2-(benzylsulphonyl)tropone at 200° reveal extensive cross-over of benzyl substitution positions.[307] Gas-phase cyclization of 1,5-diaryl-1,5-diazapentadienes gives quinolines;[308] flash vacuum pyrolysis of cinnamaldehyde phenylhydrazones at 600° gives equal amounts of cinnamonitriles and quinolines *via* species such as ArCH=CH=CH—N·.[309]

Biradicals

Spectroscopic and kinetic studies of conjugated biradicals[310] have been reviewed, as have σ-conjugation effects[311] and the MO characteristics of 1,3-dipolar species.[312] Electron repulsion can cause allylic resonance effects to be important in biradicals even when absent in the corresponding monoradicals.[313] The ground state of Firestone's extended 1,5-biradical corresponds to a non-symmetric transition state for concerted 1,3-dipolar cycloadditions.[314] Relative stabilities of biradical conformers and interconversion barriers of *gem*-di- and -tetra-fluoromethylenecyclopropanes have been calculated.[315]

Triplet 1,4-biradicals from hexan-2-one and its 5-methyl derivative have been monitored by optical absorption.[316] Triplet biradicals such as $Ph_2\dot{C}$—CH_2—(CH_2)—$\dot{C}$=O have life-times (*ca* 50 ns) similar to twisted olefin triplets and Norrish II 1,4-biradicals, but three times longer than Norrish I 1,8-biradicals.[317] The life-time of the *gauche*-1,4-biradical (**75**) at 242 ns in MeOH is longer than that of the *anti*-form (**76**) which, with other evidence, indicates a lack of generality in the relation of life-time to inter-radical terminal distance and orbital orientation.[318] Biradical character is imposed on the radical pair from photolysis of dibenzyl ketone by the constraints of sodium alkyl sulphate micelles.[319]

A new synthesis of *m*-quinodimethanes and their efficient trapping by conjugated dienes has been described.[320] A distance of 0.36 nm exists between the radical centres in (**77**), generated in singlet and triplet forms at 77 K.[321] Step-wise elimination of N_2 from compounds such as (**78**), or SO_2 from analogous sulphones, to generate an octamethyltetramethyleneethane biradical, appears to be thwarted by rapid cyclization before the second N_2 molecule is lost.[322]

Photochemical deazetation of fluorine-substituted 4-methylenepyrazolines yields "hot" trimethylenemethane biradicals, even in solution, whose cyclization products occur in statistical ratios, whereas thermally produced species are selective due to incomplete conformational equilibration.[323] The 1,3-biradical (**79**) is very short-lived relative to (**80**) but can be intercepted at high pressures of O_2 (Scheme 5).[324]

Under high-power excimer laser photolysis at 308 nm, the 1,5-biradical (**81**) ejects a 1,3-biradical (**82**).[325]

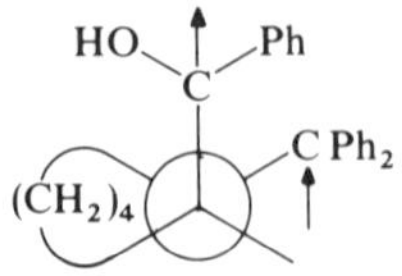

(75) 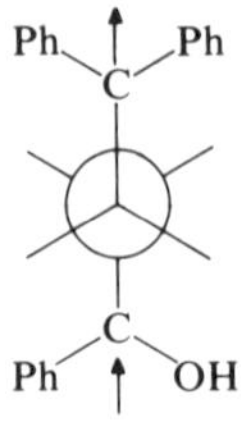 (76)

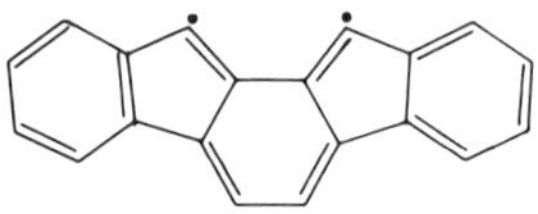

(77) 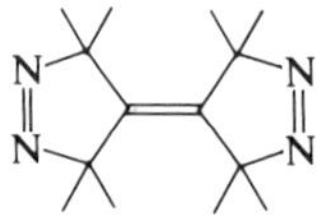 (78)

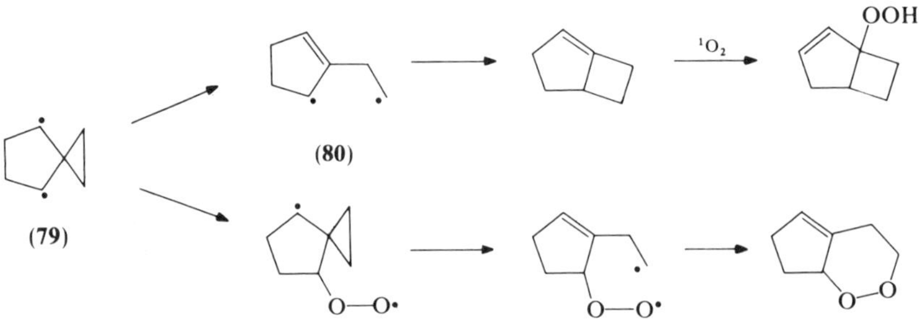

(79) (80)

SCHEME 5

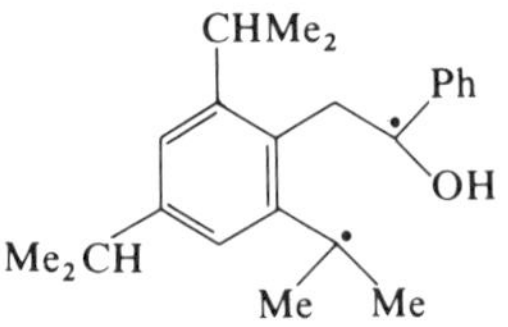

(81) 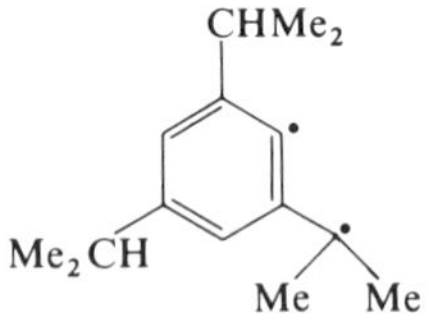 (82) 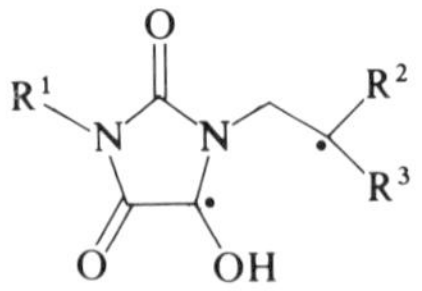 (83)

Energy barriers to ring-closure remain nearly constant but those for internal rotation and cleavage decrease with increase in strain in biradicals formed during allene cycloaddition reactions.[326] Concerted 1,2-addition of Me_2Si: to 1,3-butadienes is followed by non-concerted rearrangements involving biradical intermediates with some unusual C—C bond-cleavages.[327] Norrish Type II processes are generally unfavourable for biradicals (**83**), the exception being (**83**; R^1 = Me, $R^2 = R^3$ = H) since conjugation of one of the radical centres with the adjacent carbonyl holds the orbital perpendicular to the potentially fissionable N—CH_2 bond.[328] Attempts to influence asymmetric induction *via* the size of an ester grouping E have proved to be unsuccessful in the intramolecular 1,3-diyl trapping reaction of (**84**). The reaction is non-synchronous but equilibration of *cis,anti*- and *cis,syn*-biradicals may be occurring.[329]

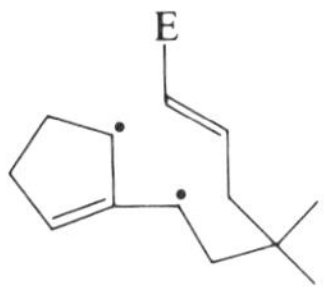

(**84**) 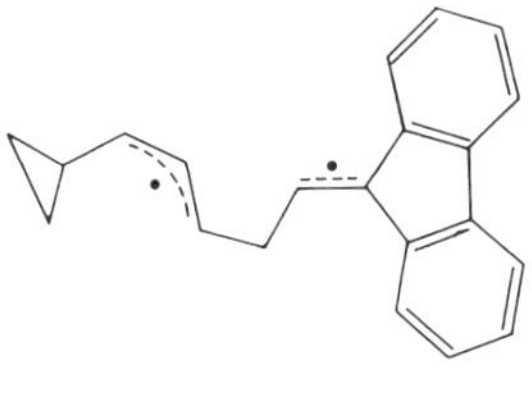 (**85**)

The cyclopropylmethyl → 3-butenyl-type rearrangement of (**85**) is unimportant below 140° compared to the intramolecular radical coupling.[330] Photoreaction of 1-acetyl-2-ethylidene-1-methoxycyclohexane to 2-ethylidene-1-methoxy-6-methylbicyclo[3.1.1]heptan-6-ol involves ring-closure of the 1,4-biradical of least stability.[331] Production of the π,*p*-state biradical cation (**87**), and thence non-stereospecific products, competes with the concerted Wagner–Meerwein rearrangement of the zwitterionic biradical (**86**) formed on photolysis in glacial acetic acid.[332]

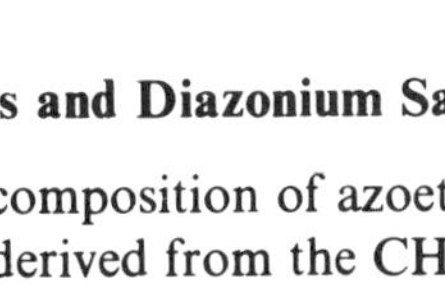

(**86**)  (**87**)

Azo Compounds and Diazonium Salts

The thermal decomposition of azoethane at 508–598 K involves a short chain and some products derived from the $CH_3CHNNCH_2CH_3$ radical.[333] Primary[334] and secondary[335] decomposition processes of azoisopropane have been evaluated. The pyrolysis of aryl azides has been reviewed.[336] Oxidation of tetramethyltetrazine by N_2O_4 has been monitored by ESR.[337]

Metal-ion-catalysed reaction of arenediazonium salts with acetylacetone provides a novel and selective synthesis of 3-arylpentane-2,4-diones.[338] Electron-withdrawing substituents in the aryl ring increase the reactivity of aryldiazonium chlorides towards radical chlorodediazoniation induced by reducing cations such as Sn^{2+}, Cu^+, and Fe^{2+}, and indicate a rate-determining electron-transfer step.[339] Dediazoniation is initiated by one-electron transfer from *N*-benzyl-1,4-dihydronicotinamide, a NAD(P)H model compound, to *p*-$XC_6H_4N_2^+$ BF_4^-, the rate increasing in the order X = Me < Br < NO_2.[340] The photoinduced reversible electron-transfer reaction:

$$\text{pyrene}^{*} + ArN_2^{+} \rightleftharpoons (\text{pyrene})^{+\cdot} + ArN_2\cdot$$

can be examined in three time regimes by CIDNP intensities of ^{1}H (90–100 ns), ^{13}C (15–20 ns), and ^{15}N (3–5 ns) nuclei.[341]

Pulse radiolytic studies of 9-diazofluorene, FlN_2, show that it reacts with $Me_2\dot{C}OH$ radicals in neutral aqueous solution to give $FlN_2H\cdot$ radicals, the rate constant being 4.6×10^{8} $M^{-1}s^{-1}$; $FlN_2^{-\cdot}$ is formed in alkaline solution and is intrinsically long-lived but electron transfer to O_2 is rapid. The rate constant for the oxidation of FlN_2 to $FlN_2^{+\cdot}$ by $Cl_3COO\cdot$ radicals is 1.2×10^{7} $M^{-1}s^{-1}$.[342] Oxidation of FlN_2 at a platinum anode or by one-electron reagents gives high yields of bifluorenylidene and fluorenone azine but the chain carrier is a poorer oxidant than $FlN_2^{+\cdot}$. The ESR of the carrier does not allow unequivocal identification of the species but a 1,1-diazene-type radical cation is most likely.[343] In the 9,10-dicyanoanthracene(DCA) photosensitized reation of $PhCH{=}N_2$, the radical cation $Ph\dot{C}HN_2^{+}$ adds rapidly to the parent molecule to form (**88**) and (**89**), the *cis*-form (**88**) being more stable than the *trans*-form (**89**), which then are reduced by $DCA^{-\cdot}$ to the stilbenes *via* the intermediate (**90**) in the case of (**88**).[344]

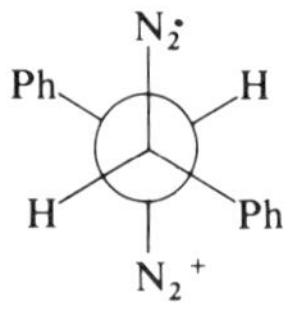

(**88**)

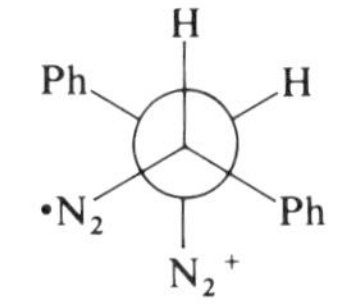

(**89**)

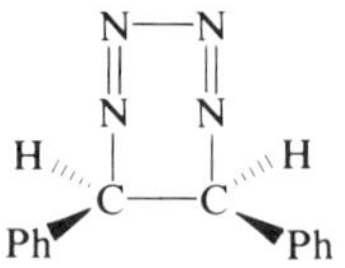

(**90**)

Electron-transfer chain catalytic reduction of FlN_2 in MeCN does not involve $FlN_2^{-\cdot}$ as the chain carrier.[345] Reduction of FlN_2 and fluoren-9-one occurs in DMSO–MeOH mixtures containing KOH.[346]

Peroxides

Free-radical decomposition mechanisms of peroxides in solution have been reviewed.[347] A correlation of the catalytic activity of d^{0}molybdenum and vanadium complexes in the oxidation of hydrocarbons with organic hydroperoxides and the oxidizing properties of the metal peroxides has been attempted since the triangular structure of bound peroxo and alkylperoxidic groups is a common characteristic.[348] Decomposition of propyl and butyl peroxides of cadmium in toluene proceeds in two stages.[349] The system Fe^{2+}/4-$HOCMe_2C_6H_4CMe_2OOCMe_3$ could prove useful as a free-radical initiator.[350]

Endoperoxides of naphthalene and anthracene photoisomerize and biradical intermediates can be trapped by *N*-methylmaleimide.[351] Biradical (**91**), derived from photolysis of the corresponding 1,4-tetrahydroaromatic endoperoxide, cyclizes to (**92**) or re-closes to a diepoxide. The biradical (**92**) undergoes hydrogen-atom transfer followed by elimination of water to give the final product (**93**).[352]

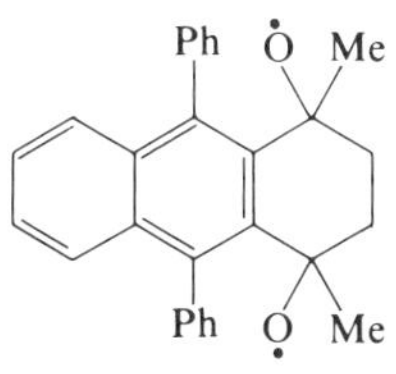

(91) 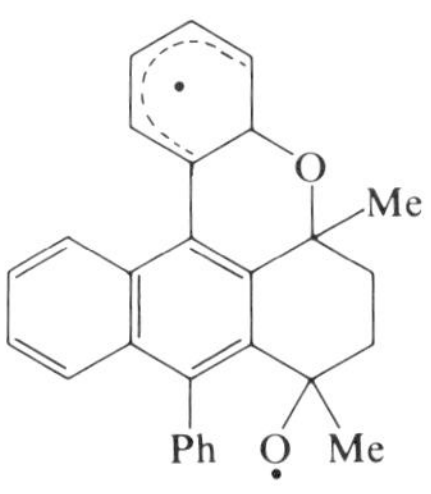 (92) 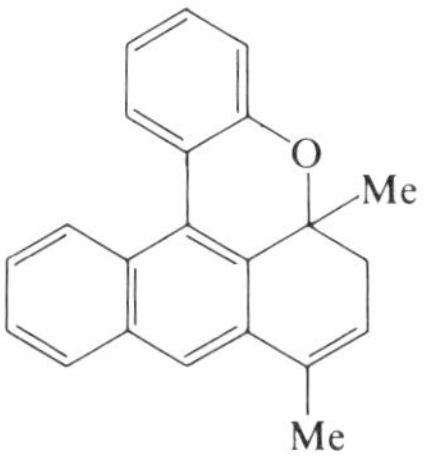 (93)

An inverse k.i.e. is found for β-deuterium substitution in the thermolysis of tetramethyl-1,2-dioxetane.[353] A step-wise mechanism for the thermal decomposition of 3,3,6,6-tetramethyl-1,2,4,5-tetraoxane in benzene at 135–165° is indicated; the activation entropy is unusually low at 0.0 e.u. compared to 9–10 e.u. for most other peroxide decompositions.[354]

The effects of toluene[355] and of benzene and chlorobenzene[356] on the gas-phase thermolysis of Bu^{t}OOH have been fully investigated. The initiation properties of 2-cyano-2-propylhydroperoxide in oxidation processes have been studied.[357] Marked isomerization during photolysis of methyl-13-hydroperoxy-(9Z,11E)-octadecadienoate in cyclohexane showed that hydrogen abstraction from OOH groups was favoured. In methanol solvent, alkoxy radicals attack the solvent and the product radicals in turn attack the substrate unsaturation.[358] Salt formation precedes the decomposition of α-primary hydroperoxides of *sec*-butyltoluene in aqueous sodium hydroxide at 70°.[359] Metal stearate-induced decomposition[360] and inhibition/acceleration by oxidation products of 2-nonylhydroperoxide[361] have received study. The thermal decomposition of $Ph_2C(Me)OOH$ has been monitored by a pulse chromatography method.[362]

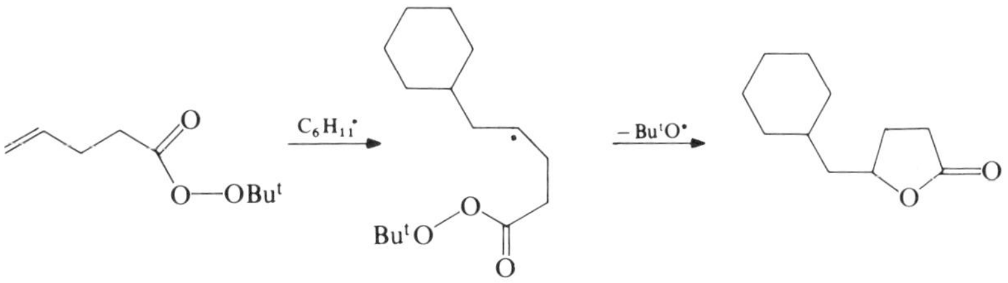

(94) (95)

Thermolysis of **(94)** in cyclohexane at 120° gives **(95)**.[363] Kinetic isotope effects, from deuterium substitution in dialkyl and aryl alkyl sulphides, provide a more reliable probe of a free-radical mechanism in the reaction of the nucleophiles with substituted *tert*-butylperoxybenzoates.[364] Electron-transfer reactions of $(C_3F_7CO_2)_2$ with methoxybenzenes rapidly give high yields of ring-substituted products even at 0°.[365] Chain depolymerization *via* β-scission of alkoxyfluoroalkyl radicals, formed by homolysis of peroxidic perfluoro polyethers, has been followed by ESR.[366] The kinetics of the photochemical breakdown of HSO_5^- in the presence/absence of propan-2-ol indicates a chain mechanism, independent of pH.[367]

The kinetics of the gas-phase reaction of $\cdot CCl_3$ radicals with O_2 and with NO at 295 K have been determined.[368] Absolute propagation and termination rate constants for $RO_2\cdot$ radicals in solution have been determined and the literature reviewed.[369] Alkylperoxy radicals react with α-tocopherol in solution at 50° with rate constants in excess of 2×10^5 M^{-1} s^{-1}.[370] Thermolysis of 4-methyl-4-phenylmalonyl peroxide is an oxygen-dpendent chemiluminescent reaction. Conjugation with a phenyl ring in the biradical $Ph(Me)\dot{C}—CO_2\cdot$ does not slow down the intramolecular cyclization sufficiently to allow trapping of the biradical by O_2. In the presence of an electron donor like perylene, the reaction proceeds down a more exothermic route but the light yield is still very low.[371] Chemiluminescence yields for hyponitrite esters increase with increasing (negative) enthalpy of disproportionation of the alkoxy radical pair involved, the pair being equivalent to the biradical intermediate in the decomposition of dioxetanes.[372] Bimolecular termination of peroxy radicals from di-*tert*-butyl peroxyoxalate in the presence of O_2 gives rise to a strong chemiluminescence.[373]

Autoxidation

Oxidation of alkanes photochemically in air/NO/CH_3ONO mixtures has provided relative rate constants for the reaction of alkylnitrates with $\cdot OH$ radicals.[374] The reactions

$$PhCH_2\cdot + O \rightarrow PhCHO + H\cdot$$

and

$$PhCH_2\cdot + HO_2\cdot \rightarrow PhCH_2\cdot + \cdot OH \rightarrow PhCHO + H\cdot + \cdot OH$$

play an important rôle in the high-temperature oxidation of toluene in a flow reactor at 1200 K.[375]

Oxidation of butyraldehyde by O_2 is initiated by the dimeric complex $Pd_2(OAc)_4Cl_2^{2-}$ $(Bu_4N^+)_2$.[376] Photoepoxidation of olefins with α-diketones and molecular oxygen involves acylperoxy radicals, $RO_3\cdot$, as key intermediates. Such radicals are strongly electrophilic and unreactive towards sulphides, sulphoxides, or pyridine, and thus permit selective epoxidation of carbon–carbon double bonds.[377, 378] Ozonolysis of (**96**) does not follow the Criegee mechanism at −80° but proceeds *via* the 1,6-biradical (**97**).[379]

(**96**)

$n = 3, 4, 5$

(**97**)

Reluctance towards deprotonation or dimerization of the radical cation of adamantylideneadamantane and rather addition of O_2 leads to good yields of the corresponding dioxetane on autoxidation in CH_2Cl_2/CF_3CO_2H.[380] Liquid-phase autoxidation of retinyl acetate proceeds *via* a highly stabilized polyenyl radical formed by hydrogen abstraction at the C(15) site. The radical shows little tendency to form the corresponding peroxy radical with O_2. Intramolecular isomerization of **(98)** to **(99)** competes successfully with intermolecular reactions.[381]

OAc O O• OAc O O

(**98**) (**99**)

A vinylogous ene reaction of 2,5-dimethyl-2,4-hexadiene with singlet oxygen forms the hydroperoxide (**101**) from the biradical (**100**).[382]

$^1O_2$ O O• O_2H

(**100**) (**101**)

Autoxidation of diphenylketene in CCl_4 at 0° gives mainly polybenzylic acid.[383] The kinetics and product analyses for the early stages of the autoxidation of $Me_2C(CH_2OAc)_2$ and $EtC(CH_2OAc)_3$ have been determined.[384] Model studies of the oxidation of associating thiols having dipeptide binding sites and various aryl "recognition sites" have been made.[385] Electron-transfer oxidation (by O_2) of 1,5-dihydroflavins has been reviewed and further explored.[386] Carbon radicals generated from esters of *N*-hydroxypyridine-2-thione react smoothly with oxygen, in the presence of *tert*-butyl thiol as chain restarter, to form nor-hydroperoxides and thence nor-alcohols and nor-oxo derivatives, so providing a convenient conversion of aliphatic and alicyclic carboxylic acids.[387] A related system facilitates deoxygenation of tertiary alcohols.[388]

References

1 Baciocchi, E., Rol, E., Sebastiani, G. V., and Serena, B., *Tetrahedron Lett.*, **25**, 1945 (1984).
2 Baciocchi, E., Rol, C., and Ruzziconi, R., *J. Chem. Res. (S)*, **1984**, 334.
3 Baciocchi, E., and Ruzziconi, R., *J. Chem. Soc., Chem. Commun.*, **1984**, 445.
4 Schlesener, C. J., and Kochi, J. K., *J. Org. Chem.*, **49**, 3142 (1984).
5 Kimura, M., and Kaneko, Y., *J. Chem. Soc., Dalton Trans.*, **1984**, 341.
6 Colter, A. K., Plank, P., Bergsma, J. P., Lahti, R., Quesnel, A. A., and Parsons, A. G., *Can. J. Chem.*, **62**, 1780 (1984).

[7] Goldstein, S. L., and McNelis, E., *J. Org. Chem.*, **49,** 1613 (1984).
[8] Lee, D. G., Congson, L. N., Spitzer, U. A., and Olson, M. E., *Can. J. Chem.*, **62,** 1835 (1984).
[9] Galliani, G., Rindone, B., and Beltrame, P. L., *Nouv. J. Chim.*, **7,** 639 (1983).
[10] Baizer, M. M., *Tetrahedron,* **40,** 935 (1984).
[11] Schlegel, G., and Schafer, H. J., *Chem. Ber.*, **117,** 1400 (1984).
[12] Howell, J. O., Goncalves, J. M., Amatore, C., Klasinc, L., Wightman, R. M., and Kochi, J. K., *J. Am. Chem. Soc.*, **106,** 3968 (1984).
[13] Vasil'eva, N. V., Starichenko, V. F., Zelinskii, A. G., and Koptyug, V. A., *Zh. Org. Khim.*, **20,** 348 (1984); *Chem. Abs.*, **101,** 6392 (1984).
[14] Pettersson, T., Parker, V. D., and Ronlan, A., *Acta Chem. Scand.*, **37B,** 675 (1983).
[15] Callot, H. J., Lovati, A., and Gross. M., *Bull. Soc. Chim. Fr. II,* **1983,** 317.
[16] Porter, Q. N., Utley, J. H. P., Machion, P. D., Pardini, V. L., Schumacher, P. R., and Viertler, H., *J. Chem. Soc., Perkin Trans. 1,* **1984,** 973.
[17] Koppenol, W. H., and Liebman, J. F., *J. Phys. Chem.*, **88,** 99 (1984).
[18] Davies, M. J., Gilbert, B. C., McCleland, C. W., Thomas, C. B., and Young, J., *J. Chem. Soc., Chem. Commun.*, **1984,** 966.
[19] Kim, S. S., *Kich'o Kwahak Yonguso Nonmunjip,* **5,** 93 (1984); *Chem. Abs.*, **101,** 38072 (1984).
[20] Fristad, W. E., and Peterson, J. R., *Tetrahedron,* **40,** 1469 (1984).
[21] Walling, C., Zhao, C., and El-Taliawi, G. M., *J. Org. Chem.*, **48,** 4910 (1983).
[22] Walling, C., El-Taliawi, G. M., and Zhao, C., *Prepr.-Am. Chem. Soc., Div. Pet. Chem.*, **29,** 348 (1984); *Chem. Abs.*, **101,** 72035 (1984).
[23] Walling, C., El-Taliawi, G. M., and Zhao, C., *J. Org. Chem.*, **48,** 4914 (1983).
[24] Abdeen, S. Z., Ohag, M. I., and Vasudeva, W. C., *Gazz. Chim. Ital.*, **114,** 197 (1984); *Chem. Abs.*, **101,** 151216 (1984).
[25] Rasmussen, J. K., Heilmann, S. M., Toren, P. E., Pocius, A. V., and Kotnour, T. A., *J. Am. Chem. Soc.*, **105,** 6845 (1983).
[26] Davies, M. J., Gilbert, B. C., and Norman, R. O. C., *J. Chem. Soc., Perkin Trans. 2,* **1984,** 503.
[27] Eberhardt, M. K., *Tetrahedron Lett.*, **25,** 3663 (1984).
[28] Jonsson, L., and Wistrand, L.-G., *J. Org. Chem.*, **49,** 3340 (1984).
[29] Koshechko, V. G., Inozemtsev, A. N., and Pokhodenko, V. D., *Teor. Eksp. Khim.*, **20,** 178 (1984); *Chem. Abs.*, **101,** 190757 (1984).
[30] Omura, K., *J. Org. Chem.*, **49,** 3046 (1984).
[31] Bauer, W., Daub, J., Eibler, E., Gieten, A., Lamm, V., and Lotter, W., *Chem. Ber.*, **117,** 809 (1984).
[32] Wuest, J. D., and Zacharie, B., *J. Org. Chem.*, **49,** 166 (1984).
[33] Wuest, J. D., and Zacharie, B., *J. Org. Chem.*, **49,** 163 (1984).
[34] Fukunaga, K., and Kimura, M., *Bull. Chem. Soc. Jpn.*, **56,** 3191 (1983).
[35] Garrard, W. N. C., and Thomas, F. G., *Aust. J. Chem.*, **36,** 1983 (1983).
[36] Shiotani, M., Nagata, Y., and Sohma, J., *J. Am. Chem. Soc.*, **106,** 4640 (1984).
[37] Toriyama, K., Nunome, K., and Iwasaki, M., *Chem. Phys. Lett.*, **107,** 86 (1984); *Chem. Abs.*, **101,** 54428 (1984).
[38] Shiotani, M., Nagata, Y., and Sohma, J., *J. Phys. Chem.*, **88,** 4078 (1984); *Chem. Abs.*, **101,** 110216 (1984).
[39] Kalcher, J. and Olbrich, G. *Theochem.*, **13,** 341 (1983); *Chem. Abs.*, **100,** 67480 (1984).
[40] Eastland, G. W., Maj, S. P., Symons, M. C. R., Hasegawa, A., Glidewell, C., Hayashi, M., and Wakabayashi, T., *J. Chem. Soc., Perkin Trans. 2,* **1984,** 1439.
[41] Maj. S. P., Rao, D. N. R., and Symons, M. C. R., *J. Chem. Soc., Faraday Trans. 1,* **80,** 2767 (1984).
[42] Rideout, J. and Symons, M. C. R., *J. Chem. Res. (S),* **1984,** 268.
[43] Rao, D. N. R., Chandra, H., and Symons, M. C. R., *J. Chem. Soc., Perkin Trans. 2,* **1984,** 1201.
[44] Rao, D. N. R., Eastland, G. W., and Symons, M. C. R., *J. Chem. Soc., Faraday Trans. 1,* **80,** 2803 (1984).
[45] Shiotani, M., Kawazoe, H., and Sohma, J., *J. Phys. Chem.*, **88,** 2220 (1984).
[46] Symons, M. C. R., and Wren, B. W., *J. Chem. Soc., Perkin Trans. 2,* **1984,** 511.
[47] Snow, L. D., and Williams, F., *Chem. Phys. Lett.*, **100,** 198 (1983); *Chem. Abs.*, **100,** 5692 (1984).
[48] Boon, P. J., Symons, M. C. R., Ushida, K., and Shida, T., *J. Chem. Soc., Perkin Trans. 2,* **1984,** 1213.
[49] Symons, M. C. R., and Boon, P. J., *Chem. Phys. Lett.*, **100,** 20 (1983); *Chem. Abs.*, **100,** 5693 (1984).
[50] Sevilla, M. D., Becker, D., Sevilla, C. L., and Swarts, S., *J. Phys. Chem.*, **88,** 1701 (1984).
[51] Rao, D. N. R., Rideout, J., and Symons, M. C. R., *J. Chem. Soc., Perkin Trans. 2,* **1984,** 1221.
[52] Eastland, G. W., Rao, D. N. R., and Symons, M. C. R., *J. Chem. Soc., Perkin Trans. 2,* **1984,** 1551.
[53] Rao, D. N. R., Symons, M. C. R., and Wren, B. W., *J. Chem. Soc., Perkin Trans. 2,* **1984,** 1681.
[54] Qin, X.-Z., Walther, B. W., and Williams, F., *J. Chem. Soc., Chem Commun.*, **1984,** 1667.

[55] Schwartz, H., *Shitsuryo Bunseki*, **32,** 3 (1984); *Chem. Abs.*, **101,** 129829 (1984).
[56] Snow, L. D., Wang, J. T., and Williams, F., *Chem. Phys. Lett.*, **100,** 193 (1983); *Chem. Abs.*, **100,** 34036 (1984).
[57] Bouma, W. J., Poppinger, D., Saebo, S., MacLeod, J. K., and Radom, L., *Chem. Phys. Lett.*, **104,** 198 (1984); *Chem. Abs.*, **100,** 208672 (1984).
[58] Clark, T., *J. Chem. Soc., Chem. Commun.*, **1984,** 666.
[59] Apeloig, Y., Karni, M., Ciommer, B., Depke, G., Frenking, G., Meyn, S., Schmidt, J., and Schwarz, H., *J. Chem. Soc., Chem. Commun.*, **1983,** 1497.
[60] Terlouw, J. K., Heerma, W., Burgers, P. C., and Holmes, J. L., *Can. J. Chem.*, **62,** 289 (1984).
[61] Hammerum, S., Kuck, D., and Derrick, P. J., *Tetrahedron Lett.*, **25,** 893 (1984).
[62] Lien, M. H., and Hopkinson, A. C., *Can. J. Chem.*, **62,** 922 (1984).
[63] Cacace, F., Ciranni, G., and Di Marzio, A., *J. Chem. Soc., Perkin Trans. 2,* **1984,** 775.
[64] Schmitt, R. J., Buttrill, S. E., and Ross, D. S., *J. Am. Chem. Soc.*, **106,** 926 (1984).
[65] Franke, H., Halim, H., and Schwarz, H., *Chem. Ber.*, **117,** 3194 (1984).
[66] Eierdanz, H., Potthoff, S., Bolze, R., and Berndt, A., *Angew. Chem. Int. Ed.*, **23,** 526 (1984).
[67] Broxterman, Q. B., Hogeveen, H., and Kingma, R. F., *Tetrahedron Lett.*, **25,** 2043 (1984).
[68] Courtneidge, J. L., Davies, A. G., Lusztyk, E., and Lusztyk, J., *J. Chem. Soc., Perkin Trans. 2,* **1984,** 155.
[69] Bock, H., Roth, B., and Maier, G., *Chem. Ber.*, **117,** 172 (1984).
[70] Courtneidge, J. L., and Davies, A. G., *J. Chem. Soc., Chem. Commun.*, **1984,** 136.
[71] Courtneidge, J. L., Davies, A. G., and Yazdi, S. N., *J. Chem. Soc., Chem. Commun.*, **1984,** 570.
[72] Courtneidge, J. L., Davies, A. G., Clark, T., and Wilhelm, D., *J. Chem. Soc., Perkin Trnas. 2,* **1984,** 1197.
[73] Glidewell, C., *J. Chem. Soc., Perkin Trans. 2,* **1984,** 1175.
[74] Gerson, F., Huber, W., and Lopez, J., *J. Am. Chem. Soc.*, **106,** 5808 (1984).
[75] Smith, I. M., Sutcliffe, L. H., Wiesner, S., Lubitz, W., and Kurreck, W., *J. Chem. Soc., Faraday Trans. 1,* **80,** 3021 (1984).
[76] Nelsen, S. F., Teasley, M. F., Kapp, D. L., and Wilson, R. M., *J. Org. Chem.*, **49,** 1843 (1984).
[77] Scaiano, J. C., Stewart, L. C., Livant, P., and Majors, A. W., *Can. J. Chem.*, **62,** 1339 (1984).
[78] Nelsen, S. F., Buschek, J. M., Göbl, M., and Asmus, K.-D., *J. Chem. Soc., Perkin Trans. 2,* **1984,** 11.
[79] Kaim, W., *J. Chem. Soc., Perkin Trans. 2,* **1984,** 1357.
[80] Minaev, B. F., Tikhomirav, V. A., and Dausheev, Zh. K., *Khim. Fiz.*, **3,** 615 (1984); *Chem. Abs.*, **101,** 71966 (1984).
[81] Chen, C. C., and Fox, M. A., *J. Comput. Chem.*, **4,** 488 (1983); *Chem. Abs.*, **100,** 120159 (1984).
[82] Weiske, T., van der Wel, H., Nibbering, N. M. M., and Schwarz, H., *Angew. Chem. Int. Ed.*, **23,** 733 (1984).
[83] Nunome, K., Toriyama, K., and Iwasaki, M., *Chem. Phys. Lett.*, **105,** 414 (1984); *Chem. Abs.*, **101,** 54413 (1984).
[84] Green, M. M., Mielke, S. L., and Mukhopadhyay, T., *J. Org. Chem.*, **49,** 1276 (1984).
[85] Kelsall, B. J. and Andrews, L., *J. Phys. Chem.*, **88,** 2723 (1984).
[86] Dass, C., Sack, T. M., and Gross, M. L., *J. Am. Chem. Soc.*, **106,** 5780 (1984).
[87] Dass, C., and Gross, M. L., *J. Am. Chem. Soc.*, **106,** 5775 (1984).
[88] Pancir, F. and Turecek, F., *Chem. Phys.*, **87,** 223 (1984); *Chem. Abs.*, **101,** 110031 (1984).
[89] Chan, W., Courtneidge, J. L., Davies, A. G., Djap, W. H., Gregory, P. S., and Yazdi, S. N., *J. Chem. Soc., Chem. Commun.*, **1984,** 1541.
[90] Camioni, D. M., and Franz, J. A., *J. Org. Chem.*, **49,** 1607 (1984).
[91] King, P. F., and O'Malley, R. F., *J. Org. Chem.*, **49,** 2803 (1984).
[92] Parker, V. D., *Acc. Chem. Res.*, **17,** 243 (1984).
[93] Alberti, A., Pedulli, G. F., Tiecco, M., Testaferri, L., and Tingoli, M., *J. Chem. Soc., Perkin Trans. 2,* **1984,** 975.
[94] Bethell, D., Clare, P. N., and Hare, G. J., *J. Chem. Soc., Perkin Trans. 2,* **1983,** 1889.
[95] Maruyama, K., and Ogawa, T., *Bull. Chem. Soc. Jpn.*, **56,** 3525 (1983).
[96] Saeva, F. D., Doney, J. J., and Morgan, D. P., *J. Org. Chem.*, **49,** 2027 (1984).
[97] Roth, H. D., and Schilling, M. L. M., *J. Am. Chem. Soc.*, **105,** 6805 (1983).
[98] Takahashi, Y., Miyashi, T., and Mukai, T., *J. Am. Chem. Soc.*, **105,** 6511 (1983).
[99] Roth, H. D., and Schilling, M. L. M., *J. Am. Chem. Soc.*, **106,** 2711 (1984).
[100] Roth, H. D., Schilling, M. L. M., and Raghavachari, K., *J. Am. Chem. Soc.*, **106,** 253 (1984).
[101] Roth, H. D., Schilling, M. L. M., and Wamser, C. C., *J. Am. Chem. Soc.*, **106,** 5023 (1984).
[102] Roth, H. D., Schilling, M. L. M., Mukai, T., and Miyashi, T., *Tetrahedron Lett.*, **24,** 5815 (1983).
[103] Pabon, R. A., and Bauld, N. L., *J. Am. Chem. Soc.*, **106,** 1145 (1984).

[104] Groenewold, G. S., Chess, E. K., and Gross, M. L., *J. Am. Chem. Soc.*, **106,** 539 (1984).
[105] Mattes, S. L., Luss, H. R., and Farld, S., *J. Phys. Chem.*, **87,** 4779 (1983).
[106] Pabon, R. A., Bellville, D. J., and Bauld, N. L., *J. Am. Chem. Soc.*, **106,** 2730 (1984).
[107] Citterio, A., Gentile, A., Minisci, F., Serravalle, M., and Ventura, S., *J. Org. Chem.*, **49,** 3364 (1984).
[108] Auricchio, S., Bianca, M., Citterio, A., Minisci, F., and Ventura, S., *Tetrahedron Lett.*, **25,** 3373 (1984).
[109] Langhals, E., Langhals, H., and Rüchardt, C., *Chem. Ber.*, **117,** 1436 (1984).
[110] Pankratov, A. N., Mushtakova, S. P., and Gribov, L. A., *Teor. Eksp. Khim.*, **20,** 202 (1984); *Chem. Abs.*, **101,** 190680 (1984).
[111] Pokhodenko, V. D., Koshechko, V. G., Titov, V. E., Colovatyi, V. G., and Shabel'nikov, V. P., *Teor. Eksp. Khim.*, **20,** 25 (1984); *Chem. Abs.*, **100,** 208873 (1984).
[112] Frolov, A. N., and Klokova, E. M., *Zh. Org. Khim.*, **19,** 1490 (1983); *Chem. Abs.*, **99,** 211849 (1983).
[113] Lobachev, V. L., Rudakov, E. S., and Zaev, E. E., *Khim. Fiz.*, **3,** 120 (1984); *Chem. Abs.*, **100,** 156053 (1984).
[114] Depew, M. C., Liv, Z., and Wan, J. K. S., *Spectrosc. Lett.*, **16,** 451 (1983) *Chem. Abs.*, **99,** 212044 (1983).
[115] Bock, H., Rittmeyer, P., Krebs, A., Schuetz, K., Voss, J., and Koepke, B., *Phosphorus Sulfur*, **19,** 131 (1984); *Chem. Abs.*, **101,** 22769 (1984).
[116] Shida, T., Haselbach, E., and Bally, T., *Acc. Chem. Res.*, **17,** 180 (1984).
[117] Gerson, F., *Top. Curr. Chem.*, **115,** 57 (1983); *Chem. Abs.*, **99,** 193997 (1983).
[118] McDonald, R. N., Chowdhury, A. K., and McGhee, W. D., *J. Am. Chem. Soc.*, **106,** 4112 (1984).
[119] Shchegoleva, L. N., Bilkis, I. I., and Schastnev, P. V., *Chem. Phys.*, **82,** 343 (1983); *Chem. Abs.*, **100,** 120157 (1984).
[120] Symons, M. C. R., and Bowman, W. R., *J. Chem. Soc., Chem. Commun.*, **1984,** 1445.
[121] Banerjee, S., Chorai, D. K., and Sen, S. P., *Indian J. Chem.*, **22B,** 598 (1983); *Chem. Abs.*, **100,** 5736 (1984).
[122] Yuknovski, I., and Andreev, G., *Izv. Khim.*, **16,** 389 (1983); *Chem. Abs.*, **101,** 6470 (1984).
[123] Wilhelm, D., Courtneidge, J. L., Clark, T., and Davies, A. G., *J. Chem. Soc., Chem. Commun.*, **1984,** 810.
[124] Walther, B. W., Williams, F., and Lemal, D. M., *J. Am. Chem. Soc.*, **106,** 548 (1984).
[125] Stevenson, G. R., Reiter, R. C., and Sedgwick, J. B., *J. Am. Chem. Soc.*, **105,** 6521 (1983).
[126] Irmen, W., Huber, W., Lex, J., and Müllen, K., *Angew. Chem. Int. Ed.*, **23,** 818 (1984).
[127] Huber, W., *Helv. Chim. Acta*, **67,** 625 (1984).
[128] Gerson, F., Knöbel, J., Metzger, A., Murata, I., and Bakasuji, K., *Helv. Chim. Acta*, **67,** 934 (1984).
[129] Screttas, C. G., and Micha-Screttas, M., *J. Phys. Chem.*, **87,** 3844 (1983).
[130] Wilhelm, D., Clark, T., Schleyer, P. von R., Courtneidge, J. L., and Davies, A. G., *J. Organomet. Chem.*, **273,** C1 (1984).
[131] Stevenson, G. R., Schock, L. E., and Reiter, R. C., *J. Phys. Chem.*, **87,** 4004 (1983).
[132] Parker, V. D., *Acta Chem. Scand.*, **38B,** 125 (1984).
[133] Parker, V. D., *Acta Chem. Scand.*, **38B,** 189 (1984).
[134] Bagdasar'yan, Kh. S., *Usp. Khim.*, **53,** 1073 (1984); *Chem. Abs.*, **101,** 151044 (1984).
[135] Karoli, L. L., Kuzmin, V. A., and Khudyakov, I. V., *Int. J. Chem. Kinet.*, **16,** 379 (1984).
[136] Boche, G., Schneider, D. R., and Wernicke, K., *Tetrahedron Lett.*, **25,** 2961 (1984).
[137] Bethell, D., McDowall, L. J., and Parker, V. D., *J. Chem. Soc., Chem. Commun.*, **1984,** 308.
[138] Butenschön, H., and de Meijere, A., *Tetrahedron Lett.*, **25,** 1693 (1984).
[139] Eisch, J. J., and Smith, L. E., *J. Organomet. Chem.*, **271,** 83 (1984).
[140] Andrieux, C. P., Saveant, J. M., and Zann, D., *Nouv. J. Chim.*, **8,** 107 (1984).
[141] Villar, H. O., Castro, E. A. and Rossi, R. A., *Z. Naturforsch.*, **39a,** 49 (1984); *Chem. Abs.*, **100,** 173922 (1984).
[142] Clark, T., *J. Chem. Soc., Chem. Commun.*, **1984,** 93.
[143] Taggart, D. L., Peppercorn, W., and Anderson, L. B., *J. Phys. Chem.*, **88,** 2875 (1984).
[144] Sosonkin, I. M., and Strogov, G. N., *Elektrokhimiya*, **20,** 481 (1984); *Chem. Abs.*, **101,** 37975 (1984).
[145] Norris, R. K., Barker, S. D., and Neta, P., *J. Am. Chem. Soc.*, **106,** 3140 (1984).
[146] Klima, J., Prousek, J., Ludvik, J., and Volke, J., *Collect. Czech. Chem. Commun.*, **49,** 1627 (1984).
[147] Gavars, R., Stradins, J., Baumane, L., and Baider, L., *J. Mol. Struct.*, **102,** 183 (1983); *Chem. Abs.*, **100,** 34044 (1984).
[148] Andrieux, C. P., Merz, A., Saveant, J.-M., and Tomahogh, R., *J. Am. Chem. Soc.*, **106,** 1957 (1984).
[149] Hammerich, O., and Parker, V. D., *Acta Chem. Scand.*, **37B,** 851 (1983).
[150] Novi, M., Garbarino, G., and Dell'Erba, C., *J. Org. Chem.*, **49,** 2799 (1984).
[151] Anne, A., and Moiroux, J., *Nouv. J. Chim.*, **8,** 259 (1984).

[152] Roberts, J. L., and Sawyer, D. T., *Isr. J. Chem.*, **23,** 430 (1983).
[153] Liebman, J. F., and Valentine, J. S., *Isr. J. Chem.* **23,** 439 (1983).
[154] Nagano, T., Yokoohji, K., and Hirobe, M., *Tetrahedron Lett.*, **25,** 965 (1984).
[155] Vidril-Robert, D., Maurette, M.-T., Oliveros, E., Hocquaux, M., and Jacquet, B., *Tetrahedron Lett.*, **25,** 529 (1984).
[156] Hocquaux, M., Jacquet, B., Vidril-Robert, D., Maurette, M.-T., and Oliveros, E., *Tetrahedron Lett.*, **25,** 533 (1984).
[157] Inoue, H., Bagano, T., and Hirobe, M., *Tetrahedron Lett.*, **25,** 317 (1984).
[158] Kim, Y. H., Yon, C. H., and Kim, H. J., *Chem. Lett.*, **1984,** 309.
[159] Ballester, M., Castañer, J., Riera, J., and Pascual, I., *J. Am. Chem. Soc.*, **106,** 3365 (1984).
[160] Ziyaev, A. A., Otroshchenko, O. S., Khalilova, Kh. D., and Tyshchenko, A. A., *Sint. Reakts. Sposobn. Org. Soedin.* **1983,** 49; *Chem. Abs.*, **101,** 110057 (1984).
[161] Baral-Tosh, S., Chattopadhyay, S. K., and Das, P. K., *J. Phys. Chem.*, **88,** 1404 (1984).
[162] Kim-Thuan, N., and Scaiano, J. C., *Int. J. Chem. Kinet.*, **16,** 371 (1984).
[163] Miller, J. R., Calcaterra, L. T., and Closs, G. L., *J. Am. Chem. Soc.*, **106,** 3047 (1984).
[164] Miller, J. R., Beitz, J. V., and Huddleston, R. K., *J. Am. Chem. Soc.*, **106,** 5057 (1984).
[165] Gerson, F., Huber, W., Martin, W. B., Caluwe, P., Pepper, T., and Szwarc, M., *Helv. Chim. Acta,* **67,** 416 (1984).
[166] Zusman, L. D., *Teor. Eksp. Khim.*, **20,** 362 (1984); *Chem. Abs.*, **101,** 151194 (1984).
[167] Yamamoto, Y., Nishida, S., Yabe, K., Hayashi, K., Takeda, S., and Tsumori, K., *J. Phys. Chem.*, **88,** 2368 (1984).
[168] Grampp, G., and Jaenicke, W., *Ber. Bunsen-Ges. Phys. Chem.*, **88,** 335 (1984); *Chem. Abs.*, **101,** 90123 (1984).
[169] Masuda, T., Kimura, T., Kadoi, E., Takahashi, T., Yamauchi, K., and Kondo, M., *J. Radiat. Res.*, **24,** 165 (1983); *Chem. Abs.*, **99,** 212083 (1983).
[170] Merenyi, G., Lind, J., and Eriksen, T. E., *J. Phys. Chem.*, **88,** 2320 (1984).
[171] Platen, M., and Steckhan, E., *Chem. Ber.*, **117,** 1679 (1984).
[172] Degtyarev, L. S., and Stetsenko, A. A., *Teor. Eksp. Khim.*, **19,** 419 (1983); *Chem. Abs.*, **100,** 5580 (1984).
[173] Bogillo, V. I., Atamanyuk, V. Yu., and Gragerov, I. P., *Teor. Eksp. Khim.*, 20, 375 (1984); *Chem. Abs.*, **101,** 151195 (1984).
[174] Bogillo, V. I., Levit, A. F., Kondratenko, N. V., Maletina, I. L., and Gragerov, I. P., *Teor. Eksp. Khim.*, **19,** 508 (1983); *Chem. Abs.*, **100,** 5581 (1984).
[175] Kaim, W., *Angew. Chem. Int. Ed.*, **23,** 613 (1984).
[176] Kurov, G. N., Svyatkina, L. I., Pal'chuk, E. G., Naumova, I. P., and Skvartsova, G. G., *Zh. Obshch. Khim.*, **54,** 178 (1984); *Chem. Abs.*, **100,** 120231 (1984).
[177] Yano, Y., Sakaguchi, T., and Nakazato, M., *J. Chem. Soc., Perkin Trans. 2,* **1984,** 595.
[178] Ohno, A., Goto, H. K. T., and Oka, S., *Bull. Chem. Soc. Jpn.*, **57,** 1279 (1984).
[179] Cotsaris, E., and Paddon-Row, M. N., *J. Chem. Soc., Perkin Trans. 2,* **1984,** 1487.
[180] Chau, D. D., Paddon-Row, M. N., and Patney, H. K., *Aust. J. Chem.*, **36,** 2423 (1983).
[181] Gurudutt, K. N., Pasha, M. A., Ravindranath, B., and Srinivas, P., *Tetrahedron,* **40,** 1629 (1984).
[182] Deshayes, H., and Pete, J.-P., *Can. J. Chem.*, **62,** 2063 (1984).
[183] Stamm, H., Assithianakis, P., Weiss, R., Bentz, G., and Buchholz, B., *J. Chem. Soc., Chem. Commun.*, **1984,** 753.
[184] Zhao, C., and Jiang, X., *Youji Huaxue,* **1983,** 466; *Chem. Abs.*, **100,** 173999 (1984).
[185] Kaim, W., *Nachr. Chem. Tech. Lab.*, **32,** 436 (1984); *Chem. Abs.*, **101,** 54070 (1984).
[186] Chu, S.-Y., Lee, T., and Lee, S.-L., *J. Phys. Chem.*, **88,** 2809 (1984).
[187] Eberson, L., *Acta Chem. Scand.*, **38B,** 439 (1984).
[188] Holm, T. *Acta Chem. Scand.*, **37B,** 567 (1983).
[189] Berberova, N. T., and Okhlobystin, O. Yu., *Khim. Geterotsikl. Soedin.*, **1984,** 1011; *Chem. Abs.*, **101,** 190665 (1984).
[190] Provsek, J., *Chem. Listy,* **78,** 284 (1984); *Chem. Abs.*, **100,** 208652 (1984).
[191] Lindley, J., *Tetrahedron,* **40,** 1433 (1984).
[192] Bowman, W. S., Rakshit, D., and Valmas, M. D., *J. Chem. Soc., Perkin Trans. 1,* **1984,** 2327.
[193] Kornblum, N., Singh, H. K., and Boyd, S. D., *J. Org. Chem.*, **49,** 358 (1984).
[194] Meijs, G. F., *J. Org. Chem.*, **49,** 3863 (1984).
[195] Swartz, J. E., and Stenzel, T. T., *J. Am. Chem. Soc.*, **106,** 2520 (1984).
[196] Tolbert, L. M., and Siddiqui, S., *J. Org. Chem.*, **49,** 1744 (1984).
[197] Bailey, W. F., Gagnier, R. P., and Patricia, J. J., *J. Org. Chem.*, **49,** 2098 (1984).

198 Liu, Y., Dang, H., Yang, D., Xu, G., and Hou, G., *Youji Huaxue*, **1983,** 363; *Chem. Abs.*, **100,** 120203 (1984).
199 Miller, B., and Haggerty, J. C., *J. Chem. Soc., Chem. Commun.*, **1984,** 1617.
200 Eisch, J. J., Behrooz, M., and Galle, J. E., *Tetrahedron Lett.*, **25,** 4851 (1984).
201 Eberson, L., and Greci, L., *J. Org. Chem.*, **49,** 2135 (1984).
202 Ashby, E. C., DePriest, R. N., Goel, A. B., Wenderoth, B., and Pham, T. N., *J. Org. Chem.*, **49,** 3545 (1984).
203 Kaim, W., *J. Am. Chem. Soc.*, **106,** 1712 (1984).
204 Kaim, W., and Lubitz, W., *Angew. Chem. Int. Ed.*, **22,** 892 (1983).
205 Nasipuri, D., Gupta, M. D., and Banerjee, S., *Tetrahedron Lett.*, **25,** 5551 (1984).
206 Newcomb, M., and Burchill, M. T., *J. Am. Chem. Soc.*, **105,** 7759 (1983).
207 Newcomb, M., and Williams, W. C., *Tetrahedron Lett.*, **25,** 2723 (1984).
208 Eleveld, M. B., and Hogeveen, H., *Tetrahedron Lett.*, **25,** 785 (1984).
209 Ashby, E. C., and Argyropoulos, J. N., *Tetrahedron Lett.*, **25,** 7 (1984).
210 Ashby, E. C., Bae, D. -H., Park, W. -S., DePriest, R. N., and Su, W. -Y., *Tetrahedron.*, **25,** 5107 (1984).
211 Magnus, P., and Albaugh-Robertson, P., *J. Chem. Soc., Chem. Commun.*, **1984,** 804.
212 Pierini, A. B., Penenory, A. B., and Rossi, R. A., *J. Org. Chem.*, **49,** 486 (1984).
213 Julliard, M., and Chanon, M., *Chem. Rev.*, **83,** 425 (1983).
214 Fox, M. A., *Acc. Chem. Res.*, **16,** 314 (1983).
215 Eaton, D. F., *Pure Appl. Chem.*, **56,** 1191 (1984).
216 Perrin, C. L., *J. Phys. Chem.*, **88,** 3611 (1984).
217 Truong, T. B., *J. Phys. Chem.*, **88,** 3906 (1984).
218 Tsuchida, A., Yamamoto, M., and Nishijima, Y., *J. Phys. Chem.*, **88,** 5062 (1984).
219 Smith, D. W., and Pienta, N. J., *Tetrahedron Lett.*, **25,** 915 (1984).
220 Kiryukhin, Yu. I., Borovkova, V. A., Sinitsyna, V. A., and Bagdasar'yan, Kh. S., *Dokl. Akad. Nauk SSSR*, **274,** 114 (1984); *Chem. Abs.*, **100,** 191291 (1984).
221 Hub. W., Schneider, S., Dörr, F., Oxman, J. D., and Lewis, F. D., *J. Phys. Chem.*, **87,** 4351 (1983).
222 Delaire, J. A., Castella, M., Faure, J., and De-Schryver, F. C., *Nouv. J. Chim.*, **8,** 231 (1984).
223 Lee, G. A., and Israel, S. H., *J. Org. Chem.*, **48,** 4557 (1983).
224 Chow, Y. L., Marciniak, B., and Mishra, P., *J. Org. Chem.*, **49,** 1457 (1984).
225 Das, P. K., Muller, A. J., and Griffin, G. W., *J. Org. Chem.*, **49,** 1977 (1984).
226 Kumar, C. V., Chattopadhyay, S. K., and Das, P. K., *J. Chem. Soc., Chem. Commun.*, **1984,** 1107
227 Mattay, J., Gersdorf, J., Leismann, H., and Steenken, S., *Angew. Chem. Int. Ed.*, **23,** 249 (1984).
228 Borg, R. M., Arnold, D. R., and Cameron, T. S., *Can. J. Chem.*, **62,** 1785 (1984).
229 Mattay, J., Gersdorf, J., and Freudenberg, U., *Tetrahedron Lett.*, **25,** 817 (1984).
230 Mazzocchi, P. H., Somich, C., and Ammon, H., *Tetrahedron Lett.*, **25,** 3551 (1984).
231 Ohga, K., Yoon, U. C., and Mariano, P. S., *J. Org. Chem.*, **49,** 213 (1984)
232 Ullrich, J. W., Chiu, F. -T., Tiner-Harding, T., and Mariano, P. S., *J. Org. Chem.*, **49,** 220 (1984).
233 Chiu, F. -T., Ullrich, J. W., and Mariano, P. S., *J. Org. Chem.*, **49,** 228 (1984).
234 Pac, C., Miyauchi, Y., Ishitani, O., Ihama, M., Yasuda, M., and Sakurai, H., *J. Org. Chem.*, **49,** 26 (1984).
235 Hashimoto, S., Bandow, H., Akimoto, H., Wang, J. -H., and Tang, X. -Y., *Int. J. Chem. Kinet.*, **16,** 1385 (1984).
236 Shepson, P. B., Edney, E. O., and Corse, E. W., *J. Phys. Chem.*, **88,** 4122 (1984).
237 Reichel, L. W., Griffin, G. W., Muller, A. J., Das, P. K., and Ege, S. N., *Can. J. Chem.*, **62,** 424 (1984).
238 Futamura, S., Kusunose, S., Ohta, H., and Kamiya, Y., *J. Chem. Soc., Perkin Trans. 1*, **1984,** 15.
239 Schaap, A. P., Siddiqui, S., Balakrishnan, P., Lopez, L., and Gagnon, S. D., *Isr. J. Chem.*, **23,** 415 (1983).
240 Gollnick, K., and Schnatterer, A., *Tetrahedron Lett.*, **25,** 185 (1984).
241 Gollnick, K., and Schnatterer, A., *Tetrahedron Lett.*, **25,** 2735 (1984).
242 Manring, L. E., Kramer, M. K., and Foote, C. S., *Tetrahedron Lett.*, **25,** 2523 (1984).
243 Yamaguchi, K., *Oxygen Radicals Chem. Bio., Proc., Int. Conf., 3rd 1983*, **1984,** 65; *Chem. Abs.*, **101,** 90109 (1984).
244 Santamaria, J., Gabillet, P., and Bokobza, L., *Tetrahedron Lett.*, **25,** 2139 (1984).
245 Davidson, R. S., and Pratt, J. E., *Tetrahedron*, **40,** 999 (1984).
246 Sawaki, Y., *Bull. Chem. Soc. Jpn.*, **56,** 3464 (1983).
247 Munger, K., and Fischer, H., *Int. J. Chem. Kinet.*, **16,** 1213 (1984).
248 Kropp, P. J., *Acc. Chem. Res.*, **17,** 131 (1984).

[249] Kropp, P. J., McNeely, S. A., and Davis, S. D., *J. Am. Chem. Soc.*, **105**, 6907 (1983).
[250] Reinholdt, K., and Margaretha, P., *Helv. Chim. Acta*, **66**, 2534 (1983).
[251] Lilie, J., and Koskikallio, J., *Acta Chem. Scand.*, **38A**, 41 (1984); *Chem. Abs.*, **101**, 54334 (1984).
[252] Hilinski, E. F., Huppert, D., Kelley, D. F., Milton, S. V., and Rentzepis, P. M., *J. Am. Chem. Soc.*, **106**, 1951 (1984).
[253] Slocum, G. H., and Schuster, G. B., *J. Org. Chem.*, **49**, 2177 (1984).
[254] Cristol, S. J., Seapy, D. G., and Aeling, E. O. *J. Am. Chem. Soc.*, **105**, 7337 (1983).
[255] Manring, L. E., and Peters, K. S., *J. Phys. Chem.*, **88**, 3516 (1984); *Chem. Abs.*, **101**, 72212 (1984).
[256] Baumann, H., and Timpe, H. -J., *Z. Chem.*, **24**, 19 (1984).
[257] Baumann, H., and Timpe, H. -J., *Z. Chem.*, **24**, 18 (1984).
[258] Encinas, M. V., Lissi, E. A., Zanocco, A., Stewart, L. C., and Scaiano, J. C., *Can. J. Chem.*, **62**, 386 (1984).
[259] Hasebe, M., Kogawa, K., and Tsuchiya, T., *Tetrahedron Lett.*, **25**, 3887 (1984).
[260] Hutton, R. S., Roth, H. D., and Bertz, S. H., *J. Am. Chem. Soc.*, **105**, 6371 (1983).
[261] Defoin, A., Defoin-Straatmann, R., and Kuhn, H. J., *Tetrahedron*, **40**, 2651 (1984).
[262] Vermeersch, G., Marko, J., Febvay-Garot, N., Caplain, S., Lablache-Cambier, A., and Jacquignon, P., *J. Chem. Soc., Perkin Trans. 1*, **1984**, 39.
[263] Vermeersch, G., Marko, J., Febvay-Garot, N., and Lablache-Combier, A., *J. Chem. Soc., Perkin, Trans. 2*, **1984**, 43.
[264] Jacobson, K. A., Muszkat, K. A., and Khait, I., *J. Chem. Soc., Chem. Commun.*, **1983**, 1384.
[265] Adeleke, B. B., Weir, D., Depew, M. C., and Wan, J. K. S., *Can. J. Chem.*, **62**, 117 (1984).
[266] McLauchlan, K. A., and Ritchie, A. J. D., *J. Chem. Soc., Perkin Trans. 2*, **1984**, 275.
[267] Yang, N. C., Chiang, W. -L., and Langan, J. R., *Tetrahedron Lett.*, **25**, 2855 (1984).
[268] Lewis, F. D., Zebrowski, B. E., and Correa, P. E., *J. Am. Chem. Soc.*, **106**, 187 (1984).
[269] Lewis, F. D., and Correa, P. E., *J. Am. Chem. Soc.*, **106**, 194 (1984).
[270] Fisch, M. H., Dannenberg, J. J., Pereyre, M., Anderson, W. G. Rens, J., and Grossman, W. E. L., *Tetrahedron*, **40**, 293 (1984).
[271] Padwa, A., Pulwer, M. J., and Rosenthal, R. J., *J. Org. Chem.*, **49**, 856 (1984).
[272] Ithoh, K., Yamada, H., and Sera, A., *Bull. Chem. Soc. Jpn.*, **57**, 2140 (1984).
[273] Swindell, C. S., deSolms, S. J., and Springer, J. P., *Tetrahedron Lett.*, **25**, 3797 (1984).
[274] Beyer, M., Ghaffari-Tabrizi, R., Jung, M., and Margaretha, P., *Helv. Chim. Acta*, **67**, 1535 (1984).
[275] Weisz, A., Kaftory, M., Vidavsky, I., and Mandelbaum, A., *J. Chem. Soc., Chem. Commun.*, **1984**, 18.
[276] Gassman, P. G., and Carroll, G. T., *J. Org. Chem.*, **49** 2074 (1984).
[277] Horowitz, A., *Chem. Halides, Pseudo-Halides, Azides*, **1**, 405 (1983); *Chem. Abs.*, **100**, 208645 (1984).
[278] Greenstock, C. L., *Isr. J. Chem.*, **24**, 1 (1984).
[279] Tilquin, B., Gourdin-Serveniere, C., Miyazaki, T., and Fueki, K., *Bull. Chem. Soc. Jpn.*, **57**, 2029 (1984).
[280] Piesiak, A., Schuchmann, M. N., Segota, H., and von Sonntag, C., *Z. Naturforsch.*, **39B**, 1262 (1984); *Chem. Abs.*, **101**, 190961 (1984).
[281] Bors, W., Tait, D., Michel, C., Saran, M., and Erben-Russ, M., *Isr. J. Chem.*, **24**, 17 (1984).
[282] Neta, P., Dizdaroglu, M., and Simic, M. G., *Isr. J. Chem.*, **24**, 25 (1984).
[283] Al-Sheikhly, M. I., Hissung, A., Schuchmann, H. -P., Schuchmann, M. N., von Sonntag, C., Garner, A., and Scholes, G., *J. Chem. Soc., Perkin Trans. 2*, **1984**, 601.
[284] Rivera, E., and Schuler, R. H. *J. Phys. Chem.*, **87**, 3966 (1983).
[285] Sevilla, M. D., Swarts, S., Riederer, H., and Hüttermann, J., *J. Phys. Chem.*, **88**, 1601 (1984).
[286] Chedekel, M. R., Land, E. J., Thompson, A., and Truscott, T. G., *J. Chem. Soc. Chem. Commun.*, **1984**, 1170.
[287] Grodkowski, J., and Neta, P., *J. Phys. Chem.*, **88**, 1205 (1984).
[288] Anderson, R. F., Packer, J. E., and Denny, W. A., *J. Chem. Soc., Perkin Trans. 2*, **1984**, 49.
[289] Clarke, E. D., Wardman, P., Wilson, I., and Tatsumi, K., *J. Chem. Soc., Perkin Trans. 2*, **1984**, 1155.
[290] Chaudhri, S. A., Göbl, M., Freyholdt, T., Asmus, K. -D., *J. Am. Chem. Soc.*, **106**, 5988 (1984).
[291] Göbl, M., Bonifacić, M., and Asmus, K. -D., *J. Am. Chem. Soc.*, **106**, 5984 (1984).
[292] Ahmad, R., and Armstrong, D. A., *Can. J. Chem.*, **62**, 171 (1984).
[293] Surdhar, P. S., Ahmad, R., and Armstrong, D. A., *Can. J. Chem.*, **62**, 580 (1984).
[294] Gilbert, B. C., Kelsall, P. A., Sexton, M. D., McConnacchie, G. D. G., and Symons, M. C. R., *J. Chem. Soc., Perkin Trans., 2*, **1984**, 629.
[295] Kiefer, J. H., and Budach, K. A., *Int. J. Chem. Kinet.*, **16**, 697 (1984).
[296] Seres, L., Gorgenyi, M., and Farkas, J., *Int. J. Chem. Kinet.*, **15**, 1133 (1983).
[297] Bernlöhr, W., Beckhaus, H. -D., and Rüchardt, C., *Chem. Ber.*, **117**, 1026 (1984).

[298] Kratt, G., Beckhaus, H. -D., and Rüchardt, C., *Chem. Ber.*, **117,** 1748 (1984).
[299] Bernlöhr, W., Beckhaus, H. -D., Lindner, H. -J., and Rüchardt, C., *Chem. Ber.*, **117,** 3303 (1984).
[300] Metzger, J. O., *Angew. Chem. Int. Ed.*, **22,** 889 (1983).
[301] Kaufmann, D., and de Meijere, A., *Chem. Ber.*, **117**, 3134 (1984).
[302] Gilbert, K. E., *J. Org. Chem.* **49,** 6 (1984).
[303] Franz, J. A., Camaioni, D. M., Beishline, R. R., and Dalling, D. K., *J. Org. Chem.*, **49,** 3563 (1984).
[304] Lamberts, J. J. M., and Laarhoven, W. H., *J. Org. Chem.*, **49,** 100 (1984).
[305] dos Santos Afonso, M., and Schumacher, H. J., *Int. J. Chem. Kinet.*, **16,** 103 (1984).
[306] Manuel, G., Faucher, A., and Mazerolles, P., *J. Organomet. Chem.*, **264,** 127 (1984).
[307] Takeshita, H., Matsuo, N., and Mametsuka, H., *Bull. Chem. Soc. Jpn.*, **57,** 2321 (1984).
[308] McNab, H., and Murray, M. E. -A., *J. Chem. Soc., Perkin Trans. 1,* **1984,** 1565.
[309] Hickson, C. L., and McNab, H., *J. Chem. Soc., Perkin Trans. 1,* **1984,** 1569.
[310] Wirz, J., *Pure Appl. Chem.*, **56,** 1289 (1984).
[311] Dewar, M. J. S., *J. Am. Chem. Soc.*, **106,** 669 (1984).
[312] Yamaguchi, K., *Theochem.*, **12,** 101 (1983); *Chem. Abs.*, **99,** 174931 (1983).
[313] Feller, B., Davidson, E. R., and Borden, W. T., *J. Am. Chem. Soc.*, **106,** 2513 (1984).
[314] Harcourt, R. D., and Little, R. D., *J. Am. Chem. Soc.*, **106,** 41 (1984).
[315] Shancke, A., and Wahlgren, U., *Acta Chem. Scand.*, **37A,** 771 (1983); *Chem. Abs.*, **100,** 208666 (1984).
[316] Naito, I., *Bull. Chem. Soc. Jpn.*, **56,** 2851 (1983).
[317] Caldwell, R. A., Sakuragi, H., and Majima, T., *J. Am. Chem. Soc.*, **106,** 2471 (1984).
[318] Caldwell, R. A., Dhawan, S. N., and Majima, T., *J. Am. Chem. Soc.*, **106,** 6454 (1984).
[319] Zimmt, M. B., Doubleday, C., and Turro, N. J., *J. Am. Chem. Soc.*, **106,** 3363 (1984).
[320] Goodman, J. L., and Berson, J. A., *J. Am Chem. Soc.*, **106,** 1867 (1984).
[321] Sugawara, T., Bethell, D., and Iwamura, H., *Tetrahedron Lett.*, **25,** 2375 (1984).
[322] Bushby, R. J., Mann, S., and Jesudason, M. V., *J. Chem. Soc., Perkin Trans. 1,* **1984,** 2457.
[323] Dolbier, W. R., and Burkholder, C. R., *J. Am. Chem. Soc.*, **106,** 2139 (1984).
[324] Adam, W., Hannemann, K., and Hössel, P., *Tetrahedron Lett.*, **25,** 181 (1984).
[325] Scaiano, J. C., and Wagner, P. J., *J. Am. Chem. Soc.*, **106,** 4626 (1984).
[326] Paste, D. J., and Yang, S. -H., *J. Am. Chem. Soc.*, **106,** 152 (1984).
[327] Lei, D., Hwang, R. -J., and Gaspar, P. P., *J. Organomet. Chem.*, **271,** 1 (1984).
[328] Aoyama, H., Ohnota, M., Sakamoto, M., and Omote, Y., *Tetrahedron Lett.*, **25,** 3327 (1984).
[329] Little, R. D., and Moeller, K. D., *J. Org. Chem.*, **48,** 4487 (1983).
[330] Nishida, S., Komiya, Z., Mizuno, T., Mikuni, A., Fukui, T., Tsuji, T., Murakami, M., and Shimizu, N., *J. Org. Chem.*, **49,** 495 (1984).
[331] Blondeel, G., de Bruyn, A., and De Keukeleire, D., *Tetrahedron Lett.*, **25,** 2055 (1984).
[332] Cristol, S. J., and Ali, M. Z., *Tetrahedron Lett.*, **24,** 5839 (1983).
[333] Peter, A., Acs, G., and Huhn, P., *Int. J. Chem. Kinet.*, **16,** 753 (1984).
[334] Peter, A., Acs, G., and Huhn, P., *Int. J. Chem. Kinet.*, **16,** 767 (1984).
[335] Peter, A., Acs, G., and Huhn, P., *Int. J. Chem. Kinet.*, **16,** 781 (1984).
[336] Dyall, L. K., *Chem. Halides, Pseudo-Halides, Azides,* **1,** 287 (1983); *Chem. Abs.*, **101,** 6186 (1984).
[337] Khusidman, M. B., Grigor'eva, N. V., Vyatkin, V. P., and Dobychin, S. L., *Zh. Obshch. Khim.*, **53,** 1568 (1983); *Chem. Abs.*, **99,** 212049 (1983).
[338] Citterio, A., and Ferrario, F., *J. Chem. Res. (S),* **1983,** 308.
[339] Galli, C., *J. Chem. Soc., Perkin Trans. 2,* **1984,** 897.
[340] Yasui, S., Nakamura, K., and Ohno, A., *J. Org. Chem.*, **49,** 878 (1984).
[341] Becker, H. G. O., Pfeifer, D., and Radeglia, R., *Z. Naturforsch.*, **38b,** 1591 (1983); *Chem. Abs.*, **100,** 156083 (1984).
[342] Packer, J. E. and Willson, R. L., *J. Chem. Soc., Perkin Trans. 2,* **1984,** 1415.
[343] Ahmad, I., Bethell, D., and Parker, V. D., *J. Chem. Soc., Perkin Trans. 2,* **1984,** 1527.
[344] Sawaki, Y., Ishiguro, K., and Kimura, M., *Tetrahedron Lett.*, **25,** 1367 (1984).
[345] Bethell, D., McDowall, L. J., and Parker, V. D., *J. Chem. Soc., Perkin Trans. 2,* **1984,** 1531.
[346] Handoo, K. L., Gadru, K., and Jain, C. K., *Indian J. Chem.*, **22B,** 526 (1983); *Chem. Abs.*, **100,** 5566 (1984).
[347] Howard, J. A., *Chem. Peroxides,* **1983,** 235; *Chem. Abs.*, **100,** 155899 (1984).
[348] Mimoun, H., *Isr. J. Chem.*, **23,** 451 (1983).
[349] Aleksandrov, Yv. A., Lebedev, S. A., Kuznetsova, N. V., and Tarakanova, N. V., *Zh. Obshch. Khim.*, **53,** 1564 (1983); *Chem. Abs.*, **99,** 211990 (1983).
[350] Min'ko, S. S., Puchin, V. A., Voronov, S. A., and Dikii, M. A., *Kinet. Katal.*, **25,** 744 (1984); *Chem. Abs.*, **101,** 151189 (1984).

[351] Rigaudy, J., Baranne-Lafont, J., Ranjon, A., and Caspar, A., *Bull. Soc. Chim. Fr. II*, **1984,** 187.
[352] Rigaudy, J., Caspar, A., Baranne-Lafont, J., and Chassagnard, C., *Bull. Soc. Chim. Fr. II*, **1984,** 195.
[353] Baumstark, A. L., and Vasquez, P. C., *J. Org. Chem.*, **49,** 2640 (1984).
[354] Cafferata, L. F. R., Eyler, G. N., and Mirifico, M. V., *J. Org. Chem.*, **49,** 2107 (1984).
[355] Mulder, P., and Louw, R., *Recl. Trav. Chim. Pays-Bas*, **103,** 148 (1984).
[356] Mulder, P., and Louw, R., *Recl. Trav. Chim. Pays-Bas*, **103,** 282 (1984).
[357] Burghardt, A., and Kulicki, Z., *Monatsh. Chem.*, **115,** 87 (1984).
[358] Schieberle, P., Grosch, W., and Firl, J., *Oxygen Radicals Chem. Biol., Proc. Int. Conf., 3rd, 1983*, **1984,** 257; *Chem. Abs.*, **101,** 130021 (1984).
[359] Kulsrestha, G. N., Shanker, U., and Sharma, J. S., *J. Chem., Technol. Biotechnol. Chem. Technol.*, **34A,** 103 (1984); *Chem. Abs.*, **101,** 90212 (1984).
[360] Lyavinets, A. S., and Chervinskii, K. A., *Deposited Doc.*, **1982,** SPSTL 317 Khp. -D82, 22pp.; *Chem. Abs.*, **101,** 72064 (1984).
[361] Lyavinets, A. S., and Chervinskii, K. A., *Deposited Doc.*, **1982,** SPSTL 318 D82, 17 pp.; *Chem. Abs.*, **101,** 72065 (1984).
[362] Shushunova, A. F., and Prokhorova, C. Yu., *J. Chromatogr.*, **283,** 365 (1984); *Chem. Abs.*, **100,** 156088 (1984).
[363] Maillard, B., Kharrat, A., and Gardrat, C., *Tetrahedron*, **40,** 3531 (1984).
[364] Pryor, W. A., and Hendrickson, W. H., *J. Am. Chem. Soc.*, **105,** 7114 (1983).
[365] Zhao, C., El-Taliawi, G. M., and Walling, C., *J. Org. Chem.*, **48,** 4908 (1983).
[366] Faucitano, A., Buttafava, A., Caporiccio, G., and Viola, C. T., *J. Am. Chem. Soc.*, **106,** 4172 (1984).
[367] Kanakaraj, P., and Maruthamuthu, P., *Int. J. Chem. Kinet.*, **15,** 1301 (1983).
[368] Ryan, K. R., and Plumb, I. C., *Int. J. Chem. Kinet.*, **16,** 591 (1984).
[369] Howard, J. A., *Isr. J. Chem.*, **24,** 33 (1984).
[370] Winterle, J., Dulin, D., and Mill, T., *J. Org. Chem.*, **49,** 491 (1984).
[371] Porter, J. E., and Schuster, G. B., *J. Org. Chem.*, **48,** 4944 (1983).
[372] Quinga, E. M. Y., and Mendenhall, G. D., *J. Am. Chem. Soc.*, **105,** 6520 (1983).
[373] Niki, E., Tanimura, R., Kamiya, Y., Takyu, C., and Inaba, R., *Sekiyu Gakkaishi*, **27,** 15 (1984); *Chem. Abs.*, **100,** 138412 (1984).
[374] Atkinson, R., Aschmann, S. M., Carter, W. P. L., Winer, A. M., and Pitts, J. N., *Int. J. Chem. Kinet.*, **16,** 1085 (1984).
[375] Brezinsky, K., Litzinger, T. A., and Glassman, I., *Int. J. Chem. Kinet.*, **16,** 1053 (1984).
[376] Onsager, O. -T., Swensen, H. C. A., and Johansen, J. E., *Acta Chem. Scand.*, **38B,** 567 (1984).
[377] Sawaki, Y., and Foote, C. S., *J. Org. Chem.*, **48,** 4934 (1983).
[378] Sawaki, Y., and Ogata, Y., *J. Org. Chem.*, **49,** 3344 (1984).
[379] van den Heuvel, C. J. M., Hofland, A., van Velzen, J. C., Steinberg, H., and de Boer, Th. J., *Recl. Trav. Chim. Pays-Bas*, **103,** 233 (1984).
[380] Akaba, R., Sakuragi, H., and Tokumaru, K., *Tetrahedron Lett.*, **25,** 665 (1984).
[381] Finkelshtein, E. I., Rubchinskaya, Yu. M., and Kozlov, E. I. *Int. J. Chem. Kinet.*, **16**, 513 (1984).
[382] Gollnick, K., and Griesbeck, A., *Tetrahedron*, **40,** 3235 (1984).
[383] Bartlett, P. D., and McCluney, R. E., *J. Org. Chem.*, **48,** 4165 (1983).
[384] Martem'yanov, V. S., Khlebnikov, V. N., Lapkin, L. M., Sharafutdinova, Z. F., and Gaisina, L. N., *Kinet. Katal.*, **25,** 781 (1984); *Chem. Abs.*, **101,** 190909 (1984).
[385] Endo, T., Hashimoto, M., Orii, T., and Ito, M. M., *Bull. Chem. Soc. Jpn.*, **57,** 1562 (1984).
[386] Bruice, T. C., *Isr. J. Chem.*, **24,** 54 (1984).
[387] Barton, D. H. R., Crich, D., and Motherwell, W. B., *J. Chem. Soc., Chem. Commun.*, **1984,** 242.
[388] Barton, D. H. R., and Crich, D., *J. Chem. Soc., Chem. Commun.*, **1984,** 774.

Organic Reaction Mechanisms 1984
Edited by A. C. Knipe and W. E. Watts

CHAPTER 5

Oxidation and Reduction

G. W. J. Fleet

Dyson Perrins Laboratory, South Parks, Road, Oxford

Oxidation by Metal Ions and Related Species

The homogenous catalysis of oxidation reactions with phosphine complexes,[1] metal nitro complexes as oxygen-transfer agents for the selective oxidation of organic substrates by oxygen,[2] the heterolytic and homolytic transfers of oxygen from d^0 metal peroxides to hydrocarbons,[3] and the rôle of transition-metal salts in one-electron-transfer organic reactions have been reviewed.[4] A discussion of transition-metal-catalysed oxidations classifies reactions into oxidations (where the oxidant acts only as an electron acceptor) and oxygenations (where one or two of the oxygen atoms of the oxidant are incorporated into the organic product).[5] It is asserted that thc oxidative cleavage of C—H bonds by solutions of metal complexes and oxidants generally proceeds *via* cyclic transition states; probable structures for these transition states have been analysed.[6]

Chromium and Manganese

Good kinetic data have been obtained for the oxidation of benzyl alcohol by chromic acid in a two-phase system by preventing further oxidation of benzaldehyde to benzoic aicd.[7] The oxidation of alkyl mercaptans by acid dichromate involves rate-determining fragmentation of a complex formed between the thiol and chromic acid.[8] Oxidation of mesidine (**1**) with chromate in methanol gives (**2**) as the principal product, with a small amount of (**3**), in which an alkyl migration has occurred.[9] 1,10-Phenanthroline catalyses the hydroxylation of anisole by chromic acid; the active oxidant is a chromium(VI)–phenanthroline complex.[10] Kinetic studies on the chromic acid oxidation of succinic acid have been reported.[11] The oxidations of diols[12] and aryl methyl sulphides[13] by pyridinium chlorochromate and of alcohols by pyridinium fluorochromate[14, 15] have been investigated. An attempt has been made to separate electronic and steric effects on the rates of oxidation of aryl methyl sulphides by chromium(VI) reagents.[16]

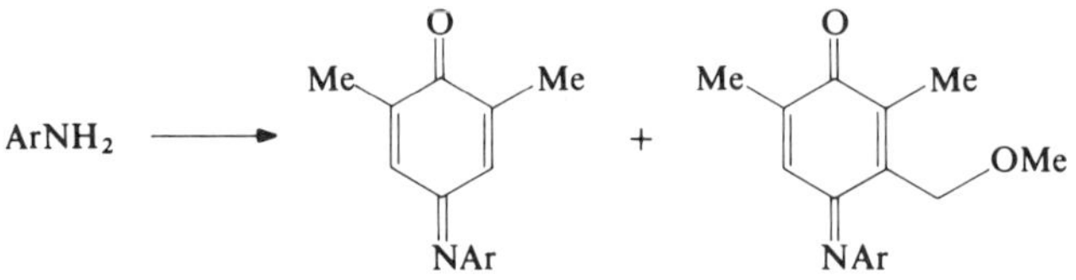

ArNH$_2$ → (**2**) + (**3**)

(**1**) (**2**) (**3**)

Ar = 2,4,6-trimethylphenyl (Me, Me, Me)

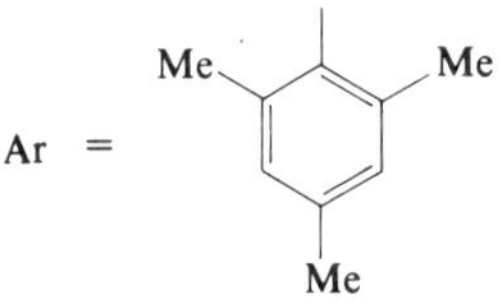

The reaction of $ClCrO_2^+$ with ethene in a selected-ion flow tube gives $C_2H_3O^+$ and $ClCrO^+$; reaction of $ClCrO^+$ with ethylene oxide gives $C_2H_3O^+$ as the only observed ionic product. This is the first study of the gas-phase reactivity of a d^0 transition-metal oxo species with an alkene; the results may be interpreted in terms of the Sharpless mechanism of alkene oxidation involving oxametallocyclobutane formation followed by reductive elimination of an epoxide (oxygen transfer) or metathesis to give formaldehyde and a chromium(VI) oxo-carbene.[17] There is considerable similarity in the oxidations of organic sulphides by chromium(VI) oxide diperoxide and the corresponding molybdenum species.[18] Kinetic studies on the oxidation of formic acid by chromium peroxydichromate have been reported.[19]

Zinc permanganate is a surprisingly powerful oxidant for organic substrates, reacting instantaneously (with fires in some cases) with such common solvents as tetrahydrofuran, methanol, ethanol, *tert*-butanol, acetone, and acetic acid; complexation of the organic substrates to zinc greatly enhances their reactivity towards permanganate oxidation.[20] In oxidations of pyrocatechol–metal salts with potassium permanganate under aprotic conditions to the corresponding semiquinones, the oxidation potentials and the semiquinone ESR coupling constants depend on the metal counter-ion unless a cryptand is added.[21] The formation of oxonitine by permanganate oxidation of aconitine has been studied in 20:1 acetone:water; isotopic labelling demonstrates that the *N*-formyl group of oxonitine originates from the methylene group of the *N*-ethyl group of aconitine under these conditions (Scheme 1).[22] Mechanistic studies have appeared on the permanganate oxidations of 1,2,3,4-thiatriazol-5-thioate ion[23] and of (2-^{14}C)-acetate.[24].

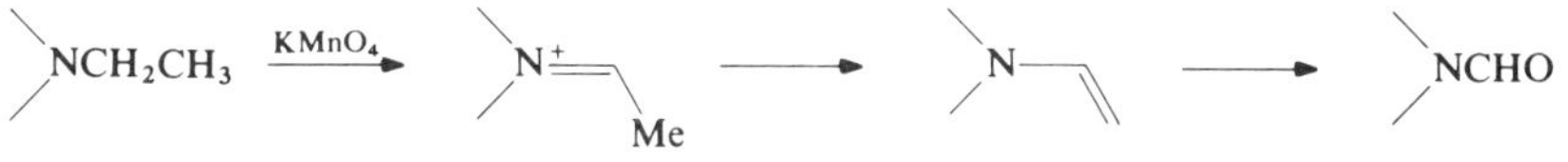

$$>NCH_2CH_3 \xrightarrow{KMnO_4} >\overset{+}{N}{=}CHMe \longrightarrow >N{-}CH{=}CH_2 \longrightarrow >NCHO$$

SCHEME 1

Manganese(III) inhibits the oxidation of 4-methylpyridine by manganese(IV) in sulphuric acid, due to the reaction of a manganese(IV)–substrate complex with manganese(III).[25] The most important factor governing the product distribution in the oxidation of alkenes by manganese(III) acetate in acetic acid–acetic anhydride is the composition of the solvent; the oxidation of 1-octene in the absence of acetic anhydride gives solely γ-decanolactone (Scheme 2), whereas the presence of even a small quantity of acetic anhydride gives other products derived from the cationic intermediate (**4**).[26] The oxidation of acetone by manganese(III) in acetic acid in the presence of alkenes and heteroaromatic bases produces the acetonyl radical which adds to the olefin; the resulting nucleophilic radical is scavenged by the bases; the radical addition reaction only competes with further manganese(III) oxidation of the alkyl adduct in the presence of strong acid.[27] The manganese(III) oxidations of mandelic acid[28, 29] and *o*-phenylenediamine[30] have been investigated. Kinetic data for the catalytic oxidation of ethylene glycol by hydrogen peroxide in the presence of bicarbonate complexes of manganese(II) indicate a radical non-chain mechanism involving bicarbonate complexes of manganese(III) and manganese(IV).[31] The rate-determining step in the epoxidation of cyclohexene by manganese(II)–prophyrin and

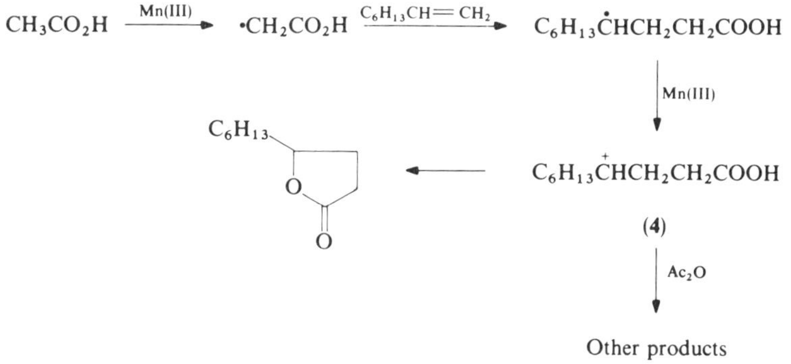

$$CH_3CO_2H \xrightarrow{Mn(III)} \cdot CH_2CO_2H \xrightarrow{C_6H_{13}CH{=}CH_2} C_6H_{13}\dot{C}HCH_2CH_2COOH$$

$$\downarrow Mn(III)$$

$$C_6H_{13}\overset{+}{C}HCH_2CH_2COOH \; (\mathbf{4}) \longrightarrow \gamma\text{-lactone } (C_6H_{13}\text{-substituted}, C{=}O)$$

$$(\mathbf{4}) \xrightarrow{Ac_2O} \text{Other products}$$

SCHEME 2

sodium hypochlorite is the conversion of a manganese(III) species into an oxo-manganese(V) species.[32] A mechanism has been proposed to account for the formation of alcohols and ketones in the oxidation of alkenes by oxygen activated by a manganese(II)–porphyrin complex.[33] The product pattern of the metal-ion oxidations of aromatic hydrocarbons and arylacetic acids by hetero-polyanions containing nickel(IV), manganese(IV), and cobalt(III) ions as central atoms is consistent with an outer-sphere electron-transfer mechanism.[34]

Gold, Silver, and Copper

The selective catalytic oxidation of thio-ethers to sulphoxides by gold(III) halides has been acheived by a phase-transfer process.[35] Radical mechanisms are involved in the silver(II) oxidations of pinacol[36] and benzamide.[37] Linear-free-energy studies on the copper(III) oxidation of aromatic aldehydes indicate that an electron-rich site develops in the transition state.[38] Copper(I) alkoxides are oxidized by molecular oxygen in acetonitrile to carbonyl compounds; a mechanism involving copper(III) has been proposed (Scheme 3).[39] The catalytic activity of copper(I) in the autoxidation of *para*-disubstituted benzoins has been investigated; a redox shuttle mechanism has been proposed where the rate-determining step is autoxidation of copper(I), followed by rapid oxidation of the benzoin by an oxocupric species.[40] Mechanistic aspects of the copper(II) oxidations of lactose and maltose,[41] and of dihydroxyfumaric acid[42] have been discussed.

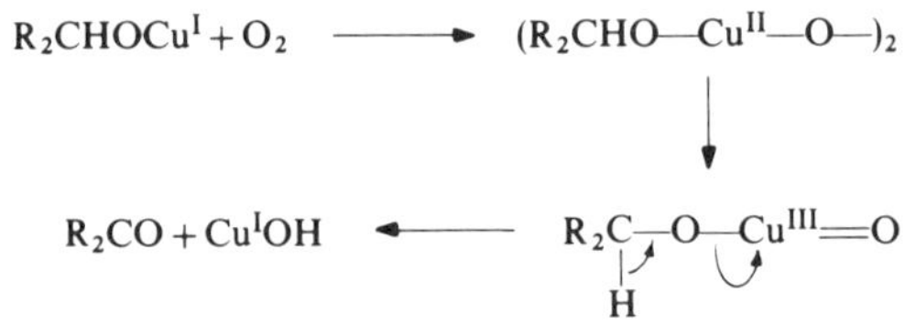

$$R_2CHOCu^I + O_2 \longrightarrow (R_2CHO{-}Cu^{II}{-}O{-})_2$$

$$\downarrow$$

$$R_2CO + Cu^IOH \longleftarrow R_2C(H){-}O{-}Cu^{III}{=}O$$

SCHEME 3

1,2-Cyclohexandiones are dioxygenated by molecular oxygen in the presence of copper(II) to give 1,5-keto-acids and carbon monoxide, probably *via* an endoperoxide (**5**) intermediate.[43] Alcohols are oxidized to aldehydes by oxygen in the presence of copper(II) and the nitrosyl radical (**6**) (Scheme 4).[44] Desoxybenzoin is converted by copper(II)–tertiary base–methanol and oxygen to a mixture of benzil,

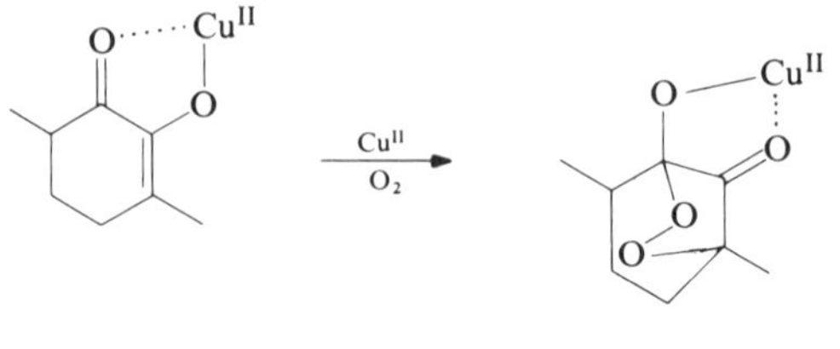

Cu^{II}, O_2

(**5**)

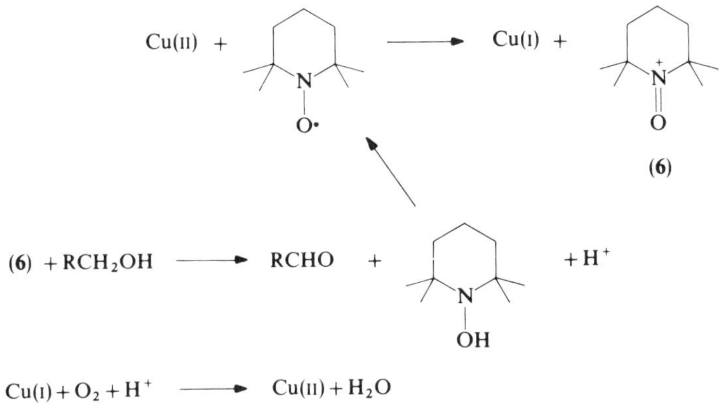

$Cu(I) + O_2 + H^+ \longrightarrow Cu(II) + H_2O$

SCHEME 4

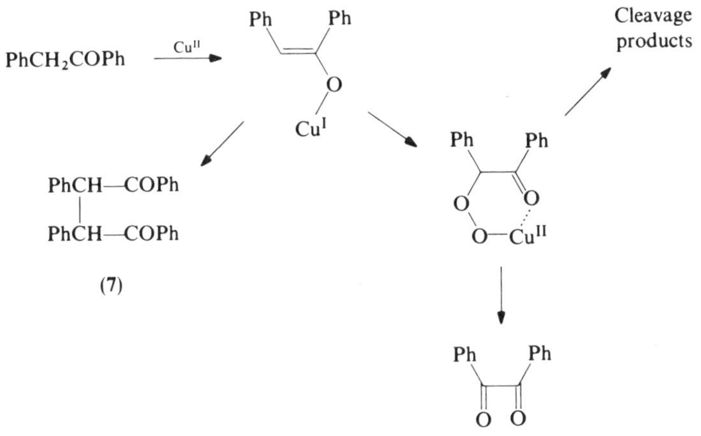

bidesyl (7), and cleavage products.[45] Mechanistic aspects of oxidations of anthraquinone,[46] dioxosuccinic acid,[47] methylhydrazines,[48] and hydroquinone[49] by molecular oxygen catalysed by copper(II) have been investigated.

Lead, Thallium, and Mercury

Oxidation of biphenylene with lead(IV) acetate gives 2-substituted biphenylenes and 2,3-biphenylenedione; oxidation of 2-methoxybiphenylene gives the ring-expanded ketone (**8**) (Scheme 5).[50] Radical cations and enamines have been demonstrated to be intermediates in the lead tetraacetate oxidation of tertiary amines.[51] Factors affecting the competition between intramolecular cyclization and β-fragmentation in the lead tetraacetate oxidation of cycloalkylmethanols include the size of the ring, transannular interactions, and steric and conformational effects.[52] A mechanism has been proposed to rationalize the products of lead tetraacetate oxidation of

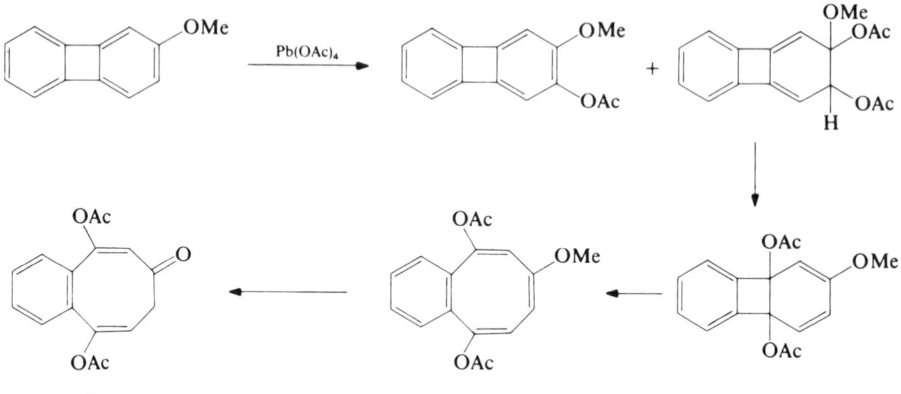

OMe Pb(OAc)$_4$ OMe OAc + OMe OAc OAc H

OAc O OAc OMe OAc OMe OAc OAc OMe OAc

(8)

SCHEME 5

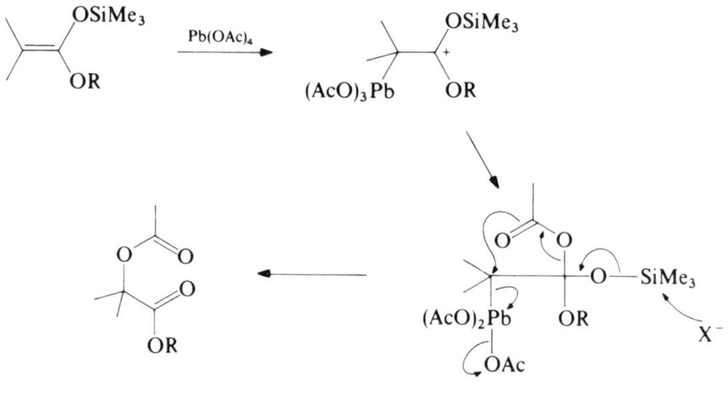

OSiMe$_3$ OR Pb(OAc)$_4$ OSiMe$_3$ + (AcO)$_3$Pb OR

O O O O OR O O O—SiMe$_3$ (AcO)$_2$Pb OR X$^-$ OAc

SCHEME 6

cyclohexylethanol.[53] Lead(IV) carboxylate oxidation of alkyltrimethylsilylketene acetals leads to α-carboyloxy-esters and -lactones (Scheme 6).[54]

An intermediate diazonium salt was detected in the oxidation of 4-ethylquinoline-2-hydrazine by thallium(III).[55] A radical mechanism has been proposed for the thallium(III) oxidation of benzaldehyde semicarbazone.[56] An acylthallium species may be an intermediate in the oxidations of maltose and lactose by thallium(III).[57] Micelle formation has been shown to increase the rate of thallium(III) oxidative cyclization of 1,3,5-triarylformazan.[58] A linear-free-energy relationship treatment has been used to separate polar and steric effects in the oxidation of aliphatic ketones by thallium(III) sulphate.[59] The kinetics of the oxidations of lactic and mandelic acids,[60] of catechol[61] by thallium(II), and of quinol by mercury(II)[62] have been investigated.

Cerium, Titanium, Vanadium, Molybdenum, and Tungsten

Studies on the Belousov–Zhabotinskii reaction are reported in the section on bromate oxidation. Ceric sulphate oxidations of lactic and atrolactic acids are complex, proceeding through three simultaneous pathways involving intermediate complexes with cerium(IV) species.[63] Polyhydroxylated benzenes are intermediates in the oxidation of nitrophenols by cerium(IV).[64] The rate-determining step in the oxidation of allyl alcohol by cerium(IV) sulphate is decomposition of a complex of cerium(IV) and allyl alcohol.[65] The reaction of cerium(IV) with 1,2-diphenylethane proceeds *via* radical cations,[66] whereas that of furfural involves free radicals.[67] Kinetic studies on the cerium(IV) oxidations of 4-methylmorpholine,[68] cyclic ketones,[69] and 3-substituted 1-aminoguanidines[70, 71] have been reported.

The mechanism of the titanium-tartrate-catalysed asymmetric epoxidation is discussed in terms of kinetic resolution of racemic secondary allylic alcohols as well as the enantioselectivity of epoxidation of prochiral substrates. A dimeric structure with C_2 symmetry has been proposed for the catalyst; the alignment of a lone pair of the reactive alkylperoxo oxygen atom with the olefin π-orbital is postulated as an important interaction in the transition state.[72] A review of the uses of the Sharpless epoxidation has appeared.[73] A mechanism, based on a study of the stoichiometry and products of the reaction, has been proposed for the oxidation of amino-alcohols by vanadium(V) (Scheme 7).[74] The vanadium(V) oxidations of *N*-phenylbenzohydroxamic acid,[75] alcohols,[76] and diols[77] have been investigated.

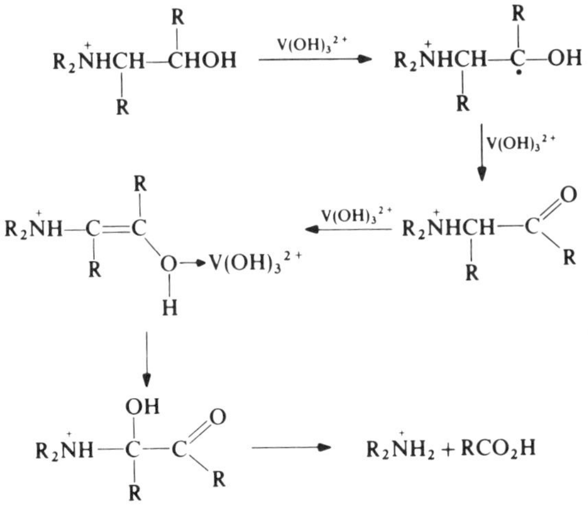

$$R_2\overset{+}{N}HCH(R)\text{—}CH(R)OH \xrightarrow{V(OH)_3^{2+}} R_2\overset{+}{N}HCH(R)\text{—}\dot{C}(R)\text{—}OH$$

$$\xrightarrow{V(OH)_3^{2+}} R_2\overset{+}{N}HCH(R)\text{—}C(=O)R \xrightarrow{V(OH)_3^{2+}} R_2\overset{+}{N}H\text{—}C(R)=C(R)\text{—}O(H)\rightarrow V(OH)_3^{2+}$$

$$\longrightarrow R_2\overset{+}{N}H\text{—}C(OH)(R)\text{—}C(=O)R \longrightarrow R_2\overset{+}{N}H_2 + RCO_2H$$

SCHEME 7

Molybdenum porphyrins are stereoselective catalysts for the epoxidation of alkenes by *tert*-butyl hydroperoxide; molybdenum–porphyrin complexes are the active epoxidizing species.[78] Other studies on molybdenum-catalysed epoxidations have appeared.[79, 80] Bis(acetylacetonato)dioxomolybdenum is an effective catalyst for the *tert*-butyl hydroperoxide oxidation of alkenes resulting in cleavage of the double bond; initial epoxidation is followed by cleavage of the epoxide *via* a five-membered metallocycle.[81] Benzyltrimethylammonium tetrabromooxomolybdate (**9**) selectively oxidizes alcohols catalytically, with *tert*-butyl hydroperoxide as the oxidant.[82] Kinetic studies on the hydrogen peroxide oxidations of allyl chloride[83] and acrylic acid[84] catalysed by sodium tungstate have been reported.

$$PhCH_2\overset{+}{N}Me_3\ {}^{-}OMoBr_4$$

(**9**)

Group VIII Metals

Oxidative substitution of methylarenes by iron(III) complexes proceeds by slow initial electron transfer (Scheme 8); Marcus theory enables the intrinsic barrier for electron transfer from arenes to be quantitatively described in terms of solvation and Jahn–Teller distortion of arene cation radicals.[85] In contrast, hydrogen atom abstraction has been claimed to be rate-determining in the oxidation of nitro-toluenes[86] and of methoxytoluenes[87] by hexacyanoferrate(III). Radicals have been identified by spectroscopic and other studies in the hexacyanoferrate oxidations of resorcinols,[88] phloroglucinol[89] pyrogallol,[90] hydroxylamine,[91] trimethylamine,[92] phenanthrene,[93] and glyceraldehyde phosphate.[94] Mechanisms have been proposed for the hexacyannoferrate(III) oxidations of quaternary pyridinium[95] and quin-olinium[96] salts.

$$ArMe + Fe(III) \xrightarrow{slow} Fe(II) + [ArCH_3]^{+\bullet} \longrightarrow products$$

SCHEME 8

Three distinct pathways occur in the ferricyanide oxidation of the 10-methyl-acridinium cation in base, depending on the pH of the reaction; for pH up to 9.7, rate-determining attack of ferricyanide on the neutral pseudo-base predominates, while at pH higher than 13 the major pathway involves the reaction of ferricyanide with the pseudobase alkoxide ion. Between pH 9.7 and 13, initial disproportionation of the acridinium cation followed by ferricyanide oxidation of the resulting 9,10-dihydro-10-methylacridine is the principal oxidation route.[97] Hexacyanoferrate(II) inhibits the oxidation of dihydropyridines and related compounds by hexacyanoferrate(III); it was not possible to differentiate between one-electron and hydride-transfer pathways for these oxidations on the basis of linear-free-energy relationship studies.[98] The effects of base on iron(III) oxidations of NADH models have been investigated.[99]

Iron(III)-catalysed oxygenation of 2-halocyclohexanones in methanol (Scheme 9) leads to the formation of adipic acid dimethyl esters.[100] Selectivity in the oxidation of saturated hydrocarbons by oxygen in pyridine–acetic acid in the presence of an iron catalyst and zinc is strongly dependent on the oxygenation of the reaction mixture; it is possible, by using air and slow stirring, to attack only the secondary positions of adamantane and *trans*-1,4-dimethylcyclohexane.[101] Iron(III) chloride

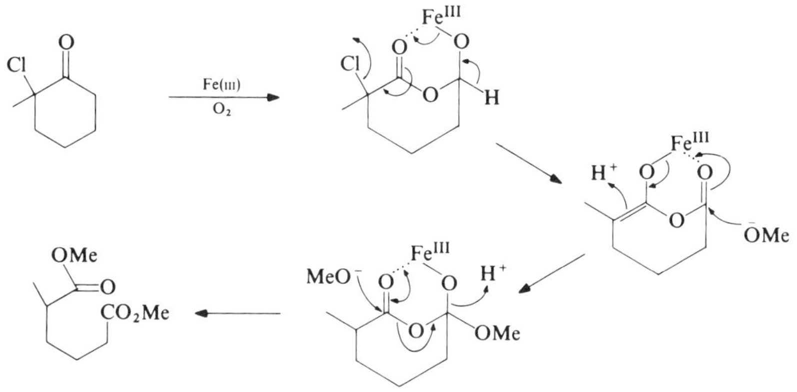

Scheme 9

and (acetylacetonato)iron(III), like (tetraphenylporphyrinato)iron, catalyse the epoxidation of styrene, stilbenes, and 1-nonene by iodosobenzene; however, several characteristics, such as stereospecificity, are different for non-porphyrin complexes, indicating the importance of the porphyin ligand in cytochrome P450 and heme model reactions.[102] The regioselectivity of olefin oxidations by iodosobenzene catalysed by metalloporphyrins has been studied. Besides epoxides, allylic alcohols and aldehydes are also formed; the latter products are not derived from the epoxides. Competition between epoxidation and allylic oxidation can be controlled by non-bonded interactions between the olefin and porphyrin substituents.[103] A peroxidase model system, with various model hemin compounds as catalysts, *m*-chloroperbenzoic acid as oxidant and tri-*tert*-butylphenol as substrate, has been investigated.[104] Mechanistic studies on the iron(III)-catalysed peracetic acid oxidation of catechol to *cis, cis*-muconic acid have been used as a model of iron(III)-containing pyrocatechase.[105] Mechanistic aspects of the hydrogen peroxide oxidations of dihydroxyfumaric acid[106] and indole-3-acetic acid[107] catalysed by iron(II), and of indigo carmine[108] and 5-methylphenazine[109] catalysed by iron(III), have been discussed.

Oxidation of aryl-conjugated olefins by cobalt(III) in wet acetic acid gives the corresponding glycol monoacetates *via* one-electron-transfer processes.[110] In the reaction of valeric acid with cobalt(III), an inner-sphere mechanism occurs involving rapid reversible formation of a cobalt(III) complex with the acid, followed by slow fragmentation to alkyl radicals.[111] Cobalt ion oxidations of methylnaphthalenes[112] and benzoin[113] have been investigated.

Among other studies on the oscillating reaction, oxidation of benzaldehyde by oxygen in the presence of cobalt ions,[114] it has been shown that the benzoyl radical (**10**) undergoes primarily reaction with oxygen in one stage while being oxidized by cobalt(III) in the second stage (Scheme 10).[115] Regioselectivity, but not stereoselectivity, is observed in the conversion of aryl olefins to benzyl alcohols by molecular oxygen and sodium borohydride in the presence of a catalytic amount of cobalto-tetraphenylporphyrin (Scheme 11).[116] Mechanistic aspects of cobalt-catalysed

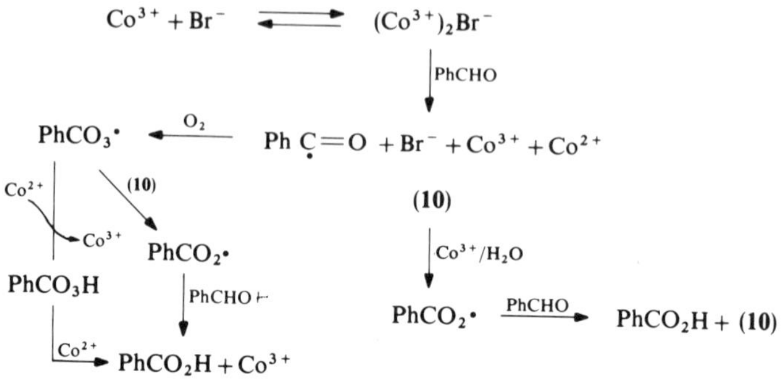

$$Co^{3+} + Br^- \rightleftharpoons (Co^{3+})_2Br^-$$

$$(Co^{3+})_2Br^- \xrightarrow{PhCHO} Ph\dot{C}{=}O + Br^- + Co^{3+} + Co^{2+}$$

(10)

$$Ph\dot{C}{=}O \xrightarrow{O_2} PhCO_3^{\bullet}$$

$$PhCO_3^{\bullet} \xrightarrow{Co^{2+} \rightarrow Co^{3+}} PhCO_3H \xrightarrow{Co^{2+}} PhCO_2H + Co^{3+}$$

$$PhCO_3^{\bullet} \xrightarrow{(10)} PhCO_2^{\bullet} \xrightarrow{PhCHO} PhCO_2H + Co^{3+}$$

$$(10) \xrightarrow{Co^{3+}/H_2O} PhCO_2^{\bullet} \xrightarrow{PhCHO} PhCO_2H + (10)$$

SCHEME 10

$$ArCH{=}CH_2 + HCoL \longrightarrow ArCH(Me)Co^{III}L \rightleftharpoons Ar\dot{C}HMe \quad Co^{II}L$$

$$Ar\dot{C}HMe \xrightarrow{O_2} ArCH(Me)OOCo^{II}L \xrightarrow{BH_4^-} ArCH(Me)OOH \longrightarrow ArCH_2OH + ArCOMe$$

$$ArCOMe \xrightarrow{BH_4^-} ArCH_2OH$$

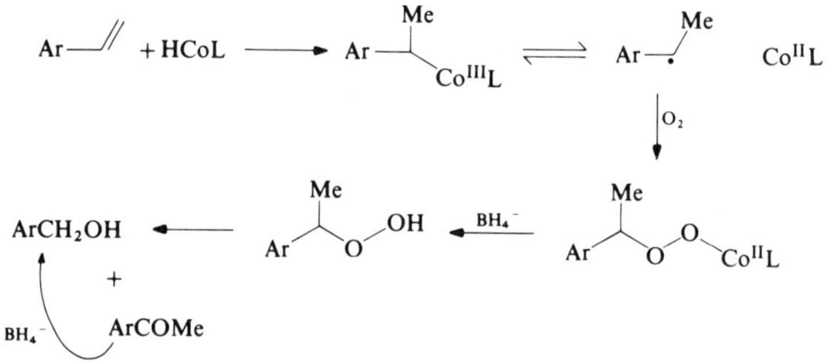

SCHEME 11

autoxidations of phenols,[117] allyl ethers,[118] cyclohexanone,[119] xylenes,[120] cyclohexane[121] and thiols[122] have been investigated.

The oxidation of 2-propanol and of cyclobutanol by stoichiometric amounts of sodium ruthenate has been examined (Scheme 12); contrary to earlier suggestions, these reactions are not initiated by traces of perruthenate present as a contaminant.[123] Cyclic sulphite diesters are oxidized with ruthenium tetroxide to the corresponding cyclic sulphate diesters with retention of configuration at sulphur.[124]

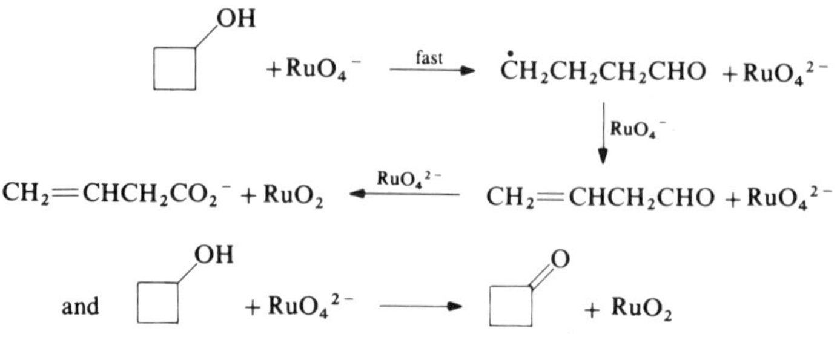

$$\text{cyclobutanol} + RuO_4^- \xrightarrow{fast} \dot{C}H_2CH_2CH_2CHO + RuO_4^{2-}$$

$$\dot{C}H_2CH_2CH_2CHO \xrightarrow{RuO_4^-} CH_2{=}CHCH_2CHO + RuO_4^{2-}$$

$$CH_2{=}CHCH_2CHO \xrightarrow{RuO_4^{2-}} CH_2{=}CHCH_2CO_2^- + RuO_2$$

and

$$\text{cyclobutanol} + RuO_4^{2-} \longrightarrow \text{cyclobutanone} + RuO_2$$

SCHEME 12

The oxidation of alcohols by oxo complexes of ruthenium (VI) and (VII) has been reported.[125] An intermediate in the oxidation of *trans*-cinnamic acid by perruthenate contains a stabilizing triply bonded spectator oxo group.[126] The kinetics of some ruthenium-catalysed oxidations by iron(III),[127-129] cerium(IV),[130] iodate,[131] periodate,[132] bromamine-T,[133,134] and phenyliodosoacetate[135,136] have been studied.

The mechanism for the selective oxygenation of sulphides to sulphoxides catalysed by dihaloruthenium(II) complexes involves oxidation of ruthenium(II) to give an oxidized metal species, probably ruthenium(IV), and peroxide; the active sulphide oxidant is peroxide; the reductant for the metal species is the solvent alcohol, thus completing the catalytic cycle.[137] A ruthenium(II)-polypyridyl complex is a catalyst for the photo-oxidation of some carbinols to aldehydes.[138] The oxidation of some amines complexed to ruthenium-bipyridyls has been studied.[139]

The use of palladium in selective oxidation reactions has been reviewed.[140] The oxidative carbonylation of methanol in the presence of palladium complexes and *p*-benzoquinone involves both palladium(I) and palladium(II) catalysts.[141] Thienylpalladium(II) is a common intermediate in both substitution and coupling reactions of thiophene by oxygen in the presence of palladium salts.[142] Kinetics studies on palladium(II)-catalysed oxygenations of alkenes,[143,144] and hydrocarbons[145,146] have been reported. The reactivity of di-μ-chloro-3-(nitrosooxy)bicyclo[2.2.1]hept-2-yl-(*C*, *N*)- dipalladium, an intermediate complex formed during the palladium-catalysed oxygenation of norbornene has been investigated.[147] Palladium(0)-catalysed oxidation of alcohols, *via* their allyl carbonates under neutral conditions, gives the corresponding carbonyl compounds in the absence of triphenylphosphine (A) whereas allyl coupling occurs (B) with triphenylphosphine present (Scheme 13).[148] The interaction of bis(*tert*-butylisocyanoperoxonickel(II) and bis(triphenylphosphine oxide)dibromodioxomolybdenum(VI) in the presence of cyclohexene yields 1,2-dibromocyclohexane as the major product; this result may be explained by initial epoxidation followed by subsequent conversion into the dibromide.[149]

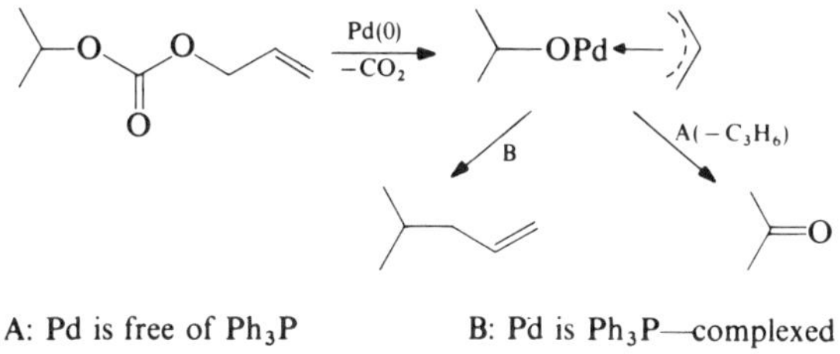

SCHEME 13

Deuterium-labelling studies have shown that, in the rhodium-catalysed co-oxygenation of *cis*-cyclooctene and triphenylphosphine, formation of the allyl alcohol product involves oxygenation at a vinylic centre; this observation may have general consequences for the mechanisms of Group VIII metal-catalysed co-oxygenations.[150] Isotopic-labelling studies have been used to elucidate the mechanism for the rhodium-catalysed oxygenation of terminal olefins to methyl ketones.[151]

An empirical formulation to predict the osmylation of allylic alcohols and their derivatives is that, for both cyclic and acyclic systems, osmium tetroxide approaches preferentially to the face of the olefin opposite to that of the pre-existing hydroxyl or alkoxyl group; *e.g.* cyclohexenol gives the cyclohexanetriol (**11**). Preliminary experiments indicate that permanganate oxidation and bromohydrin formation also seem to follow this empirical formulation.[152] Kinetic studies on the osmium-tetroxide-catalysed oxidations of sulphoxides[153,154] and the iridium(IV)-catalysed oxidation of nickel(II) dimethylglyoximate[155] have appeared. A comparison between different mechanistic pathways for the variable-valence metal-ion oxidation of furfural has been made for a number of metals.[156]

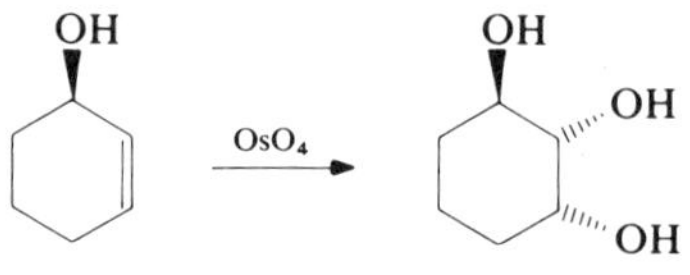

OH → (OsO_4) → OH, OH, OH

(**11**)

Oxidation by Compounds of Non-metallic Elements

Selenium and Sulphur

Bis(*p*-methoxyphenyl) selenoxide is an excellent co-oxidant for the oxidation of benzyl alcohols by a catalytic amount of selenium dioxide (Scheme 14).[157]

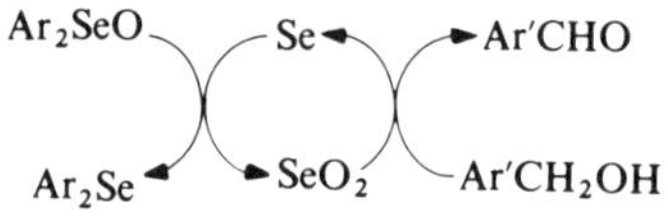

Ar_2SeO → Ar_2Se; Se → SeO_2; $Ar'CH_2OH$ → $Ar'CHO$

$Ar = p\text{-}MeOC_6H_4$

SCHEME 14

Competitive oxidation of two alcohols by less than the equivalent amount of oxidant under Swern conditions ($Me_2SCl^+Cl^-$) shows significant selectivity, with alcohols with electron-withdrawing groups or steric crowding being oxidized more slowly.[158] Diisopropyl sulphide and *N*-chlorosuccinimide at 0° oxidizes primary alcohols to aldehydes, but does not affect secondary alcohols; in contrast, at −78° the same reagent oxidizes secondary alcohols to ketones but does not affect primary alcohols (Scheme 15).[159] An NMR study of the kinetics and equilibrium for the oxidation of penicillamine and *N*-acetylpenicillamine by glutathione disulphide has been described.[160]

Chiral dioxiranes, such as (**12**), are formed from the reaction of potassium peroxomonosulphate with chiral ketones and allow epoxidation of simple prochiral

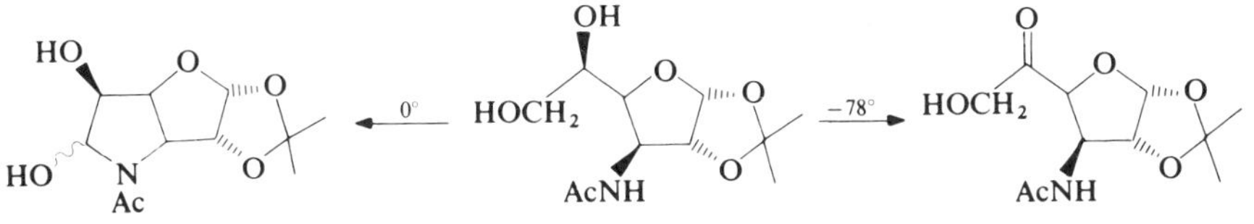

Reagent: $(Me_2CH)_2S$, *N*-chlorosuccinimide

SCHEME 15

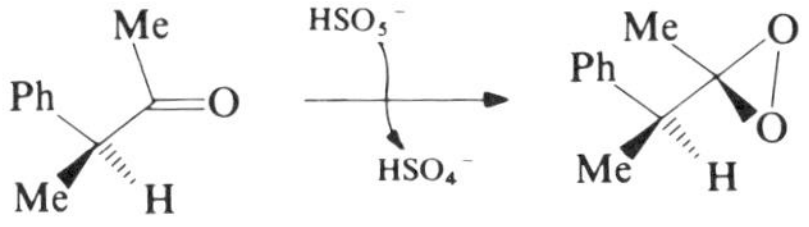

(12)

olefins to the corresponding epoxides with enantiomeric excesses in the range of 9–12%.[161] The mechanisms of the oxidations of amino-acids by peroxomonosulphate in the presence and absence of formaldehyde have been elucidated.[162–164] The ^{18}O isotope effect on ^{31}P-NMR has been used to identify the origin of the oxygen in the oxidation of triphenylphosphine by potassium persulphate.[165]

Peroxydisulphate oxidations in organic chemistry have been reviewed.[166] The oxidation of sulphides by peroxydisulphate anion proceeds through dimeric radical cations of the sulphide; sulphoxides may undergo oxidation by sulphate radical anions to form aliphatic radicals which are directly oxidized by peroxydisulphate anion (Scheme 16).[167] In contrast, the lack of observation of ESR signals in the

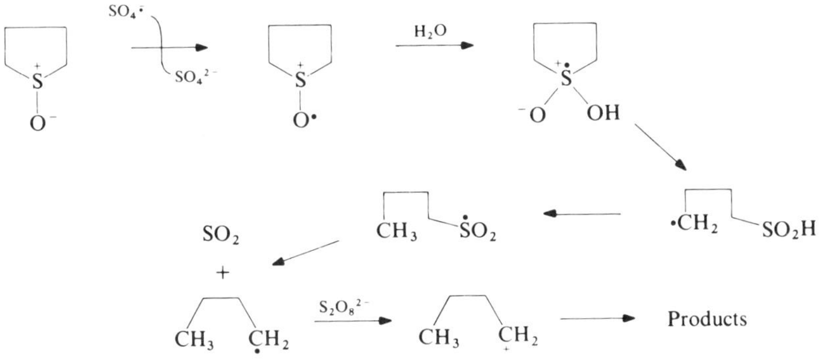

SCHEME 16

peroxydisulphate oxidation of diethyl and diphenyl sulphoxides has been taken to indicate that radicals are not formed in the reaction.[168] Radicals may be involved in the uncatalysed oxidations of phenylacetic acid[169] and acrylic acid[170] by peroxydisulphate. The mechanism for the oxidation of alkenes to vicinal *trans*-

diacetates with iron(II)–peroxydisulphate proceeds by initial addition of a sulphate radical anion to the alkene, followed by acetic acid solvolysis (Scheme 17).[171] Aliphatic ketones are oxidized by iron(II)–peroxydisulphate to diketones (Scheme 18).[172] The peroxydisulphate oxidation of aromatic side-chains catalysed by copper (II) proceeds *via* radical cations.[173–175] Kinetic studies on peroxydisulphate oxidations of cyclohexanecarboxylic acid and its methyl ester,[176] of alkanenitriles[177] catalysed by copper(II), and of dioxan[178] and acrylic acid[179] catalysed by silver(I), have been reported.

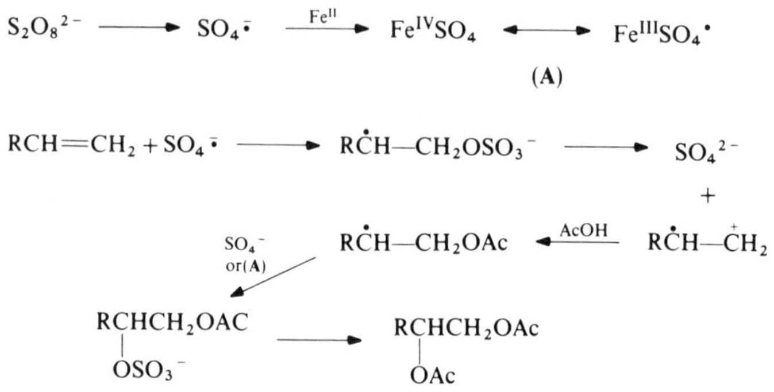

$$S_2O_8^{2-} \longrightarrow SO_4^{-\bullet} \xrightarrow{Fe^{II}} Fe^{IV}SO_4 \longleftrightarrow Fe^{III}SO_4^{\bullet}$$
(**A**)

$$RCH{=}CH_2 + SO_4^{-\bullet} \longrightarrow R\dot{C}H{-}CH_2OSO_3^{-} \longrightarrow SO_4^{2-} + R\dot{C}H{-}\overset{+}{C}H_2$$

$$R\overset{+}{C}H{-}\overset{+}{C}H_2 \xrightarrow{AcOH} R\dot{C}H{-}CH_2OAc \xrightarrow{SO_4^{-}\text{ or }(\mathbf{A})} RCH(OSO_3^{-})CH_2OAC \longrightarrow RCH(OAc)CH_2OAc$$

SCHEME 17

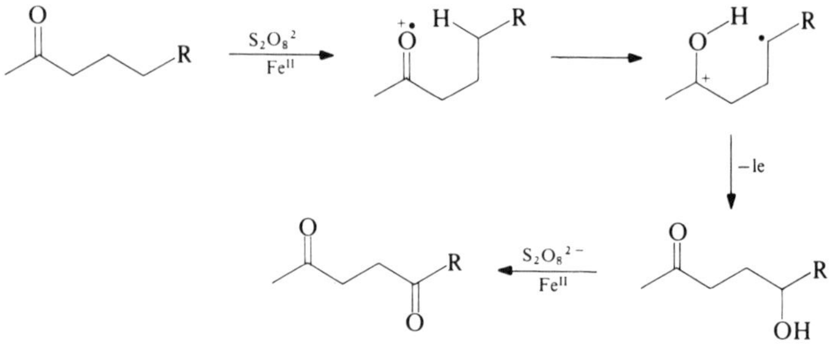

SCHEME 18

Halogens, including Bromate

Positive[180] and hypervalent[181] halogen compounds have been reviewed. A mechanism involving hydride abstraction has been proposed for the oxidation of alcohols by chlorine dioxide.[182,183] The oxidative decarboxylation of propiolic acids by iodine and iodine pentoxide in methanol leads to the corresponding ketal esters containing one less carbon atom; iodoacetylenic compounds are intermediates (Scheme 19).[184] Neighbouring-group participation by nitrogen is observed in the oxidative cleavage of disulphide (**13**) by iodine.[185] The oxidation of 5-methyl-1-thia-5-azacyclooctane by the dibromine radical anion (**14**) gives the transannular-stabilized three-electron-bonded thiaazoniacyclooctane (**16**) as an intermediate.[186] The oxidation of cysteine by the radical anions (**14**) and (**15**) has been re-investigated;

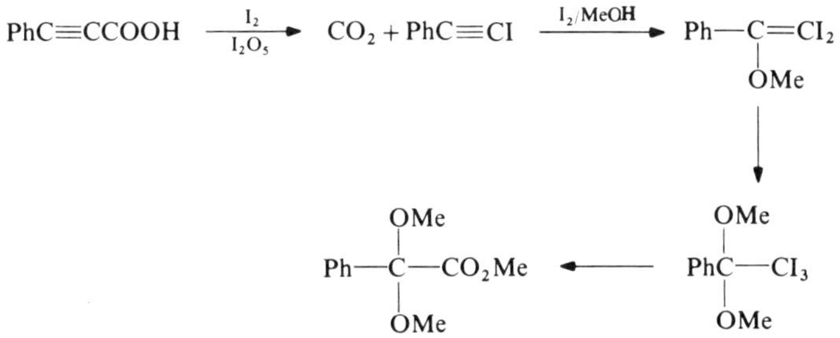

$$PhC{\equiv}CCOOH \xrightarrow[I_2O_5]{I_2} CO_2 + PhC{\equiv}CI \xrightarrow{I_2/MeOH} Ph{-}C(OMe){=}CI_2 \rightarrow PhC(OMe)_2{-}CI_3 \rightarrow Ph{-}C(OMe)_2{-}CO_2Me$$

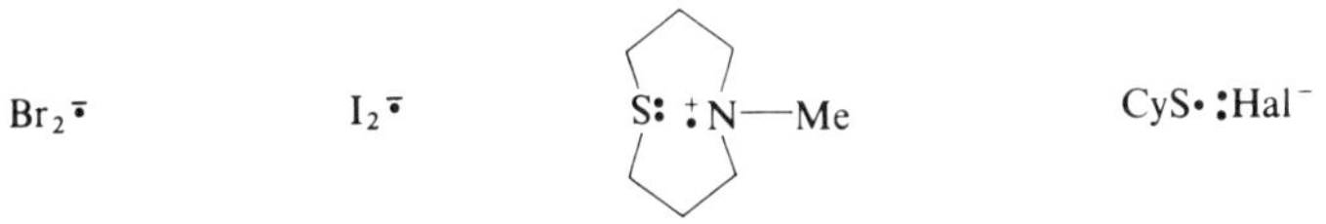

SCHEME 19

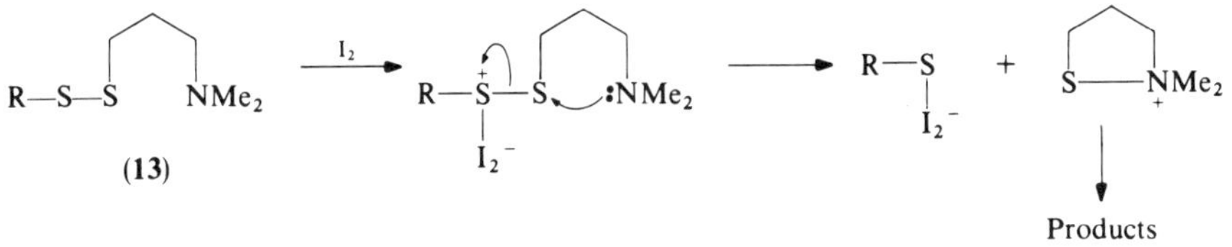

R—S—S(CH2)3NMe2 (13) —I_2→ R—S⁺(I₂⁻)—S…:NMe₂ → R—S—I₂⁻ + cyclic S—N⁺Me₂ → Products

$Br_2^{\bar{\cdot}}$ (14) $I_2^{\bar{\cdot}}$ (15) S∴⁺N—Me (16) CyS• :Hal⁻ (17)

the existence of complexes between the thiyl radical from cysteine and bromide and iodide ions has been demonstrated, and the properties of these intermediates (**17**) discussed in relation to other sulphur and halogen species with a three-electron half-order bond.[187] The products obtained from reaction of bromine with bistrimethylsilylenol ether (**18**) depend on the temperature (Scheme 20).[188] The oxidation of N-

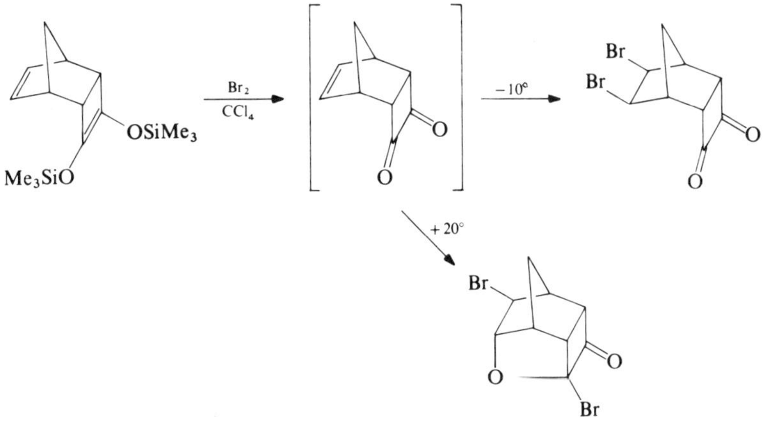

Me3SiO / OSiMe3 —Br_2, CCl_4→ [dione] —$-10°$→ Br, Br dione; —$+20°$→ Br, O, O, Br

SCHEME 20

alkyl-*N*′-tosylhydrazones with bromine gives alkyl bromides and vicinal dibromides, together with a trace of the corresponding alcohols.[189] The oxidation of furan by bromine in aqueous solution[190] and the oxidative chlorination of trichloroethylene[191] have been investigated.

o-Iodosylbenzoic acid in basic methanol converts ketones to α-hydroxydimethylacetals (Scheme 21);[192] intramolecular participation by a phenolic

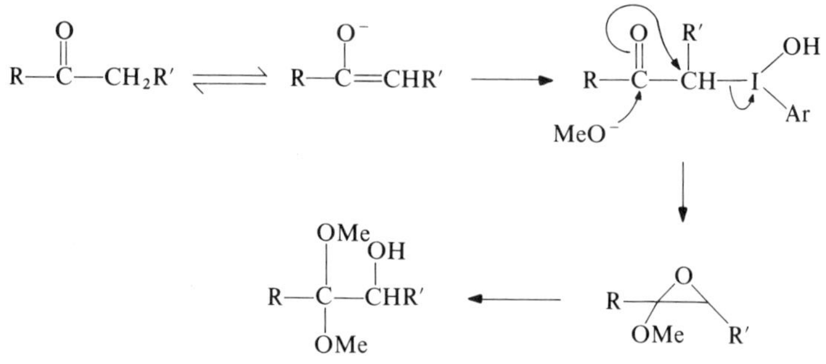

$Ar = o\text{-}C_6H_4CO_2^-$

SCHEME 21

hydroxyl group is observed in the oxidation of *o*-hydroxyphenyl alkyl ketones by phenyl iodosoacetate.[193] Oxidation of 3-cholestanone with hypervalent iodine species leads to ring-contraction products (Scheme 22).[194] Mechanistic proposals

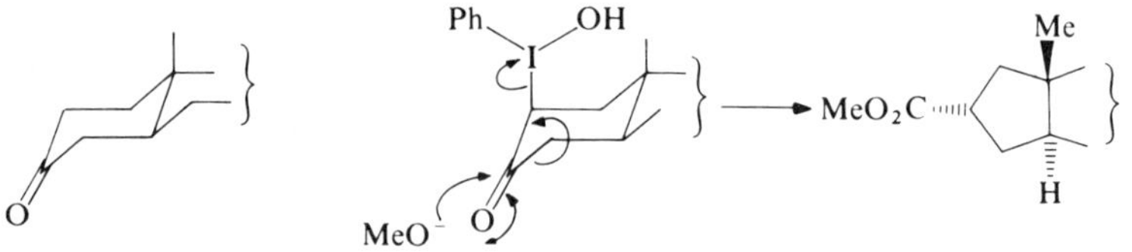

SCHEME 22

have been made for the oxidations of α,β-unsaturated ketones,[195] β-diketones,[196] and polycyclic aromatic compounds[197] by phenyl iodosoacetate. Iodoxybenzene (**18**) is isoelectronic with ozone; the reaction of (**18**) with phenanthrene proceeds initially in a similar manner to the reaction of ozone with phenanthrene (Scheme 23).[198]

Inorganic hypochlorite in the presence of a quaternary ammonium salt as a phase-transfer catalyst not only epoxidizes several arenes to arene oxides but also converts toluene into benzyl chloride and causes ring chlorination of anisole; alkenes are converted into complex mixtures of products arising from competing chlorination and oxidation. It is suggested that these reactions proceed by a free-radical pathway involving Cl_2O and $ClO^{\bullet}$ as reactive species.[199] Mechanistic studies have been

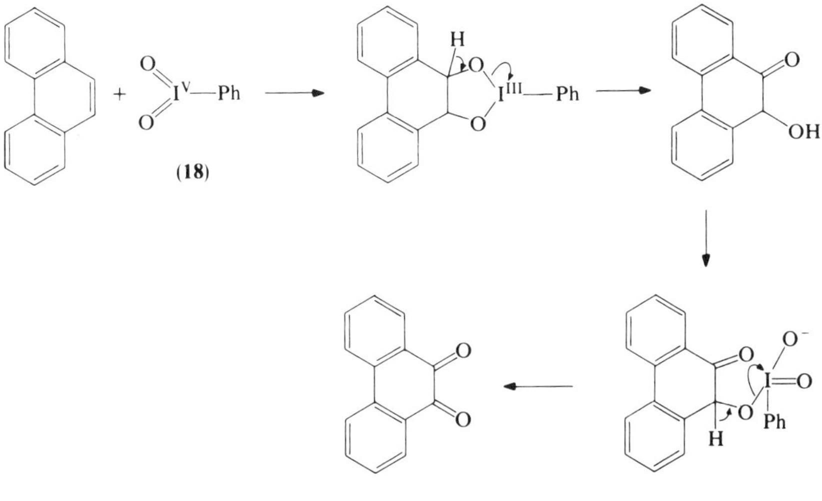

(18)

SCHEME 23

reported on the reactions of hypochlorite with α-hydroxy-acids[200] and ethanolamines.[201,202] The reactions of 3-sulpholene[203] and of quinone and ascorbic acid[204] with hypochlorous acid have been investigated.

α-Bromoketones are formed in the reaction of olefins with sodium bromite (Scheme 24); this is in contrast to the behaviour with sodium hypobromite (which gives bromohydrins) and of sodium bromate (which does not react).[205] In the acid bromate oxidation of *para*-substituted phenyl methyl sulphides, the active oxidant is $HBrO_3$.[206] Kinetics of metal-catalysed bromate oxidations of oxalic acid[207] and phenols[208] have been studied. Among other investigations of the Belousov–Zhabotinskii and related reactions,[209-216] a comparative study of uncatalysed and catalysed oscillatory behaviour of mono- and di-hydroxybenzoic acids in acidic bromate has been reported.[217] Chaos in the non-stirred Belousov–Zhabotinskii reaction is induced by interaction of waves and stationary dissipative structures. No chemical turbulence occurs and the wave pattern remains regular when physical conditions are inappropriate for the formation of dissipative structures; chaos appears in the reaction as a result of structure formation.[218]

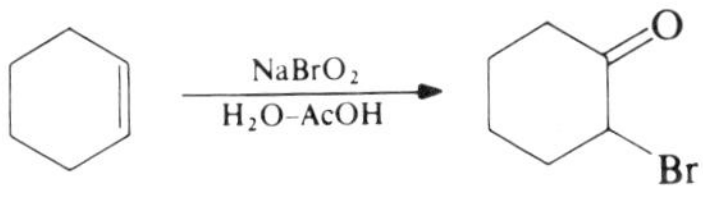

$NaBrO_2$, H_2O–AcOH

SCHEME 24

The oxidation of acetophenones by acid iodate may involve rate-determining electrophilic attack of IO_2^+ on the methyl carbon of the acetophenone.[219] Mechanistic aspects of the periodate oxidations of anilines,[220] glycols,[221] and levoglucosenone[222] have been investigated. Mechanistic studies on *N*-halogeno oxidants have been reported for the following organic substrates: fluoren-9-ols,[223] benzylic alcohols,[224,225] ethanolamines,[226] diols,[227] aliphatic ketones,[228] ketoglutaric acids,[229] pyridoxine,[230] and dimethyl sulphoxide.[231]

Miscellaneous Oxidations

Studies on the oxidizing ability of the oxoammonium chloride (**19**) have established that phenols, enolizable ketones, amines, and anilines are oxidized but that olefins, aromatic ethers, and sulphides are not affected by the reagent.[232] The oxidation of adamantane by a mixture of nitric and sulphuric acids has been investigated.[233]

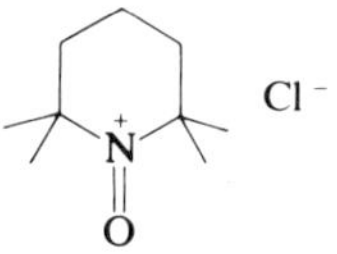

Cl^-

(**19**)

A mechanism for the transfer of oxygen from oxaziridine to ethylene has been proposed on the basis of *ab initio* orbital calculations in which the significance of the interaction between the ethylene π-bond and the Walsh orbital containing the oxygen lone pair is considered.[234] The ability of aryl oxaziridines to transfer oxygen to phenolates has been examined and an electron-transfer mechanism proposed on the basis of ESR and other studies (Scheme 25); these reactions may serve as models for the suggested flavin-based oxaziridine (**20**) in enzyme-mediated mono-oxygenations.[235] The low values of the NIH shift and the nearly 1 : 1 ratio of *ortho*:*para* hydroxylation suggest that arene oxides are not intermediates in the hydroxylation of anisole by 2-sulphonyloxaziridines.[236]

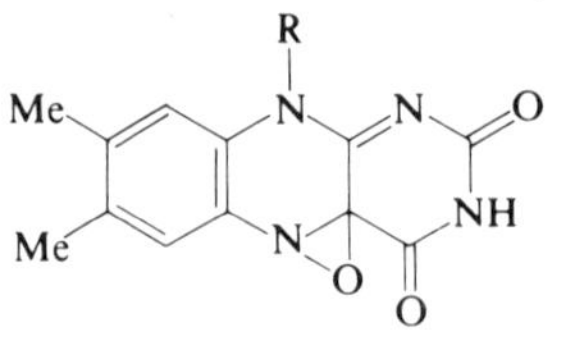

(**20**)

Complete intramolecular transfer of a central chiral element to an axial chiral element has been observed in a hydride transfer to dichlorodicyano-*p*-benzoquinone.[237] Studies on the interconversion by hydrogen transfer of unsymmetrically substituted quinhydrones in the solid state have been reported.[238] The dehydrogenation of 1,4-cyclohexadiene by dichlorodicyano-*p*-benzoquinone is stereospecifically *cis*.[239]

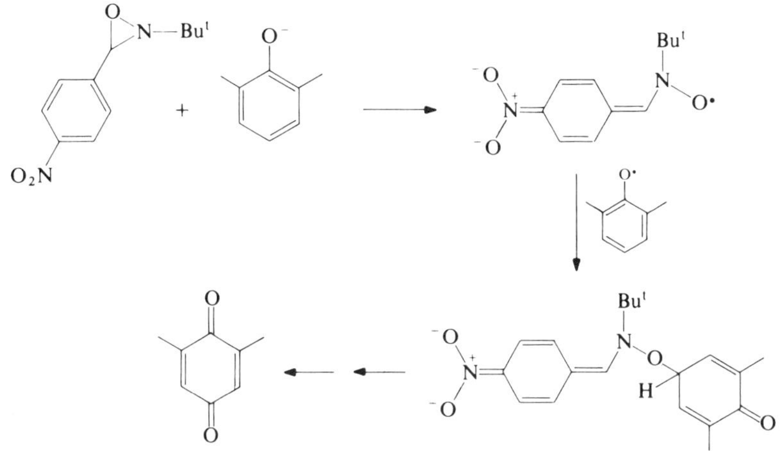

SCHEME 25

Ozonolysis and Ozonation

The rate of reaction of ozone with alkenes is sensitive to electronic effects, with electron-deficient alkenes reacting several orders of magnitude more slowly than electron-rich alkenes. The relative rates of reaction with various halogenated ethenes are consistent with a 1, 3-dipolar cycloaddition; however, it is possible that a change in reaction mechanism occurs with more electron-rich olefins.[240] 1,1-Dichloroethyl hydroperoxide and the corresponding anion (**21**) are intermediates in the ozonolysis of 2,3-dichloro-2-butene (Scheme 26).[241] Other studies on the ozonolysis of haloalkenes in solution,[242, 243] and by Fourier-transform infrared spectroscopy in the gas phase,[244] have been reported. The effect of solvent polarity on the rate of reaction of ozone with unsaturated compounds has been investigated.[245] The inverse kinetic secondary isotope effect observed in the ozonolysis of differently

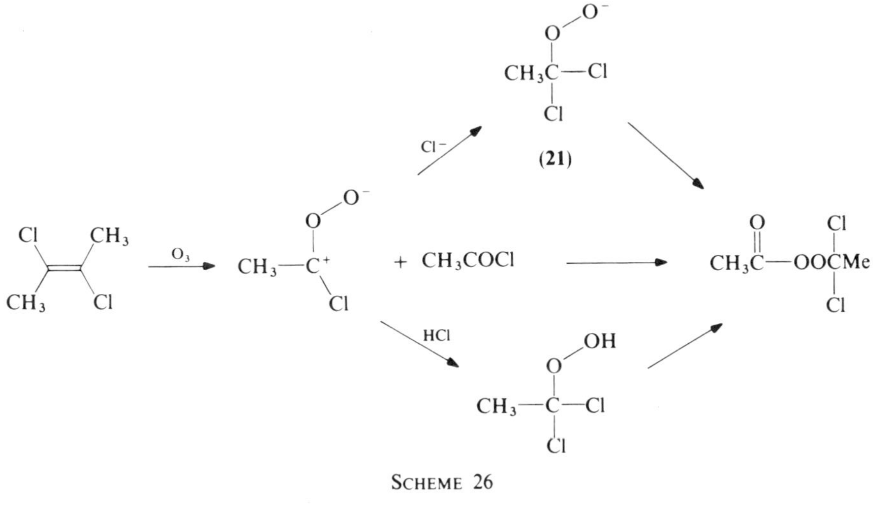

SCHEME 26

deuterated ethylenes in the presence of acetaldehyde is consistent with a concerted Criegee mechanism.[246] Ozonolysis of cycloalkenes with one equivalent of ozone in an ether-type solvent, followed by hydrogenation over the Lindlar catalyst does not yield dicarboxylic acids directly as has previously been claimed; an excess of ozone is required. The peroxidic ozonolysis products obtained by this procedure have been characterized and are different from those previously reported.[247] The kinetics of the reactions of ozone with a number of alkenes in the presence of oxygen[248] and hydroxyl radicals,[249] and of alkynes in the presence of hydroxyl radicals,[250] have been studied. Aspects of the ozonation of cyclopropylidenecycloalkanes,[251] benzenesulphonic acid,[252] and phenylazonapthalenesulphonic acids[253] have been investigated.

Linear-free-energy relationships between the rates of reaction of ozone with a number of polycyclic hydrocarbons and calculated molecular orbital parameters are best correlated by models based on rate determining π- or σ-complex formation, rather than simultaneous addition of ozone to two carbon atoms.[254] Ozonolysis of phenanthrene in HCl–methanol gives unsubstituted and chloro-substituted 6*H*-dibenzo[*b*, *d*]pyran-6-ones in addition to dimethyl 2,2′-biphenyldicarboxylate.[255] The ozonolysis of a series of di- and tri-substituted indenes in carbon tetrachloride leads to the formation of the corresponding bicyclic ozonides, usually as a mixture of *exo*- and *endo*-isomers; the results may be explained in terms of the direction of approach of ozone to the substrate, since the product stereochemistry shows little dependence on either the relative stability of the isomers or on the substituent electronic effects.[256] Ozonolysis of 1-methyl-2,3,-diphenylindene in methylene chloride-methanol gives (**22**) in an unprecedented solvolytic diversion of the

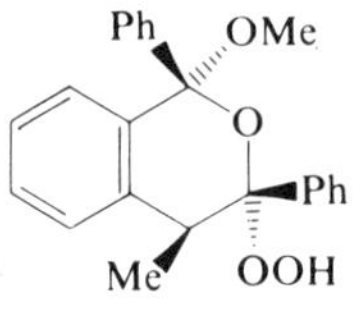

Ph OMe O Ph Me OOH

(**22**)

carbonyl oxide intermediates.[257] The abnormal product (**24**) derived from the treatment of 1-methyl-3-styryl[3.3] paracyclophane (**23**) on ozonolysis is a reflection of the high nucleophilicity of the cyclophanyl group, explaining its high migratory aptitude.[258] Alkylbenzenes with ozone undergo competitive ring ozonolysis and benzylic substitution (to give benzyl hydrotrioxides); introduction of methyl or phenyl groups at the exocyclic position of toluene markedly enhances benzylic substitution.[259] The ozonation of *p*-nitrotoluene in acetic anhydride gives *p*-$O_2NC_6H_4CH(OAc)_2$ as the major product.[260] The ozonation of single bonds has been reviewed.[261] Studies on the liquid-phase autoxidation of cumene initiated by ozone have been reported.[262] The reaction of sydnones with ozone leads to an overall deamination (Scheme 27).[263] Ozone is produced on the irradiation of cyclopentadienone *O*-oxide.[264]

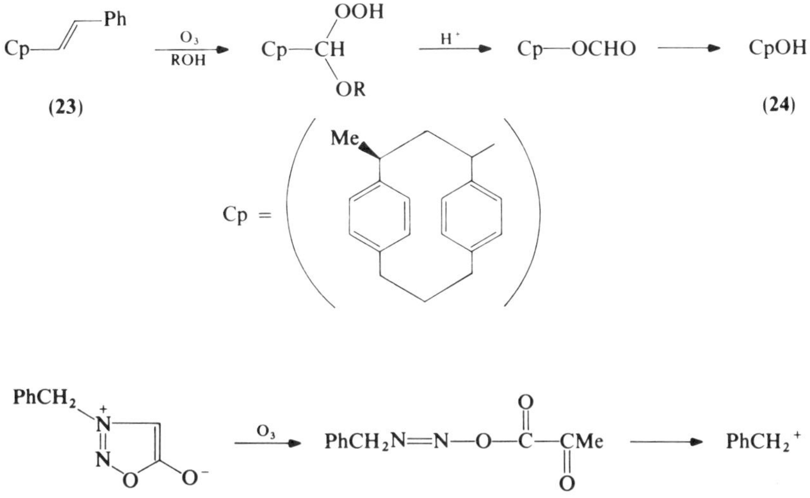

(23) (24)

Cp =

SCHEME 27

Peracids, Peroxides, and Superoxide

Several aspects of the chemistry of the O—O bond have been reviewed including organic sulphur and phosphorus peroxides,[265] pyrolysis of peroxides in the gas phase,[266] diacyl peroxides, peroxycarboxylic acids and peroxy-esters,[267] alkylhydroperoxides and dialkyl peroxides,[268] endoperoxides,[269] oxirenes,[270] the photochemistry and radiation chemistry of peroxides,[271] polar reaction mechanisms involving peroxides in solution,[272] and the preparation and uses of isotopically labelled peroxides.[273]

Conformational and steric effects on the regioselectivity in the Baeyer–Villiger oxidation of six β,β- and γ-substituted cyclopentanones and cyclohexanones have been studied. Selectivity for the migration of α-carbon relative to that of the α'-carbon in (**25**) is determined by steric factors; this preference is ascribed to greater relief of non-bonded interactions as the α–β bond lengthens in going towards the transition state; thus, in the tetrahedral intermediate (**26**), 1,3-diaxial-like interactions between the hydroxyl group and the closest γ-carbon are diminished to a greater extent when the α; rather than the α'-carbon, migrates.[274] The observed regioselectivity in the Baeyer–Villiger oxidation of a series of tetracyclic ketones has

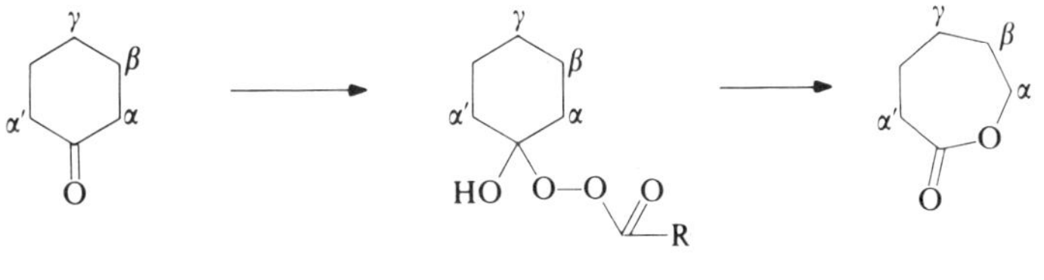

(**25**) (**26**)

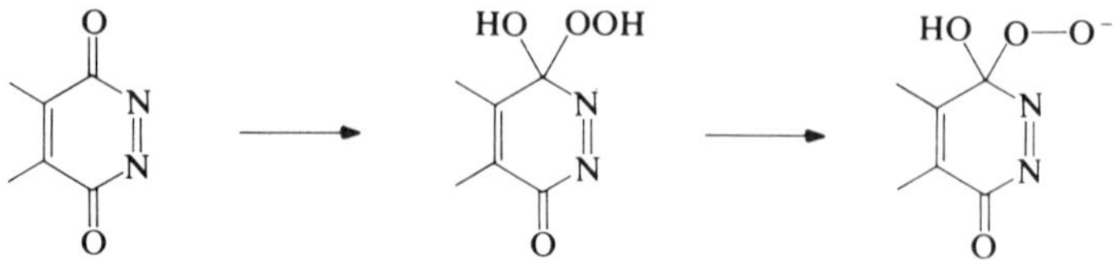

(27) (28)

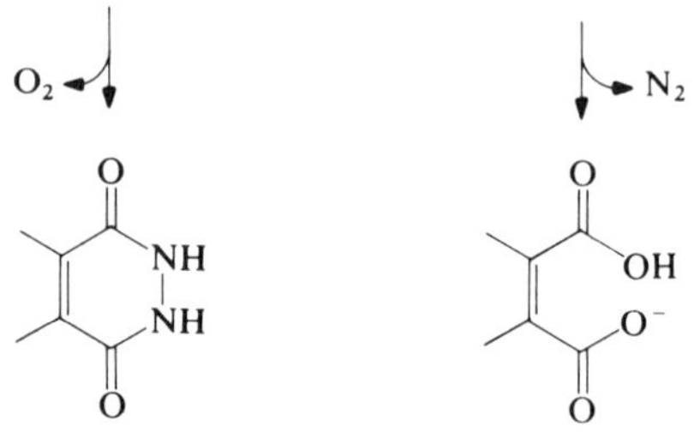

(30) (29)

been interpreted in terms of torsional effects.[275] The first critical intermediate in the chemiluminescent hydrogen peroxide oxidation of luminol is the hydroperoxide (**27**): at a pH above the pK_a of (**27**) the corresponding anion (**28**) fragments with loss of nitrogen to form the monoprotonated dicarboxylate (**29**); at a lower pH (**27**) decomposes into oxygen and the parent hydrazide (**30**).[276] The peracid oxidation of 1-pyrroline 1-oxides proceeds *via* initial step-wise epoxidation of the carbon–nitrogen double bond (Scheme 28).[277]

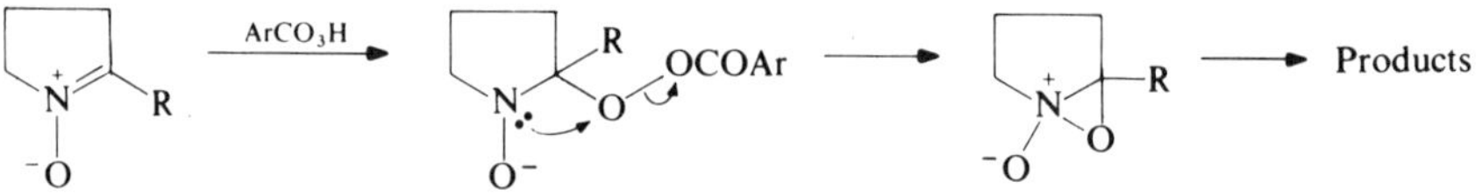

$ArCO_3H$ OCOAr R Products

SCHEME 28

The ease of formation of the epoxides of a series of bicylic olefins where distant sulphonyl groups are present is variable; there is some correlation between the relative reactivity of the vinyl groups and their π-ionization potential due to orbital interactions through bonds or space between the sulphone and the double bond.[278] In order to demonstrate an earlier structural misassignment in the literature,[279] it has been found that peracid oxidation of the acylindole (**31**) gives the equilibrium mixture (**32**) of the dihydroxyacylindoline and the keto-amide.[280] Kinetic studies on the epoxidation of α,β-unsaturated aldehydes[281] and of α,β- and γ,δ-unsaturated sulphones[282] by alkaline hydrogen peroxide, and of olefins by peroxypropionic acid,[283] have been reported. The initial step in the *m*-chloroperbenzoic acid oxidation of trialkyl-substituted furans is the formation of a monoepoxide which sequentially rearranges to an enedione intermediate which undergoes a

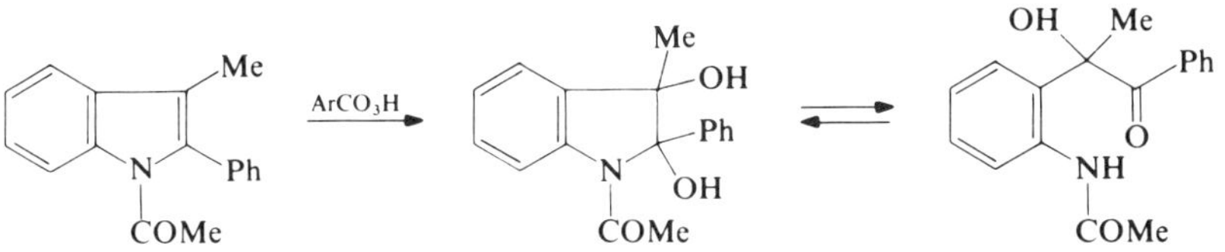

(31) $\xrightarrow{ArCO_3H}$ ⇌ (32)

Baeyer–Villiger oxidation.[284] *O*-Alkylperoxycarbonic acids[285] and α-azohydroperoxides[286] have been used to epoxidize carbon–carbon double bonds.

A dioxirane intermediate may be involved in the oxidation of sulphides with α-azohydroperoxide.[287] The possibility of dioxirane formation from the decomposition of α-*tert*-butyldimethylsilyl peroxybenzoates (**33**) has been examined.[288] Thianthrene 5-oxide (**34**) can be used as a probe for distinguishing carbonyl oxides

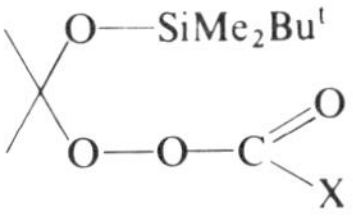

(**33**)

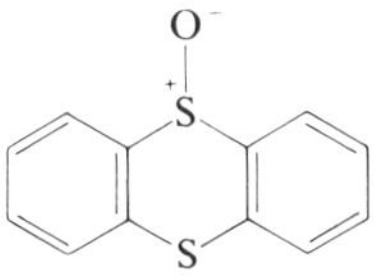

(**34**)

and dioxiranes in oxygen-transfer reactions.[289] Silylated thiirans react with peracids to give the corresponding *S*-oxides and some products derived from subsequent ring-opening (Scheme 29).[290] Oxidation of alkyl phenyl selenides with *m*-chloroperbenzoic acid in methanol gives the corresponding methyl ethers (Scheme 30).[291] Oxidative ether cleavage with *p*-nitroperbenzoic acid proceeds *via* selective α-(C—H) bond fragmentation to give hemiacetals.[292] A radical anion intermediate is invoked to explain the acceleration of the formation of *o*-nitrophenol from *o*-dinitrobenzene by hydrogen peroxide.[293] Aromatic hydroxylation by peracid oxidation of a cyclopalladated azoxybenzene has been described.[294]

Organic reactions involving the superoxide anion,[295] the chemical reactivity of superoxide with organic substrates in aprotic media,[296] the activation of superoxide by reaction with a number of organic substrates,[297] and the nucleophilic substitution of aryl halides by superoxides[298] have been reviewed. The initial step in the reaction of superoxide with a number of organic substrates containing the trichloromethyl group is transfer of an electron to the CCl_3 group followed by nucleophilic displacement of chloride by superoxide.[299] Superoxide in aprotic

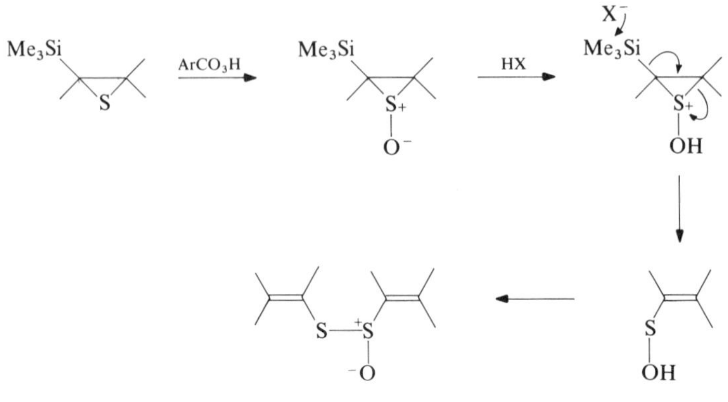

SCHEME 29

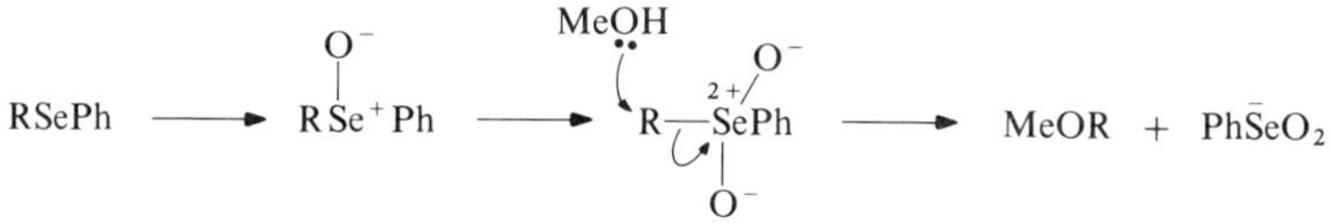

SCHEME 30

media and in the gas phase reacts with 1,2-disubstituted hydrazines to produce the anion radical of the corresponding 1,2-disubstituted azo compound which with oxygen gives the azo compound (Scheme 31); this combination of reactions represents a superoxide-catalysed autoxidation and parallels the xanthine oxidase-catalysed autoxidation of reduced flavins.[300] Oxidative cleavage of pyrocatechol by superoxide at the *meta*-position may provide a model reaction for pyrocatechol-2,3-dioxygenase.[301] Quantum-mechanical studies of the interaction of superoxide ion with a model of the active site of superoxide mutase have suggested a new mechanism of action for the enzyme.[302] Vanadate and molybdate stimulate the oxidation of NADH by superoxide.[303] The mechanism of inhibition of catalase by the superoxide ion has been investigated.[304] Glutathione is oxidized by superoxide to the disulphide and the sulphonate, yielding singlet oxygen.[305] Superoxide has been trapped by *p*-benzoquinone in some dicyanoanthracene-sensitized photo-oxidations.[306] Superoxide is formed by one-electron transfer from organic electron

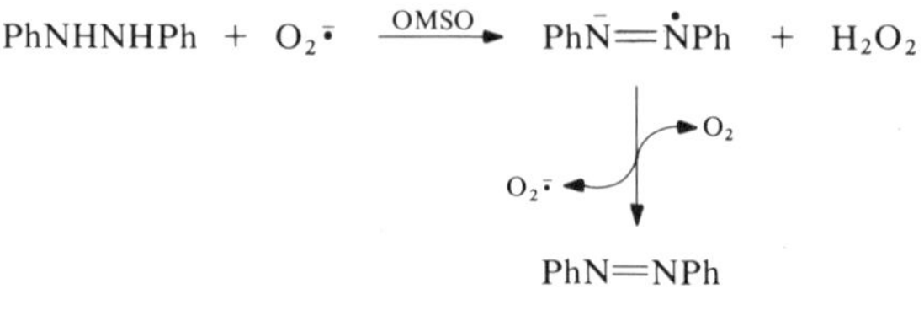

SCHEME 31

donors, such as anilines, to singlet oxygen.[307] A mechanism has been proposed for the production of superoxide in reaction mixtures containing NADH, phenazine methosulphate, and nitro blue tetrazolium.[308]

Atomic Oxygen and Singlet Oxygen

The reaction of atomic oxygen with isomeric butanols shows that, although the α-(C—H) bond in these alcohols is the most reactive, appreciable attack also occurs with other C—H bonds.[309] The primary processes in the reaction of atomic oxygen with ethene have been investigated.[310] The absolute rate constants for the reaction of oxygen atoms with a number of olefins have been determined using a laser photolysis–chemiluminescence technique.[311]

Reviews have appeared on singlet oxygen in peroxide chemistry,[312] photochemical formation and reactivity of singlet oxygen,[313] and other aspects of the chemistry of singlet oxygen.[314] Solvent isotope effects have demonstrated that the direct photo-oxidation of highly fluorescent anthracenes involves the production of singlet oxygen from the quenching of anthracene excited singlet states by oxygen.[315] Singlet oxygen is generated from molecular oxygen quenching of both the lowest excited singlet and triplet states of 9,10-diphenylanthracene.[316] The use of juglone formation as an indicator for singlet oxygen has been described.[317] Mesodiphenylhelianthrene (**35**) has been shown to be an exceptionally efficient acceptor of singlet oxygen.[318] The rate constants for the quenching of singlet oxygen by a number of organic sulphur compounds have been determined in a discharge flow system in the absence of oxygen atoms.[319] A mechanism has been proposed for the catalysed oxidation of singlet substrates by singlet oxygen and the kinetic implications for the catalytic and photocatalytic action of metal ions discussed.[320]

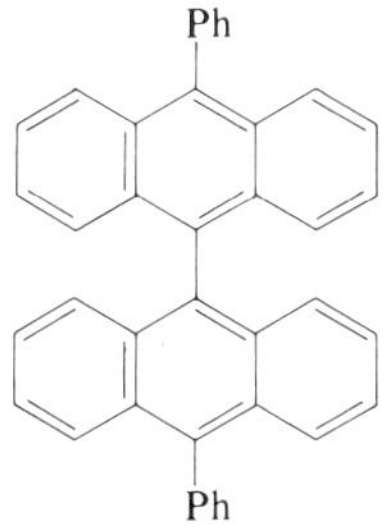

Ph

Ph

(**35**)

MINDO/3 Calculations of the reactions of ethene with singlet and triplet oxygen suggest that both processes involve $\cdot CH_2CH_2OO\cdot$ as an intermediate. The singlet biradical can easily close to the dioxetane, whereas the triplet biradical may initiate other radical processes.[321] The 9,10-dicyanoanthracene-sensitized photo-oxygenation of a number of alkyl-substituted olefins has been shown to proceed *via* singlet oxygen.[322] An easy "one-pot" synthesis of cyclopentenones and related compounds from the corresponding cycloalkenes has been reported (Scheme 32).[323]

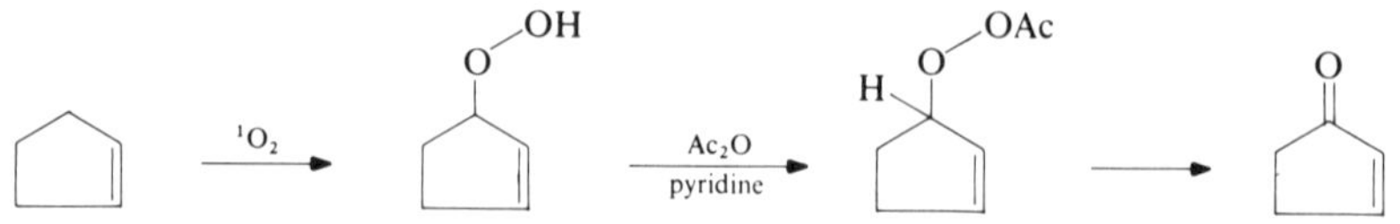

OH O $^1O_2$ Ac_2O pyridine H O OAc O

SCHEME 32

The possibility of an exciplex intermediate in the addition of singlet oxygen to substituted furans has been explored.[324] 2,5-Dimethyl-2,4-hexadiene undergoes three concomitant reactions with singlet oxygen, *viz.* the ene reaction to give allylic hydroperoxide, 4+2-cycloaddition to give an endoperoxide, and 2+2-cycloaddition to form a dioxetane; a vinylogous ene reaction occurs to give the diene hydroperoxide (**36**).[325] Singlet oxygen reacts stereospecifically with the dihydroazulene (**37**) to give the endoperoxides (**38**) and (**39**).[326] Substituent, temperature, and solvent effects have been studied for the singlet oxygenation of 7-alkyl- and

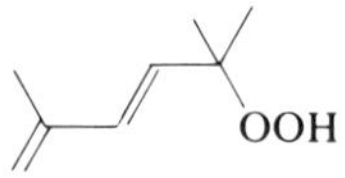

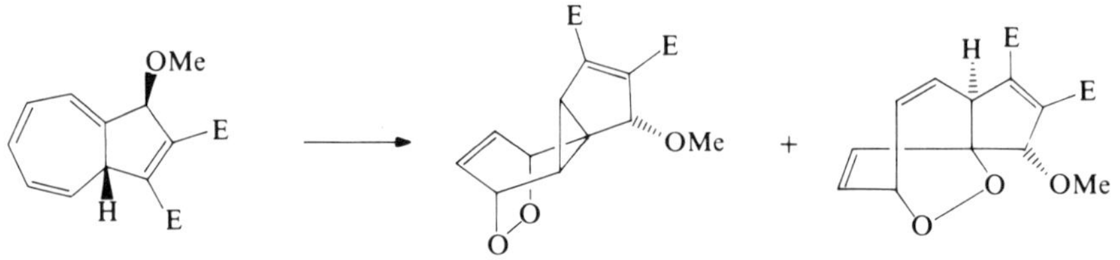

OOH

(**36**)

OMe E H E

(**37**)

E E OMe O O

(**38**)

H E E O OMe O

(**39**)

$E = CO_2Me$

7-aryl-1,3,5-cycloheptatrienes.[327] The initial products of the photosensitized oxygenations of monocyclic 1*H*-1,3-diazepines are 4,7-endoperoxides and 4,5-dioxetanes.[328] 1,2,4-Trioxanes are formed in the reaction of singlet oxygen with 1,3-dimethylindole in the presence of acetaldehyde (Scheme 33).[329] The singlet oxygenations of psoralens and related benzofurans,[330] chalcones,[331] and bilirubin[332] have been studied. The dicyanoanthracene-sensitized photo-oxidation of *cis*- and *trans*-2,3-diphenylaziridine gives only the *cis*-dioxazolidine whereas the photo-oxidation of *N*-alkyl-substituted 2,3-diphenylaziridines gives both isomers of the cyclic peroxide; it has been proposed that the mechanism involves singlet oxygen addition to an intermediate azomethine ylid (Scheme 34).[333] Air oxidation of 1-phenyl-1-diazopent-4-ene (**40**) under basic conditions leads to a carbonyl oxide which performs an intramolecular epoxidation to (**41**), while photo-oxygenation of

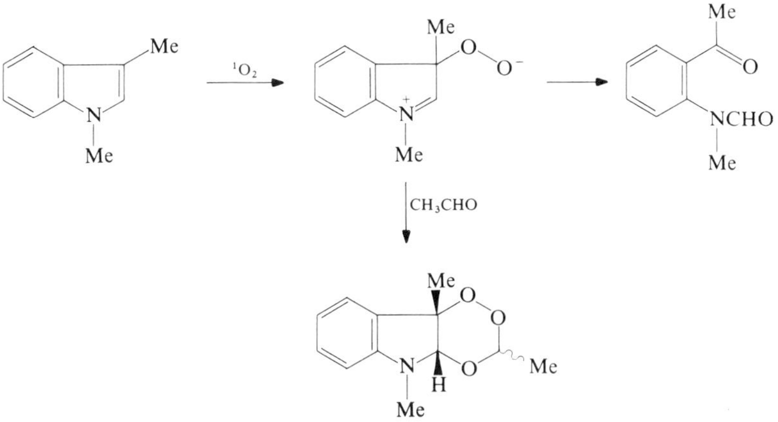

SCHEME 33

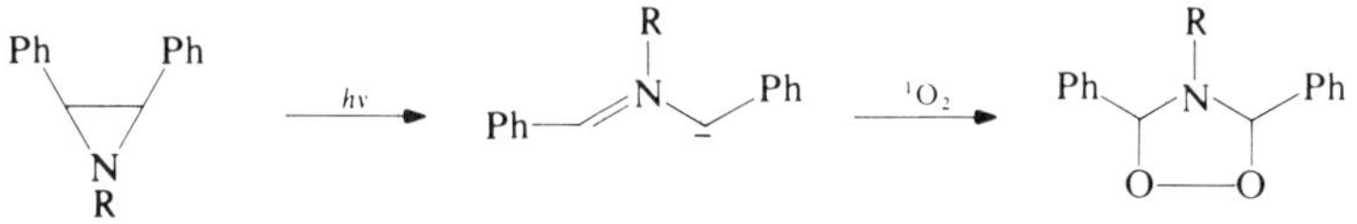

SCHEME 34

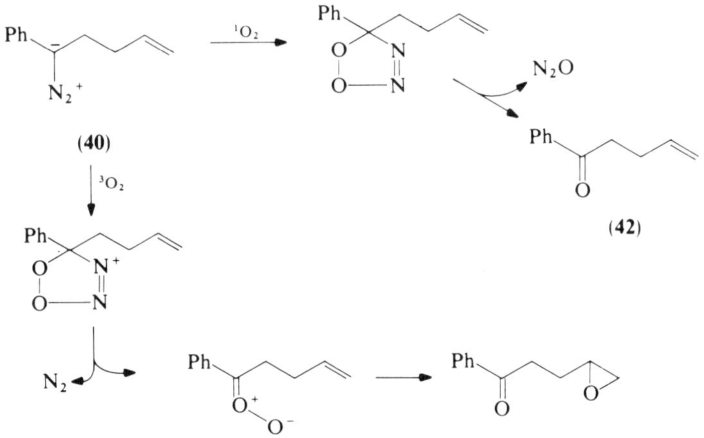

(**40**) produces nitrous oxide and the ketone (**42**).[334] Singlet oxygen with adamantanone azine affords, in addition to adamantanone, products derived from a dioxirane intermediate.[335] The oxidations of diphenyl sulphide by singlet oxygen-generated carbonyl oxides,[336] and of thioketones[337] and 6-mercaptopurine[338] by singlet oxygen, have been investigated. The rôle of singlet oxygen in the liquid-phase oxidation of ethylbenzene has been elucidated.[339] Singlet sulphur, the analogue of singlet oxygen, has been generated and trapped in the Diels–Alder reaction with a number of dienes.[340]

Other Reactions of Oxygen, including Autoxidation

The oxidation of olefins using molecular oxygen has been reviewed.[341] The mechanism of the photo-oxidation of olefins with α-diketones involves addition of oxygen to triplet diketone to give an acylperoxy radical, which is a key epoxidizing species.[342,343] In a study on the mechanism of olefin oxygenations by electron-transfer processes, the electronic and geometrical structures of the ethylene peroxy cation and anion radicals have been investigated.[344] An electron-transfer complex between the substrate and oxygen is probably involved in the photo-oxygenation of 4-flavanols and benzylic alcohols.[345] A co-sensitized electron-transfer photo-oxygenation of epoxides provides a new route to ozonides.[346] Furan ring-formation is observed in hydroxyl-radical-initiated photo-oxidation of 1,3-dienes.[347] Up to 500 K, the primary process in the reaction of ethyl radical with molecular oxygen is the formation of EtOO· whereas a second route to form ethene and H_2O· is of major importance above 600 K.[348] Microwave discharge of oxygen and carbon dioxide as a synthetic tool for the oxidations of unsaturated compounds on solid support has been described.[349] Studies on the oxidation of acetaldehyde in a field of ultrasonic waves have been described.[350–352] Diphenyl(dimethylamino)phosphine, Ph_2PNMe_2, reacts with sulphur dioxide to give $Ph_2P(O)NMe_2$ and $Ph_2P(S)NMe_2$; the reaction proceeds *via* an intermediate singlet sulphurane which fragments to triplet oxygen in an intersystem-crossing process.[353]

The reaction of oxygen with unsaturated fatty acids has been reviewed.[354] The autoxidation of four isomeric methyl 9,12-octadienoates has been investigated in benzene–cyclo-1,4-hexadiene mixtures. With no cyclohexadiene present, the distribution of product hydroperoxides from the various isomers is virtually identical; with increasing cyclohexadiene concentrations, the product hydroperoxide mixtures becomes different and reflects the stereochemistry of particular diene precursors. A scheme for diene fatty acid autoxidation involving reversible oxygen addition to intermediate pentadienyl carbon radicals has been proposed.[355] A synergistic mechanism for the effect of glutathione on the inhibiting power of vitamin E during the autoxidation of methyl linolenate has been described.[356] Rate and product studies have been reported on the oxidation of methyl linolenate dispersed in aqueous solution.[357] Other studies on the autoxidation of olefins have appeared.[358–360] Oxirenes have been shown to be intermediates in the autoxidation of 1-phenylalkynes (Scheme 35).[361] Oxidation rates for ascorbic acid and its derivatives do not support the contention that bicyclic ring-formation is required for its autoxidation.[362] The mechanism for the autoxidation of diphenylketene to polybenzilic acid involves diphenylacetolactone (**43**) as an intermediate; under appropriate conditions, the lactone (**43**) can be trapped by diphenylketene to give (**44**).[363]

Autoxidation of 1,3-disubstituted thioureas with oxygen in the presence of *tert*-

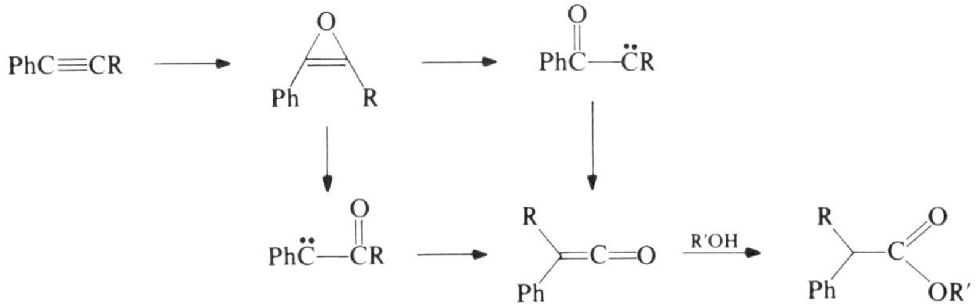

SCHEME 35

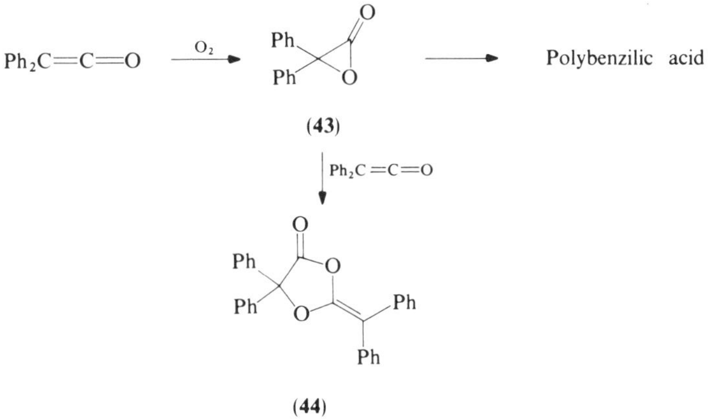

(43)

(44)

butoxide gives ureas, together with a small quantity of the corresponding guanidines (Scheme 36).[364] Several autoxidations of stabilized carbanions have been reported.[365–367] Among other studies on the autoxidation of aromatic hydrocarbons,[368–372] the mechanism for the atmospheric oxidation of toluene has been discussed.[373] *Gem*-Hydroxy(hydroperoxy) intermediates have been identified in the oxidation of alcohols.[374, 375] Kinetic investigations of the oxidations by oxygen of sodium pentamethylenedithiocarbamate,[376] of hexafluoropropylene at low temperature,[377] and of *n*-decane in the presence of sulphur dioxide,[378] have appeared. Quantitative kinetic measurements on the autoxidation of aqueous dispersions of oxidizable organic substrates can be made by using either water- or lipid-soluble initiators with water- or lipid-soluble chain-breaking antioxidants.[379]

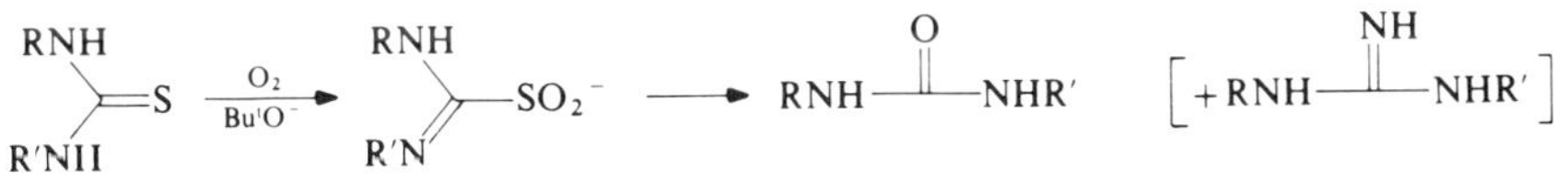

SCHEME 36

Reduction by Complex Metal Hydrides

Reductions with chiral boron reagents[380] and chiral modifications of lithium aluminium hydride[381] have been reviewed. Much interest has centred on stereochemical aspects of the hydride reduction of carbonyl groups. High optical yields have been obtained in the asymmetric reduction of prochiral ketones by a reagent derived from lithium aluminium hydride partially decomposed by (−)-N-methylephedrin and 2-(alkylamino)pyridine.[382] Essentially complete optical induction is found in the reduction of α-keto-esters with *B*-(3-pinanyl)-9-borabicyclo[3.3.1]nonane.[383] A decrease in the significance of product-stability control has been used to rationalize the different diastereoselectivities of lithium trimethoxyaluminium hydride in comparison to the unmodified lithium aluminium hydride.[384] Highly diastereocontrolled reductions of 2-(dimethylamino)ketones and 2-hydroxyketones by hydrosilanes occur in the presence of a catalytic amount of tetrabutylammonium fluoride in HMPA; the *threo*-selectivity may be explained according to the Felkin transition-state model by considering attack by the bulky $[R_3SiHF]^-$ anion in the reduction.[385] Diastereoselectivity in the reduction of carbonyl compounds by zinc borohydride[386] and other metal complex hydrides[387] has been discussed. The stereochemistry of the hydride reduction of 3,*N*-diphenylbutane-2-imine has been accounted for by consideration of Felkin models.[388]

The crystal structures of some *p*-nitrobenzoates of some boat 4-hydroxycyclohexanones have been determined in order to examine the reaction path for the 1,4-transfer of hydride in the corresponding alkoxide ions.[389] The reaction between borohydride ion and formaldehyde has been investigated in the gas phase using ion cyclotron resonance spectroscopy. No hydride transfer is observed and a novel reaction between enolate ions and diborane has been discovered.[390] For the borohydride reduction of ketones, micro-emulsion catalysis has been found to be more effective than either phase-transfer catalysis or the use of a tetraalkylammonium borohydride in a hydrocarbon solvent.[391] Tin formates reduce carbonyl compounds by a radical process in which tin hydrides are not intermediates.[392]

α-Oxoketene dithiacetals are regio- and stereo-specifically reduced to *threo*-β-alkyl-γ-bis(methylthio)-alcohols (Scheme 37).[393] Stereocontrol of a Michael hy-

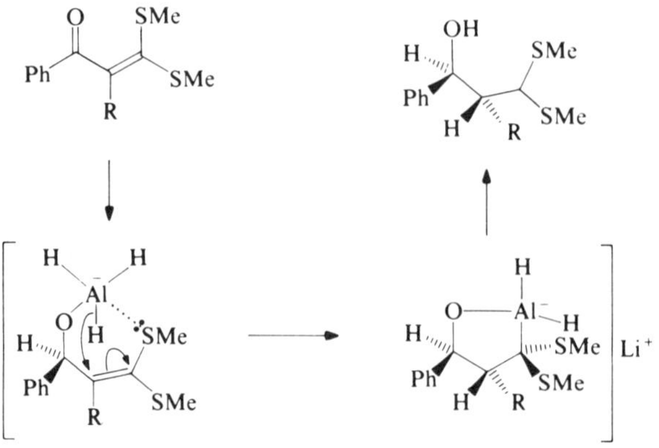

O SMe Ph SMe R

OH H SMe Ph SMe H R

H H Al O H H SMe Ph SMe R

H O Al H H SMe Ph SMe H R Li+

SCHEME 37

dride reduction by a remote hydroxyl group is observed in the reduction of the α,β-unsaturated ester (Scheme 38).[394] The regio- and stereo-selectivity found in the lithium aluminium hydride reductions of 2-cyclopropenyl esters and alcohols has been rationalized by initial formation of an alkoxyaluminium hydride, followed by intramolecular reduction of the double bond.[395]

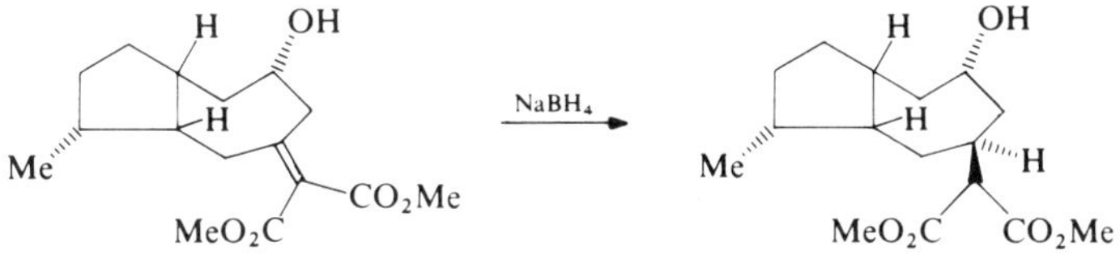

SCHEME 38

Triethylborane greatly facilitates the reductions of epoxides[396] and esters[397] by lithium borohydride. The monochloroborane–ether complex is a good reagent for the reductive cleavage of ketals and acetals.[398] Amination of aromatic and heteroaromatic organometallics is achieved by sequential treatment with diphenyl phosphorazidate and sodium bis(2-methoxyethoxy)aluminum hydride.[399]

Reduction by Metals, Metal Ions, and Metal Complexes

The stereochemistry of the reduction of ketones by metals in liquid ammonia has been studied as a function of the metal and the proton donor.[400] Factors which determine the diastereoselectivity and the ratio of pinacol to carbinol in the reduction of prochiral alkyl aryl ketones with alkali-metal amalgams have been investigated.[401] The formation of the tricyclic ketone (**47**) by reduction of the α,β-unsaturated ketone (**45**) by lithium in liquid ammonia involves intramolecular Michael addition by the vinyl anion (**46**).[402] Among other studies on the Birch reduction,[403–405] it has been found that remotely connected, but spatially proximate, hydroxyl groups have considerable effects on rates and the regiochemistry of the Birch reduction of aromatic rings and double bonds; the

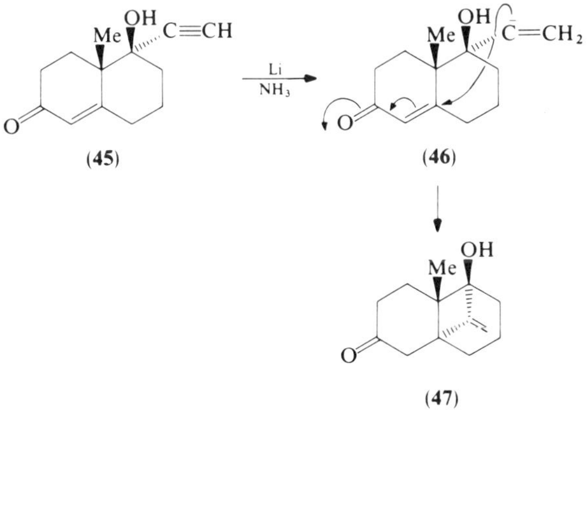

(**45**)

(**46**)

(**47**)

presence of intramolecular OH . . . . π-bonding can markedly influence both the regio- and stereo-chemistry of the Birch reduction, as well as give rise to dramatic rate enhancements.[406] Chemiluminescence is observed in the Birch reduction of norbornenes such as (**48**); the chemiluminescence arises from dissociation of the vinylic carbon–chlorine bond in the radical anion corresponding to (**48**) which gives rise to $^2\Pi$-excited vinyl radicals which fluoresce to their $^2\Sigma$ ground state.[407]

The rôle of low-valent titanium complexes in the deoxygenation of aromatic nitroso compounds,[408] and the selective reduction of aromatic nitro compounds to the corresponding amines,[409] have been studied.

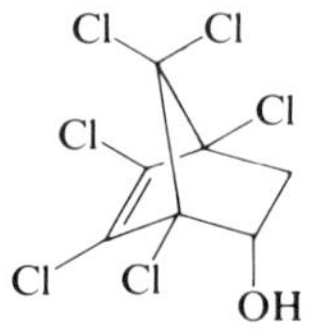

(**48**)

Miscellaneous Reductions

Reviews have appeared on complex reducing agents[410] and the reduction of sulphoxides.[411] Dibromocarbene deoxygenates aldehydes and ketones *via* an intermediate dibromocarbonyl ylid.[412] The reduction of the nitroazobenzene (**49**) by rongalite to the triazole (**50**) involves initial fragmentation of rongalite to

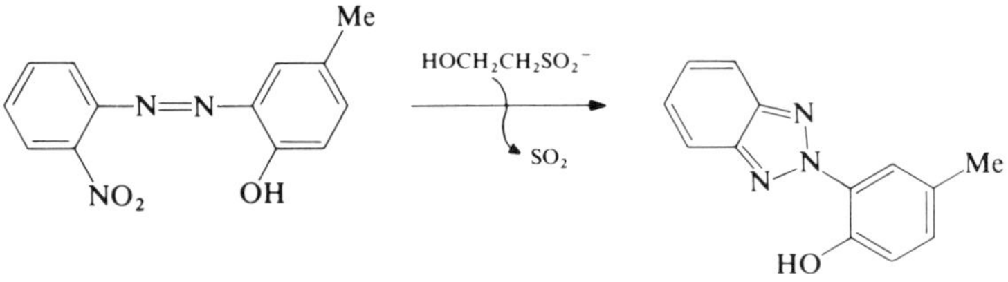

$HOCH_2CH_2SO_2^-$, $-SO_2$

(**49**) (**50**)

formaldehyde and SO_2^{2-}.[413] Kinetic studies on the reaction of aqueous sulphite with *N*-chloroalanylalanylalanine[414] and the reduction of 3-pyridinecarboxaldehyde by thiourea dioxide[415] have been reported. Tributylphosphine–diphenyl disulphide reduces oximes to imines (Scheme 39); reduction under these condition of nitro compounds with a suitable carbonyl group leads to the formation of pyrroles (Scheme 40).[416] Triphenylphosphine–iodine is a good reagent for the deoxygenation of epoxides (Scheme 41).[417] Chlorotrimethylsilane–sodium iodide effects the reductive removal of tertiary hydroxyl groups in α,β-unsaturated-γ-*tert*-hydroxyketones.[418] The reduction of azides to amines by sodium hydrogen telluride proceeds by initial nucleophilic attack of the telluride ion on the azide group.[419] In the zeolite process for the conversion of methanol into hydrocarbons, an oxygen ylid is a probable intermediate.[420]

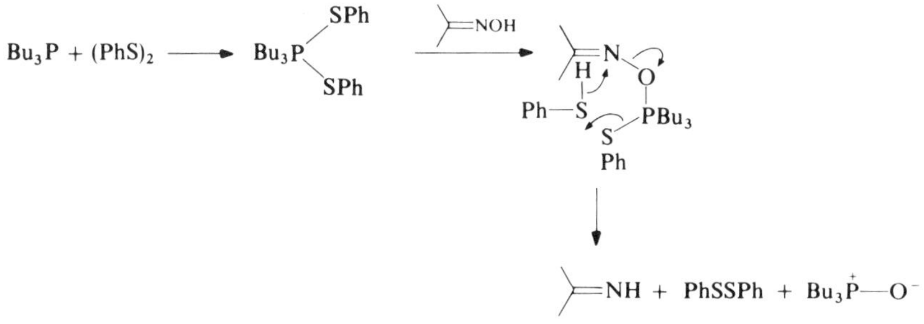

SCHEME 39

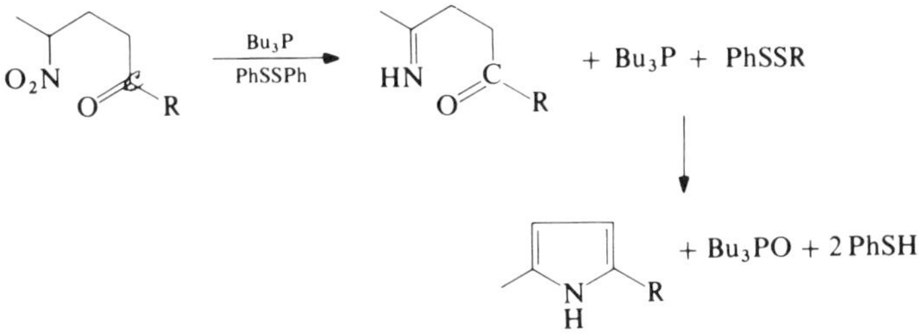

SCHEME 40

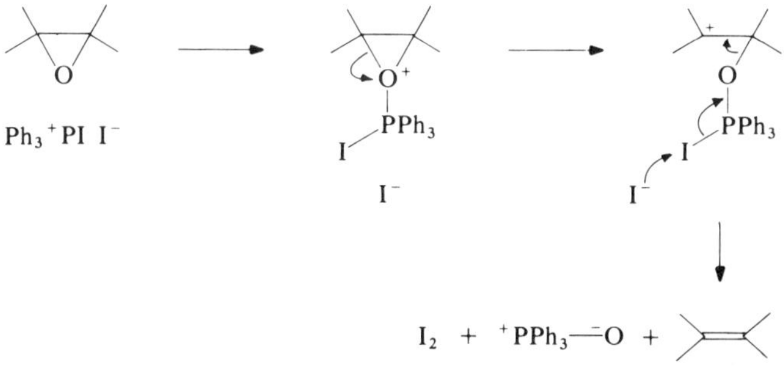

SCHEME 41

Hydrogenation

Modern catalytic studies over metal single crystals, including hydrogenation reactions, have been discussed.[421] A mechanistic comparison has been undertaken of noble-metal catalysts in olefin hydrogenation reactions in terms of product stereochemistry, double-bond isomerization and hydrogen-uptake data. The platinum-catalysed reactions give data in agreement with previously reported results explained by the Horiuto–Polanyi mechanism. Rhodium, although similar to platinum, is more susceptible to steric effects. Olefin hydrogenations over palladium are quite different, taking place by way of a π-allyl absorbed species rather than the 1,2-σ entity in the Horiuto–Polanyi process. Ruthenium gives anomalous results while iridium is mechanistically similar to platinum.[422] Other comparisons of catalysts for hydrogenation reactions include the reduction of 2,4-dinitroaniline[423] and of aromatic and heteroaromatic nitro compounds by transition-metal carbonyl hydrides.[424]

The hydrogenation of prochiral methyl ketones occurs in a high optical yield with Raney nickel asymmetrically modified by tartaric acid.[425] The selective hydrogenation of acrolein to allyl alcohol has been achieved with a Raney-type catalyst.[426] The use of pure palladium foil or powder as Lindlar and Rosenmund catalysts has shown that selective poisons in these syntheses are unnecessary; the action of the poisons involves the reorganization of high-surface-area palladium to a lower-surface-area metallic palladium *via* labile hydride formation.[428] The regioselectivities of palladium-catalysed hydrogenations of cyclopentadiene[427] and of substituted cyclopropanes (where Masso's rule generally predicts the ring-opening correctly)[429] have been investigated. A selective palladium-catalysed hydrogenation of bromocyanamides has been used to develop a synthesis of vicinal diamines.[430] Five sequential hydrogenation steps occurred in the palladium-catalysed reduction of the enal (**51**) to the acetonide of swainsonine.[431]

O O N_3 $PhCH_2O$ O H CHO —Pd, 5 H_2→ O O N H HO

(**51**) (**52**)

Dienes derived from 1,3-[432] and 1,4-[433] di-*tert*-butylbenzenes with a rhodium catalyst and hydrogen have been identified; although little or no diene is desorbed during the hydrogenation, the results indicate that the preferred reaction pathway proceeds *via* the addition of the first hydrogen to a carbon bearing a *tert*-butyl roup. The results of a theoretical study of the homogeneous hydrogenation of olefins catalysed by Wilkinson's catalyst have been analysed by comparison to some experimental data.[434] Kinetic studies on the hydrogenations of unsaturated fatty and 2-oxo-acids,[435] acrylonitrile,[436] and allyl alcohol,[437] catalysed by homogeneous rhodium complexes, have been reported. The mechanism of rhodium-promoted triphenylphosphine reactions in hydroformylation processes has been investigated.[438] The axially dissymmetric bis(triaryl)phosphine (**53**) is an excellent ligand for rhodium(I)-catalysed asymmetric hydrogenation of α-acylaminoacrylic acids and esters; factors controlling the enantioselectivity and mechanistic aspects of the reaction have been probed by ^{31}P-NMR experiments.[439] Other studies on rhodium-catalysed enantioselective reductions of α-acylaminoacrylic acids[440] and dehydropeptides[441] have appeared. The reduction of alkyl-substituted cyclohexanones by

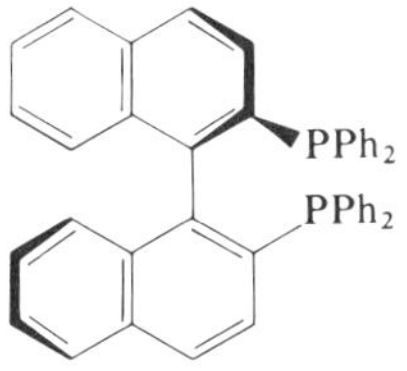

(53)

hydrogen-transfer reactions catalysed by rhodium(I) complexes gives mainly *cis*-substituted cyclohexanols;[442] the effect of the basicity of the phosphine ligand on the stereoselectivity of this reaction has been investigated.[443] The mechanisms of the homogeneous hydrogenations of cyclohexene[444] and α,β-unsaturated aldehydes,[445] acids, and esters,[446] catalysed by ruthenium complexes, have been studied by several techniques including ^{31}P-NMR spectroscopy and kinetic isotope effects.

In the presence of 2,2′-bipyridyl, acetylenes are hydrogenated by cyanocobaltate, reflecting the formation of the highly active hydride anionic complex $[Co(CN)_3(bipyridyl)H]^-$.[447] The rate enhancement of hydrogenation of olefins catalysed by pentacyanohydrocobaltate caused by micelles has been attributed to the concentration of the substrate and catalyst on the micelle surface.[448,449] Homogeneous catalysis of the hydrogenation of olefins by an iridium(II) catalyst[450] and hydrogen transfer from formic acid to olefins catalysed by polymer-bound iridium[451] have been investigated. Photolysis of *cis*-$HMn(CO)_4PPh_3$ in the presence of hydrogen and a terminal olefin results in catalytic hydrogenation and isomerization of the alkene.[452] Iron pentacarbonyl in the presence of water, carbon monoxide, base, and a phase-transfer agent catalyses the regiospecific reduction of the nitrogen-containing ring in quinoline; an electron-transfer process is involved.[453]

Reductions and Oxidations of Biological Interest

A theoretical study of linear *versus* bent transition states for hydride transfers mediated by NADH models has been described.[454] The semi-classical (*i.e.* in the absence of tunnelling) kinetic isotope effects for reduced NADH model hydride-transfer reactions are in the range 2–3, consistent with rate-determining but non-linear hydride transfer through a charge-transfer type of complex.[455] In a study of rate and equilibrium constants of hydride transfer between NAD analogues, the data are not consistent with a multi-step (electron-transfer)–(proton-transfer)–(electron-transfer) mechanism;[456] kinetic isotope effects indicate a one-step mechanism.[457] The mechanisms of hydride transfer from *N*-benzylacridans[458] and NADH models[459] to *p*-benzoquinone have been studied; charge-transfer complexes between an NADH model and quinones have been isolated and it is postulated that the radical ion pair formed by one-electron transfer is closer to a transition state than to an intermediate. Ternary complexes have been identified in the reaction of 4-*trans*-(*N*,*N*-dimethylamino) cinnamaldehyde with oxidized NAD and horse liver alcohol dehydrogenase[460] and in the reduction of a *p*-benzoquinone derivative by NADH in the presence of magnesium ions.[461] Metal-ion catalysis of the reduction of carbonyl compounds by 1,4-dihydropyridines has attracted much attention. Zinc ion produces a remarkable acceleration in the rate of reduction of 2-pyridinecarbaldehyde with dihydroquinolines in aqueous solution; it has been found that the kinetic complexities disappear when the reaction is conducted in the dark and also that a hydroxyl group on the dihydroquinoline accelerates the reaction considerably.[462] Simultaneous general acid–base catalysis involving an enzyme and an external catalyst has been observed in the lactate dehydrogenase-catalysed formation of the NAD–pyruvate adduct.[463] A rare example of an acid-catalysed NADH model has been observed in the reduction of an arylnitroso compound to the corresponding hydroxylamine.[464] The reductive dediazoniation of arenediazonium salts by *N*-benzyl-1,4-dihydronicotinamide proceeds by a radical-chain pathway initiated by one-electron transfer from the NADH model.[465] A radical-chain mechanism is also involved in the oxidation of, and addition reactions to, *N*-benzyl-1,4-dihydronicotinamide catalysed by benzoyl peroxide.[466] Among other aspects of stereoselectivity in the transfer of hydrogen from NADH model compounds,[467,468] it has been pointed out that a re-evaluation of the currently accepted theories is required.[469] Other mechanistic studies on enzymic[470,471] and non-enzymic[472–474] NADH reactions have been reported.

The crown ether flavin mimic (**54**), containing both the flavin as a catalytic site and the crown ether as a recognition site, has been prepared; pseudo-intramolecular fluorescence quenching of (**54**) by tryptamine hydrochloride is due to recognition of the ammonium group by the crown ether ring.[475] Evidence has been presented for

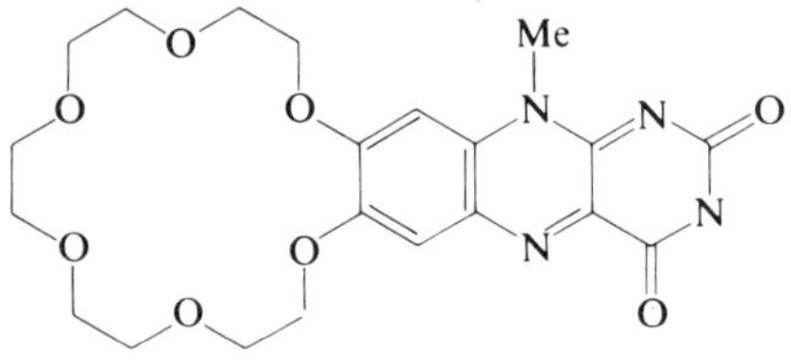

(**54**)

the transfer of the β-hydrogen to the flavin N(5) position as a hydride species in acyl-CoA dehydrogenases.[476] There are marked differences in the efficiency of electron transfer from reduced flavin to aromatic nitro compounds as a function of the metal ion.[477] The interconversion of the reduced, radical, and oxidized forms of 1,10-ethano-5-ethyllumiflavin have been studied;[478] the mechanism of proton and one-electron transfer steps in the reaction of dihydroflavins with oxygen has bee investigated.[479] Implications derived from the reduction of maleimides by a 1,5-dihydroflavin for dehydrogenating flavoenzymes such as succinic acid dehydrogenase and acyl-CoA oxidase have been considered.[480] Oxygen flavin chemistry has been reviewed[481] and several other aspects of flavin oxidations discussed.[482–484]

The active oxidant in the epoxidation of olefins catalysed by manganese(III)–porphyrin is a high-valent oxomanganese complex which reacts with the olefin to produce a relatively stable intermediate; the decomposition of this intermediate to manganese(III)–porphyrin and the product epoxide is the rate-determining step of the catalytic cycle with lithium hypochlorite as the oxidizing agent.[485] Evidence for group migration in intermediates formed in the oxidations of vinylidene chloride and *trans*-1-phenyl-1-butene by cytochrome P450 has been obtained.[486] Other mechanistic studies on oxidations by cytochrome P450 have been published,[487–493] as well as a unified view of the mechanisms of reactions of cytochrome P450.[494] Several other mechanistic studies on enzyme-catalysed oxidation–reduction reactions in biological systems have been reported.[495–510]

References

1 Roundhill, D. M., *Homogeneous Catal. Met. Phosphine Complexes,* **1983,** 377; *Chem. Abs.,* **100,** 120104 (1984).
2 Mares, F., and Diamond, S. E., *Fundam. Res. Homogeneous Catal.,* **4,** 55 (1984); *Chem. Abs.,* **100,** 138223 (1984).
3 Mimoun, H., *Isr. J. Chem.,* **23,** 451 (1983).
4 Minisci, F., *Fundam. Res. Homogeneous Catal.,* **4,** 173 (1984); *Chem. Abs.,* **100,** 138225 (1984).
5 Meunier, B., *Bull. Soc. Chim. Fr. II,* **1983,** 345
6 Rudakov, E. S., *React. Kinet. Catal. Lett.,* **22,** 319 (1983); *Chem. Abs.,* **100,** 67653 (1984).
7 Thyrion, F. C., *Bull. Soc. Chim. Belg.,* **93,** 281 (1984).
8 Reddy, K. N., Devi, P. S. U., and Saiprakash, P. K., *Acta Cienc. Indica,* [*ser.*] *Chem.,* **9,** 112 (1983); *Chem. Abs.,* **100,** 208915 (1984).
9 Goldstein, S. L., and McNelis, E., *J. Org. Chem.,* **49,** 1613 (1984).
10 Reddy, P. M., Jagannadham, V., Sethuram, B., and Rao, T. N., *Pol. J. Chem.,* **56,** 865 (1982); *Chem. Abs.,* **100,** 67616 (1984).
11 Mata, F., and Sancho-Orcajo, A., *An. Quim.,* **80A,** 23 (1984); *Chem. Abs.,* **101,** 90143 (1984).
12 Banerji, K. K., *Indian J. Chem.,* **22B,** 650 (1983); *Chem. Abs.,* **100,** 33934 (1984).
13 Rajasekaran, K., Baskaran, T., and Gnanasekaran, C., *J. Chem. Soc., Perkin Trans. 2,* **1984,** 1183.
14 Nonaka, T., Kanemoto, S., Oshima, K., and Nozaki, H., *Bull. Chem. Soc. Jpn.,* **57,** 2019 (1984).
15 Bhattacharjee, M. N., Chaudhuri, M. K., and Dasgupta, H. S., *Bull. Chem. Soc. Jpn.,* **57,** 258 (1984).
16 Srinivasan, C., Kuthalingam, P., Chellamani, A., Rajagopal, S., and Arumugam, N., *Proc. Indian Acad. Sci., Chem. Sci.,* **93,** 157 (1984); *Chem. Abs.,* **101,** 90110 (1984).
17 Walba, D. M., DePuy, C. H., Grabowski, J. J., and Bierbaum, V. M., *Organometallics,* **3,** 498 (1984).
18 Curci, R., Giannattasio, S., Sciacovelli, O., and Troisi, L., *Tetrahedron,* **40,** 2763 (1984).
19 Sharma, V. K., and Raj, R. C., *J. Indian Chem. Soc.,* **60,** 745 (1983); *Chem. Abs.,* **100,** 102565 (1984).
20 Wolfe, S., and Ingold, C. F., *J. Am. Chem. Soc.,* **105,** 7755 (1983).
21 Bock, H., and Jaculi, D., *Angew. Chem. Int. Ed.,* **23,** 305 (1984).
22 Amiya, T., Kanaiwa, Y., Bando, H., and Suginome, H., *Chem. Lett.,* **1984,** 859.
23 Polito, W. L., and Neves, E. F. A., *An. Simp. Bras. Electroquim. Electroanal., 4th,* **1984,** 7; *Chem. Abs.,* **101,** 90133 (1984).
24 Zielinski, M., *Zesz. Nauk Uniw, Jagiellon., Pr. Chem.,* **28,** 79, (1983); *Chem. Abs.,* **101,** 72049 (1984).
25 Tyupalo, N. F., and Zamashchikov, V. V., *Kinet. Katal.,* **24,** 994 (1983); *Chem. Abs.,* **99,** 175072 (1983).
26 Midgley, G., and Thomas, C. B., *J. Chem. Soc., Perkin Trans. 2,* **1984,** 1537.
27 Citterio, A., Gentile, A., Minisci, F., and Serravalle, M., *Gazz. Chim. Ital.,* **113,** 443 (1983); *Chem. Abs.,* **100,** 120301 (1984).
28 Rao, B. M., and Gandhi, P. K., *Z. Phys. Chem. (Leipzig),* **265,** 782 (1984).
29 Oswal, S. L., and Pathak, K. G., *J. Indian Chem. Soc.,* **61,** 89 (1984); *Chem. Abs.,* **101,** 90146 (1984).
30 Ramachandran, M. S., Vivekanandam, T. S., Subbaratnam, N. R. and Rajaram, N., *Indian J. Chem., Sect. A.,* **22A,** 895 (1983); *Chem. Abs.,* **100,** 138393 (1984).
31 Sychev, A. Y., Mai Hui Hiem, and Isak, V. G., *Zh. Fiz. Khim.,* **58,** 906 (1984); *Chem. Abs.,* **101,** 72026 (1984).
32 Razenberg, J. A. S. J., Nolte, R. J. M., and Drenth, W., *Tetrahedron Lett.,* **25,** 789 (1984).
33 Solov'eva, A. B., Mel'nikova, V. I., Pivnitskii, K. K., Karakozova, E. I., Bogdanova, K. A.,

Karmilova, L. V., Nikiforov, G. A., and Enikolopov, N. S., *Izv. Akad. Nauk SSSR, Ser. Khim.*, **10,** 2327 (1983); *Chem. Abs.*, **100,** 67626 (1984).
[34] Joensson, L., *Acta Chem. Scand.*, **37B,** 761 (1983).
[35] Gasparrini, F., Giovannoli, M., Natile, G., and Palmieri, G., *Congr. Naz. Chim. Inorg.*, [*Atti*], *15th*, **1982,** 178; *Chem. Abs.*, **101,** 6358 (1984).
[36] Heyward, M. P., and Wells, C. F., *J. Chem. Soc., Faraday Trans. 1*, **80,** 2155 (1984).
[37] Ahmad, F., Sabir, S., and Baswani, V. S., *J. Indian Chem. Soc.*, **60,** 1052 (1983); *Chem. Abs.*, **101,** 72017 (1984).
[38] Reddy, K. B., Sethuram, B., and Rao, T. N., *Indian J. Chem.*, **23A,** 593 (1984); *Chem. Abs.*, **101,** 90196 (1984).
[39] Capdevielle, P., Audebert, P., and Maumy, M., *Tetrahedron Lett.*, **25,** 4397 (1984).
[40] Bampp, H., Haspra, P., Spieler, W., and Zvberbuhler, A. D., *Helv. Chim. Acta*, **67,** 1019 (1984).
[41] Gupta, K. C., and Pandey, S. P., *Z. Phys. Chem. (Leipzig)* **265,** 365 (1984).
[42] Sychev, A. Y., and Duka, G. G., *Khim. Fiz.*, **4,** 521 (1982); *Chem. Abs.*, **100,** 33976 (1984).
[43] Utaka, M., Hojo, M., Fujii, Y., and Takeda, A., *Chem. Lett.*, **1984,** 635.
[44] Semmelhack, M. F., Schmid, C. R., Dortez, D. A., and Chou, C. S., *J. Am. Chem. Soc.*, **106,** 3374 (1984).
[45] Sayre, L. M., and Jin, S.-J., *J. Org. Chem.*, **49,** 3498 (1984).
[46] Morozov, V. P. and Smelova, I. V., *Izv. Vyssh. Uchebn. Zaved., Khim. Khim. Tekhnol.*, **26,** 761 (1983); *Chem. Abs.*, **99,** 194171 (1983).
[47] Sychev, A. Y., Duka, G. G., and Chub, L. S., *Izv. Akad. Nauk Mold. SSR, Ser. Biol. Khim. Nauk*, **1983,** 65; *Chem. Abs.*, **100,** 102586 (1984).
[48] Banerjee, S., Pack, E. J., Jr., Sikka, H. and Kelly, C. M., *Chemosphere*, **13,** 549 (1984); *Chem. Abs.*, **101,** 22804 (1984).
[49] Panova, G. V., Solozhenko, E. G., Garbar, A. V., Astanina, A. N., and Potapov, V. M., *Zh. Obshch. Khim.*, **54,** 1391 (1984); *Chem. Abs.*, **101,** 151197 (1984).
[50] Kurosawa, K., Takamura, T., Ueno, Y., McOmie, J. F. W., and Pearson, N. D., *Bull. Chem. Soc. Jpn.*, **57,** 1914 (1984).
[51] Galliani, G., Rindone, B., and Beltrame, P. L., *Nouv. J. Chim.*, **7,** 639 (1983).
[52] Cekovic, Z., Musicki, B., Bosnjak, J., and Mihailovic, M. L., *Glas. Hem. Drus. Beograd*, **48,** 681 (1983); *Chem. Abs.*, **101,** 90103 (1984).
[53] Mihailovic, M. L., Andrejevic, V., Gojkovic, S., Milosavljevic, S., and Konstantinovic, S.; *Glas. Hem. Drus. Beograd*, **48,** 283 (1983); *Chem. Abs.*, **100,** 138379 (1984).
[54] Rubottom, G. M., Gruber, J. M., Marrero, R., Juve, H. D., and Kim, C. W., *J. Org. Chem.*, **48,** 4940 (1983).
[55] Banthia, U. S., Joshi, B. C., and Gupta, Y. K., *Indian J. Chem.*, **23A,** 413 (1984); *Chem. Abs.*, **101,** 129999 (1984).
[56] Balakrishnan, R., and Vangalur, S., *Proc. Indian Acad. Sci., Chem. Sci.*, **93,** 171 (1984); *Chem. Abs.*, **101,** 54261 (1984).
[57] Shukla, S. N., and Kesarwani, R. N., *Carbohydr. Res.*, **133,** 319 (1984).
[58] Balakrishnan, R., Raghavan, P. S., and Srinivasan, V. S., *Proc. Indian Acad. Sci.*, [*Ser.*]*: Chem. Sci.*, **92,** 283 (1983); *Chem. Abs.*, **100,** 138369 (1984).
[59] Pechal, M., and Strasak, M., *Collect. Czech. Chem. Commun.*, **48,** 2819 (1983).
[60] Gupta, P., Sharma, P. D., and Gupta, Y. K., *Indian J. Chem.*, **23A,** 392 (1984); *Chem. Abs.*, **101,** 90184 (1984).
[61] Menard, R., and Zador, M., *Inorg. Chim. Acta*, **45,** L217 (1980); *Chem. Abs.*, **100,** 208912 (1948).
[62] Gupta, K. C., Misra, V. D., and Gupta, K., *Monatsh. Chem.*, **115,** 405 (1984).
[63] Calvaruso, G., Cavasino, F. P., and Sbriziolo, C., *Int. J. Chem. Kinet.*, **16,** 1201 (1984).
[64] Ignaczak, M., Deka, M., and Dziegiec, J., *Pol. J. Chem.*, **56,** 229 (1982); *Chem. Abs.*, **100,** 156034 (1984).
[65] Bhatt, K., and Nand, K. C., *Z. Phys. Chem. (Leipzig)*, **264,** 1195 (1983).
[66] Baciocchi, E., and Ruzziconi, R., *J. Chem. Soc., Chem. Commun.*, **1984,** 445.
[67] Gopalan, R., and Kannamma, E., *Indian J. Chem.*, **23A,** 518 (1984); *Chem. Abs.*, **101,** 13009 (1984).
[68] Singh, B., and Seth, G. K., *J. Indian Chem. Soc.*, **60,** 1107 (1983); *Chem. Abs.*, **101,** 37970 (1984).
[69] Ignaczak, M., Dziegiec, J., Skowronski, R., and Turala, L., *Pol. J. Chem.*, **56,** 887 (1982); *Chem. Abs.*, **100,** 102581 (1984).
[70] Kramer, C. R., Schelenz, T., and Stein, J., *Z. Phys. Chem. (Leipzig)*, **265,** 849 (1984).
[71] Schelenz, T., Kramer, C. R., and Stein, J., *Z. Phys. Chem. (Leipzig)*, **265,** 865 (1984).
[72] Sharpless, K. B., Woodard, S. S., and Finn, M. G., *Pure Appl. Chem.*, **55,** 1823 (1983).

[73] Behrens, C. H., and Sharpless, K. B., *Aldrichimica Acta,* **16,** 67 (1983); *Chem. Abs.,* **100,** 120106 (1984).
[74] Hutchinson, R. J., Smith, J. R. L., and Twigg, M. V., *J. Chem. Soc., Perkin Trans. 2,* **1984,** 1583.
[75] Mishra, A. K., Sharma, S. D., Sarpal, N. S., and Agnihotri, V. K., *Vijnana Parishad Anusandhan Patrika,* **27,** 93 (1984); *Chem. Abs.,* **101,** 151220 (1984).
[76] Virtanen, P. O. I., and Sammalkivi, R., *Finn. Chem. Lett.,* **1983,** 151.
[77] Srivastava, M., Singh, R. K., and Tripathi, S. R., *Natl. Acad. Sci. Lett. (India),* **6,** 191 (1983); *Chem. Abs.,* **100,** 174081 (1984).
[78] Ledon, H. J., Varescon, F., and Durbut, P., *Chem. Uses Molybdenum, Proc. Int. Conf., 4th,* **1982,** 319; *Chem. Abs.,* **100,** 120262 (1984).
[79] Vretsena, N. B., Nikipanchuk, M. V., and Chernyak, B. I., *Ukr. Khim. Zh. (Russ. Ed.),* **50,** 370 (1984); *Chem. Abs.,* **101,** 54308 (1984).
[80] Chang, B., and Lin, F., *Gaodeng Xuexiao Huaxue Xuebao,* **4,** 775 (1983); *Chem. Abs.,* **100,** 191125 (1984).
[81] Jitsukawa, J., Kaneda, K., and Teranishi, S., *J. Org. Chem.,* **49,** 199 (1984).
[82] Masuyama, Y., Takahashi, M., and Kuruse, Y., *Tetrahedron Lett.,* **25,** 4417 (1984).
[83] Khare, A. P., and Agrawal, G. L., *J. Inst. Chem. (India),* **55,** 137 (1983); *Chem. Abs.,* **100,** 33972 (1984).
[84] Khare, P., and Agrawal, G. L., *React. Kinet. Catal. Lett.,* **23,** 207 (1983); *Chem. Abs.,* **101,** 71998 (1984).
[85] Schlesener, C. J., Amatore, C., and Kochi, J. K., *J. Am. Chem. Soc.,* **106,** 3567 (1984).
[86] Bhattacharjee, A. K., and Mahanti, M. K., *Bull. Korean Chem. Soc.,* **4,** 120 (1983); *Chem. Abs.,* **99,** 211967 (1983).
[87] Bhattacharjee, A. K., and Mahanti, M. K., *React. Kinet. Catal. Lett.,* **23,** 361 (1983); *Chem. Abs.,* **101,** 6394 (1984).
[88] Bhattacharjee, M., and Mahanti, M. K., *Indian J. Chem.,* **22A,** 634 (1983); *Chem. Abs.,* **99,** 211956 (1983).
[89] Bhattacharjee, M., and Mahanti, M. K., *Bull. Soc. Chim. Fr. II,* **1983,** 225.
[90] Bhattacharjee, M., and Mahanti, M. K., *React. Kinet. Catal. Lett.,* **22,** 445 (1983); *Chem. Abs.,* **100,** 67651 (1984).
[91] Vital, A. S. P., Rao, P. V. K., and Rao, K. J. M., *React. Kinet. Catal. Lett.,* **23,** 175 (1983); *Chem. Abs.,* **100,** 138417 (1984).
[92] Dasgupta, G., and Mahanti, M. K., *React. Kinet. Catal. Lett.,* **23,** 393 (1983); *Chem. Abs.,* **101,** 6396 (1984).
[93] Bhattacharjee, A. K., and Mahanti, M. K., *React. Kinet. Catal. Lett.,* **22,** 227 (1983); *Chem. Abs.,* **100,** 33931 (1984).
[94] Nikolaenko, T. K., Mel'nichenko, I. V., and Yasnikov, A. A., *Dopov. Akad. Nauk Ukr. RSR, Ser. B: Geol., Khim. Biol. Nauki,* **1984,** 47; *Chem. Abs.,* **100,** 208850 (1984).
[95] Nesvadba, P., and Kuthan, J., *Collect. Czech. Chem. Commun.,* **49,** 543 (1984).
[96] Nesvadba, P. and Kuthan, J., *Collect. Czech. Chem. Commun.,* **48,** 2965 (1983).
[97] Bunting, J. W., and Kauffman, G. M., *Can. J. Chem.,* **62,** 729 (1984).
[98] Powell, M. F., Wu, J. C., and Bruice, T. C., *J. Am. Chem. Soc.,* **106,** 3850 (1984).
[99] Fukuzumi, S., Kondo, Y., and Tanaka, T., *J. Chem. Soc., Perkin Trans. 2,* **1984,** 673.
[100] Ito, S., Aihara, K., and Matsumoto, M., *Tetrahedron Lett.,* **25,** 3891 (1984).
[101] Barton, D. H. R., Boivin, J., Ozbalik, N., and Schwartzentruber, K. M., *Tetrahedron Lett.,* **25,** 4219 (1984).
[102] Fontecave, M., and Mansuy, D., *J. Chem. Soc., Chem. Commun.,* **1984,** 879.
[103] Mansuy, D., Leclaire, J., Fontecave, M., and Dansette, P., *Tetrahedron,* **40,** 2847 (1984).
[104] Traylor, T. G., Lee, W. A., and Stynes, D. V., *J. Am. Chem. Soc.,* **106,** 755 (1984).
[105] Pandell, A. J., *J. Org. Chem.,* **48,** 3908 (1983).
[106] Sychev, A. Y., and Duka, G. G., *Zh. Fiz. Khim.,* **58,** 1382 (1984); *Chem. Abs.,* **101,** 90175 (1984).
[107] Brennan, T., and Frenkel, C., *Bot. Gaz. (Chicago),* **144,** 32 (1983); *Chem. Abs.,* **99,** 194157 (1983).
[108] Sychev, A. Y., Suen, N. V., and Isak, V. G., *Khim. Fiz.,* **1983,** 1565; *Chem. Abs.,* **100,** 50892 (1984).
[109] Richter, H. W., and Waddell, W. H., *Proc. Int. Conf. Superoxide Superoxide Dismutase, 3rd,* **1,** 89 (1982); *Chem. Abs.,* **100,** 67615 (1984).
[110] Hirano, M., and Morimoto, T., *J. Chem. Soc., Perkin Trans. 2,* **1984,** 1033.
[111] Budholia, D. C., and Sthapak, J. K., *J. Indian Chem. Soc.,* **60,** 404 (1983); *Chem. Abs.,* **99,** 194169 (1983).
[112] Kashiyae, M., and Yoshitomi, S., *Nippon Kagaku Kaishi,* **1983,** 685; *Chem. Abs.,* **99,** 211918 (1983).

[113] Chaplin, R. P., Walpole, A. S., Zadro, S., Vorlow, S., and Wainwright, M. S., *J. Mol. Catal.*, **22,** 269 (1984); *Chem. Abs.*, **100,** 156045 (1984).

[114] Rastogi, R. P., and Das, I., *Indian J. Chem.*, **23A,** 363 (1984); *Chem. Abs.*, **101,** 90182 (1984).

[115] Roelofs, M. G., Wasserman, E., Jensen, J. H., and Nader, A. E., *J. Am. Chem. Soc.*, **105,** 6329 (1983).

[116] Okamoto, T., and Oka, S., *J. Org. Chem.*, **49,** 1589 (1984).

[117] Wang, X. Y., Motekaitis, R. J., and Martell, A. E., *Inorg. Chem.*, **23,** 271 (1984); *Chem. Abs.*, **100,** 67628 (1984).

[118] Toivonen, H., *Acta Chem. Scand.*, **38B,** 37 (1984).

[119] Rudakov, E. S., Tret'yakov, V. P., Chudaev, V. V., Zimtseva, G. P., and Simonov, M. A., *Kinet. Katal.*, **24,** 1081 (1983); *Chem. Abs.*, **100,** 33973 (1984).

[120] Ariko, N. G., Samtsevich, V. S., and Mitskevich, N. I., *Vestsi Akad. Navuk BSSR, Ser. Khim. Navuk*, **1984,** 14; *Chem. Abs.*, **101,** 110097 (1984).

[121] Barkanova, S. V., Zheltukhin, I. A., and Kaliya, O. L., *Zh. Org. Khim.*, **19,** 2212 (1983); *Chem. Abs.*, **100,** 33967 (1984).

[122] Brouwer, W. M., Piet, P., and German, A. L., *J. Mol. Catal.*, **22,** 297 (1984); *Chem. Abs.*, **100,** 156046 (1984).

[123] Lee, D. G., Cowgson, L. N., Spitzer, U. A., and Olson, M. E., *Can. J. Chem.*, **62,** 1835 (1984).

[124] Lowe, G., and Salamone, S. J., *J. Chem. Soc., Chem. Commun.*, **1983,** 1392.

[125] Green, G., Griffith, W. P., Hollinshead, D. M. M., Ley, S. V., and Schroder, M., *J. Chem. Soc., Perkin Trans. 1*, **1984,** 681.

[126] Lee, D. G., and Helliwell, S., *Can. J. Chem.*, **62,** 1085 (1984).

[127] Singh, H. S., and Singh, P., *Ann. Soc. Sci. Bruxelles, Ser. 1*, **97,** 21 (1983); *Chem. Abs.*, **100,** 67656 (1984).

[128] Behari, K., Narayan, H., Shukla, R. S., and Gupta, K. C., *Int. J. Chem. Kinet.*, **16,** 195 (1984).

[129] Singh, B., Singh, K., and Singh, J. P., *J. Indian Chem. Soc.*, **60,** 704 (1983); *Chem. Abs.*, **100,** 50894 (1984).

[130] Tandon, P. K., Manibala, K., Singha, H. S., and Krishna, B., *Z. Phys. Chem. (Leipzig)*, **265,** 609 (1984).

[131] Rao, M. D. P., Ahmad, M., and Kanungo, T. K., *React. Kinet. Catal. Lett.*, **24,** 141 (1984); *Chem. Abs.*, **101,** 90101 (1984).

[132] Swarnalakshmi, N., Uma, V., Sethuram, B., and Rao, T. N., *Indian J. Chem.*, **23A,** 386 (1984); *Chem. Abs.*, **101,** 90183 (1984).

[133] Singh, B., Singh, N. B., and Saxena, B. B. L., *J. Indian Chem. Soc.*, **61,** 319 (1984); *Chem. Abs.*, **101,** 190902 (1984).

[134] Singh, B., Singh, A. K., Singh, R. K., Singh, N. B., and Saxena, B. B. L., *Natl. Acad. Sci. Lett. (India)*, **6,** 265 (1983); *Chem. Abs.*, **101,** 190869 (1984).

[135] Pati, S. C., and Dev, B. R., *Curr. Sci.*, **53,** 477 (1984); *Chem. Abs.*, **101,** 72043 (1984).

[136] Pati, S. C., Dev, B. R., Behera, N., and Mishra, M., *Proc. Indian Natl. Sci. Acad., Part A*, **49,** 538 (1984); *Chem. Abs.*, **100,** 138396 (1984).

[137] Riley, D. P., and Shumate, R. E., *J. Am. Chem. Soc.*, **106,** 3179 (1984).

[138] Cano-Yelo, M., and Deronzier, A., *Tetrahedron Lett.*, **25,** 5517 (1984).

[139] Keene, F. R., Ridd, M. J., and Snow, M. R., *J. Am. Chem. Soc.*, **105,** 7075 (1983).

[140] Backvall, J.-E., *Acc. Chem. Res.*, **16,** 335 (1983).

[141] Zhir-Lebed, L. N., and Temkin, O. N., *Deposited Doc.*, **1982,** SPSTL 828; *Chem. Abs.*, **101,** 6399 (1984).

[142] Rudakov, E. S., and Ignatenko, V. M., *React. Kinet. Catal. Lett.*, **22,** 75 (1983); *Chem. Abs.*, **99,** 194176 (1983).

[143] Shlapak, M. S., Brailovskii, S. M., and Temkin, O. N., *Kinet. Katal.*, **24,** 1380 (1983); *Chem. Abs.*, **100,** 120244 (1984).

[144] Kuznetsova, N. I., Likholobov, V. A., and Ermakov, Y. I., *React. Kinet. Catal. Lett.*, **22,** 139 (1983); *Chem. Abs.*, **99,** 194177 (1983).

[145] Rudakov, E. S., Lutsyk, A. I., and Yaroshenko, A. P., *Ukr. Khim. Zh. (Russ. Ed.)*, **49,** 1083 (1983); *Chem. Abs.*, **99,** 194215 (1983).

[146] Ivanov, S., and Tanielyan, S., *Geterog. Katal., 5th Pt.* **1,** 99 (1983); *Chem. Abs.*, **101,** 54254 (1984).

[147] Chauvet, F., Heumann, A., and Waegell, B., *Tetrahedron Lett.*, **25,** 4393 (1984).

[148] Jsuji, J., Minami, I., and Shimizu, I., *Tetrahedron Lett.*, **25,** 2791 (1984).

[149] Arzoumanian, H., Lai, R., Metzger, J., and Petrignani, J. F., *J. Organomet. Chem.*, **267,** 207 (1984).

[150] Read, G., and Shaw, J., *J. Chem. Soc., Chem. Commun.*, **1984,** 1313.

[151] Bortolini, O., Di Furia, F., Modena, G., and Seraglia, R., *J. Mol. Catal.*, **22,** 313 (1984); *Chem. Abs.*, **100,** 156047 (1984).
[152] Cha, J., Christ, W. J., and Kishi, Y., *Tetrahedron,* **40,** 2247 (1984).
[153] Mahadevappa, D. S., Ananda, S., Murthy, A. S. A., and Rangappa, K. S., *Tetrahedron,* **40,** 1673 (1984).
[154] Ganapathy, K., and Jayagandhi, P., *Proc. Indian Acad. Sci., Chem. Sci.*, **93,** 23 (1984); *Chem. Abs.*, **100,** 191145 (1984).
[155] Kalinina, V. E., Zhukov, Y. A., and Koroleva, T. N., *Vopr. Kinet, i Kataliza, Ivanovo,* **1982,** 25; *Chem. Abs.*, **100,** 50872 (1984).
[156] Krupenskii, V. I., *Izv. Vyssh. Uchebn. Zaved., Khim. Khim. Tekhnol.*, **26,** 1163 (1983); *Chem. Abs.*, **100,** 33977 (1984).
[157] Ogura, F., Otsubo, T., Ariyoshi, K., and Yamaguchi, H., *Chem. Lett.*, **1983,** 1833.
[158] Marx, M. and Tidwell, T. T., *J. Org. Chem.*, **49,** 788 (1984).
[159] Kim, K. S., Cho, I. H., Yoo, B. K., Song, Y. H., and Hahn, C. S., *J. Chem. Soc., Chem. Commun.*, **1984,** 762.
[160] Rabenstein, D. L., and Theriault, Y., *Can. J. Chem.*, **62,** 1672 (1984).
[161] Curci, R., Fiorentino, M., and Serio, M. R., *J. Chem. Soc., Chem. Commun.*, **1984,** 155.
[162] Ramachandran, M. S., and Vivekanandam, T. S., *Tetrahedron,* **40,** 4929 (1984).
[163] Ramachandran, M. S., and Vivekanandam, T. S., *J. Chem. Soc., Perkin Trans. 2,* **1984,** 1341.
[164] Ramachandran, M. S., Vivekanandam, T. S., and Raj, R. P. M. M., *J. Chem. Soc., Perkin Trans. 2,* **1984,** 1345.
[165] Risley, J. M., and Van Etten, R. L., *Int. J. Chem. Kinet.*, **16,** 1167 (1984).
[166] Ogawa, K., and Nomura, Y., *Yuki Gosei Kagaku Kyokaishi,* **42,** 98 (1984); *Chem. Abs.*, **100,** 138201 (1984).
[167] Davies, M. J., Gilbert, B. C., and Norman, R. O. C., *J. Chem. Soc., Perkin Trans. 2,* **1984,** 503.
[168] Levitt, L. S., and Pytcher, J. R., *J. Phys. Chem.*, **88,** 1177 (1984).
[169] Abdeen, S. Z., Ohag, M. I., and Vasudeva, W. C., *Gazz. Chim. Ital.*, **114,** 197 (1984); *Chem. Abs.*, **101,** 151216 (1984).
[170] Khan, G. M., and Hamid, A., *Pak. J. Sci. Ind. Res.*, **26,** 16 (1983); *Chem. Abs.*, **100,** 102571 (1984).
[171] Fristad, W. E., and Peterson, J. R., *Tetrahedron,* **40,** 1469 (1984).
[172] Nikishin, G. I., Troyansky, E. I., and Lazareva, M. I., *Tetrahedron Lett.*, **25,** 4987 (1984).
[173] Walling, C., El Taliawi, G. M., and Zhao, C., *Prepr. Am. Chem. Soc., Div. Pet. Chem.*, **29,** 348 (1984); *Chem. Abs.*, **101,** 72035 (1984).
[174] Walling, C., Zhao, C., and El-Taliawi, G. M., *J. Org. Chem.*, **48,** 4910 (1983).
[175] Walling, C., El-Taliawi, G. M., and Zhao, C., *J. Org. Chem.*, **48,** 4914 (1983).
[176] Troyanskii, E. I., Svitan'ko, O. S., and Nikishin, G. I., *Izv. Akad. Nauk SSSR, Ser. Khim.*, **1983,** 1537; *Chem. Abs.*, **99,** 194046 (1983).
[177] Troyanskii, E. I., Svitan'ko, I. V., Ogibin, Y. N., and Nikishin, G. I., *Izv. Akad. Nauk SSSR, Ser. Khim.*, **1983,** 2316; *Chem. Abs.*, **100,** 67562 (1984).
[178] Srivastava, S. P., Gupta, V. K., Sharma, R. G., and Singh, K. P., *React. Kinet. Catal. Lett.*, **24,** 167 (1984); *Chem. Abs.*, **100,** 191176 (1984).
[179] Agrawal, G. L., *Z. Phys. Chem. (Leipzig)*, **265,** 691 (1984).
[180] Foucaud, A., *Chem. Halides, Pseudo-Halides, Azides,* **1,** 441 (1983); *Chem. Abs.*, **100,** 138192 (1984).
[181] Koser, G. F., *Chem. Halides, Pseudo-Halides, Azides,* **1,** 721 (1983); *Chem. Abs.*, **100,** 138196 (1984).
[182] Sharma, C. B., *Rev. Roum. Chim.*, **28,** 967 (1983).
[183] Kudesia, V. P., and Charma, C. B., *Rev. Roum. Chim.*, **28,** 263 (1983).
[184] Cohen, M. J., and McNelis, E., *J. Org. Chem.*, **49,** 515 (1984).
[185] Doi, J. T., and Musker, W. K., *J. Am. Chem. Soc.*, **106,** 1887 (1984).
[186] Musker, W. K., Surdhar, P. S., Ahmad, R., and Armstrong, D. A., *Can. J. Chem.*, **72,** 1874 (1984).
[187] Packer, J. E., *J. Chem. Soc., Perkin Trans. 2,* **1984,** 1015.
[188] Ried, W., and Bellinger, O., *Liebigs Ann. Chem.*, **1984,** 1778.
[189] Palmieri, G., *Tetrahedron,* **39,** 4097 (1983).
[190] Tee, O. S., and Swedlund, B. E., *Can. J. Chem.*, **61,** 2171 (1983).
[191] Paraskevov, V. G., Pimenov, I. F., Treger, Y. A., and Dasaeva, G. S., *Kinet. Katal.*, **24,** 1007 (1983); *Chem. Abs.*, **99,** 194107 (1983).
[192] Moriarty, R. M., and Hov, K.-C., *Tetrahedron Lett.*, **25,** 691 (1948).
[193] Moriarty, R. M., Prakash, O., Prakash, I., and Musallam, H. A., *J. Chem. Soc., Chem. Commun.*, **1984,** 1342.
[194] Moriarty, R. M., Prakash, I., and Musallam, H. A., *Tetrahedron Lett.*, **25,** 5867 (1984).

[195] Moriarty, R. M., Prakash, O., and Freeman, W. A., *J. Chem. Soc., Chem. Commun.*, **1984,** 927.
[196] Podolesov, B., *J. Org. Chem.*, **49,** 2644 (1984).
[197] Radhakrishnamurti, P. S., and Panda, B. K., *React. Kinet. Catal. Lett.*, **22,** 379 (1983); *Chem. Abs.*, **100,** 85058 (1984).
[198] Ranganathan, S., Ranganathan, D., and Ramachandran, P. V., *Tetrahedron,* **40,** 3145 (1984).
[199] Fonouni, H. E., Krishnan, S., Kuhn, D. G., and Hamilton, G. A., *J. Am. Chem. Soc.*, **105,** 6772 (1983).
[200] Carlsen, P. H., *Acta Chem. Scand.*, **38B,** 343 (1984).
[201] Antelo, J. M., Arce, F., Armesto, J. L., Casado, J., and Varela, A., *An. Quim.*, **79A,** 196 (1983); *Chem. Abs.*, **99,** 175073 (1983).
[202] Antelo, J. M., Arce, F., Armesto, J. L., Casado, J., Penedo, F. J., and Varela, A., *Bull. Soc. Chim. Fr.*, **1984,** 101; *Chem. Abs.*, **101,** 129924 (1984).
[203] Shkaraputa, L. N., Sklyar, V. T., Kononov, A. V., and Danilenko, V. V., *Neftepererab. Neftekhim. (Kiev)*, **25,** 40 (1983); *Chem. Abs.*, **99,** 194117 (1983).
[204] Uskov, A. M., Kozlov, Y. N., and Purmal, A. P., *Zh. Fiz. Khim.*, **58,** 1677 (1984); *Chem. Abs.*, **101,** 130020 (1984).
[205] Kageyama, T., Tobito, Y., Katoh, A., Ueno, Y., and Okawawa, M., *Chem. Lett.*, **1984,** 1481.
[206] Rajasekaran, K., Baskaran, T., and Gnanasekaran, C., *Indian J. Chem.*, **22A,** 1041 (1983); *Chem. Abs.*, **101,** 90164 (1984).
[207] Sevcik, P., Adamcikova, L., Gunarova, D., and Kovacikova, D., *Acta Fac. Rerum Nat. Univ. Comenianae, Chim.*, **31,** 17 (1983); *Chem. Abs.*, **100,** 191152 (1984).
[208] Pati, S. C., and Mishra, M., *Bull. Pure Appl. Sci.*, **1,** 106 (1982); *Chem. Abs.*, **101,** 151198 (1984).
[209] Bruss, M. A., and Colussi, A. J., *Int. J. Chem. Kinet.*, **15,** 1335 (1983).
[210] Reindl, L., and Zvac, V., *React. Kinet. Catal. Lett.*, **22,** 451 (1983); *Chem. Abs.*, **100,** 67652 (1984).
[211] Begar, V. A., Livshits, V. A., and Kuznetsov, A. N., *Deposited Doc.*, **1983,** VINITI 3139; *Chem. Abs.*, **101,** 130018 (1984).
[212] Zhao, X., Zhang, Y., Zhao, H., Wang, S., Du, Z., and Zang, Y., *Cuihua Xuebao*, **5,** 69 (1984); *Chem. Abs.*, **101,** 22653 (1984).
[213] Jiang, L., *Beijing Shifan Daxue Xuebao, Ziran Kexueban*, **1983,** 33; *Chem. Abs.*, **101,** 22787 (1984).
[214] Agulova, L. P., and Opalinskaya, A. M., *Deposited Doc.*, **1982,** VINITI 5415; *Chem. Abs.*, **99,** 211975 (1983).
[215] Tikhonova, L. P., and Kovalenko, A. S., *Teor. Eksp. Khim.*, **19,** 565 (1983); *Chem. Abs.*, **100,** 22242 (1984).
[216] Adamcikova, L., and Knappova, O., *Collect. Czech. Chem. Commun.*, **48,** 2335 (1983).
[217] Gupta, V. K., and Srinivasulu, K., *React. Kinet. Catal. Lett.*, **25,** 87 (1984); *Chem. Abs.*, **101,** 90179 (1984).
[218] Agladze, K. I., Krinsky, V. I., and Pertsov, A. M., *Nature (London)*, **308,** 834 (1984); *Chem. Abs.*, **101,** 151171 (1984).
[219] Manikyamba, P., Rao, P. R., and Sundaram, E. V., *J. Indian Chem. Soc.*, **60,** 652 (1983); *Chem. Abs.*, **100,** 67636 (1948).
[220] Srivastava, S. P., Gupta, V. K., Jain, M. C., Ansari, M. N., and Kaushik, R. D., *Thermochim. Acta*, **68,** 27 (1983); *Chem. Abs.*, **100,** 5597 (1984).
[221] Swarnalakshmi, N., Uma, V., Sethuram, B., and Rao, T. N., *Indian J. Chem.*, **23A,** 646 (1984); *Chem. Abs.*, **101,** 110119 (1984).
[222] Essig, M. G., and Shafizadeh, F., *Carbohydr. Res.*, **127,** 235 (1984).
[223] Gunasekaran, S., and Venkatasubramanian, N., *Indian J. Chem.*, **22A,** 774 (1983); *Chem. Abs.*, **100,** 50884 (1984).
[224] Manoharan, V., and Venkatasubramanian, N., *Indian J. Chem.*, **23A,** 389 (1984); *Chem. Abs.*, **101,** 110109 (1984).
[225] Gunasekaran, S., and Venkatasubramanian, N., *Proc. Indian Acad. Sci., Chem. Sci.*, **92,** 107 (1983); *Chem. Abs.*, **100,** 156069 (1984).
[226] Antelo, J. M., Arce, F., Casado, J., Castro, R., Sanchez, M. E., and Varela, A., *Environ. Sci. Technol.*, **18,** 97 (1984); *Chem. Abs.*, **100,** 50911 (1984).
[227] Shah, B., Jain, A. L., and Banerji, K. K., *Indian J. Chem.*, **22B,** 720 (1983); *Chem. Abs.*, **100,** 5593 (1984).
[228] Singh, B., Saxena, B. B. L., and Samant, A. K., *Tetrahedron,* **40,** 3321 (1984).
[229] Singh, B., Samant, A. K., and Saxena, B. B. L., *Proc. Indian Natl. Sci. Acad.*, **49A,** 550 (1983); *Chem. Abs.*, **100,** 138398 (1984).
[230] Jayaram, B., and Mayanna, S. M., *Bull. Chem. Soc. Jpn.*, **57,** 1439 (1984).

[231] Mahadevappa, D. S., Madegowda, M. B., Ananda, S., and Rangappa, K. S., *Indian J. Chem.*, **23A**, 325 (1984); *Chem. Abs.*, **101**, 72059 (1984).
[232] Hunter, D. H., Barton, D. H. R., and Motherwell, W. J., *Tetrahedron Lett.*, **25**, 603 (1984).
[233] Butenko, L. N., and Gorelov, V. I., *Khimiya i Tekhnol. Elementoorgan. Poluproduktov i Polimerov, Volgograd*, **1982**, 22; *Chem. Abs.*, **100**, 33966 (1984).
[234] Bach, R. D., and Wolber, G. J., *J. Am. Chem. Soc.*, **106**, 1410 (1984).
[235] Wagner, W. R., Spero, D. M., and Rastetter, W. H., *J. Am. Chem. Soc.*, **106**, 1476 (1984).
[236] Davis, F. A., and Abdul-Malik, N. F., *J. Org. Chem.*, **48**, 5128 (1983).
[237] Meyers, A. I., and Wettlaufer, D. G., *J. Am. Chem. Soc.*, **106**, 1135 (1984).
[238] Patil, A. O., Curtin, D. Y., and Paul, I. C., *J. Am. Chem. Soc.*, **106**, 4010 (1984).
[239] Muller, P., Joly, D., and Mermond, F., *Helv. Chim. Acta*, **67**, 105 (1984).
[240] Pryor, W. A., Giamalva, D., and Church, D. F., *J. Am. Chem. Soc.*, **105**, 6858 (1983).
[241] Gab, S., and Turner, W. V., *J. Org. Chem.*, **49**, 2711 (1984).
[242] Abramyan, Z. I., Sevoyan, T. S., and Avetisyan, D. P., *Deposited Doc.*, **1982**, VINITI 3923; *Chem. Abs.*, **99**, 194216 (1983).
[243] Gillies, C. W., and Kuczkowski, R. L., *Isr. J. Chem.*, **23**, 446 (1983).
[244] Niki, H., Maker, P. D., Savage, C. M., Breitenbach, L. P., and Martinez, R. I., *J. Phys. Chem.*, **88**, 766 (1984).
[245] Gomes, M., Razumovskii, S. D., and Zaikov, G. E., *Int. J. Chem. Kinet.*, **16**, 1 (1984).
[246] Choe, J. I., Painter, M. K., and Kuczkowski, R. L., *J. Am. Chem. Soc.*, **106**, 2891 (1984).
[247] Habib, R. M., Chiang, C.-Y., and Bailey, P. S., *J. Org. Chem.*, **49**, 2780 (1984).
[248] Bahta, A., Simonaitis, R., and Heicklein, J., *Int. J. Chem. Kinet.*, **16**, 1227 (1984).
[249] Atkinson, R., Aschmann, S. M., and Carter, W. P. L., *Int. J. Chem. Kinet.*, **16**, 967 (1984).
[250] Atkinson, R. and Aschmann, S. M., *Int. J. Chem. Kinet.*, **16**, 259 (1984).
[251] van der Heuvel, C. J. M., Hofland, A., van Velzen, J. C., Steinberg, H., and de Boer, T. J., *Recl. Trav. Chim. Pays-Bas*, **103**, 233 (1984).
[252] Insola, A., and Lignola, P. G., *Oxid. Commun.*, **3**, 355 (1983); *Chem. Abs.*, **100**, 138403 (1984).
[253] Onari, Y., *Nippon Kagaku Kaishi*, **1983**, 1526; *Chem. Abs.*, **100**, 120265 (1984).
[254] Pryor, W. A., Gleicher, G. J., and Church, D. F., *J. Org. Chem.*, **48**, 4198 (1983).
[255] Neumeister, J., and Griesbaum, K., *Chem. Ber.*, **117**, 1640 (1984).
[256] Miura, M., Nojima, M., Kusabayashi, S., and McCullogh, K. J., *J. Am. Chem. Soc.*, **106**, 2932 (1984).
[257] McCullough, K. J., Nojima, M., Miura, M., Fujisaka, T., and Kusabayashi, S., *J. Chem. Soc., Chem. Commun.*, **1984**, 35.
[258] Nishimura, J., Okada, Y., Nakajima, H., Hashimoto, K., and Oku, A., *Chem. Lett.*, **1984**, 187.
[259] Prior, W. A., Gleicher, G. J., and Church, D. F., *J. Org. Chem.*, **49**, 2574 (1984).
[260] Galstyan, G. A., Vedernikov, V. V., and Gudym, A. N., *Izv. Vyssh. Uchebn. Zaved., Khim. Kjim. Tekhnol.*, **26**, 1041 (1983); *Chem. Abs.*, **100**, 50890 (1984).
[261] Keinan, E., and Varkony, H. T., *Chem. Peroxides*, **1983**, 649; *Chem. Abs.*, **100**, 138189 (1984).
[262] Sotelo, J. L., Torregross, J., and Beltran, F. J., *An. Quim.*, **79A**, 755 (1983).
[263] White, E. H., and Egger, N., *J. Am. Chem. Soc.*, **106**, 3701 (1984).
[264] Chapman, O. L., and Hess, T. C., *J. Am. Chem. Soc.*, **106**, 1842 (1984).
[265] Hoffman, R. V., *Chem. Peroxides*, **1983**, 259; *Chem. Abs.*, **100**, 138185 (1984).
[266] Batt, L., and Liu, M. T. H., *Chem. Peroxides.*, **1983**, 685; *Chem. Abs.*, **100**, 138190 (1984).
[267] Bouillon, G., Lick, C., and Schank, K., *Chem. Peroxides*, **1983**, 279; *Chem. Abs.*, **100**, 138186 (1984).
[268] Sheldon, R. A., *Chem. Peroxides*, **1983**, 161; *Chem. Abs.*, **100**, 138183 (1984).
[269] Saito, I., and Nittala, S. S., *Chem. Peroxides*, **1983**, 311; *Chem. Abs.*, **100**, 138187 (1984).
[270] Lewars, E. G., *Chem. Rev.*, **83**, 519 (1983).
[271] Ogata, Y., Tomizawa, K., and Furuta, K., *Chem. Peroxides*, **1983**, 711; *Chem. Abs.*, **100**, 138232 (1984).
[272] Plesnicar, B., *Chem. Peroxides*, **1983**, 521; *Chem. Abs.*, **100**, 138231 (1984).
[273] Oae, S., and Fujimori, K., *Chem. Peroxides*, **1983**, 585; *Chem. Abs.*, **100**, 138188 (1984).
[274] Dave, V., Stothers, J. B., and Warnhoff, E. W., *Can. J. Chem.*, **62**, 1965 (1984).
[275] Suryawanshi, S. N., Swenson, C. J., Jorgensen, W. L., and Fuchs, P. L., *Tetrahedron Lett.*, **25**, 1859 (1984).
[276] Lind, J., Merenyi, G., and Eriksen, T. E., *J. Am. Chem. Soc.*, **105**, 7655 (1983).
[277] Bapat, J. B., and Durie, A., *Aust. J. Chem.*, **37**, 211 (1984).
[278] Aitken, R. A., Gosney, F. H., Palmer, M. H., Simpson, I., Cadogan, J. I. G., and Tinley, E. J., *Tetrahedron*, **40**, 2487 (1984).

[279] Kershaw, J. W., and Taylor, A., *J. Chem. Soc.*, **1964,** 4320.
[280] Harrison, D. M., *Tetrahedron Lett.*, **25,** 6063 (1984).
[281] Tishchenko, I. G., Revinskii, I. F., Burd, V. N., and Nahar, P., *Dokl. Akad. Nauk BSSR*, **27,** 1095 (1983); *Chem. Abs.*, **100,** 102562 (1984).
[282] Rao, D. S. R., *J. Indian Chem. Soc.*, **60,** 557 (1983); *Chem. Abs.*, **100,** 138375 (1984).
[283] Lusparyan, A. P., Paronikyan, D. G., and Vardanyan, I. A., *Arm. Khim. Zh.*, **37,** 329 (1984); *Chem. Abs.*, **101,** 110098 (1984).
[284] Gingerich, S. B., and Jennings, P. W., *J. Org. Chem.*, **49,** 1284 (1984).
[285] Tsunokawa, Y., Iwasaki, S., and Okuda, S., *Chem. Pharm. Bull.*, **31,** 4578 (1983).
[286] Tezuka, T., and Iwaki, M., *Heterocycles*, **22,** 725 (1984).
[287] Tezuka, T., Iwaki, M., and Haga, Y., *J. Chem. Soc., Chem. Commun.*, **1984,** 325.
[288] Saito, I., Nagata, R., and Matsuvra, T., *Tetrahedron Lett.*, **25,** 2687 (1984).
[289] Adam, W., Haas, W., and Sieker, G., *J. Am. Chem. Soc.*, **106,** 5020 (1984).
[290] Bonini, B. F., Foresti, E., Leardini, R., Maccagnani, G., and Mazzanti, G., *Tetrahedron Lett.*, **25,** 445 (1984).
[291] Uemura, S., Fukuzawa, S., and Toshimitsu, A., *J. Chem. Soc., Chem. Commun.*, **1983,** 1501.
[292] Schneider, H.-J., Ahlhelm, A., and Muller, W., *Chem. Ber.*, **117,** 3297 (1984).
[293] Heller, R. A., Fewster, M., and Lambert, T., *Can. J. Chem.*, **61,** 2455 (1983).
[294] Mahapatra, A. K., Bandyopadhyay, D., Bandyopadhyay, P., and Chakravorty, A., *J. Chem. Soc., Chem. Commun.*, **1984,** 999.
[295] Frimer, A. A., *Chem. Peroxides*, **1983,** 429; *Chem. Abs.*, **100,** 138229 (1984).
[296] Sawyer, D. T., Roberts, J. L., Calderwood, T. S., Tsuchiya, T., and Stamp, J. J., *Proc. Int. Conf. Superoxide Superoxide Dismutase, 3rd*, **1,** 8 (1982); *Chem. Abs.*, **99,** 193995 (1983).
[297] Roberts, J. L., and Sawyer, D. T., *Isr. J. Chem.*, **23,** 430 (1983).
[298] Gareil, M., Pinson, J., and Saveant, J. M., *Electr. Fr., Bull, Dir. Etud. et Rech., Ser. A, Nucl., Hydraul., Therm.*, **1983,** 79; *Chem. Abs.*, **101,** 22627 (1984).
[299] Roberts, R. L., Calderwood, T. S., and Sawyer, D. T., *J. Am. Chem. Soc.*, **105,** 7691 (1983).
[300] Calderwood, T. S., Johlman, C. L., Roberts, J. L., Wilkins, C. L., and Sawyer, D. T., *J. Am. Chem. Soc.*, **106,** 4683 (1984).
[301] Muller, R., and Lingens, F., *Angew. Chem. Int. Ed.*, **23,** 79 (1984).
[302] Osman, R., and Basch, H., *J. Am. Chem. Soc.*, **106,** 5710 (1984).
[303] Darr, D., and Fridovich, I., *Arch. Biochem. Biophys.*, **232,** 562 (1984).
[304] Shimizu, N., Kobayashi, K., and Hayashi, K., *J. Biol. Chem.*, **259,** 4414 (1983).
[305] Wefers, H., and Sies, H., *Eur. J. Biochem.*, **137,** 29 (1983).
[306] Manring, L. E., Kramer, M. K., and Foote, C. S., *Tetrahedron Lett.*, **25,** 2523 (1984).
[307] Saito, I., Matsuura, T., and Inoue, K., *Oxygen Radicals Chem. Biol., Proc., Int. Conf., 3rd*, **1984,** 535; *Chem. Abs.*, **101,** 72005 (1984).
[308] Picker, S. D., and Fridovich, I., *Arch. Biochem. Biophys.*, **228,** 155 (1984).
[309] Roscoe, J. M., *Can. J. Chem.*, **61,** 2716 (1983).
[310] Sridharan, U. C., and Kaufman, F., *Chem. Phys. Lett.*, **102,** 45 (1983).
[311] Perry, R. A., *J. Chem. Phys.*, **80,** 153 (1984); *Chem. Abs.*, **100,** 156039 (1984).
[312] Frimer, A. A., *Chem. Peroxides*, **1983,** 201; *Chem. Abs.*, **100,** 138184 (1984).
[313] Yun, M. J., *Hwahak Kwa Kongop Ui Chinbo*, **24,** 305 (1984); *Chem. Abs.*, **101,** 89911 (1984).
[314] Foote, C. S., Gu, C. L., Manring, L., Liang, J. J., Kanner, R., Boyd, J., Kacher, M. L., Kramer, M., and Ogilby, P., *Methods Stereochem. Anal.*, **1983,** 3; *Chem. Abs.*, **101,** 89925 (1984).
[315] Davidson, R. S., Goodwin, D., and Pratt, J. E., *Proc., Int. Conf. 3rd*, **1983,** 473; *Chem. Abs.*, **101,** 72004 (1984).
[316] Drews, W., Schmidt, R., and Brauer, H. D., *Chem. Phys. Lett.*, **100,** 466 (1983); *Chem. Abs.*, **99,** 194285 (1983).
[317] Duchstein, H. J., and Wurm, G., *Arch. Pharm. (Weinheim, Ger.)*, **317,** 809 (1984).
[318] Acs, A., Schmidt, R., and Brauer, H. D., *Photochem. Photobiol.*, **38,** 527 (1983); *Chem. Abs.*, **100,** 138378 (1984).
[319] Raja, N., Chatha, J. P. S., Arora, P. K., and Vohra, K. G., *Int. J. Chem. Kinet.*, **16,** 205 (1984).
[320] Mach, I., and Veprek-Siska, J., *Oxid. Commun.*, **3,** 171 (1983); *Chem. Abs.*, **100,** 156042 (1984).
[321] Minaev, B. F., and Tikhomirov, V. A., *Zh. Fiz. Khim.*, **58,** 646 (1984); *Chem. Abs.*, **101,** 6400 (1984).
[322] Araki, Y., Dobrowolski, D. C., Goyne, T. E., Hanson, D. C., Jiang, Z. Q., Lee, K. J., and Foote, C. S., *J. Am. Chem. Soc.*, **106,** 4570 (1984).
[323] Mihelich, E. D., and Eickhoff, D. J., *J. Org. Chem.*, **48,** 4135 (1983).
[324] Clennan, E. L., and Mehrsheikh-Mohammadi, M. E., *J. Org. Chem.*, **49,** 1321 (1984).

[325] Gollnick, K., and Griesbeck, A., *Tetrahedron,* **40,** 3235 (1984).
[326] Daub, J., and Knochel, T., *Liebigs Ann. Chem.,* **1984,** 773.
[327] Adam, W., and Rebollo, H., *Isr. J. Chem.,* **23,** 399 (1984).
[328] Kurita, J., Kojima, H., and Tsuchiya, T., *Heterocycles,* **22,** 721 (1984).
[329] Jefford, C. W., Jaggi, D., Boukouvalas, J., and Kohmoto. S., *Helv. Chim. Acta,* **67,** 1104 (1984).
[330] Berdahl, D. R., and Wasserman, H. H., *Isr. J. Chem.,* **23,** 409 (1983).
[331] Chawla, H. M., and Chakrabarty, K., *J. Chem. Soc., Perkin Trans 1,* **1984,** 1511.
[332] Galliani, G., Manitto, P., and Monti, D., *Isr. J. Chem.,* **23,** 219 (1983); *Chem. Abs.,* **100,** 138376 (1984).
[333] Schaap, A. P., Prasad, G., and Siddiqui, S., *Tetrahedron Lett.,* **25,** 3035 (1984).
[334] Miyashi, T., Kamata, M., Nishizawa, Y., and Mukai, T., *J. Chem. Soc., Chem. Commun.,* **1984,** 147.
[335] Ando, W., Sato, R., Sonobe, H., and Akasaka, T., *Tetrahedron Lett.,* **25,** 853 (1984).
[336] Agarwal, S. K., and Murray, R. W., *Isr. J. Chem.,* **23,** 405 (1983).
[337] Rao, T. J., Ramamurthy, V., Schaumann, E., and Nimmesgern, H., *J. Org. Chem.,* **49,** 615 (1984).
[338] Hemmens, V. J., and Moore, D. E., *J. Chem. Soc., Perkin Trans. 2,* **1984,** 209.
[339] Darmanyan, A. P., Vidoczy, T., Irinyi, G., and Gal, D., *Izv. Akad. Nauk SSSR, Ser. Khim.,* **1983,** 2702; *Chem. Abs.,* **101,** 54246 (1984).
[340] Steliou, K., Gareau, Y., and Harpp, D. N., *J. Am. Chem. Soc.,* **106,** 799 (1984).
[341] Pritzkow, W., and Schnurpfeil, D., *Mitteilungsbl.—Chem. Ges. DDR,* **29,** 145 (1982); *Chem. Abs.,* **101,** 109865 (1984).
[342] Sawaki, Y., and Foote, C. S., *J. Org. Chem.,* **48,** 4934 (1983).
[343] Sawaki, Y., *Bull. Chem. Soc. Jpn.,* **56,** 3464 (1983).
[344] Yamaguchi, K., *Proc. Int. Conf., 3rd,* **1983,** 65; *Chem. Abs.,* **101,** 90109 (1984).
[345] Suzuki, M., Yamazaki, E., Takabe, F., Morioka, M., Mizuno, H., and Matsushima, R., *Bull. Chem. Soc. Jpn.,* **57,** 1870 (1984).
[346] Schaap, A. P., Siddiqui, S., Balakrishnan, P., Lopez, L., and Gagnon, S. D., *Isr. J. Chem.,* **23,** 415 (1984).
[347] Ohta, T., *Bull. Chem. Soc. Jpn.,* **57,** 960 (1984).
[348] Slagle, I. R., Feng, Q., and Gutman, D., *J. Phys. Chem.,* **88,** 3648 (1984).
[349] Zadok, E., Rubinraut, S., and Mazur, Y., *Isr. J. Chem.,* **23,** 457 (1983).
[350] Vasilina, T. V., Starchevskii, V. L., and Mokryi, E. N., *Zh. Fiz. Khim.,* **58,** 1926 (1984); *Chem. Abs.,* **101,** 190891 (1984).
[351] Vasilina, T. V., Starchevskii, V. L., and Mokryi, E. N., *Visn, L'viv. Politekh. Inst.,* **181,** 115 (1984); *Chem. Abs.,* **101,** 72034 (1984).
[352] Starchevskii, V. L., Vasilina, T. V., Grindel, L. M., Margulis, M. A., and Mokryi, E. N., *Zh. Fiz. Khim.,* **58,** 1940 (1984); *Chem. Abs.,* **101,** 190890 (1984).
[353] Burgess, E. M., Zoller, U., and Burger, R. L., *J. Am. Chem. Soc.,* **106,** 1128 (1984).
[354] Gunstone, F. D., *J. Am. Oil Chem. Soc.,* **61,** 441 (1984); *Chem. Abs.,* **100,** 138220 (1984).
[355] Porter, N. A., and Wujek, D. G., *J. Am. Chem. Soc.,* **106,** 2626 (1984).
[356] Rousseau, C., Richard, C., and Martin, R., *J. Chim. Phys. Phys.-Chim. Biol.,* **81,** 137 (1984); *Chem. Abs.,* **101,** 110077 (1984).
[357] Yamamoto, Y., Haga, S., Niki, E., and Kamiya, Y., *Bull. Chem. Soc. Jpn.,* **57,** 1260 (1984).
[358] Futu-Tangu, M., Lawson, E., Pritzkow, W., and Voerckel, V., *J. Prakt. Chem.,* **325,** 545 (1983); *Chem. Abs.,* **100,** 5613 (1984).
[359] Biela, R., Bilas, W., Ihsan, U., Pritzkow, W., and Schmidt-Renner, W., *J. Prakt. Chem.,* **325,** 893 (1983).
[360] Finkelshtein, E. I., Rubchinskaya, Y. M., and Kozlov, E. I., *Int. J. Chem. Kinet.,* **16,** 513 (1984).
[361] Dao, L. T. A., Blau, K., Pritzkow, W., Schmidt-Renner, W., Voerckel, V., and Willecke, L., *J. Prakt. Chem.,* **326,** 73 (1984); *Chem. Abs.,* **101,** 6373 (1984).
[362] Fleming, J. E., Miyashita, K., Quay, S. C., and Bensch, K. G., *Biochem. Biophys. Res. Commun.,* **115,** 531 (1983); *Chem. Abs.,* **100,** 5586 (1984).
[363] Bartlett, P. D., and McCluney, R. E., *J. Org. Chem.,* **48,** 4165 (1983).
[364] Kim, Y. H., Kim, H. J., and Yon, G. H., *J. Chem. Soc., Chem. Commun.,* **1984,** 1064.
[365] Schirmann, P. J., Matthews, R. S., and Dittmer, D. C., *J. Org. Chem.,* **48,** 4426 (1983).
[366] Lehmann, J., Ghoneim, K. M., and El-Gendy, A. A., *Arch. Pharm. (Winheim, Ger.)* **317,** 188 (1984).
[367] Neumann, R., and Sasson, Y., *J. Org. Chem.,* **49,** 1282 (1984).
[368] Dimitrov, D. and Kalinkova, I., *Chem. Tech. Leipzig,* **35,** 401 (1983); *Chem. Abs.,* **99,** 211936 (1983).
[369] Kulsrestha, G. N., Pathania, B. S., Sharma, K. G., Sharma, J. S., Negi, J., and Bhattacharyya, K. K., *Indian J. Chem.,* **22A,** 636 (1983); *Chem. Abs.,* **99,** 211957 (1983).

[370] Burghardt, A., and Kulicki, Z., *Monatsh. Chem.*, **115,** 87 (1984).
[371] Brezinsky, K., Litzinger, T. A., and Glassman, I., *Int. J. Chem. Kinet.*, **16,** 1053 (1984).
[372] Howard, J. A., *Isr. J. Chem.*, **24,** 33 (1984).
[373] Leone, J. A., and Seinfeld, J. H., *Int. J. Chem. Kinet.*, **16,** 159 (1984).
[374] Petryaev, E. P., Maslovskaya, L. A., and Shadyro, O. I., *Zh. Org. Khim.*, **19,** 2263 (1983); *Chem. Abs.*, **100,** 67642 (1984).
[375] Camacho, R. F., Diaz, R. F., Torres, S. M., and Gonzalez, M. J., *An. Quim.*, **79A,** 271 (1983); *Chem. Abs.*, **100,** 5587 (1984).
[376] Titova, G. F., Anan'eva, T. A., and Al'yanov, M. I., *Vopr. Kinet. i Kataliza, Ivanovo*, **1982,** 21; *Chem. Abs.*, **100,** 50871 (1984).
[377] Shapovalov, V. V., Poluektov, V. A., and Ryabinin, N. A., *Kinet. Katal.*, **25,** 540 (1984); *Chem. Abs.*, **101,** 90185 (1984).
[378] Komissarov, V. D., Safiullin, R. L., and Zaripov, R. N., *Izv. Akad. Nauk SSSR, Ser. Khim.*, **1984,** 1673; *Chem. Abs.*, **101,** 151218 (1984).
[379] Barclay, L. R. C., Locke, S. J., MacNeil, J. M., VanKessel, J., Burton, G. W., and Ingold, K. U., *J. Am. Chem. Soc.*, **106,** 2479 (1984).
[380] Midland, M. M., *Asymmetric Synth.*, **2,** 45 (1983); *Chem. Abs.*, **101,** 37794 (1984).
[381] Grandbois, E. R., Howard, S. I., and Morrison, J. D., *Asymmetric Synth.*, **2,** 71, (1983); *Chem. Abs.*, **101,** 37795 (1984).
[382] Kawasaki, M., Suzuki, Y., and Terashima, S., *Chem. Lett.*, **1984,** 239.
[383] Brown, H. C., Pai, G. G., and Jadhav, P. K., *J. Am. Chem. Soc.*, **106,** 1531 (1984).
[384] Rei, M. H., *J. Org. Chem.*, **48,** 5386 (1983).
[385] Fujita, M., and Hiyama, T., *J. Am. Chem. Soc.*, **106,** 4629 (1984).
[386] Oishi, T., and Nakata, T., *Acc. Chem. Res.*, **17,** 338 (1984).
[387] Caro, B., Boyer, B., Lamaty , G., and Jaouen, G., *Bull. Soc. Chim. Fr. II,* **1983,** 281.
[388] Alvarez-Ibarra, C., Arjona, O., Perez-Ossorio, R., Perez-Rubalcaba, A., Quiroga, M. L., and Valdes, F., *J. Chem. Res. Synop.*, **1984,** 224.
[389] Cernik, R. V., Craze, G.-A., Mills, O. S., Watt, I., and Whittleton, S. N., *J. Chem. Soc., Perkin Trans. 2,* **1984,** 685.
[390] Kayser, M. M., and McMahon, T. B., *Tetrahedron Lett.*, **25,** 3379 (1984).
[391] Jaeger, D. A., Ward, M. D., and Martin, C. A., *Tetrahedron,* **40,** 2691 (1984).
[392] Wuest, J. D., and Zacharie, B., *J. Org. Chem.*, **49,** 163 (1984).
[393] Gamill, R. B., Bell, L. T., and Nash, S. A., *J. Org. Chem.*, **49,** 3039 (1984).
[394] Salomon, R. G., Sachinvala, N. D., Raychaudhuri, S. R., and Miller, D. B., *J. Am. Chem. Soc.*, **106,** 2211 (1984).
[395] Vincens, M., Dumont, C., Vidal, M., and Domnin, I. N., *Tetrahedron,* **39,** 4281 (1983).
[396] Yoon, N. M., Oh, I. H., Choi, K. I., and Lee, H. J., *Heterocycles,* **22,** 39 (1984).
[397] Yoon, N. M., Park, H. M., Cho, B. T., and Oh, I. H., *Bull. Korean Chem. Soc.*, **4,** 287 (1983); *Chem. Abs.*, **100,** 208837 (1984).
[398] Borders, R. J., and Bryson, T. A., *Chem. Lett.*, **1984,** 9.
[399] Mori, M., Aoyama, T., and Shioiri, T., *Tetrahedron Lett.*, **25,** 429 (1984).
[400] J. W. Huffman, *Acc. Chem. Res.*, **16,** 399 (1983).
[401] Horner, L., and Dickerhof, K., *Liebigs Ann. Chem.*, **1984,** 1240.
[402] Kannan, R., Geetha, P., and Swaminathan, S., *Tetrahedron Lett.*, **25,** 1601 (1984).
[403] Bhattacharyya, S., Basi, B., and Mukherjee, D., *Tetrahedron,* **39,** 4221 (1983).
[404] Kariv-Miller, E., Swenson, K. E., and Zemach, D., *J. Org. Chem.*, **48,** 4210 (1983).
[405] Rabideau, P. W., and Huser, D. L., *J. Org. Chem.*, **48,** 4266 (1983).
[406] Cotsaris, E., and Paddon-Row, M. N., *J. Chem. Soc., Perkin Trans. 2,* **1984,** 1487.
[407] Cotsaris, E., and Paddon-Row, M. N., *J. Chem. Soc., Chem. Commun.*, **1984,** 95.
[408] Fochi, G., and Floriani, C., *J. Chem. Soc., Dalton Trans.*, **1984,** 2577.
[409] Bellamy, F. D., and Ou, K., *Tetrahedron Lett.*, **25,** 839 (1984).
[410] Che, R., *Huaxue Tongbao,* **1984,** 38; *Chem. Abs.*, **101,** 190620 (1984).
[411] Drabowicz, J., Togo, H., Mikolajczyk, M., and Oae, S., *Org. Prep. Proced. Int.*, **16,** 171 (1984); *Chem. Abs.*, **101,** 89881 (1984).
[412] Huan, Z., Landgrebe, J. A., and Peterson, K., *J. Org. Chem.*, **48,** 4519 (1983).
[413] Sokolovı, I. N., Budanov, V. V., Polenov, Y. V., and Polyakova, I. R., *Izv. Vyssh. Uchebn. Saved., Khim. Khim. Tekhnol.*, **26,** 822 (1983); *Chem. Abs.*, **99,** 175059 (1983).
[414] Stanbro, W. D., and Lenkevich, M. J., *Int. J. Chem. Kinet.*, **16,** 251 (1984).
[415] Jansone, D., Leitis, L., Shimanskaya, M. V., *Latv. PSR Zinat. Akad. Vestis, Kim. Ser.*, **1983,** 470; *Chem. Abs.*, **99,** 194191 (1983).

[416] Barton, D. H. R., Motherwell, W. B., Simon, E. S., and Zard, S. Z., *J. Chem. Soc., Chem. Commun.*, **1984,** 337; Barton, D. H. R., Motherwell, W. B., and Zard, S. Z., *Tetrahedron Lett.*, **25,** 3707 (1984).
[417] Paryzek, Z., and Wydra, K., *Tetrahedron Lett.*, **25,** 2601 (1984).
[418] Serma, D. N., Sarma, J. C., Barua, N. C., and Sharma, R. P., *J. Chem. Soc., Chem. Commun.*, **1984,** 813.
[419] Suzuki, H., and Takaoka, K., *Chem. Lett.*, **1984,** 1733.
[420] Rimmelin, P., Taghavi, H., and Sommer, J., *J. Chem. Soc., Chem. Commun.*, **1984,** 1210.
[421] Goodman, D. W., *Acc. Chem. Res.*, **17,** 194 (1984).
[422] Augustine, R. L., Yaghmaie, F., and van Peppen, J. F., *J. Org. Chem.*, **49,** 1865 (1984).
[423] Avanesova, K. M., Koval'chuk, I. N., Petrova, T. Y., Dul'tseva, Z. A., and Shcherbakova, S. S., *Tekhnol. Sinteza Organ. Soedin., L.*, **1983,** 21; *Chem. Abs.*, **101,** 72055 (1984).
[424] Fish, R. H., *Ann. N. Y. Acad. Sci.*, **415,** 292 (1983); *Chem. Abs.*, **101,** 90107 (1984).
[425] Osawa, T., and Harada, T., *Bull. Chem. Soc. Jpn.*, **57,** 1518 (1984).
[426] Negase, Y., Hattori, H., and Tanable, K., *Chem. Lett.*, **1984,** 1615.
[427] McEwen, A. B., Guttieri, M. J., Maier, W. F., and Laine, R. M., *J. Org. Chem.*, **48,** 4436 (1983).
[428] Hirai, H., Komatsuzaki, S., and Toshima, N., *Bull. Chem. Soc. Jpn.*, **57,** 488 (1984).
[429] Neumann, N., and Boldt, P., *Chem. Ber.*, **117,** 1935 (1984).
[430] Jung, S.-H., and Kohn, H., *Tetrahedron Lett.*, **25,** 399 (1984).
[431] Fleet, G. W. J., Gough, M. J., and Smith, P. W., *Tetrahedron Lett.*, **25,** 1853 (1984).
[432] Cozort, J. R., Outlaw, J. F., Hawkins, A., and Siegel, S., *J. Org. Chem.*, **48,** 4190 (1983).
[433] Outlaw, J. F., Cozort, J. R., Garti, N., and Siegel, S., *J. Org. Chem.*, **48,** 4186 (1983).
[434] Climent, M. A., Esteban, A. L., and Perez, J. M., *An. Quim.*, **79A,** 700 (1983); *Chem. Abs.*, **101,** 90114 (1984).
[435] Joo, F., Somsak, L., and Beck, M. T., *J. Mol. Catal.*, **24,** 71 (1984); *Chem. Abs.*, **101,** 72010 (1984).
[436] Woo, J. C., and Chin, C. S., *Bull. Korean Chem. Soc.*, **4,** 169 (1983); *Chem. Abs.*, **99,** 211960 (1983).
[437] Wadkar, J. G., and Chaudhari, R. V., *J. Mol. Catal.*, **22,** 103 (1983); *Chem. Abs.*, **100,** 120277 (1984).
[438] Abatjoglou, A. G., Billig, E., and Bryant, D. R., *Organometallics*, **3,** 923 (1984).
[439] Miyashita, A., Takaya, H., Souchi, T., and Noyori, R., *Tetrahedron*, **40,** 1245 (1984).
[440] Horner, L., and Simons, G., *Z. Naturforsch.*, **39B,** 512 (1984); *Chem. Abs.*, **101,** 37974 (1984).
[441] Ojima, I., Yoda, N., Yatabe, M., Tanaka, T., and Kogure, T., *Tetrahedron*, **40,** 1255 (1984).
[442] Felfoldi, K., Kapocsi, I., and Bartok, M., *J. Organomet. Chem.*, **277,** 439 (1984).
[443] Felfoldi, K., Kapocsi, I., and Bartok, M., *J. Organomet. Chem.*, **277,** 443 (1984).
[444] Patil, S. R., Chaudhari, R. V., and Sen, D. N., *J. Mol. Catal.*, **23,** 51 (1984); *Chem. Abs.*, **100,** 156059 (1984).
[445] Sanchez-Delgado, R. A., Andriollo, A., and Valencia, N., *J. Mol. Catal.*, **24,** 217 (1984); *Chem. Abs.*, **101,** 72020 (1984).
[446] Yoshinaga, K., Taketoshi, K., and Katsutoshi, O., *J. Chem. Soc., Perkin Trans. 2*, **1984,** 469.
[447] Funabiki, T., Yamazaki, Y., Sato, Y., and Yoshida, S., *J. Chem. Soc., Perkin Trans. 2*, **1983,** 1915.
[448] Yamashita, K., and Ohkubo, K., *Nippon Kagaku Kaishi*, **1984,** 505; *Chem. Abs.*, **101,** 90102 (1984).
[449] Ohkubo, K., Kawabe, T., Yamashita, K., and Sakaki, S., *J. Mol. Catal.*, **24,** 83 (1984); *Chem. Abs.*, **101,** 72011 (1984).
[450] Moon, C. J., and Chin, C. S., *Bull. Korean Chem. Soc.*, **4,** 180 (1983); *Chem. Abs.*, **99,** 211961 (1983).
[451] Azran, J., Buchman, O., Orchin, M., and Blum, J., *J. Org. Chem.*, **49,** 1327 (1984).
[452] Bogden, P. L., Sullivan, P. J., Donovan, T. A., and Atwood, J. D., *J. Organomet. Chem.*, **269,** C51 (1984).
[453] Lynch, T. J., Banah, M., Kaesz, H. D., and Porter, C. R., *J. Org. Chem.*, **49,** 1266 (1984).
[454] van der Kerk, S. M., van Gerresheim, W., and Verhoeven, J. W., *Recl. Trav. Chim. Pays-Bas*, **103,** 143 (1984).
[455] Powell, M. F., and Bruice, T. C., *J. Am. Chem. Soc.*, **105,** 7139 (1983).
[456] Roberts, R. M. G., Ostovic, D., and Kreevoy, M. M., *Faraday Discuss. Chem. Soc.*, **1982,** 257; *Chem. Abs.*, **99,** 175027 (1983).
[457] Ostovic, D., Roberts, R. M. G., and Kreevoy, M., *J. Am. Chem. Soc.*, **105,** 7629 (1983).
[458] Colter, A. K., Lai, C. C., Williamson, T. W., and Berry, R. E., *Can. J. Chem.*, **61,** 2544 (1983).
[459] Fukuzumi, S., Nishizawa, N., and Tanaka, T., *J. Org. Chem.*, **49,** 3571 (1984).
[460] Dahl, K. H., and Dunn, M. F., *Biochemistry*, **23,** 4094 (1984).
[461] Fukuzumi, S., Nishizawa, N., and Tanaka, T., *Chem. Lett.*, **1983,** 1755.
[462] Tabushi, I., Kuroda, Y., and Mizutani, T., *J. Am. Chem. Soc.*, **106,** 3377 (1984).
[463] Burgner, J. W., and Ray, W. J., *Biochemistry*, **23,** 3626 (1984).
[464] Awano, H., Hirabayashi, T., and Tagaki, W., *Tetrahedron Lett.*, **25,** 2005 (1984).
[465] Yasui, S., Nakamura, K., and Ohno, A., *J. Org. Chem.*, **49,** 878 (1984).

[466] Boiko, T. S., Uzienko, A. B., and Yasnikov, A. A., *Dopov. Akad. Nauk Ukr. RSR, Ser. B: Geol., Khim. Biol. Nauki,* **1983,** 30; *Chem. Abs.,* **100,** 67644 (1984).
[467] Rob, F., van Ramesdonk, H. J., van Gerresheim, W., Bosma, P., Scheele, J. J., and Verhoeven, J. W., *J. Am. Chem. Soc.,* **106,** 3826 (1984).
[468] Baba, N., Amano, M., Oda, J., and Inouye, Y., *J. Am. Chem. Soc.,* **106,** 1481 (1984).
[469] Oppenheimer, N. J., *J. Am. Chem. Soc.,* **106,** 3032 (1984).
[470] Vanoni, M. A., and Matthews, R. G., *Biochemistry,* **23,** 5272 (1984).
[471] Bertini, I., Gerber, M., Lanini, G., Maret, W., Rawer, S., and Zeppezauler, M., *J. Am. Chem. Soc.,* **106,** 1826 (1984).
[472] Murkami, Y., Kikuchi, J., and Nishida, K., *Chem. Lett.,* **1983,** 1565.
[473] Shinkai, S., Tsuno, T., and Manabe, O., *J. Chem. Soc., Perkin Trans 2,* **1984,** 661.
[474] Ohno, A., Ushida, S., and Oka, S., *Bull. Chem. Soc. Jpn.,* **57,** 506 (1984).
[475] Shinkai, S., Ishikawa, Y., Shinkai, H., Tsuno, T., Makishima, H., Ueda, K., and Manabe, O., *J. Am. Chem. Soc.,* **106,** 1801 (1984).
[476] Ghisla, S., Thorpe, C., and Massey, V., *Biochemistry,* **23,** 3154 (1984).
[477] Yano, Y., Sakaguchi, T., and Nakazato, M., *J. Chem. Soc., Perkin Trans. 2,* **1984,** 595.
[478] Eberlein, G., and Bruice, T. C., *J. Am. Chem. Soc.,* **105,** 6679 (1983).
[479] Eberlein, G., and Bruice, T. C., *J. Am. Chem. Soc.,* **105,** 6685 (1983).
[480] Venkataram, U. V., and Bruice, T. C., *J. Am. Chem. Soc.,* **106,** 5703 (1984).
[481] Bruice, T. C., *Isr. J. Chem.,* **24,** 54 (1984).
[482] Yano, Y., and Ohya, E., *J. Chem. Soc., Perkin Trans. 2,* **1984,** 1227.
[483] Yano, Y., Ohshima, M., Sutoh, S., and Nakazato, M., *J. Chem. Soc., Chem. Commun.,* **1984,** 1031.
[484] Yano, Y., Ohshima, M., and Sutoh, S., *J. Chem. Soc., Chem. Commun.,* **1984,** 695.
[485] Collman, J. P., Brauman, J. I., Meunier, B., Raybuck, S. A., and Kodadek, T., *Proc. Natl. Acad. Sci. U. S. A.,* **81,** 3245 (1984).
[486] Liebler, D. C., and Guengerich, F. P., *Biochemistry,* **22,** 5482 (1983).
[487] Miwa, G. T., Walsh, J. S., and Lu, A. Y. H., *J. Biol. Chem.,* **259,** 3000 (1983).
[488] Harada, N., Miwa, G. T., Walsh, J. S., and Lu, A. Y. H., *J. Biol. Chem.,* **259,** 3005 (1983).
[489] Khenkin, A. M., and Shteinman, A. A., *J. Chem. Soc., Chem. Commun.,* **1984,** 1219.
[490] Smith, J. R. L., Nee, M. W., Noar, J. B., and Bruice, T. C., *J. Chem. Soc., Perkin Trans. 2,* **1984,** 255.
[491] Heimbrook, D. C., Murray, R. I., Egeberg, K. D., and Sligar, S. G., *J. Am. Chem. Soc.,* **106,** 1514 (1984).
[492] Reid, L. S., Mauk, M. R., and Mauk, A. G., *J. Am. Chem. Soc.,* **106,** 2182 (1984).
[493] Guengerich, F. P., Willard, R. J., Shea, J. P., Richards, L. E., and Macdonald, T. L., *J. Am. Chem. Soc.,* **106,** 6446 (1984).
[494] Guengerich, F. P., and Macdonald, T. L., *Acc. Chem. Res.,* **17,** 9 (1984).
[495] Walsh, T. A., Ballou, D. P., Mayer, R., and Que, L. J., *J. Biol. Chem.,* **258,** 14422 (1983).
[496] Kalyanaraman, B., Felix, C. C., and Sealy, R. C., *J. Biol. Chem.,* **259,** 7584 (1983).
[497] Burgner, J. W., and Ray, W. J., *Biochemistry,* **23,** 3636 (1984).
[498] Waskiewicz, D. E., and Hammes, G. G., *Biochemistry,* **23,** 3136 (1984).
[499] Pascal, R. A., and Walsh, C. T., *Biochemistry,* **23,** 2745 (1984).
[500] Mangold, J. B., and Klinman, J. P., *J. Biol. Chem.,* **259,** 7772 (1983).
[501] Lambeir, A. M., and Dunford, H. B., *J. Biol. Chem.,* **258,** 3558 (1983).
[502] Kovar, J., Simek, K., Kucera, I., and Matyska, L., *Eur. J. Biochem.,* **139,** 585 (1984).
[503] Hornby, D. P., and Engel, P. C., *Eur. J. Biochem.,* **143,** 557 (1984).
[504] Ting, H.-H., and Crabbe, M. J. C., *Biochem. J.,* **215,** 361 (1984).
[505] Viola, R. E., *Arch. Biochem. Biophys.,* **228,** 415 (1984).
[506] Matos, J. R., Smith, M. B., and Wong, C.-H., *Bioorg. Chem.,* **12,** 121 (1984).
[507] Magnusson, R. P., Taurog, A., and Dorris, M. L., *J. Biol. Chem.,* **259,** 197 (1983).
[508] Whittaker, J. W., and Lipscomb, J. D., *J. Biol. Chem.,* **259,** 4476 (1983).
[509] Blanchard, J. S., and Englard, S., *Biochemistry,* **22,** 5922 (1983).
[510] Chauncey, T. R., and Westley, J., *J. Biol. Chem.,* **258,** 5037 (1983).

Organic Reaction Mechanisms 1984
Edited by A. C. Knipe and W. E. Watts

CHAPTER 6

Carbenes and Nitrenes

C. J. Moody

Department of Chemistry, Imperial College of Science and Technology, London SW7

The year has seen the publication of a new book on reactive intermediates,[1] together with the usual annual review.[2] More specialized reviews have appeared on the flash photolysis studies of carbenes and their reaction kinetics,[3,4] and on the gas-phase chemistry of novel cation radicals with carbene-type structures.[5] The chemistry of unsaturated carbenes has been reviewed,[6] as has the chemistry of carbynes[7] and the theoretical aspects of olefin metathesis.[8]

Structure and Reactivity

The structure and reactivity of carbenes has been reviewed.[9]

The rate of intersystem crossing of singlet to triplet diphenylcarbene is faster in polar solvents despite the fact that the singlet state should be stabilized by such solvents. The reasons for this apparent anomaly have been discussed.[10] The effect of the angle at the carbenic centre has been investigated in several cases. ESR work on the unusual diarylcarbenes (**1**; n = 8–12) of specific steric structure shows that one consequence of the cyclophane structure is a high concentration of triplet electron density on the carbenic carbon due to restricted delocalization into the aromatic

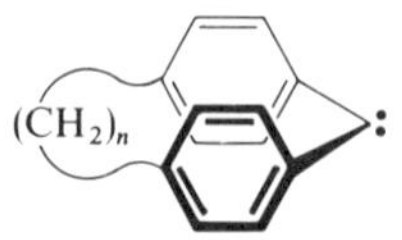

$(CH_2)_n$

(**1**) 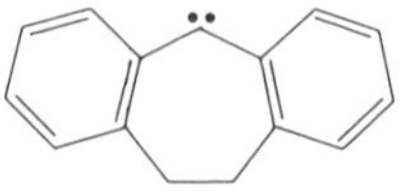 (**2**) 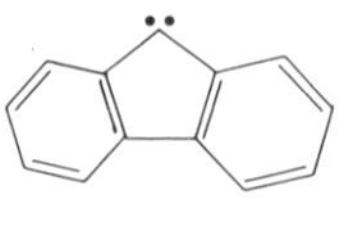 (**3**) 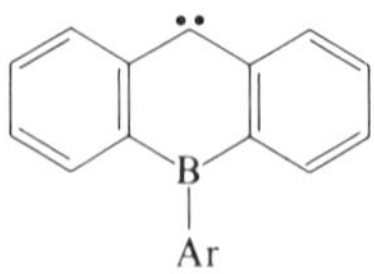 (**4**)

B
Ar

rings.[11] Further studies on dimesitylcarbene confirm that it is considerably less bent than diphenylcarbene, and the increased singlet–triplet gap results in the two species exhibiting quite different chemistry.[12,13] The effect of bond angle on the singlet–triplet gap is also apparent in other carbenes. Thus the larger angled carbene (**2**) has a considerably larger singlet–triplet gap than fluorenylidene (**3**) as evidenced by ESR studies in a methanol matrix at 77 K.[14] Studies on 9-mesityl-9,10-dihydro-9-boraanthracenylidene (**4**; Ar = 2,4,6-$Me_3C_6H_2$) also show that equilibration between the singlet and triplet states is slower than in fluorenylidene,[15] where the latest estimate of the singlet–triplet gap is 1.1 kcal mol^{-1}.[16] Full papers on the laser flash photolytic generation of fluorenylidene (**3**)[17] and on its reaction with *E*- and *Z*-1,2-dichloroethene[18] confirm the rapid interconversion of the singlet and triplet spin states of this carbene. The possibility of conjugation between the carbene centres in (**5**) has been investigated by ESR spectroscopy, which concludes that the 1,4-σ-biradical tautomer makes an important contribution.[19] The magnetic behaviour of *m*-phenylenebis[(diphenylmethylen-3-yl)methylene] (**6**) suggests that this unusual carbene has nonet multiplicity.[20]

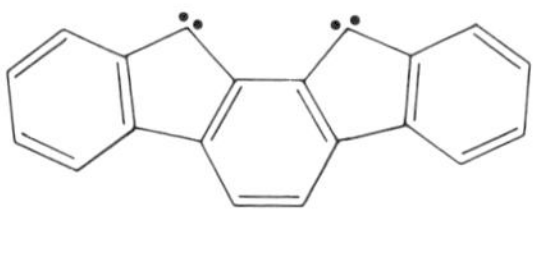

(**5**) 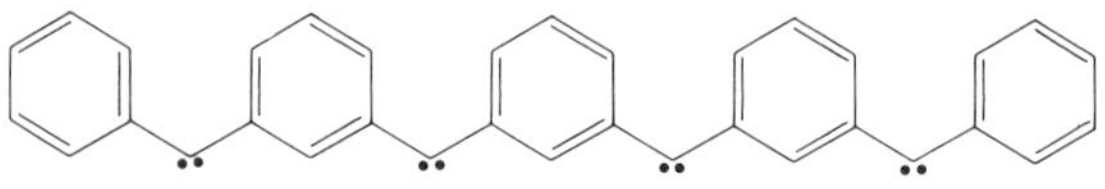 (**6**)

Further studies *o*-(9-fluorenyl)arylnitrenes confirm that the two conformers (**7**) and (**8**) undergo different chemistry.[21] The first evidence for the generation of a nitrilo-λ^5-phosphane (**9**), a valence tautomer of the nitrene (**10**), has been obtained in the photolysis of $(Pr^i_2N)_2PN_3$.[22,23]

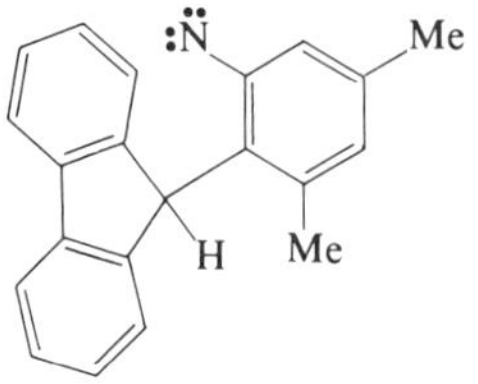

:N̈ Me H Me

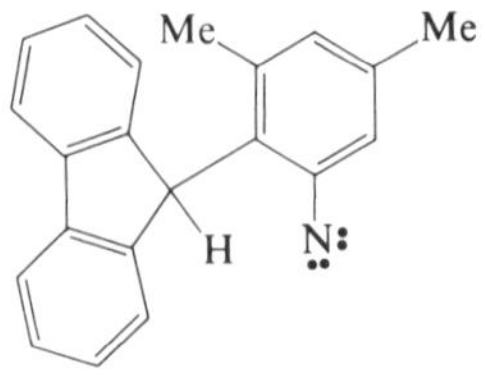

Me Me H N̈:

$(Pr^i_2N)_2P{\equiv}N$ $(Pr^i_2N)_2$—PN̈:

(**9**) (**10**)

(**7**) (**8**)

Generation

Carbenes

Singlet methylene is probably generated in the atomic carbon mediated deoxygenation of formaldehyde,[24] whilst cycloalkylidenes can be formed by the radiolysis of C_7–C_{10} cycloalkanes by unimolecular hydrogen elimination.[25] The distribution of products is similar to that obtained by photochemical decomposition of the sodium salts of the tosylhydrazones of the corresponding cycloalkanones. The photochemical generation of carbenes from alkyl halides has been reviewed,[26] and deuterium labelling studies have shown that the alkenes which result from such reactions are formed substantially although not exclusively by the carbene pathway.[27]

The 1,2-hydrogen shift from 90° twisted propene to give a carbene has been examined *ab initio* calculations,[28] and ^{13}C-labelling has shown that in the pyrolysis of the alkynone (**11**) to generate the unsaturated carbene (**12**), it is the hydrogen rather than the acyl group which migrates.[29] Although pyrolysis of benzobicyclo[3.1.0]hex-2-ene (**13**) and its phenyl-substituted derivatives leads mainly to products derived by cleavage of the internal bond, a minor pathway does involve generation of the carbene (**14**) by retro-1,2-addition.[30] In contrast to direct irradiation which gives dimesitylsilylene and 1,1,3,3-tetramethylindan-2-one, photolysis of (**15**; Ar = 2,4,6-$Me_3C_6H_2$) in the presence of an electron-transfer reagent such as tetracyanoethene gives the carbene (**16**) and dimesitylsilanone. It is thought

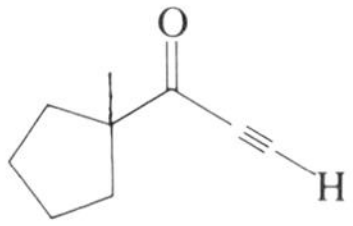

(11) 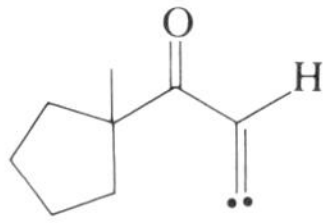 (12) 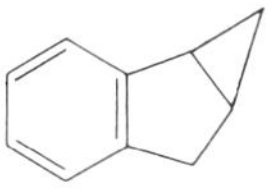 (13)

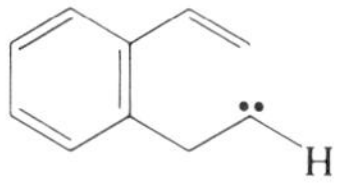

(14) 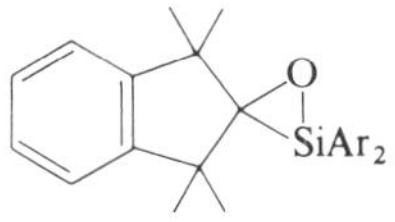 (15) 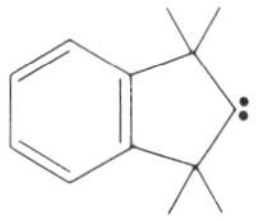 (16)

that the carbene arises from the radical cation of the starting material.[31] Initial attempts to generate the Dewar furan (**18**) by photolysis of (**17**) have resulted in further decomposition to the carbene (**19**), and formation of 3-acetyl-1,2,3-trimethylcyclopropene (Scheme 1).[32]

Cyclopropenes also act as precursors to carbenes. Thus vapour-phase pyrolysis of spiro[2.4]hept-1-ene (**20**) at 225° results in ring-opening to give the vinylcarbene (**21**). The corresponding photochemical ring-opening is believed to proceed from an excited state of the cyclopropene and gives the S_1 excited state of the carbene.[33] Treatment of 1,1-dimethyl-2,2,3,3-tetrachlorocyclopropane with methyllithium at −70° leads to 1,2-dichloro-3,3-dimethylcyclopropene which ring-opens to the

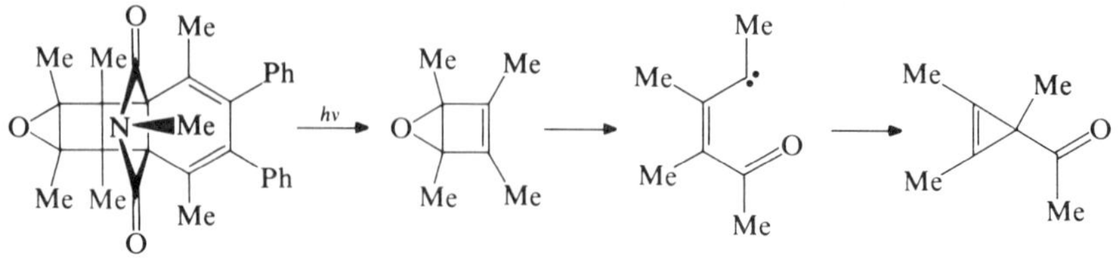

(17) **(18)** **(19)**

SCHEME 1

vinylcarbene (**22**) at about 0°. Attempts to intercept the cyclopropene have, however, been unsuccessful.[34] The vinylcarbene (**23**) is generated by irradiation of the cyclopropene (**24**).[35]

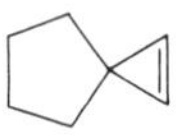

(20) 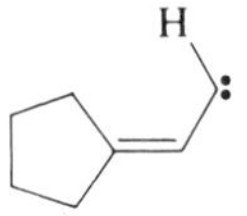 **(21)** 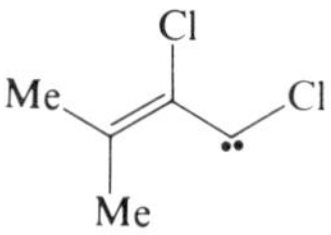 **(22)**

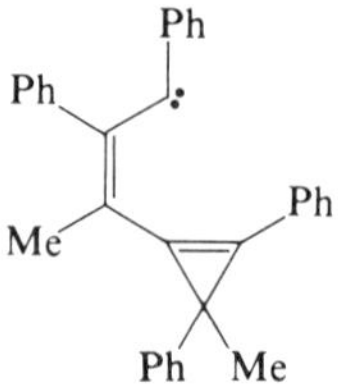

(23) 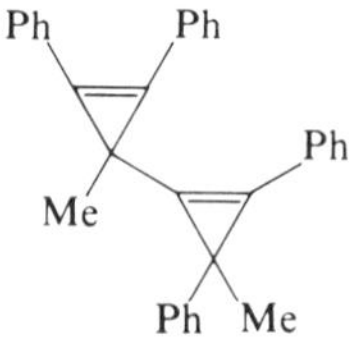 **(24)**

Decomposition of the tosyl azo compound (**25**) at room temperature under neutral or basic conditions generates the unsaturated carbenes, $R_2C{=}C$:, *via* the corresponding diazo compounds.[36] A vinylidene carbene is also produced by treatment of the optically active vinyl bromide (**26**) with *tert*-butyllithium. Since the resulting products are also optically active it is assumed that the leaving halide blocks one enantio-face, directing incoming nucleophiles to the opposite face.[37] The unsaturated carbene (**27**) is formed by rearrangement of the highly strained cyclobutyne intermediate (**28**) in the reverse of a known rearrangement (vinylidene carbene to alkyne).[38]

$R_2C{=}CHN{=}NTs$

(25) **(26)** 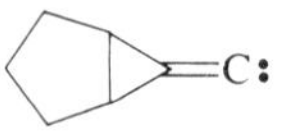 **(27)** 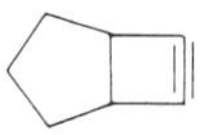 **(28)**

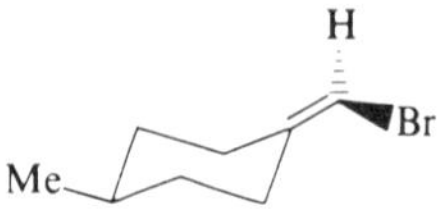

A new method of dichlorocarbene generation involves treatment of carbon tetrachloride with a mixture of dimethyl sulphone, potassium hydroxide and *tert*-butanol.[39] Extrusion of difluorocarbene from perfluoronorbornadiene with concomitant formation of hexafluorobenzene does not occur until about 240°. These surprisingly vigourous conditions contrast with the formation of the same carbene from perfluorocyclopropane which occurs at 160° without aromatization as the driving force, and highlight the strain energy associated with highly fluorinated cyclopropanes.[40] Nucleophilic attack by iodide on $PhHgCF_3$ followed by decomposition of the intermediate is the key step in the generation of difluorocarbene from the organomercurial,[41] and MINDO/3 calculations implicate difluorocarbene as the most likely intermediate in the reaction of 1,1-difluoroalkenes with oxygen (^3P) atoms.[42] Fluoro(phenyl)carbene can now be generated photochemically from the corresponding diazirine. The selectivities of the carbene towards various alkenes are not identical to the species generated by treatment of PhCHBrF with potassium *tert*-butoxide. On the other hand the photochemically generated carbene is very similar to the species generated by the α-elimination route in the presence of 18-crown-6, suggesting that the "free" carbene is important in both cases.[43]

Nitrovinylcarbenes (**29**) are generated by photolysis of the 3*H*-pyrazoles (**30**),[44] and the formation of the oxacarbenes (**31**) by irradiation of the ketones (**32**) has been shown to proceed by a diradical mechanism, rather than by the concerted pathway postulated for the formation of oxacarbenes from cyclobutanones.[45]

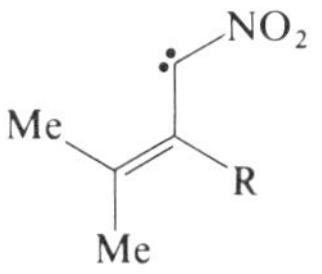

(**29**) 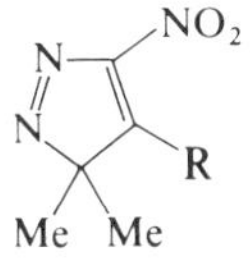 (**30**) 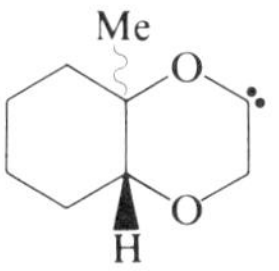 (**31**) 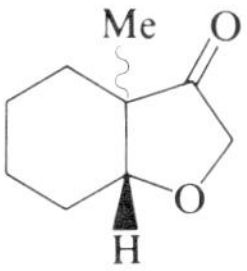 (**32**)

Epoxyenones (**33**; R = H or Me, X = Y = H or XY = CH_2O) undergo photochemical ring-opening to give products derived from the carbenes (**34**).[46, 47] The scrambling of ^{13}C-labelled benz[*a*]anthracene (**35**) at high temperatures, related to the known α–β scrambling in naphthalene, is postulated to occur by generation of the carbene (**36**).[48]

Nitrenes

Treatment of *N*-chloro-2-nitroanilines with base leads to a transient red colour associated with the corresponding nitrogen anion, which then gives the singlet arylnitrene by irreversible loss of chloride.[49] The first-order rate constants for the decomposition of 25 sulphonyl azides, RSO_2N_3, have been obtained and are in accord with rate-limiting sulphonylnitrene formation. The kinetics of a variety of sulphonyl azides bearing reactive neighbouring groups has failed to reveal any significant anchimeric assistance in their decomposition.[50] The decomposition of phenylsulphinyl azide also follows first-order kinetics in the presence of various

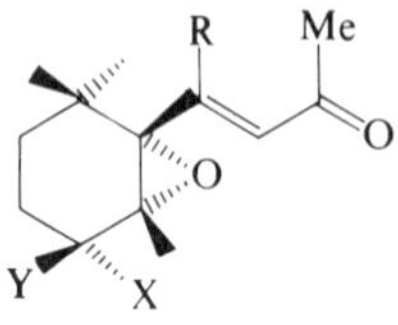

(33) 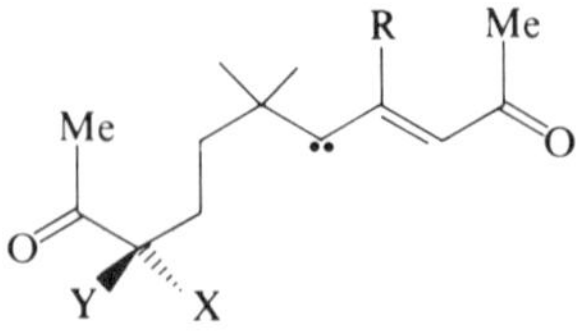 (34) 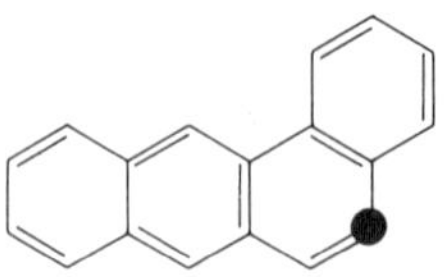 (35)

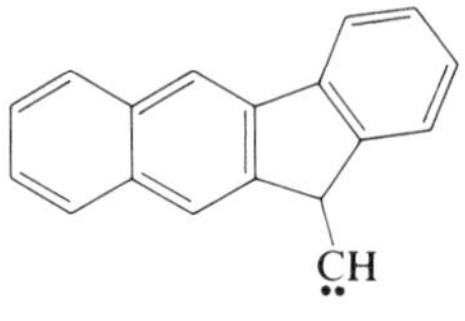

(36)

substrates, consistent with the generation of phenylsulphinylnitrene.[51] Arylsulphenylnitrenes can be generated by thermolysis of the bicyclic compounds (**37**).[52]

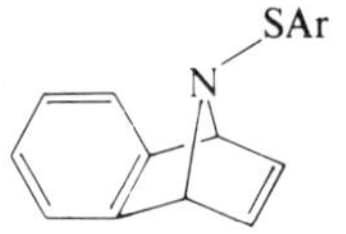

(37) 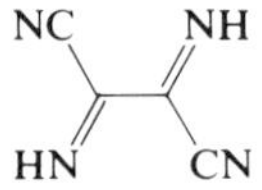 (38) 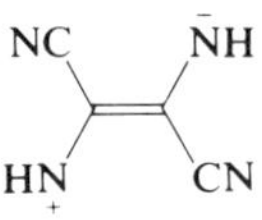 (39) 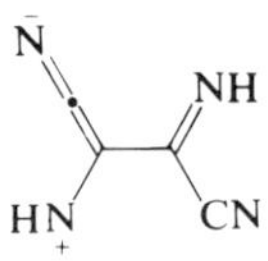 (40)

Excited halonitrenes are generated from HN_3 and halogens in a discharge flow apparatus.[53] Diiminosuccinonitrile (**38**) is a latent nitrenium ion source. It is postulated that imines exhibit nitrenium ion reactivity if the C=N bond can be polarized in the opposite direction to normal. In this case one C=N can induce the reverse polarization in the other, *i.e.* (**39**), further stabilized by the nitrile groups, *i.e.* (**40**).[54]

Addition

MINDO/3 calculations of the addition of methylene to ethene suggest that an *endo* asynchronous approach is most favourable.[55] Similar calculations on the transition state for the reaction of the carbenoid $LiCH_2F$ with ethene suggest that the CH_2 moiety is in a plane nearly parallel to the alkene plane, with the LiF substantially decomplexed.[56] Methylene, generated photochemically from ketene, adds to vinyl chloride to give chlorocyclopropane in a chemically activated state.[57] Decomposition of ethyl diazoacetate in the presence of copper (II) complexes of chiral Schiff bases results in asymmetric cyclopropanation of alkenes.[58]

Triplet diphenylcarbene adds to styrenes as an ambiphile, reacting more rapidly with both electron-rich and electron-poor styrenes than with styrene itself.[59] Electrophilic chloro(phenyl)carbene reacts with electrophilic alkenes such as diethyl maleate to give cyclopropanes in good yield *via* a dipolar intermediate (**41**).[60]

1,3-Diphenylisoindenylidene (**42**), a highly electrophilic carbene, adds readily to electron-rich alkenes such as vinyl ethers.[61] In a new example of a concerted reaction of a carbene with two σ-bonds, careful examination of the reaction between di(methoxycarbonyl)carbene and quadricyclane has shown that the primary product is the cyclopropane (**43**).[62]

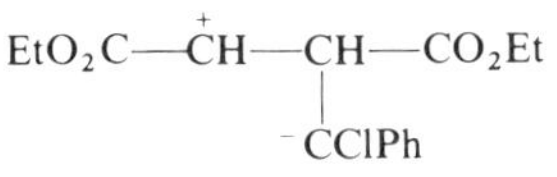

$EtO_2C—\overset{+}{C}H—CH(^{-}CClPh)—CO_2Et$

(**41**) 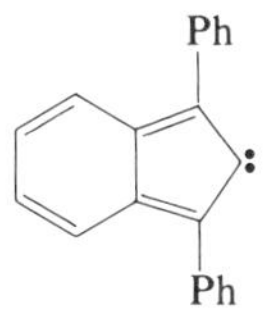 (**42**) 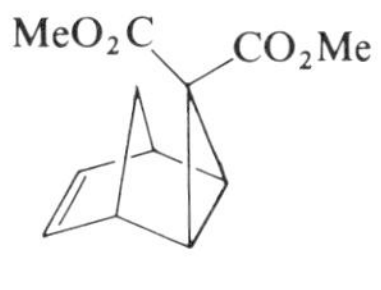 (**43**)

Several studies on the addition reactions of dihalocarbenes have been published. *Ab initio* calculations on the reaction of difluorocarbene with propene reveal two transition states of identical energy with the fluorines *syn* and *anti* to the methyl group. The overall activation energy is calculated to be 1.3 kcal mol^{-1} lower than the corresponding reaction of ethene, in line with the expected donor effect of a methyl substituent.[63] The relative rates of the addition of dichlorocarbene to the double bond in various bicyclic systems have been interpreted in terms of steric hindrance at the double bond.[64] In the reaction of E- and Z-1,3,5-hexatriene and cycloheptatriene with both dichloro- and dibromo-carbene, the cyclopropanation occurs regioselectively at the 1,2-double bond in each case.[65] The minor product of the reaction of dibromocarbene with E-2,2,2′,2′-tetramethylcyclopropylidene (**44**) is thought to arise from an unusual addition reaction to give (**45**), which subsequently reacts with more dibromocarbene to give the observed 1:2 adduct.[66] The observation that the product ratio of the reaction of alkenes with dibromocarbene, generated from $PhHgCBr_3$, is proportional to the concentration of $PhHgCBr_3$ has been attributed to formation of a complex between the carbene and its precursor; full details of this work have now appeared.[67] Addition of dichlorocarbene to 2,5-dihydro-3-trimethylsilylfuran gives the expected cycloadduct together with C—H insertion products. The corresponding adduct from 4,5-dihydro-2-trimethylsilylfuran, however, cannot be isolated, rearranging rapidly to the pyran (**46**).[68] The formation of the dihalostyrene (**47**) from dihalocarbene and 1,3-diaryliso-benzofurans is thought to proceed by initial 1,2-addition; the alternative 1,4-addition is considered to be much less likely.[69]

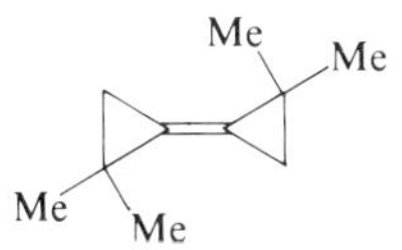

(**44**) 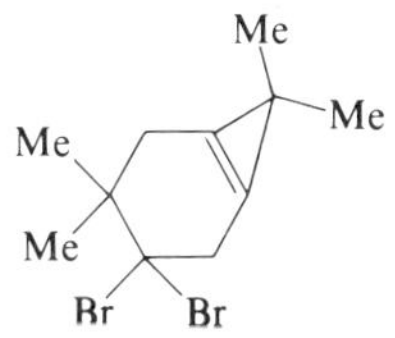 (**45**) 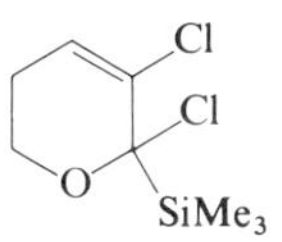 (**46**) 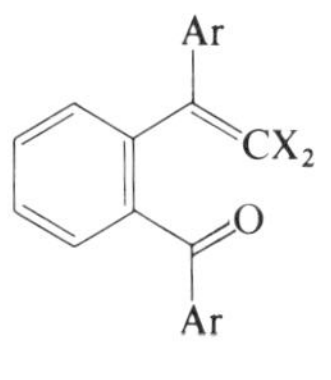 (**47**)

Perchlorovinylcarbene (**48**), a bulky electrophile similar in reactivity to chloro(ethoxycarbonyl)carbene, generated by thermolysis of perchlorocyclopropene at 180°, reacts readily with alkenes, although the rates of addition are more sensitive to steric hindrance at the double bond than the corresponding reactions of dichlorocarbene.[70] The related carbenes (or carbenoids) (**49**; X = H, Y = Cl and X = Cl, Y = H) can also be generated by α-elimination from the vinyl halides (**50**) using lithium hexamethyldisilazide in ether.[71] The corresponding reaction of (**50**; X = H, Y = Cl) with potassium *tert*-butoxide in pentane, however, gives the carbene (**51**).[72] Both types of carbene react with alkenes to give cyclopropanes.

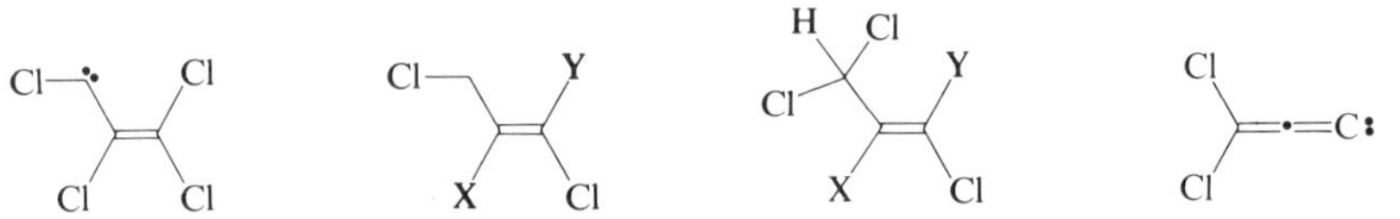

(**48**) (**49**) (**50**) (**51**)

The possible rôle of carbene–alkene complexes as intermediates in the addition reaction of carbenes to alkenes continues to be a subject of intense interest. On one hand, the stereochemical results obtained in the addition of dihalocarbenes to E- and Z-cyclooctene are claimed to be only consistent with the formation of such carbene–alkene complexes,[73] whilst on the other hand, *ab initio* calculations on the reaction of dichloro- and difluoro-carbenes with ethene reveal that possible π-complexes are not energy minima and are unlikely to be involved as intermediates.[74] The negative activation energies which are sometimes found for such reactions are quite consistent with fast reactions having no inherent potential energy barrier.[75] Benzyl(chloro)carbenes undergo addition to alkenes in competition with a 1,2-hydrogen shift to give chlorostyrenes. The kinetics of these reactions are also taken as evidence for the intervention of carbene–alkene complexes.[76] However, another author[77] regards these claims as unsubstantiated, the kinetics having been misinterpreted. It would seem that all papers postulating reversible carbene–alkene complex formation should be viewed critically.

Ethoxycarbonylnitrene adds readily to the double bond of allyl ethers,[78] but the remote SO_2 group in cyclic unsaturated sulphones such as (**52**) drastically diminishes the susceptibility of the double bond to attack by the same nitrene, generated by treatment of $EtO_2CNHOSO_2Ar$ with base. This effect is explained by through-space orbital interactions between the SO_2 group and the π-bond, and the explanation is supported by UV and PES measurements. Interestingly, the alkene (**52**) does react with ethoxycarbonylnitrene generated from ethyl azidoformate, or from $EtO_2CNHOSO_2Ar$ under phase-transfer conditions.[79, 80]

The key step in the synthesis of triannulanes, *e.g.* (**53**), involves intramolecular addition of the carbene derived from benzophenone-sensitized photolysis of the diazo compound (**54**).[81] Irradiation of the diazo-ketone (**55**) in a matrix at 10 K leads to the ketone (**56**) by intramolecular cycloaddition of the corresponding carbene.[82] Full details of the intramolecular additions of *N*-nitrenes derived from 3-

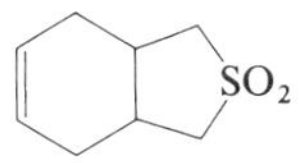

(52) 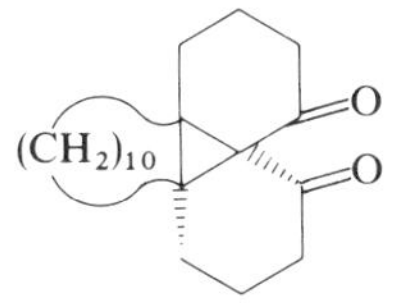 (53) 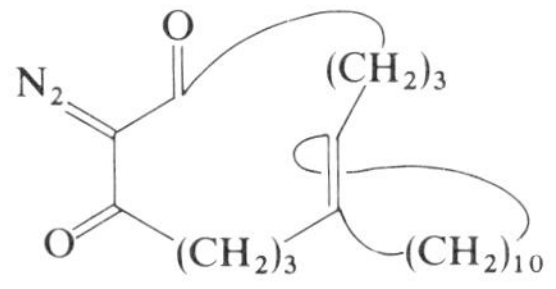 (54)

aminoquinazolin-4-ones (**57**) have been published. The results suggest that the reaction is non-concerted and proceeds by dipolar intermediates reached by a seven-membered transition state, depicted as (**58**).[83]

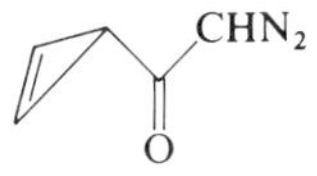

(55) 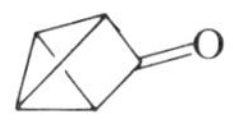 (56) 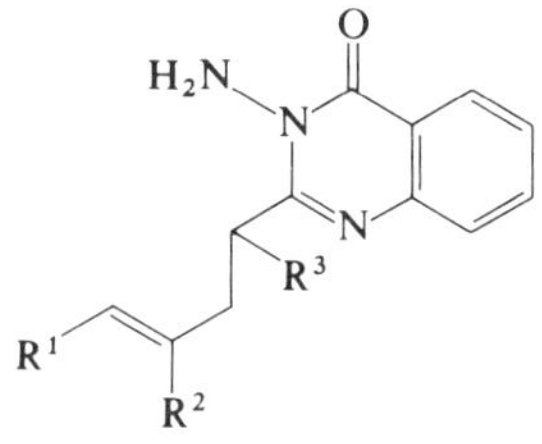 (57) 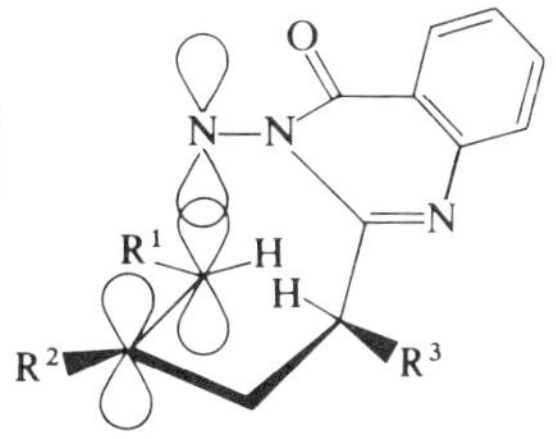 (58)

Insertion and Abstraction

Intermolecular

Several insertion and abstraction reactions have been the subject of theoretical calculations. Methylene is shown to have a lower barrier to insertion into the Si—H bond of silane than into the C—H bond of methane.[84, 85] A corresponding calculation on the insertion of methylene, fluorocarbene, and difluorocarbene into hydrogen shows that there is a dramatic increase in barrier height with fluorine substitution. The effect of the fluorine is two-fold: it lowers the electron-pair energy, and it raises the empty *p*-orbital energy by anti-bonding interactions with its own filled π-type lone pairs.[86] A MINDO/3 comparison of the hydrogen insertion reactions of cyclopropenylidene and cyclopropylidene suggests that the reaction of the former carbene has the higher activation energy.[87]

The abstraction of hydrogen from cyclohexane by triplet diphenylcarbene can be monitored in the laser flash photolysis of diphenyldiazomethane. A kinetic isotope effect of 2.6 for cyclohexane *vs.* d_{12}-cyclohexane is observed.[88] Photolysis of diphenyldiazomethane in piperylene oligomer at 77 K produces radical pairs formed by hydrogen abstraction from tertiary C—H bonds of the oligomer by diphenylcarbene.[89] The carbene (**59**) inserts into cyclohexane and into cumene, although it does not undergo cycloaddition to cyclohexene.[90] Decomposition of ethyl diazoacetate in *o*-carborane gives four products resulting from insertion into B—H bonds, with no products from C—H insertion being isolated.[91] Isopropylidene inserts into Se—Se and Te—Te bonds to give moderate yields of ketene seleno- and telluro-ketals (**60**; M = Se or Te).[92]

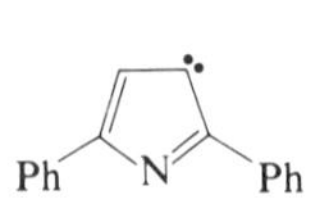

(59) 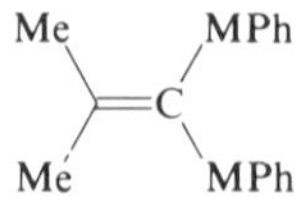 (60) 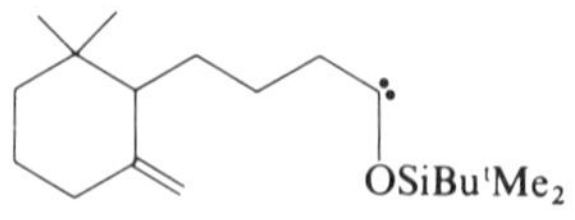 (61)

A recent paper on the reaction of triplet diphenylcarbene with methanol, as well as reporting the measurement of activation energies for the O—H insertion reaction in various solvents by flash photolysis of diphenyldiazomethane, reviews the published data for this reaction. It has been concluded that if these data are correct, then the triplet carbene must react to give Ph_2CHOMe in a process that must access or modify the singlet state at some point. However, if the reported value for singlet–triplet intersystem crossing is in error, being too high, then the data may be consistent with a mechanism involving a singlet–triplet equilibrium, followed by reaction of the singlet with methanol.[93] Kinetic studies of the reaction of benzyl(chloro)carbene with methanol show that it is second order in methanol, the amount of the insertion product, $PhCH_2CH(OMe)_2$, increasing with increasing methanol concentration, at the expense of chlorostyrene formed by rearrangement of the carbene. The results are interpreted in terms of reversible complex formation between the carbene and two molecules of methanol.[94] The related reaction with ethanol is also temperature-dependent, the yield of the insertion product, $PhCH_2CH(OEt)_2$, increasing with decreasing temperature, although the trends are reversed in a solid matrix. No competing C—H insertion is observed.[95] The carbene (**61**) reacts preferentially by intermolecular insertion into O—H bonds. Products arising from intramolecular addition to the double bond, or from 1,2-rearrangement are only minor.[96]

Intramolecular

The cyclopropylidene (**62**), or its lithium carbenoid, undergoes selective intramolecular insertion into the OCH_3 group to give the bicyclic compound (**63**) possibly as a result of coordination of the lithium by the oxygen atom.[97] The *syn*-stereochemical requirement for intramolecular insertion of cyclopropylidene into adjacent C—H bonds has been established in the carbene (**64**). The epimeric alcohol does not undergo intramolecular insertion.[98] The tricyclic hydrocarbon (**65**) is a minor product from generation of the carbene (**66**).[99]

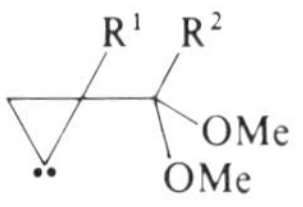

(62) 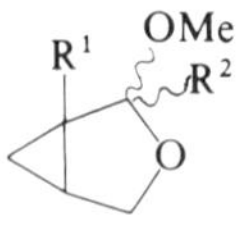 (63) 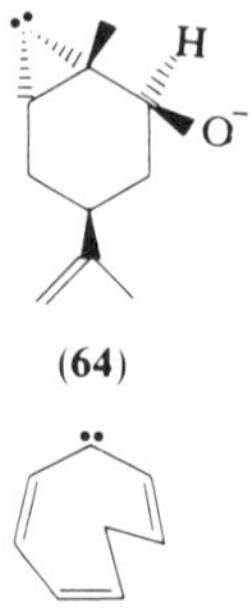 (64)

(65) (66) (67)

Attempts to generate homocycloheptatrienylidene (**67**) result in the formation of cyclooctatetraene as the major product *via* intramolecular insertion of the carbene into the transannular C—H bond, followed by electrocyclic ring-opening of the resulting bicyclooctatriene.[100]

Cyclopentenes can be formed in reasonable yield by intramolecular insertions of the alkylidene carbenes (**68**).[101] Amino (phenyl) carbenes (**69**; X = 0 or NMe) also give five-membered rings by intramolecular insertion into O—H or N—H bonds.[102]

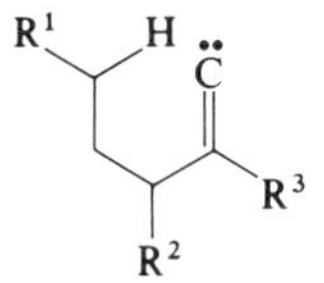

(**68**) 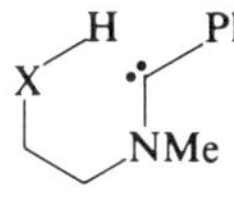 (**69**) 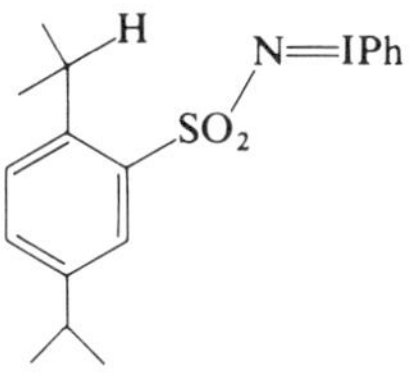 (**70**)

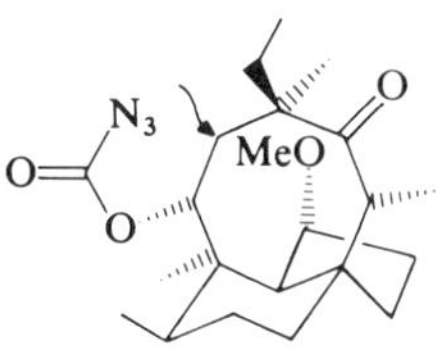

(**71**) 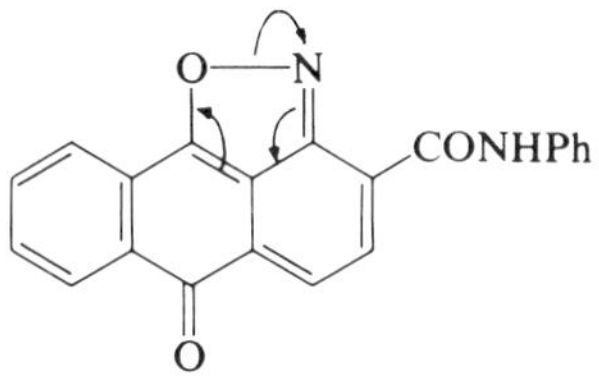 (**72**) 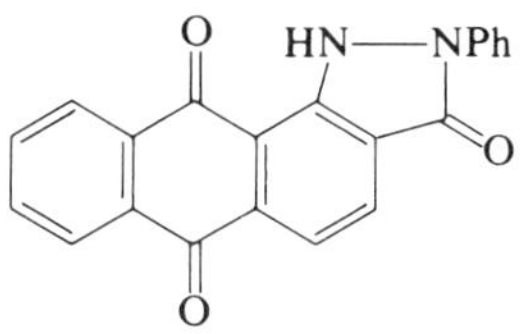 (**73**)

Intramolecular nitrene insertions have been observed in the iron(III)-mediated nitrene generation from (**70**),[103] and in the decomposition of the natural product derived azide (**71**) into the C—H bond indicated.[104] The pyrolysis of the isoxazole (**72**) in 1,2-dichlorobenzene gives the anthraquinone (**73**) by ring-opening as shown followed by insertion of the resulting nitrene into the adjacent amide N—H bond.[105]

Rearrangement

MNDO calculations on the rearrangement of carbenes and nitrenes by 1,2-shift reveal a preferred transition state in which the C—R bond of the migrating group, R, is eclipsed to the empty *p*-orbital of the reactive intermediate.[106]

Deuterium labelling experiments have established that in the rearrangement of hydroxy(methyl)carbene to give acetaldehyde, it is the hydrogen which is attached to oxygen which undergoes the 1,2-shift.[107] Treatment of 4-hydroxy-5,5,5-tris(phenylthio)pent-2-ene with *sec*-butyllithium gives 5-phenylthiopent-2-en-4-one *via* a 1,2-hydrogen shift in the intermediate carbene (**74**).[108]

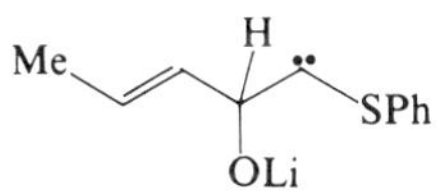

(**74**) 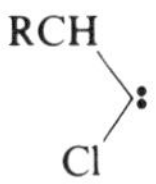 (**75**) 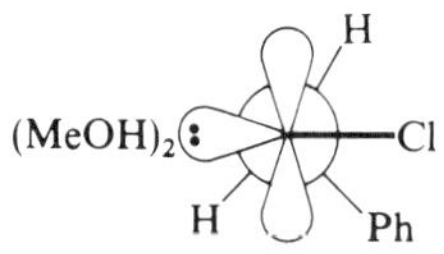 (**76**) 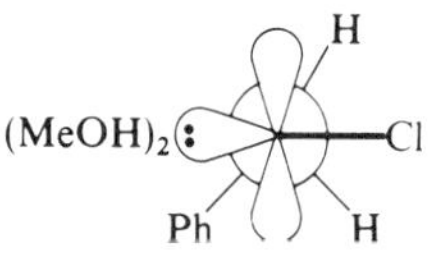 (**77**)

1,2-Hydrogen shifts in several chlorocarbenes have been studied. The shift is accelerated by a methoxy and decelerated by a phenyl group at the migration origin. Thus the carbene (**75**; R = OMe) rearranges more readily than the carbene (**75**; R = Ph).[109] In the benzyl(chloro)carbenes (**75**; R = Ar) the stereochemistry of the chlorostyrene resulting from the 1,2-hydrogen shift is dependent on the nature of the aryl group, the amount of E-alkene decreasing as the substituents in the aromatic ring are changed from electron-donating to electron-withdrawing.[110] When the same carbenes (**75**; R = Ar) are generated in methanol solution, the E/Z ratio of the chlorostyrenes increases with increasing methanol concentration. This is taken as further evidence for a carbene–methanol complex, the complex (**76**) leading to the Z-alkene being more stable than the complex (**77**) leading to the E-isomer.[94] The Z/E ratio of alkenes is also dependent on the temperature, increasing as the phase is changed to a solid matrix at low temperature.[95] When the rearrangement of the carbene (**75**; R = Ph) is carried out in the presence of AcOD, some deuterium is found in the chlorostyrene product, indicating that some of the alkene results from the $PhCH_2CClD$ cation as well as from the carbene. Evidence that the carbene is generated even in acetic acid comes from its interception with 2,3-dimethylbut-2-ene.[111]

Generation of carbene (**78**) by photochemical reaction of 2,3-dimethylbut-2-ene with pent-2-yn-4-one results in a 1,4-hydrogen shift to give the diene (**79**). Labelling experiments prove that the reaction involves a genuine 1,4-hydrogen shift.[112] The possible involvement of an intermediate such as (**80**) in the 1,2-benzoyl shift in the benzoyloxycarbene (**81**) to give PhCOCOMe has been ruled out by oxygen labelling experiments.[113]

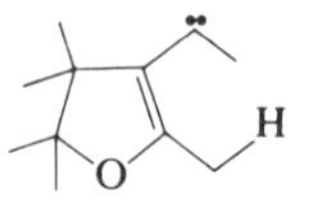

H O (**78**) O (**79**) Ph O Me O (**80**) O PhC O Me (**81**)

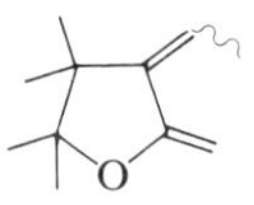

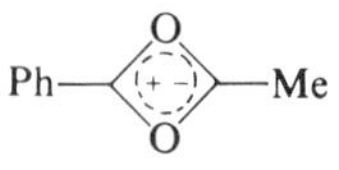

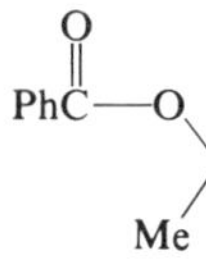

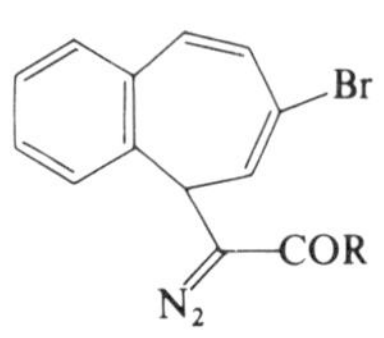

Br COR N_2 (**82**) Br COR (**83**) $SiMe_3$ Me_2Si O Ad (**84**) Me_3Si $SiMe_2$ O Ad (**85**)

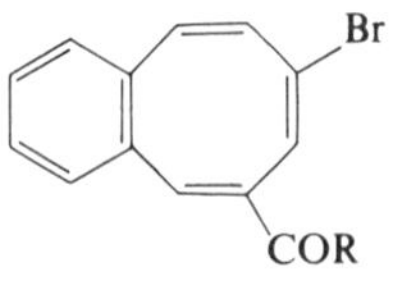

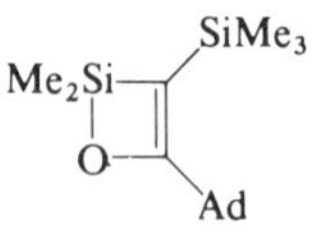

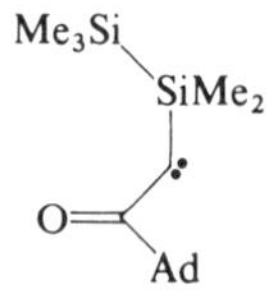

Copper-catalysed decomposition of the diazo compound (**82**) results in exclusive vinyl migration to give the benzocyclooctatetraene (**83**). No competing Wolff rearrangement is observed.[114] The key step in the formation of the first 1,2-silaoxetene (**84**) involves a 1,2-shift of the trimethylsilyl group in the carbene (**85**; Ad = 1-adamantyl), followed by electrocyclic ring-closure of the resulting acylsilene.[115]

A comprehensive review has appeared on oxirenes, and discusses their possible involvement in the Wolff rearrangement of acylcarbenes.[116]. The formation of cyclobutadiene by low-temperature irradiation of the diazo compound (**55**) is rationalized by a sequence involving Wolff rearrangement of the initial carbene, loss of CO from the resulting ketene to give cyclopropenylcarbene, followed by 1,2-rearrangement to give the four-membered ring.[82] In studies designed to probe the effects of strain in the transition state for Wolff rearrangement, the photochemical decomposition of the diazo-ketone (**86**) was investigated under various conditions. It was concluded that the triplet carbene was the initial and only photo-product from the singlet excited diazo compound, and that under these conditions all the Wolff rearrangement proceeded *via* the triplet state.[117] Irradiation of the 4-diazopyrazoline-3,5-diones (**87**) in the presence of a nucleophile (NuH) gives the aza-β-lactams (**88**) by Wolff rearrangement of the intermediate carbenes with migration of a nitrogen substituent. In the related bicyclic carbene (**89**), a competing fragmentation (arrows, although not necessarily concerted) to give carbon suboxide and 3,4,5,6-tetrahydropyridazine is the major pathway.[118]

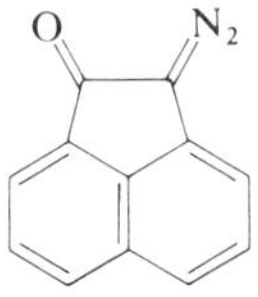

(**86**) 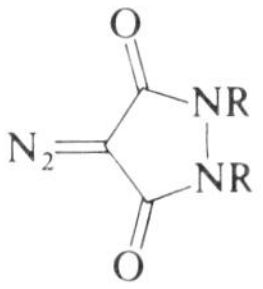 (**87**) 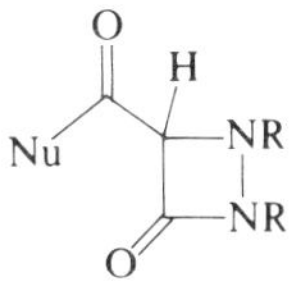 (**88**) 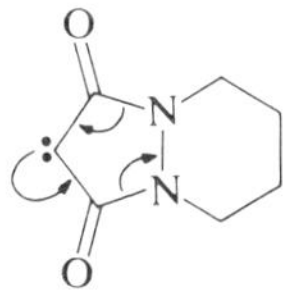 (**89**)

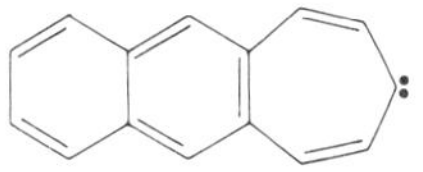

(**90**) 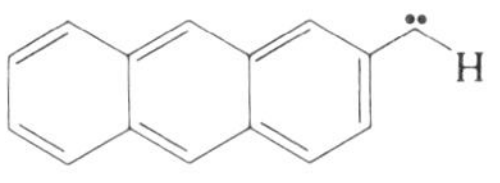 (**91**) 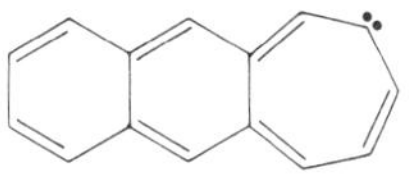 (**92**) 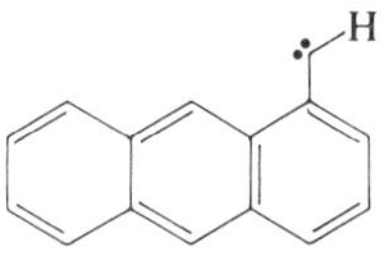 (**93**)

Several carbene–carbene rearrangements have been reported. Generation of the carbene (**90**) from the corresponding tosylhydrazone gives among other products both 1- and 2-methyl-anthracene by mechanisms involving rearrangement to the carbenes (**91**), (**92**), and (**93**).[119] Similarly, generation of the fluorenylcarbene (**94**) gives the hydrocarbon (**95**) as a minor product, by initial rearrangement to the isomeric carbene (**96**) followed by intramolecular insertion. The major product from the carbene (**94**) is benz[*a*]azulene.[120] Cyclopropenylidene, which can be isolated in a matrix at 10 K, rearranges to the carbene (**97**) on irradiation.[121]

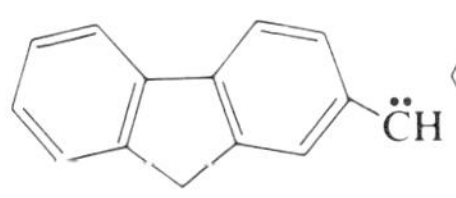

$HC{\equiv}C{-}\ddot{C}H$

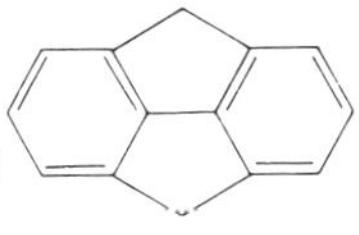

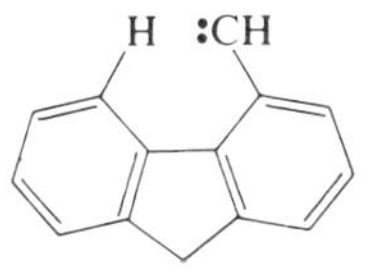

(**97**)

(**94**) (**95**) (**96**)

The mechanism of the vinylcyclopropylidene – cyclopentenylidene rearrangement has been discussed in detail. Decomposition of (**98**), a possible precursor of the carbene (**99**), gives products formally derived from the rearranged carbene (**100**). However, it has been concluded that the butyllithium-mediated reaction proceeds through carbenoids rather than free carbenes, whilst the purely thermal reaction proceeds by an ionic mechnism.[122, 123] In the case of the homologous rearrangement, (**101**) to (**102**), free carbenes are involved when the precursor is the *N*-nitrosourea (**103**).[124]

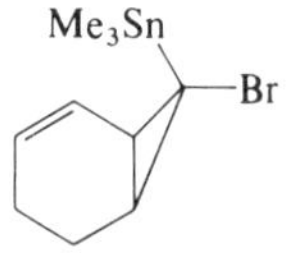

Me_3Sn Br

(**98**)

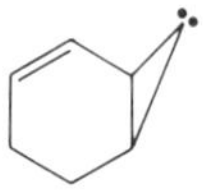

(**99**)

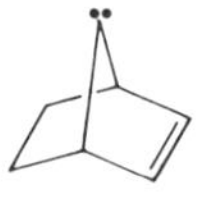

(**100**)

H NCONH$_2$ NO

(**101**)

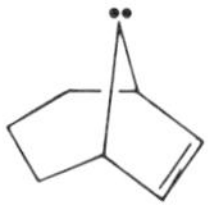

(**102**)

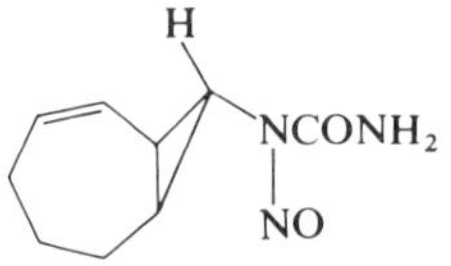

(**103**)

Ab initio calculations on the formation of azirine from vinyl azide suggest that a "concerted" pathway is preferred to one involving vinylnitrene.[125] Although the singlet nitrene is undoubtedly generated in the photolysis of the azides (**104**), it has been shown that the "Curtius rearrangement" to give (**105**), which is trapped by methanol, proceeds by a non-nitrene concerted pathway. In the azide (**104**; R^1 = Ph, R^2 = Bu^t) the *tert*-butyl group appears to have a higher migratory aptitude than phenyl.[126] The formation of methoxyazepines in the photolysis of arylazides in methanol is promoted by an electron-withdrawing ester group *para* to the azide. Thus the azide (**106**; R = CO_2Me) gives a good yield of the azepine, in contrast to the azide (**106**; R = OMe) which only gives the corresponding azo compound upon irradiation.[127]

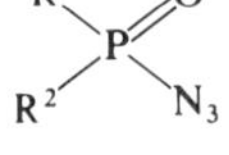

R^1 O P R^2 N_3

(**104**)

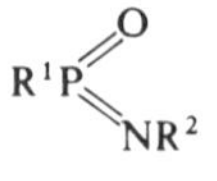

R^1P O NR^2

(**105**)

OMe R N_3

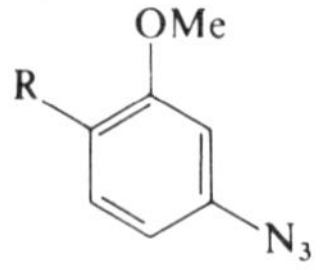

(**106**)

Aromatics

The carbene (**107**) gives cycloheptatrienes on reaction with aromatic compounds; full details of this work have appeared.[128] Generation of the carbene (**59**) in benzene leads to the novel cyclooctapyrrole (**108**) by rearrangement of the initially formed norcaradiene intermediate.[90]

Intramolecular carbene attack on the thiophen ring to give (**109**) occurs on decomposition of the tosylhydrazone (**110**), no attack on the thiophen sulphur being observed.[129]

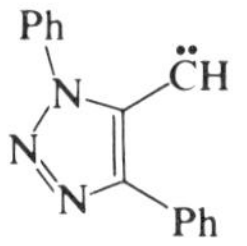

(**107**) 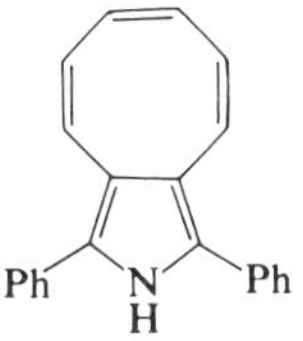 (**108**) 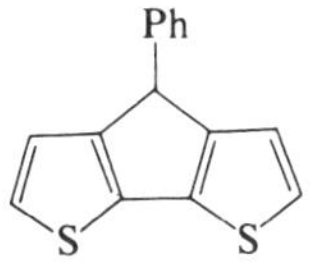 (**109**) 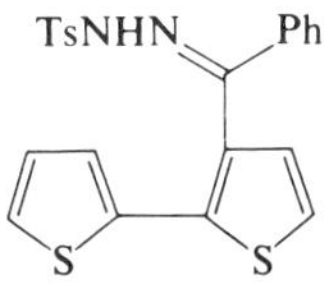 (**110**)

Thermal decomposition of ethyl azidoformate in aromatic solvents in the presence of trifluoroacetic acid gives much better yields of the urethanes $ArNHCO_2Et$ than in the simple thermal reaction, suggesting that in the presence of the acid the mechanism involves electrophilic aromatic substitution by the ethoxycarbonylnitrenium ion.[130] *N*-Nitrenes derived from oxidation of 3-aminoquinazolin-4-ones (**111**) add intramolecularly to the aromatic ring, the product depending on the reaction solvent.[131]

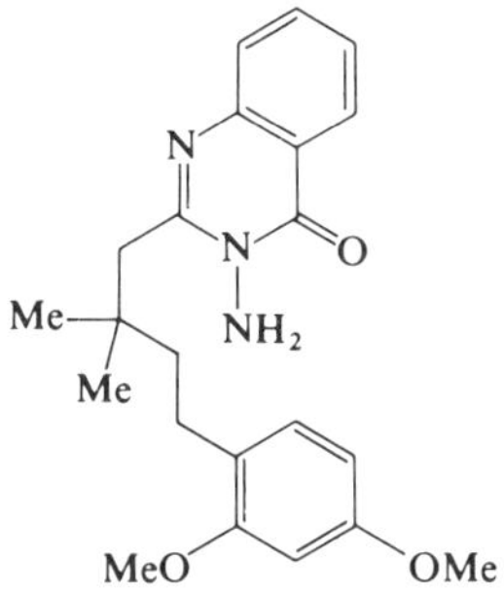

(**111**) 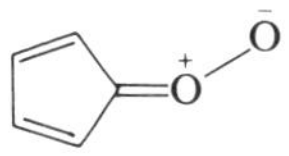 (**112**)

Nucleophiles and Electrophiles

The reaction of methylene with molecular oxygen has been studied by laser magnetic resonance, and the rate constant obtained.[132,133] A second research group has investigated the photolytic decomposition of diazocyclopentadiene in an argon matrix in the presence of oxygen, and confirmed that the initial product is cyclopentadienone-*O*-oxide (**112**).[134]

The deoxygenation of aldehydes and ketones by dibromocarbene, generated from $PhHgCBr_3$, has been further studied. In the reaction of benzaldehyde labelled at the carbonyl carbon, the label is retained in the $PhCHBr_2$ product, consistent with a mechanism involving attack by the electrophilic carbene on the carbonyl oxygen to give a carbonyl ylid as the initial product.[135] The lithium carbenoids, LiCRMeBr, also react with aldehydes and ketones, but the products in this case are epoxides which are formed in good yield.[136] Intramolecular nucleophilic attack by a thioamide occurs on generation of the carbene (**113**) to give the mesoionic compound (**114**) in excellent yield.[137]

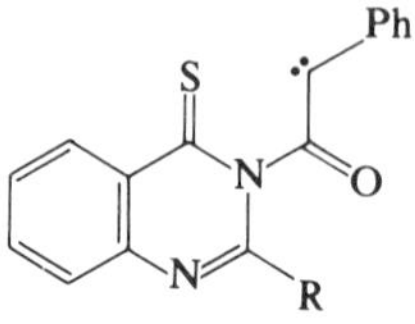

(**113**) 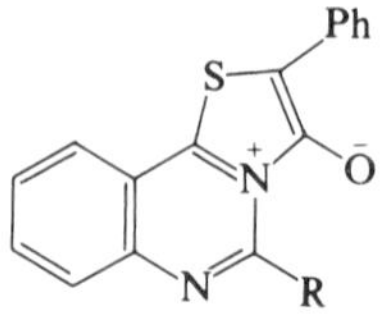 (**114**) 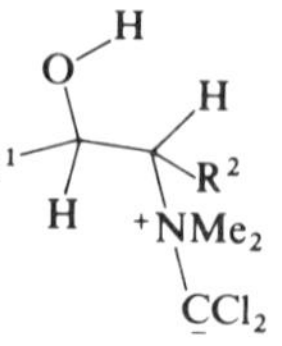 (**115**)

Several reactions involving nucleophilic attack on carbenes by nitrogen compounds have been reported. The quenching of the photoexcited triplet state of diphenylcarbene by amines has been studied by picosecond laser experiments,[138] and related experiments on the reaction of triplet carbon atoms with trimethylamine have also been carried out.[139] Treatment of vicinal amino-alcohols, $R^1CH(OH)CHR^2NMe_2$, with phase-transfer generated dichlorocarbene gives epoxides by ring-closure of the initially formed ylids (**115**).[140] Attempts to form bicyclic aza-β-lactams by "insertion" of the rhodium carbenoid derived from the diazo compound (**116**) into the adjacent N—H bond are frustrated by the greater nucleophilicity of N(1) which results in ring-closure to give the spiro-ylid (**117**). The final products arise from further rearrangement of this ylid.[141] Copper (I) bromide catalysed decomposition of phenyldiazomethane in the presence of imines, $ArCR^1{=}NR^2$, has been shown to proceed by the *trans*-azomethine ylid (**118**), which, depending on the substituents, either undergoes conrotatory ring-closure to the *cis*-aziridine, or cycloaddition to a second molecule of the imine.[142] Alkoxycarbenes, generated by thermolysis of the 1,3,4-oxadiazolines (**119**), react with the precursor by electrophilic attack on the azo group to give products derived from the adduct (**120**), and its N(4) isomer.[143] Isopropylidene reacts with azobenzenes to give isoindazoles (**121**). The mechanism involves nucleophilic attack on the carbene to give (**122**), followed by cyclization, rather than direct 1,4-addition.[144]

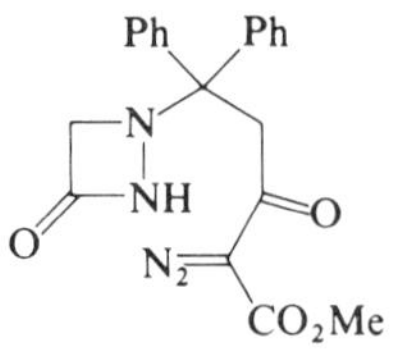

(**116**) 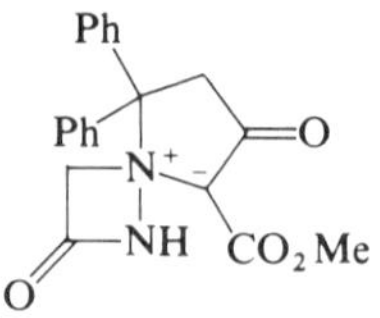 (**117**) 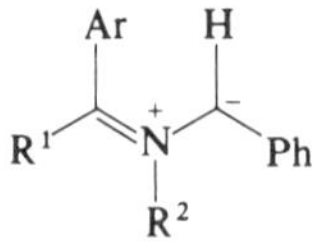 (**118**) 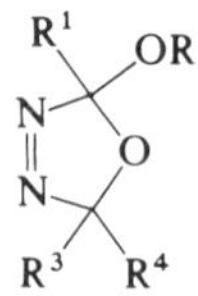 (**119**)

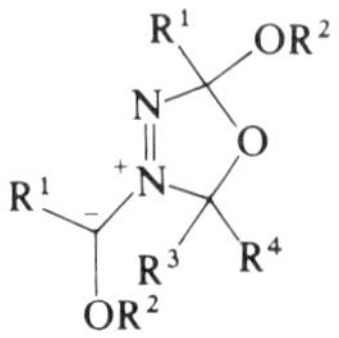

(**120**) 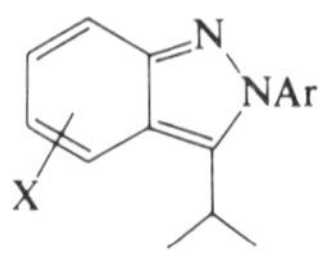 (**121**) 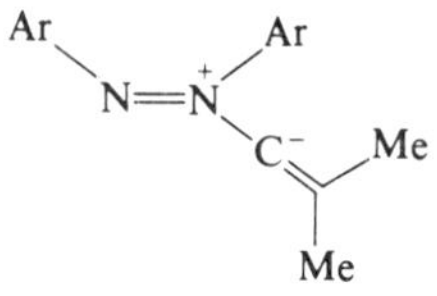 (**122**)

MO calculations on the electrophilic nitrenoids $LiNR^1OR^2$, which aminate organolithium reagents, R^3Li, show that the facile replacement of OR^2 by R^3 is due to the formation of the "dimeric" species (**123**), and to an exceptionally long N—O bond in the original nitrenoid.[145] Thermal decomposition of *para*-substituted arylazides, ArN_3, in phenyl isocyanate can give products derived from phenylnitrene and the aryl isocyanate. One possible explanation for this "exchange" reaction involves direct electrophilic attack on the isocyanate nitrogen to give (**124**), which can then fragment to phenylnitrene and aryl isocyanate, presumably *via* the diaziridinone (**125**).[146]

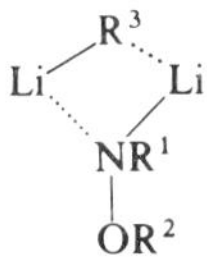

(**123**) 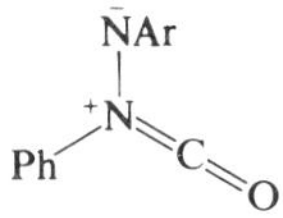 (**124**) 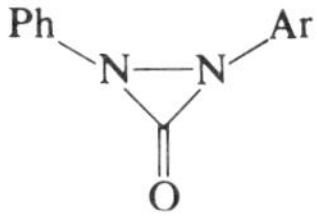 (**125**)

Further details of the decomposition of aryl azides in the presence of sulphides have appeared. The results are entirely consistent with singlet arylnitrene attack on sulphur to give sulphimides (**126**), which then undergo [2,3]-sigmatropic rearrangement under the reaction conditions.[147] Decomposition of azidoformates or tosyl azide in tetrachlorothiophen leads to the ylids (**127**; R = CO_2R' or Ts), the first examples of thiophen S=N ylids. Photolysis of the ylids regenerates the nitrenes by cleavage of the S=N bond.[148] An intramolecular attack on a thiophen sulphur is thought to account for the unprecedented ring-cleavage of the 3-azidothiophen (**128**) with extrusion of acetylene to form ethyl 5-cyanoisothiazole-3-carboxylate; full details of this work have appeared.[149] This intramolecular nitrene attack on a thiophen sulphur is in contrast to the carbene derived from (**110**), which exclusively attacks the thiophen-3-position rather than the sulphur.[129]

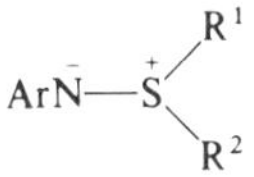

(**126**) 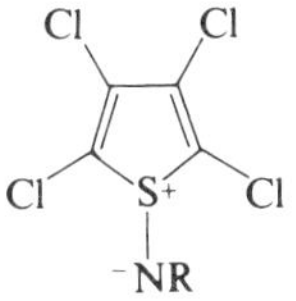 (**127**) 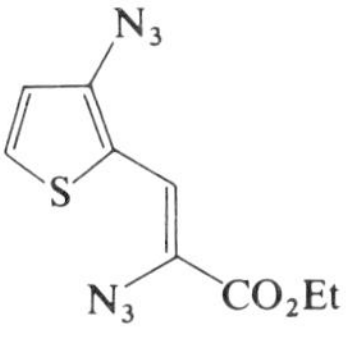 (**128**)

Silylenes and Germylenes

The chemistry of disilenes has been reviewed.[150]

The results of a detailed study of the gas-phase decomposition of alkylsilanes, $RSiH_3$, and their deuterated analogues, $RSiD_3$, have been published.[151–155] The mechanism of decomposition involves loss of H_2 (or D_2) to give the corresponding silylene. Singlet silylene itself can be generated from high-energy recoiling silicon atoms in the presence of phosphine or silane.[156] Photochemical decomposition of the disilacyclopropane (**129**; Ar = 2,6-$Me_2C_6H_3$) results in retro-1,2-addition and generation of the diarylsilylene, which can be trapped with methanol or 2,3-dimethylbuta-1,3-diene, and the silene $Ar_2Si{=}CH_2$.[157]

Ab initio calculations have been carried out for the insertion reactions of silylene into NH_3, H_2O, HF, PH_3, H_2S, and HCl. The results suggest that all the reactions involve initial formation of a donor–acceptor complex, followed by hydrogen shift *via* an unsymmetrical high-energy transition state.[158] Similar calculations on the insertion of silylene into hydrogen, methane, and silane show that the barrier is higher than for the corresponding insertions of methylene,[84,85] and in common with the carbene case, the barrier height is dramatically raised by fluorine substitution.[86]

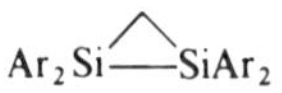

(**129**) 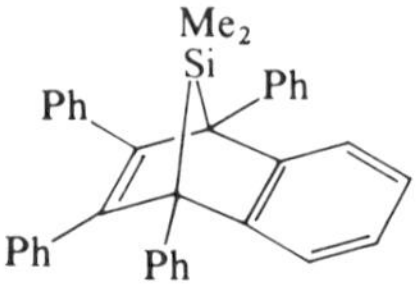 (**130**) 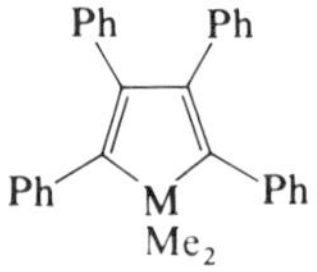 (**131**)

Irradiation of the dimethylsilylene precursor (**130**) in the presence of 1,1-dimethyl-2,3,4,5-tetraphenyl-1-germacyclopentadiene (**131**; M = Ge) leads to silylene–germylene exchange, and the formation of the silacyclopentadiene (**131**; M = Si) and products derived from dimethylgermylene.[159] Dimethylgermylene exhibits inverse electron demand in its reactions with 1,3-dienes, reacting more rapidly with electron-deficient dienes.[160]

Transition-metal Complexes

The use of metal–carbene complexes in organic synthesis has been covered in a review article.[161]

The chromium carbene complex (**132**) is a highly reactive dienophile in Diels–Alder reactions, exhibiting rate accelerations of approximately 10^4 over methyl acrylate with various dienes.[162] Similar chromium complexes have already found wide use in synthesis, and recent examples include a new benzannelation procedure,[163] and a new route to cyclohexadienones.[164] The reaction of such complexes with imines gives β-lactams, and this work has been reviewed.[165]

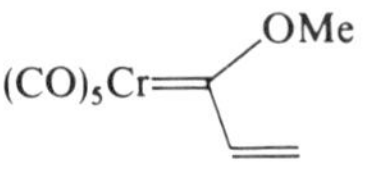

(**132**) $Ar_2CHCHCl_2$ (**133**) (**134**)

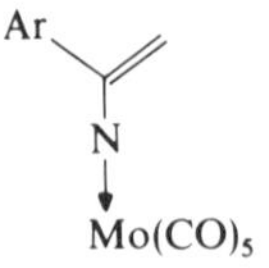

Further examples of the cobaloxime-mediated formation of *trans*-stilbenes from the dichlorides (**133**) have appeared. The mechanism is thought to involve a cobalt carbenoid intermediate formed by displacement of chloride by cobalt (I), followed by α-elimination.[166] Thermal decomposition of α-azidostyrenes catalysed by hexacarbonylmolybdenum gives 2,5- and 2,4-diarylpyrroles *via* the initially formed complex (**134**).[167]

References

1 Wentrup, C., *Reactive Molecules. The Neutral Reactive Intermediates in Organic Chemistry*, John Wiley and Sons, New York, 1984.
2 Baird, M. S., *Ann. Rep. Prog. Chem., Sect. B*, **1983,** 83.
3 Griller, D., Nazran, A. S., and Scaiano, J. C., *Acc. Chem. Res.*, **17,** 283 (1984).
4 Scaiano, J. C., *Contrib. Cient. Tecnol.*, **13** (Numero Espec.), 9 (1983); *Chem. Abs.*, **101,** 72062 (1984).
5 Schwarz, H., *Shitsuryo Bunseki*, **32,** 3 (1984); *Chem. Abs.*, **101,** 129829 (1984).
6 Xu, L., Tao, F., and Wu, S., *Huaxue Tongbao*, **1984,** 1; *Chem. Abs.*, **101,** 190619 (1984).
7 Yuan, Y., and Cao, J., *Huaxue Tongbao*, **1983,** 31; *Chem. Abs.*, **100,** 67464 (1984).
8 Rasch, G., and Remmler, K. D., *Wiss. Z. Tech. Hochsch. "Carl Schorlemmer" Leuna-Merseburg*, **25,** 190 (1983); *Chem. Abs.*, **99,** 193990 (1983).
9 Murahashi, S., *Kagaku, Zokan (Kyoto)*, **1983,** 31; *Chem. Abs.*, **100,** 33793 (1984).
10 Sitzmann, E. V., Langan, J., and Eisenthal, K. B., *J. Am. Chem. Soc.*, **106,** 1868 (1984).
11 Alt, R., Staab, H. A., Reisenauer, H. P., and Maier, G., *Tetrahedron Lett.*, **25,** 633 (1984).
12 Nazran, A. S., and Griller, D., *J. Am. Chem. Soc.*, **106,** 543 (1984).
13 Nazran, A. S., Lee, F. L., Gabe, E. J., Lepage, Y., Northcott, D. J., Park, J. M., and Griller, D., *J. Phys. Chem.*, **88,** 5251 (1984).
14 Wright, B. B., and Platz, M. S., *J. Am. Chem. Soc.*, **106,** 4175 (1984).
15 Lapin, S. C., Brauer, B.-E., and Schuster, G. B., *J. Am. Chem. Soc.*, **106,** 2092 (1984).
16 Grasse, P. B., Brauer, B.-E., Zupancic, J. J., Kaufmann, K. J., and Schuster, G. B., *J. Am. Chem. Soc.*, **105,** 6833 (1983).
17 Griller, D., Hadel, L., Nazran, A. S., Platz, M. S., Wong, P. C., Savino, T. G., and Scaiano, J. C., *J. Am. Chem. Soc.*, **106,** 2227 (1984).
18 Gaspar, P. P., Lin, C.-T., Whitsel Dunbar, B. L., Mack, D. P., and Balasubramanian, P., *J. Am. Chem. Soc.*, **106,** 2128 (1984).
19 Suguwara, T., Bethell, D., and Iwamura, H., *Tetrahedron Lett.*, **25,** 2375 (1984).
20 Suguwara, T., Bandow, S., Kimura, K., Iwamura, H., and Itoh, K., *J. Am. Chem. Soc.*, **106,** 6449 (1984).
21 Murata, S., Suguwara, T., Nakashima, N., Yoshihara, K., and Iwamura, H., *Tetrahedron Lett.*, **25,** 1933 (1984).
22 Sicard, G., Baceiredo, A., Bertrand, G., and Majoral, J.-P., *Angew, Chem. Int. Ed.*, **23,** 459 (1984).
23 Baceiredo, A., Bertrand, G., Majoral, J.-P., Sicard, G., Jaud, J., and Galy, J., *J. Am. Chem. Soc.*, **106,** 6088 (1984).
24 Ahmed, S. N., and Shevlin, P. B., *J. Am. Chem. Soc.*, **105,** 6488 (1983).
25 Wojnarovits, L., *J. Chem. Soc., Perkin Trans. 2*, **1984,** 1449.
26 Kropp, P. J., *Acc. Chem. Res.*, **17,** 131 (1984).
27 Kropp, P. J., Sawyer, J. A., and Snyder, J. J., *J. Org. Chem.*, **49,** 1583 (1984).
28 Kikuchi, O., *Bull. Chem. Soc. Jpn.*, **57,** 2007 (1984).
29 Koller, M., Karpf, M., and Dreiding, A. S., *Helv. Chim. Acta*, **66,** 2760 (1983).
30 Lamberts, J. J. M., and Laarhoven, W. H., *J. Org. Chem.*, **49,** 100 (1984).
31 Ando, W., Hamada, Y., and Sekiguchi, A., *Tetrahedron Lett.*, **25,** 5057 (1984).
32 Warrener, R. N., Pitt, I. G., and Russell, R. A., *J. Chem. Soc., Chem. Commun.*, **1984,** 1464.
33 Steinmetz, M. G., Yen, Y.-P., and Poch, G. K., *J. Chem. Soc., Chem. Commun.*, **1983,** 1504.
34 Baird, M. S., Buxton, S. R., and Whitley, J. S., *Tetrahedron Lett.*, **25,** 1509 (1984).
35 Padwa, A., Pulwer, M. J., and Rosenthal, R. J., *J. Org. Chem.*, **49,** 856 (1984).
36 Fox, D. P., Bjork, J. A., and Stang, P. J., *J. Org. Chem.*, **48,** 3994 (1984).
37 Duraisamy, M., and Walborsky, H. M., *J. Am. Chem. Soc.*, **106,** 5035 (1984).
38 Baumgart, K.-D., and Szeimies, G., *Tetrahedron Lett.*, **25,** 737 (1984).
39 Poon, C.-D., Yuen, P.-W., Man, T.-O., Li, C.-S., and Chan, T.-L., *J. Chem. Soc., Perkin Trans. 1*, **1984,** 1561.
40 Dailey, W. P., and Lemal, D. M., *J. Am. Chem. Soc.*, **106,** 1169 (1984).
41 Koroniak, H., *J. Fluorine Chem.*, **24,** 503 (1984); *Chem. Abs.*, **101,** 54478 (1984).
42 Goldstein, E., Hammond, B. L., Sadri, B., and Hsia, Y. P., *THEOCHEM*, **14,** 315 (1983); *Chem. Abs.*, **100,** 138245 (1984).

[43] Moss, R. A., and Lawrynowicz, W., *J. Org. Chem.*, **49**, 3828 (1984).
[44] Franck-Neumann, M., and Miesch, M., *Tetrahedron Lett.*, **25**, 2909 (1984).
[45] Grewel, R. S., Burnell, D. J., and Yates, P., *J. Chem. Soc., Chem. Commun.*, **1984**, 759.
[46] Siewinski, A., Henggeler, B., Wolf, H. R., Frei, B., and Jeger, O., *Helv. Chim. Acta*, **67**, 120 (1984).
[47] O'Sullivan, A., Frei, B., and Jeger, O., *Helv. Chim. Acta*, **67**, 815 (1984).
[48] Scott, L. T., Tsang, T.-H., and Levy, L. A., *Tetrahedron Lett.*, **25**, 1661 (1984).
[49] Chapman, K. J., Dyall, L. K., and Frith, L. K., *Aust. J. Chem.*, **37**, 341 (1984).
[50] McManus, S. P., Smith, M. R., Abramovitch, R. A., and Offor, M. N., *J. Org. Chem.*, **49**, 683 (1984).
[51] Maricich, T. J., Angeletakis, C. N., and Mjanger, R., *J, Org. Chem.*, **49**, 1928 (1984).
[52] Atkinson, R. S., Lee, M., and Malpass, J. R., *J. Chem. Soc., Chem. Commun.*, **1984**, 919.
[53] Pritt, A. T., Patel, D., and Coombe, R. D., *Int. J. Chem. Kinet.*, **16**, 977 (1984).
[54] Fukunaga, T., and Begland, R. W., *J. Org. Chem.*, **49**, 813 (1984).
[55] Moreno, M., Lluch, J. M., Oliva, A., and Bertran, J., *THEOCHEM*, **16**, 227 (1984); *Chem. Abs.*, **100**, 191057 (1984).
[56] Mireda, J., Rondan, N. G., Houk, K. N., Clark, T., and Schleyer, P. von R., *J. Am. Chem. Soc.*, **105**, 6997 (1983).
[57] Eichler, K., and Heydtmann, H., *Int. J. Chem. Kinet.*, **16**, 835 (1984).
[58] Laidler, D. A., and Milner, D. J., *J. Organomet. Chem.*, **270**, 121 (1984).
[59] Tomioka, H., Ohno, K., Izawa, Y., Moss, R. A., and Munjal, R. C., *Tetrahedron Lett.*, **25**, 5415 (1984).
[60] Doyle, M. P., Terpstra, J. W., and Winter, C. H., *Tetrahedron Lett.*, **25**, 901 (1984).
[61] Tolbert, L. M., and Siddiqui, S., *J. Am. Chem. Soc.*, **106**, 5538 (1984).
[62] Tetef, M. L., and Jones, M., *Tetrahedron Lett.*, **25**, 161 (1984).
[63] Rondan, N. G., and Houk, K. N., *Tetrahedron Lett.*, **25**, 5965 (1984).
[64] Kostikov, R. R., and Menchikov, L. G., *Dokl. Akad. Nauk SSSR*, **276**, 123 (1984); *Chem. Abs.*, **101**, 151157 (1984).
[65] Kostikov, R. R., and Molchanov, A. P., *Zh. Org. Khim.*, **20**, 1003 (1984); *Chem. Abs.*, **101**, 129944 (1984).
[66] Tubul, A., and Bertrand, M., *Tetrahedron Lett.*, **25**, 2219 (1984).
[67] Lambert, J. B., Bosch, R. J., Mueller, P. H., and Kobayashi, K., *J. Am. Chem. Soc.*, **106**, 3584 (1984).
[68] Lukevics, E., Gevorgyan, V. N., Goldberg, Y. S., Gaukhman, A. P., Gavars, M. P., Popelis, J. J., and Shymanska, M. V., *J. Organomet. Chem.*, **265**, 237 (1984).
[69] Dehmlow, E. V., Broda, W., and Schladerbeck, N. H., *Tetrahedron Lett.*, **25**, 5509 (1984).
[70] Kostikov, R., and de Meijere, A., *J. Chem. Soc., Chem. Commun.*, **1984**, 1528.
[71] Göthling, W., Keyaniyan, S., and de Meijere, A., *Tetrahedron Lett.*, **25**, 4101 (1984).
[72] Keyaniyan, S., Göthling, W., and de Meijere, A., *Tetrahedron Lett.*, **25**, 4105 (1984).
[73] Dehmlow, E. V., and Kramer, R., *Angew. Chem. Int. Ed.*, **23**, 706 (1984).
[74] Houk, K. N., Rondan, N. G., and Mareda, J., *J. Am. Chem. Soc.*, **106**, 4291 (1984).
[75] Houk, K. N., and Rondan, N. G., *J. Am. Chem. Soc.*, **106**, 4293 (1984).
[76] Tomioka, H., Hayashi, N., Izawa, Y., and Liu, M. T. H., *J. Am. Chem. Soc.*, **106**, 454 (1984).
[77] Warner, P. M., *Tetrahedron Lett.*, **25**, 4211 (1984).
[78] Loreto, M. A., Pellacani, L., Tardella, P. A., and Toniato, E., *Tetrahedron.*, **25**, 4271 (1984).
[79] Aitken, R. A., Cadogan, J. I. G., Farries, H., Gosney, I., Palmer, M. H., Simpson, I., and Tinley, E. J., *J. Chem. Soc., Chem. Commun.*, **1984**, 791.
[80] Aitken, R. A., Gosney, I., Farries, M. H., Simpson, I., Cadogan, J. I. G., and Tinley, E. J., *Tetrahedron*, **40**, 2487 (1984).
[81] Marshall, J. A., Peterson, J. C., and Lebioda, L., *J. Am. Chem. Soc.*, **106**, 6006 (1984).
[82] Maier, G., Hoppe, M., Lanz, K., and Reisenauer, H. P., *Tetrahedron Lett.*, **25**, 5645 (1984).
[83] Atkinson, R. S., Malpass, J. R., Skinner, K. L., and Woodthorpe, K. L., *J. Chem. Soc., Perkin Trans. 1*, **1984**, 1905.
[84] Gordon, M. S., *J. Am. Chem. Soc.*, **106**, 4054 (1984).
[85] Gordon, M. S., and Gano, D. R., *J. Am. Chem. Soc.*, **106**, 5421 (1984).
[86] Sosa, C., and Schlegel, H. B., *J. Am. Chem. Soc.*, **106**, 5847 (1984).
[87] Tsang, H. T., and Li, W. K., *THEOCHEM*, **13**, 95 (1983); *Chem. Abs.*, **100**, 22179 (1984).
[88] Hadel, L. M., Platz, M. S., and Scaiano, J. C., *J. Am. Chem. Soc.*, **106**, 283 (1984).
[89] Vorotnikov, A. P., Davydov, E. Ya., and Toptygin, D. Ya., *Izv. Akad. Nauk SSSR, Ser. Khim.*, **1983**, 1499; *Chem. Abs.*, **99**, 194348 (1983).
[90] Nagarajan, M., and Shechter, H., *J. Org. Chem.*, **49**, 62 (1984).
[91] Zheng, G., and Jones, M., *J. Am. Chem. Soc.*, **105**, 6487 (1983).
[92] Stang, P. J., Roberts, K. A., and Lynch, L. F., *J. Org. Chem.*, **49**, 1653 (1984).

[93] Griller, D., Nazran, A. S., and Scaiano, J. C., *J. Am. Chem. Soc.*, **106**, 198 (1984).
[94] Liu, M. T. H., and Subramanian, R., *J. Chem. Soc., Chem. Commun.*, **1984**, 1062.
[95] Tomioka, H., Hayashi, N., Izawa, Y., and Liu, M. T. H., *Tetrahedron Lett.*, **25**, 4413 (1984).
[96] Scheller, M. E., and Frei, B., *Helv. Chim. Acta*, **67**, 1734 (1984).
[97] Arct, J., Skattebøl, L., and Stenstrom, Y., *Acta Chem. Scand.*, **37B**, 681 (1983).
[98] Baird, M. S., Buxton, S. R., and Sadler, P., *J. Chem. Soc., Perkin Trans. 1*, **1984**, 1379.
[99] Stadler, H., Rey, M., and Dreiding, A. S., *Helv. Chim. Acta*, **67**, 1379 (1984).
[100] Parker, R. H., and Jones, W. M., *Tetrahedron Lett.*, **25**, 1245 (1984).
[101] Gilbert, J. C., Giamalva, D. H., and Weerasooriya, U., *J. Org. Chem.*, **48**, 5251 (1983).
[102] Moss, R. A., Cox, D. P., and Tomioka, H., *Tetrahedron Lett.*, **25**, 1023 (1984).
[103] Breslow, R., and Gellman, S. H., *J. Am. Chem. Soc.*, **105**, 6728 (1983).
[104] Berner, H., Vyplel, H., Schulz, G., and Stuchlik, P., *Tetrahedron*, **40**, 919 (1984).
[105] Gornostaev, L. M., and Lavrikova, T. I., *Zh. Org. Khim.*, **20**, 874 (1984); *Chem. Abs.*, **101**, 110020 (1984).
[106] Frenking, G., and Schmidt, J., *Tetrahedron*, **40**, 2123 (1984).
[107] Weiner, B. R., and Rosenfeld, R. N., *J. Org. Chem.*, **48**, 5362 (1983).
[108] Cohen, T., and Yu, L.-C., *J. Org. Chem.*, **49**, 605 (1984).
[109] Liu, M. T. H., and Tencer, M., *Tetrahedron Lett.*, **24**, 5713 (1983).
[110] Tomioka, H., Hayashi, N., Izawa, Y., and Liu, M. T. H., *J. Chem. Soc., Chem. Commun.*, **1984**, 476.
[111] Liu, M. T. H., Chishti, N. H., Tencer, M., Tomioka, H., and Izawa, Y., *Tetrahedron*, **40**, 887 (1984).
[112] Saba, S., Wolff, S., Schröder, C., Margaretha, P., and Agosta, W. C., *J. Am. Chem. Soc.*, **105**, 6902 (1983).
[113] Brown, R. F. C., Browne, N. R., and Eastwood, F. W., *Aust. J. Chem.*, **36**, 2355 (1983).
[114] Böhshar, M., Heydt, H., and Regitz, M., *Chem. Ber.*, **117**, 3093 (1984).
[115] Sekiguchi, A., and Ando, W., *J. Am. Chem. Soc.*, **106**, 1486 (1984).
[116] Lewars, E. G., *Chem. Rev.*, **83**, 519 (1983).
[117] Hayes, R. A., Hess, T. C., McMahon, R. J., and Chapman, O. L., *J, Am. Chem. Soc.*, **105**, 7786 (1983).
[118] Lawton, G., Moody, C. J., and Pearson, C. J., *J. Chem. Soc., Chem. Commun.*, **1984**, 754.
[119] Hackenburger, A., and Dürr, H., *Chem. Ber.*, **117**, 2644 (1984).
[120] Wentrup, C., and Becker, J., *J. Am. Chem. Soc.*, **106**, 3705 (1984).
[121] Reisenauer, H. P., Maier, G., Riemann, A., and Hoffmann, R. W., *Angew. Chem. Int. Ed.*, **23**, 641 (1984).
[122] Warner, P. M., and Herold, R. D., *J. Org. Chem.*, **48**, 5411 (1983).
[123] Warner, P. M., and Herold, R. D., *Tetrahedron Lett.*, **25**, 4897 (1984).
[124] Warner, P. M., and Chu, I.-S., *J. Org. Chem.*, **49**, 3666 (1984).
[125] Yamabe, T., Kaminoyama, M., Minato, T., Hori, K., Isomura, K., and Taniguchi, H., *Tetrahedron*, **40**, 2095 (1984).
[126] Harger, M. J. P., and Westlake, S., *J. Chem. Soc., Perkin Trans. 1*, **1984**, 2351.
[127] Mustill, R. A., and Rees, A. H., *J. Org. Chem.*, **48**, 5041 (1983).
[128] Bedford, C. D., and Smith, P. A. S., *J. Org. Chem.*, **48**, 4002 (1983).
[129] Skramstad, J., and Storflor, H., *Acta Chem. Scand.*, **38B**, 533 (1984).
[130] Takeuchi, H., and Mastubara, E., *J. Chem. Soc., Perkin, Trans. 1*, **1984**, 981.
[131] Atkinson, R. S., Fawcett, J., Gawad, N. A., Russell, D. R., and Sherry, L. J. S., *J. Chem. Soc., Chem. Commun.*, **1984**, 1072.
[132] Boehland, T., Temps, F., and Wagner, H. G., *Ber. Bunsenges. Phys. Chem.*, **88**, 455 (1984); *Chem. Abs.*, **101**, 72045 (1984).
[133] Boehland, T., *Ber.-Max-Planck-Inst. Stroemungsforsch.*, **1984**, 53; *Chem. Abs.*, **101**, 110319 (1984).
[134] Chapman, O. L., and Hess, T. C., *J. Am. Chem. Soc.*, **106**, 1842 (1984).
[135] Huan, Z., Landgrebe, J. A., and Peterson, K., *J. Org. Chem.*, **48**, 4519 (1983).
[136] Tarhouni, R., Kirschleger, B., and Villieras, J., *J. Organomet. Chem.*, **272**, C1 (1984).
[137] Potts, K. T., and Murphy, P., *J. Chem. Soc., Chem. Commun.*, **1984**, 1348.
[138] Sitzmann, E. V., Langan, J., and Eisenthal, K. B., *Chem. Phys. Lett.*, **102**, 446 (1983); *Chem. Abs.*, **100**, 138310 (1984).
[139] McPherson, D. W., McKee, M. L., and Shevlin, P. B., *J. Am. Chem. Soc.*, **106**, 2712 (1984).
[140] Castedo, L., Castro, J. L., and Riguera, R., *Tetrahedron Lett.*, **25**, 1205 (1984).
[141] Taylor, E. C., and Davies, H. M. L., *J. Org. Chem.*, **49**, 113 (1984).
[142] Bartnik, R., and Mloston, G., *Tetrahedron*, **40**, 2569 (1984).
[143] Keus, D., Kaminski, M., and Warkentin, J., *J. Org. Chem.*, **49**, 343 (1984).
[144] Krageloh, K., Anderson, G. H., and Stang, P. J., *J. Am. Chem. Soc.*, **106**, 6015 (1984).

[145] Boche, G., and Wagner, H.-U., *J. Chem. Soc., Chem. Commun.*, **1984,** 1591.
[146] Patel, D. I., and Smalley, R. K., *J. Chem. Soc., Perkin Trans. 1*, **1984,** 2587.
[147] Benati, L., Montevecchi, P. C., and Spagnolo, P., *J. Chem. Soc., Perkin Trans. 1*, **1984,** 625.
[148] Meth-Cohn, O., and van Vuuren, G., *J. Chem. Soc., Chem. Commun.*, **1984,** 190.
[149] Moody, C. J., Rees, C. W., and Tsoi, S. C., *J. Chem. Soc., Perkin Trans. 1*, **1984,** 915.
[150] West, R., *Pure Appl. Chem.*, **56,** 163 (1984).
[151] Sawrey, B. A., O'Neal, H. E., Ring, M. A., and Coffey, D., *Int. J. Chem. Kinet.*, **16,** 7 (1984).
[152] Sawrey, B. A., O'Neal, H. E., and Ring, M. A., *Int. J. Chem. Kinet.*, **16,** 23 (1984).
[153] Sawrey, B. A., O'Neal, H. E., Ring, M. A., and Coffey, D., *Int. J. Chem. Kinet.*, **16,** 31 (1984).
[154] Rickborn, S. F., Ring, M. A., O'Neal, H. E., and Coffey, D., *Int. J. Chem. Kinet.*, **16,** 289 (1984).
[155] Rickborn, S. F., Ring, M. A., and O'Neal, H. E., *Int. J. Chem. Kinet.*, **16,** 1371 (1984).
[156] Gaspar, P. P., Konieczny, S., and Mo, S. H., *J. Am. Chem. Soc.*, **106,** 424 (1984).
[157] Masamune, S., Murakami, S., Tobita, H., and Williams, D. J., *J. Am. Chem. Soc.*, **105,** 7776 (1983).
[158] Raghavachari, K., Chandrasekhar, J., Gordon, M. S., and Dykema, K. J., *J. Am. Chem. Soc.*, **106,** 5853 (1984).
[159] Hawari, J. A., and Griller, D., *J. Chem. Soc., Chem. Commun.*, **1984,** 1160.
[160] Köcher, J., and Neumann, W. P., *J. Am. Chem. Soc.*, **106,** 3861 (1984).
[161] Dötz, K. H., *Angew. Chem. Int. Ed.*, **23,** 587 (1984).
[162] Wulff, W. D., and Yang, D. C., *J. Am. Chem. Soc.*, **105,** 6726 (1983).
[163] Wulff, W. D., Chan, K.-S., and Tang, P.-C., *J. Org. Chem.*, **49,** 2293 (1984).
[164] Tang, P.-C., and Wulff, W. D., *J. Am. Chem. Soc.*, **106,** 1132 (1984).
[165] Hegedus, L. S., *Pure Appl. Chem.*, **55,** 1745 (1983).
[166] Nome, F., Rezende, C., and deSouza, N. S., *J. Org. Chem.*, **48,** 5357 (1983).
[167] Nitta, M., and Kobayashi, T., *Chem. Lett.*, **1983,** 1715.

Organic Reaction Mechanisms 1984
Edited by A. C. Knipe and W. E. Watts

CHAPTER 7

Nucleophilic Aromatic Substitution

MICHAEL R. CRAMPTON

Department of Chemistry, Durham University

General

The decomposition of benzenediazonium ions in methanol may occur by competing ionic or free-radical pathways. The volume of activation for the homolytic process has a value of 5.4 $cm^3\ mol^{-1}$ which is much lower than that for the heterolytic process.[1] An ionic mechanism is likely for the reaction of 2,6-dimethylbenzenediazonium ions with tropone which yields an aryloxytropylium salt.[2] Phase-transfer catalysis has been reported in the reaction of anthraquinonediazonium ions with benzene,[3] and there is spectroscopic evidence for intramolecular complexing of a diazonium group with a 21-crown-7 macro-ring in a biphenyl derivative.[4] Evidence has been presented[5] for rate-determining electron transfer in the Sandmeyer reaction—the chlorodediazoniation of arenediazonium ions in the presence of reducing agents such as Sn^{2+}, Fe^{2+} or Cu^{+}. Radical chain processes are likely in the reaction of arenediazonium ions with acetylacetone induced by reducing metal species,[6] and in the reduction of arenediazonium ions with *N*-benzyl-1,4-dihydronicotinamide, a NAD(P)H model.[7]

Interest in aromatic substitution by the radical chain $S_{RN}1$ mechanism continues to expand. Reactions by this mechanism have been the subject of an excellent monograph[8] and of a review.[9] Competition experiments in liquid ammonia have been used to determine the relative rates of addition of nucleophiles to aryl radicals produced in $S_{RN}1$ reactions. In group VA the reactivity order is $NH_2^- < Ph_2P^- \sim Ph_2As^-$ and it is likely that the latter two nucleophiles react at the diffusion-controlled rate.[10] Similar experiments with Group VIA give an order PhO^- (0.0) PhS^- (1.0), $PhSe^-$ (5.8), $PhTe^-$ (28) which increases with the "softness" of the nucleophile.[11] The formation of the reduction product, benzene, in the irradiated reaction of benzenesilenate ions with halobenzenes has been shown[12] to result from photostimulated electron transfer from the nucleophile, $PhSe^-$, to the reaction product, Ph_2Se. Competition experiments of a different type using hetaryl radicals have been used[13] to compare the relative rates of reaction by addition, to benzenethiolate ions, or proton abstraction, from methanol or methoxide ions; 3-pyridyl, 4-isoquinolyl, and 3-quinolyl radicals have been used and in each case the thiolate ion is more reactive (2–5 times) than methoxide ion which is more reactive (*ca.* 39 times) than is methanol. It has been shown[14] that the photostimulated reactions of *o*-bis(phenylsulphonyl)benzene derivatives with arenethiolate ions in dimethyl sulphoxide may yield reduction products or addition products from a common radical intermediate. Further work has been reported[15,16] on the fragmentation of radical anions $[ArSR]^{\cdot-}$ formed in the $S_{RN}1$ reactions of substituted aryl halides with sulphur nucleophiles; with heterocyclic substrates carrying electron-withdrawing groups the S_NAr mechanism may supplant the $S_{RN}1$ pathway. Other studies have shown that the regiospecificity of the phenylation of indenyl anions can be explained by attack at the most basic site,[17] and that the reaction between *o*-halogenobenzylamines with ketone enolate ions yields the isoquinoline ring system.[18]

The reactions of ketone enolate ions and of diethyl phosphite ions with aryl halides are catalysed by iron(II) salts, apparently *via* the $S_{RN}1$ mechanism.[19] It is possible that iron catalysis may be an important new method of initiating $S_{RN}1$ reactions. There have been two reports of the use of redox catalysts in electrochemically initiated $S_{RN}1$ reactions. Thus benzonitrile has been used as redox catalyst in the reaction of bromobenzene with benzenethiolate ions,[16] and 4-cyanopyridine in the reaction of 2-chlorobenzonitrile with benzenethiolate.[20] Further work[21] on the oxidatively catalysed $S_{ON}2$ mechanism has shown that 1- and 2-fluoronaphthalene form 1- and 2-acetoxynaphthalene with acetate ions in the presence of chemical oxidants such as benzoyl peroxide or copper(III), while electrochemical oxidation results mainly in the substitution of hydrogen.

There has been a general review of the participation of radicals in the reactions of aromatic compounds with nucleophiles,[22] and a specific one on the reactions of aryl halides with superoxide ions.[23] The formation of *o*-nitrophenol from *o*-dinitrobenzene and hydroxide ions is accelerated by the presence of hydrogen peroxide; a possible mechanism involves electron transfer from the hydroperoxide anion to the substrate to give the *o*-dinitrobenzene radical anion as an intermediate.[24]

The S_NAr Mechanism

Reactions of nucleophiles in the gas phase have been reviewed.[25] Use of ^{18}O- and ^2H-labelling has shown that the reactions of anions with alkyl phenyl ethers may proceed either by *ipso*-substitution or by S_N2 substitution and are accompanied by extensive inter- and intra-molecular hydrogen exchange.[26] The main reaction pathways in the gas-phase reactions between pentafluorophenyl ethers and anions are formation of the $C_6F_5O^-$ ion and S_NAr attack at fluorine-substituted ring atoms. The latter leads to the formation of an F^- ion–molecule complex whose lifetime is long enough to allow secondary reactions to occur.[27]

There has been a theoretical discussion of aromatic substitution based on ionic π-electron structure.[28] *Ab initio* MO calculations indicate the importance of σ^*–π^* orbital mixing for nucleophilic displacements; when the substrate has low-lying σ^*-orbitals, as in 4-nitrochlorobenzene, then mixing with π^*-orbitals may occur facilitating the concerted displacement which has been postulated.[29] Calculations by the CNDO, INDO, and MNDO semi-empirical methods indicate the stronger interaction of amide ion than of ammonia with halonitrobenzenes,[30] and in the case of 2-nitrofluorobenzene suggest the intermediacy of hydrogen-bonded structure (**1**). A further study of photostimulated reactions with amine nucleophiles indicates that the *meta*-directing properties usually associated with the nitro group are limited to small amines while with large, and thus softer, nucleophiles *para*-orientation may be observed; calculations here indicate a change from charge-controlled to frontier-orbital-controlled reaction.[31]

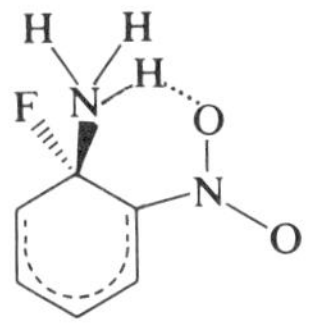

H H H F N O N O

(**1**)

A study has been reported[32] of the variation with solvent composition of the solvent isotope effect $k(CH_3OD)/k(CH_3OH)$ for the methoxydechlorination of 1-chloro-2,4-dinitrobenzene; the data suggest a transition state involving methoxide-promoted attack of a solvating methanol molecule.

Base catalysis in substitution reactions may be observed (Scheme 1) when $k_2 + k_3[B] \leqslant k_{-1}$. There is increasing evidence that the mechanism of the base-catalysed step is solvent-dependent. Thus there have been two independent reports of reactions occurring in protic solvents by rate-limiting deprotonation of the zwitterionic intermediate followed by spontaneous leaving-group expulsion. One of these involves the reaction of N-(2,4-dinitrophenyl)imidazole with piperidine *via* the intermediate (**2**) and provides further evidence for steric hindrance associated with this nucleophile.[33] The other involves the intramolecular reactions (Smiles rearrangements) of 2-(p-nitrophenoxy)ethylamine[34] and its N-alkyl-substituted derivatives[35] *via* the intermediates (**3**). In the intramolecular reaction of N-(2-(p-

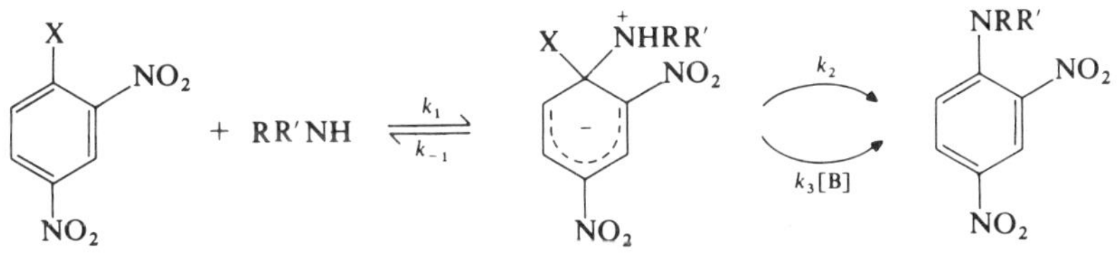

X NO₂ NO₂ + RR′NH $\underset{k_{-1}}{\overset{k_1}{\rightleftharpoons}}$ X $\overset{+}{N}$HRR′ NO₂ NO₂ k_2 $k_3[B]$ NRR′ NO₂ NO₂

SCHEME 1

nitrophenoxy)ethyl)ethylenediamine[36] in water the catalytic effect of the neighbouring amino group is dominant in promoting rate-determining deprotonation of the spiro-intermediate (**4**). In contrast, substitutions in the dipolar aprotic solvent

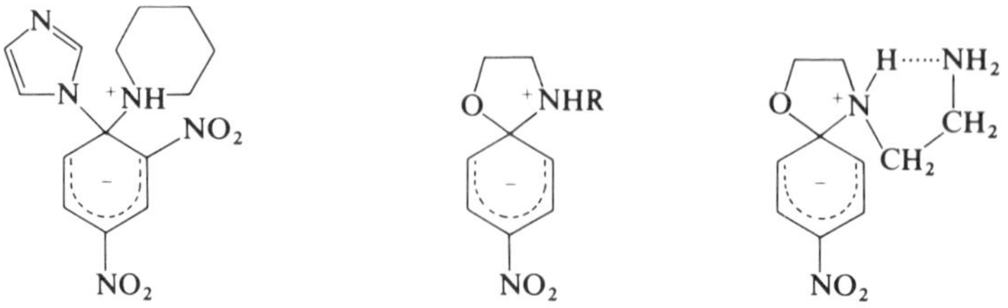

(**2**) (**3**) (**4**)

dimethyl sulphoxide (DMSO) have been reported in which the zwitterionic intermediate and its conjugate base are in rapid equilibrium and the rate-determining step is general-acid-catalysed departure of the leaving group. These studies involve the reaction of methyl 4-methoxy-3,5-dinitrobenzoate with *n*-butylamine where (**5**) is the observable anionic intermediate,[37] and the reactions of 2,4,6-trinitrophenetole with aliphatic amines where adducts (**6**; NRR′ = NHBu, $NHCH_2Ph$, NC_5H_{10}) are observed.[38] In the latter reaction there is evidence for severe steric crowding at the reaction centre when piperidine is the nucleophile and this results in large reductions in both the rate of proton transfer from zwitterionic intermediate to amine catalyst and the rate of leaving-group expulsion. Specific effects of chloride ions on σ-adduct formation in DMSO have been attributed[38,166,168] to association of the type ($RR'NH_2^+$ ---Cl^-). Halide ions have also been shown to influence the course of substitutions involving aniline and its *N*-methyl derivative in acetonitrile. Thus the reaction of 1-chloro-2,4-dinitrobenzene with aniline, where the first step (k_1 in Scheme 1) is rate-determining, is catalysed by halide due to hydrogen bonding in the transition state (**7**) leading to

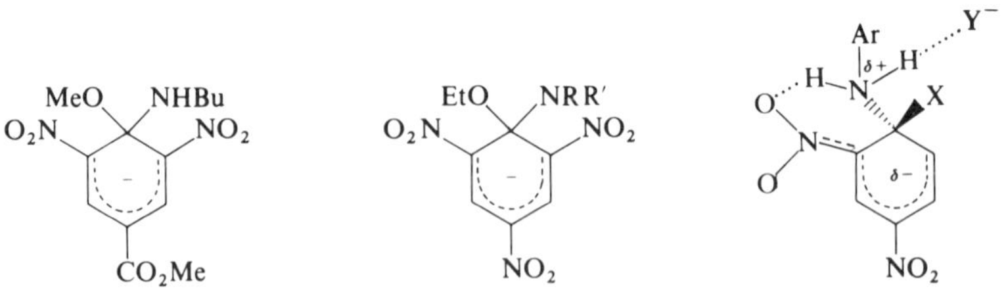

(**5**) (**6**) (**7**)

the zwitterionic adduct.[39] The absence of an unbound hydrogen prevents such catalysis in the analogous reaction with *N*-methylaniline. However, the observed chloride ion catalysis of the reaction of 1-fluoro-2,4-dinitrobenzene with *N*-methylaniline is believed to be due to base-catalysed decomposition of the zwitterionic intermediate.[40]

Solvent effects have been reported[41] for the reaction of 1-chloro-2,4-dinitrobenzene with piperidine in aprotic solvents where the initial nucleophilic attack is rate-limiting. It is known that in non-polar solvents substitutions involving amines often show a high kinetic dependence on the concentration of nucleophile. Further evidence that dimers of the amine are the main nucleophilic reagent has been reported for the reaction of 2,6-dinitroanisole with cyclohexylamine.[42] An alternative mechanism[43] invoked for the reaction of 1-fluoro-2,4-dinitrobenzene with aromatic amines proposes that molecular complexes formed from substrate and nucleophile[44] undergo nucleophilic substitution.

The synthesis of compounds carrying the strongly electron-withdrawing substituents $C_6H_5I^+$ and $C_6F_5I^+$ has been reported. The 2-methoxy-3,5-dinitrophenyl(phenyl)iodonium cation (**8**) is solvolysed by water or alcohols at room temperature to yield the zwitterion (**9**) which reacts readily with nucleophiles.[45] An

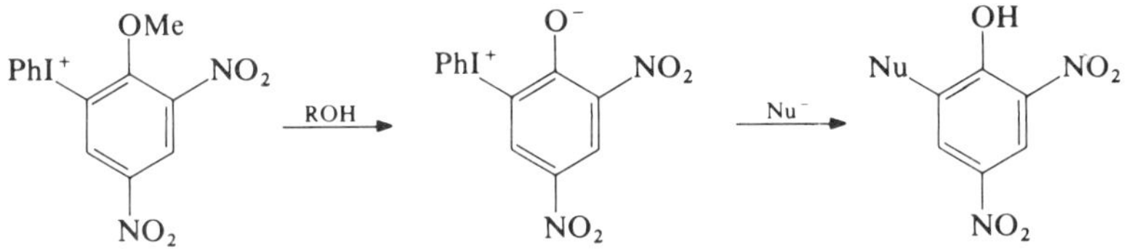

(**8**) (**9**)

unusual substitution reaction has been observed between 8,11-dichloro [5]metacyclophane and *tert*-butyllithium;[46] reaction at the hindered 11-position is attributed to bending of the "bow" of the aromatic plane leading to some pyramidalization at C(11) and an addition–elimination mechanism *via* (**10**) is likely. There have been reports of the activating effects of a *para*-immonium group ($-CH{=}\overset{+}{N}Me_2$) in substitutions of alkoxy and halogeno groups by amines,[47] and of the reactions of anhydrides and imides of 4-substituted naphthalic acid with dialkylaminopropionitriles to give 4-dialkylamino-substituted products.[48]

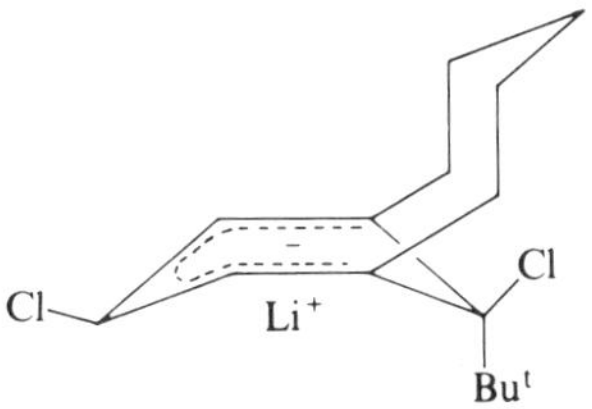

(**10**)

In dimethylformamide unactivated aryl halides react with lithium methyl selenide to give aryl methyl selenides; dealkylation of the products yields the synthetically useful aryl selenide ions.[49] Studies of the biological transformation of pentachloronitrobenzene into pentachloro(methylthio)benzene by *Tetrahymena thermophila* show that pentachlorobenzenethiol is an intermediate.[50] The use of the phase-transfer catalysis has been demonstrated in the reactions of octafluorotoluene and pentafluoropyridine with phenolic and alcoholic functions in steroids; the reversibility of the reactions allows use of these perfluoro compounds as protecting groups.[51] Phase-transfer catalysis has also been used in the regioselective replacement of chloride by alkyl thiolate ions in di-, tri-, and tetra-chlorobenzenes,[52] and in the reactions of aryl halides with phenoxides and alkoxides.[53] It has been shown that *N*-alkyl-4-(*N*′,*N*′-dialkylamino)pyridinium salts are useful phase-transfer catalysts for displacements by phenoxide ions.[54]

There have been several reports of micellar catalysis. The reactions of 1-chloro-2,4-dinitro-benzene and -naphthalene with hydroxide in the presence of the functional surfactant hexadecyl(2-hydroxyethyl)dimethylammonium hydroxide involve the initial formation of ethers and their subsequent reaction with micellar bound hydroxide ions.[55] The pseudo-phase model of micellar kinetics has been successfully applied to reactions of chlorodinitroarenes with hydroxide ions in the presence of didodecyldimethylammonium hydroxide, a twin-tailed surfactant,[56] and with azide ions in micelles of CTAB.[57] Kinetic studies[58, 59] of the effects of the macrocyclic ammonium salt (**11**; X = —$(CH_2)_8$—) on substitutions by anionic nucleophiles indicate that rate accelerations are caused by desolvation of the nucleophiles on their inclusion in the molecular cavity.

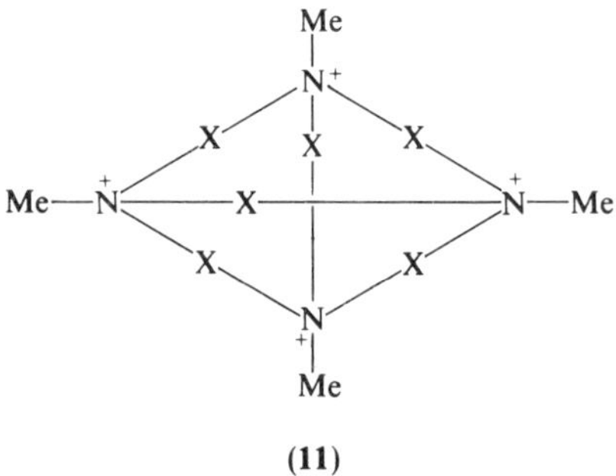

(**11**)

The use of ^{18}O-labelled base indicates that the hydroxylation of 4-substituted nitrobenzenes in liquid ammonia containing hydroxide and oxygen may involve S_NAr reaction.[60] The reactions of the isomeric fluoronitrobenzenes with alkoxide ions in liquid ammonia are rapid.[61] Other studies have been reported of solvent effects on the rate of reaction of 1-fluoro-2,4-dinitrobenzene with the ethyl ester of tyrosine,[62] and of the effects of the cation and of solvent on the reactions of picryl derivatives with the ethyl acetoacetate enolate ion which yield *C*- or *O*-substituted products.[63] Kinetic studies of the hydrolysis of aminobenzenesulphonic acids in aqueous hydrochloric acid indicate that reaction involves the protonated substrates.[64]

Detailed kinetic studies have been reported[34–36] of the Smiles rearrangements of 2-(*p*-nitrophenoxy)ethylamine and its *N*-alkyl derivatives *via* the intermediates (**3**). There is evidence for the involvement of the spiro-adduct (**12**) during the Smiles rearrangement of 4-nitrophenyl-*N*-hydroxycarbamate[65] and for the intermediacy of (**13**) during the reactions with base of both 2-(acetylamino)ethyl-2,6-dinitrophenyl ether and *N*-acetyl-*N*-(2-hydroxyethyl)-2,6-dinitroaniline.[66] It has been shown[67] that Smiles rearrangements activated by a chlorine substituent occur during the syntheses of phenoxazines from *o*-aminophenyl phenyl ethers. Coulombic repulsion between the negative centres in (**14**) is thought to be the driving

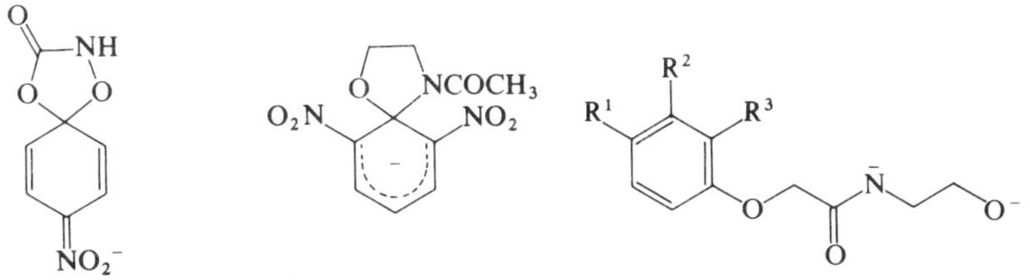

(12) (13) (14)

force for the Smiles rearrangements of *N*-(2-hydroxyethyl)aryloxyacetamides to *N*-(2-hydroxyethyl)anilines.[68] Intramolecular elimination of hydrogen fluoride from methyl and fluorine substituents lying *ortho* to a N=N linkage leads to a synthesis of diarene-1,2-diazepines.[69] An example of intramolecular substitution of an aryl hydrogen by the diazo group occurs in the formation of benzo-1,2-diazepines from 1-aryl-3-diazoalkenes,[70] while the photoinduced intramolecular substitution of ω-anilinoalkyl-*m*-nitrophenyl ethers results in displacement of a hydrogen or methoxyl group situated *para* to the nitro group.[71]

There have been studies of the effects of ring substituents on the vicarious substitution of hydrogen by carbanions containing a potential leaving group (Scheme 2). Reactions of carbanions of α-haloalkyl phenyl sulphones with

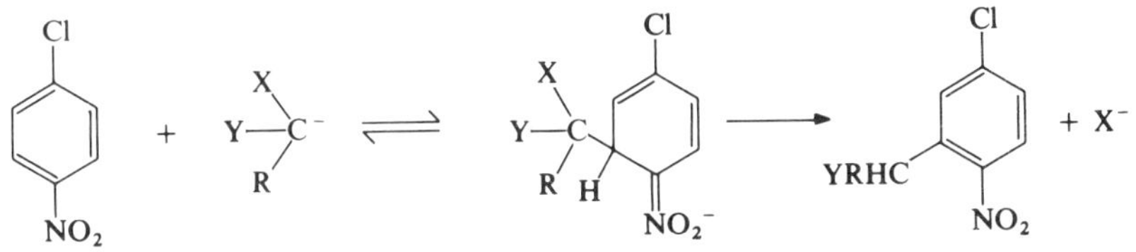

SCHEME 2

substituted nitrobenzenes result in replacement of hydrogen *ortho*- or *para* to the nitro group rather than replacement of good leaving groups such as halogen, methoxy, or phenoxy.[72] Similarly carbanions generated from alkanenitriles bearing α-chloro, α-OR, or α-SR groups and from aliphatic esters bearing α-SR groups react with substituted nitroarenes to preferentially replace hydrogen.[73] These results are in agreement with the general rule[155] that nucleophilic attack is faster at unsubstituted ring positions than at similarly activated, but substituted, ring positions. The specific *ortho*-orientation observed in the vicarious substitution of hydrogen in nitroarenes by the carbanion of chloromethyl phenyl sulphone may be attributed to the ion-pairing expected in this non-polar solvent.[74] Examples of intramolecular vicarious substitutions have been reported in the base-catalysed cyclizations of *N*-methyl-*N*-3-nitrophenyl-(and 3-nitrobenzyl-)chloromethane-sulphonamides.[75] Vicarious substitutions have been used to promote ring-closure in a synthesis of an aklavinone precursor from chrysazin[76] and in a synthesis of substituted indole derivatives from *m*-nitrophenyl isocyanides.[77] There has been a report of the formal displacement of a methoxyl group by a carbaldehyde entity during the Reimer–Tiemann reaction.[78]

Activation of aromatic compounds to nucleophilic substitution may be achieved by π-complexing with transition-metal ligands.[79] Regioselectivity in the addition of carbanions to the 1,4-dimethoxynaphthalene–tricarbonylchromium complex has been examined and a new synthesis of anthracyclinones reported.[80] The catalysis by the (η^5-ethyltetramethylcyclopentadienyl) (η^6-benzene)rhodium(III) cation of the intramolecular cyclization of 3-(2-fluorophenyl)propanols to give chromans is believed to involve π-bonding of the aryl fluoride with the metal cation.[81] The use of organoiron complexes has been reviewed[82] and the regioselectivity of carbanion addition to hydride adducts of $(arene)_2Fe^{2+}$ complexes examined.[83]

There has been a review, bringing out the diversity of possible mechanisms, of copper-assisted substitutions of aryl halogen.[84] The potential for use of Cu(I) catalysis as an alternative to $S_{RN}1$ reaction has been examined and a mechanism involving inner-sphere electron transfer and a Cu(III) intermediate proposed.[85] The Ullman-type reaction of aryl halides with thiophenols has been shown to involve cuprous thiophenolates as intermediates.[86] There have been studies of the copper-promoted intramolecular coupling of aryl rings attached to a chiral 1,1′-binaphthyl residue, which may provide a method of synthesis of chiral biaryls,[87] and of the copper-mediated hydroxylation of an arene as a model system for the action of copper mono-oxygenases.[88]

The formation of arylpalladium intermediates has been postulated in the palladium-catalysed reaction of aryl iodides with substituted acetylenes,[89] and also in the cyclization of 2-substituted halogenoarenes to yield heterocyclic derivatives.[90] Ni(0)–trialkylphosphine complexes have been shown to be effective catalysts for the coupling of aryl halides in the presence of zinc,[91] and nickel catalysts have been used to effect cross-coupling of chlorobenzenes with long-chain *n*-alkylmagnesium bromides.[92]

There has been a survey[93] of the conjugate addition of alkyl Grignard reagents to mononitroarenes leading to ring-alkylated products (Scheme 3); and the reaction of trialkylaluminium reagents with *N*-arylhydroxylamine derivatives, leading to ring-alkylated products, has been applied to the synthesis of indoles.[94]

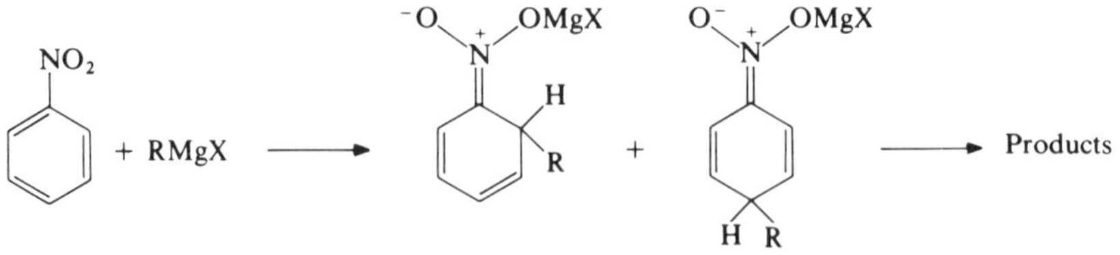

NO_2 + RMgX ⟶ ^-O, $\overset{+}{N}$, OMgX, H, R + O^-, $\overset{+}{N}$, OMgX, H R ⟶ Products

SCHEME 3

Heterocyclic Systems

Substitutions in the pyrrole ring have been compared with those in the benzene ring, and changes in the relative activation of methoxydenitration reactions by *o*-nitro groups and *o*-cyano groups have been explained by the different ring geometries; the bulkier nitro group has a larger steric effect in the more crowded benzene ring.[95]

Changes, in the two ring systems, in the relative rates of methoxide attack at ring carbon atoms carrying nitro groups or halogen atoms may be due to the increased conjugation of the nitro groups with the pyrrole ring leading to greater electron density at the nitro group and hence greater repulsion with the incoming nucleophile.[96] 1-Alkyl-2,4-dinitropyrroles have been found[97] to react with piperidine in dipolar aprotic solvents to give *trans*-pyrrolines (**15**; R = alkyl).

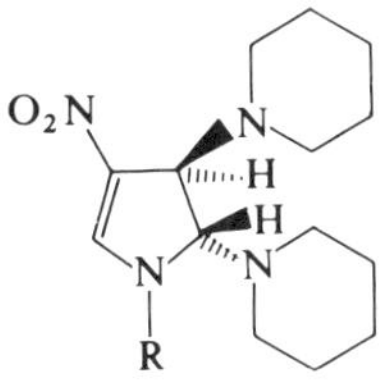

(**15**) 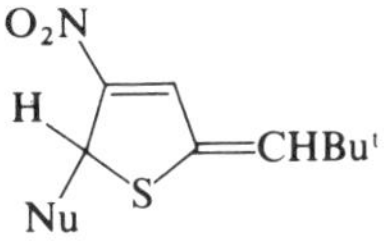 (**16**)

The effects of substituents and solvent on the rates of the chlorine isotope exchange between $Li^{36}Cl$ and 2-chloro-3-nitro-5-X-thiophenes have been determined. The results show that the charge developed in the transition state is strongly affected by the electronic effects of substituents.[98] A further example of the S_N(AEAE) mechanism has been reported[99] in the reaction of (4-nitro-2-thienyl)neopentyl chloride with nucleophiles involving the intermediate (**16**). A kinetic study has been reported of the reactions of 2-methoxy-3-nitro-5-X-thiophenes with piperidine and with *n*-butylamine in benzene; the reactions with piperidine are catalysed by piperidine, a result which was interpreted by the SB-GA mechanism.[100] The kinetics of the reaction of 2,4,6-triphenylpyrilium ion with amines in methanol to give ring-opened products indicate the mechanism of Scheme 4; with primary amines formation of the intermediate (**17**) is rate-determining while with secondary amines the slow step is proton transfer from (**17**) to amine.[101] Base

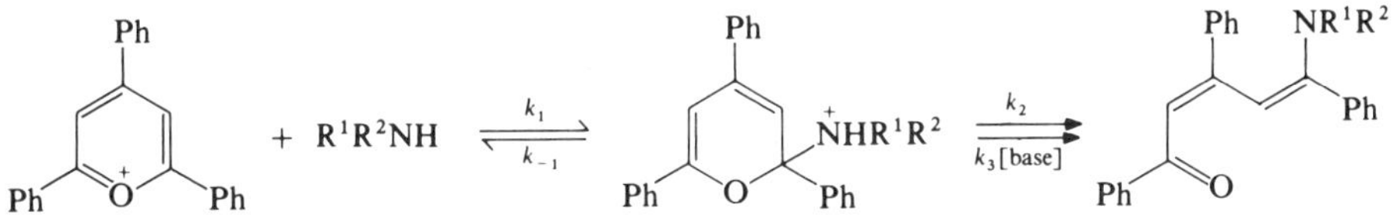

(**17**)

Scheme 4

catalysis has also been observed in the reactions of 1-, 2-, 3-, and 4-nitro-9-chloroacridines with *N,N*-dimethylpropane-1,3-diamine.[102] The steric and electronic effects of the nitro group on these reactions have been discussed;[102] and the effects of ring substituents on the reaction of 9-chloroacridines with sodium methoxide have been determined.[103] Kinetic studies show that the reactions between 5-bromothiazole and benzenethiols in methanol are autocatalytic and indicate that the rate-determining step involves charged nucleophiles.[104] Autocatalysis, involving acidic species, has been observed in the substitution reactions of 3-chloro-4-methyl-6-phenylpyridazine with aliphatic amines.[105]

The facile *ipso*-substitution of sulphinyl or sulphonyl groups at the 2- or 4-positions of pyridine rings has been used in the synthesis of pyridine macrocycles.[106] Reaction in ether of the pyridine derivative (**18**) with organolithium compounds gives predominantly 3,4-disubstituted products; however, with tetramethylethylenediamine as solvent, reaction occurs mainly at the 6-position.[107] The reactions of 1-ethoxycarbonylpyridinium chloride and of 1-*tert*-butyldimethylsilylpyridinium triflate with organometallic reagents are highly regioselective and yield 4-alkyl(aryl)pyridines in good yield.[108] Addition of carbanions to *N*-ethoxycarbonylthiazolium chloride[109] occurs at the 2-position to give thiazolines (19).

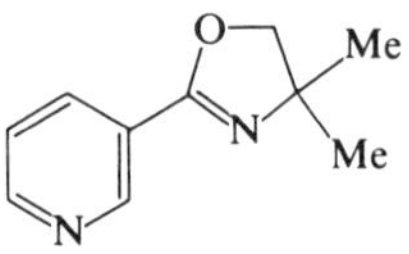

(**18**) 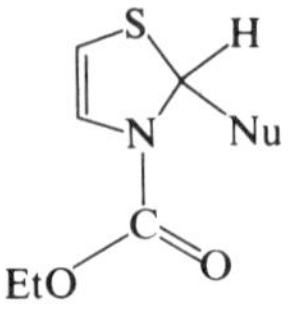 (**19**)

A study of the mechanism of the chlorination of ring-substituted pyrazine-*N*-oxides by phosphoryl chloride indicates that reaction is likely to involve nucleophilic attack by chloride ions on the initially formed dichlorophosphate esters.[110] The high reactivity of the benzofuroxan ring system is shown in a kinetic study of the reaction of 5-chloro-4,6-dinitrobenzofuroxan with aniline.[111] There have also been kinetic studies of the displacement of halogen by ethoxide in methyl-substituted 2-halogenopicoline *N*-oxides,[112] and of the displacement of bromine and nitro groups by phenoxide in dibenzothiophene-5,5-dioxides.[113] Other studies have been reported of the displacement of halogen by 3-amino-2-butenoates in various heterocycles to yield vinylogous amines,[114] and of the displacement of chloride by nucleophiles in 3-alkyl-5-chloroisoxazoles.[115] Charge-transfer complexes of 3,5-dinitropyridine with aromatic donors have been characterized.[116]

^{13}C-NMR spectroscopy has been used to show that nucleophilic substitution occurs preferentially at the 7-position in 5,7-dichlorothieno[3,2-*b*]pyridines,[117] and to examine the regiospecificity of mono-substitution in dibromoimidazo[1,2-*a*]pyrazines.[118] Interesting reactions have been reported of 3-trichloromethylpyridine and its α-chlorinated derivatives with methoxide, where nucleophilic attack occurs both at ring positions and at the trichloromethyl group,[119] and of 1,2,4-triazines with nitronate anions where oximes (**20**; R = H, alkyl) are produced.[120]

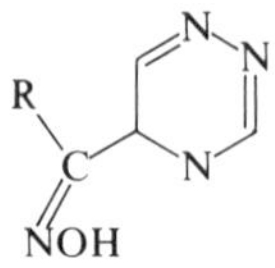

(**20**) 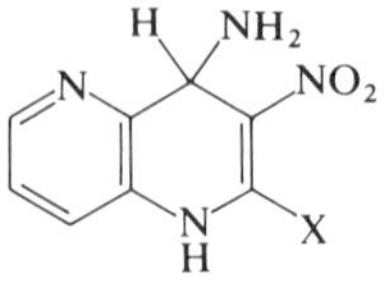 (**21**)

Further kinetic studies involving 2,4,6-trichloro-*s*-triazines have shown that the successive replacement of chlorine by substituted alkoxide ions, $RCH_2CH_2O^-$, becomes progressively more difficult, and is accompanied by hydrolysis of the substrate.[121–124] The reaction with 2,4,6-trichloro-*s*-triazine has been used to examine the effects of substituents, X, in 4′-X-4-aminoazobenzenes.[125]

A review of the reactions of naphthyridines with nitrogen nucleophiles[126] includes Chichibabin aminations (replacement of a ring hydrogen by an amino group). There is ^{1}H-NMR evidence for the formation of the adducts (**21**) when 3-nitro-2-X-1,5-naphthyridines are dissolved in liquid ammonia; oxidation with potassium permanganate yields the 4-amino derivatives.[127] Unless the steric effect of the 1-substituent is important covalent amination of 1-substituted 3-carbamoyl-pyridinium ions occurs at the 6-position[128] to give adducts (**22**), which may be oxidized by potassium permanganate[129] to the 6-imino derivatives (**23**).

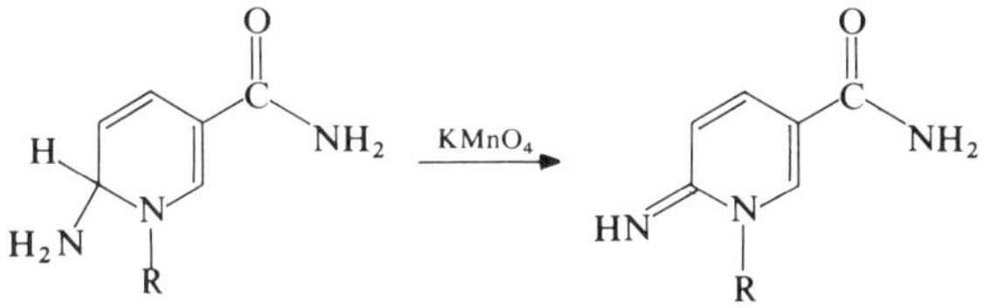

O C NH2 H H2N N R —KMnO4→ O C NH2 HN N R

(**22**) (**23**)

The reaction of 2-chloro-5-nitropyridine with sodium deuterioxide yields the salt of 2-hydroxy-5-nitropyridine. However the isolation of the intermediate (**24**) during this reaction indicates the operation of the S_N(ANRORC) mechanism (addition of nucleophile, ring-opening, ring-closure), with initial base attack occurring at the 6-position.[130] The conversion of 2-phenylamino-1,3,4-oxadiazole into 4-phenyl-1,2,4-triazolin-5-one is likely to involve initial base attack at the 5-position to give (**25**) followed by ring-opening and -reclosure.[131] Carbanion addition at the 2-position of the pyrimidine ring in azolopyrido[2,3-*d*]pyrimidines is followed by ring-opening,[132] and there has been a report of the cyanide-induced transformation of 1,3-oxazines into pyrroles.[133]

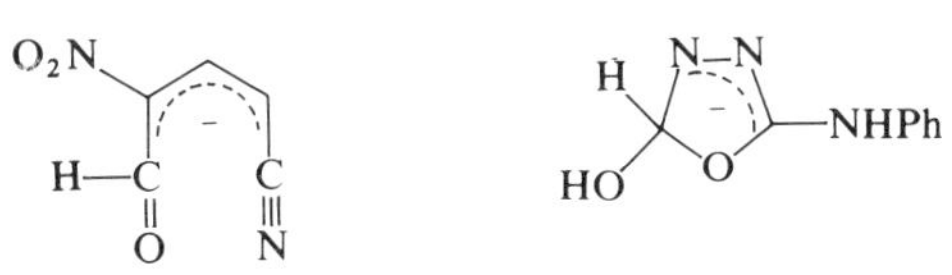

O2N H—C O C N H N–N NHPh HO O

(**24**) (**25**)

Experimental[134] and theoretical[135] studies of the reactions of substituted pyrilium and thiopyrilium cations with the azide anion show that covalent azides are formed only from sterically hindered cations while unhindered cations yield donor–acceptor complexes. The covalent azidopyrans rearrange readily to 1,3-oxazepins while the azidothiopyrans give, on heating, unstable thiazepins which may decompose to pyridines after extrusion of sulphur or to thiophenes after elimination of benzonitrile.[136] The reaction of azide ions with substituted chromylium salts may yield azido-2*H*- and azido-4*H*-chromenes which, on heating, give good yields of benz(*f*)oxazepins.[137]

Kinetic and equilibrium studies have been reported[138] for the reaction of 4-(4-methoxy-3-sulphophenyl)-2,6-bis(4-sulphophenyl)pyrilium perchlorate, a water-soluble pyrilium salt, with water to give the enedione pseudo-base (**26**) and its anion (**27**). Reaction of this pyrilium salt with amines[139] yields the divinylogous amide (**28**) which cyclizes to pyridinium ion (**29**). The use of this, and related water-soluble pyrilium salts, has been reported[140–144] in the transformations of natural products containing amino groups.

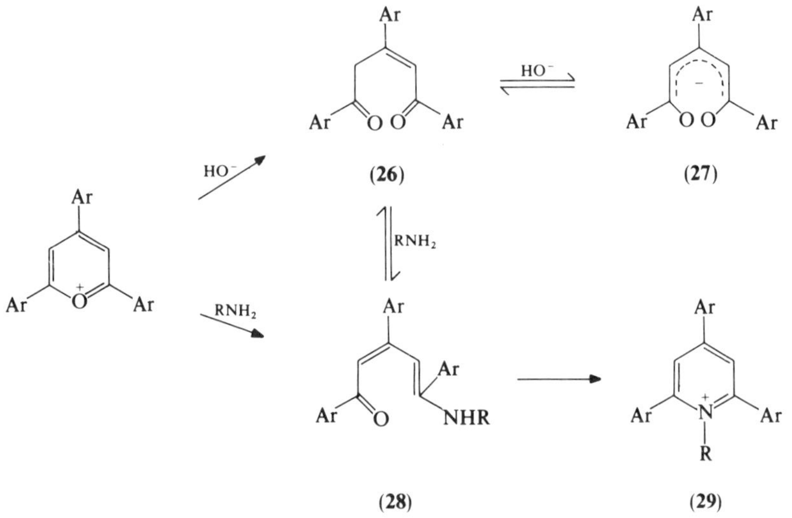

(**26**) (**27**)

(**28**) (**29**)

Studies of pseudo-base formation from 3-substituted 1-methylquinolinium cations show that the C(2) pseudo-bases are kinetically favoured in all cases while the thermodynamic preference for attack at C(2) or C(4) depends on the nature of the 3-substituent.[145] The effects of the substituent on the rates and equilibria of reaction with hydroxide of 9-substituted 10-methylacridinium cations to give the pseudo-bases (**30**) have been examined.[146] The pH dependence of the disproportionation reaction of the 10-methylacridinium ion in basic solution indicates rate-determining reaction between the pseudo-base alkoxide anion and acridinum cation; pathways for the ferricyanide ion oxidation have been discussed.[147] Evidence from ^{13}C-NMR suggests that thiolate adducts from *N*-methylpyridinium salts derived from pyridine-3,5-dicarboxylic acids may exist as ion-pairs (**31**) in polar solvents.[148] Rate and equilibrium data have been reported[149] for the hydration of the 1,3-benzodithiolylium ion to give (**32**), and the addition of nucleophiles at the 3-position of anthranilium salts has been shown to yield adducts which may undergo ring-opening reactions.[150]

There is continuing interest in the mechanisms of hydride-transfer processes involving NADH model compounds. A theoretical study (MNDO) of hydride transfer between 1,4-dihydropyridine and the pyridinium ion indicates that a linear transition state is strongly preferred.[151] The invariance of the value of the primary

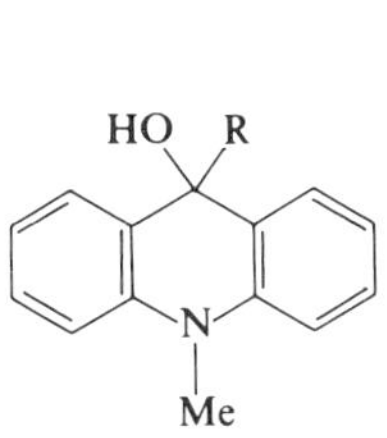

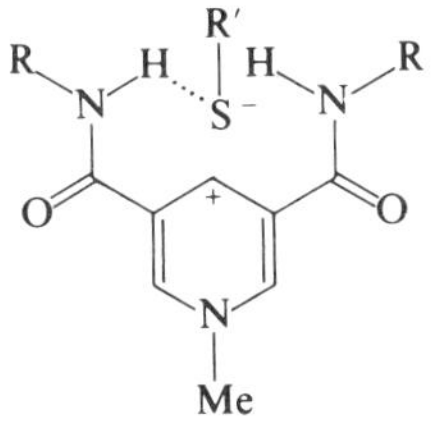

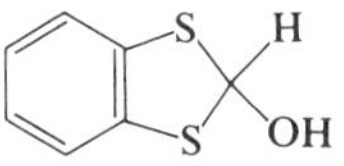

(30) (31) (32)

kinetic isotope effect for hydride trasfers between NAD^+ analogues has been used as evidence that here the mechanism involves a single step.[152] However for the reaction of 1-benzyl-1,4-dihydronicotinamide with a series of *p*-benzoquinone derivatives there is a large variation in the kinetic isotope effect and there is evidence for charge-transfer interaction, suggesting a mechanism involving sequential electron and proton transfers.[153] A kinetic study of hydride transfer involving 2- and 3-methoxy-*N*-methylacridans and various electron acceptors has been used to investigate transition-state structures.[154]

Meisenheimer and Related Adducts

The chemistry and properties of anionic σ-adducts derived from aromatic and heteroaromatic substrates has been comprehensively summarized in a book,[155] and there has been a review of the reactions of aryl trifluoromethyl sulphones with nucleophiles.[156]

Kinetic data for the reactions of ethoxide ions in ethanol with a series of alkyl 2,4,6-trinitrophenyl ethers indicate the importance of steric effects; increasing the size of the alkyl substituent causes decreases in the values of k_3, the rate coefficient, and K_3, the equilibrium constant for reaction at the unsubstituted 3-position, and also in the values of k_1 for reaction at the substituted position.[157] Kinetic and equilibrium data[158] for reaction of alkali-metal methoxides with 2,4-disubstituted 1-methoxy-naphthalenes to give, 1,1-dimethoxy adducts indicate the importance of an *o*-nitro group in the observation of anion–cation association (**33**). A study[159] of the reaction of 1-phenyl-2,4,6-trinitrobenzene with methoxide ions indicates kinetic preference for attack at the unsubstituted 3-position with the 1,1-adduct thermodynamically preferred; for reaction of this substrate with phenoxide ions in DMSO a 1,3-adduct (**34**) is observed in which phenoxide is bound via the *para*-carbon atom.

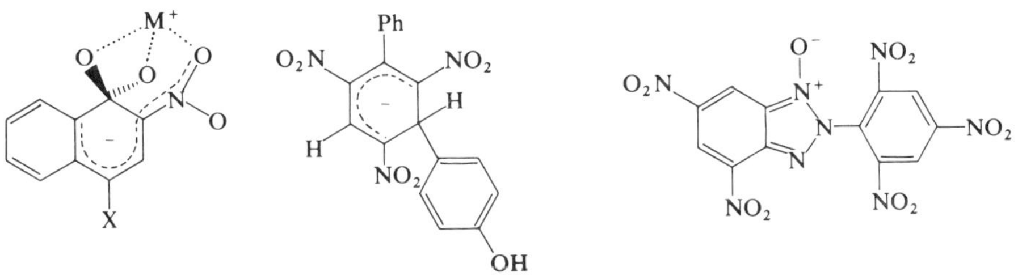

(33) (34) (35)

Ambient behaviour of nucleophiles is also observed in the reaction of 4,6-dinitrobenzofuroxan with aromatic amines[160] where carbon- and nitrogen-bonded adducts have been observed. A further report has appeared[161] of the reaction with nucleophiles of the highly electrophilic reagent (**35**), 4,6-dinitro-2-(2,4,6-trinitrophenyl) benzotriazole-1-oxide; attack may occur at the benzotriazole ring to give σ-adducts or at the picryl ring to give nucleophilic displacement.

Formation of the adduct (**36**) from 1,3,5-trinitrobenzene (TNB) and liberated trichloromethyl anions has been used to monitor the rate of decarboxylation of trichloroacetic acid in DMSO,[162] and competition for trihalogenomethyl anions between TNB and a series of aldehydes has been used in comparisons of reactivity.[163] In a related study[164] equilibrium constants for cyanide addition to TNB and a series of substituted benzaldehydes in DMSO have been reported.

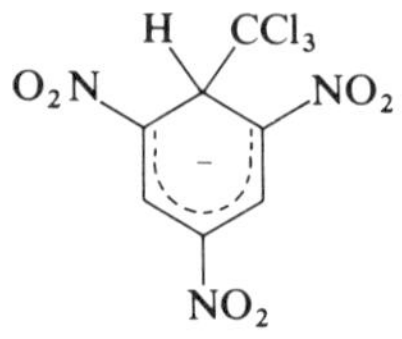

(**36**) 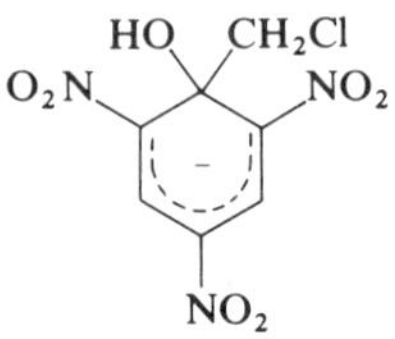 (**37**)

Kinetic and equilibrium data have been reported[165] for hydroxide attack at the 1-position of 2,4,6-trinitrobenzyl chloride (TNBCl) to give (**37**). Three processes observable in the reactions of TNBCl with primary aliphatic amines in DMSO are σ-adduct formation by attack at the 3- or 1-position or transfer of a side-chain proton; with secondary amines σ-adduct formation at the 1-position is disfavoured for steric reasons. Rate studies of these reactions show that in the formation of σ-adducts proton transfers from zwitterionic intermediates to base may be kinetically significant. Reductions of rate constants for proton transfer below the values expected for diffusion-controlled reaction are attributed to steric effects which are increased when reaction involves secondary amines and when the CH_2Cl group is at the reaction site.[166] Similar effects have been observed in the reactions with amines of 2,2′,4,4′,6,6′-hexanitrobibenzyl[167, 168] and of 2,2′,4,4′,6,6′-hexanitrostilbene;[169] the presence of two aromatic rings in these compounds encourages the formation of dianionic species.

Kinetic studies have been reported of substitution reactions involving the σ-adduct intermediates (**5**)[37] and (**6**),[38] and the spiro-intermediate (**13**).[66, 170]

There have been studies of the Janovsky adducts formed from polynitrodiphenyl sulphides with acetone in the presence of potassium hydroxide,[171] and kinetic data have been reported.[172] Carbanion addition is involved here, as it is in the Jaffé reaction of picrate ions with creatinine in alkaline solution. Further kinetic studies have been reported[173] for the latter reaction and ^{1}H- and ^{13}C-NMR have been used to examine the conformation of the adducts formed.[174] There is NMR and crystallographic evidence for the structures (**38**; X = S, NMe) for the adducts formed from 2,4-bis-dimethylamino-thiazoles and -imidazoles with trinitrobenzene.[175]

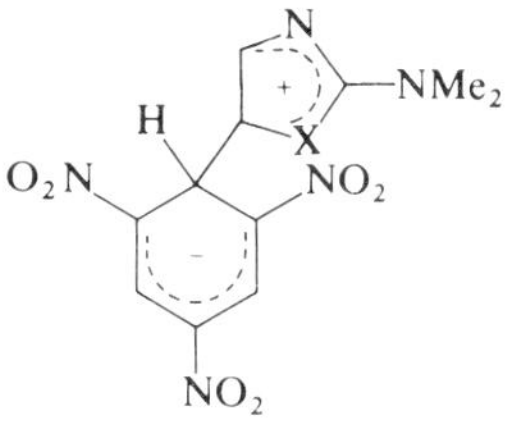

(38) 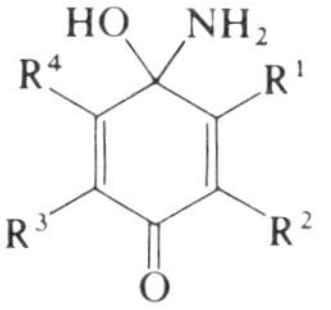 (39) 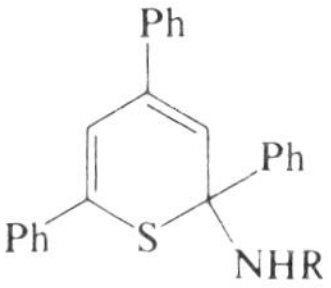 (40)

Kinetic, equilibrium, and ^{13}C-NMR data have been reported for the formation of 2,2-dimethoxy adducts from 2-methoxy-3-nitro-5-substituted-thiophenes and provide further evidence for hyper-*ortho* interaction in the thiophene ring.[176] NMR studies of the reactions of substituted 1,4-benzoquinones with liquid ammonia show that reversible addition occurs at the carbonyl-substituted positions to give 1:1 adducts (**39**) and also 1:2 adducts.[177] A kinetic study[178] of the reactions of 2,4,6-triphenylthiopyrilium cations with amines in DMSO indicates initial attack at the 2- and 4-positions with the 2-adducts (**40**) having the greater thermodynamic stability; the rate-determining step in the formation of these adducts is base addition rather than proton transfer (*cf.* Refs. 166–169). Several other studies[134–150] involve the formation of neutral adducts by reaction of heterocyclic cations with nucleophiles. The strong colour observed in the reactions of transition-metal complexes of 5-nitro-1,10-phenanthroline with amines in DMSO indicate the likely presence of Meisenheimer adducts.[179]

Other studies have involved the reaction of picryl chloride with sodium glycolates,[180] and solvent effects on the reactions of methoxide with nitroanilines[181] and 1,3-dinitrobenzene with amines.[182]

Benzyne and Related Intermediates

There have been reports of the thermal decomposition of polymeric diaryliodonium-2-carboxylates to give mono-substituted *o*-benzynes,[183] of fluoride-ion-induced 1,2-elimination from 2-trimethylsilyl-3-triflylbenzamide to give (**41**) which may be trapped by cycloaddition or nucleophilic addition,[184] and of the preparation of 1,4-dilithiotetrahaloarenes such as (**42**) which eliminate lithium bromide to give arynes.[185] Regiospecific substitutions of 2-bromo-*p*-xylene and 4-bromoveratrole by nucleophiles occur via aryne intermediates.[186].

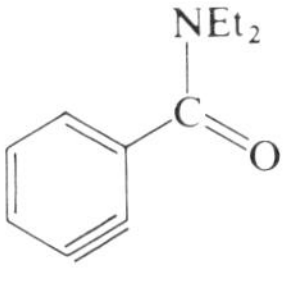

(41) 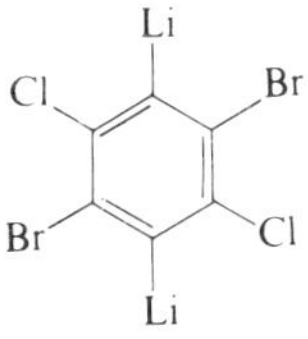 (42)

Examination of [$^{13}C_2$]biphenylene formed by gas-phase pyrolysis of doubly labelled phthalic anhydride or benzocyclobutenedione provides evidence for 1,2- → 1,3-rearrangement of the C_6H_4 carbon skeleton. However, whether this rea: rangement occurs before or after benzyne formation is open to question.[187]

The reaction of benzynes with organic sulphides often results in initial attack at the sulphur atom. Thus the reaction with thiiranes has been used[188] in a stereospecific synthesis of phenyl vinyl sulphides. However reaction of benzynes with 9-*tert*-butylthioanthracene has been shown to involve Diels–Alder addition to give 9-*tert*-butylthiotriptycenes.[189]

The course of the reaction of benzynes derived from aryloxazolines with lithioalkyl nitriles has been shown to involve addition followed by cyclization to benzocyclobutanone imines which on fragmentation may give 3-cyano-2-alkylbenzoic acids and -benzaldehydes.[190] Reports of other synthetically important reactions have included the condensation of benzyne with ketone enolates to give benzocyclobutenols,[191] and the addition of benzyne to benzofuran to give a precursor of benz[*e*]acephenanthrylene.[192]

References

1 Kuokkanen, T., *Finn. Chem. Lett.*, **1984,** 38.
2 Hancock, R. A., and Hill, H., *J. Chem. Res.*, **1984,** (S) 68.
3 Pushkina, L. L., Ponomareva, L. A., Shelyapin, O. P., and Shein, S. M., *Zh. Org. Khim.*, **19,** 1486 (1983); *Chem. Abs.*, **99,** 212258 (1983).
4 Beadle, J. R., and Hill, H., *J. Chem. Res.*, **1984,** (S) 68.
5 Galli, C., *J. Chem. Soc., Perkin Trans. 2*, **1984,** 897.
6 Citterio, A., and Ferrario, F., *J. Chem. Res.*, **1983,** (S) 308, (M) 2656.
7 Yasui, S., Nakamura, K., and Ohno, A., *J. Org. Chem.*, **49,** 878 (1984).
8 Rossi, R. A., and de Rossi, R. H., *Aromatic Substitution by the $S_{RN}1$ Mechanism*, A.C.S. Monograph, 1983.
9 Norris, R. K., *Chemistry of the Halides, Pseudo-halides and Azides*, Vol. 1, Wiley–Interscience, New York–London, 1983, p. 681.
10 Alonso, R. A., Bardón, A., and Rossi, R. A., *J. Org. Chem.*, **49,** 3584 (1984).
11 Pierini, A. B., Peñéñory, A. B., and Rossi, R. A., *J. Org. Chem.*, **49,** 486 (1984).
12 Peñéñory, A. B., Pierini, A. B., and Rossi, R. A., *J. Org. Chem.*, **49,** 3834 (1984).
13 Zoltewicz, J. A., and Locko, G. A., *J. Org. Chem.*, **48,** 4214 (1983).
14 Novi, M., Garbarino, G., and Dell'Erba, C., *J. Org. Chem.*, **49,** 2977 (1984).
15 Beugelmans, R., Bois-Choussy, M., and Boudet, B., *Tetrahedron*, **39,** 4153 (1983).
16 Swartz, J. E., and Stenzel, T. T., *J. Am. Chem. Soc.*, **106,** 2520 (1984).
17 Tolbert, L. M., and Siddiqui, S., *J. Org. Chem.*, **49,** 1744 (1984).
18 Beugelmans, R., Chastanet, J., and Roussi, G., *Tetrahedron*, **40,** 311 (1984).
19 Galli, C., and Bunnett, J. F., *J. Org. Chem.*, **49,** 3041 (1984).
20 Amatore, C., Oturan, M. A., Pinson, J., Savéant, J.-M., and Thiébault, A., *J. Am. Chem. Soc.*, **106,** 6318 (1984).
21 Jönsson, L., and Nistrand, L.-G., *J. Org. Chem.*, **49,** 3340 (1984).
22 Shein, S. M., *Izv. Sib. Otd. Akad. Nauk SSSR, Ser. Khim. Nauk.*, **1983,** 20; *Chem. Abs.*, **99,** 174913 (1983).
23 Gareil, M., Pinson, J., Savéant, J.-M., *Electr. Fr., Bull. Dir. Etud. et Rech., Ser. A: Nucl., Hydraul., Therm.*, **1983,** 79; *Chem. Abs.*, **101,** 22627 (1984).
24 Heller, R. A., Fewster, M., and Lambert, T., *Can. J. Chem.*, **61,** 2455 (1983).
25 Bowie, J. H., *Mass. Spectrom. Rev.*, **3,** 1 (1984).
26 Kleingeld, J. C., and Nibbering, N. M. M., *Tetrahedron*, **39,** 4193 (1983).
27 Ingemann, S., and Nibbering, N. M. M., *Nouv. J. Chim.*, **8,** 299 (1984).
28 Van Hooydonk, G., and De Keukeleire, D., *Bull. Soc. Chim. Belg.*, **92,** 673 (1983).
29 Yamabe, S., Minato, T., and Kawakata, Y., *Can. J. Chem.*, **62,** 235 (1984).

[30] Nudelman, N. S., and MacCormack, P., *Tetrahedron*, **40,** 4227 (1984).
[31] Cervello, J., Figueredo, M., Moreno-Mañas, M., Bertran, J., and Lluch, J. M., *Tetrahedron Lett.*, **25,** 4147 (1984).
[32] Gopalakrishnan, G., and Hogg, J. L., *J. Org. Chem.*, **49,** 1191 (1984).
[33] de Vargas, E. B., and de Rossi, R. H., *J. Org. Chem.*, **49,** 3978 (1984).
[34] Knipe, A. C., Lound-Keast, J., and Sridhar, N., *J. Chem. Soc., Perkin Trans. 2,* **1984,** 1885.
[35] Knipe, A. C., Sridhar, N., and Lound-Keast, J., *J. Chem. Soc., Perkin Trans. 2,* **1984,** 1893.
[36] Knipe, A. C., Sridhar, N., and Lound-Keast, J., *J. Chem. Soc., Perkin Trans. 2,* **1984,** 1901.
[37] Hasegawa, Y., *J. Chem. Soc., Perkin Trans. 2,* **1984,** 547.
[38] Crampton, M. R., and Routledge, P. J., *J. Chem. Soc., Perkin Trans. 2,* **1984,** 573.
[39] Hirst, J., and Onyido, I., *J. Chem. Soc., Perkin Trans. 2,* **1984,** 711.
[40] Bamkole, T. O., Hirst, J., and Hussain, G., *J. Chem. Soc., Perkin Trans. 2,* **1984,** 681.
[41] Mancini, P. M. E., Martinez, R. D., Vottero, L. R., and Nudelman, N. S., *J. Chem. Soc., Perkin Trans. 2,* **1984,** 1133.
[42] Nudelman, N. S., and Palleros, D., *J. Chem. Soc., Perkin Trans. 2,* **1984,** 246.
[43] Forlani, L., *J. Chem. Res.*, **1984,** (S) 260, (M) 2379.
[44] Golle, N. N., *Fiz.-Khim. Metody Issled i Analiya, Tyumen,* **1982,** 107; *Chem. Abs.*, **100,** 120432 (1984).
[45] Spyroudis, S., and Varvoglis, A., *J. Chem. Soc., Perkin Trans. 1,* **1984,** 135.
[46] Jenneskens, L. W., Klamer, J. C., de Wolf, W. H., and Bickelhaupt, F., *J. Chem. Soc., Chem. Commun.*, **1984,** 733.
[47] Yudin, L. G., Blokhin, A. V., Bundel, Yu. G., Simkin, B. Ya., and Terenin, V. I., *Zh. Org. Khim.*, **19,** 2361 (1983); *Chem. Abs.*, **100,** 102468 (1984).
[48] Plakidin, V. L., and Vostrova, V. N., *Zh. Org. Khim.*, **19,** 2591 (1983); *Chem. Abs.*, **100,** 138317 (1984).
[49] Tiecco, M., Testaferri, L., Tingoli, M., Chianelli, D., and Montanucci, M., *J. Org. Chem.*, **48,** 4289 (1983).
[50] Fall, R., and Murphy, S. E., *J. Am. Chem. Soc.*, **106,** 3033 (1984).
[51] Jarman, M., and McCague, R., *J. Chem. Soc., Chem. Commun.*, **1984,** 125.
[52] Brunelle, D. J., *J. Org. Chem.*, **49,** 1309 (1984).
[53] Cho, B. R., and Park, S. D., *Bull. Korean Chem. Soc.*, **5,** 126 (1984); *Chem. Abs.*, **101,** 90061 (1984).
[54] Brunelle, D. J., and Singleton, D. A., *Tetrahedron Lett.*, **25,** 3383 (1984).
[55] Bunton, C. A., Gan, L. H., and Savelli, G., *J. Phys. Chem.*, **87,** 5491 (1983).
[56] Cipiciani, A., Germani, R., Savelli, G., and Bunton, C. A., *Tetrahedron Lett.*, **25,** 3765 (1984).
[57] Rodenas, C. E., *An. Quim., Ser. A,* **79,** 638 (1983); *Chem. Abs.*, **101,** 22721 (1984).
[58] Schmidtchen, F. P., *Chem. Ber.*, **117,** 725 (1984).
[59] Schmidtchen, F. P., *Chem. Ber.*, **117,** 1287 (1984).
[60] Molykhin, E. V., and Shteingarts, V. D., *Zh. Org. Khim.*, **19,** 2211 (1983); *Chem. Abs.*, **100,** 50797 (1983).
[61] Kizner, J. A., and Shteingarts, V. D., *Zh. Org. Khim.*, **20,** 1089 (1984); *Chem. Abs.*, **101,** 71949 (1984).
[62] Zhu, Z., *Kexue Tongbao (Foreign Lang Ed.),* **29,** 567 (1984); *Chem. Abs.*, **101,** 90041 (1984).
[63] Kurts, A. L., Davydov, D. V., and Bundel, Yu G., *Vestn. Mosk. Univ., Ser. 2: Khim.*, **24,** 385 (1983); *Chem. Abs.*, **99,** 174995 (1983).
[64] Khelevin, R. N., *Zh. Obshch. Khim.*, **53,** 2352 (1983); *Chem. Abs.*, **100,** 50925 (1984).
[65] Fitzgerald, L. R., Blakeley, R. L., and Zerner, B., *Chem. Lett.*, **1984,** 29.
[66] Sekiquchi, S., Hirai, M., and Tomoto, N., *J. Org. Chem.*, **49,** 2378 (1984).
[67] Schmidt, D. M., and Bonvicino, G. E., *J. Org. Chem.*, **49,** 1664 (1984).
[68] Baker, W. R., *J. Org. Chem.*, **48,** 5140 (1983).
[69] Alty, A. C., Banks, R. E., Fishwick, B. R., Pritchard, R. G., and Thompson, A. P., *J. Chem. Soc., Chem. Commun.*, **1984,** 832.
[70] Miller, T. K., Sharp, J. T., Sood, H. R., and Stefaniuk, E., *J. Chem. Soc., Perkin Trans. 2,* **1984,** 823.
[71] Mutai, K., Kobayashi, K., and Yokoyama, K., *Tetrahedron*, **40,** 1755 (1984).
[72] Makosza, M., Golinski, J., and Baran, J., *J. Org. Chem.*, **49,** 1488 (1984).
[73] Makosza, M., and Winiarski, J., *J. Org. Chem.*, **49,** 1494 (1984).
[74] Makosza, M., Glinka, T., and Kinowski, A., *Tetrahedron*, **40,** 1863 (1984).
[75] Makosza, M., and Wojciechowski, K., *Tetrahedron Lett.*, **25,** 4791 (1984).
[76] Murphy, R. A., and Cava, M. P., *Tetrahedron Lett.*, **25,** 803 (1984).
[77] Wojciechowski, K., and Makosza, M., *Tetrahedron Lett.*, **25,** 4793 (1984).
[78] Bird, C. W., and Brown, A. L., *Chem. Ind. (London),* **1983,** 827.

[79] Litvak, V. V., and Shteingarts, V. D., *Izv. Sib. Otd. Akad. Nauk SSSR, Ser. Khim. Nauk,* **1983,** 59; *Chem. Abs.,* **99,** 174915 (1983).
[80] Kundig, E. P., Desobry, V., and Simmons, D. P., *J. Am. Chem. Soc.,* **105,** 6962 (1983).
[81] Houghton, R. P., Voyle, M., and Price, R., *J. Chem. Soc., Perkin Trans. 1,* **1984,** 825.
[82] Astruc, D., *Tetrahedron,* **39,** 4027 (1983).
[83] Madonik, A. M., Mandon, D., Michaud, P., Lapinte, C., and Astruc, D., *J. Am. Chem. Soc.,* **106,** 3381 (1984).
[84] Lindley, J., *Tetrahedron,* **40,** 1433 (1984).
[85] Bowman, W. R., Heaney, H., and Smith, P. H. G., *Tetrahedron Lett.,* **25,** 5821 (1984).
[86] Yamamoto, T., and Sekine, Y., *Can. J. Chem.,* **62,** 1544 (1984).
[87] Miyano, S., Handa, S., Shimizu, K., Tagami, K., and Hashimoto, H., *Bull. Chem. Soc. Jpn.,* **57,** 1943 (1984).
[88] Karlin, K. D., Hayes, J. C., Gultneh, Y., Cruse, R. W., McKown, J. W., Hutchinson, J. P., and Zubieta, J., *J. Am. Chem. Soc.,* **106,** 2121 (1984).
[89] Cacchi, S., Felici, M., and Pietroni, B., *Tetrahedron Lett.,* **25,** 3137 (1984).
[90] Ames, D. E., and Opalko, A., *Tetrahedron,* **40,** 1919 (1984).
[91] Takagi, K., Hayama, N., and Sasaki, K., *Bull. Chem. Soc. Jpn.,* **57,** 1887 (1984).
[92] Eapen, K. C., Dua, S. S., and Tamborski, C., *J. Org. Chem.,* **49,** 478 (1984).
[93] Bartoli, G., *Acc. Chem. Res.,* **17,** 109 (1984).
[94] Fujiwara, J., Fukutani, Y., Sano, H., Maruoka, K., and Yamamoto, H., *J. Am. Chem. Soc.,* **105,** 7177 (1983).
[95] Bazzano, F., Mencarelli, P., and Stegel, F., *J. Org. Chem.,* **49,** 2375 (1984).
[96] Annulli, A., Mencarelli, P., and Stegel, F., *J. Org. Chem.,* **49,** 4065 (1984).
[97] Mencarelli, P., and Stegel, F., *J. Chem. Res.,* **1984,** (S) 18.
[98] Attia, M., Davé, D., Gore, P. N., Ikejiani, A. O. O., Morris, D. F. C., Short, E. L., Consiglio, G., Spinelli, D., and Frenna, V., *J. Chem. Soc., Perkin Trans. 2,* **1984,** 1637.
[99] Flower, F. I., Newcombe, P. J., and Norris, R. K., *J. Org. Chem.,* **48,** 4202 (1983).
[100] Consiglio, G., Arnone, C., Spinelli, D., Noto, R., and Frenna, V., *J. Chem. Soc., Perkin Trans. 2,* **1984,** 781.
[101] Doddi, G., Illuminati, G., Mecozzi, M., and Nunziante, P., *J. Org. Chem.,* **48,** 5268 (1983).
[102] Kunikowski, A., and Ledochowski, A., *Pol. J. Chem.,* **57,** 435 (1983); *Chem. Abs.,* **101,** 109974 (1984).
[103] Gridukevich, A. N., Chernykh, V. I., Makurina, V. I., and Kravchenko, A. A., *Org. React. (Tartu),* **20,** 112 (1983).
[104] Forlani, L., *Gazz. Chim. Ital.,* **114,** 279 (1984); *Chem. Abs.,* **101,** 151145 (1984).
[105] Maghioros, G., Schlewer, G., Wermuth, C. G., Lagrange, J., and Lagrange, Ph., *Nouv. J. Chem.,* **7,** 667 (1983).
[106] Furukawa, N., Ogawa, S., Kawai, T., and Oae, S., *J. Chem. Soc., Perkin Trans 1,* **1984,** 1839.
[107] Hauck, A. E., and Giam, C. S., *J. Chem. Soc., Perkin Trans. 1,* **1984,** 2227.
[108] Akiba, K., Iseki, Y., and Wada, M., *Bull. Chem. Soc. Jpn.,* **57,** 1994 (1984).
[109] Dondoni, A., Dall'Occo, T., Fantin, G., Fognagnolo, M., and Medici, A., *Tetrahedron Lett.,* **25,** 3633 (1984).
[110] Sato, N., *J. Chem. Res.,* **1984,** (S) 318, (M) 2860.
[111] Sharnin, G. P., and Mukharlyamov, R. I., *Zh. Org. Khim.,* **19,** 2358 (1983); *Chem. Abs.,* **100,** 138303 (1984).
[112] Puszko, A., *Pr. Nauk. Akad. Ekon. im. Oskara Langego Wroclawia,* **238,** 113 (1983); *Chem. Abs.,* **101,** 109947 (1984).
[113] Kazin, V. N., Plakhtinskii, V. V., Dorogov, M. V., Ustinov, V. A., Mironov, G. S., and Nizhnikova, N. V., *Deposited Doc.* **1982,** SPSTL 1318 Khp-D82; *Chem. Abs.,* **101,** 109966 (1984).
[114] Hartman, G. D., Hartman, R. D., and Cochran, D. W., *J. Org. Chem.,* **48,** 4119 (1983).
[115] Stevens, R. K., and Albizati, K. F., *Tetrahedron Lett.,* **25,** 4587 (1984).
[116] Pietrzycki, W., Tomasik, P., and Koziol, J., *Chem. Scr.,* **22,** 159 (1983).
[117] Barker, J. M., Huddleston, P. R., Holmes, D., Keenan, G. J., and Wright, B., *J. Chem. Res.,* **1984,** (S) 84, (M) 771.
[118] Bonnet, P. A., Sablayrolles, C., and Chapat, J. P., *Aust. J. Chem.,* **37,** 1357 (1984).
[119] Dainter, R. S., Suschitzky, H., Wakefield, B. J., Hughes, N., and Nelson, A. J., *Tetrahedron Lett.,* **25,** 5693 (1984).
[120] Rykowski, A., and Makosza, M., *Tetrahedron Lett.,* **25,** 4795 (1984).
[121] Marchukov, V. A., Nikitin, O. A., V'yunov, K. A., and Ginak, A. I., *Zh. Org. Khim.,* **19,** 2198 (1983); *Chem. Abs.,* **100,** 50796 (1984).

[122] Marchukov, V. A., Nikitin, O. A., V'yunov, K. A., and Ginak, A. I., *Zh. Org. Khim.*, **19,** 2600 (1983); *Chem. Abs.*, **100,** 102502 (1984).
[123] Marchukov, V. A., Nikitin, O. A., V'yunov, K. A., and Ginak, A. I., *Zh. Org. Khim.*, **20,** 585 (1984); *Chem. Abs.*, **101,** 38797 (1984).
[124] Marchukov, V. A., Nikitin, O. A., V'yunov, K. A., and Ginak, A. I., *Zh. Org. Khim.*, **20,** 647 (1984); *Chem. Abs.*, **101,** 37899 (1984).
[125] Havlik, I., and Bacaloglu, R., *J. Prakt. Chem.*, **325,** 936 (1983); *Chem. Abs.*, **100,** 22212 (1984).
[126] van der Plas, H. C., Wozniak, M., and van der Haak, H. J. W., *Adv. Het. Chem.*, **31,** 95 (1983).
[127] Wozniak, M., van der Plas, H. C., Tomula, M., and van Veldhuizen, A., *Recl. Trav. Chim. Pays-Bas*, **102,** 511 (1983).
[128] Angelino, S. A. G. F., van Veldhuizen, A., Buurman, D. J., and van der Plas, H. C., *Tetrahedron*, **40,** 433 (1984).
[129] van der Plas, H. C., and Buurman, D. J., *Tetrahedron Lett.*, **25,** 3763 (1984).
[130] Reinheimer, J. D., Sourbatis, N., Lavallee, R. L., Goodwin, D., and Gould, G. L., *Can. J. Chem.*, **62,** 1120 (1984).
[131] Noto, R., Buccheri, F., Werber, G., Consiglio, G., and Spinelli, D., *J. Chem. Soc., Perkin Trans. 2*, **1984,** 537.
[132] Petrič, A., Stanovnik, B., and Tišler, M., *J. Org. Chem.*, **48,** 4132 (1983).
[133] Yogo, M., Hirota, K., and Maki, Y., *J. Chem. Soc., Chem. Commun.*, **1984,** 332.
[134] Desbene, P. -L., Cherton, J. -C., Le Roux, J. -P., and Basselier, J. -J., *Tetrahedron*, **40,** 3539 (1984).
[135] Desbene, P. -L., Richard, D., Cherton, J. -C., and Chaquin, P., *Tetrahedron*, **40,** 3549 (1984).
[136] Desbene, P. -L., and Cherton, J. -C., *Tetrahedron*, **40,** 3559 (1984).
[137] Desbene, P. -L., and Cherton, J. -C., *Tetrahedron*, **40,** 3567 (1984).
[138] Katritzky, A. R., De Rosa, M., and Grzeskowiak, N. E., *J. Chem. Soc., Perkin Trans. 2*, **1984,** 841.
[139] Katritzky, A. R., Mokrosz, J. L., and De Rosa, M., *J. Chem. Soc., Perkin Trans. 2*, **1984,** 849.
[140] Katritzky, A. R., Yang, Y. -K., Gabrielsen, B., and Marquet, J., *J. Chem. Soc., Perkin Trans. 2*, **1984,** 857.
[141] Katritzky, A. R., and Leahy, D. E., *J. Chem. Soc., Perkin Trans. 2*, **1984,** 867.
[142] Katritzky, A. R., Mokrosz, J. L., and Lopez-Rodriguez, M. L., *J. Chem. Soc., Perkin Trans. 2*, **1984,** 875.
[143] Katritzky, A. R., Yang, Y. -K., Ellison, J., and Marquet, J., *J. Chem. Soc., Perkin Trans. 2*, **1984,** 879.
[144] Katritzky, A. R., and Yang, Y. -K., *J. Chem. Soc., Perkin Trans. 2*, **1984,** 885.
[145] Bunting, J. W., and Fitzgerald, N. P., *Can. J. Chem.*, **62,** 1301 (1984).
[146] Bunting, J. W., Chew, V. S. F., Abhyankar, S. B., and Goda, Y., *Can. J. Chem.*, **62,** 351 (1984).
[147] Bunting, J. W., and Kauffman, G. M., *Can. J. Chem.*, **62,** 729 (1984).
[148] van Keulen, B. J., and Kellogg, R. M., *J. Am. Chem. Soc.*, **106,** 6029 (1984).
[149] Okuyama, T., and Fueno, T., *Bull. Chem. Soc. Jpn.*, **57,** 1128 (1984).
[150] Van der Meer, R. K., and Olofson, R. A., *J. Org. Chem.*, **49,** 3373 (1984).
[151] van der Kerk, S. M., van Gerresheim, N., and Verhoeven, J. W., *Recl. Trav. Chim. Pays-Bas*, **103,** 143 (1984).
[152] Ostović, D., Roberts, R. M. G., and Kreevoy, M. M., *J. Am. Chem. Soc.*, **105,** 7629 (1983).
[153] Fukuzumi, S., Nishizawa, N., and Tanaka, T., *J. Org. Chem.*, **49,** 3571 (1984).
[154] Colter, A. K., Plank, P., Bergsma, J. P., Lahti, R., Quesnel, A. A., and Parsons, A. G., *Can. J. Chem.*, **62,** 1780 (1984).
[155] Buncel, E., Crampton, M. R., Strauss, M. J., and Terrier, F., *Electron Deficient Aromatic – and Heteroaromatic–Base Interactions*, Elsevier, Amsterdam, 1984.
[156] Boiko, V. N., and Yagupol'skii, L. M., *Izv. Sib. Otd. Akad. Nauk SSSR, Ser. Khim. Nauk*, **1983,** 50; *Chem. Abs.*, **99,** 174914 (1983).
[157] Cooney, A., and Crampton, M. R., *J. Chem. Soc., Perkin Trans. 2*, **1984,** 1793.
[158] Sekiguchi, S., Aizawa, T., and Tomoto, N., *J. Org. Chem.*, **49,** 93 (1984).
[159] Buncel, E., Murarka, S. K., and Norris, A. R., *Can. J. Chem.*, **62,** 534 (1984).
[160] Read, R. W., Spear, R. J., and Norris, W. P., *Aust. J. Chem.*, **37,** 985 (1984).
[161] Renfrow, R. A., Strauss, M. J., Cohen, S., and Buncel, E., *Aust. J. Chem.*, **36,** 1843 (1983).
[162] Atkins, P. J., Gold, V., and Marsh, R., *J. Chem. Soc., Perkin Trans. 2*, **1984,** 1239.
[163] Atkins, P. J., Gold, V., and Wassef, W. N., *J. Chem. Soc., Perkin Trans 2.*, **1984,** 1247.
[164] Gold, V., and Wassef, W. N., *J. Chem. Soc., Perkin Trans. 2*, **1984,** 1431.
[165] Crampton, M. R., Routledge, P. J., and Golding, P., *J. Chem. Soc., Perkin Trans. 2*, **1984,** 939.
[166] Crampton, M. R., Routledge, P. J., and Golding, P., *J. Chem. Soc., Perkin Trans. 2*, **1984,** 329.
[167] Crampton, M. R., Routledge, P. J., and Golding, P., *J. Chem. Res.*, **1983,** (S) 314.

[168] Crampton, M. R., Routledge, P. J., and Golding, P., *J. Chem. Soc., Perkin Trans. 2,* **1984,** 1421.
[169] Crampton, M. R., Routledge, P. J., and Golding, P., *J. Chem. Soc., Perkin Trans. 2,* **1984,** 1785.
[170] Sekiguchi, S., Hirai, M., and Tomoto, N., *Bull. Chem. Soc. Jpn.,* **56,** 2752 (1983).
[171] Alekhina, N. N., Gitis, S. S., Grudtsyn, Yu. D., and Kaminskii, A. Ya., *Zh. Org. Khim.,* **20,** 1045 (1984); *Chem. Abs.,* **101,** 110294 (1984).
[172] Alekhina, N. N., Kaminskii, A. Ya., Gitis, S. S., Ivanov, A. V., and Kuznetsov, Yu. M., *Org. React. (Tartu),* **20,** 561 (1983).
[173] Diamandis, E. P., and Hadjiioannou, T. P., *Microchem. J.,* **28,** 399 (1983); *Chem. Abs.,* **99,** 174979 (1983).
[174] Kovar, K. -A., and Rupp, K. -P., *Arch. Pharm. (Weinheim, Ger.),* **317,** 571 (1984).
[175] Gompper, R., Kruck, P., and Schelble, J., *Tetrahedron Lett.,* **24,** 3563 (1983).
[176] Consiglio, G., Spinelli, D., Arnone, C., Sancassan, F., Dell'Erba, C., Noto, R., and Terrier, F., *J. Chem. Soc., Perkin Trans. 2,* **1984,** 317.
[177] Chudek, J. A., Foster, R., and Reid, F. J., *J. Chem. Soc., Perkin Trans. 2,* **1984,** 287.
[178] Doddi, G., and Ercolani, G., *J. Org. Chem.,* **49,** 1806 (1984).
[179] Bartolotta, A., Cusumano, M., Di Marco, G., Giannetto, A., and Guglielmo, G., *Polyhedron,* **3,** 701 (1984).
[180] Mel'nikov, A. I., Gitis, S. S., and Kaminskii, A. Ya., *Zh. Org. Khim.,* **19,** 2217 (1983); *Chem. Abs.,* **100,** 102851 (1984).
[181] Ohsawa, S., and Takeda, M., *Nippon Kagaku Kaishi,* **1983,** 1111; *Chem. Abs.,* **100,** 22261 (1984).
[182] Sosonkin, I. M., Kaminskii, A. Ya., Gershkovich, I. M., and Ponomareva, T. K., *Khim. Fiz.,* **1983,** 1288; *Chem. Abs.,* **99,** 194208 (1983).
[183] Gaviña, F., Luis, S. V., Gil, P., and Costero, A. M., *Tetrahedron Lett.,* **25,** 779 (1984).
[184] Shankaran, K., and Snieckus, V., *Tetrahedron Lett.,* **25,** 2827 (1984).
[185] Hart, H., and Nwokogu, G. C., *Tetrahedron Lett.,* **24,** 5721 (1983).
[186] Xin, H. Y., and Biehl, E. R., *J. Org. Chem.,* **48,** 4397 (1983).
[187] Barry, M., Brown, R. F. C., Eastwood, F. W., Gunawardana, D. A., and Vogel, C., *Aust. J. Chem.,* **37,** 1643 (1984).
[188] Nakayama, J., Takene, S., and Hoshino, M., *Tetrahedron Lett.,* **25,** 2679 (1984).
[189] Nakamura, N., *Chem. Lett.,* **1983,** 1795.
[190] Meyers, A. I., and Pansegrau, P. D., *Tetrahedron Lett.,* **25,** 2941 (1984).
[191] Carre, M. -C., Gregoire, B., and Caubere, P., *J. Org. Chem.,* **49,** 2050 (1984).
[192] Anthony, I. J., and Wege, D., *Aust. J. Chem.,* **37,** 1283 (1984).

Organic Reaction Mechanisms 1984
Edited by A. C. Knipe and W. E. Watts

CHAPTER 8

Electrophilic Aromatic Substitution

R. B. Moodie

University of Exeter

General

A full report of the electrophilic substitution reactions of (**1**) has appeared.[1] The cross-conjugated π-electron system of (**2**) is probably best represented by the structure shown; aromatic character has been demonstrated through six electrophilic substitution reactions.[2]

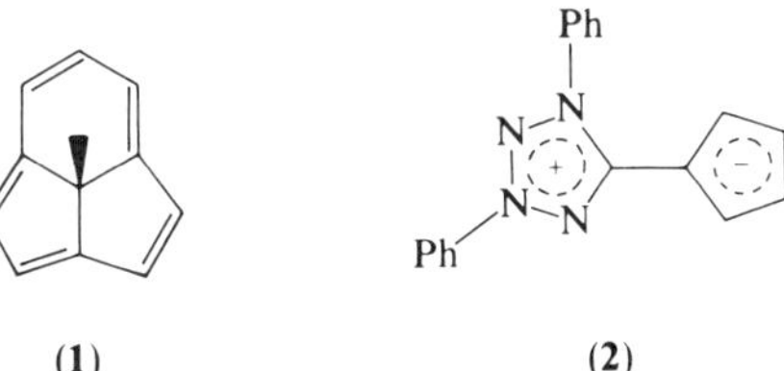

(**1**) (**2**)

Calculations using MNDO and MINDO/3 and including the influence of solvent, indicate that the preference for β-substitution shown by benzo[1,3]dioxole may change in favour of α-substitution in solvents of high dielectric constant.[3]

Reactivity in five-membered ring heteroaromatic compounds has been probed with CNDO/2 calculations.[4] Localization energies calculated by CNDO/2 for furan, thiophene and pyrrole have been considered in relation to σ^+ constants, and to the parameters of the Yukawa–Tsuno, Swain–Lipton, and Dewar–Grisedale equations.[5] Electrophilic substitution of (**3**), (**4**), and (**5**) gives in each case the expected 2-substituted product.[6] With (**6**),[7] (**7**),[7] and (**8**),[8] substitution occurs at the positions indicated, in accord with electron densities calculated by the CNDO method.

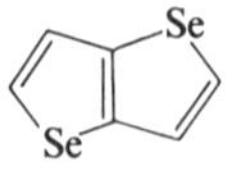

(3) 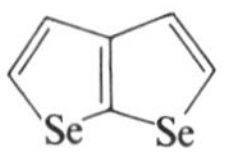 (4) 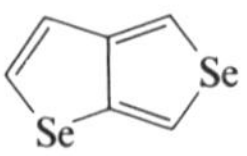 (5)

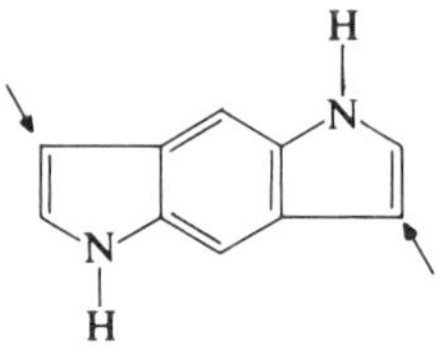

(6) 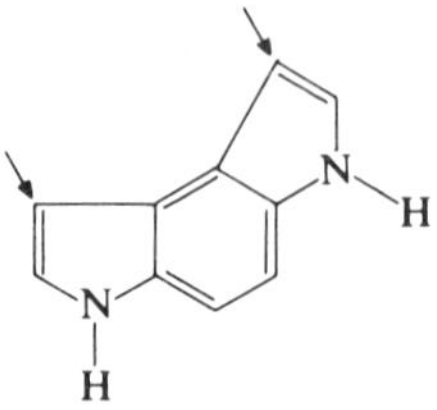 (7) 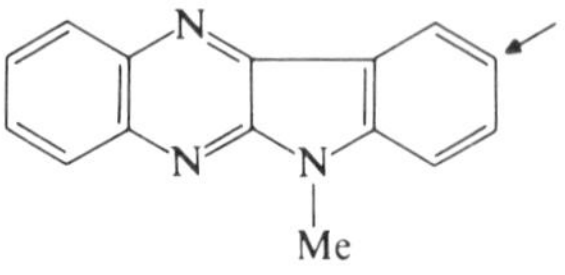 (8)

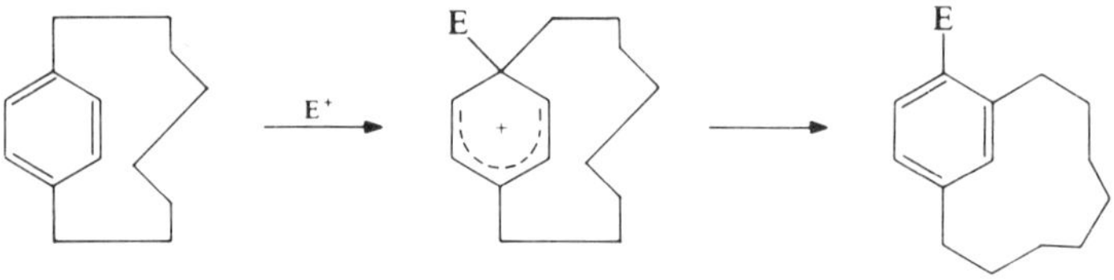

(9) (10)

Initial electrophilic attack on (**9**) is largely at the *ipso*-position, and followed by a 1,2-alkyl shift to give (**10**).[9]

Hydrogen Exchange

The isomerization of bromoxylenes, following protonation by H_3^+ in the gas phase, has been investigated. Energetic considerations, and the failure to detect Br^+ by chemical ionization mass spectrometry, suggest that the intramolecular bromine migration involves the intermediacy of a radical cation–bromine atom pair.[10]

Further evidence for 1,2-hydrogen shifts in the *p*-tritiophenyl cation (a process for which a rather high energy barrier is predicated by semi-empirical calculations) has come from studies of the products of its gas-phase reaction with methyl halides. An alternative explanation for scrambling of the label when the *p*-tritiophenyl cation reacts with methanol to give tritioanisoles (see *Organic Reaction Mechanisms 1982*) is held to be not tenable.[11]

Gas-phase tritiation of five-membered heteroaromatics by $^3HeT^+$, generated by β-decay of molecular tritium, gives an intramolecular tritium distribution which may be controlled by electrostatic interactions within the encounter pair. In pyrrole for instance β-tritiation is slightly preferred, in accord with theoretical estimates of charge distribution. In contrast, protonation in solutions occurs predominantly at the α-position.[12] The question of α *vs.* β-protonation of pyrrole and indole has been discussed.[13]

Hydrogen isotope exchange rates for *o*-, *m*-, and *p*-hydroxybenzoic acid have been measured, and correlated with H_0.[14]

Crystal structures of (**11**; R = Ph or CF_3) have been determined in an attempt to understand why the former is protonated predominantly on oxygen but the latter predominantly at the acyl-bearing ring carbon.[15]

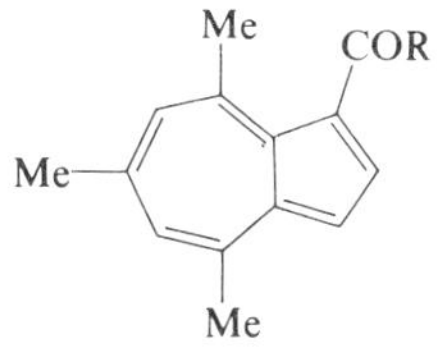

(**11**)

Halogenation

Toluene, naphthalene, and nitrobenzene react with fluoroxysulphate ion, SO_4F^-, to give the products of electrophilic fluorination and of oxidative side-reactions. Fluorination yields are enhanced by acid catalysts.[16]

Activation parameters have been evaluated for the chlorination of *p*-haloanisoles.[17] In the chlorination of anisole, *p*-cresol, and some related compounds by *N*-chlorosuccinimide in the presence of LiCl the rate-determining step is the formation of molecular chlorine and does not involve the aromatic.[18] Isomer proportions from the hypochlorite chlorination of anisole are independent of acidity, but with phenol the *ortho*:*para* ratio increases markedly with pH, reaching 4.3 at pH 10. The importance of the ionizable proton of phenol is clear, and phenyl hypochlorite may be an intermediate.[19] Regioselective *ortho*-chlorination of phenol may be effected with *N*,*N*-dichloro-*tert*-butylamine in CCl_4.[20] 2(1H)-quinoxalinone is chlorinated or iodinated exclusively at C(6), when treated with Cl_2 or ICl in H_2SO_4 containing Ag_2SO_4.[21] The kinetics and mechanisms of chlorination of *p*-aminobenzoic acid by chloramine-B have been investigated.[22]

Rate constants for bromination of disubstituted benzenes have been successfully correlated by an interactive free energy relationship of the form:

$$\log(k_{xy}/k_{HH}) = \rho_1^H\sigma_x + \rho_2^H\sigma_y + q\sigma_x\sigma_y$$

Comparison of substituent interaction terms, q, for this and other reactions suggests that q reflects at least in part the change in the transition-state position induced by the substituent.[23]

The kinetics of the reactions of 4-pyridone and some substituted derivatives with bromine in water show the reaction to occur through the predominant 4-pyridone tautomer at pH values less than 6, and through the conjugate base at higher pH.[24]

The kinetics of bromination of anisole in the presence of α-cyclodextrin indicate complexation of Br_3^- by α-cyclodextrin.[25]

Bromination of substituted anisoles by *N*-bromosuccinimide is catalysed by Ag^+; a value of $\rho = -7.85$ has been reported.[26] A similar ρ value governs bromination by *N*-bromosaccharin, for which acetyl hypobromite, formed in a pre-equilibrium step, is believed to be the reagent.[27] Bromination of anisole by $Br_2/FeCl_3$ is first order in each reactant and in the catalyst.[28] The bromination of phenol, anisole, and their derivatives by IBr in acetic acid catalysed by mercuric acetate is zeroth order in IBr and exhibits a rather low ρ^+ value; rate-determining initial mercuration is suggested.[29] There has been further work on the kinetics of bromination of substituted acetanilides, using bromine[30] and *N*-bromosuccinimide.[31] The latter reagent has also been used in studies of the bromination of aromatic azo compounds.[32] The kinetics of bromination of some substituted triphenylamines by Br_2 indicate that reactions start with a one-electron oxidation to give a cation radical.[33]

There is evidence that intramolecular hydrogen bonds in compounds such as (**12**) decrease the reactivity towards bromination of the unsubstituted position *para* to hydroxyl.[34]

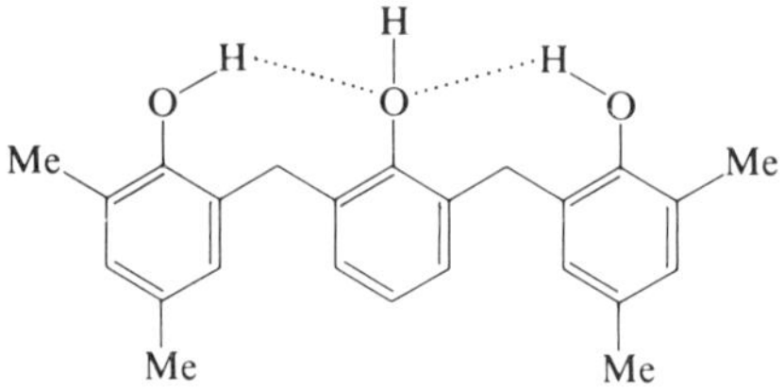

(**12**)

Nitration

Rate coefficients for nitration of nitrobenzene and methyl phenyl sulphone in 80–98% H_2SO_4 have received a quantitative interpretation based on the M_C function and the fraction of nitric acid ionized to nitronium ion as determined by Raman spectroscopy.[35] Benzene and more reactive aromatics are nitrated in phosphoric acid at their rate of encounter with nitronium ion; this is slightly faster for bicyclic than for monocyclic compounds. Isomer ratios indicate that viscosity can affect rates of translation within encounter pairs.[36]

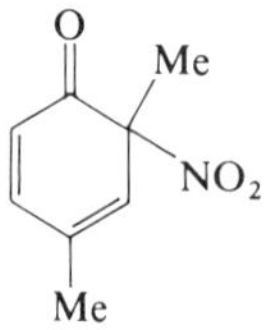

(**13**) 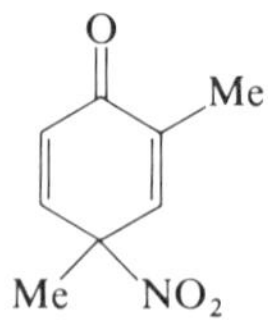 (**14**) 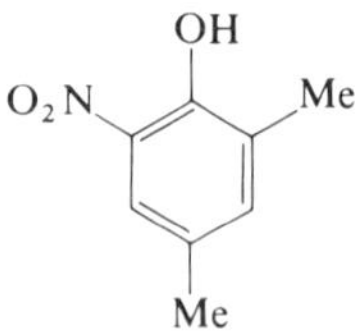 (**15**)

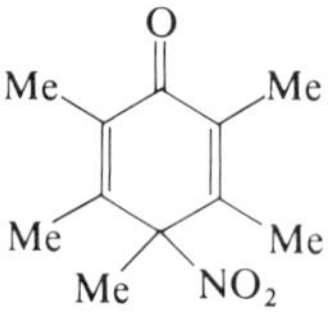

(**16**) 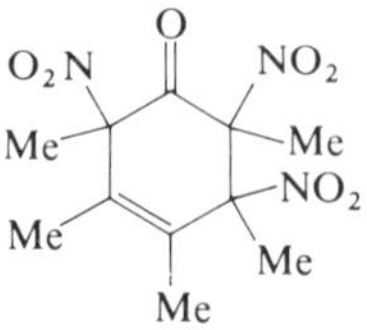 (**17**) 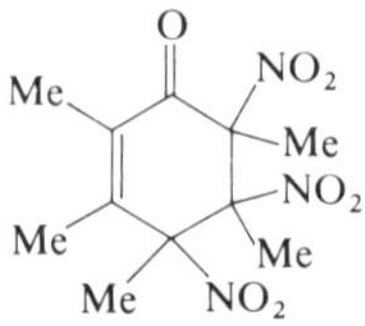 (**18**) 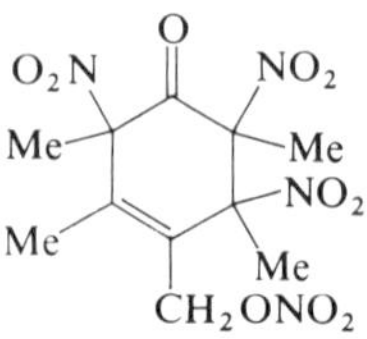 (**19**)

In the nitration of *o*-cresol in acetic anhydride, the preferred site of attack is *ipso* to methyl, and 6-methyl-6-nitrocyclohexa-2,4-dienone so formed rearranges regiospecifically to 6-nitro-*o*-cresol. 2,4-dimethylphenol gives rise to (**13**) and (**14**) in approximately equal proportions when nitration is effected in acetic anhydride at $-60°C$; on warming first (**13**) then (**14**) rearranges to (**15**).[37] Amongst the products of nitration of pentamethylphenol with nitrogen dioxide in benzene are (**16**), all four isomers of (**17**), and three of the four isomers of (**18**). Nitration with fuming nitric acid in dichloromethane gives rise additionally to all four isomers of (**19**).[38] Six adducts (Scheme 1) have been isolated in the yields shown from the reactions of methyl-2-furoate with acetyl nitrate in acetic anhydride. The lack of sterospecificity clearly indicates step-wise rather than concerted addition, and it seems reasonable to suppose that this is initiated by nitronium ion attack.[39]

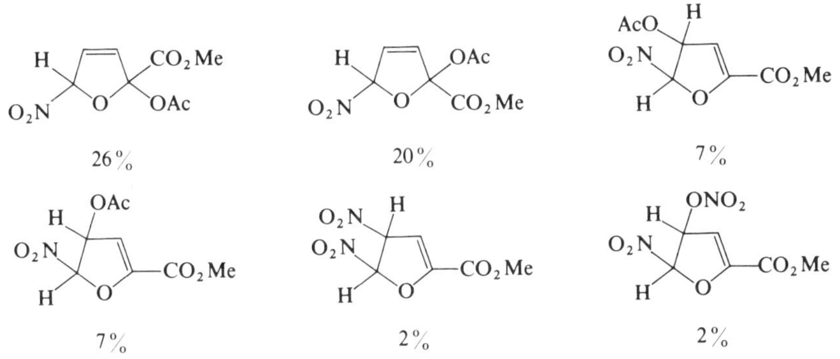

26% 20% 7%

7% 2% 2%

SCHEME 1

Nitrous acid catalysed nitration of 2-methoxyphenol gives 2-methoxy-4,6-dinitrophenol. The first but not the second (more rapid) nitration step proceeds *via* an intermediate nitroso compound. Independent investigation of the kinetics of the second nitration step suggests that either the formation of NO_2 from nitrous and nitric acids, or loss of a proton from a dinitrocyclohexadienone intermediate, can be rate-limiting.[40] 2-Methoxy-4,6-dinitrophenol (**21**) is also produced when (**20**; R=CH_2OH, CHO, or COOH) is treated with nitric acid in the presence of nitrous acid, and the kinetics and suggested mechanism are similar.[41]

Ceric ammonium nitrate has been used for the decarboxylative nitration of 4-hydroxybenzoic acid.[42]

Reactions of 3-substituted indoles, 4-substituted *N,N*-dimethylanilines, and 1- and 3-substituted indolizines with nitrous acid in acetic acid give rise to the products of *ipso*-substitution by the nitro group. A mechanism of electron transfer, followed by *ipso*-substitution and then oxidation of the nitroso compound to the nitro compound, has been suggested.[43]

The kinetics of nitration of anthraquinone in sulphuric acid,[44] and of carbazole in acetic acid,[45] have been studied. The latter reaction gives 1- and 3-nitro derivatives by a two-stage process involving formation and rearrangement of 9-nitrocarbazole.[46]

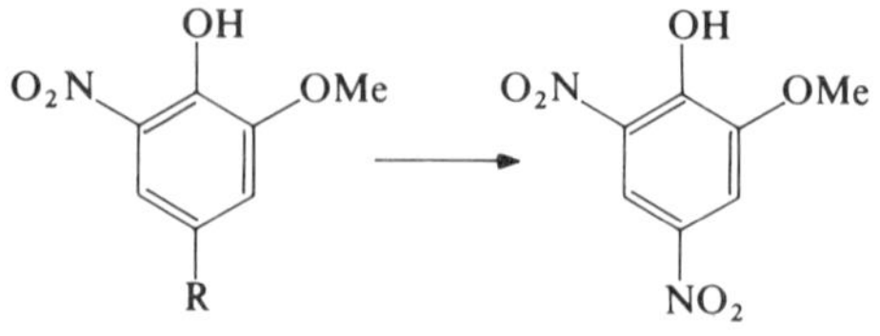

(20) (21)

Treatment of (**22**) with nitric acid in sulphuric acid gives (**23**) after aqueous work-up. The high yield and the absence of substituted nitrothiophenes amongst the products precludes a mechanism involving nitrodebromination followed by electrophilic bromination of starting material; Scheme 2 depicts the suggested mechanism.[47] An alternative possibility is that hydrolysis of the *gem*-dibromide (**22**) liberates bromide ion, which is oxidized by nitric acid to an electrophilic brominating agent and that this reacts with (**22**) or its hydrolysis product to give (**23**).

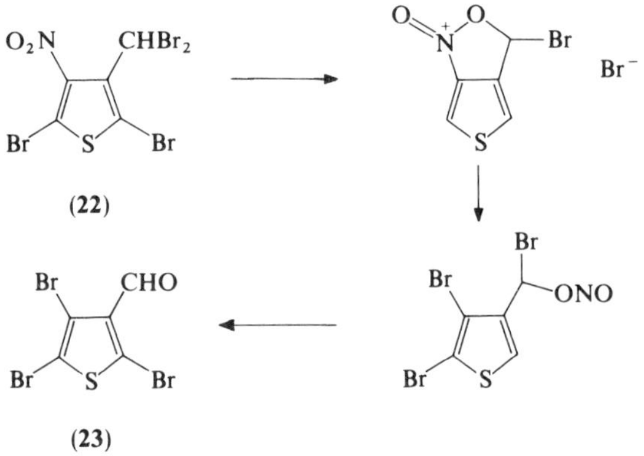

(22)

(23)

SCHEME 2

Sulphonation

A review, in Russian, of the kinetics and mechanisms of aromatic sulphonation has appeared.[48] The kinetics and products of sulphonation of aniline and its *N*-substituted derivatives in 80–100% H_2SO_4 have been interpreted in terms of concurrent reactions of protonated and unprotonated forms.[49] Both the *ortho*- and *para*-isomers of aminobenzenesulphonic acid isomerize in sulphuric acid in favour of the *meta*-isomer, by proto-desulphonation and resulphonation of the protonated amine. This process is aided by Hg^{2+} and it has been suggested that *meta*-mercuration of the protonated amine is an intermediate step.[50] The kinetics of isomerization of *p*-alkylbenzenesulphonic acids to the *meta*-isomer in mixtures of benzenesulphonic acid, water, and sulphuric acid have been investigated.[51]

(**24**) is more reactive than naphthalene towards SO_3 in dioxane; sulphonation occurs at the position indicated. The deactivating effect of fluorine substitution at the bridgehead has also been investigated.[52]

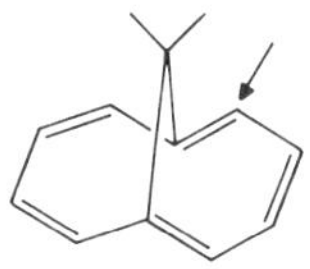

(24)

Friedel–Crafts and Related Reactions

A book on Friedel–Crafts alkylation chemistry has appeared.[53] Alkylations of anisole and toluene have been compared. Significant *meta*-alkylation of anisole is observed only with "swamping" catalyst concentrations, possibly indicating substitution of the ether–catalyst complex. With toluene there can be isomerization towards the thermodynamic mixture of products, and mechanisms for this have been discussed.[54] Alkylations of phenol and chlorophenols have been investigated.[55]

A study of the kinetics of benzylation of toluene and benzene, by $C_6H_5CH_2Cl/TiCl_4$, has revealed that when water is rigorously excluded the results are in good agreement with Brown's selectivity relationship, and show no evidence for a π-complex; a transition state close to the σ-complex is implied.[56] Benzylation by $C_6H_5CH_2SO_2Cl/AlCl_3$ is zeroth order in aromatic substrate and probably involves rate-determining loss of SO_2 from a complex formed from the reagent and $AlCl_3$.[57]

An observable mixture of isomeric σ-complexes (**25**) and (**26**) is formed from thiophene, *tert*-butyl chloride, and $AlCl_3$, or alternatively from 2- or 3-*tert*-butylthiophene and $HCl/AlCl_3$.[58] In each case the ratio (**25**):(**26**) approaches 97:3 over two days at room temperature.

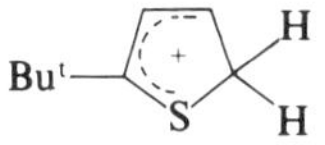

(25) 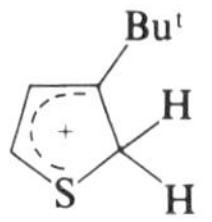 (26)

A radiochemical method has been used to investigate the ionic composition of catalytic complexes of HCl, $AlCl_3$, and mono-, di-, or tri-ethylbenzenes, and rate constants for the alkylation of benzene using these complexes and ethene have been determined.[59] The intramolecular alkylative rearrangement of (**27**) to (**29**), catalysed by $AlCl_3$, probably proceeds through the cation (**28**), even though this cannot be diverted by anisole to give the product of intermolecular alkylation.[60] Intramolecular alkylations such as that in Scheme 3 have been investigated.[61] Methanesulphonamides of *N*-(2,2-diethoxyethyl)anilines in the presence of $TiCl_4$ cyclize to indoles; steroelectronic effects in this intramolecular electrophilic substitution have been discussed.[62]

The Friedel–Crafts reactions depicted in Scheme 4 may be effected starting with either (**30**) or (**31**). If (**30**) is used the proportions of (**32**) and (**33**) depend in part

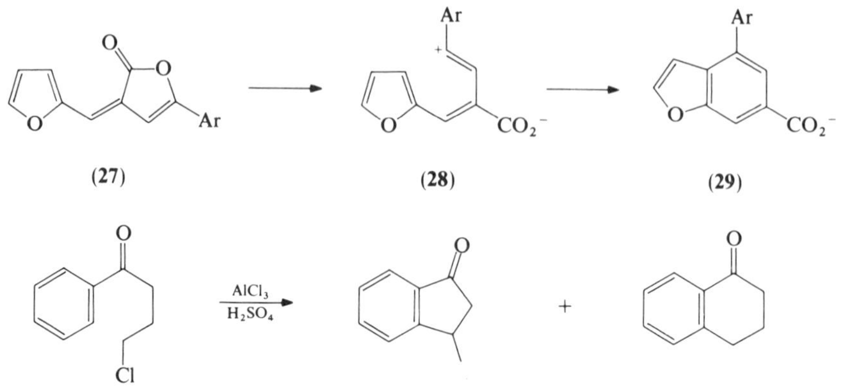

(27) (28) (29)

SCHEME 3

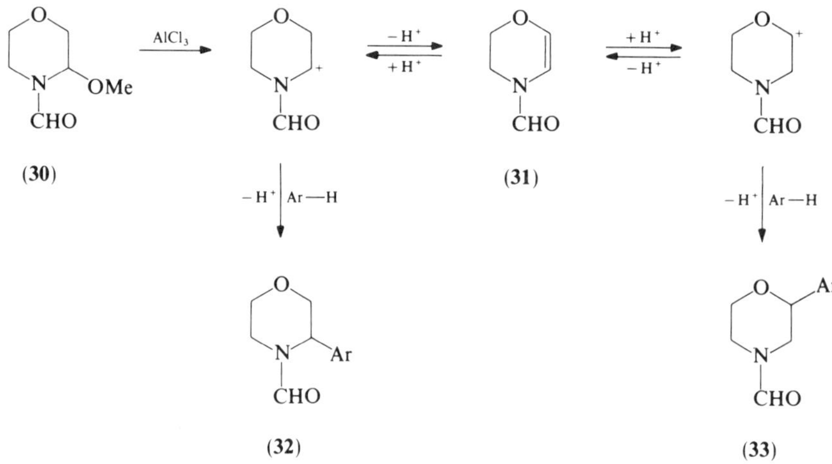

(30) (31) (32) (33)

SCHEME 4

upon the substrate; for instance, 1,3,5-trimethoxybenzene is sufficiently reactive to give only **(32)**.[63]

Rate constants for isopropylation of benzene with isopropyl acetate in the presence of various Lewis acids have been determined, and correlated with previously determined σ^* constants for the catalysts.[64] There has been a study of the kinetics of isopropylation of 1,2,3,4-tetrahydroquinoline with isopropanol in 90% H_2SO_4.[65]

The complex kinetics of the alkylation of *o*- and *p*-chlorophenol by cyclohexene in the presence of H_2SO_4 in an inert solvent have been probed.[66] Exclusively *para*-carboxylation of phenol has been achieved using tetrachloromethane, copper powder, and β-cyclodextrin; to explain the high stereoselectivity it has been suggested that the trichloromethyl cation formed from tetrachloromethane and copper is trapped in the cavity of β-cyclodextrin **(34)**.[67]

(**34**)

α-Metallated oxime ethers can act as the synthetic equivalent of α-acylcarbonium ions ($RCO—\overset{+}{C}H_2$); their aromatic substitution reactions have been reported.[68] Decay of tritium in *n*-butane labelled at C(1) in the presence of benzene in the gas, liquid, and solid phases gives rise to *sec*-butylbenzene as the main product; only from the gas phase reaction is there a significant amount of *tert*-butylbenzene formed.[69]

Benzoyl cation, generated by decay of multi-tritiated benzene in the presence of CO, reacts in the gas phase with the *n*- but not the π-type nucleophilic centres of phenol and aniline, and is unreactive towards pentamethylbenzene.[70] The kinetics of $AlCl_3$-catalysed 2,4-dichlorobenzoylations of toluene and benzene in nitromethane have been studied. The rate expression

$$\text{rate} = k_3[\text{Ar}][\text{2,4-dichlorobenzoyl chloride}][\text{AlCl}_3]$$

applies ($AlCl_3$ is consumed by complexation with the product) but the values of k_3 for both substrates, and their ratio, decrease with increasing initial concentrations of $AlCl_3$. This highlights the difficulty of interpreting observed intermolecular selectivities in such reactions.[71] Formylation with 1,1-dichloromethyl methyl ether and $TiCl_4$/or $AlCl_3$ in nitromethane has been studied by initial rates at low temperatures, to avoid problems arising from decomposition of the ether. The results are consistent with Brown's selectivity relationship.[72]

The reaction of mesitylene with acetyl chloride in the presence of $AlCl_3$ results in the formation of 1,1-dimesitylethene, as shown in Scheme 5.[73] Further examples of this unusual reaction have been reported.[74] Treatment of dichloroacetyl chloride with excess anisole in the presence of $AlCl_3$ [3] gives rise to some (**35**), possibly by an acylation-alkylation sequence.[75]

Friedel–Crafts acylation of 1,2,3,4-tetramethyldibenzofuran gives rise to the 7- and 8-acyl derivatives in a ratio which depends upon whether a nitro-hydrocarbon or a chloro-hydrocarbon solvent is used. An explanation based on regiospecific solvation of the σ-complex by the nitro-hydrocarbon, leading to the 8-isomer has been suggested.[76]

The Friedel–Crafts acetylation of 2,3-methylenedithiophene occurs mainly at C(5).[77] An unusual regiospecific *ortho*-acylation has been reported; bromomagnesium salts of phenols react with oxalyl chloride as shown in Scheme 6.[78]

The kinetics of reaction of some substituted pyrroles with the Vilsmeier–Haack reagent derived from *N*,*N*-dimethylbenzamide and phosphoryl chloride have been investigated; unusually large steric retardations by 1-alkyl substituents in the pyrrole have been rationalized in terms of a transition-state structure which places the aromatic ring of the benzamide above the heteroatom, which is at the positive end of

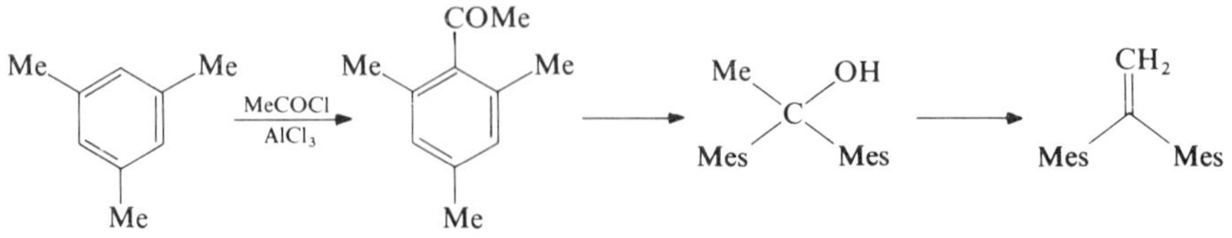

SCHEME 5

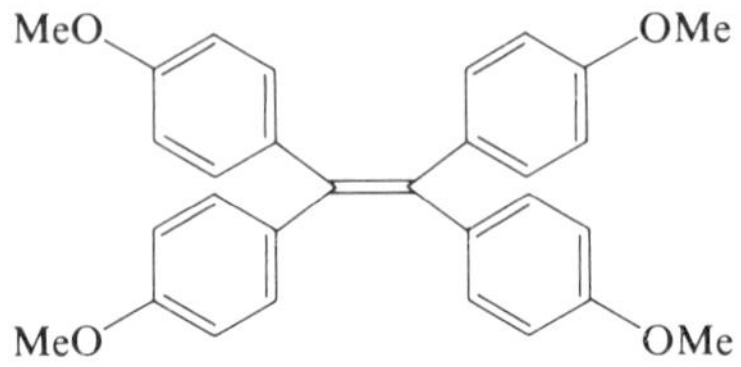

(35)

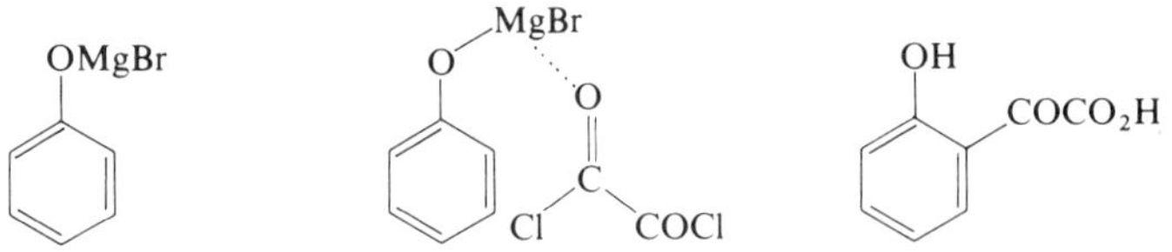

SCHEME 6

the pyrrole dipole.[79] Effects of aryl substituents in the benzamide have also been investigated.[80] Acylations with alkanoic acids and trifluoroacetic anhydride, in which the reactive intermediate is a mixed anhydride, are aided by the presence of phosphoric acid.[81]

Miscellaneous Reactions

Kinetic studies of the transmetallation of $PhSnEt_3$ with Hg(II) have established the reactivity order $HgI_3^- \ll HgI_2 \ll HgCl_2$.[82] The oxidative coupling of monosubstituted arenes with a mixture of $Pd(CH_3CO_2)_2$ and $Tl(CF_3CO_2)_3$ dissolved in the arenes gives mainly the 4,4′-biaryl; competition experiments indicate a ρ^+ value of -6.2.[83]

Aminophenols are the main products of the reaction of anilines with H_2O_2 and SbF_5–HF. The predominance of the *meta*-isomer indicates reaction of the protonated aniline.[84]

Ethoxycarbonyl nitrenium ion, generated thermally from ethyl azidoformate in the presence of TFA, gives rise to ethyl *N*-arylcarbamates by electrophilic substitution and displays low selectivity ($\rho^+ = -1.7$).[85] A nitrenium ion intermediate is likely in a novel synthetic route to 1-methoxy-2-oxindoles (Scheme 7).[86]

Products from the cationic phenylation of anisole give evidence of electrophilic attack at oxygen followed by intermolecular methylation.[87] Dialkylphenyloxonium

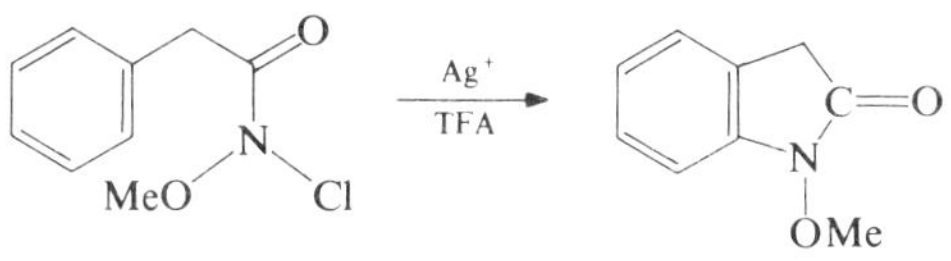

SCHEME 7

ions are the main initial products of the gas-phase reaction between tritiated phenyl cations and dialkyl ethers; their fragmentation and isomerization reactions have been studied.[88] The degenerate 1,2-shift of the phenyl group in the 1-phenyl-1,2,3,4,5,6-hexamethylbenzenonium ion occurs more slowly in the crystalline state than in solution, due to a large decrease in the entropy of activation.[89]

References

1 McCague, R., Moody, C. J., and Rees, C. W., *J. Chem. Soc., Perkin Trans. 1*, **1984,** 175.
2 Araki, S., and Butsugan, Y., *Tetrahedron Lett.*, **25,** 441 (1984).
3 Raabe, G., and Fleischhauer, J., *Z. Naturforsch.*, **39A,** 381 (1984); *Chem. Abs.*, **101,** 89935 (1984).
4 Gorb, L. G., Morozova, I. M., Belen'kii, L. I., and Abronin, I. A., *Izv. Akad. Nauk SSSR, Ser. Khim.*, **1983,** 828; *Chem. Abs.*, **99,** 4894 (1983).
5 Gorb, L. G., Morozova, I. M., Belen'kii, L. I., and Abronin, I. A., *Izv. Akad. Nauk SSSR, Ser Khim.*, **1983,** 1981; *Chem. Abs.*, **99,** 194002 (1983).
6 Konar, A., *Chem. Scr.*, **22,** 177 (1983).
7 Kadzhrishvili, D. O., Samsoniya, Sh. A., Gordeev, E. N., Kurkovskaya, L. N., Zhigachev, V. E., and Suvorov, N. N., *Khim. Geterotsikl. Soedin.*, **8,** 1086 (1983); *Chem. Abs.*, **100,** 22196 (1984).
8 Grachev, V. T., Ivashchenko, S. P., Ivanova, O. A., Gerasimov, B. G., Sarkisyan, A. Ts, Vasil'eva, A. D., Lisyutenko, V. N., Ivashchenko, A. V., and Mikhailova, T. A., *Khim. Geterotsikl. Soedin.*, **1984,** 840; *Chem. Abs.*, **101,** 151123 (1984).
9 Noble, K.-L., Hopf, H., and Ernst, L., *Chem. Ber.*, **117,** 455 (1984).
10 Cacace, F., Ciranni, G., and Di Marzio, A., *J. Chem. Soc., Perkin Trans. 2*, **1984,** 775.
11 Speranza, M., Keheyan, Y., and Angelini, G., *J. Am. Chem. Soc.*, **105,** 6377 (1983).
12 Angelini, G., Laguzzi, G., Sparapani, C., and Speranza, M., *J. Am. Chem. Soc.*, **106,** 37 (1984).
13 Catalán, J., and Yánez, M., *J. Am. Chem. Soc.*, **106,** 421 (1984).
14 Murano, Y., and Yoshihara, K., *Radiochim. Acta*, **33,** 159 (1983); *Chem. Abs.*, **100,** 138313 (1984).
15 Tafeenko, V. A., Porshnev, Yu. N., Polyakova, I. N., Gerasimov, B. G., Cherkashin, M. I., and Dyumev, K. M., *Dokl. Akad. Nauk SSSR*, **273,** 899 (1983); *Chem. Abs.*, **100,** 174186 (1984).
16 Appelman, E. H., Basile, L. J., Hayatsu, R., *Tetrahedron*, **40,** 189 (1984).
17 Jayaraman, S., and Ganesan, R., *Indian J. Chem.*, **22A,** 980 (1983); *Chem. Abs.*, **101,** 6278 (1984).
18 Mohamed Farook, S., Sivakamasundari, S., and Viswanathan, S., *Indian J. Chem.*, **23A,** 239 (1984); *Chem. Abs.*, **101,** 54208 (1984).
19 Ogata, Y., Kimura, M., Kondo, Y., Katoh, H., and Chen, F.-A., *J. Chem. Soc., Perkin Trans. 2*, **1984,** 451.
20 Ogata, Y., Takagi, K., Kondo, Y., Hsin, S.-C., Woo, W.-I., and Chen, F.-C., *J. Chin. Chem. Soc. (Taipei)*, **30,** 261 (1983); *Chem. Abs.*, **100,** 50813 (1984).
21 Sakata, G., and Makino, K., *Chem. Lett.*, **1984,** 323.
22 Jayaram, B., and Mayanna, S. M., *React. Kinet. Catal. Lett.*, **24,** 379 (1984); *Chem. Abs.*, **101,** 22712 (1984).
23 Dubois, J.-E., Ruasse, M.-F., and Argile, A., *J. Am. Chem. Soc.*, **106,** 4840 (1984).
24 Tee, O. S., and Paventi, M., *Can. J. Chem.*, **61,** 2556 (1983).
25 Tee, O. S., and Bennett, J. M., *Can. J. Chem.*, **62,** 1585 (1984).
26 Srinivasan, S. P., and Gnanapragasam, N. S., *J. Indian. Chem. Soc.*, **60,** 953 (1983); *Chem. Abs.*, **101,** 71958 (1984).
27 Srinivasan, S. P., and Gnanapragasam, N. S., *J. Indian Chem. Soc.*, **60,** 1106 (1983); *Chem. Abs.*, **101,** 22726 (1984).
28 Srinivasan, S. P., and Gnanapragasam, N. S., *J. Indian Chem. Soc.*, **60,** 1104 (1983); *Chem. Abs.*, **101,** 6315 (1984).

[29] Prasada Rao, M. D., Ahmad, M., and Kanungo, T. K., *Indian J. Chem.*, **22A,** 699 (1983); *Chem. Abs.*, **99,** 194114 (1983).
[30] Dangat, V. T., Bonde, S. L., and Rohokale, G. Y., *Indian J. Chem.*, **23A,** 237 (1984); *Chem. Abs.*, **101,** 54207 (1984).
[31] Rao, T. S., and Dalve, S. P., *Curr. Sci.*, **53,** 258 (1984); *Chem. Abs.*, **100,** 138334 (1984).
[32] Vijayasree, M., Srinivas, K., and Subba Rao, P. V., *Indian J. Chem.*, **22A,** 806 (1983); *Chem. Abs.*, **100,** 138298 (1984).
[33] Koshechko, V. G., Inozemtsev, A. N., and Pokhodenko, V. D. *Teor. Eksp. Khim.*, **20,** 178 (1984); *Chem. Abs.*, **101,** 190757 (1984).
[34] Böhmer, V., Stotz, D., Beismann, K., and Vogt, W. *Monatsh. Chem.*, **115,** 65 (1984).
[35] Marziano, N. C., Sampolini, M., Pinna, F., and Passerini, A., *J. Chem. Soc., Perkin Trans. 2,* **1984,** 1163.
[36] Moodie, R. B., Schofield, K., and Wait, A. R., *J. Chem. Soc., Perkin Trans. 2,* **1984,** 921.
[37] Cross, G. G., Fischer, A., Henderson, G. N., and Smyth, T. A., *Can. J. Chem.*, **62,** 1446 (1984).
[38] Hartshorn, M. P., Robinson, W. T., Vaughan, J., White, J. N., and Whyte, A. R., *Aust. J. Chem.*, **37,** 1489 (1984).
[39] Kolb, V. M., Darling, S. D., Koster, D. F., and Meyers, C. Y., *J. Org. Chem.*, **49,** 1636 (1984).
[40] Bazanova, G. V., and Stotskii, A. A., *Zh. Org. Khim.*, **19,** 780 (1983); *Chem. Abs.*, **99,** 52734 (1983).
[41] Bazanova, G. V., and Stotskii, A. A., *Zh. Org. Khim.*, **19** 2124 (1983); *Chem. Abs.*, **100,** 33890 (1984).
[42] Chawla, H. M., and Mittal, R. S., *Indian J. Chem.*, **22B,** 1129 (1983); *Chem. Abs.*, **101,** 37895 (1984).
[43] Colonna, M., Greci, L., and Poloni, M., *J. Chem. Soc., Perkin Trans. 2,* **1984,** 165.
[44] Adamek, M., Cervinka, M., Remes, M., Vavrickova, M., *Chem. Prum.*, **33,** 581 (1983); *Chem. Abs.*, **100,** 67574 (1984).
[45] Novikova, G. M., and Shishkina, V. I., *Vopr. Kinet. i Kataliza, Ivanovo* **1982,** 107; *Chem. Abs.*, **100,** 50782 (1984).
[46] Kyziol, J. B., and Daszkiewicz, Z., *Tetrahedron,* **40,** 1857 (1984).
[47] Coyle, J. D., *Heterocycles,* **22,** 1175 (1984).
[48] Radyshevskaya, O. N., and Gordeev, L. S., *Deposited Doc.*, **1983,** VINITI 3067-83; *Chem. Abs.*, **101,** 89909 (1984).
[49] Khalevin, R. N., *Zh. Org. Khim.*, **20,** 379 (1984); *Chem. Abs.*, **101,** 6293 (1984).
[50] Khelevin, R. N., *Zh. Org. Khim.*, **20,** 791 (1984); *Chem. Abs.*, **101,** 110038 (1984).
[51] Popkova, I. A., and Kozlov, V. A., *Izv. Vyssh. Uchebn. Zaved., Khim. Khim. Tekhnol.*, **27,** 35 (1984); *Chem. Abs.*, **100,** 174028 (1984).
[52] Cerfontain, H., Gossens, H., Koeberg-Telder, A., Kruk, C., and Lambrechts, H. J. A., *J. Org. Chem.*, **49,** 3097 (1984).
[53] Roberts, R. M., and Khalaf, A. A., *Friedel–Crafts Alkylation Chemistry: A Century of Discovery,* Marcel Dekker, New York, 1984; *Chem. Abs.*, **101,** 109843 (1984).
[54] Olah, G. A., Olah, J. A., and Ohyama, T., *J. Am. Chem. Soc.*, **106,** 5284 (1984).
[55] Alieva, M. K., and Akhmedov, K. N., *Zh. Org. Khim.*, **19,** 2131 (1983); *Chem. Abs.*, **100,** 138679 (1984).
[56] DeHaan, F. P., Covey, W. D., Ezelle, R. L., Margetan, J. E., Pace, S. A., Sollenberger, M. J., and Wolf, D. S., *J. Org. Chem.*, **49,** 3954 (1984).
[57] Covey, W. D., DeHaan, F. P., Delker, G. L., Dawson, S. F., Kilpatrick, P. K., Rattinger, G. B., and Read, W. G., *J. Org. Chem.*, **49,** 3967 (1984).
[58] Belen'kii, L. I., and Yakabov, A. P., *Tetrahedron,* **40,** 2471 (1984).
[59] Shut'ko, A. P., and Basov, V. P., *Ukr. Khim. Zh. (Russ. Ed.),* **49,** 1108 (1983); *Chem. Abs.*, **99,** 194126 (1983).
[60] Hashem, A. I., and Abd E1-Mottaleb, M. S. A., *Egypt. J. Chem.*, **25,** 541 (1982); *Chem. Abs.*, **99,** 211859 (1983).
[61] Khalaf, A. A., Abde l-Wahab, A.-M. A., El-Khawaga, A. M., and El-Zohry, M. F., *Bull. Soc. Chim. Fr. II,* **1984,** 285.
[62] Sundberg, R. J., and Laurino, J. P., *J. Org. Chem.*, **49,** 249 (1984).
[63] Eberson, L., Malmberg, M., and Nyberg, K., *Acta Chem. Scand.*, **38B,** 345 (1984).
[64] Lysenko, Yu. A., and Troshina, E. A., *Zh. Obshch. Khim.*, **54,** 402 (1984); *Chem. Abs.*, **100,** 191069 (1984).
[65] Okhrimenko, Z. A., Chekhuta, V. G., and Kachurin, O. I., *Khim. Geterotsikl. Soedin.*, **1984,** 506; *Chem. Abs.*, **101,** 71921 (1984).
[66] Kas'yanov, V. V., and Muganlinskii, F. F., *Kinet. Katal.*, **25,** 13 (1984); *Chem. Abs.*, **101,** 6286 (1984).
[67] Komiyama, M., and Hirai, H., *J. Am. Chem. Soc.*, **106,** 174 (1984).
[68] Shatzmiller, S., Lidor, R., Shalom, E., and Bahar, E. *J. Chem. Soc., Chem. Commun.*, **1984,** 795.

[69] Nefedov, V. D., Sinotova, E. N., Arkhipov, Yu.M., and Gomzina, N. A., *Radiokhimiya*, **25,** 557 (1983); *Chem. Abs.,* **100,** 5471 (1984).
[70] Occhiucci, G., Speranza, M., and Cacace, F., *J. Chem. Soc., Chem. Commun.,* **1984,** 723.
[71] DeHaan, F. P., Covey, W. D., Delker, G. L., Baker, N. J., Feigon, J. F., Ono, D., Miller, K. D., and Stelter, E. D., *J. Org. Chem.,* **49,** 3959 (1984).
[72] DeHaan, F. P., Delker, G. L., Covey, W. D., Bellomo, A. F., Brown, J. A., Ferrara, D. M., Haubrich, R. H., Lander, E. B., MacArthur, C. J., Meinhold, R. W., Neddenriep, D., Schubert, D. M., and Stewart, R. G., *J. Org. Chem.,* **49,** 3963 (1984).
[73] Roberts, R. M., El-Khawaga, A. M., and Roengsumran, S., *J. Org. Chem.,* **49,** 3180 (1984).
[74] El-Khawaga, A. M., and Roberts, R. M., *J. Org. Chem.,* **49,** 3832 (1984).
[75] Mahato, S. B., Mandal, N. B., Pal, A. K., and Maitra, S. K., *J. Org. Chem.,* **49,** 718 (1984).
[76] Keumi, T., Yago, Y., Kato, Y., Taniguchi, R., Temporin, M., and Kitajima, H., *J. Chem. Soc., Perkin Trans. 2,* **1984,** 799.
[77] Barker, J. M., Huddleston, P. R., and Smith R., *J. Chem. Res. Synop.,* **1984,** 280.
[78] Bigi, F., Casiraghi, G., Casnati, G., and Sartori, G., *J. Chem. Soc., Perkin Trans. 1,* **1984,** 2655.
[79] White, J., and McGillivray, G., *J. Chem. Soc., Perkin Trans. 2,* **1984,** 1179.
[80] White, J., *J. Chem. Soc., Perkin Trans. 2,* **1984,** 1607.
[81] Galli, C., *J. Chem. Res. Synop.,* **1984,** 272.
[82] Sedaghat-Herati, M. R., and Nahid, P., *J. Organomet. Chem.,* **239,** 307 (1982).
[83] Deiko, S. A., Ryabov, A. D., Yatsimirskii, A. K., and Berezin, I. B., *Dokl. Akad. Nauk SSSR,* **266,** 874 (1982); *Chem. Abs.,* **98,** 71396 (1983).
[84] Jacquesy, J. C., Jouannetaud, M.-P., Morellet, G., and Vidal, Y., *Tetrahedron Lett.,* **25,** 1479 (1984).
[85] Takeuchi, H., and Mastubara, E., *J. Chem. Soc., Perkin Trans. 1,* **1984,** 981.
[86] Kikugawa, Y., and Kawase, M., *J. Am. Chem. Soc.,* **106,** 5728 (1984).
[87] Eustathopoulos, H., Court, J., and Bonnier, J. M., *J. Chem. Soc., Perkin Trans. 2,* **1983,** 803.
[88] Fornarini, S., and Speranza, M., *J. Chem. Soc., Perkin Trans. 2,* **1984,** 171.
[89] Borodkin, G. I., Nagi, Sh. M., Mamatyuk, V. I., Shakirov, M. M., and Shubin, V. G., *Zh. Org. Khim.,* **20,** 552 (1984); *Chem. Abs.,* **101,** 54236 (1984).

Organic Reaction Mechanisms 1984
Edited by A. C. Knipe and W. E. Watts

CHAPTER 9

Carbocations

R. A. Cox

Department of Chemistry, University of Toronto, Canada

Introduction

Several areas have been the subject of general reviews this year. These include the generation and reactivity of carbocations,[1] the whole subject of non-classical carbocations,[2] halonium ions,[3] and the structure and reactivity of arenium ions.[4] Carbocation rearrangements by 1,2-shifts have been the subject of an extensive review,[5] and alkyl cation rearrangements in solution have been discussed.[6] The useful isotope perturbation method for determining the structures and energy surfaces of carbocations is the subject of a short review.[7] Of interest from the synthetic point of view is an article on the cycloaddition of allyl cations (**1**) to 1,3-dienes (**2**) as a general method for seven-membered carbocycles (**3**).[8]

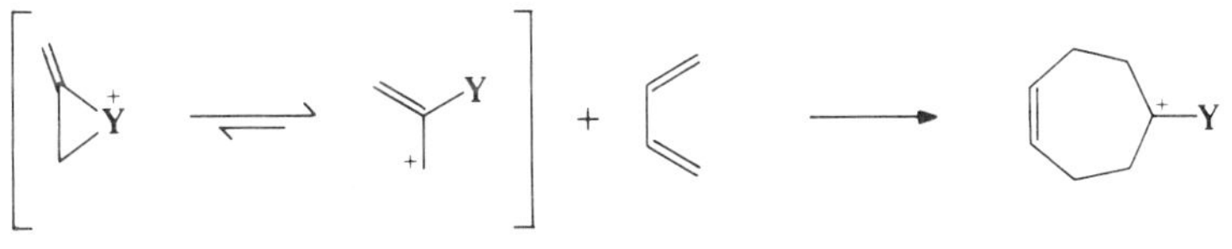

(**1**) (**2**) (**3**)

2-Norbornyl and Related Systems

Brown has reviewed the energetics of the 2-norbornyl cation and the search for non-classical stabilization.[9] Olah and Prakash maintain that NMR data for the long-lived ion support the σ-bridged, non-classical structure with two-electron three-centre bonds, and that the differentiation of this ion from classical trivalent structures can serve as a prototype in defining the limiting classical carbocations.[10]

A ^{13}C-NMR study of the model classical 2-norbornyl ions formed by (**4**) in $SbF_5/FSO_3H/SO_2ClF$ at $-80°$, in which allylic resonance stabilizes the carbocation centre with no need for σ-participation, shows that the deviations from linearity observed in plots of $\Delta\delta C^+$ *vs.* σ^{C^+} for these systems have no predictive value, and cannot be used to test for the presence or absence of σ-bridging in individual systems.[11] Another study finds no enhanced electron release from the *exo*- as compared to the *endo*-position in 2-norbornyl cations.[12] It is found that the energy barrier to rotation of the 2-aryl ring in (**5**), a measure of the extent of π-delocalization, can be measured by studying the separation of the *o*-C signals in the ^{13}C-NMR spectrum under slow rotation conditions, and determining the coalescence temperature.[13] These energy barriers increase markedly when electron-donating substituents are present, which is held to explain the less negative ρ^{C^+} values found for them in $\Delta\delta C^+$ *vs.* σ^{C^+} plots.[13]

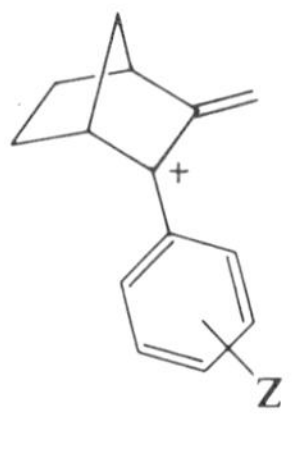

(**4**) 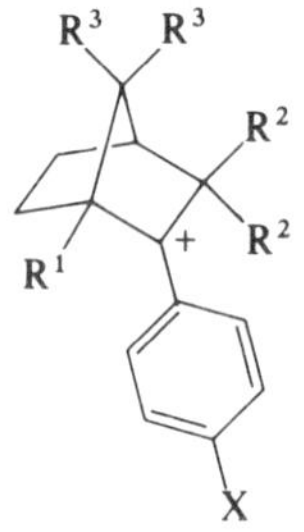 (**5**) (**6**)

X = H, Me, OMe
a: $R^1 = R^2 = R^3 = H$
b: $R^2 = H$, $R^1 = R^3 = Me$
c: $R^3 = H$, $R^1 = R^2 = Me$

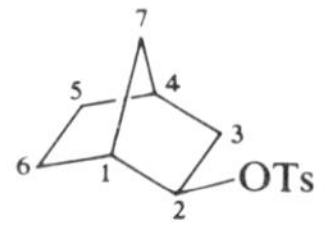

The effects of substituents at the C(5), C(6), and C(7) positions on the solvolysis rates of 2-norbornyl *p*-tosylate (**6**), in 80% EtOH at 70°, have been studied.[14] It is found that inductive interaction between C(2) and C(6) is stronger than that between C(2) and the equidistant C(7), refuting the contention that the direct-field effect should lead to comparable interactions, and supporting the view that the 2-norbornyl cation is anisotropic to the transmission of polar effects.[14]

Double labelling experiments have demonstrated that the observed products on the (+)-camphenilone route to (−)-albene depend on *exo*-3,2-methyl and -alkyl shifts, *endo*-6,2-hydride shifts and Wagner–Meerwein rearrangements, but not on *endo*-1,2-methyl shifts.[15] It has been found that the high degree of strain associated

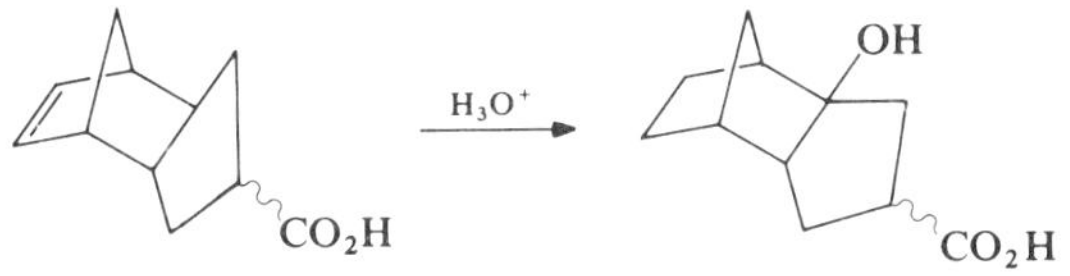

H$_3$O$^+$

OH

CO_2H CO_2H

(7) (8) (9)

with the *trans*-trimethylene bridge in (7) permits rearrangement to (**8**) by a 6,2-type hydride shift only, as determined by deuterium labelling; Wagner–Meerwein rearrangement and 3,2-type hydride shifts are completely suppressed.[16] The products of the solvolysis of (**9**) in carboxylic acids are found to depend on the acid polarity.[17] Wagner–Meerwein rearrangements and 6,2(6,1)-hydride shifts are found; the results have been interpreted in terms of asymmetric ion-pairs.[17] Trifluoromethyl group substitution in (**10**) results in dramatic reductions in the rates of solvolysis with CF_3CH_2OH buffered with 2,6-lutidine; the rate ratio $10^{12}:10^6:1$ observed for (**10**):(**11**):(**12**) is held to be overwhelming evidence for a symmetrical (non-classical) transition state for the double-bond-assisted solvolysis.[18]

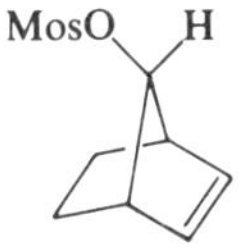

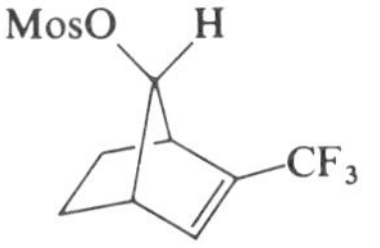

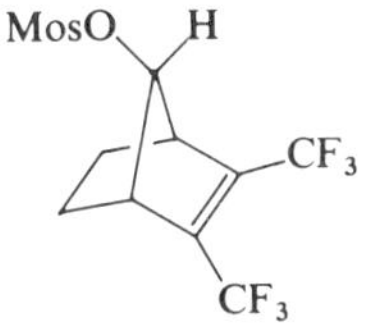

MosO H MosO H MosO H

CF_3 CF_3 CF_3

(**10**) (**11**) (**12**)

Mos = 4-methoxybenzenesulphonate

The decomposition of 3,3-, 4,4- and 5,5-dimethyl-2-norbornanediazonium ions leads to *gem*-dimethylnorbornyl cations in which 6,1- and 6,2-hydride shifts can be detected by structural isomerization and racemization, respectively; the results can reasonably be explained in terms of unsymmetrical ion-pairs, but are incompatible with equilibrating open ions.[19] The reaction of (**13**) and (**14**) with magnesium and organomagnesium salts in ether has been studied.[20] The silver-ion-assisted acetolysis of (**15**) gives only *exo*-product.[21] Calculations on (**16**) and (**17**) (*ab initio*,

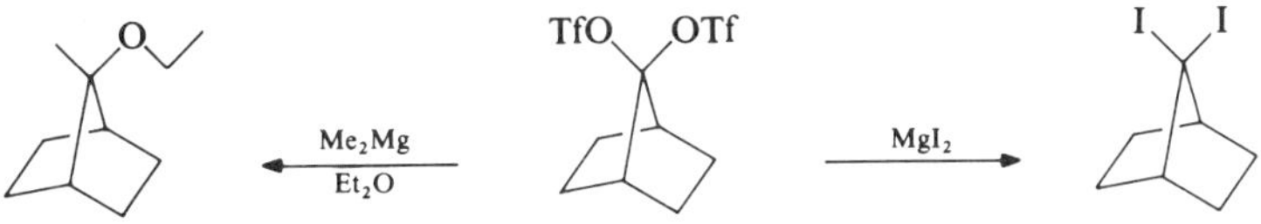

O TfO OTf I I

Me_2Mg / Et_2O MgI_2

65% (**13**) 79%

TfO OTf I I

MgI_2

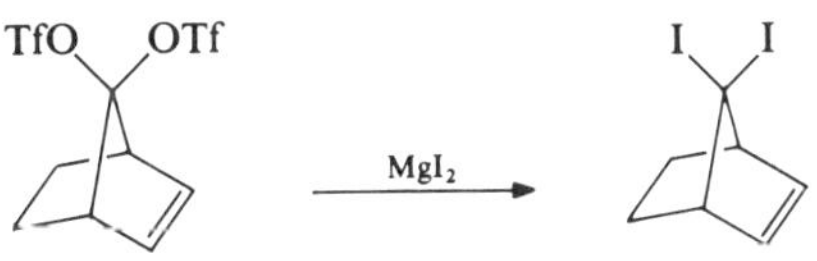

(**14**) 71%

AgOAc/HOAc, CH_3CN/Δ

(15)

(16)

(17)

STO-3G) suggest that homoconjugated carbonyl groups and related ($-M$, $-I$) functions can be remote electron donors.[22]

Other Bicyclic and Bridged Systems

The totally degenerate $(CH)_9{}^+$ barbaralyl cation has been found to have the structure (**18**; R = H), according to ^{13}C-labelling, isotopic perturbation by eight deuteriums, and *ab initio* calculations; according to the latter structure (**20**) is too high in energy to be an intermediate or transition state in the partially degenerate rearrangement of (**18**).[23] On the other hand the ^{19}F-NMR spectrum of (**18**; R = F) suggests structure (**19**).[24] Cations (**18**; R = vinyl) and (**18**; R = aryl) have been found to be in the charge range where the degenerate rearrangement changes from a Cope to a divinylcyclopropylmethyl cation mechanism.[25]

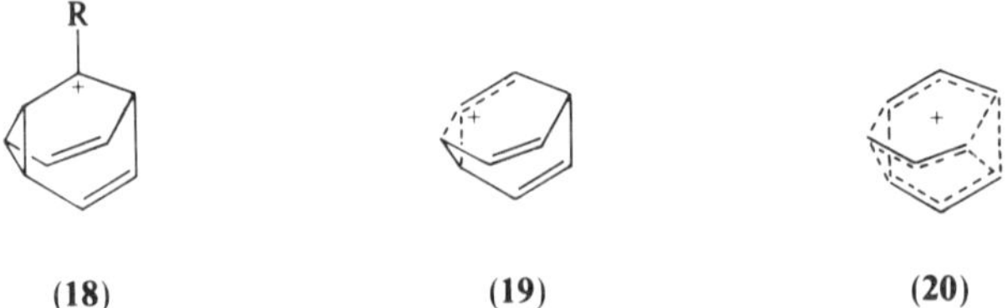

(18) (19) (20)

Low-temperature ^{13}C-NMR spectroscopy and line-shape analysis show that (**22**) is an intermediate in the automerization of the armilenyl cation (**21**), but that (**23**) is not.[26] Compound (**24**) has been synthesized as a test for the elusive phenalenyl cation (**25**); ^{13}C- and ^{1}H-NMR spectra suggest some charge delocalization into the etheno bridges.[27] Evidence for classical precursors of non-classical carbocations results from the deamination of bicyclo[2.2.2]octan-2-yl- and bicyclo[3.2.1]octan-2-yl-amines;[28] the deamination of *endo*- and *exo*-bicyclo[3.2.1]octan-3-ylamines and their derivatives has also been studied.[29] The novel cation (**26**) is suggested as an intermediate in the decomposition of bicyclo[3.2.0]heptane-*endo*-2-diazonium

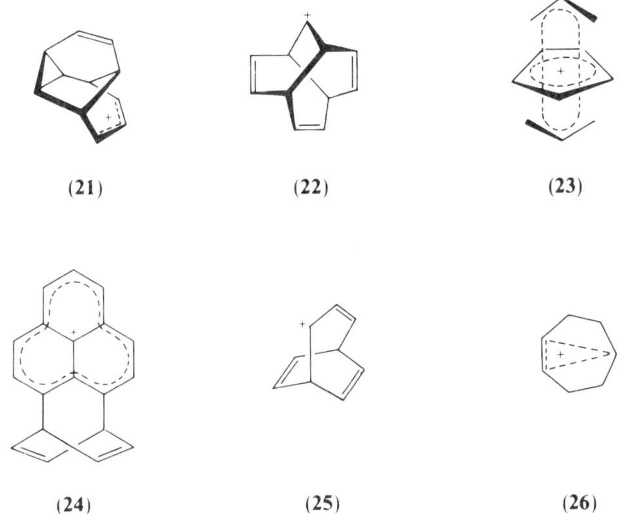

(21) (22) (23)

(24) (25) (26)

ion.[30] Different products result when methanol is used to quench the carbocations resulting from the ionization of (**27**) in super-acid, and from the ionization of (**28**) or the protonation of (**29**), apparently because of the formation of different carbocation species in the two cases.[31] A reversal of product configuration is observed during the solvolyses of (**30**) and (**31**) as R^1 is changed from electron-donating to electron-withdrawing, and is attributed to non-classical carbocation intermediates.[32] A detailed NMR investigation of the homotropylium ion (**32**) has been published; surprisingly it is found that both *exo*- and *endo*-protons are shielded in this ion.[33]

(27) (28) (29)

(30) (31) (32)

Vinyl and Aryl Cations

A novel way of generating vinyl cations from azibenzils (**33**) *via* vinyldiazonium ions (**34**) has been published.[34] Vinyl cations can also be generated photochemically, by the photolysis of triarylvinyl bromides in a two-phase system, as demonstrated by trapping them as isoquinolines with cyanate ion, or as isothioquinolines or vinyl isothiocyanates with thiocyanate.[35] Degenerate β-aryl rearrangements in these photochemically generated triarylvinyl cations have also been studied.[36] *Ab initio* calculations using the improved virtual orbital formalism show that the bridged and classical vinyl cation structures are very similar in the ground state, but that either can predominate in various excited states.[37]

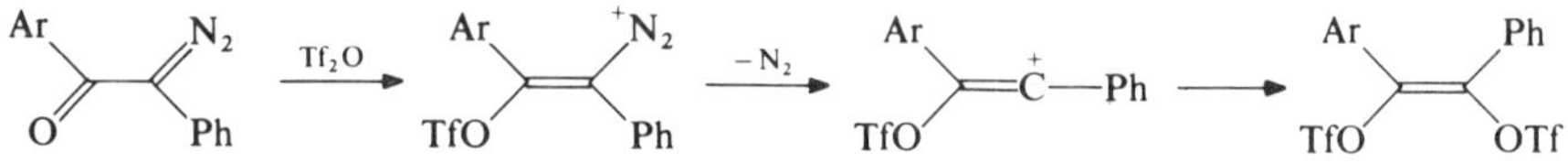

(33) (34)

The stable vinyl cation (**35**) has been observed by ^{13}C-NMR at $-100°$ in $SbF_5/SO_2ClF/SO_2F_2$,[38] and the isomeric ion (**36**) has had its rotational barrier measured.[39] The first-formed cyclopropylvinyl cation (**38**), presumably generated when (**37**) is treated with $AgSbF_6$, rearranges to the allyl cation (**39**) before reacting with alkenes.[40] Cyclobutanones result when (**40**) is solvolysed.[41] Ion (**41**) can be trapped when 9-((α-bromo)- and 9-((α-tosyloxy)-*p*-methoxybenzylidene)xanthenes are solvolysed.[42] Despite the strain in the system, (**42**) solvolyses 10^5 times faster than cyclohexenyl triflate and 11 times faster than the unstrained *cis*-2-butenyl triflate, a result explained by an unusual aryl stabilization of the incipient vinyl cation (**43**).[43] The trifluoroethanolyses of (**44**), (**45**), and (**46**) involve the classical vinyl cations (**47**) and (**48**) which equilibrate *via* the bridged ion (**49**).[44] The ethanolysis and trifluoroethanolysis of a number of vinyl triflates gives rate constants which are linear in σ^*, enabling a k_C mechanism to be deduced.[45] The product distribution in the degenerate rearrangement of doubly labelled triarylvinyl cations can be explained if complete equilibration prevails.[46]

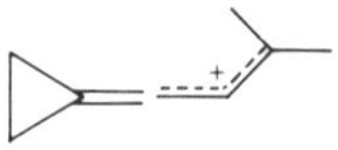

(35) 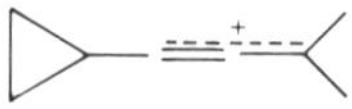 (36)

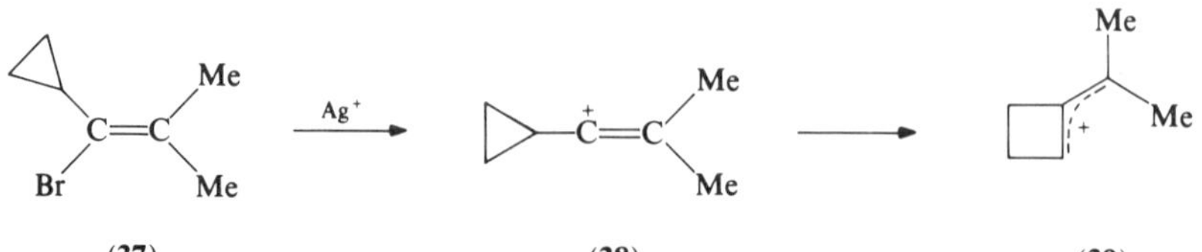

(37) (38) (39)

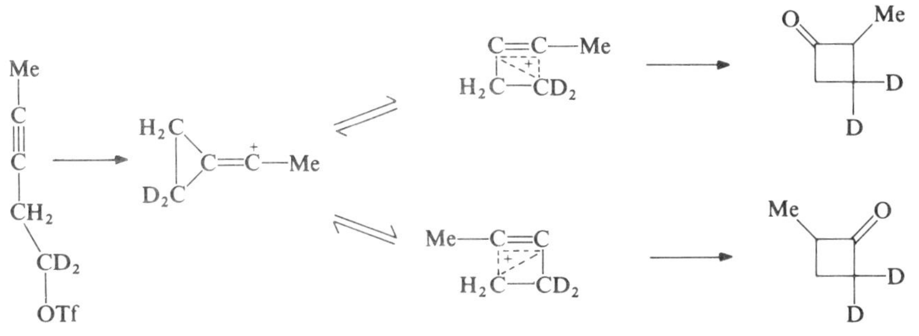

(40)

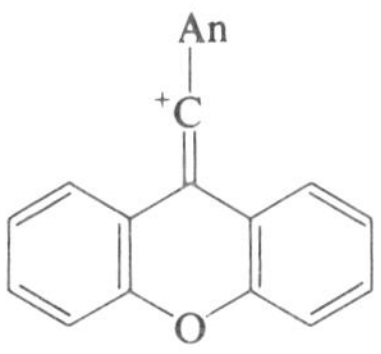

(41)

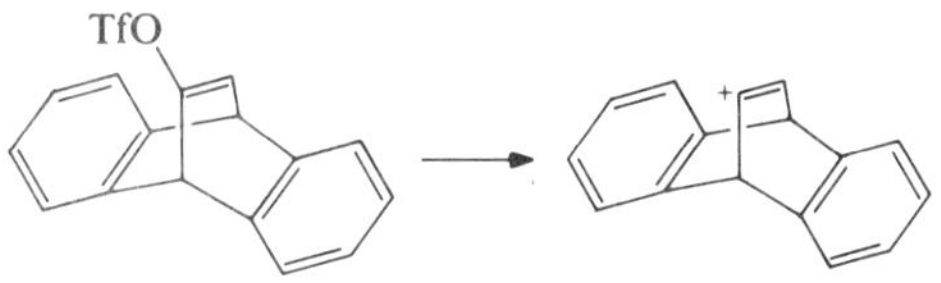

(42)

(43)

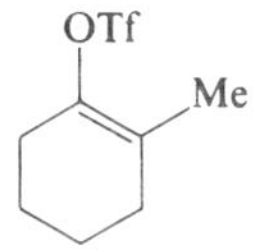

(44)

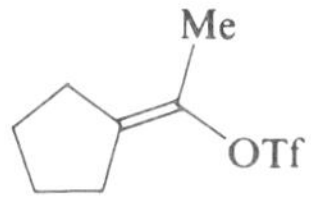

(45)

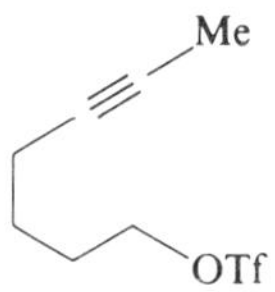

(46)

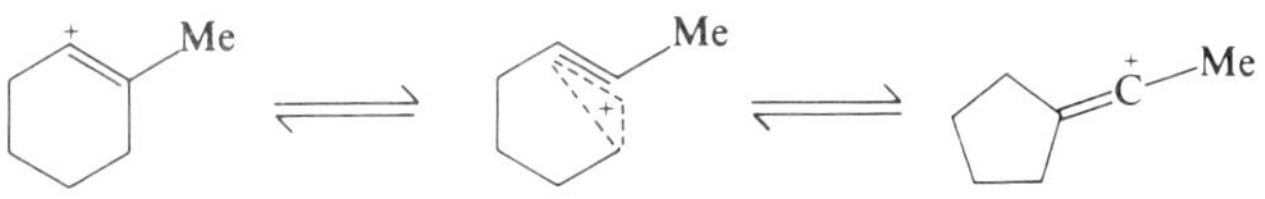

(47)

(49)

(48)

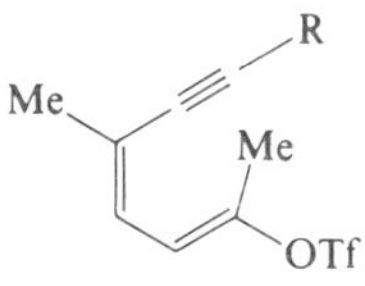

(50)

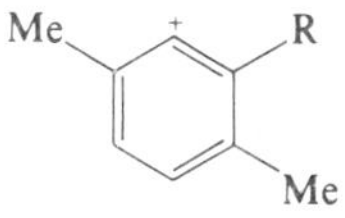

(51)

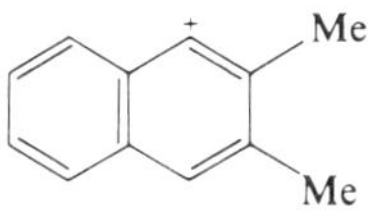

(52)

R = H, Me, $SiMe_3$

Aryl cations cannot be detected directly during the laser photolysis of diazonium fluoroborate salts, even those with ground-state triplets.[47] The life-time of Ph^+ in water has been determined to be some 500 ps.[47] The (E)-isomer (**50**), but not the (Z)-isomer (Me and OTf transposed), solvolyses to give phenyl ethers by way of the intermediate phenyl cation (**51**).[48] (E)-1-Methyl-2-(2-(propyn-1-yl)-phenyl)vinyl triflate reacts similarly by way of (**52**).[49] Evidence against 1,2-migrations in aryl cations is provided by the lack of detection of hydroxy migration product during the photolytic methoxydediazoniation of 2-diazo-4- and 2-diazo-5-methylphenol in 50% $MeOH/H_2SO_4$.[50]

Destabilized Carbocations

This area has been reviewed by Tidwell.[51] Carbocation stability can be reduced by structural effects, as in reducing the bond angle at the cationic centre, or by anti-aromatic character, as in the cyclopentadienyl cation, but the greatest study has been of destabilization caused by the presence of electron-withdrawing substituents.[51] The reactivity of α-acyl carbocations in non-nucleophilic media has been reviewed.[52] Solvolysis studies on a number of α-trifluoromethyl carbocations indicate that carbocation formation is rate-limiting.[53] The effect of an α-cyano substituent is demonstrated by comparing the solvolysis kinetics of (**53**; R = CN) with (**53**; R = H); the cyano group magnifies neighbouring-group participation.[54] This is also one of the effects found in a detailed study of the solvolysis of some α-keto norbornyl trifluoroacetates and triflates.[55] Tertiary carbocations such as (**54**) can be intermediates, but secondary α-carbonyl carbocations cannot.[55] The stable α-keto carbocations R_2C^+COR (R = Ph, *p*-$MeOC_6H_4$) have been prepared and their rates of electrophilic substitution studied, reacting with anisole or intramolecularly.[56] The reduction peak potentials of these and related carbocations have also been determined, and the relation between oxidizing ability and electrophilicity discussed.[56]

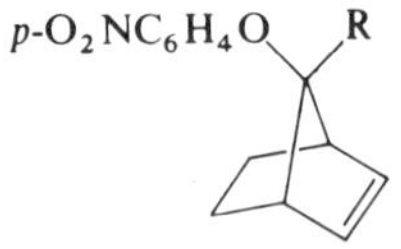

p-$O_2NC_6H_4O$ R

(**53**)

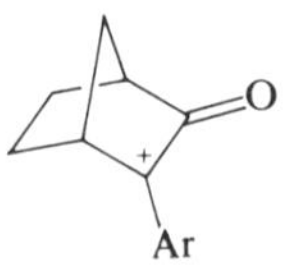

O Ar

(**54**)

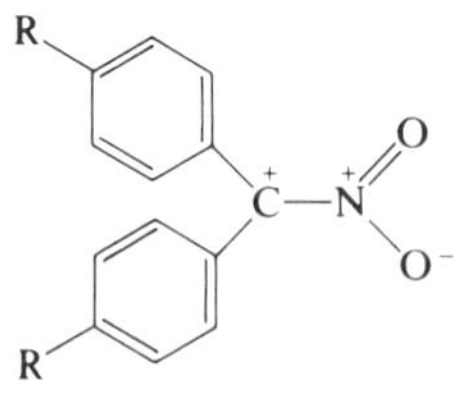

R R O O⁻ C N

(**55**)

R = H, Me

Treating the corresponding α,α-dinitrodiarylmethanes with FSO_3H/SO_2ClF at $-78°$ causes a loss of HNO_2 and the formation of the stable α-nitrodiarylmethyl cations (**55**), characterized by ^1H-, ^{13}C-, and ^{15}N-NMR spectroscopy.[57] Warming these solutions gives the protonated ketone; this is the only product when 9,9-dinitrofluorene is similarly treated, even at $-120°$.[57] The ion $^+C(CN)_3$ has been prepared from $ClC(CN)_3$ and $SbCl_5$.[58] The rearrangement of α-benzoyl carbocations in the gas phase has been studied.[59] Theoretical calculations show that α-keto or α-cyano groups actually provide conjugative stabilization, by charge delocalization to the O or N atoms.[60] However, another study shows that α-CN groups in carbocations are destabilizing, overall, at all levels of calculation, but that the α-NC group is stabilizing, better than α-F.[61]

Multiply Charged Systems

The topic of multiply charged carbocations and related species in solution has been reviewed,[62] as has that of carbodication structure.[63] Ion (**56**) forms when 1-nitronaphthalenes are diprotonated in triflic acid.[64] The stereomutation of (**57**)

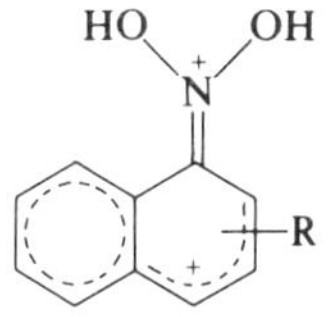

(**56**) 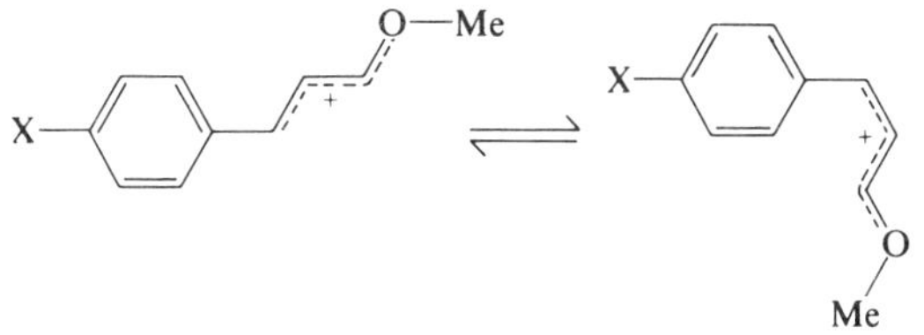 (**57**)

is catalysed by a number of super-acids, presumably by way of the oxygen-protonated dication.[65] Stable alkoxycarbenium and oxonium dications, such as $[CH_2O(CH_2)_3OH_2]^{2+}$, $[CH_2O(CH_2)_3OCH_2]^{2+}$, and $[H_2O(CH_2)_3OH_2]^{2+}$, form when 1,3-dioxanes are protonated in $HSO_3F/SbF_5/SO_2$ media.[66] Trimethylenemethane dications do not show the expected "Y-aromatic" stabilization in solution; for instance (**58**) only gives (**59**) with magic acid at $-115°$, and (**60**) is not protonated again.[67] However, (**61**) is stable to 0°.[67]

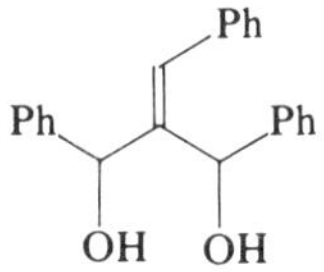

(**58**) 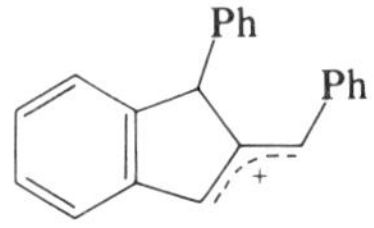 (**59**) 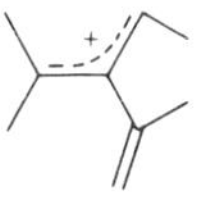 (**60**) 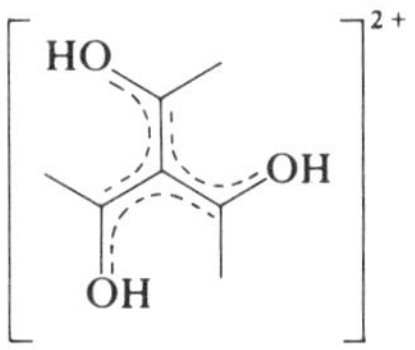 (**61**)

Open-chain, Monocyclic, and Heterocyclic Systems

The formation of trityl cations from trityl trifluoroacetates has been studied by NMR, and by relaxation and conductivity measurements, as a function of π-donor substituents and medium acidity.[68] It has been found that anion solvation by acid favours ion formation in acid media; the ion recombination rate is drastically reduced even at low acid concentrations, and the dissociation rate is increased at high ones.[68] The thermodynamics and kinetics of complex formation between trityl cations and DMF has been studied by NMR line-shape analysis.[69]

Arylalkene brominations involving benzylic carbocations, $XC_6H_4{}^+CRCHBrR'$, have been studied as a function of R and R′ (H, Me, OMe, and Ar′).[70] It has been found that ρ values are dependent on the reactivity of the parent alkene.[70] A bridged carbocation intermediate is believed to occur in the synthesis of γ-lactones from (E)-β-alkylcinnamic acids in concentrated H_2SO_4.[71] The solid-state ^{13}C-NMR CPMAS spectra of the rapidly equilibrating 2,3-dimethyl-2-butyl, 2,3,3-trimethyl-2-butyl, cyclopentyl, and 1,2-dimethylcyclopentyl cations at 128–193 K in solid SbF_5 reveal the essentially classical nature of these ions; peaks due to bridging methyl groups have not been detected.[72]

α-Seleno carbocations have been prepared and characterized by ^{1}H- and ^{13}C-NMR spectroscopy, and by X-ray structure determination.[73] In the reaction of 2-(phenylseleno)allyl cations with furan, thiophene, and pyrrole it is theorized that the cations have the bridged structure (**62**) rather than the open form (**63**).[74] The mechanism of the sulphohaloform reaction, which involves intermediate thionium ions (**64**),[75] has been revised and extended following a study of the aqueous chlorinolysis of a series of benzylic dithioacetals and α-chlorobenzylic sulphides.[76]

Ph Se$^+$ SePh

(**62**) (**63**)

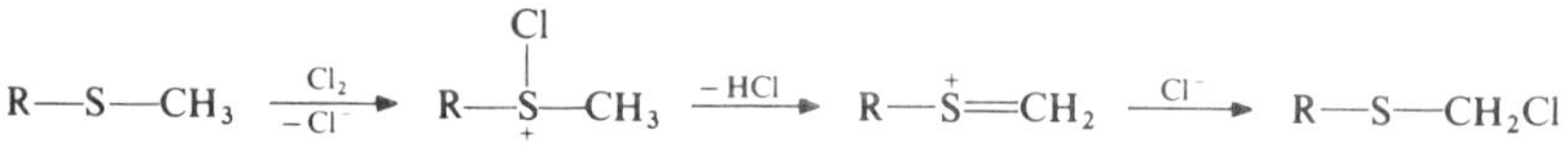

$$R{-}S{-}CH_3 \xrightarrow[-Cl^-]{Cl_2} R{-}\overset{Cl}{\underset{+}{S}}{-}CH_3 \xrightarrow{-HCl} R{-}\overset{+}{S}{=}CH_2 \xrightarrow{Cl^-} R{-}S{-}CH_2Cl$$

(**64**)

The bromination of cyclopropanes to 1,3-dibromides proceeds *via* corner-bromination.[77] Deuterated cyclopropane (**65**) yields (**67**) and (**68**); more that 85 % of the (**67**) is *erythro*, which can only be explained by invoking the intermediacy of (**66**).[77] Proton loss from (**69**) has unexpectedly been found to compete appreciably with capture by solvent.[78] Phenoxenium ions (**70**) have been identified as intermediates in the acid-catalysed solvolysis of *N*-tosyl-*O*-arylhydroxylamines, and in the thermolysis of *N*-aryloxypyridinium salts; in the presence of aromatic compounds they are captured by electrophilic attack.[79]

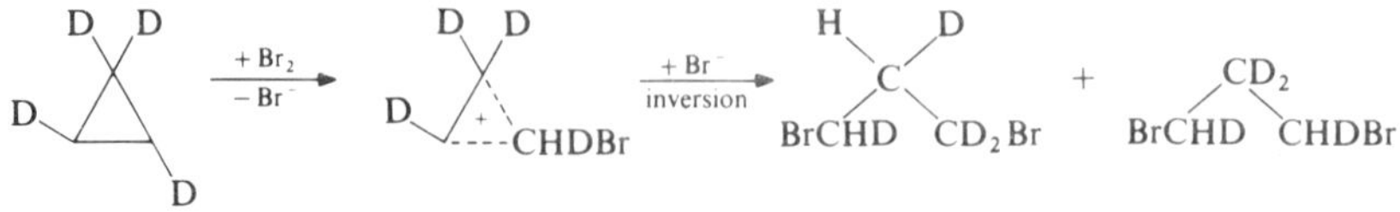

D D D D; $\xrightarrow[-Br^-]{+Br_2}$ D D D CHDBr; $\xrightarrow[\text{inversion}]{+Br^-}$ H D C BrCHD CD_2Br + CD_2 BrCHD CHDBr

(**65**) (**66**) (**67**) (**68**)

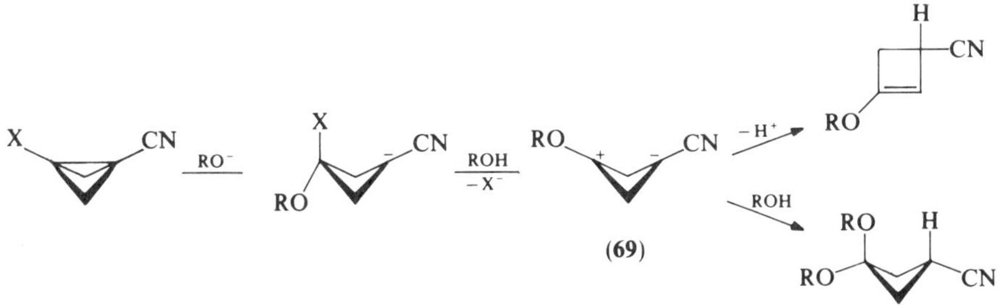

(69)

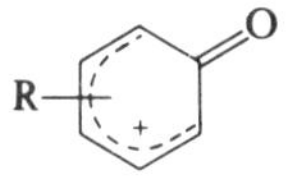

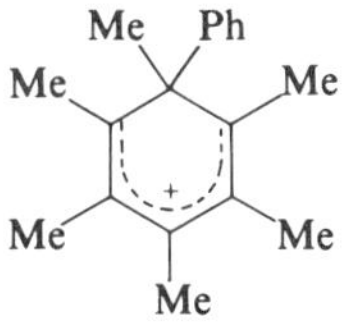

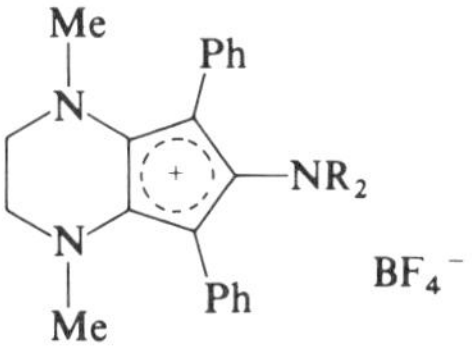

(70) (71) (72)

In the solid phase ion (**71**) undergoes degenerate 1,2-phenyl shifts more slowly than it does in solution; 1,2-methyl shifts are not seen.[80] NMR isotope shifts have been used to determine that the 1-(*p*-fluorophenyl)cyclopentyl cation adopts the twist conformation, similar to that found for cyclopentanone.[81] The regio- and stereo-chemistry of the addition of CF_3CO_2D to *cis*-cyclooctene do not support an open carbocation intermediate; the reaction involves a 1,5-hydride-bridged one.[82] Thermally stable cyclopentadienylium salts (**72**) have been prepared and studied,[83] as have the novel doubly cross-conjugated cations (**73**) and (**74**).[84]

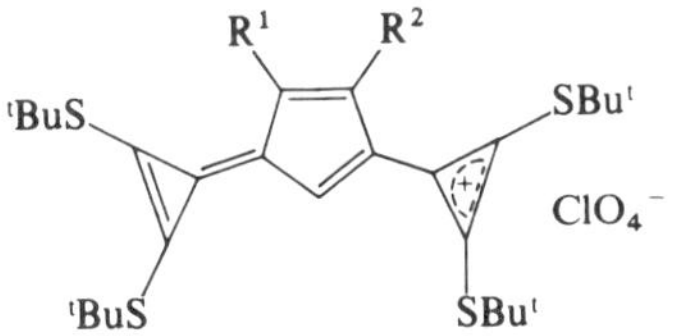

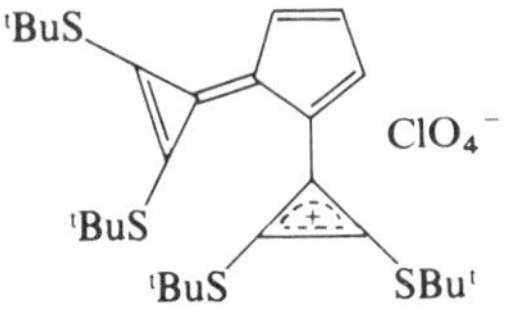

(73) (74)

$R^1 = R^2 = H$; $R^1, R^2 = (CH{=}CH)_2$

The reactions of the variously substituted (cycloheptatrienyl)diphenylmethyl carbocations differ according to the position of substitution.[85] The 1- and 2-cycloheptatrienyl ions electrocyclize to dihydrobenz[*a*]azulenes through the cycloheptatriene form, while the 3-cycloheptatrienyl ion gives the same product through the norcaradiene form. The 7-cycloheptatrienyl ion rearranges, giving triphenylethylene.[85] The mechanism of the reaction between 9-arylfluoren-9-yl cations and polymethylbenzenes in TFA involves a reduction to the 9-arylfluorene, probably by hydride transfer; at the transition state electron transfer may have progressed further than nuclear motion of the migrating hydrogen.[86]

The reaction of azide ion with pyrylium, thiopyrylium, and chromylium ions has been extensively investigated.[87–89] For example, the reaction of chromylium ions (**75**) with azide constitutes an easy preparation of benz[*f*]oxazepins (**76**).[87] Only sterically hindered pyrylium or thiopyrylium cations give covalent azide intermediates, non-hindered ones giving a dead-end donor–acceptor complex.[88] This reactivity pattern is consistent with that predicted by EHMO calculations.[89] The reaction of the 2-(*p*-methoxyphenyl)-1,3-dithiolan-2-yl cation with water and other nucleophiles has been the subject of a kinetic study.[90] Hydration is unusually slow compared to the oxygen analogue; it is rate-determining at pH > 6, it is general-base-catalysed, and other nucleophiles may compete with water. At higher acidity decomposition of the 2-hydroxythiolane intermediate to a thiol-ester becomes the slow step.[90] The thermodynamics of the equilibria (**77**)$\rightleftharpoons$(**78**)$\rightleftharpoons$(**79**) in the phthalan (X = 0) and thiophthalan (X = S) systems in aqueous buffer have been investigated.[91]

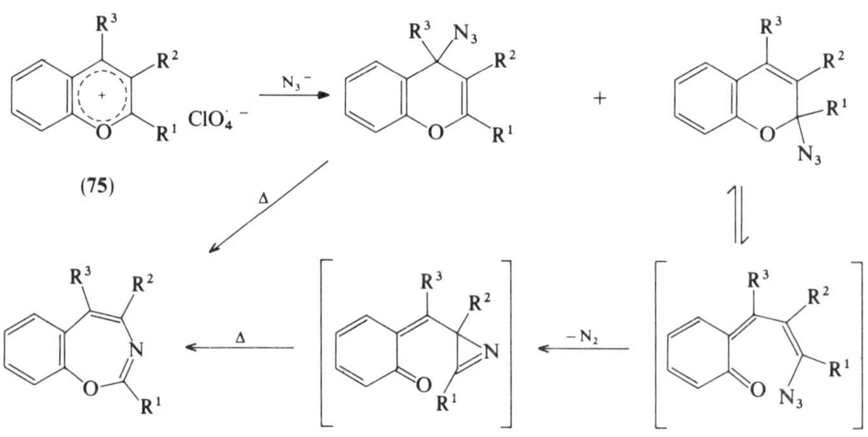

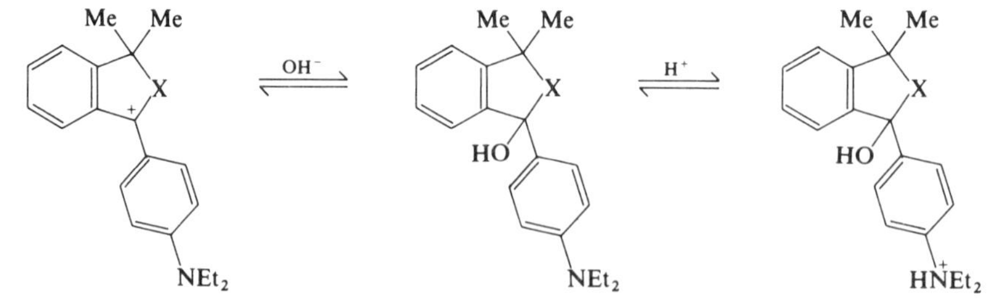

Aqueous Media, Solvolyses

The formation, stability, and reactions of ring-substituted 1-phenylethyl carbocations in 50:50 v/v TFE/H_2O,[92] and in 50:50 v/v TFE/ROH,[93] have been the subject of an extensive study, and the conditions under which general base catalysis applies to the addition of hydroxylic reagents to these ions have been delineated.[94] Very weak nucleophiles like ClO_4^-, $CF_3SO_3^-$, FSO_3^-, and TsO^- may give modest to high yields of products of covalent binding with carbocations derived from processes like electrophilic alkene additions, amine deaminations, electrophilic epoxide ring-openings, and oxidative alkyl iodide deiodinations, despite the presence of other nucleophiles such as AcOH, H_2O, and halide ions.[95] The acidic solvolyses of allylic halids and acetates of methylenecyclopropanes and cyclopropylmethanes do not result in ring-opening when an ethoxycarbonyl substituent is present on the ring methylene, only the cyclopropenyl derivatives being obtained.[96] Pseudo-base formation from 9-substituted 10-methylacridinium cations in aqueous solution has been investigated,[97] and internal return of carbocations has been probed by ^{17}O-NMR.[98]

The hydrolysis of neopentyl iodide with electrophilic cooperation by silver and mercury ions has been shown to proceed by the S_N1 mechanism without appreciable anchimeric assistance.[99] Equilibrium and rate constants for the reactions of methyl-substituted phenyltropylium ions in aqueous solution at 25° have been determined; the 2′,6′-di- and 2′,4′,6′-tri-substituted compounds show evidence of steric hindrance to coplanarity.[100] A rate constant of more than 1×10^8 s^{-1} is found for the reaction of the benzyl carbocation with water.[101] Various allylic *p*-nitrobenzoates (**80**; R = Et, $CH{=}CH_2$, $CH{=}CMe_2$, OMe, SMe) have had their solvolysis rate constants and secondary deuterium isotope effects determined; a step-wise mechanism with no neighbouring-group participation is suggested.[102]

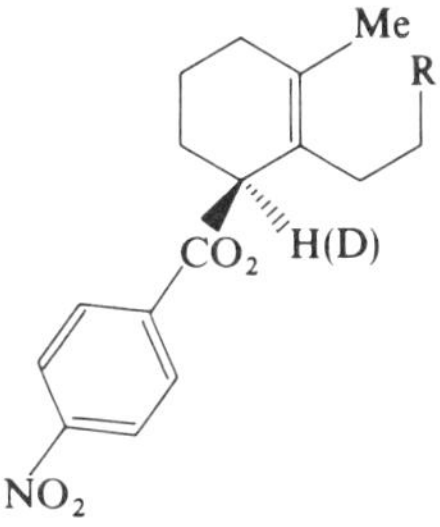

(**80**)

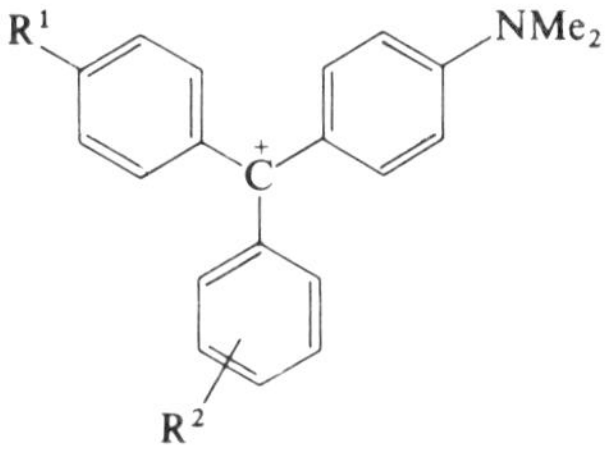

(**81**)

Kinetic data for the reaction of (**81**; $R^1 = R^2 = H$; $R^1 = NMe_2$, $R^2 = 4\text{-}NO_2$; $R^1 = OMe$, $R^2 = 3\text{-}Br$) with hydroxide or azide in aqueous acetonitrile have been used to probe the applicability of electrostatic theory to these processes.[103] Salt-effect studies on the ionization of 1-adamantyl iodide in acetonitrile suggest that the rate-determining step is the conversion from contact to solvent-separated ion-pairs.[104] The reactions of Crystal Violet, Malachite Green and 2,4-dinitrochlorobenzene with hydroxide ion are much faster in aqueous micro-emulsions than they are in aqueous solution alone;[105] Methyl Violet reacts with hydroxide ion faster in surfactant micelles.[106]

Super-acid Media

An extended thermochemical scale of carbocation stabilities in $SbF_5/FSO_3H/SO_2ClF$ has been published.[107] The variation of the cationic carbon ^{13}C-

NMR chemical shift with increasing electron demand in a series of 1,1-diaryl-1-ethyl carbocations in $SbF_5/FSO_3H/SO_2ClF$ or FSO_3H/SO_2ClF at $-70°$ has been interpreted as indicative of the importance of inductive localized π-polarization effects;[108] this also applies to a series of benzhydryl carbocations under similar conditions.[109] The effects of *p*-alkyl and *p*-cycloalkyl groups in a series of *p*-alkyl-*tert*-cumyl cations have also been studied.[12]

Cation (**82**) has been prepared and characterized;[110] α-ethylenehaloarenium ions (**83**) are stable at $-60°$ in $HF:SbF_5/SO_2ClF$.[111] The *cis*-1-methylcyclopropylcarbinyl cation (**84**) has been prepared, from *cis*-chloromethylcyclobutane precursors and $SbF_5/SO_2ClF/SO_2F_2$ at $-135°$, and its isomerization to the stable *trans*-isomer (**85**) at $-100°$ studied.[112] This is not a direct rotation but a complex process involving at least seven intermediates, one of which, the 1-ethylallyl cation, is observable.[112] The 2,3-dimethyl-3-fluoro-2-butyl cation undergoes exclusive methyl exchange in SbF_5/SO_2 or SbF_5/SO_2ClF at $-90°$; fluorine shifts through a bridged fluoronium ion do not occur.[113] The conformations of benzobicyclo[4.1.0]heptyl cations in super-acid have been studied.[114] Rearrangement of the first-formed ion results in (**86**) as a stable species; on raising the temperature dialkylnaphthalenium ions are obtained.[114]

R = H, Me (starting material, with CH_2CH_2F): SbF_5/SO_2ClF, $-90°$ → (**82**) ← SbF_5/SO_2ClF, $-80°$ to $-90°$ (starting alcohol, R = H, Me, Ph)

(**83**) X = F, Cl, Br

(**84**) → (**85**)

R = H, Me, Ph: FSO_3H/SO_2ClF, $-130°$ → [intermediate cation] → (**86**)

$R-C_6H_4-C(=O)-N_3$ $\xrightarrow{FSO_3H/SbF_5/SO_2,\ -78°}$ $R-C_6H_4-C^+(OH)-N_3$ (**87**) $\xrightarrow{\text{warm},\ -HN_3}$ $R-C_6H_4-\overset{+}{C}{=}O$

Azidocarboxonium ions (**87**) have been studied by ^{1}H-, ^{13}C-, and ^{15}N-NMR; significant charge delocalization into both the aryl ring and the N_3 group is found.[115] Protonation of 2,7-di-*tert*-butylthiepins in FSO_3H/SO_2 at $-70°$, or in concentrated H_2SO_4, results in homothiopyrylium ions.[116] The transformations undergone by tetrahydrofurfural acetals in FSO_3H have been characterized.[117] The "spin-charge exchange" process undergone in a stable radical carbocation has been studied.[118] The use of $ClSO_3H$ as a solvent for mono- and di-cations is the subject of a continuing investigation.[119]

Organometallic Systems

The use of iron-stabilized carbocations as intermediates in organic synthesis has been reviewed.[120] A vinyl ether/iron complex is an effective α-acrylic ester equivalent in the synthesis of α-methylene-γ-lactones.[121] The synthesis and some reactions of vinyl cations stabilized by the 3-η^5-$C_5H_5Fe^{II}$-η^5-(3)-1,2-$C_2B_9H_{10}$-1 group have been described; deprotonation to acetylene derivatives is a common reaction.[122] π-Allylpalladium cations react with stabilized nucleophiles at the less hindered position, but with PhZnCl at the more crowded site.[123]

The hydrolysis of ferrocenyliminium ions has been the subject of a detailed kinetic study.[124] Tricarbonyltropylium manganese cations react with enolate ions to give ring-substituted complexes; decomplexation provides convenient regio- and stereo-controlled functionalization of the seven-membered ring.[125] The reaction of the equivalent Mo and W complexes with alkoxide ions has been investigated spectroscopically at low temperatures; the intermediates observed provide evidence for both metal and carbonyl attack.[126] Extended Hückel calculations provide a method for predicting the position of attack in these cases.[127]

The Gas Phase

Structurally complex organic ions, including some carbocations, in the gas phase have been reviewed from the point of view of thermochemistry and non-covalent interactions.[128] Also reviewed is the subject of carbon skeletal rearrangements *via* pyrimidal carbocations.[129] The heats of formation of a number of simple carbocations have been obtained from the photoelectron spectra of the radicals.[130] A photoelectron spectroscopic study of the ground states of CH_2Cl^+, $CHCl_2{}^+$, and $CHFCl^+$ is also reported.[131]

Free phenylium cations, formed in the gas phase by the β-decay of pertritiated benzene followed by helium loss, attack both the C—H and C—C bonds of propene, giving tritiated cyclopropylbenzene, and various isomeric tritiated phenylpropenes and isopropylbenzenes, respectively.[132] Attack on the π-bond is preferred. Cyclopropane is also attacked, giving tritiated indane as a product.[132] The *p*-tritiophenylium ions formed in the β-decay of 1,4-ditritiobenzene react with dialkyl ethers to give, predominantly, tritiated aryl ethers by way of dialkylphenyloxonium ions.[133] Their reaction with gaseous methane, ethane, and propane has also been studied.[134] With gaseous methyl halides halobenzenes with significant scrambling of the tritium label are obtained; the results are consistent with consecutive 1,2-hydrogen shifts, with a rate constant of about 10^7 s^{-1}, in the phenyl cation, not in the first-formed products.[135]

The structures of the $C_3H_5^+$ ions obtained by protonating allene and propyne with radiolytically generated H_3^+, CH_5^+, or $C_2H_5^+$ in the gas phase have been investigated by trapping them with benzene and 1,4-dibromobutane.[136] It is found that the 2-propenyl cation rather than the allyl cation is formed almost exclusively,[136] a result similar to that found in aqueous solution.[137] The gas-phase pyrolysis of protonated glyme and diglyme involves methanol loss and oxycarbonium ion formation; rate constants and activation parameters have been determined.[138] The reaction of $C_2H_3^+$ with CH_4 has been studied,[139] and the mechanism of the interaction of Ph_3C^+ with THF investigated.[140] The reactions of the $FeCH_3^+$ and $CoCH_3^+$ ions with aliphatic alkanes[141] and cycloalkanes[142] involve insertion into C—H bonds; ring-cleavage is observed if the rings are small.[142]

Theoretical Calculations

The reality of the carbocation three-centre, two-electron bond has been reviewed.[143] Results in good agreement with *ab initio* calculations can be obtained by the SINDO method for simple carbocations.[144] Relative carbocation stabilities can be obtained from the substituent equivalent, which is a sum of all the electronic effects of all the substituents around the C^+ centre.[145] Activation barriers for 1,2-shifts of a number of groups in carbocations have been calculated using MINDO/3; they can be correlated with the degree of LUMO delocalization.[146]

The $C_4H_7^+$ potential surface has been explored using MINDO/3.[147] This method shows that both the cyclopropylcarbinyl and cyclobutyl cations are minima, in contrast to the *ab initio* method but in agreement with experiment; the latter ion has a 1-protonated bicyclobutane non-classical structure.[147] The mono-substituted allyl cation has been studied by *ab initio* calculations;[148] the canonical structures $Me_2\overset{+}{C}—C{\equiv}CH \leftrightarrow Me_2C{=}C{=}\overset{+}{C}H$ do not adequately describe the properties of the 1,1-dimethylpropargyl carbocation.[149] A computational verification of Markownikoff's rule has been provided.[150] A quantum-chemical study of the fragmentation of complexes of alcohols with H^+, Me^+, Me_3C^+, and $AlCl_3$ shows that as the acid strength increases, so does the tendency of the alcohol to fragment to the carbocation.[151] The ease of formation of carbocations from non-alternate polycyclic aromatic hydrocarbons has been discussed.[152]

A new class of ions, cation-substituted methonium ions (**88**), has been postulated as a result of *ab initio* calculations on the interaction of CH_4^{2+} with CO, NH_3, N_2, and H_2O.[153] The same method has been used on the 17 $C_2H_4N^+$ isomers; the most

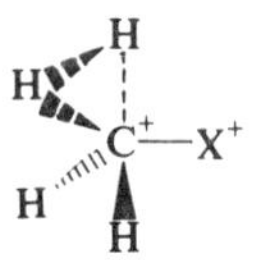

(**88**) 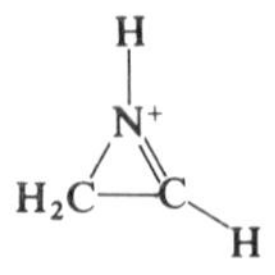 (**89**) 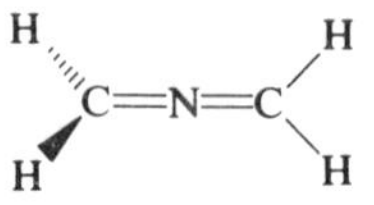 (**90**) 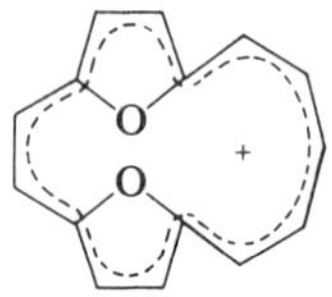 (**91**)

stable ion with C—N—C topology is $H\overset{+}{C}{=}N{-}CH_3$, that with C—C—N topology is $H_3C{-}\overset{+}{C}{=}NH$ (the best of the 17), and the most stable cyclic structure is (**89**).[154] The 2-azaallenium ion (**90**) is predicted to be experimentally accessible.[154] Successive substitution of vinyl groups for H preferentially stabilizes Si^+ relative to C^+ in $CH_3{}^+$ and $SiH_3{}^+$ [155]. The H_2SiX^+ ion has been compared to the analogous carbocations; with X = CHO or CN the ion is destabilized, but X = C≡CH is stabilizing.[156] $H_2Si^+CF_3$ and H_3SiCO^+ have also been examined.[156] The dioxido[15]annulenyl cation (**91**) has been studied,[157] and the effect of a cyano substituent on the cyclopropyl cation investigated.[158]

References

1. Kirmse, W., *Chem. Future, Proc. IUPAC Congr., 29th.*, **1983** (publ. 1984), 225; *Chem. Abs.*, **100,** 155894 (1984).
2. Barkhash, V. A., *Top. Curr. Chem.*, **116,** 1 (1984).
3. Koser, G. F., in *The Chemistry of Halides, Pseudo-Halides and Azides,* (Eds. Patai, S., and Rappoport, Z.), Wiley, Chichester, 1983, p. 1265; *Chem. Abs.*, **100,** 138199 (1984).
4. Koptyug, V. A., *Top. Curr. Chem.*, **122,** 1 (1984).
5. Shubin, V. G., *Top. Curr. Chem.*, **117,** 267 (1984).
6. Reutov, O. A., *Usp. Khim.*, **53,** 462 (1984); *Chem. Abs.*, **101,** 71818 (1984).
7. Saunders, M., Kates, M. R., and Walker, G. E., *Prepr.—Am. Chem. Soc., Div. Pet. Chem.*, **28,** 410 (1983); *Chem. Abs.*, **100,** 120132 (1984).
8. Hoffmann, H. M. R., *Angew. Chem. Int. Ed.*, **23,** 1 (1984).
9. Brown, H. C., *Prepr.—Am. Chem. Soc., Div. Pet. Chem.*, **28,** 374 (1983); *Chem. Abs.*, **100,** 155889 (1984).
10. Olah, G. A., and Prakash, G. K. S., *Prepr.—Am. Chem. Soc., Div. Pet. Chem.*, **28,** 366 (1983); *Chem. Abs.*, **100,** 120409 (1984).
11. Brown, H. C., Periasamy, M., Perumal, P. T., and Kelly, D. P., *J. Am. Chem. Soc.*, **106,** 2359 (1984).
12. Brown, H. C., Periasamy, M., and Perumal, P. T., *J. Org. Chem.*, **49,** 2754 (1984).
13. Coxon, J. M., and Steel, P. J., *J. Chem. Soc., Chem. Commun.*, **1984,** 344.
14. Grob, C. A., and Sawlewicz, P., *Tetrahedron Lett.*, **25,** 2973 (1984).
15. Baldwin, J. E., and Barden, T. C., *J. Am. Chem. Soc.*, **105,** 6656 (1983).
16. Clemans, G. B., Samaritoni, J. G., Holloway, R. J., and Edinger, W., *J. Org. Chem.*, **49,** 3457 (1984).
17. Kirmse, W., and Brandt, S., *Chem. Ber.*, **117,** 2524 (1984).
18. Gassman, P. G., and Hall, J. B., *J. Am. Chem. Soc.*, **106,** 4267 (1984).
19. Kirmse, W., and Brandt, S., *Chem. Ber.*, **117,** 2510 (1984).
20. Martínez, A. G., Rios, I. E., Barcina, J. O., and Hernando, M. M., *Chem. Ber.*, **117,** 982 (1984).
21. Wilt, J. W., and George, C., *J. Org. Chem.*, **49,** 2297 (1984).
22. Carrupt, P.-A., and Vogel, P., *Tetrahedron Lett.*, **25,** 2879 (1984).
23. Ahlberg, P., Jonsall, G., Engdahl, C., Huang, M. B., and Goscinski, O., *Prepr.—Am. Chem. Soc., Div. Pet. Chem.*, **28,** 357 (1983); *Chem. Abs.*, **100,** 174167 (1984).
24. Shin, J. H., *Chayon Kwahak Taehak Nomunjip (Soul Taehakkyo)*, **8,** 91 (1983); *Chem. Abs.*, **101,** 130123 (1984).
25. Jonsäll, G., and Ahlberg, P., *J. Chem. Soc., Chem. Commun.*, **1984,** 1125.
26. Goldstein, M. J., and Dinnocenzo, J. P., *J. Am. Chem. Soc.*, **106,** 2473 (1984).
27. Sugihara, Y., Hashimoto, K., and Murata, I., *J. Org. Chem.*, **49,** 1146 (1984).
28. Maskill, H., and Wilson, A. A., *J. Chem. Soc., Perkin Trans. 2,* **1984,** 119.
29. Maskill, H., and Wilson, A. A., *J. Chem. Soc., Perkin Trans. 2,* **1984,** 1369.
30. Kirmse, W., Siegfried, R., and Streu, J., *J. Am. Chem. Soc.*, **106,** 2465 (1984).
31. Nisnevich, G. A., Vyalkov, A. I., Komshii, G. T., Mamatyuk, V. I., and Barkhash, V. A., *Zh. Org. Khim.*, **19,** 2081 (1983); *Chem. Abs.*, **100,** 138351 (1984).
32. Nisnevich, G. A., Mamatyuk, V. I., and Barkhash, V. A., *Zh. Org. Khim.*, **19,** 2551 (1983); *Chem. Abs.*, **100,** 208707 (1984).
33. Childs, R. F., McGlinchey, M. J., and Varadarajan, A., *J. Am. Chem. Soc.*, **106,** 5974 (1984).
34. Mass, G., and Lorenz, W., *J. Org. Chem.*, **49,** 2273 (1984).

[35] Kitamura, T., Kobayashi, S., and Taniguchi, H., *Chem. Lett.*, **1984,** 1523.
[36] Kitamura, T., Kobayashi, S., and Taniguchi, H., *J. Org. Chem.*, **49,** 3167 (1984).
[37] Winkelhofer, G., Janoschek, R., Fratev, F., and Schleyer, P. von R., *Croat. Chem. Acta*, **56,** 509 (1983); *Chem. Abs.*, **100,** 67483 (1984).
[38] Siehl, H.-U., and Koch, E.-W., *J. Org. Chem.*, **49,** 575 (1984).
[39] Siehl, H.-U., *J. Chem. Soc., Chem. Commun.*, **1984,** 635.
[40] Brennenstuhl, W., and Hanack, M., *Tetrahedron Lett.*, **25,** 3437 (1984).
[41] Collins, C. J., Hanack, M., Stutz, H., Auchter, G., and Schoberth, W., *J. Org. Chem.*, **48,** 5260 (1983).
[42] Rappoport, Z., Kaspi, J., and Tsidoni, D., *J. Org. Chem.*, **49,** 80 (1984).
[43] Ladika, M., Stang, P. J., Schiavelli, M. D., and Kowalski, M. H., *J. Am. Chem. Soc.*, **106,** 6372 (1984).
[44] Collins, C. J., Fuchs, K. A., and Hanack, M., *Prepr.—Am. Chem. Soc., Div. Pet. Chem.*, **28,** 352 (1983); *Chem. Abs.*, **101,** 22661 (1984).
[45] Collins, C. J., Martínez, A. G., Alvarez, R. M., and Aguirre, J. A., *Chem. Ber.*, **117,** 2815 (1984).
[46] Rappoport, Z., Fiakpui, C. Y., Yu, X.-D., and Lee, C. C., *J. Org. Chem.*, **49,** 570 (1984).
[47] Scaiano, J. C., and Nguyen, K. T., *J. Photochem.*, **23,** 269 (1983).
[48] Holweger, W., and Hanack, M., *Chem. Ber.*, **117,** 3004 (1984).
[49] Bleckmann, W., and Hanack, M., *Chem. Ber.*, **117,** 3021 (1984).
[50] Haberfield, P., Cardona, C., and Bell, M., *J. Org. Chem.*, **49,** 78 (1984).
[51] Tidwell, T. T., *Angew. Chem. Int. Ed.*, **23,** 20 (1984).
[52] Charpentier-Morize, M. G., *Prepr.—Am. Chem. Soc., Div. Pet. Chem.*, **28,** 297 (1983); *Chem. Abs.*, **100,** 155888 (1984).
[53] Allen, A.D., Ambidge, I. C., Kanagasabapathy, V. M., and Tidwell, T. T., *Prepr.—Am. Chem. Soc., Div. Pet Chem.*, **28,** 339 (1983); *Chem. Abs.*, **101,** 71906 (1984).
[54] Gassman, P. G., Talley, J. J., Saito, K., Guggenheim, T. L., Doherty, M. M., and Dixon, D. A., *Prepr.—Am. Chem. Soc., Div. Pet. Chem.*, **28,** 334 (1983); *Chem. Abs.*, **101,** 37888 (1984).
[55] Creary, X., and Geiger, C. C., *J. Am. Chem. Soc.*, **105,** 7123 (1983).
[56] Okamoto, K., Takeuchi, K., and Kitagawa, T., *Prepr.—Am. Chem. Soc., Div. Pet. Chem.*, **28,** 347 (1983); *Chem. Abs.*, **101,** 110053 (1984).
[57] Olah, G. A., Prakash, G. K. S., Arvanaghi, M., Krishnamurthy, V. V., and Narang, S. C., *J. Am. Chem. Soc.*, **106,** 2378 (1984).
[58] Beaumont, R. C., Aspin, K. B., Demas, T. J., Hoggatt, J. H., and Potter, G. E., *Inorg. Chim. Acta*, **84,** 141 (1984).
[59] Gruetzmacher, H. F., and Dommroese, A. M., *Org. Mass Spectrom.*, **18,** 601 (1983); *Chem. Abs.*, **101,** 22806 (1984).
[60] Dixon, D. A., Eades, R. A., Frey, R., Gassman, P. G., Hendewerk, M. L., Paddon-Row, M. N., and Houk, K. N., *J. Am. Chem. Soc.*, **106,** 3885 (1984).
[61] Lien, M. H., Hopkinson, A. C., and McKinney, M. A., *THEOCHEM*, **14,** 37 (1983).
[62] Pagni, R. M., *Tetrahedron*, **40,** 4161 (1984).
[63] Schleyer, P. von R., *Prepr.—Am. Chem. Soc., Div. Pet. Chem.*, **28,** 413 (1983); *Chem. Abs.*, **100,** 155921 (1984).
[64] Ohta, T., Shudo, K., and Okamoto, T., *Tetrahedron Lett.*, **25,** 325 (1984).
[65] Blackburn, C., and Childs, R. F., *J. Chem. Soc., Chem. Commun.*, **1984,** 812.
[66] Akhmatdinov, R. T., Kudasheva, I. A., Kantor, E. A., and Rakhmankulov, D. L., *Zh. Org. Khim.*, **19,** 1965 (1983); *Chem. Abs.*, **100,** 50818 (1984).
[67] Schötz, K., Clark, T., Schaller, H., and Schleyer, P. von R., *J. Org. Chem.*, **49,** 733 (1984).
[68] Blumenstock, H., Dickert, F., Fackler, H., and Hammerschmidt, A., *Z. Phys. Chem. (Weisbaden)*, **135,** 157 (1983).
[69] Blumenstock, H., Dickert, F. L., and Hammerschmidt, A., *Z. Phys. Chem. (Weisbaden)*, **139,** 123 (1984).
[70] Ruasse, M.-F., Argile, A., and Dubois, J.-E., *J. Am. Chem. Soc.*, **106,** 4846 (1984).
[71] Jalander, L., *Tetrahedron Lett.*, **25,** 457 (1984).
[72] Myhre, P. C., Kruger, J. D., Hammond, B. L., Lok, S. M., Yannoni, C. S., Macho, V., Limbach, H. H., and Vieth, H. M., *J. Am. Chem. Soc.*, **106,** 6079 (1984).
[73] Hevesi, L., Desauvage, S., Georges, B., Evrard, G., Blanpain, P., Michel, A., Harkema, S., and van Hummel, G. J., *J. Am. Chem. Soc.*, **106,** 3784 (1984).
[74] Halazy, S., and Hevesi, L., *J. Org. Chem.*, **48,** 5242 (1983).
[75] Grossert, J. S., and Langler, R. F., *Can. J. Chem.*, **55,** 407 (1977).
[76] Baum, J. C., Hardstaff, W. R., Langler, R. F., and Makkinje, A., *Can. J. Chem.*, **62,** 1687 (1984).

[77] Lambert, J. B., Schulz, W. J., Mueller, P. H., and Kobayashi, K., *J. Am. Chem. Soc.*, **106**, 792 (1984).
[78] Hoz, S., and Aurbach, D., *J. Chem. Soc., Chem. Commun.*, **1984**, 364.
[79] Iijima, H., Endo, Y., Shudo, K., and Okamoto, T., *Tetrahedron*, **40**, 4981 (1984).
[80] Borodkin, G. I., Nagi, Sh. M., Mamatyuk, V. I., Shakirov, M. M., and Shubin, V. G., *Zh. Org. Khim.*, **20**, 552 (1984); *Chem. Abs.*, **101**, 54236 (1984).
[81] Forsyth, D. A., and Botkin, J. H., *J. Am. Chem. Soc.*, **106**, 4296 (1984).
[82] Nordlander, J. E., Kotian, K. D., Raff, D. E., and Njoroge, G. F., *Prepr.—Am. Chem. Soc., Div. Pet. Chem.*, **28**, 285 (1983); *Chem. Abs.*, **101**, 22660 (1984).
[83] Gompper, R., and Glöckner, H., *Angew. Chem. Int. Ed.*, **23**, 53 (1984).
[84] Yoshida, Z., Shibata, M., Kida, S., Miki, S., Sugimoto, T., and Yoneda, S., *Tetrahedron Lett.*, **25**, 345 (1984).
[85] Mizumoto, K., Okada, K., and Oda, M., *Tetrahedron Lett.*, **25**, 2999 (1984).
[86] Bethell, D., Clare, P. N., and Hare, G. J., *J. Chem. Soc., Perkin Trans. 2*, **1983**, 1889.
[87] Desbene, P.-L., and Cherton, J.-C., *Tetrahedron*, **40**, 3567 (1984).
[88] Desbene, P.-L., Cherton, J.-C., Le Roux, J.-P., and Basselier, J.-J., *Tetrahedron*, **40**, 3539 (1984).
[89] Desbene, P.-L., Richard, D., Cherton, J.-C., and Chaquin, P., *Tetrahedron*, **40**, 3549 (1984).
[90] Okuyama, T., Fujiwara, W., and Fueno, T., *J. Am. Chem. Soc.*, **106**, 657 (1984).
[91] Oparin, D. A., *Vestsi Akad. Navuk BSSR, Ser. Khim. Navuk*, **1984**, 58; *Chem. Abs.*, **100**, 191261 (1984).
[92] Richard, J. P., Rothenberg, M. E., and Jencks, W. P., *J. Am. Chem. Soc.*, **106**, 1361 (1984).
[93] Richard, J. P., and Jencks, W. P., *J. Am. Chem. Soc.*, **106**, 1373 (1984).
[94] Richard, J. P., and Jencks, W. P., *J. Am. Chem. Soc.*, **106**, 1396 (1984).
[95] Zefirov, N. S., Koz'min, A. S., Zhdankin, V. V., Kirin, V. N., Yur'eva, N. M., and Sorokin, V. D., *Chem. Scr.*, **22**, 195 (1983).
[96] Arnaud, R., Dussauge, A., Faucher, H., Subra, R., Vidal, M., and Vincens, M., *Tetrahedron*, **40**, 315 (1984).
[97] Bunting, J. W., Chew, V. S. F., Abhyankar, S. B., and Goda, Y., *Can. J. Chem.*, **62**, 351 (1984).
[98] Chang, S., *Diss. Abstr. Int. B*, **44**, 2158 (1984); *Chem. Abs.*, **100**, 120137 (1984).
[99] Zamashchikov, V. V., Rudakov, E. S., Bezbozhnaya, T. V., and Matveev, A. A., *Zh. Org. Khim.*, **20**, 14 (1984); *Chem. Abs.*, **100**, 156002 (1984).
[100] Vaskuri, J., and Virtanen, P. O. I., *Finn. Chem. Lett.*, **1983**, 104.
[101] Lilie, J., and Koskikallio, J., *Acta Chem. Scand.*, **A38**, 41 (1984).
[102] Ladika, M., and Sunko, D. E., *Croat. Chem. Acta*, **57**, 179 (1984); *Chem. Abs.*, **101**, 54170 (1984).
[103] Sinev, V. V., and Nikolova, T. A., *Zh. Org. Khim.*, **20**, 583 (1984); *Chem. Abs.*, **101**, 54473 (1984).
[104] Ponomareva, E. A., Tarasenko, P. V., Yurchenko, A. G., and Dvorko, G. F., *Zh. Org. Khim.*, **19**, 2503 (1983); *Chem. Abs.*, **100**, 120438 (1984).
[105] Blandamer, M. J., Clark, B., and Burgess, J., *J. Chem. Soc., Faraday Trans. 1*, **80**, 1651 (1984).
[106] Malaviya, S., and Katiyar, S. S., *Z. Phys. Chem. (Leipzig)*, **265**, 26 (1984).
[107] Arnett, E. M., and Hofelich, T. C., *Prepr.—Am. Chem. Soc., Div. Pet. Chem.*, **28**, 406 (1983); *Chem. Abs.*, **100**, 174185 (1984).
[108] Brown, H. C., Periasamy, M., Perumal, P. T., Kelly, D. P., and Giansiracusa, J. J., *J. Am. Chem. Soc.*, **105**, 6300 (1983).
[109] Kelly, D. P., and Jenkins, M. J., *J. Org. Chem.*, **49**, 409 (1984).
[110] Olah, G. A., and Singh, B. P., *J. Am. Chem. Soc.*, **106**, 3265 (1984).
[111] Olah, G. A., Singh, B. P., and Liang, G., *J. Org. Chem.*, **49**, 2922 (1984).
[112] Falkenberg-Andersen, C., Ranganayakulu, K., Schmitz, L. R., and Sorensen, T. S., *J. Am. Chem. Soc.*, **106**, 178 (1984).
[113] Olah, G. A., Prakash, G. K. S., and Krishnamurthy, V. V., *J. Org. Chem.*, **48**, 5116 (1983).
[114] Kelly, D. P., Leslie, D. R., and Smith, B. D., *J. Am. Chem. Soc.*, **106**, 687 (1984).
[115] Mertens, A., Arvanaghi, M., and Olah, G. A., *Chem. Ber.*, **116**, 3926 (1983).
[116] Yamamoto, K., Yamazaki, S., Matsukawa, A., and Murata, I., *J. Chem. Soc., Chem. Commun.*, **1984**, 604.
[117] Karakhanov, R. A., Skurko, M. R., Ramazanov, O. M., Kantor, E. A., Bartok, M., and Bucsi, I., *Acta Phys. Chem.*, **29**, 181 (1983); *Chem. Abs.*, **101**, 22826 (1984).
[118] Ballester, M., Castañer, J., Riera, J., and Pascual, I., *J. Am. Chem. Soc.*, **106**, 3365 (1984).
[119] Puri, J. K., and Dhillon, D. S., *Inorg. Chim. Acta*, **90**, 165 (1984).
[120] Pearson, A. J., *Science (Washington, D. C., 1883–)*, **223**, 895 (1984); *Chem. Abs.*, **100**, 138179 (1984).
[121] Chang, T. C. T., and Rosenblum, M., *Isr. J. Chem.*, **24**, 99 (1984).
[122] Zakharkin, L. I., and Kobak, V. V., *J. Organomet. Chem.*, **270**, 229 (1984).

[123] Kienan, E., and Sahai, M., *J. Chem. Soc., Chem. Commun.*, **1984,** 648.
[124] Bunton, C. A., Davoudzadeh, F., Jagdale, M. H., and Watts, W. E., *J. Chem. Soc., Perkin Trans. 2*, **1984,** 395.
[125] Pearson, A. J., Bruhn, P., and Richards, I. C., *Tetrahedron Lett.*, **25,** 387 (1984).
[126] Brown, D. A., Fitzpatrick, N. J., Glass, W. K., and Taylor, T. H., *J. Organomet. Chem.*, **275,** C9 (1984).
[127] Brown, D. A., Fitzpatrick, N. J., and McGinn, M. A., *J. Organomet. Chem.*, **275,** C5 (1984).
[128] Meot-Ner (Mautner), M., *Acc. Chem. Res.*, **17,** 186 (1984).
[129] Schwarz, H., Thies, H., and Franke, W., *NATO ASI Ser., Ser. C*, **118** (Ionic Processes Gas Phase), 267 (1984); *Chem. Abs.*, **100,** 173910 (1984).
[130] Schultz, J. C., Houle, F. A., and Beauchamp, J. L., *J. Am. Chem. Soc.*, **106,** 3917 (1984).
[131] Andrews, L., Dyke, J. M., Jonathan, N., Keddar, N., and Morris, A., *J. Am. Chem. Soc.*, **106,** 299 (1984).
[132] Colosimo, M., Speranza, M., Cacace, F., and Ciranni, G., *Tetrahedron*, **40,** 4873 (1984).
[133] Fornarini, S., and Speranza, M., *J. Chem. Soc., Perkin Trans. 2*, **1984,** 171.
[134] Angelini, G., Sparapani, C., and Speranza, M., *Tetrahedron*, **40,** 4865 (1984).
[135] Speranza, M., Keheyan, Y., and Angelini, G., *J. Am. Chem. Soc.*, **105,** 6377 (1983).
[136] Fornarini, S., Speranza, M., Attinà, M., Cacace, F., and Giacomello, P., *J. Am. Chem. Soc.*, **106,** 2498 (1984).
[137] Cramer, P., and Tidwell, T. T., *J. Org. Chem.*, **46,** 2683 (1981).
[138] Sieck, L. W., and Meot-Ner (Mautner), M., *J. Phys. Chem.*, **88,** 5324 (1984).
[139] Senzer, S. N., Lim, K. P., and Lampe, F. W., *J. Phys. Chem.*, **88,** 5314 (1984).
[140] Bekhli, E. Yu., Fomina, M. V., and Entelis, S. G., *Khim. Fiz.*, **1982,** 649; *Chem. Abs.*, **101,** 130137 (1984).
[141] Jacobson, D. B., and Freiser, B. S., *J. Am. Chem. Soc.*, **106,** 3891 (1984).
[142] Jacobson, D. B., and Freiser, B. S., *J. Am. Chem. Soc.*, **106,** 3900 (1984).
[143] Lenoir, D., *Nachr. Chem., Tech. Lab.*, **31,** 889 (1983); *Chem. Abs.*, **100,** 33795 (1984).
[144] Dwivedi, C. P. D., *Indian J. Phys.*, **58B,** 69 (1984); *Chem. Abs.*, **101,** 190683 (1984).
[145] Lin, Z., *Huaxue Tongbao*, **1983,** 46; *Chem. Abs.*, **99,** 174934 (1983).
[146] Frenking, G., *Tetrahedron*, **40,** 377 (1984).
[147] Dewar, M. J. S., and Reynolds, C. H., *J. Am. Chem. Soc.*, **106,** 6388 (1984).
[148] Lien, M. H., and Hopkinson, A. C., *J. Phys. Chem.*, **88,** 1513 (1984).
[149] Andres, J., Silla, E., Bertran, J., and Tapia, O., *THEOCHEM*, **16,** 211 (1984).
[150] Hu, J., and Wei, T., *Xibei Shifan Xueyuan Xuebao, Ziran Kexueban*, **1982,** 67; *Chem. Abs.*, **100,** 67564 (1984).
[151] Sangalov, Yu. A., Babkin, V. A., Nel'kenbaum, Yu. Ya., and Minsker, K. S., *Teor. Eksp. Khim.*, **19,** 615 (1983); *Chem. Abs.*, **100,** 33820 (1984).
[152] Lowe, J. P., and Silverman, B. D., *J. Am. Chem. Soc.*, **106,** 5955 (1984).
[153] Lammertsma, K., *J. Am. Chem. Soc.*, **106,** 4619 (1984).
[154] Würthwein, E.-U., *J. Org. Chem.*, **49,** 2971 (1984).
[155] Truong, T., Gordon, M. S., and Boudjouk, P., *Organometallics*, **3,** 484 (1984).
[156] Hopkinson, A. C., and Lien, M. H., *THEOCHEM*, **13,** 303 (1984).
[157] Banerjee, P., and Das Gupta, N. K., *Indian J. Chem.*, **22B,** 905 (1983); *Chem. Abs.*, **100,** 173919 (1984).
[158] Moffat, J. B., *THEOCHEM*, **17,** 93 (1984).

Organic Reaction Mechanisms 1984
Edited by A. C. Knipe and W. E. Watts

CHAPTER 10

Nucleophilic Aliphatic Substitution

J. SHORTER

Department of Chemistry, The University, Hull.

Vinylic Systems

Rappoport's extensive studies of vinylic substitution have continued with a study of the solvolysis of (**1a, b**) in various media.[1] The high selectivity of the derived vinyl cation (**2**) and the k_{OTs}/k_{Br} values have been discussed in terms of steric effects on the approach of the nucleophiles to (**2**) and the relative steric acceleration of the solvolysis of (**1a**). In the substitution of tricyanovinyl halides by the *para*-position of three dialkylanilines, $k_{Br}/k_{Cl} = 2.37 \pm 0.18$, indicating an early transition state for the expulsion of Hal^- from the intermediate zwitterion.[2] Rappoport and the original authors have also offered a reinterpretation of the distribution of products in the degenerate rearrangement of doubly labelled triarylvinyl cations (see *Organic Reaction Mechanisms* 1983, Chap. 10, Refs. 3 and 4).[3]

In the solvolyses of the stereoisomeric 1,4-dimethyl-1,3-hexadien-5-yn-1-yl triflates (**3**) and (**4**), the (E)-isomers (**4**), in contrast to the (Z)-isomers (**3**) react mainly *via* the intermediate phenyl cation (**5**) to give phenyl ethers (**6**).[4] In the solvolyses of a further (E)–(Z) pair of related triflates, a naphthyl cation has been found to be intermediate for the (E)-isomer.[5]

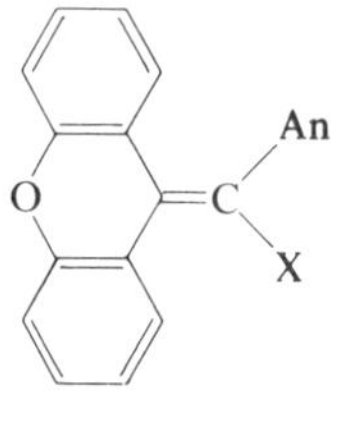

(**1**)

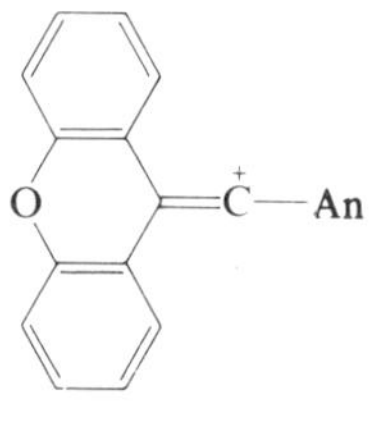

(**2**)

a: X = Br
b: X = OTs
An = *p*-$MeOC_6H_4$

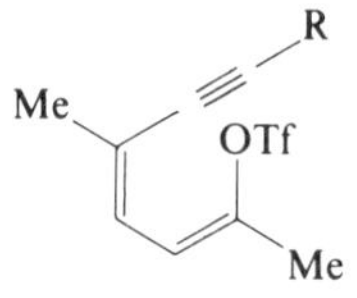

(**3**)

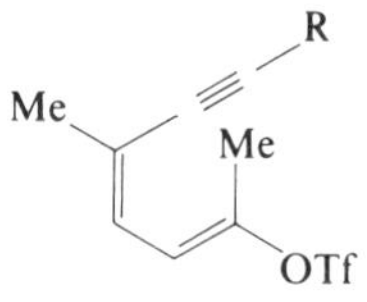

(**4**)

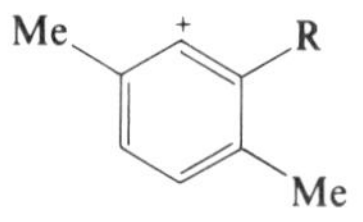

(**5**)

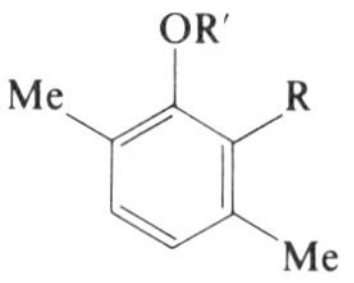

(**6**)

For a series of cyclic vinyl triflates, solvolysis products in TFE and rates of solvolysis in 50 % ethanol have been determined.[6] From a successful correlation with Taft's σ^* values it has been deduced that the solvolyses of most of the substrates occur by a k_c mechanism.

Unactivated vinyl halides react with the sodium salts of alkanethiols in HMPT giving high yields of vinyl alkyl sulphides with complete retention of configuration.[7] Reaction with alkyl and aryl selenide anions in various dipolar aprotic solvents is similarly stereospecific.[8]

The application of frontier MO theory and *ab initio* MO calculations to nucleophilic addition to alkenes leads to the conclusion that nucleophilic substitution with simple alkenes involves a concerted one-step mechanism, proceeding with retention of configuration.[9] If a poor leaving group is involved, however, or a stabilized carbanionic intermediate is formed, stereoconvergence results. HOMO–LUMO interaction is very important in vinylic substitution.

The displacement of halide ion from 2-halo-1,4-quinones may involve attack at the *ipso* or vicinal carbon atom, depending on the nature of the nucleophile and the solvent.[10] For example, the hard nucleophiles F^- and MeO^- attack predominantly *ipso*, while the soft 1,1-dimethoxyethene prefers vicinal attack; with amines the favoured mode of attack is vicinal in benzene and acetonitrile, but *ipso* in methanol.

The kinetics and mechanism of reactions of *trans*-aryl β-chlorovinyl sulphoxides with amines,[11] and of $\beta\beta$-dibromo- and β-chloro-β-iodo-vinyl sulphones with NaOMe and 4-RC_6H_4SNa (various groups R)[12] have been studied. Kinetic studies of nucleophilic vinylic substitution of furan and thiophene derivatives have also been made.[13]

Allylic and Various Unsaturated Systems

The hydrolysis of β-methyl-1- and -3-phenylallyl chlorides, (**7a**) and (**8a**), in aqueous dioxan proceeds *via* the same mechanism as previously reported for (**7b**) and (**8b**), respectively, with different behaviour by the first intermediate;[14] for (**7a**) this rearranges to the corresponding conjugated compound through a transition state rather than a further intermediate. The application of the extended Grunwald–Winstein equation to the rates of solvolysis of allyl arenesulphonates indicates a marked dependence on both the solvent nucleophilicity and solvent ionizing power.[15] This sensitivity diminishes as charge delocalization in the leaving group increases. The solvolysis in acid medium of allylic halides and acetates of methylenecyclopropanes (**9**) or their cyclopropenyl isomers (**10**) does not lead to

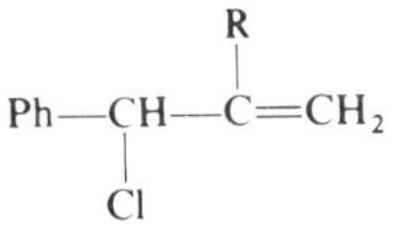

Ph—CH(Cl)—C(R)=CH_2 (**7**)

Ph—CH=C(R)—CH_2Cl (**8**)

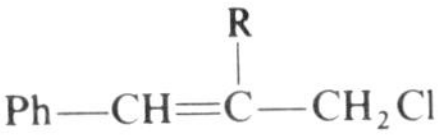

a: R = Me
b: R = H

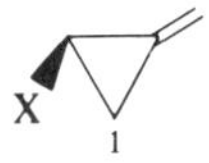

(**9**) 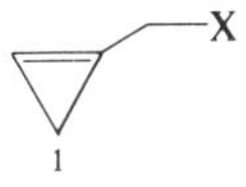 (**10**)

ring-opening when there is a CO_2Et group at position 1 of the ring.[16] Only the cyclopropenyl derivative is obtained. A MNDO study of the possible allylic cation was carried out. The introduction of a CF_3 group in the 1-position of allylic sulphonates retards solvolysis by a factor of 2×10^6, compared with 4×10^4 for a CF_3 group in the 3-position.[17]

For the reactions of substituted sodium phenoxides with allyl chloride or bromide in mixtures of aprotic solvents, kinetic studies support a concerted one-step S_N2 mechanism for the formation of *ortho*-allylated product;[18] reaction is retarded in protic solvents.

S_N2' reactions (bimolecular nucleophilic substitution with allylic rearrangement) continue to excite interest. Their preferred *syn*-stereochemistry has been ascribed to a multiplicity of conformational effects in the transition-state structures.[19] *Ab initio* MO calculations show that preferred *trans*-bending upon nucleophilic attack, the rotation of allylic groups into staggered conformations, and the preferred *anti*-periplanar arrangement of the developing lone-pair and the leaving group lead to the *syn*-preference. HMO theory has been used to study the stereoselectivity of S_N2' and other reactions.[20] Such reactions have also been featured in a rather general theoretical article under the title "Multibond reactions cannot normally be synchronous".[21] The S_N2' reaction is currently thought to involve a synchronous multi-bond mechanism, but on reviewing the evidence the author concludes that there is no clear support for synchronicity. MNDO studies support a step-wise process for the S_N2' reaction of chloride ion with allyl chloride. Further similar studies on S_N2 and S_N2' reactions involving anionic nucleophiles have confirmed the determining rôle of the desolvation of the anion.[22] Thus the S_N2' reaction of chloride ion with allyl chloride should take place without activation in the gas phase; its slowness in solution must be due to the energy required to remove solvent from the anion so that the allyl derivative can approach.

The rôle of copper compounds in the S_N2' reaction has received further attention. Thus, in the reaction of 2-allylthiobenzothiazoles with Grignard reagents in the presence of Cu(I) bromide, the S_N2' product is favoured by a low ratio of Grignard reagent to CuBr, and the S_N2 product by a high one.[23] S_N2' reactions with a high *anti*-selectivity are favoured in the reactions of a variety of cuprate reagents with optically active 1,3-dibromoallenes.[24] In related work, the *anti*-stereochemistry observed in organocuprate S_N2' displacements has been ascribed to a stereoelectronic effect arising from "bidentate" binding involving a *d*-orbital of nucleophilic copper and π^* and σ^*-orbitals of the substrate.[25]

trans-6-Chloro-4-decene, 3-chloro-1-heptene, and *trans*-1-chloro-2-heptene react with a molar equivalent of methyltin tris(methanethiolate) in benzene–dichloromethane at 70° to form allylic methyl sulphides in very high yield.[26] Considerable amounts of allylic rearrangement occur, and this has been explained in terms of a possible mechanism involving the discharge of unsymmetrical (allylic cation)–$MeSn(SMe)_nCl_{4-n}^-$ ion pairs ($n = 1-3$) *via* MeS ligand transfer. Reactions of allylic esters with nucleophiles such as benzenesulphinate and diethylmalonate anions (usually in the presence of triphenylphosphine) are catalysed by palladium–graphite and related heterogeneous catalysts.[27] Rearrangement products may occur.

For the solvolysis of four 4-substituted homopropargyl tosylates and triflates, $RC{\equiv}CCH_2CH_2X$, and several related compounds, the ratios of C(3) ring to C(4) ring-closure have been measured in various solvents.[28] As expected, ring-closure (k_Δ) increases and solvent displacement (k_s, S_N2) decreases with decreasing nucleophilicity of solvent.

Norbornyl and Closely Related Systems

Non-classical carbocations have been comprehensively reviewed in 265 pages with 790 references.[29] There is considerable attention to kinetics, stereochemistry, *etc.* of the norbornyl and related systems.

The most substantial experimental work on these systems continues to come from C.A. Grob's group.[30–35] Solvolysis rates [80 % (v/v) ethanol] and products of the 6-substituted 2-methyl-2-*exo*- and 2-methyl-2-*endo*-norbornyl 2,4-dinitrophenyl ethers, (**11**) and (**12**), have been determined.[30,31] The reaction constants ρ_I are -1.30 for (**11**) and -0.74 for (**12**). These differing sensitivities to the inductive effects of substituents at C(6) indicate that graded bridging of C(2) by C(6) occurs in the ionization of the *exo*-(**11**) but not of the *endo*-(**12**). Substitution is with retention at C(2) in (**11**) but with inversion at C(2) in (**12**). It has been concluded that stereoelectronic and polar effects, rather than steric bulk effects, lead to the high *exo*/*endo* rate ratios of the parent norbornyl derivatives (R = H).

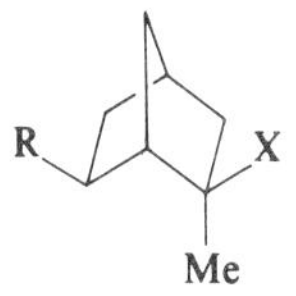

(**11**) 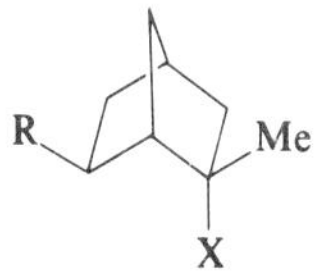 (**12**) 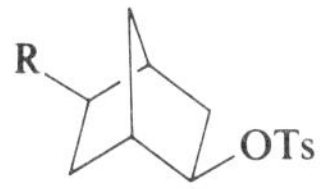 (**13**) 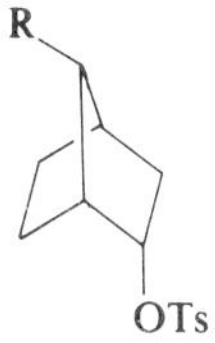 (**14**)

$X = 2{,}4(NO_2)_2C_6H_3O$

Solvolysis rates and products have been reported for *exo*- and *endo*-tosylates in five bicyclic systems closely related to 2-norbornyl.[32] A great range of *exo*/*endo* rate ratios and relative rates was found, and the stereochemistry of substitution was largely as stated for (**11**) and (**12**) above. Once again it appears that rates and products are mainly governed by the degree of bridging between the cationic centre and a dorsal carbon atom in the transition state and in the resulting ion pairs.

Rate constants for the solvolysis of a series of 5-*exo*-substituted 2-*exo*-norbornyl tosylates (**13**) in 80 % (v/v) ethanol at 70° have been measured.[33] The reaction constant ρ_I is -0.96, *i.e.* higher than that for 7-*anti*-substituted 2-*endo*-analogues (**14**) (-0.72) but lower than that for 6-*exo*-substituted 2-*exo*-analogues (**15**) (-2.0), previously reported. These results "confirm that through-space induction is directional and depends on distance and bridging strain".

Related work from the same group concerns the bicyclo[2.2.2]octyl system.[34,35] The solvolysis rates and products of several 6-substituted 2-*exo*- and 2-*endo*-bicyclo[2.2.2]octyl tosylates (**16**) and (**17**), respectively, have been determined. The reaction constant ρ_I is -1.50 for the *exo*-series (**16**), considerably less negative than -2.0 for the corresponding 6-*exo*-substituted 2-*exo*-norbornyl tosylate (**15**), as referred to above. It has been suggested that for geometrical reasons bridging is not so strong in the bicyclooctane series (**16**) as in the corresponding norbornane series (**15**). For the *endo*-series (**17**), however, ρ_I is -1.0, compared with -0.78 in the corresponding 2-*endo*-norbornane series (**18**), probably because in the former the bridging of C(6) is less hindered by the departing anion. The relative yields of *exo*- and *endo*-substitution products from (**16**) and (**17**) are in accord with graded bridging of C(6) in the incipient bicyclooctyl cations.

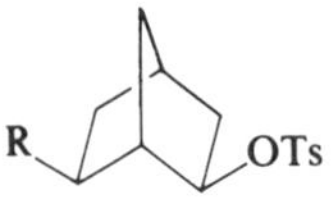

R OTs R OTs R OTs R OTs

(**15**) 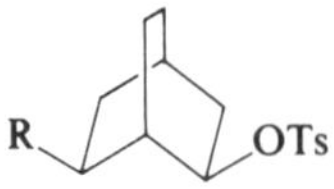 (**16**) 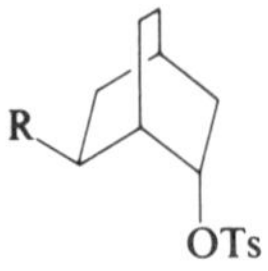 (**17**) 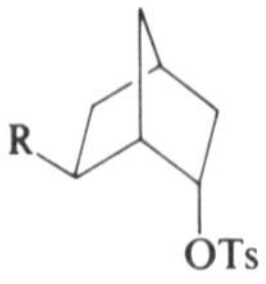 (**18**)

Studies of the effect of the cyano group in the norbornyl system have continued with solvolysis of all possible stereoisomers of 3-cyano-2-norbornyl tosylate.[36] A semi-quantitative analysis of the rates was made and bridged rather than classical structures were supported for the *exo*-2-norbornyl cations. Oxygen scrambling does not occur during the ethanolysis of [ether-^{17}O]-*endo*-2-norbornyl mesylate.[37] This observation and related facts have led to several conclusions, the most notable being that "the *exo*- and *endo*-2-norbornyl substrates clearly have different rate-controlling steps in solvolysis: formation of an ion pair in *endo* and dissociation of the pair in *exo*".

The study of long-range aryl migration combined with electrocyclic ring-opening (LRAMERO) has continued with norbornyl-related substrates designed to test the need for the presence of a second ("stationary") aryl group.[38] A second aryl group is not necessary for LRAMERO, although proximate positioning of the migrating aryl group and the leaving-group centre remain critical.

Studies of the solvolysis of α-keto-substituted 2-norbornyl systems have continued.[39] The effect of a cyano group at an incipient carbocation centre has also been studied.[40]

Miscellaneous Polycyclic Systems

Product selectivities in solvolysis of 1-adamantyl bromide in several binary protic solvent mixtures reveal the relative importance of solvent acidity and bulk.[41] Rates of solvolysis of 2-adamantyl azoxytosylate in various protic media have been measured at several temperatures;[42] isotope effects, salt effects, and product composition have also been studied. The suggested mechanism involves initial unimolecular, rate-determining synchronous fragmentation of the substrate to give nitrous oxide and an ion-pair; the latter is either captured by solvent to give a solvolysis product or, in the less polar media, undergoes ion-pair combination to yield covalent 2-adamantyl tosylate.

The stannylation of 1-bromo- and 1-chloro-4-iodobicyclo[2.2.2]octane with $(Me_3Sn)Li$ indicates a leaving-group mobility order $Br > I > Cl$.[43] Usually bridgehead iodides are more reactive than bromides, and chlorides are unreactive; the intervention of a carbanionic intermediate has been suggested tentatively to explain the above abnormal order.

Solvolysis rates in methanol for four β-methyl and β-phenyl bicyclic tosylates indicate solvent nucleophilic assistance, but in 97% hexafluoroisopropanol three of the mentioned compounds show π or σ neighbouring group assistance.[44] In related work, solvolysis rates and products have been investigated for hindered secondary substrates.[45] Rate-determining and product-determining steps are not identical and the results support the "S_N2(intermediate)" mechanism.

Hydrolysis rates and products have been determined for the *endo*- and *exo*-2-methylcycloprop[2,3]inden-1-yl 3,5-dinitrobenzoates, (**19a**) and (**20a**), in 80% aqueous acetone.[46] Comparison with the unsubstituted esters (**19b**) and (**20b**) finds a mild retarding effect of 2-Me in the *endo*-substrate, which contrasts with a reported 250-fold rate decrease in acid-catalysed epimerization of *endo*-cycloprop[2,3]inden-1-ol.

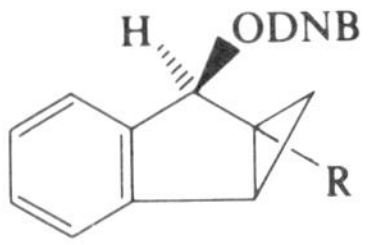

H ODNB R

(**19**)

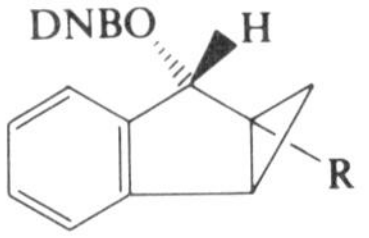

DNBO H R

(**20**)

a: R = Me
b: R = H

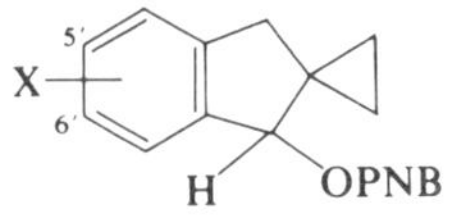

X 5′ 6′ H OPNB

(**21**)

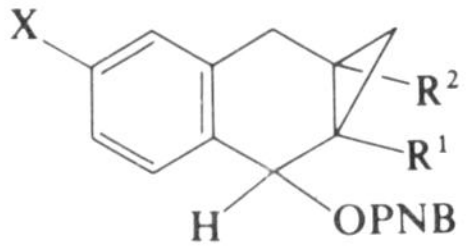

X R² R¹ H OPNB

(**22**)

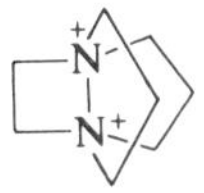

N⁺ N⁺

(**23**)

The solvolysis of 5′- or 6′-substituted spiro[cyclopropan-1,2′-indan]-1′-yl *p*-nitrobenzoate (**21**) in 80% aqueous acetone gives a ρ value (σ^+ correlation) of -3.33.[47] This contrasts with a value of -2.11 for the related substrate (**22**).

1,5-Diazoniatricyclo[3.3.2.0]decane (**23**) undergoes S_N2 attack on the four-membered ring with ring-opening (C—N$^+$ cleavage).[48] Various similar related substrates showed only deprotonation at α-C, with concomitant N$^+$—N$^+$ cleavage (an *E*2 reaction).

Epoxide Reactions

Relative rate studies have been carried out for the reactions of protected amino-acids with (R)- and (S)-methyloxirane.[49] Their relevance to the toxicology of methyl-oxiranes has been discussed. The action of proton donor additives on the reaction of octyl β-hydroxyethyl sulphide with propylene oxide has been studied quantitatively and the results correlated with the dissociation constants of the proton donors.[50]

Rate constants for chain transfer and chain propagation have been determined for the reaction of caesium perfluoroalkoxides with hexafluoropropylene oxide in tetrahydrofuran, in presence of aprotic solvent additives.[51] The kinetics of the reaction of 1-chloro-2,3-epoxypropane with methanol in the presence of Sn(IV) chloride have been studied.[52]

In the reaction of 1-phenyl-7-oxabicyclo[4.1.0]heptane (**24a**) with ammonia, or mono- or di-methylamine, *trans*-ring-opening always occurs, with the main products being derived by nucleophilic attack on C(2).[53] Some regioisomer is also formed, in yield depending on the amine and solvent. In the reactions of 1,2-epoxy-1-trimethylsilylcyclohexane (**24b**) with several nucleophiles under acidic conditions, ring-opening again occurs in *trans*-fashion, but usually with attack by the nucleophile on C(1).[54] The only exception is the reaction with HCl when 2-chloro-1-trimethylsilylcyclohexanol is also formed as a minor product, in amount which depends on the solvent.

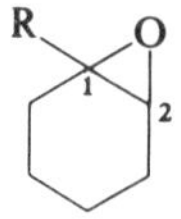

(**24**)

a: R = Ph
b: R = $SiMe_3$

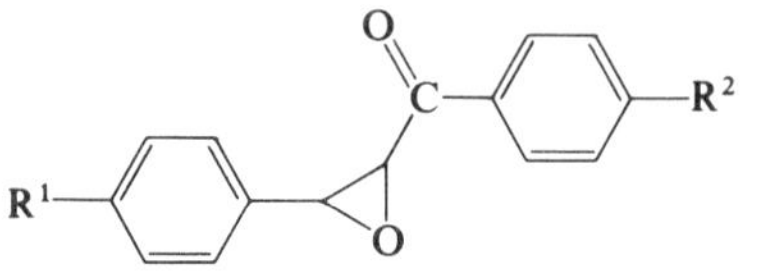

(**25**)

Rate constants for the nucleophilic ring-cleavage of several *para*-substituted epoxybenzalacetophenones (**25**) by piperidine in mixed dipolar aprotic and polar protic solvents have been determined.[55] Correlations with σ^+ and σ^0 for R^1 and R^2, respectively, are found. In reactions of oxiranes with sulphur nucleophiles under conditions in which the observed rate constant does not depend on pH, aryloxiranes do not show the expected enhancement of reactivity.[56] This result is explained by conformational restriction of orthogonal substitution.

The reactions of (dimethylamino)phenyloxosulphonium methylide, $Me_2N\overset{+}{S}O(Ph)(CH_2^-)$, with aromatic epoxides yield cyclopropyl sulphones and oxetanes, but with aliphatic epoxides the only products are the cyclopropyl sulphones.[57] The ylide acts as a methylene-transfer reagent to give the corresponding oxetanes, while the formation of cyclopropyl sulphones is attributed to an intramolecular S_N2-type reaction of the intermediate betaine. In the opening of epoxides by trimethylsilyl cyanide–zinc iodide, adjacent hydroxyl, trimethylsiloxyl, and acetoxyl groups provide complete control of the regiospecificity.[58] The method has been applied to the establishment of three contiguous chiral centres.

Transition-state structures for ring-opening of oxirane initiated by attack of a fluoride ion have been examined by *ab initio* MO calculations.[59] The preference for ring-opening of oxiranes with inversion rather than retention is explained by disfavouring of the latter through strong overlap repulsion between the oxirane ring and the attacking nucleophile and through the forced distortion of the three-membered ring in the transition state.

Rigid, bisected oxiranylcarbinyl mesylates solvolyse *via* an apparent oxiranylcarbinyl cation, which rearranges to a 2-oxahomoallyl cation, while perpendicular isomers give displacement with inversion.[60]

The only epoxide paper relevant to carcinogenesis is a kinetic study of the hydronium-ion-catalysed and pH-independent hydrolyses of several benzo ring diol epoxides, derived from polycyclic aromatic hydrocarbons, that possess bay-region *trans*-diol groups.[61] A detailed conformational interpretation has been given.

Other Small Rings

The effects of ester and carboxylic acid groups on the regio- and stereo-selectivities of Hg(II)-induced cleavage of unactivated cyclopropanes have been investigated.[62] Three types of mechanism have been distinguished for cyclopropane ring-openings by organocuprates.[63] The simplest mechanism is direct nucleophilic displacement, with an enolate leaving group.

The reaction of the spiro-activated cyclopropane (**26**) with pyridine in acetonitrile gives a zwitterionic addition product; rate and equilibrium constants can be measured.[64] An extended Brønsted treatment gives $\beta_{nuc} = 0.26$ for the reactions with substituted pyridines.

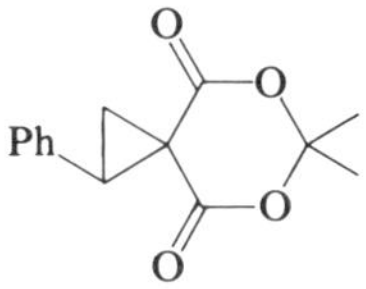

Ph O O O O

(**26**)

X CN

(**27**)

X = Cl, Br

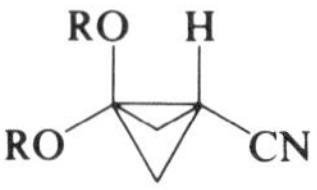

RO H RO CN

(**28**)

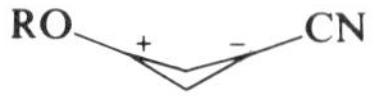

RO + − CN

(**29**)

In the reaction of 3-halobicyclobutanecarbonitrile (**27**) with several nucleophiles, the main product is the ketal 3,3-dialkoxycyclobutanecarbonitrile (**28**).[65] Kinetic studies have found an element effect $k_{Cl}/k_{Br} \approx 4$ and this indicates that the first step is the nucleophilic cleavage of the central bond of (**27**). The actual displacement of the leaving group occurs with the formation of a zwitterionic intermediate, involving an ionic central bond (**29**). A similar mechanism and intermediate is found for the reaction of (**27**) with PhS^- in methanol but the mechanism is altered in DME.[66]

The MINDO/3 method has been used to study the reaction of protonated ethylenimine with ethylamine to give protonated $EtNHCH_2CH_2NH_2$.[67] The same method has also been applied to the reaction of ammonia with aziridine, protonated aziridine, and a aziridine–ammonium complex ion.[68]

The addition of hydrogen fluoride to 1-phenyl-9-azabicyclo[6.1.0]nonane gives two fluoroamines;[69] the likely intermediate is a hydrido-bridged cation.

Substitution at Elements Other than Carbon

The hydrolysis (in H_2O–MeCN or H_2O–Me_2SO) and methanolysis of $(Me_3Si)_3CSiR_2I$ (R = Ph, Et) give only unrearranged products $(Me_3Si)_3CSiR_2Y$ (Y = OH,OMe).[70] The methanolyses of $(Me_3Si)_3CSiMe_2I$ and $(Me_3Si)_3CSiPhHI$ therefore seem more likely to be S_N2(intermediate) processes rather than the S_N1 processes previously suggested.

Detailed kinetic studies of the hydrolyses of 2-trimethylsilyl-*N*-methylimidazole, 2-trimethylsilylpyridine, and 2-((trimethylsilyl)methyl)-*N*-methylimidazole have been reported.[71] The dominant mechanism involves nucleophilic attack of OH^- on the trimethylsilyl unit of the *N*-protonated base, the leaving group being the zwitterion (ylide) of the heterocycle.

In the hydrolysis or methanolysis of the oxasilacyclopentane (**30a**) there is a large acceleration (10^3–10^4-fold) relative to the six-membered ring (**30b**) and open-chain analogues, the reaction occurring with inversion.[72] This great rate enhancement seems to occur when an oxa leaving group is displaced by an oxa nucleophile, for it is less pronounced with analogous thia electrophiles and with carbon nucleophiles, but it is also found with analogous phosphorus electrophiles.

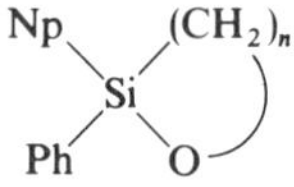

Np, $(CH_2)_n$, Si, Ph, O

(30)

a: $n = 3$
b: $n = 4$
Np = 1-Naphthyl

$Et_n\overset{+}{P}Ph_{4-n}I^-$

(31)

Reactions of various nucleophiles with ring-substituted 1,3,2-dioxaphospholanes, 1,3,2-oxazaphospholanes, 1,2-oxaphospholes, and phosphetanes bearing the leaving groups Cl, OR, or NR_2 on phosphorus have been studied.[73] In the main the actual substitution step proceeds with inversion and the lack of stereoselectivity sometimes observed is due to competing isomerization of products or reactants. Third-order rate constants have been measured for the hydrolyses of ethyl(phenyl)phosphonium iodides (**31**) in 70% aqueous methanol or aqueous tetrahydrofuran.[74] For the substrates with $n = 0$ or 1 there is a large acceleration by THF, but this drops almost to zero for the substrates with $n = 3$ or 4, to greater delocalization of the positive charge by ethyl than by phenyl groups.

Nucleophilic substitution of perfluoroalkyl 3-oxaperfluoroalkanesulphonates occurs exclusively *via* S—O cleavage, S_N2 attack on α-C of the alcohol moiety being prevented by the shielding effect of geminal F atoms.[75] The kinetics of the S—S cleavage (at pH 8) of Ellman's reagent in vesicular $(n\text{-}C_{16}H_{33})_2NMe_2^+ Br^-$ have been studied.[76] This is a two-site vesicular reaction.

Thiolates react with 2-halomethyl-5-nitrofurans to yield 5-nitrofurfuryl sulphides by an S_N2(C) mechanism, and disulphides and 2-methyl-5-nitrofuran by an S_N2(X) mechanism.[77] Single-electron transfer mechanisms are unlikely.

Intramolecular Substitution

A series of papers on ring-closure reactions has continued with a kinetic study of cyclization of diethyl (ω-bromoalkyl)malonates (as anions) in dimethyl sulphoxide at 25°.[78] Rate constants and effective molarities (*EM*) have been obtained for ring sizes 4–13, 17, and 21, the reactivity data covering nine powers of ten. The transition-state strain energies tend to parallel cycloalkane strain energies for the seven-membered and larger rings, but not for the smaller rings. For the latter, the ring product no longer appears to be a proper model for understanding the transition state. In a further study for the six-membered ring, it has been shown that the rate is depressed by alkali-metal cations.[79] The results have been analysed in terms of independent kinetic contributions from free reactant anion and from cation-paired reactant anion.

From a study of the reactions of selected 2-azido-alcohols, their pivalates, and methanesulphonates with triphenylphosphine, it was shown that the formation of aziridines proceeds *via* 1,3,2 λ^5-oxazaphospholidines.[80] A mechanistic duality in the cyclialkylation of alkenylmetal derivatives has been confirmed; σ-type or π-type processes may be involved.[81]

Anchimeric Assistance

Study of the solvolysis of β-methyl and β-phenyl tosylates of two cycloalkyl systems finds evidence for considerable phenyl assistance.[82] A competitive solvent-assisted pathway is of minor importance.

Ab initio MO calculations of isotope effects have been made for model processes of neopentyl ester solvolysis.[83] Calculated ^{14}C kinetic isotope effects of the migrating methyl group are normal, *i.e.* > 1, for all model processes, and so the experimentally observed normal isotype effect is not decisive evidence for the k_Δ mechanism. However, ^{2}H isotope effects are calculated to be normal for the k_Δ but inverse for the k_c; thus, since the inverse effect is not compatible with experiment, the k_Δ mechanism for solvolysis of neopentyl esters is supported.

Neopentyl iodide hydrolysis with electrophilic cooperation by silver or mercury ions proceeds by the S_N1 process without appreciable anchimeric assistance.[84] *p*-Nitrobenzal chloride is solvolysed in aqueous ethanol faster than the *ortho*-isomer, indicating that solvolysis of the latter does not involve participation by the *ortho*-nitro group.[85] Kinetic data for the reactions of *N*-substituted benzimidoyl chlorides with aniline in benzene have been discussed in relation to possible participation of the C=N bond.[86]

A Taft σ^* plot for the acetolysis of cyclopentenylmethyl and 2-cyclopentenylethyl *p*-nitrobenzenesulphonates and related substrates indicates that σ- and π-assisted paths, as well as solvent-assisted nucleophilic substitution, may variously occur in members of this series.[87]

Comparison of rate ratios suggests that O(3) participation in the solvolysis of 2-oxaadamantanes provides anchimeric assistance of the order of 10^4-fold.[88] Through use of D- and ^{13}C-labelled substrates, the transformation of certain (2-hydroxyethyl)pyrroles [*e.g.* (**32**)] into corresponding (2-haloethyl)pyrroles (by action of thionyl chloride/pyridine or triphenyl phosphine/carbon tetrabromide) has been shown to proceed with scrambling of the two carbons in the side-chain.[89] The proposed mechanism involves neighbouring-group participation by the pyrrole nucleus to give an ethylenepyrrolonium ion, *e.g.* (**33**). Neighbouring-group participation by pyridine rings and *N*-oxides has been examined in the solvolyses of 5,8-dihydro-5,8-methanoiso-quinoline derivatives.[90]

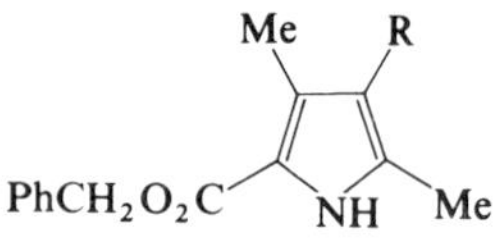

Me R PhCH$_2$O$_2$C NH Me

(**32**)

a: $R = CD_2CH_2OH$
b: $R = CH_2{}^{13}CH_2OH$

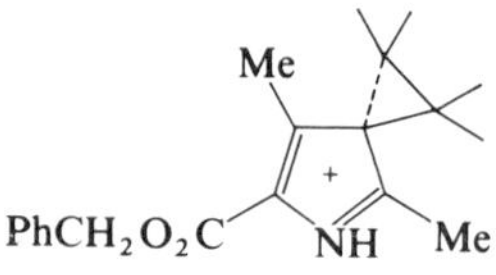

Me PhCH$_2$O$_2$C NH Me

(**33**)

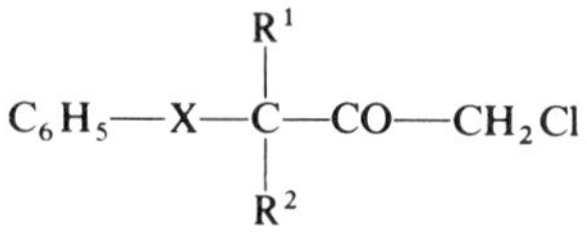

$C_6H_5—X—C(R^1)(R^2)—CO—CH_2Cl$

(**34**)

a: X = O
b: X = S

H H Me Me H

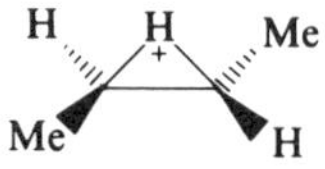

(**35**)

The acetolyses of a series of α'-phenoxy-α-chloro-ketones (**34a**) have been compared with results previously reported for the corresponding α'-phenylthio-α-chloro-ketones (**34b**), and have been discussed in terms of enolization–solvolysis mechanisms.[91] The easier solvolysis of several thio compounds has been ascribed to anchimeric assistance by sulphur in the ionization of enol allylic chlorides.

Two studies of the trifluoroacetolysis of optically active 2-butyl tosylate agree in finding the striking result of a low percentage of net inversion.[92,93] There is disagreement, however, in the interpretation of this and related results. One group considers that the mechanism involves "... attack by solvent both upon an intermediate where the leaving group determines the stereochemistry of the reaction and upon free solvated hydrogen-bridged 2-butyl cation" (**35**).[92] The other group favours ". . . . formation of an ion pair consisting of an open 2-butyl cation and a tosylate anion . . ." and considers that the proposal that ". . . . a hydrogen-bridged butyl cation is the predominant intermediate in this reaction does not give a simple prediction of these results".[93]

Ambident Nucleophiles

The stereo- and regio-selectivity of the alkylation of benzophenone–sodium by means of chiral reagents depends on the leaving group of the alkylating agent and on the sodium/benzophenone ratio.[94] *C*-Alkylation of the ketyl with inversion of configuration or racemization, and also alkylation of the aromatic nucleus, are variously found. *O*-Alkylation does not occur in these reactions, and the hard–soft alkylation theory (Pearson) is not applicable to the ketyl dianion or radical anion alkylations.

N-Monoalkylations of ditertiary diamines succeed best in nitromethane or ethanol at low temperatures.[95] Under these conditions unsymmetrical 1,3-diaminopropanes as model compounds react selectively at the sterically less-hindered nitrogen atom.

Alkali salts of 1,3,3-triphenylpropyne $[Ph_2CC{\equiv}CPh]^- M^+$ (M = Na, Cs) react with EtX (X = Cl, Br, I) in THF to give $Ph_2CEtC{\equiv}CPh$ and $Ph_2C{=}C{=}CEtPh$ in ratios between 2.2 and 5.4.[96] Factors favouring contact ion-pair formation enhance the amount of the former, while factors favouring solvent-separated ion-pairs enhance the amount of the latter.

Isotope Effects

Contrary to an earlier report, the α-D kinetic isotope effect (KIE) for the trifluoroacetolysis of 2-propyl naphthalene-1-sulphonate is smaller than 1.22 at 25°, suggesting that the mechanism is not a limiting S_N1.[97]

Extensive studies have been made of primary ^{14}C and secondary α-T KIEs for the Menschutkin-type reactions of (substituted)benzyl (substituted)benzenesulphonates with substituted *N,N*-dimethylanilines (*i.e.* heavy isotope in methylene group).[98] Large ^{14}C (1.117–1.151) and small α-T (1.026–1.041) KIEs are consistent with the S_N2 mechanism. Discussion of the transition-state structure is based on the variation of KIE with substituents.

Solvent KIEs $[k(H_2O)/k(D_2O)]$ have been measured for the solvolyses of benzyl nitrate, 4-chlorobutan-1-ol, and piperidylsulphamoyl chloride at several temperatures.[99] The standard activation parameters are discussed in relation to the reaction mechanism.

Solvent deuterium KIEs in the hydrolysis of $N_2CHCO(CH_2)_nCOCHN_2$ ($n = 0$, 2, 3, 4) in aqueous perchloric acid indicate that the rate-determining step is not protonation of the substrate.[100] Reactivity differences are governed by different rates of decomposition of the diazonium ions formed.

Gas-phase Reactions

Reactions of nucleophiles in the gas phase have been reviewed at length.[101]

Bare and mono-solvated hydroxide ions (ICR spectrometer) and bulk-solvated hydroxide ions react with acetonitrile to give three different sets of products.[102] Bare OH^- deprotonates CH_3CN in the gas phase, while $HOH\cdots{}^-OH$ reacts to form $CH_3OH\cdots{}^-CN$. In aqueous solution the first intermediate is $HOC(N^-)CH_3$.

Rate constants for the gas-phase reactions:

$$Cl^- + RBr \rightarrow ClR + Br^-$$

have been determined for temperatures between 25° and 390° with a pulsed electron-beam, high-ion source, pressure mass spectrometer (R = alkyl).[103] Analysis of the results yields approximate values for ΔE_0, the energy of the trigonal bipyramidal transition state relative to the energy of the reactants, as follows: Me, −2.5; Et, 0.8; Bu^n, −0.5; Pr^i, +5.1; Bu^i, +5.7 kcal mol^{-1}. The trend of these values agrees fairly well with the trends of activation energies in solution and of calculated strain energies.

Rate constants in the gas phase at room temperature (flowing afterglow) have been measured for nucleophilic displacement reactions of CH_3Cl and CH_3Br with the solvated homoconjugate anions $A^-(AH)_n$, where n = 0–3, and AH = H_2O, D_2O, CH_3OH, C_2H_5OH, CH_3COCH_3, HCO_2H, and CH_3CO_2H.[104] There is comparison with the kinetics of the analogous reactions in solution. The step-wise addition of solvent molecules greatly diminishes the reactivity of the nucleophiles.

An ICR study has shown that alkanol–alkoxide negative ions $[R^1O\cdots H\cdots OR^2]^-$ react with alkoxysilanes Me_3SiOR^3 (M) to produce both $[M+R^1O^-]$ and $[M+R^2O^-]$ ions.[105] When $R^1 < R^2$ and either R^1 or $R^2 \geqslant$ Pr the following four-centre reaction occurs:

$$[R^1O\cdots H\cdots OR^2]^- + Me_3SiOR^3 \rightarrow [R^2O\cdots H\cdots OR^3]^- + Me_3SiOR^1$$

Activation barriers of the S_N2 reactions of F^- + MeF and H^- + MeF have been estimated by fourth-order MB RSPT calculations.[106] For H^- + MeF the barrier is calculated to be 12 kJ mol^{-1}, in good agreement with experiment. Sophisticated *ab initio* calculations have been made for S_N2 reactions involving cationic substrates, *e.g.* $H_2O + CH_3OH_2^+$ and $CH_3OH + CH_3OH_2^+$.[107] The most stable structures correspond to a front-side complex but a back-side S_N2 pathway appears feasible since it requires little or no activation.

Radical Processes

The photo-behaviour of alkyl halides in solution has been reviewed.[108] The available data are most consistent with a mechanism involving initial light-induced homolytic cleavage of the C—Hal bond, followed by electron transfer within the resulting caged radical-pair to give an ion pair (Scheme 1). Iodides tend to show ionic behaviour, with the resultant carbocations displaying their typical behaviour, *i.e.* undergoing nucleophilic trapping, rearrangement, and deprotonation. Bromides exhibit substantially more radical behaviour.

Study of the reactions of nucleophiles with organic peroxides has continued in an examination of products and kinetics for the reaction of dimethyl sulphide and a

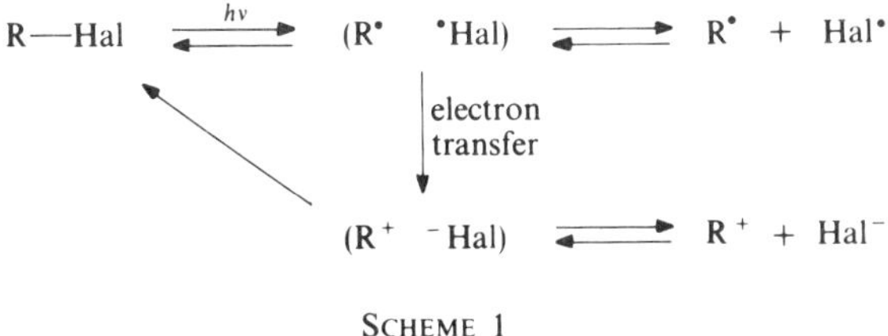

SCHEME 1

series of ring-substituted aryl methyl sulphides with *tert*-butyl peroxybenzoate (TBP) and with four ring-substituted TBPs.[109] Products are consistent with, but not conclusive for, a radical process, while solvent and substituent effects are not useful for distinguishing S_N2 and ET (electron-transfer) pathways. However, kinetic isotope effects for $(CD_3)_2S$ in reaction with TBP or $3,5(NO_2)_2$TBP indicate an ET process. A related study of *N,N*-dimethylaniline-d_6 with benzoyl peroxide has found an S_N2 reaction.

Medium Effects

From a factor analysis of data matrices containing 35 empirical and non-empirical parameters for 85 solvents (or a limited matrix of 20 parameters for 51 solvents), an orthogonal set of parameters with normed normal distribution has been suggested for characterizing the influence of the solvents on chemical processes.[110] These are to be applied in a four-parameter multiple regression, and have been tested on 22 examples from the literature including several nucleophilic aliphatic substitutions. This procedure has been compared with earlier multi-parameter treatments of solvent effects.

In some "Cautionary Comments concerning the Application of the Swain *A* and *B* Parameters to Solvolytic Displacement Reactions" evidence is presented that *A* and *B* are inferior in correlation analysis to the commonly used *N* and *Y* parameters, particularly for non-polar and dipolar aprotic solvents.[111]

Solvation effects and mechanism of heterolysis of *tert*-butyl halides have been reviewed extensively.[112] Hexafluoropropan-2-ol and the role of the solvent in the study of solvolytic reactions have been reviewed (in Serbo-Croatian).[113]

The work of Katritzky's group on the "Kinetics and Mechanism of Nucleophilic Displacements with Heterocycles as Leaving Groups" has continued in the measurement of ρ^* values for solvolyses of a series of 1-(*sec*-alkyl)pyridinium cations in trifluoroethanol or hexafluoropropan-2-ol in the absence of nucleophiles.[114] Corresponding reactions with pyridine, piperidine, or morpholine present as nucleophiles in several aprotic or protic solvents are, in fact, kinetically first order, and again ρ^* values have been measured. Values of ρ^* correlate with E_T values for the solvents, with certain strongly deviating points. Product analyses permit various mechanistic conclusions.

Kinetic studies have revealed that the mechanism of heterolysis of Me_3CCl or Me_3CBr changes on going from aprotic to protic solvents and so these compounds are of limited use as standards in determining the ionizing ability of solvents;[115] benzhydryl bromide is a better probe. Rates of S_N1 solvolysis of *tert*-butyl halides

(Cl, Br, or I) in methanol–acetonitrile mixtures have been correlated with Taft's solvatochromic parameter π^*, and α, a hydrogen-bond donor acidity scale.[116] Rate constants have been measured for the solvolysis of diphenylmethyl chloride in aqueous acetonitrile and other aqueous mixtures.[117] A logarithmic plot *vs.* the Grunwald–Winstein Y parameter gives considerable dispersion of lines for the various solvent systems. It has been suggested that a limitation of all treatments of this type lies in the neglect of initial-state interactions, so that it is probably not feasible to devise generally applicable kinetic solvent scales for S_N1 reactions. In related work, various parameters of this nature have been measured for aqueous acetonitrile containing 40–90% H_2O w/w.[118]

Work on perchlorate esters has continued in studies of rates and products of the solvolysis of 2-adamantyl perchlorate in a wide range of pure and aqueous organic solvents.[119] Nucleofugality of the perchlorate ion in S_N1 reactions is several orders of magnitude higher than for the more usual leaving groups, such as tosylate or bromide. A scale of Y_{OClO_3} values has been developed and compared with Y scales based on other leaving groups.

Solvolyses of 13 arenesulphonates in ethanol–water mixtures, and acetic or formic acid have been studied.[120] The complex pattern of solvent effects has been interpreted in terms of differences in the involvement of solvent in the transition state as between solvolyses assisted by bridging and those assisted by C—C σ-hyperconjugation.

The new solvent polarity scale π_0^* has been applied successfully to several types of reaction including Menschutkin reactions.[121] The rates of quaternization of benzyldimethylamine with derivatives of some oxyethylated nonylphenols increase with increasing dielectric constant in a series of solvents.[122]

For the reaction of ethyl iodide with chloride ions, rate constants and activation parameters have been measured for reaction in mixtures of methanol with acetonitrile or *N*,*N*-dimethylacetamide; enthalpies of solution of tetraethylammonium bromide and chloride in the same solvents have also been determined.[123] The thermodynamics of clustering of dimethyl sulphoxide molecules around halide ions in the gas phase have been studied and discussed in relation to corresponding data for ionic solvation in liquid DMSO and other solvents.[124] It has been concluded that "the smaller decrease of solvation with an increase of ion radius is the principal reason for the much higher rates of reactions:

$$A^- + B \rightleftarrows (AB^-)^{\neq} \rightarrow C^- + D$$

in dipolar aprotic solvents".

Electrophilic assistance to reactions at a C—X bond has been extensively reviewed.[125] Primary alkyl iodides show both M^+-S_N1 and M^+-S_N2 reactions in the presence of silver (I) or mercury(II) salts.[126] The reaction of methyl iodide with tetrachloroplatinite anion shows electrophilic assistance by mercury(II), present largely as $HgCl_4{}^{2-}$.[127] Silver(I) assistance of isopropyl iodide hydrolysis has also been studied.[128]

Solvolysis of *tert*-butyl and *tert*-amyl iodides in aqueous ethanol and aqueous hexafluoropropan-2-ol is catalysed by iodine, and certain other systems show analogous catalysis.[129] It has been suggested that the mechanism involves pre-association of iodine with alkyl halide but there are various possibilities for the rate-determining step.

A general empirical equation for kinetic salt effects has been proposed:

$$\ln k = \ln k_0 + A \ln[(C/C^*) + 1]$$

and other special forms.[130] (A and C^* are constants characteristic of the reaction and the particular salt.) The equation has been tested on a variety of processes.

The valence-bond state correlation diagram model has been used to analyse solvent effects for the identity reactions:[131]

$$X^- + CH_3X \rightarrow XCH_3 + X^-$$

"The trends in the reaction barriers are shown to arise from the interplay of intrinsic reactant properties and solvent properties in a manner which reflects *the nature of* S_N2 *as a transformation involving a single electron shift attended by bond interchange.*"

The first computation of an energy profile for an S_N2 reaction in aqueous solution has been made.[132] This is for the identity reaction of Cl^- and CH_3Cl, and goes considerably beyond previous simulations, in so far as a large number of water molecules are explicitly included. The calculations support the conventional view of a unimodal energy surface in aqueous solution. Calculated $\Delta G^{\ddagger}$ is 26.3 ± 0.5 kcal mol^{-1}, in excellent agreement with the experimental value, 26.6.

Monte Carlo calculations have been made for the effect of large numbers of water molecules on the identity reaction of F^- with CH_3F.[133] The same reaction has also been treated by *ab initio* calculations using the 3-21G basis set.[134]

The CNDO/2 method has been applied to the mechanism of solvolysis of methyl fluoride by various numbers of water molecules.[135] The process implies the transfer of one proton through a chain of water molecules. Winstein's mechanism is supported and solvent assistance to carbocation formation appears as a very important factor in the ionization part of the process.

Kinetics of organic substitution reactions in molten acetate and thiocyanate salts have been studied.[136] For example, alkyl *p*-hydroxybenzoates undergo displacements in molten acetate salts, and the reactivity order is typical of an S_N2 reaction.

Phase-transfer Catalysis and Other Intermolecular Effects

A book on *Interphase Transfer Catalysis: Applications to Organic Chemistry* has been published in Romanian.[137]

The kinetics of phase-transfer-catalysed reversible or irreversible S_N2 substitution of an organic substrate RY by aqueous X^- have been studied theoretically.[138] The poisoning effect of a foreign counter-ion Z^- of the phase-transfer catalyst has been assessed. In a related theoretical study of the irreversible S_N2 reaction of RY with X^-, the necessary conditions have been established for the application of the method of initial rates in phase-transfer catalysis.[139]

The influence of the concentration of aqueous potassium hydroxide on the hydration and hence reactivity of anions (Cl^-, Br^-, I^-, SCN^-, N_3^-) in aliphatic nucleophilic substitutions has been studied for catalysis by lipophilic [2.2.2]cryptand (**36**) and dicyclohexano-18-crown-6 (DCH 18C6) (**37**) under phase-transfer conditions.[140] Comparison has been made with the same reactions under classical liquid–liquid PTC and homogeneous anhydrous conditions. Unlike quaternary 'onium salts, even at the highest [KOH] (53 %, $a(H_2O) \approx 0$], water is not completely removed by (**36**). In consequence, rate constants are enhanced compared with those found under conventional PTC conditions, but do not reach those of anhydrous solutions.

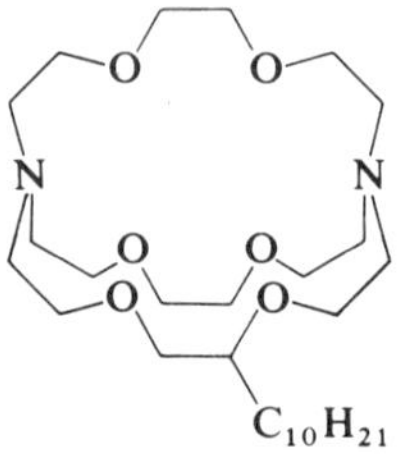

$C_{10}H_{21}$

(**36**)

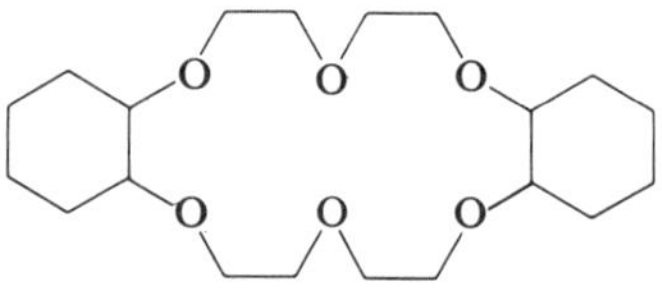

(**37**)

The action of sodium cyanide on two *gem*-diarylchloropropanones may variously produce an ester, epoxynitrile, or cyclopropane derivative when the experimental conditions are varied with respect to solvent, phase-transfer catalyst, or presence of a crown ether.[141] The reaction of sodium thiophenoxide with some perchlorofluoroethanes proceeds by an anionic chain-process and is accelerated in presence of crown ether.[142] Catalytic amounts of α- and β-cyclodextrins act as phase-transfer catalysts in nucleophilic substitution of octyl bromide by CN^-, I^-, or CNS^- in a liquid–liquid two-phase system.[143]

Several sulphoxide-substituted pyridines act as good phase-transfer catalysts for S_N2 displacements of octyl bromide with PhS^-, CN^-, CNS^-, or PhO^- in solid–liquid two-phase systems.[144] Other examples have also been investigated.

Polymer-supported crown ethers (**38**) have been tested for phase-transfer catalytic activity in anion-promoted nucleophilic aliphatic substitutions, in comparison with that of structurally similar soluble crown ethers and of polymer-supported and soluble phosphonium salts.[145] Various polymer-supported crown ethers act as phase-transfer catalysts for the reactions of *n*-octyl bromide in toluene with aqueous potassium iodide at 90° and with aqueous sodium *p*-methylthiophenolate at 80°, but not with aqueous potassium cyanide at 90°.[146]

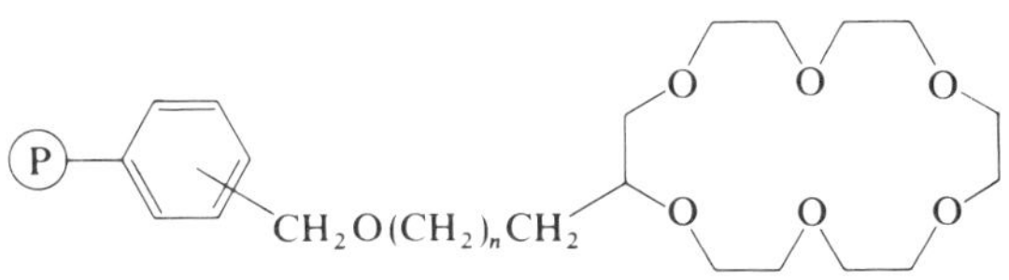

(**38**)

$n = 0,8$

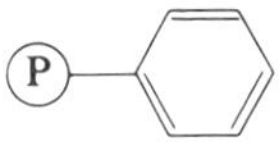

= microporous cross-linked polystyrene

Ultrasonic irradiation of a mixture of benzyl bromide, potassium cyanide, and alumina in an aromatic solvent yields benzyl cyanide, whereas mechanical agitation gives a Friedel–Crafts-type product.[147] Thus sonication changes the main reaction pathway from aromatic electrophilic to aliphatic nucleophilic substitution.

A survey of polar reaction mechanisms has included micellar reactions.[148]

Aqueous cationic micelles of cetyltrimethylammonium surfactants or dodecyltrimethylammonium bromide, and anionic micelles of sodium lauryl sulphate inhibit S_N1 hydrolysis of sterically hindered arenesulphonates and chlorides, *e.g.* 2-adamantyl brosylate, pinacolyl tosylate, and 2,2-dimethyl-1-phenylpropyl chloride.[149] Micellar inhibition increases with increasing substrate hydrophobicity, but is always larger with cationic than with anionic micelles. In the reaction between phenacyl bromide and benzoate ion in aqueous methanol or aqueous acetone in the presence of anionic surfactant aggregates of sodium dodecyl sulphate, both electrostatic and hydrophobic effects are operative.[150] For the reaction of sodium benzoate and benzyl chloride, certain quaternary ammonium salts $RNMe_3Cl$ produce micelles and the yield of benzyl benzoate is low.[151] When certain salts $R^1R^2NMe_2Cl$ are used, no micelles are formed and the quaternary salt acts as an effective phase-transfer catalyst, producing an excellent yield of the above ester.

Micellar solutions of CTAB and other surfactants producing micelles or vesicles enhance markedly (sometimes by two or more orders of magnitude) the reactivity of nucleophiles derived from the dissociation of weak acids, *e.g.* n-$C_7H_{15}SH$.[152] Analysis of the kinetics indicate that the main effect of the micelles and vesicles on the rate is due to changes in the local concentration of the reactants in the restricted environment of the pseudo-phase.

Structural Effects

A review of work by Katritzky's group during the past decade has appeared under the title "New Insights into Aliphatic Nucleophilic Substitution Reactions from the Use of Pyridines as Leaving Groups".[153] Five mechanisms, frequently in some degree of competition, have been distinguished, *viz.* radical-pair collapse (second-order kinetics), S_N2 displacement (classical, second-order kinetics), displacement on intimate molecule–ion pair (*i*, second-order kinetics; *ii*, first-order kinetics), and S_N1 reaction (first-order kinetics).

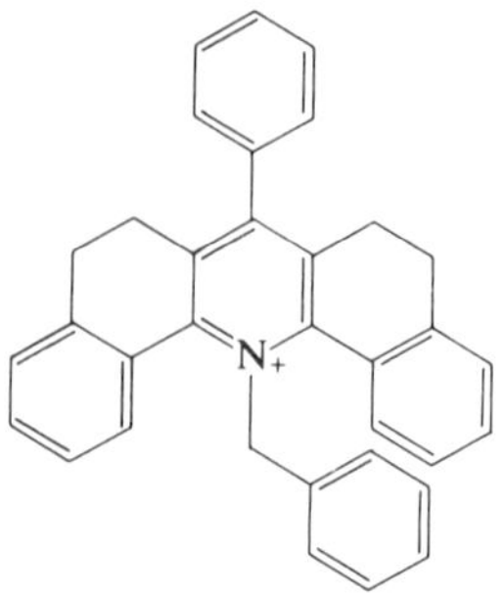

(39)

In further new work by the above group, the effect of hydrostatic pressure has been studied for the second-order displacement of pyridines from *N*-benzylpyridinium cations by piperidine.[154] The most striking result concerns the substrate (**39**), reacting as a tetrafluoroborate in chlorobenzene at 30°. The logarithmic rate *vs.* pressure plot shows a clear minimum, the low-pressure branch showing $\Delta V^{\neq}$ *ca.* $+22\ cm^3\ mol^{-1}$, while the high-pressure branch has $\Delta V^{\neq}$ *ca.* $-20\ cm^3\ mol^{-1}$. The former is considered to correspond to second-order reaction of the nucleophile on an intimate ion–molecule pair, while the latter corresponds to a normal S_N2 reaction. Part 17 of the main series of papers produced by this group deals with rates, activation parameters, and products of solvolysis of 14-(primary alkyl)-5,6,8,9-tetrahydro-7-phenyl-dibenzo[*c*, *h*]acridiniums in several protic solvents.[155] In methanol classical S_N2 substitution predominates but in acetic acid the main mechanism is nucleophilic attack on ion–molecule pairs.

Ponderal effects in substitution reactions have been reviewed (in Japanese).[156]

In a study of the effect of a perfluoroalkyl group on the reactions of iodoalkanes, the series $R_F(CH_2)_2I$ on reaction with strong base gives only $R_FCH{=}CH_2$, but $R_F(CH_2)_3I$ gives 4–10 times as much substitution as elimination.[157] Rate measurements of various series of perfluoroalkyl-substituted substrates indicate a remarkable damping by CH_2 groups of the inductive effect of R_F.

The synthesis and reactivity of α-halogenated ketones have been extensively reviewed.[158] The rates of S_N2 reactions of *meta*- and *para*-substituted phenacyl bromides with quinoline are increased by both electron-donating and electron-withdrawing substituents.[159] Bimolecular displacement of chloride at C=N leading to open-chain imidoyl azides is stereospecific: the z-compound (**40**) reacts with retention, while its isomer (**41**) reacts with inversion of configuration on reaction with azide ion, to give the same product (**42**).[160] The kinetics of the reaction of several chloromethylated phenols with substituted anilines in dimethyl sulphoxide have been studied by a conductivity method.[161] A two-step mechanism has been proposed, involving the reversible formation of an intermediate quinone methide. Reactions of alkanediazotic acids $R^1R^2CHN{=}NOH$ have been examined in $H_2{}^{18}O$ at pH 7.0–8.5 and > 14.[162] The intervention of numerous ion-pairs has been postulated to account for products, ^{18}O incorporation, and stereochemistry.

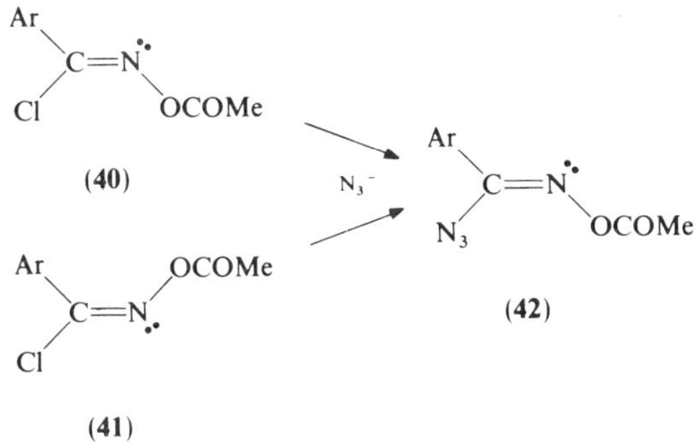

Ar(Cl)C=N–OCOMe

(40)

Ar(Cl)C=N–OCOMe

(41)

N_3^-

Ar(N_3)C=N–OCOMe

(42)

Correlation Analysis by Hammett or Brønsted Equations

Characteristic vector analysis (CVA) has been applied to the kinetic data for 1260 S_N1 solvolysis reactions of $R^1C_6H_4NHCH_2C(NO_2)_2C_6H_4R^2$, with respect to variation of R^1, R^2, temperature, and solvent composition.[163]. The $\log k$ data comprise a 42×30 matrix, *i.e.* there are 42 combinations of R^1 and R^2 and 30 combinations of temperature and solvent. Two characteristic vectors V_1 and V_2 describe 99.25% of the total variability. V_1 and V_2 are linearly related to the σ° constants for *m*- and *p*-R^1 and -R^2. V_1 and V_2 appear in the CVA equation with weightings S_1 and S_2; their scalar values after rotation of axes S'_1 and S'_2 are linearly dependent on the Grunwald–Winstein Y solvent parameter and the reciprocal of temperature, respectively.

The effect of substituents on the acetolysis of 2-arylethyl tosylates has been dissected into effects on aryl-assisted and -unassisted processes.[164] The latter is correlated by σ° constants; the former requires the Yukawa–Tsuno equation, which is also required in the correlation of the rearrangement ratios of solvolysis products of the same substrates.[165] The same research group has also studied the effect of substituents in the leaving group on the acetolysis of neophyl arenesulphonates.[166] All π-acceptor *para*-substituents appear to be considerably less electron-withdrawing than expected from their σ_p° values. In a further study, substituent effects of *p*-OMe and *p*-SMe, deactivated by additional *meta*-substituents, in solvolyses of cumyl and α-phenylethyl chlorides have been examined to provide evidence for the higher resonance demand in the latter system.[167]

The valence-bond configuration mixing model (VBCM) has been applied to the nucleophilic substitution reactions of α-carbonyl derivatives.[168] The following have been explained: the dependence of the rate-enhancing effect of the carbonyl upon the nucleophilicity of the entering group, the unusually high ρ value for reaction of $PhCOCH_2Br$ with substituted pyridines, and the mechanism of transmission of the rate-enhancing effect of the carbonyl group to the reaction centre.

A study of the kinetics of the hydrolysis of 2-bromo-4-dibromomethylphenol in 95% aqueous dioxan has led to an estimate of -1.36 for the σ^+ value of *p*-OH, compared with the usual value -0.92 based on electrophilic aromatic substitution.[169]

The kinetics of bimolecular nucleophilic substitution reactions of 1-phenylethyl derivatives have been studied in respect of Hammett, Swain–Scott, and Grunwald–Winstein correlations, and also leaving-group and salt effects.[170] It has been concluded that most of "these reactions are S_N2 displacements that proceed through an open, 'exploded' transition state that closely resembles a carbocation". In related work general base catalysis of the addition of ROH to unstable carbocations has been studied.[171] For the addition of trifluoroethanol to 1-(4-methoxyphenyl)ethyl carbocation the Brønsted coefficient β is 0.08. Variation of β with ROH and with carbocation stability has been explored.

First-order rate constants for the hydrolysis of ferrocenyl(aryl)methyl acetates FcCH(OAc) (p-XC_6H_4) in acetone–water (4:1 v/v) have been correlated with σ^+ constants and $E_{1/4}$ chronopotentiometric potentials.[172] The value $\rho = -1.4$, is remarkably low compared to the values for related systems, and this has been attributed to the high electron-releasing effect of the Fc group.

Rate constants for the reactions of 4-$R^1C_6H_4CH_2Cl$ with $R^2C_6H_4NH_2$ have been discussed in terms of Hammett and Brønsted relations and of potential energy surface and quantum-mechanical models of the transition state;[173] a related study involves the corresponding bromides and iodides.[174] In all cases a dissociative S_N2 mechanism is supported but a hypersurface treatment does not account for transition-state variations due to leaving-group effects.

The reaction of HBr with $RC_6H_4CH(OH)Me$ shows second-order kinetics and gives a concave-down Hammett plot.[175] The mechanism suggested involves protonation of the substrate followed by attack of Br^-.

A series of papers on methyl-transfer reactions has continued in a study of the identity reaction:

$$XC_6H_4{}^*SO_3{}^- + CH_3O_3SC_6H_4X \rightarrow XC_6H_4{}^*SO_3CH_3 + XC_6H_4SO_3{}^-$$

in sulpholane by ^{35}S labelling.[176] The value of ρ is 0.6. Various related systems have also been studied with respect both to rates and to equilibria, and an excellent application of the Marcus equation has been achieved.

Steric Effects

For the Menschutkin reaction of 2-alkyl-pyridines or -thiazoles and CH_3X ($X = I$, SO_3F) transition-state structures have been calculated by the Allinger (1973) force-field method.[177] Experimental differences in steric energies between the quaternary iminium ions and the activated complexes have been used as measures for the position of the transition state. There is a $22 \pm 4\,\%$ extension of the C—N distance in the transition state compared with the ground state, depending upon the model used.

Strain changes occurring during solvolysis of secondary tosylates have been calculated by a molecular mechanics program (MM2) using an empirical force-field for carbenium ions.[178] For the acetolysis of k_c substrates the rate constants correlate well with these strain energy changes, and anchimeric assistance in other substrates may be estimated from deviations.

The second-order rate coefficients for the reaction of *ortho*-substituted benzoate ions with ethyl bromoacetate in 90% aqueous acetone have been correlated in a multiple regression with σ_I, σ_R, and the steric parameter ε (Charton).[179] The regression coefficients of σ_I and σ_R are negative while that of ε is positive, the last-mentioned indicating steric acceleration by the *ortho*-substituent. In a related study of *ortho*-substituted benzoate ions reacting with *para*-substituted phenacyl bromides, a similar multi-parameter treatment has been effected.[180] The system presents a failure of the reactivity–selectivity principle.

Second-order rate constants for the reaction of 1-alkyl-2-halopyridinium salts with primary and secondary amines in acetonitrile have been correlated with inductive σ^* and steric (E_N) constants.[181]

Nucleophilicity

The rates of S_N2 reactions of 19 nitranions with $PhCH_2Cl$ and of 9 nitranions with m-$CF_3C_6H_4CH_2Cl$ in DMSO at 25° have been measured.[182] Brønsted-type plots of $\log k$ *vs.* pK_{HA} for the reactions of anion families based on carbazoles, phenothiazines, and diphenylamines with $PhCH_2Cl$ are rectilinear with slopes 0.32–0.33. These and related findings have led to the conclusion "that basicity is the primary factor in controlling nucleophilicities of nitranions, carbanions, and oxanions of diverse structural types in S_N2 reactions. Donor atom, solvation, and steric effects generally play a secondary role".

A long series of papers on the reactivity of carbanions has continued in a study of the kinetics of alkylation of 9-substituted fluorenyl carbanions and their ion-pairs (with Li^+ or Cs^+), and of closely related substrates, with butyl bromide.[183] A relative nucleophilicity scale of the carbanions has been based on the results. In further studies, rate constants have been determined for the reactions of various carbanions with β-bromocinnamic acid nitrile[184] and with butyl bromide.[185] Throughout this work the ion-pairs (with Li^+ or Cs^+) are found to react more slowly than the free carbanions.

The selectivities of a series of substituted 1-phenylethyl carbocations towards alcohols and other nucleophiles have been determined by product analysis.[186] For example, the 1-(4-(dimethylamino)phenyl)ethyl carbocation is highly selective towards alcohols, with $k_{EtOH}/k_{TFE} = 140$ and $\beta_{nuc} = 0.5$. In the case of activation-limited reactions with alcohols, the selectivity decreases progressively with increasing reactivity of the carbocation, in contrast to the behaviour expected from the N^+ scale of nucleophilicity.

As part of an ongoing study of cation–anion combination reactions free energies of bond dissociation in solution (assumed to be approximately equal to enthalpies of bond dissociation in the gas phase) have been used in a thermodynamic cycle to obtain free energies for one-electron transfer from various nucleophiles to the proton, and other properties of nucleophiles in aqueous solution.[187] The general problem of nucleophilic reactivity and the relevance of the above properties thereto has been discussed.

Rates of reaction of dichloromethane with various amines in dichloromethane as solvent do not correspond to the nucleophilicities of the amines as revealed in reactions with the usual alkyl halides.[188] The nucleophilic reactivity of a series of methyl(cyanomethyl)phosphines has been studied.[189]

Theoretical Treatments

The qualitative valence-bond approach to organic reactivity, extensively developed in recent years, has been reviewed.[190] Its main application is to S_N2 reactions. The valence-bond configuration mixing (VBCM) model has been applied to the problem of rate–equilibrium relationships.[191] The analysis suggests the types of process for which such relationships may or may not hold, and also the processes for which the Brønsted α or β may or may not provide a measure of transition-state structure.

The mechanistic difference between S_N2 and S_N1 reactions of MeCl and Bu^{t}Cl has been investigated by *ab initio* MO calculations.[192] Various models involving substrate with OH^- and/or H_2O, as governed by computational restrictions, have been examined. Among the conclusions it is stated that "The position of a substitution in the S_N1–S_N2 spectrum depends on the nucleophilicity of solvent, stability of the carbocation, and the bond strength of the C-leaving group". The same group has investigated nucleophilic substitution at acyl carbon and has shown the importance of σ^*–π^* orbital mixing in governing the relative reactivity of the substrates.[193] Two reviews, in Japanese, on the application of MO theory to nucleophilic substitution have also appeared.[194,195]

Relative activation energies $\Delta E^{\neq}$ for the methylation of pyridine, 37 alkyl-substituted pyridines, and 6 hetero-substituted pyridines have been calculated by MINDO/3.[196] Correlation equations for $\log k_{rel}$ *vs.* $\Delta E^{\neq}$ have been presented for the entire data set and for various sub-sets, *e.g.* for the entire set the correlation coefficient $r = 0.921$. This research group has used the same system as a basis for discussing various issues in structure–reactivity relationships, *e.g.* the hypothesis that ground-state geometries may be used to quantify chemical reactivity.[197]

A theoretical analysis of the reactions of Bu^{i}Br, and Pen^{neo}Br with I^- has confirmed the view (Hughes and Ingold) that the rate ratio is governed more by electronic than by steric effects.[198] The applications of a new stereochemical notation system to various examples, including the alkylation of enolates, have been reviewed.[199]

S_N2 Reactions (Miscellaneous)

The reaction of amines with methylene chloride has been reviewed.[200]

An S_N2 deprotection reaction for synthetic peptides occurs when dimethyl sulphide is used as a diluent for HF.[201]

Labelling by C(HDT) has been used to show that the enzymic transfer of a methyl group from the sulphur of *S*-adenosylmethionine to the nitrogen of histamine occurs with inversion in an S_N2 process.[202] Similarly, methyl transfer from betaine to homocysteine in homocysteine *S*-methyltransferase occurs directly in a second-order nucleophilic substitution.[203]

Kinetic Studies (Miscellaneous)

These have included the following: the solvolysis of benzyl azoxytosylate;[204] the reactions under pressure of β-phenylethyl brosylate with pyridine[205] and of phenacyl bromide with pyrimidine,[206] the quaternization of tri-*n*-propylamine by *n*-propyl bromide and iodide,[207] the thermokinetic characterization of the reaction of methyl iodide with triethylamine in nitrobenzene,[208] transalkylation of *N*-alkylarylamines,[209] the hydrobromination of *n*-propyl alcohol by hydrogen bromide in aqueous solution,[210] the acid-catalysed etherification of a hydroxybenzyl alcohol with aliphatic alcohols,[211] the de-*O*-alkylation of phenoxyacetic acids,[212] the hydrolysis of 4,4-dimethyl-1,3-dioxan in aqueous–alcoholic solutions,[213] and the chlorine isotopic exchange reaction between chloramine-T and chloride ion.[214]

References

1 Rappoport, Z., Kaspi, J., and Tsidoni, D., *J. Org. Chem.*, **49**, 80 (1984).
2 Rappoport, Z., and Rav-Acha, C., *Tetrahedron Lett.*, **25**, 117 (1984).
3 Rappoport, Z., Fiakpui, C. Y., Yu, X.-D., and Lee, C. C., *J. Org. Chem.*, **49**, 570 (1984).
4 Holweger, W., and Hanack, M., *Chem. Ber.*, **117**, 3004 (1984).
5 Bleckmann, W., and Hanack, M., *Chem. Ber.*, **117**, 3021 (1984).
6 Collins, C. J., Martinez, A. G., Alvarez, R. M., and Aguirre, J. A., *Chem. Ber.*, **117**, 2815 (1984).
7 Tiecco, M., Testaferri, L., Tingoli, M., Chianelli, D., and Montanucci, M., *J. Org. Chem.*, **48**, 4795 (1983).
8 Tiecco, M., Testaferri, L., Tingoli, M., Chianelli, D., and Montanucci, M., *Tetrahedron Lett.*, **25**, 4975 (1984).
9 Bach, R. D., and Wolber, G. J., *J. Am. Chem. Soc.*, **106**, 1401 (1984).
10 Cameron, D. W., Chalmers, P. J., and Feutrill, G. I., *Tetrahedron Lett.*, **25**, 6031 (1984).
11 Kravchenko, V. V., Popov, A. F., and Kostenko, L. I., *Zh. Org. Khim.*, **19**, 1906 (1983); *Chem. Abs.*, **100**, 33877 (1984).
12 Shainyan, B. A., and Mirskova, A. N., *Zh. Org. Khim.*, **20**, 972 (1984); *Chem. Abs.*, **101**, 109980 (1984).
13 Knoppova, V., Vegh, D., and Kovac, J., *Top. Furan Chem., Proc. Symp. Furan Chem., 4th*, **1983**, 146; *Chem. Abs.*, **101**, 110004 (1984).
14 Shandala, M. Y., Khalil, S. M., and Al-dabbagh, M. S., *Tetrahedron*, **40**, 1195 (1984).
15 Kevill, D. N., and Rissmann, T. J., *J. Chem. Soc., Perkin Trans. 2*, **1984**, 717.
16 Arnaud, R., Dussauge, A., Faucher, H., Subra, R., Vidal, M., and Vincens, M., *Tetrahedron*, **40**, 315 (1984).
17 Gassman, P. G., and Harrington, C. K., *J. Org. Chem.*, **49**, 2258 (1984).
18 Panigrahi, G. P., and Sinha, T. K., *Indian J. Chem.*, **23B**, 57 (1984); *Chem. Abs.*, **101**, 71938 (1984).
19 Houk, K. N., Paddon-Row, M. N., and Rondan, N. G., *THEOCHEM*, **12**, 197 (1983); *Chem. Abs.*, **99**, 211818 (1983).
20 Chen, N., and Zhou, B., *Youji Huaxue*, **1983**, 257; *Chem. Abs.*, **99**, 211825 (1983).
21 Dewar, M. J. S., *J. Am. Chem. Soc.*, **106**, 209 (1984).
22 Carrion, F., and Dewar, M. J. S., *J. Am. Chem. Soc.*, **106**, 3531 (1984).
23 Calò, V., Lopez, L., and Carlucci, W. F., *J. Chem. Soc., Perkin Trans. 1*, **1983**, 2953.
24 Corey, E. J., and Boaz, N. W., *Tetrahedron Lett.*, **25**, 3059 (1984).
25 Corey, E. J., and Boaz, N. W., *Tetrahedron Lett.*, **25**, 3063 (1984).
26 Katz, H. E., and Starnes, W. H., *J. Org. Chem.*, **49**, 2758 (1984).
27 Boldrini, G. P., Savoia, D., Tagliavini, E., Trombini, C., and Umani-Ronchi, A., *J. Organomet. Chem.*, **268**, 97 (1984).
28 Collins, C. J., Hanack, M., Stutz, H., Auchter, G., and Schoberth, W., *J. Org. Chem.*, **48**, 5260 (1983).
29 Barkhash, V. A., *Top. Curr. Chem.*, **116**, 1 (1984).
30 Grob, C. A., and Waldner, A., *Helv. Chim. Acta*, **66**, 2481 (1983).

[31] Grob, C. A., von Sprecher, G., and Waldner, A., *Helv. Chim. Acta,* **66,** 2656 (1983).
[32] Grob, C. A., Waldner, A., and Zutter, U., *Helv. Chim. Acta,* **67,** 717 (1984).
[33] Grob, C.A., and Sawlewicz, P., *Tetrahedron Lett.,* **25,** 2973 (1984).
[34] Grob, C. A., and Sawlewicz, P., *Helv. Chim. Acta,* **67,** 1859 (1984).
[35] Grob, C. A., and Sawlewicz, P., *Helv. Chim. Acta,* **67,** 1906 (1984).
[36] Wilcox, C. F., and Brungardt, B., *Tetrahedron Lett.,* **25,** 3403 (1984).
[37] Chang, S., and Le Noble, W. J., *J. Am. Chem. Soc.,* **106,** 810 (1984).
[38] Wilt, J. W., Curtis, V. A., Congson, L. N., and Palmer, R., *J. Org. Chem.,* **49,** 2937 (1984).
[39] Creary, X., and Geiger, C. C., *J. Am. Chem. Soc.,* **105,** 7123 (1983).
[40] Gassman, P. G., Talley, J. J., Saito, K., Guggenheim, T. L., Doherty, M. M., and Dixon, D. A., *Prepr.-Am. Chem. Soc., Div. Pet. Chem.,* **28,** 334 (1983); *Chem. Abs.,* **101,** 37888 (1984).
[41] McManus, S. P., and Zutant, S. E., *Tetrahedron Lett.,* **25,** 2859 (1984).
[42] Maskill, H., Thompson, J. T., and Wilson, A. A., *J. Chem. Soc., Perkin Trans. 2,* **1984,** 1693.
[43] Adcock, W., Iyer, V. S., Kok, G. B., and Kitching, W., *Tetrahedron Lett.,* **24,** 5901 (1983).
[44] Laureillard, J., and Casadevall, E., *Tetrahedron,* **40,** 2117 (1984).
[45] Laureillard, J., Casadevall, A., and Casadevall, E., *Tetrahedron,* **40,** 4921 (1984).
[46] Friedrich, E. C., and De Lucca, G., *J. Org. Chem.,* **48,** 4563 (1983).
[47] Ohkata, K., Akiyama, M., Wada, K., Sakaue, S., Toda, Y., and Hanafusa, T., *J. Org. Chem.,* **49,** 2517 (1984).
[48] Alder, R. W., Sessions, R. B., Gmünder, J. O., and Grob, C. A., *J. Chem. Soc., Perkin Trans. 2,* **1984,** 411.
[49] Ellis, M. K., Golding, B. T., and Watson, W. P., *J. Chem. Soc., Perkin Trans. 2,* **1984,** 1737.
[50] Vints, V. V., and Malievskii, A. D., *Kinet. Katal.,* **24,** 844 (1983); *Chem. Abs.,* **99,** 194139 (1983),
[51] Gubanov, V. A., Tyul'ga, G. M., Solodkaya, I. G., and Sherman, M. A., *Zh. Org. Khim.,* **20,** 493 (1984); *Chem. Abs.,* **101,** 22748 (1984).
[52] Drugov, M. V., Barantsevich, E. N., and Drach, V. A., *Zh. Org. Khim.,* **20,** 1151 (1984); *Chem. Abs.,* **101,** 151163 (1984).
[53] Anselmi, C., Catelani, G., and Monti, L., *Gazz. Chim. Ital.,* **114,** 205 (1984); *Chem. Abs.,* **101,** 190733 (1984).
[54] Berti, G., Canedoli, S., Crotti, P., and Macchia, F., *J. Chem. Soc., Perkin Trans. 1,* **1984,** 1183.
[55] Youssef, A. H., and Ed-Sadany, S. K., *Indian J. Chem.,* **22B,** 601 (1983); *Chem. Abs.,* **100,** 5496 (1984),
[56] Lynas-Gray, J. I., and Stirling, C. J. M., *J. Chem. Soc., Chem. Commun.,* **1984,** 483.
[57] Okuma, K., Nishimura, K., and Ohta, H., *Chem. Lett.,* **1984,** 93.
[58] Gassman, P. G., and Gremban, R. S., *Tetrahedron Lett.,* **25,** 3259 (1984).
[59] Fujimoto, H., Hataue, I., Koga, N., and Yamasaki, T., *Tetrahedron Lett.,* **25,** 5339 (1984).
[60] Clark, G. R., *Tetrahedron Lett.,* **25,** 2839 (1984).
[61] Sayer, J. M., Whalen, D. L., Friedman, S. L., Paik, A., Yagi, H., Vyas, K. P., and Jerina, D. M., *J. Am. Chem. Soc.,* **106,** 226 (1984).
[62] Collum, D. B., Mohamadi, F., and Hallock, J. S., *J. Am. Chem. Soc.,* **105,** 6882 (1983).
[63] Bertz, S. H., Dabbagh, G., Cook, J. M., and Honkan, V., *J. Org. Chem.,* **49,** 1739 (1984).
[64] McKinney, M. A., Kremer, K. G., and Aicher, T., *Tetrahedron Lett.,* **25,** 5477 (1984).
[65] Hoz, S., and Aurbach, D., *J. Am. Chem. Soc.,* **105,** 7685 (1983).
[66] Hoz, S., and Aurbach, D., *J. Org. Chem.,* **49,** 3285 (1984).
[67] Timofeeva, L. M., and Zhuk, D. S., *Izv. Akad. Nauk SSSR, Ser. Khim.,* **1984,** 442; *Chem. Abs.,* **100,** 208810 (1984).
[68] Koldobskii, S. G., Bobylev, V. A., Tereschenko, G. F., and Puzanov, Yu. V., *Zh. Obshch. Khim.,* **53,** 2356 (1983); *Chem. Abs.,* **100,** 33908 (1984).
[69] Haufe, G., Lacombe, S., Laurent, A., and Rousset, C., *Tetrahedron Lett.,* **24,** 5877 (1983).
[70] Al-Shali, S. A. I., Eaborn, C., Fattah, F. A., and Najim, S. T., *J. Chem. Soc., Chem. Commun.,* **1984,** 318.
[71] Brown, R. S., Slebocka-Tilk, H., Buschek, J. M., and Ulan, J. G., *J. Am. Chem. Soc.,* **106,** 5979 (1984).
[72] Corriu, R., Fernandez, J. M., and Guérin, C., *Nouv. J. Chim.,* **8,** 279 (1984).
[73] Nielsen, J., and Dahl, O., *J. Chem. Soc., Perkin Trans. 2,* **1984,** 553.
[74] Dawber, J. G., Tebby, J. C., and Waite, A. A. C., *J. Chem. Soc., Perkin Trans. 2,* **1983,** 1923.
[75] Chen, Q., and Zhu, S., *Huaxue Xuebao,* **41,** 1044 (1983); *Chem. Abs.,* **100,** 102491 (1984).
[76] Moss, R. A., and Schreck, R. P., *J. Am. Chem. Soc.,* **105,** 6767 (1983).
[77] Beadle, C. D., Bowman, W. R., and Prousek, J., *Tetrahedron Lett.,* **25,** 4979 (1984).

[78] Casadei, M. A., Galli, C., and Mandolini, L., *J. Am. Chem. Soc.*, **106,** 1051 (1984).
[79] Galli, C., and Mandolini, L., *J. Chem. Soc., Perkin Trans. 2,* **1984,** 1435.
[80] Pöchlauer, P., Müller, E. P., and Peringer, P., *Helv. Chim. Acta,* **67,** 1238 (1984).
[81] Boardman, L. D., Bagheri, V., Sawada, H., and Negishi, E., *J. Am. Chem. Soc.*, **106,** 6105 (1984),
[82] Laureillard, J., Casadevall, A., and Casadevall, E., *Helv. Chim. Acta,* **67,** 352 (1984).
[83] Yamataka, H., Ando, T., Nagase, S., Hanamura, M., and Morokuma, K., *J. Org. Chem.*, **49,** 631 (1984).
[84] Zamashchikov, V. V., Rudakov, E. S., Bezbozhnaya, T. V., and Matveev, A. A., *Zh. Org. Khim.*, **20,** 14 (1984); *Chem. Abs.*, **100,** 156002 (1984).
[85] Bansal, R. K., and Jain, S. K., *Indian J. Chem.*, **23A,** 235 (1984); *Chem. Abs.*, **101,** 54206 (1984).
[86] Mikhailov, V. A., Drizhd, L. P., Litvinenko, L. M., Bondarenko, L. I., and Savelova, V. A., *Zh. Org. Khim.*, **19,** 1456 (1983); *Chem. Abs.*, **99,** 211848 (1983).
[87] Sokolov, M. T., Nikolic, G. S., and Muskatirovic, M. D., *Glas. Hem. Drus. Beograd,* **49,** 145 (1984); *Chem. Abs.*, **101,** 190769 (1984).
[88] Subramaniam, R., and Fort, R. C., *J. Org. Chem.*, **49,** 2891 (1984).
[89] Smith, K. M., Martynenko, Z., Pandey, R. K., and Tabba, H. D., *J. Org. Chem.*, **48,** 4296 (1983).
[90] Tanida, H., Irie, T., and Hayashi, Y., *J. Org. Chem.*, **49,** 2527 (1984).
[91] Pusino, A., Rosnati, V., and Saba, A., *Tetrahedron,* **40,** 1893 (1984).
[92] Dannenberg, J. J., Barton, J. K., Bunch, B., Goldberg, B. J., and Kowalski, T., *J. Org. Chem.*, **48,** 4524 (1983).
[93] Allen, A. D., Ambidge, I. C., and Tidwell, T. T., *J. Org. Chem.*, **48,** 4527 (1983).
[94] Herbert, E., Mazaleyrat, J. P., and Welvart, Z., *Nouv. J. Chim.*, **7,** 55 (1983).
[95] Chung, B.-H., and Zymalkowski, F., *Arch. Pharm.* (*Weinheim, Ger.*), **317,** 307 (1984).
[96] Dem'yanov, P. I., Fedot'eva, I. B., Petrosyan, V. S., and Reutov, O. A., *Dokl. Akad. Nauk SSSR,* **274,** 602 (1984) [Chem]; *Chem. Abs.*, **101,** 22720 (1984).
[97] Yamataka, H., Tamura, S., Hanafusa, T., and Ando, T., *J. Chem. Soc., Chem. Commun.*, **1984,** 362.
[98] Ando, T., Tanabe, H., and Yamataka, H., *J. Am. Chem. Soc.*, **106,** 2084 (1984).
[99] Blandamer, M. J., Burgess, J., Robertson, R. E., Koshy, K. M., Ko, E. C. F., Golinkin, H. S., and Scott, J. M. W., *J. Chem. Soc., Faraday Trans. 1,* **80,** 2287 (1984).
[100] Kirkovski, L. I., Sipyagin, A. M., and Kartsev, V. G., *Izv. Akad. Nauk SSSR, Ser. Khim.*, **1984,** 1192; *Chem. Abs.*, **101,** 71951 (1984).
[101] Bowie, J. H., *Mass Spectrom. Rev.*, **3,** 1 (1984); *Chem. Abs.*, **100,** 155913 (1984).
[102] Caldwell, G., Rozeboom, M. D., Kiplinger, J. P., and Bartmess, J. E., *J. Am. Chem. Soc.*, **106,** 809 (1984).
[103] Caldwell, G., Magnera, T. F., and Kebarle, P., *J. Am. Chem. Soc.*, **106,** 959 (1984).
[104] Bohme, D. K., and Raksit, A. B., *J. Am. Chem. Soc.*, **106,** 3447 (1984).
[105] Hayes, R. N., Bowie, J. H., and Klass, G., *J. Chem. Soc., Perkin Trans. 2,* **1984,** 1167
[106] Urban, M., Cernusak, I., and Kello, V., *Chem. Phys. Lett.*, **105,** 625 (1984); *Chem. Abs.*, **101,** 54100 (1984).
[107] Raghavachari, K., Chandrasekhar, J., and Burnier, R. C., *J. Am. Chem. Soc.*, **106,** 3124 (1984).
[108] Kropp, P. J., *Acc. Chem. Res.*, **17,** 131 (1984).
[109] Pryor, W. A., and Hendrickson, W. H., *J. Am. Chem. Soc.*, **105,** 7114 (1983).
[110] Svoboda, P., Pytela, O., and Večeřa, M., *Collect. Czech. Chem. Commun.*, **48,** 3287 (1983).
[111] Kevill, D. N., *J. Chem. Res. Synop.*, **1984,** 86.
[112] Dvorko, G. F., Ponomareva, E. A., and Kulik, N. I., *Usp. Khim.*, **53,** 948 (1984); *Russ. Chem. Rev.*, **53,** 547 (1984); *Chem. Abs.*, **101,** 89921 (1984).
[113] Ladika, M., and Mihalic, Z., *Kem. Ind.*, **32,** 433 (1983); *Chem. Abs.*, **99,** 174922 (1983).
[114] Katritzky, A. R., Lopez-Rodriguez, M. L., and Marquet, J., *J. Chem. Soc., Perkin Trans. 2,* **1984,** 349.
[115] Ponomareva, E. A., Dvorko, G. F., Kulik, N. I., and Evtushenko, N. Yu., *Dokl. Akad. Nauk SSSR,* **272,** 373 (1983) [Chem]; *Chem. Abs.*, **100,** 51024 (1984).
[116] Lee, I., Lee, B. S., Koo, I. S., and Sohn, S. C., *Bull. Korean Chem. Soc.*, **4,** 189 (1983); *Chem. Abs.*, **99,** 211871 (1983).
[117] Bunton, C. A., Mhala, M. M., and Moffatt, J. R., *J. Org. Chem.*, **49,** 3639 (1984).
[118] Bunton, C. A., Mhala, M. M., and Moffatt, J. R., *J. Org. Chem.*, **49,** 3637 (1984).
[119] Kevill, D. N., Bahari, M. S., and Anderson, S. W., *J. Am. Chem. Soc.*, **106,** 2895 (1984).
[120] Roberts, D. D., *J. Org. Chem.*, **49,** 2521 (1984).
[121] Stolarova, M., and Bekarek, V., *Acta Univ. Palacki, Olomuc., Fac. Rerum Nat.*, **76** (Chem. 22), 131 (1983); *Chem. Abs.*, **100,** 138300 (1984).

[122] Kulic, J., Poskocil, J., and Kosla, J., *Sb. Ved. Pr., Vys. Sk. Chemickotechnol. Pardubice,* **45,** 331 (1982); *Chem. Abs.,* **100,** 191039 (1984).
[123] Kondo, Y., Shiotani, H., and Kusabayashi, S., *J. Chem. Soc., Faraday Trans. 1,* **80,** 2145 (1984).
[124] Magnera, T. F., Caldwell, G., Sunner, J., Ikuta, S., and Kebarle, P., *J. Am. Chem. Soc.,* **106,** 6140 (1984).
[125] Kevill, D. N., *Chem. Halides, Pseudo-Halides Azides,* **2,** 933 (1983); *Chem. Abs.* **100,** 155907 (1984).
[126] Zamashchikov, V. V., Rudakov, E. S., and Bezbozhnaya, T. V., *React. Kinet. Catal. Lett.,* **24,** 65 (1984); *Chem. Abs.,* **101,** 71911 (1984).
[127] Zamashchikov, V. V., Litvinenko, S. L., and Uzhik, O. N., *Kinet. Katal.,* **25,** 765 (1984); *Chem. Abs.,* **101,** 110001 (1984).
[128] Zamashchikov, V. V., Rudakov, E. S., Uzhik, O. N., and Litvinenko, S. L., *Ukr. Khim. Zh.* (*Russ. Ed.*), **50,** 585 (1984); *Chem. Abs.,* **101,** 151141 (1984).
[129] Cox, B. G., and Maskill, H., *J. Chem. Soc., Perkin Trans. 2,* **1983,** 1901.
[130] Rezende, M. C., Zanette, D., and Zucco, C., *Tetrahedron Lett.,* **25,** 3423 (1984).
[131] Shaik, S. S., *J. Am. Chem. Soc.,* **106,** 1227 (1984).
[132] Chandrasekhar, J., Smith, S. F., and Jorgensen, W. L., *J. Am. Chem. Soc.,* **106,** 3049 (1984).
[133] Černušák, I., and Urban, M., *Collect. Czech. Chem. Commun.,* **49,** 1854 (1984).
[134] Jaume, J., Lluch, J. M., Oliva, A., and Bertran, J., *Chem. Phys. Lett.,* **106,** 232 (1984); *Chem. Abs.,* **101,** 71840 (1984).
[135] Revetllat, J. A., Oliva, A., and Bertrán J., *J. Chem. Soc., Perkin Trans. 2,* **1984,** 815.
[136] Crowell, T. I., Braue, E. H., and Hillery, P. S., *Proc.-Electrochem. Soc.,* **84**-2 (Molten Salts), 265 (1984); *Chem. Abs.,* **101,** 71913 (1984).
[137] Vlassa, M., Kezdi, M., Strajescu, M., and Teodor, F., *Interphase Transfer Catalysis: Applications to Organic Chemistry,* Dacia: Cluj-Napoca, Romania, 1983; *Chem. Abs.,* **101,** 6557 (1984).
[138] Bar, R., de la Zerda, J., and Sasson, Y., *J. Chem. Soc., Perkin Trans. 2,* **1984,** 1875.
[139] Bar, R., de la Zerda, J., and Sasson, Y., *J. Chem. Soc., Perkin Trans. 2,* **1984,** 1881.
[140] Landini, D., Maia, A., and Montanari, F., *J. Am. Chem. Soc.,* **106,** 2917 (1984).
[141] Galons, H., Farnoux, C. C., and Miocque, M., *C. R. Hebd. Séances Acad. Sci., Sér. 2,* **297,** 255 (1983).
[142] Li, X., Pan, H., and Jiang, X., *Huaxue Xuebao,* **42,** 297 (1984); *Chem. Abs.,* **101,** 54178 (1984).
[143] Trifonov, A., and Nikiforov, T., *J. Mol. Catal.,* **24,** 15 (1984); *Chem. Abs.,* **101,** 22703 (1984).
[144] Furukawa, N., Ogawa, S., Kawai, T., and Oae, S., *J. Chem. Soc., Perkin Trans. 1,* **1984,** 1833.
[145] Anelli, P. L., Czech, B., Montanari, F., and Quici, S., *J. Am. Chem. Soc.,* **106,** 861 (1984).
[146] Hodge, P., Khoshdel, E., and Waterhouse, J., *J. Chem. Soc., Perkin. Trans. 1,* **1984,** 2451.
[147] Ando, T., Sumi, S., Kawate, T., Ichihara (née Yamawaki), J., and Hanafusa, T., *J. Chem. Soc., Chem. Commun.,* **1984,** 439.
[148] Morris, D. G., *Annu. Rep. Prog. Chem., Sect. B,* **79,** 51 (1983); *Chem. Abs.,* **100,** 173903 (1984).
[149] Bunton, C. A., and Ljunggren, S., *J. Chem. Soc., Perkin Trans. 2,* **1984,** 355.
[150] Behera, G. B., Mishra, S. S., Panda, D. S., and Sutar, S., *J. Indian Chem. Soc.,* **60,** 465 (1983); *Chem. Abs.,* **100,** 22197 (1984).
[151] Chang, J. R., Yeh, M. Y., and Shih, Y. P., *J. Chin. Inst. Chem. Eng.,* **14,** 457 (1983); *Chem. Abs.,* **100,** 191042 (1984).
[152] Chaimovich, H., Bonilha, J. B. S., Zanette, D., and Cuccovia, I. M., *Surfactants Solution, (Proc. 4th Int. Symp. 1982),* **2,** 1121 (1984); *Chem. Abs.,* **101,** 151103 (1984).
[153] Katritzky, A. R., and Musumarra, G., *Chem. Soc. Rev.,* **13,** 47 (1984).
[154] Katritzky, A. R., Sakizadeh, K., Gabrielsen, B., and le Noble, W. J., *J. Am. Chem. Soc.,* **106,** 1879 (1984).
[155] Katritzky, A. R., Dega-Szafran, Z., Lopez-Rodriguez, M. L., and King, R. W., *J. Am. Chem. Soc.,* **106,** 5577 (1984).
[156] Tanaka, K., *Yuki Gosei Kagaku Kyokaishi,* **41,** 904 (1983); *Chem. Abs.,* **100,** 5370 (1984).
[157] Brace, N. O., Marshall, L. W., Pinson, C. J., and van Wingerden, G., *J. Org. Chem.,* **49,** 2361 (1984).
[158] De Kimpe, N., *Chem. Halides, Pseudo-Halides Azides,* **1,** 813 (1983); *Chem. Abs.,* **100,** 138197 (1984).
[159] Kim, C. S., and Hong, S. Y., *Taehan Hwahakhoe Chi,* **28,** 265 (1984); *Chem. Abs.,* **101,** 190784 (1984).
[160] Hegarty, A. F., and Mullane, M., *J. Chem. Soc., Chem. Commun.,* **1984,** 913.
[161] Stein, G., Kämmerer, H., Böhmer, V., *J. Chem. Soc., Perkin Trans. 2,* **1984,** 1285.
[162] Gold, B., Deshpande, A., Linder, W., and Hines, L., *J. Am. Chem. Soc.,* **106,** 2072 (1984).
[163] Zalewski, R. I., and Geltz, Z., *J. Chem. Soc., Perkin Trans. 2,* **1983,** 1885.

[164] Fujio, M., Funatsu, K., Goto, M., Seki, Y., Mishima, M., and Tsuno, Y., *Mem. Fac. Sci. Kyushu Univ., Ser. C.,* **14,** 177 (1983); *Chem. Abs.,* **99,** 194102 (1983).

[165] Fujio, M., Goto, M., Seki, Y., and Tsuno, Y., *Mem. Fac. Sci., Kyushu Univ., Ser. C.,* **14,** 187 (1983); *Chem. Abs.,* **99,** 194103 (1983).

[166] Mishima, M., Funatsu, K., Isoda, M., Shibata, K., Yoshinaga, H., Fujio, M., and Tsuno, Y., *Mem. Fac. Sci., Kyushu Univ., Ser. C.,* **14,** 199 (1983); *Chem. Abs.,* **99,** 194104 (1983).

[167] Fujio, M., Adachi, T., Shibuya, Y., Murata, A., and Tsuno, Y., *Tetrahedron Lett.,* **25,** 4557 (1984).

[168] McLennan, D. J., and Pross, A., *J. Chem. Soc., Perkin Trans. 2,* **1984,** 981.

[169] de la Mare, P. B. D., and Newman, P. A., *J. Chem. Soc., Perkin Trans. 2,* **1984,** 1797.

[170] Richard, J. P., and Jencks, W. P., *J. Am. Chem. Soc.,* **106,** 1383 (1984).

[171] Richard, J. P., and Jencks, W. P., *J. Am. Chem. Soc.,* **106,** 1396 (1984).

[172] Sterzo, C. L., and Ortaggi, G., *Tetrahedron,* **40,** 593 (1984).

[173] Lee, I., Sohn, S. C., Lee, B. C., and Song, H. B., *Bull. Korean Chem. Soc.,* **4,** 208 (1983); *Chem. Abs.,* **100,** 191035 (1984).

[174] Lee, I., Sohn, S. C., Song, H. B., and Lee, B. C., *Taehan Hwahakhoe Chi,* **28,** 155 (1984); *Chem. Abs.,* **101,** 129895 (1984).

[175] Ramakrishnan, S., and Venkatasubramanian, N., *Indian J. Chem.,* **23A,** 60 (1984); *Chem. Abs.* **101,** 109955 (1984).

[176] Lewis, E. S., and Hu, D. D., *J. Am. Chem. Soc.,* **106,** 3292 (1984).

[177] Berg, U., and Gallo, R., *Acta Chem. Scand.,* **37B,** 661 (1983).

[178] Müller, P., and Mareda, J., *Tetrahedron Lett.,* **25,** 1703 (1984).

[179] Srinivasan, C., Shunmugasundaram, A., and Arumugam, N., *J. Chem. Soc., Perkin Trans. 2,* **1984,** 213.

[180] Srinivasan, C., Shunmugasundaram, A., Roja, M., and Arumugam, N., *Indian J. Chem.,* **23B,** 555 (1984); *Chem. Abs.,* **101,** 190779 (1984).

[181] Litvinenko, L. M., Titskii, G. D., and Mitchenko, E. S., *Zh. Org. Khim.,* **19,** 1970 (1983); *Chem. Abs.,* **100,** 22211 (1984).

[182] Bordwell, F. G., and Hughes, D. L., *J. Am. Chem. Soc.,* **106,** 3234 (1984).

[183] Solov'yanov, A. A., Beletskaya, I. P., and Reutov, O. A., *Zh. Org. Khim.,* **19,** 1822 (1983); *Chem. Abs.,* **100,** 22206 (1984).

[184] Solov'yanov, A. A., Shtern, M. M., Beletskaya, I. P., and Reutov, O. A., *Zh. Org. Khim.,* **19,** 1835 (1983); *Chem. Abs.,* **100,** 22207 (1984).

[185] Solov'yanov, A. A., Beletskaya, I. P., and Reutov, O. A., *Zh. Org. Khim.,* **19,** 1840 (1983); *Chem. Abs.,* **100,** 22208 (1984).

[186] Richard, J. P., and Jencks, W. P., *J. Am. Chem. Soc.,* **106,** 1373 (1984).

[187] Ritchie, C. D., *J. Am. Chem. Soc.,* **105,** 7313 (1983).

[188] Nevstad, G. O., and Songstad, J., *Acta Chem. Scand.,* **38B,** 469 (1984).

[189] Dahl, O., Henriksen, U., and Trebbien, C., *Acta Chem. Scand.,* **37B,** 639 (1983).

[190] Pross, A., and Shaik, S. S., *Acc. Chem. Res.,* **16,** 363 (1983).

[191] Pross, A., *J. Org. Chem.,* **49,** 1811 (1984).

[192] Yamabe, S., Yamabe, E., and Minato, T., *J. Chem. Soc., Perkin Trans. 2,* **1983,** 1881.

[193] Yamabe, Y., Minato, T., and Kawabata, Y., *Can. J. Chem.,* **62,** 235 (1984).

[194] Fujimoto, H., Yamabe, S., and Inagaki, S., *Kagaku* (*Kyoto*), **39,** 256 (1984); *Chem. Abs.,* **101,** 22629 (1984).

[195] Fujimoto, H., Yamabe, S., and Inagaki, S., *Kagaku* (*Kyoto*), **39,** 333 (1984); *Chem. Abs.,* **101,** 71828 (1984).

[196] Schug, J. C., Viers, J. W., and Seeman, J. I., *J. Org. Chem.,* **48,** 4892 (1983).

[197] Seeman, J. I., Viers, J. W., Schug, J. C., and Stovall, M. D., *J. Am. Chem. Soc.,* **106,** 143 (1984).

[198] Ayapbergenov, K. A., Muldakhmetov, Z. M., and Finaeva, M. G., *Deposited Doc.,* **1982,** VINITI 3954; *Chem. Abs.,* **99,** 194127 (1983).

[199] Toromanoff, E., *Actual. Chim.,* **1984,** 13; *Chem. Abs.,* **101,** 129872 (1984).

[200] Mills, J. E., Maryanoff, C. A., Cosgrove, R. M., Scott, L., and McComsey, D. F., *Org. Prep. Proced. Int.,* **16,** 97 (1984); *Chem. Abs.,* **101,** 22594 (1984).

[201] Tam, J. P., Heath, W. F., and Merrifield, R. B., *J. Am. Chem. Soc.,* **105,** 6442 (1983).

[202] Asano, Y., Woodard, R. W., Houck, D. R., and Floss, H. G., *Arch. Biochem. Biophys.,* **231,** 253 (1984).

[203] Awad, W. M., Whitney, P. L., Skiba, W. E., Mangum, J. H., and Wells, M. S., *J. Biol, Chem.,* **258,** 12790 (1983).

[204] Maskill, H., and Jencks, W. P., *J. Chem. Soc., Chem. Commun.,* **1984,** 944.

[205] Yoh, S. D., and Park, J. H., *Taehan Hwahakhoe Chi*, **28,** 143 (1984); *Chem. Abs.*, **101,** 22714 (1984).
[206] Hwang, J. U., Chung, J. J., Yoh, S. D., and Jee, J. G., *Bull. Korean Chem. Soc.*, **4,** 237 (1983); *Chem. Abs.*, **100,** 120221 (1984).
[207] Mravec, D., Palenik, L., Petro, M., and Ilavsky, J., *Petrochemia*, **23,** 69 (1983); *Chem. Abs.*, **100,** 67558 (1984).
[208] Beldie, C., Nemtoi, G., Onu, A., and Aelenei, N., *Rev. Chim.* (*Bucharest*), **35,** 128 (1984); *Chem. Abs.*, **101,** 54210 (1984).
[209] Matvienko, N. M., Kachurin, O. I., and Chekhuta, V. G., *Ukr. Khim. Zh.* (*Russ. Ed.*), **50,** 84 (1984), *Chem. Abs.*, **101,** 6538 (1984).
[210] Vinnik, M. I., Skakun, S. A., and Shelud'ko, A. B., *Izv. Akad. Nauk SSSR, Ser. Khim.*, **1983,** 1747; *Chem. Abs.*, **99,** 174997 (1983).
[211] Sorokin, M. F., Kochnova, Z. A., Fedosina, A. A., and Kupryashkina, E. I., *Deposited Doc.*, **1982,** VINITI 4307-82; *Chem. Abs.*, **100,** 67554 (1984).
[212] Adamets, Yu.D., Chimishkyan, A. L., and Krylov, I. A., *Deposited Doc.*, **1983,** VINITI 519–83; *Chem. Abs.*, **100,** 209074 (1984).
[213] Ryvkina, L. A., Gankin, V. Yu, Sinitsyn, A. V., and Polyakova, M. A., *Zh. Org. Khim.*, **19,** 1957 (1983); *Chem. Abs.*, **100,** 22219 (1984).
[214] Narayanan, S. S., and Rao, V. R. S., *Radiochim. Acta*, **32,** 211 (1983); *Chem. Abs.*, **99,** 194340 (1983).

Organic Reaction Mechanisms 1984
Edited by A. C. Knipe and W. E. Watts

CHAPTER 11

Carbanions and Electrophilic Aliphatic Substitution

I. WATT

Department of Chemistry, University of Manchester

Carbanion Structure and Stability

MO Calculations

The carbanion-stabilizing ability of —CN and —NC groups have been compared. *Ab initio* MO calculations for the substituted methanes, ethanes, and 2-propanes, and their derived carbanions, with a 3-21G basis set, show that —CN is the more effective group.[1] Similar calculations have been used in the examination of 1-substituted cyclopropyl and 2-substituted 2-propyl anions.[2] Substituents have included —CN, —NC, —H, and —F, and carbanion stabilization decreases in this order. All carbanions are pyramidal with the exception of 2-cyano-2-propyl, and out-of-plane angles and barriers to inversion increase with decreasing carbanion stabilization. Structures and energies of silyl anions of the form H_2Si^-X, with X = CHO, CN, or C_2H, have been calculated with a double-zeta basis set, and compared with their carbon analogues.[3] The silyl anions are pyramidal with large inversion barriers and, in the series studied, the lowest barrier is 19.9 kcal mol^{-1}, found in H_2Si^-CHO.

Conformation and rotational barriers of a series of functionalized fluoro-carbanions, X—CF_2—$\bar{C}$F—Y, with X = CN or N_3 and Y = F or OCF_3 have been calculated[4] with a split-valence 4-31G basis set. The carbanions are pyramidal and when Y = F, butane-like rotational profiles are obtained, with the lowest energy conformation having the lone pair *anti* to the X group. With Y = OCF_3, the conformations with the lone pair *anti* to a β-fluorine are lowered relatively and should be significantly populated. These and related fluoro-carbanions can be generated by nucleophilic addition to fluoro-olefins, and the results of the computational study are used to account for the variation of the ratio of elimination of fluoride to trapping by carbon dioxide to give β-substituted polyfluoro-propionates.

The $C_9H_9^-$ potential energy surface has been explored using semi-empirical and *ab initio* SCF methods, with agreement between MNDO and STO-3G methods as to energy ordering of the structures.[5] Both bicyclo[3.2.2]nonatrienyl (**1**) and barbaralyl (**2**) are stable structures with the MNDO placing (**2**) 16.7 kcal mol^{-1} above (**1**). Related structures with C_{2v} or higher symmetry, which might lie on the barbaralyl Cope rearrangement pathway, are all more than 35 kcal mol^{-1} above (**2**) showing that, in contrast to the situation in the cation, the rearrangement should be slow. On the same scale, the cyclononatetraenyl anion is 10.5 kcal mol^{-1} more stable than (**1**). The same semi-empirical and *ab initio* methods have been compared in calculation of the structure of the diphenylmethide anion.[6]

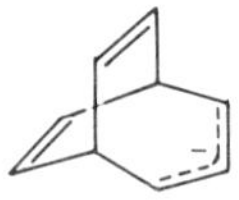

(**1**)

(**2**)

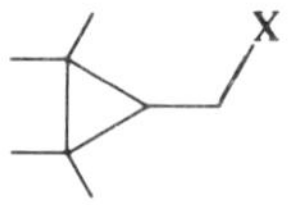

(**3**)

a: X = Li
b: X = Cl
c: X = Hg

Organolithiums

Tetramethylcyclopropylmethyllithium (**3a**) has been prepared by reaction of the chloride (**3b**) or the mercurial (**3c**) with lithium powder in ether.[7] Solvent-free crystals have been obtained, and X-ray crystallography shows that they contain hexamers, in which the lithium atoms occupy the corners of a trigonal anti-prism. The end faces are unoccupied, but each of the lateral faces is coordinated to a cyclopropylmethyl ligand.

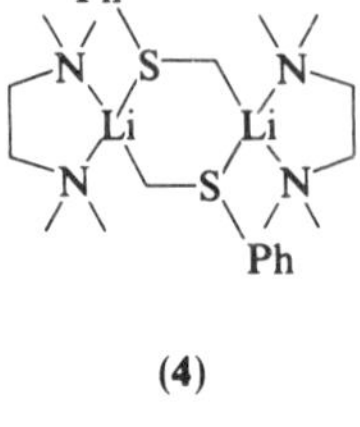

(**4**)

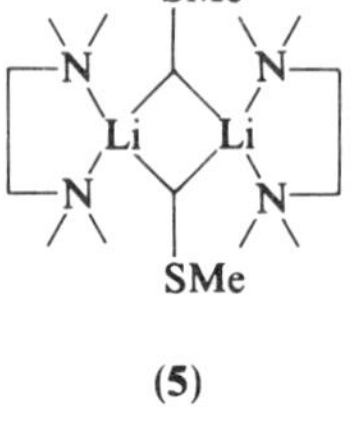

(**5**)

The low-temperature X-ray crystal structures of the TMEDA complexes of phenylthiomethyllithium (**4**) and methylthiomethyllithium (**5**) show that both exist as dimers.[8] In the case of (**4**), the lithium atoms are also coordinated to the sulphur atoms giving a chair conformation six-membered ring with equatorial phenyl groups. In contrast, (**5**) contains a four-membered ring. Earlier conclusions as to the ion-pair nature of the C-Li bond in the structure of 2-(2-phenyl-1,3-dithianyl)lithium have been revised. With corrections to the electron density map, it is now not possible to measure charge distribution because of the similar scattering behaviour of neutral lithium and its monocation. Crystalline bis(trimethylsilyl)methyllithium contains linear polymers with an unusual single alkyl bridging between lithium atoms.[9] The gas-phase (413 K) electron diffraction structure of the monomer has also been determined. Geometries found in experimentally determined structures of allyl-, benzyl-, and trityl-lithium can be reproduced by a simple hard-sphere electrostatic model of the bonding between lithium and the organic residues.[10] Lithium–carbon and lithium–hydrogen distances in phenyllithium solutions have been determined from the contribution of ^{7}Li to the spin-lattice relaxation times of C(1) and H(2).[11] In diethyl ether–cyclohexane solution, the distances are in good agreement with those in the X-ray crystal structure of the tetrameric ether solvate, but chemical shift variations suggest that, in dilute solution, the tetramer coexists with dimer. In THF solution, phenyllithium is dimeric.

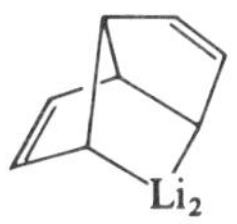

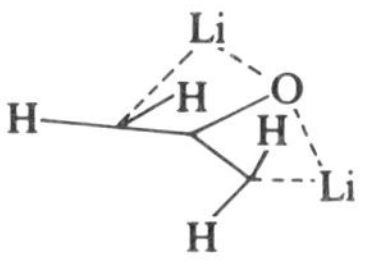

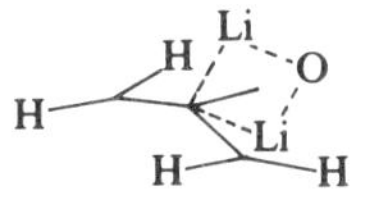

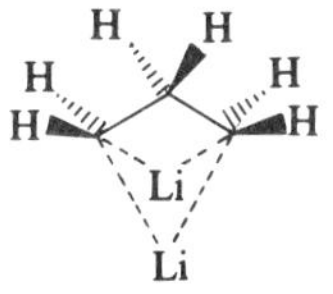

(**6**) (**7**) (**8**) (**9**)

Full details of the preparation and solution spectroscopic properties of the dilithium salts of 1,1-dibenzylethylene and of diphenylacetone have been published.[12] Diastereoisomeric clusters with C_{2h} and D_2 symmetry have been proposed to account for the Li-NMR spectra of "dilithiumsemibullvalene" (**6**)[13]. Dilithium compounds appear to crystallize as monomers. X-ray crystallography of the bis-TMEDA complex of 1,3-dilithiodiphenylacetone shows chiral monomers with near C_2 symmetry.[14] Each TMEDA-solvated lithium is coordinated to the oxygen on opposite sides of the plane of the acetone unit. The C—O bond-length (1.334 Å) is similar to that found in structures of aggregated lithium enolates. *Ab initio* MO calculations on 1,3-dilithioacetone itself, with a split valence 3-21G basis set, gratifyingly, gives the corresponding chiral C_2 structure (**7**) as energy minimum.[15] Racemization *via* a C_{2v} structure (**8**) is estimated to have a barrier of 16 kcal mol^{-1}. With the same MO methods, the lowest energy geometry for 1,3-dilithiopropane is the doubly lithium-bridged structure (**9**).[16] Although stable with respect to extended conformations, or disproportionation with propane, rearrangement to an allyllithium–lithium hydride complex is exothermic by 29.8 kcal mol^{-1}.

Reinvestigation of 1,2-dilithioethane shows that the global minimum for $C_2H_4Li_2$ is an analogous vinyllithium–lithium hydride complex. 1,4-Dilithiobutane is stable to such rearrangement. The nature of the double bridging commonly found in 1,4-dilithium compounds has been discussed.[17] Symmetrical arrangements are favoured by electrostatic considerations when charges are localized; this preference must be balanced against the effects of delocalization.[18]

Aromatic and Other Delocalized Ions

Pentafluorocyclopentadiene has been prepared and has $15.5 < pK_a < 12.2$, illustrating the relatively small effect of fluorine substitution on the aromatic nucleus.[19] THF solutions of the metal salts are not stable, but [Na^+ 18-crown-6]C_5F_5 solutions survive at room temperature for several hours. Treatment of specifically ^{13}C-labelled 1,2,3-triphenyl-3-(trimethylsilyl)cyclopropene (**10**) with tetra-*n*-butylammonium fluoride in aqueous THF results in the formation of 1,2,3-triphenylcyclopropene with the label nearly fully distributed over the cyclopropene ring sites.[20] No scrambling is found in the recovered (**10**). If the protiodesilylation involves formation of the cyclopropenyl anion, the result indicates that pseudorotation in a structure of less than D_{3h} symmetry must occur faster than protonation. Cyclopropenide itself has been reinvestigated by *ab initio* SCF theory using 3-21G and 6-31G* basis sets with electron correlation.[21] The fully optimized geometry has two long and one shorter C—C bond, and the charge is effectively localized at a unique pyramidal carbon. The highest occupied MO is a bound state. A single-point calculation (6-31G*/unrestricted Moller–Plesset) on a triplet of D_{3h} symmetry placed it 19 kcal mol^{-1} above the non-planar singlet. Squaric acid dianions, in which the oxygens have been sequentially replaced by dicyanomethylene groups have been prepared and an X-ray crystal structure of the substituted dianion determined.[22]

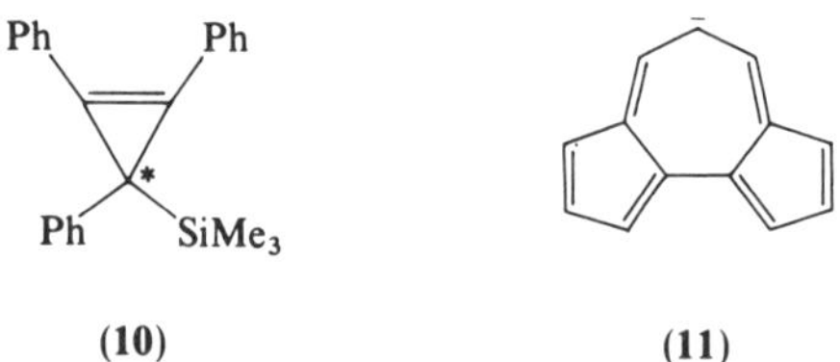

(**10**) (**11**)

A full paper on the chemistry of the cyclopent[*e*]azulenide (**11**) has appeared.[23] The NMR spectroscopic properties of the ion and the low pK_a (between 12 and 15) of the parent hydrocarbon are consistent with aromatic stabilization in the 14π-electron periphery. Deprotonation of 2,5-di-*tert*-butyl-7*H*-cyclopenta[*a*]pentalene with lithium tetramethylpiperidide gives the novel ion (**12**).[24] THF solutions are stable at $-78°$ but decompose rapidly at $-30°$. At present, neither electronic nor H-NMR spectra allow distinction between a fully delocalized ion and rapidly equilibrating localized forms, but the high pK_a of the hydrocarbon suggests an important anti-aromatic contribution from the 12π-electron perimeter. Potentially anti-aromatic α-heteronaphthalenyl anions have been prepared by KNH_2 deprotonation of precursors.[25] Only the thianaphthalenide is sufficiently stable for NMR observation at $-35°$, but rearranges on warming to 2-methylbenzothiophene.

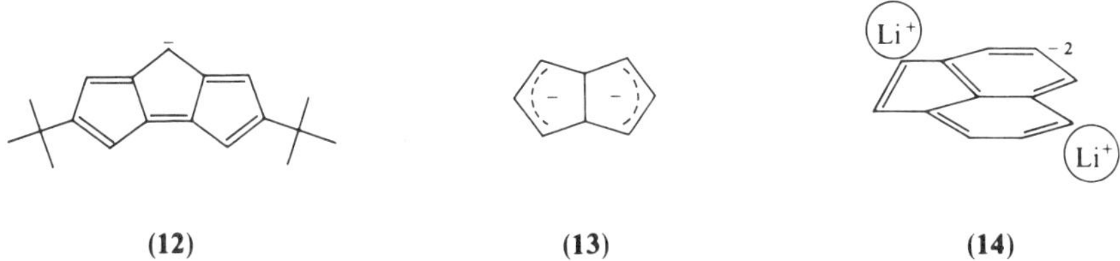

(12) (13) (14)

Potassium *tert*-alkoxide/butyllithium mixtures can form polyanions from alkylbenzenes[26,27] and metallate pyridine.[28] Reaction of tetrahydropentalene with the mixed reagent yields the cyclooctatetraene dianion (COT^{-2}).[29] Since use of less than a full equivalent of base gives no COT^{-2}, its formation presumably involves disrotatory ring-opening of the bicyclic dianion (**13**). MNDO calculations show this rearrangement to be exothermic by 27.7 kcal mol^{-1}. An inference from the observations that dianion (**13**) is not an intermediate in the reduction of semibullvalene is premature. Although reduction of semibullvalene with lithium gives dilithium semibullvalene (**6**) which is stable in THF at $-78°$ and protonates to give tetrahydropentalenes, slow rearrangement to COT^{-2} does occur at 0°.[30] Furthermore, use of potassium as the reductant yields the COT^{-2} directly at $-78°$.

Cyclophanes with unsaturated bridges have been reduced to give dianions with perimeters with 22–38 π-electrons.[31] The solvent, cation, and temperature dependence of the NMR spectra of the acenaphthalene dianion have been examined.[32] The ^{13}C-NMR spectra of the $Na_2{}^+$ and $K_2{}^+$ salts are consistent with contact ion-pairs, and show little variation with temperature. On cooling, signals from C(1) and C(2) in the $Li_2{}^+$ salt shift upfield, and those from C(5) and C(6) experience a larger downfield shift. No concentration effects are observed. A model of the ion-pair (**14**) with the metal ion occupying non-equivalent positions is proposed to account for the data, and the authors emphasize that without knowledge of the ion structure, correlation of charge densities with chemical shifts is not sound. Claims that alkali-metal reduction of pyrene (p) can give distorted di- or tetra-anions have been shown to be wrong.[33,34] The major reduction product, in THF, is the known pyrene dianion, p^{-2}, with full D_{2h} symmetry as evidenced by the simplicity of its ^{13}C- and ^{1}H-NMR spectra which show temperature-dependent line-shape changes associated with rapid electron exchange with the anion radical, $p^{-1}\cdot$. Spectra of the supposed unsymmetrical polyanions appear to have been those of the monoanion (**15**), presumably formed by protonation of the dianion by residual moisture. The ^{13}C-NMR spectra of mixtures of tetraphenylethylene and its dilithium salt in deuterobenzene[35] give a remarkably low upper limit for rates of bimolecular exchange of 100 $\text{M}^{-1}\,\text{s}^{-1}$.

The IR spectra of some nitro-stabilized carbanions have been reported.[36,37] Variable-temperature ^{13}C-NMR methods have been used to determine rotational barriers in diphenylmethyl anions stabilized by trimethylsilyl and tricarbonylchromium (0) groups.[38] Solutions of the lithium salts of 9,10-dihydroanthracenes in liquid ammonia have also been examined by ^{13}C-NMR.[39] Chemical shifts are indicative of substantial delocalization of charge and the one-bond coupling constants at the carbanionic site, $J_{C(10)-H(10)}$, are consistent with sp^2 hybridization of

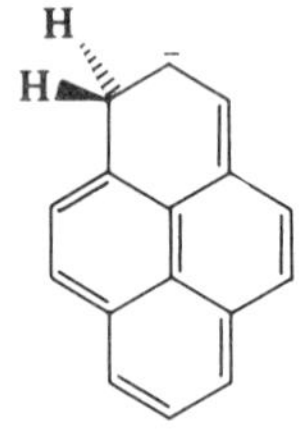

(15)

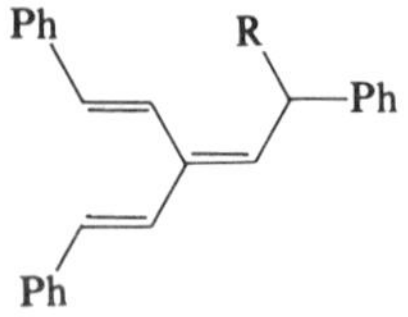

(16)

R = H, $SiMe_3$

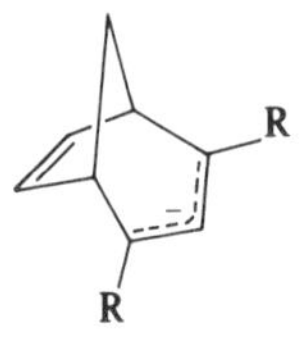

(17)

a: R = H
b: R = Ph

that carbon. CD spectra of fluorenyl-, benzyl-, and diphenylmethyl-lithium, in non-polar solvents such as toluene or heptane, show induced circular dichroism with added chiral tertiary amines.[40]

*p*K*a* *Measurements*

Tristyrylmethane has a $pK_a = 20.1$ in 9:1 DMSO:EtOH, and is thus 12 pK_a units more acidic than triphenylmethane.[41] Both ^{1}H- and ^{13}C-NMR data are reported. Shifts are similar for Li^+ and K^+ salts in THF indicating loose or solvent-separated ion-pairs. Quenching of the ion with water or trimethylsilyl chloride gives the cross-conjugated triene (**16**).

Equilibrium caesium ion-pair acidities of a range of unsaturated hydrocarbons have been determined in THF using 9-phenylfluorene as reference[42] at $pK_a = 18.49$. There is notably good agreement between the values found in this solvent and those reported for DMSO, cyclohexylamine, and dimethoxyethane. Bicyclo[3.2.1]octadiene has $pK_a = 31.4$ in cyclohexylamine/caesium cyclohexylamide.[43] By this method, the diene is more than 9 pK_a units more acidic than bicyclo[3.2.0]oct-2-ene. Bis-homoaromatic stabilization in the ion (**17a**), from introduction of the remote second double bond compares favourably with the effects of simple conjugation. It has been pointed out that the ^{13}C-NMR signals from C(6) and C(7) in the substituted ion (**17b**) show upfield shifts of 12.5 and 22.5 ppm respectively, relative to the parent hydrocarbon.[44] Since long-range π-interaction is not likely to be important in this ion, chemical shift behaviour alone is evidently not a good indicator of delocalization. Equilibria for *ortho*-lithiation of mono-substituted benzenes by lithium tetramethylpiperidide in THF are rapidly established, and concentrations have been measured by ^{13}C-NMR methods.[45] The most acidic compound is phenyl *N,N*-diethylcarbamate with $pK_a = 37.2$ (relative to tetramethylpiperidine at 37.3). These substituent effects on the equilibria are not reflected in the reactivity of the same compounds towards alkyllithiums in ether. In ethereal solvents, deprotonation of the 2-methyl group of *s*-collidine by phenyllithium is kinetically favoured.[46] Slow equilibration to the anion at the 4-position occurs in THF or DME and rates are enhanced by addition of HMPA. In liquid ammonia, with lithium or sodium amide bases, only the anion from the 4-methyl site is observed. Isomeric cyclohexadienes and dihydronaphthalenes have also been equilibrated with metal amide bases in liquid ammonia.[47] Counter-ion effects, relevant when choosing metals and conditions in Birch reductions, have been found. The pK_as of 4-substituted phenylethynes in DMSO correlate with σ or σ° constants.[48] Acidities of naphthyl- and pyridyl-ethynes have also been measured.

Gas-phase IR laser-induced decomposition of alkoxide anions $(CH_3)_2CO^-R$ with R = H, CF_3, Ph, H, or vinyl, in a pulsed ICR spectrometer, results in exclusive formation of acetone enolate by loss of RH.[49] Since the C—R bonds in this series are known to be weaker to heterolysis than to homolysis, the result is consistent with a step-wise heterolytic mechanism, where products reflect the stability of an intermediate ion–molecule complex. The decrease in isotope effect in the fragmentation of deuteromethyl alkoxides as the leaving group is altered from CF_3 ($k_H/k_D = 6$) to CH_3 ($k_H/k_D = 1.6$) is inconsistent with a concerted mechanism. A relative order of leaving-group abilities has been determined and reveals an unusual ordering of alkane acidities with $Bu^tH > MeH > Pr^iH > EtH$. A similar ordering is implied by the behaviour of product ratios in the gas-phase reaction of OH^- with substituted trimethylsilanes,[50] $(CH_3)_3SiR$, yielding either $(CH_3)_3SiO^-$ and RH or $(CH_3)_2(R)SiO^-$ and CH_4. In this case, calibration with substituents of known acidity allows quantitative estimates to be made of gas-phase acidities and radical electron affinities. The ethyl, propyl, and *tert*-butyl radicals appear to be unstable with respect to electron attachment. Deprotonation of phenylcyclopropane, 2-phenylpropene, or 3-phenylpropene in the gas phase by NH_2^- yields ions, *m/z* 117, which exhibit different reactivity patterns to N_2O, O_2, and in exchange with D_2O.[51] For example, under conditions where the ion from 2-phenylpropene is completely converted to d_4-anion, that from phenylcyclopropane still contains large amounts of its parent ion, and smaller amounts of d_1–d_5 anions. Use of labelled substrate demonstrated that these are formed by exchanges at the aromatic ring. Slower exchange of the non-benzylic cyclopropyl hydrogens is also observed, but ring-opening to the 1-phenylallyl anion does not occur within the time-scale of the flowing-afterglow experiment.

Carbanion Reactions

General Reviews

The following subjects involving carbanion reactions have been reviewed: stereoselective aldol condensations;[52–54] stereoselective alkylations of chiral metal enolates;[55] selectivity in carbon–carbon bond-forming reactions;[56] substitution and addition reactions of nitrile-stabilized carbanions;[57] β-acyl and β-vinyl carbonyl anions in organic synthesis;[58] unmasked acyl carbanions from carbonylation of alkyllithiums;[59] and the formation and transformation of allenic and acetylenic carbanions.[60] More generally, a useful review of techniques in carbanion chemistry has also been published.[61]

Enolates and Related Species

N-Fluoro-*N*-alkylsulphonamides are useful reagents for selective fluorination of a wide range of carbanions.[62] Their use does not require any of the specialized equipment usually necessary for the reaction. The lithium enolate of acetaldehyde, formed from the action of butyllithium on THF, reacts with aryl azides to give diazomethane and the corresponding aryl formamide.[63] Salt-free butylpotassium also reacts with THF to give the acetaldehyde enolate,[64] but the metallated ether can be intercepted at low temperatures by electrophiles such as chlorotrimethylsilane to give the α-substituted ether. Sodium hydride induces efficient cyclization of the isocyanate (**18**) to 1-phenyl-1,5-methanobenzazocines.[65] Carbanions from methyl nitroacetate can be generated in polymer-bound form on Amberlite IRA 400.[66] The storable polymer can be treated with alkyl halides or Michael acceptors to give good yields of α-nitro-esters.

LDA and the more hindered base, lithium *tert*-octyl-*tert*-butylamide, LOBA, have been found to be compatible with chlorotrimethylsilane in THF at −78°. Deprotonation of ketones and esters by these bases in presence of TMSCl yields silyl enol ethers from kinetically generated enolates with enhanced regio- and stereoselectivity.[67] Typically, selectivities for methyl deprotonation in a range of methylalkyl ketones are greater than 95%, and the reaction of 3-pentanone with the LOBA–TMSCl combination affords the E- and Z-silylenol ethers in a 98:2 ratio. The results have been attributed to the minimization of steric interactions in the six-centre cyclic transition state (**19**) for the deprotonation. With added HMPA rapid equilibration occurs, with formation of increased amounts of product from the more stable *Z*-enolate. Deprotonation of the labelled 3-methylbutenoates (**20a**) and (**20b**) with LDA, followed by quench with saturated aqueous ammonium chloride forms the deconjugated esters (**21a**) and (**21b**) respectively, indicating that the deprotonations occur at the methyl group *cis* to the ester.[68] It is likely that the selectivity is also associated with the coordination of the lithium to carbonyl oxygen.

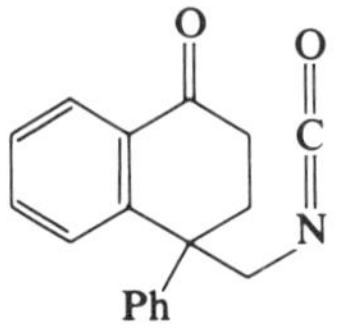

(**18**) 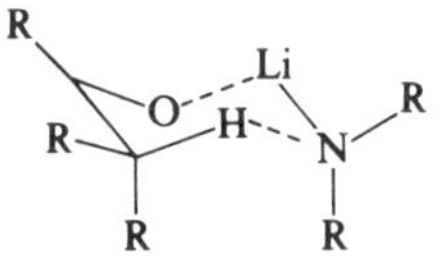 (**19**) 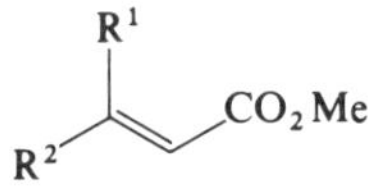 (**20**) 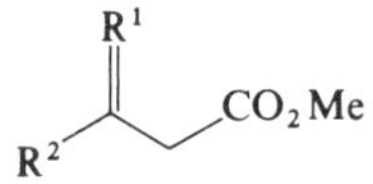 (**21**)

a: $R^1 = CH_3/CH_2$; $R^2 = CD_3$
b: $R^1 = CD_3/CD_2$; $R^2 = CH_3$

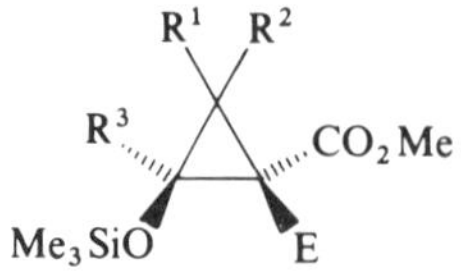

(**22**)

R = H, Ph, alkyl
E = Me, allyl, SPh

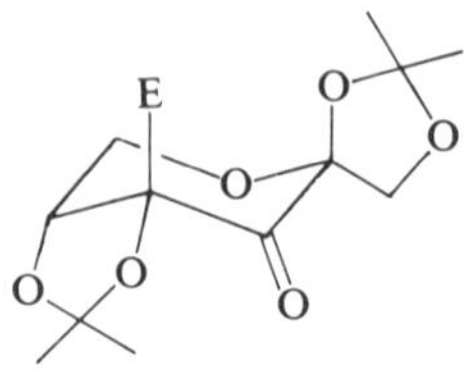

(**23**)

a: E = H
b: E = Me, CH_2OH

Lithium enolates of 2-silyloxycyclopropanecarboxylic acid esters react with alkyl halides and dimethyl disulphide to yield 1-substituted products (**22**) with high selectivity for attack of the electrophile *syn* to the silyloxy group.[69] Carbohydrate uloses such as (**23a**) have been converted to their lithium enolates by the action of LDA/HMPA[70] and converted to their methyl or hydroxymethyl (**23b**) derivatives in yields of 40–70%. Competing Cannizzaro reactions also give reduced products in the reaction with formaldehyde, but the alkylations occur with high axial selectivity. The stabilized anions from the conformationally biased β-keto-ester and β-keto-nitrile dimethylhydrazones (**24a**) react with methyl iodide to give 2-methyl derivatives with a preference for axial substitution with selectivity in the nitrile, 26 : 1, being the greater.[71] The selectivities are remarkably insensitive to variation of counter-ion and to added HMPA. Similar axial preferences have been found in the 6-methyl series (**24b**), which includes the dimethylhydrazone of 2-methylcyclohexanone itself. An X-ray crystal structure of the lithiated cyclohexanone dimethylhydrazone/TMEDA complex has been determined. There are two independent molecules in the asymmetric unit, and the six-membered rings have half-chair conformations with C(1), C(2), C(3), and C(6) coplanar. The hydrazone nitrogen atoms and C(1) and C(2) are also coplanar, with a lithium atom coordinated to these atoms in η^4-fashion and lying 1.64 Å from the plane so that the faces of the π-system are differentiated in a way not found in simple enolate structures. Alkylation of the stabilized carbanions derived from the cyclic β-keto-sulphoxides (**25a**)[72] and (**25b**)[73] has been examined. The reaction with alkyl halides is selectively *cis* to the S—O bond in the case of (**25a**) and *trans* in the case of (**25b**). The dianion from (**25b**)[74] reacts at the 1-position, also with *trans*-selectivity for alkylation, but quenching with D_2O introduces a deuteron *cis* to the S—O bond at that position. Reaction of methyl iodide with lithium enolates of camphor and its bromo derivatives yields the thermodynamically less stable *exo*-methyl compound, despite apparent steric hindrance to attack by the 7-substituents.[75] It has been pointed out that, with the exception of singlet oxygen, electrophilic reagents generally react with camphor enol derivatives to give larger proportions of *exo*-product, and that the steric effect of the 7-substituents on reagent approach is often overestimated.

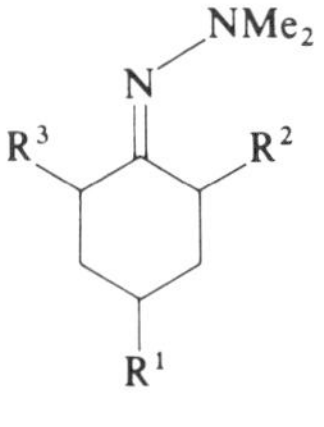

NMe$_2$ N R^3 R^2 R^1

(**24**)

a: R^1 = Bu^t; R^2 = CN, CO$_2$Me; R^3 = H
b: R^1 = H; R^2 = Me, CN, CO$_2$Me, SMe, H; R^3 = Me

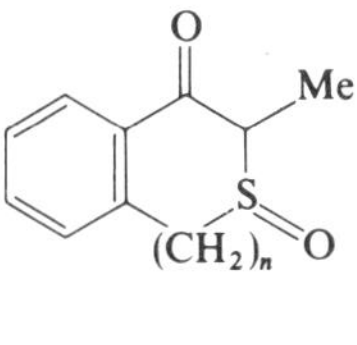

O Me S O (CH$_2$)$_n$

(**25**)

a: $n = 0$
b: $n = 1$

NMe$_2$ R SMgBr

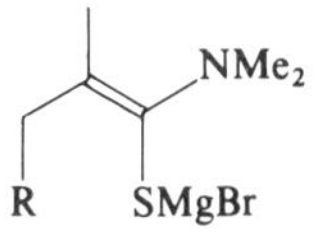

(**26**)

Tetra-substituted enolates generated by addition of Grignard reagents to *N,N*-dimethyl-α-methacrylothiamide react with simple aldehydes at low temperature to give high yields of *erythro* aldol condensation product.[76] Raised reaction times and temperatures result in reduced yields and selectivities, apparently because of reversibility in the aldol reaction. It has been suggested that the high *erythro*-selectivities under conditions of kinetic control arise, in the first instance, from high Z stereochemical purity of the enolate (**26**) formed by addition of the Grignard reagent RMgBr to the thioamide *via* a six-centre cyclic array with the magnesium coordinated to the sulphur atom. Diastereoisomer ratios have been determined in the products of conjugate addition of $BuCu.BF_3$ to E-2-methylbutenoates, after protonation of the resultant enolate.[77] Selectivities are small but protonation of an enolate conformation with an eclipsed enolate double bond and β-C—H (**27**) is favoured. Selectivities in the methylation of the lithium enolate of 3-methylheptanoate are also consistent with this model, although again selectivities are not great. A full description of the generation of stabilized vinyl carbanions by deprotonation of acrylonitrile or methyl acrylate has been published.[78] Use of specifically *cis*-β-deuterated precursors has demonstrated that the ion from acrylonitrile has some configurational stability.

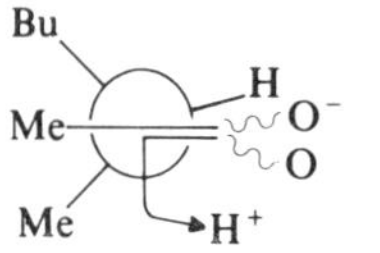

(**27**)

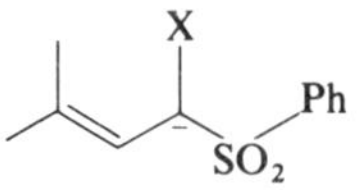

(**28**)

a: X = H
b: X = Cl

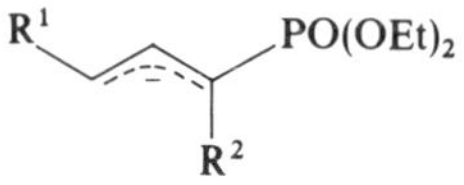

(**29**)

a: R^1 = H; R^2 = SMe, OEt, $OCOBu^t$
b: R^1 = $SiMe_3$; R^2 = $OP(OEt)_2$

Tertiary enolates react with trichloroethylene giving good yields of E-1,2-dichlorovinyl adducts which can be converted to chloroacetylenes in good yields.[79] Evidence supporting an elimination–addition mechanism, with the intermediacy of dichloroacetylene functioning as a Michael acceptor, has been presented. Similar enolate condensation with hexachlorobutadiene yields either pentachlorodienyl adducts, or trichloroenyne adducts, and both addition–elimination and elimination–addition mechanisms appear to operate in this case with the balance depending on the nature of the enolate.

Hexachloroethane efficiently α-chlorinates the allylsulphone (**28a**) under phase-transfer base catalysis.[80] Under the conditions, the α-chloroallylsulphone anion (**28b**) is formed and may react with a range of alkyl halides, Michael acceptors, and aldehydes through the α-position. 1-Phosphorylallyl anions, carrying a range of heteroatom substituents at their 1-position (**29**) have been prepared by deprotonation of allyl or vinyl phosphonates.[81] Their reactions with alkyl halides show only low regioselectivity for the 1-site. Generally, reactions with carbonyl electrophiles are not successful, but the silylated derivative give functionalized dienes in moderate yield by Wittig–Horner condensation. Conversion of substituted allyl anions (**30**) to their trialkylboron aluminium "ate" complexes enhances selectivity for the α-site in their reactions with reactive alkyl halides or carbonyl compounds.[82] The regio- and stereo-chemistry of the reactions of η^3-allyltitanium complexes with aldehydes has been studied.[83] Anions formed by deprotonation of α,β- or β,γ-unsaturated or α-aryl *N*-phenyl thioimidoesters are alkylated on the α-carbon by reactive alkyl halides.[84] Those from the saturated analogues alkylate exclusively on nitrogen.

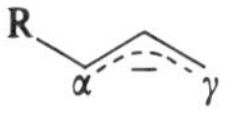

(30)

R = PhS, OPr^i, OCH_2OMe, NMe_2, SePh

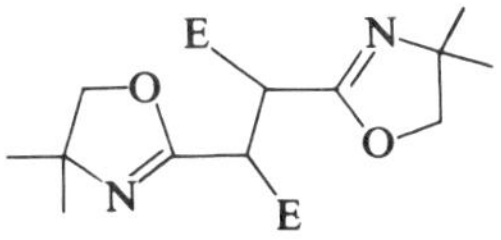

(31)

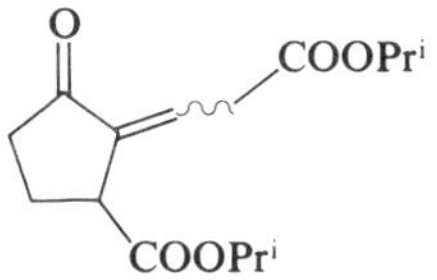

(32)

Diisopropyl succinate[85] or the 1,2-bis(oxazolinyl)ethane **(31)**[86] are cleanly deprotonated by two equivalents of LDA or butyllithium, respectively. The resulting 1,2-dianions are versatile synthetic intermediates, most importantly, giving high yields of cyclopentane-1,2-dicarboxylic acid derivatives with 1,3-dihalides or β-bromo-esters. The extended dianion from diisopropyl 3-hexenedioate also reacts with 3-bromopropionate to form five-membered rings.[87] The resulting cyclopentanone **(32)** has been transformed to sarkomycin and a range of prostaglandins. Butyllithium also doubly deprotonates aldehyde oximes.[88] Low temperature is essential to prevent nitrile formation, but the dianions can be alkylated on carbon with reactive alkyl halides. Improved techniques for preparation and alkylation of α,α'-dianions from β-keto-esters and β-keto-sulphones have been described.[89,90] Highly functionalized tetrahydroanthracenes, intermediates in the synthesis of (±)-auramycinine and (±)-13-methylaklavinone, have been prepared by "bicyclo cyclization" of dianions from trihydroxynaphthalene derivatives.[91]

Heteroatom-stabilized Carbanions

Electrolysis of 1 : 1 mixtures of methyl trichloroacetate and methyl dichloroacetate in presence of α-branched aldehydes affords diastereoisomeric 2,2-dichloro-3-hydroxyalkanoates by condensation of the dichloro(methoxycarbonyl)methyl anion with the aldehyde.[92] Yields are only moderate, but high diastereoselectivities are possible. Thus the reaction with 2,2-dimethyl-1,3-dioxolane-4-carboxaldehyde gives the adduct **(33)** with the indicated stereochemistry and over 95% selectivity. Trichloromethyl anion has been generated in an analogous electroductive chain-reaction, from a carbon tetrachloride–chloroform mixture, and reacts also with aldehydes but in less stereoselective fashion. Experiments on the ready decarboxylation of trichloroacetic acid in DMSO have been fully described.[93] The reaction, monitored by CO_2 evolution is first order in trichloroacetic acid but rates are depressed by addition of acids; this is consistent with slow decarboxylation of a trichloroacetate anion, to give trichloromethyl anion which can be trapped as a Meisenheimer complex with trinitrobenzene. The rates are unaffected by the concentration of the trapping agent and those for decomposition of tribromoacetate are between ten-fold and one hundred-fold faster. The relative reactivities of trichloromethide towards a series of 4-substituted benzaldehydes correlate with

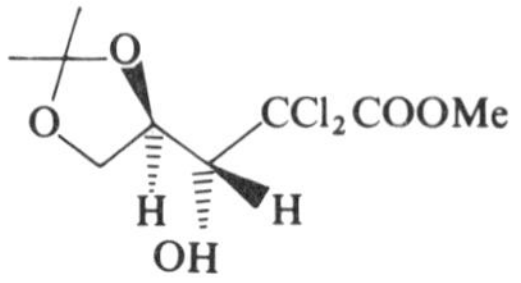

(33) 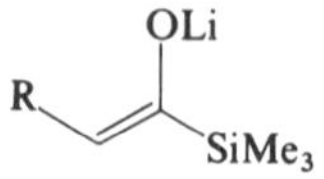 (34) 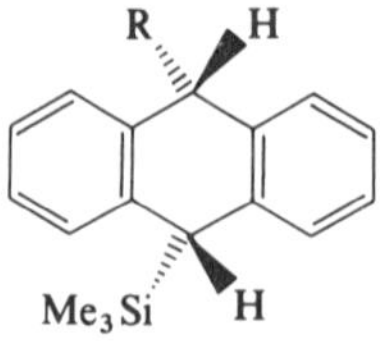 (35)

R = Me, Et, Pr^i

σ constants.[94] Interestingly, the most reactive of the aldehydes are only half as reactive to trichloromethide as protons, showing that the proton reaction is not encounter-controlled. The potassium salt of 2,4,6-trinitrophenylacetic acid also decarboxylates at room temperature in DMSO or THF, yielding solutions of the trinitrobenzyl anion.[95] The THF reactions are promoted by addition of crown ether.

Carbonylation of α-trimethylsilylalkyllithiums at $-15°$ yields acyllithiums which rearrange to E-acylsilane enolates (**34**) by intramolecular 1,2-silicon shift.[96] The enolates react with aldehydes to yield *erythro* aldol condensation products with a high degree of selectivity. When the carbonylation is carried out at lower temperatures, the acylsilane enolates are not formed cleanly because reaction of the initially formed acyllithium with a second equivalent of carbon monoxide competes with the 1,2-silicon shift. Deprotonation of 9-alkyl-9-(trimethylsilyl)-9,10-dihydroanthracenes with butyllithium is accompanied by rapid rearrangement involving intramolecular 1,4-shift of the trimethylsilyl group.[97] Protonation of the resulting anion yields preferentially the *cis*-9,10-dihydroanthracenes (**35**). Haller–Bauer cleavage of the optically active cyclopropyl ketone (**36**) with $NaNH_2$ yields the 2,2-diphenyl-1-(trimethylsilyl)cyclopropane with complete retention of configuration.[98] Evidently, the intermediate TMS-stabilized cyclopropyl-carbanion is configurationally stable.

Desilylation of bis(trimethylsilyl)phenylmethane (**37a**) by alkoxides in HMPA yields the α-silyl-carbanion $Ph\bar{C}HSiMe_3$ which reacts with benzaldehyde to give near-quantitative yields of stilbenes.[99] The isomer ratio *cis*:*trans* = 1:1.4, is insensitive to temperature, the nature of the alkoxide, addition of inert salts such as lithium perchlorate, and to variation of the solvent. This behaviour contrasts dramatically with that of the Wittig reaction, and presumably reflects the irreversible formation of the *threo*- and *erythro*-α-silylalkoxide intermediates in formation of the *trans*- and *cis*-stilbenes. Increase of the substituent size on the silyl groups (**37b**) alters the ratio in favour of the *cis*-isomer. The cleavage of 1-(phenylthio)-1-(trimethylsilyl)alkanes with lithium naphthalenide has been used to prepare α-silyl-carbanions which are inaccessible by other methods.[100] These give good yields of adducts with non-enolizable ketones or aldehydes.

Carbanions α to (alkoxy)silyl or (amino)silyl groups have been prepared by addition of alkyllithium or Grignard reagents to vinylsilanes.[101] With chiral groups on the silicon, the method affords chiral anions which may be coupled with alkyl halides.[102] Oxidative cleavage of the Si—C bond then gives optically active alcohols with modest enantiomeric excess.

(36)

$PhCH_2(SiR_3)_2$

(37)

a: R = Me
b: R = Et, Bu^t, Ph

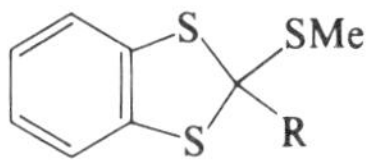

(38)

a: R = CD_3
b: R = –

Sequential addition of methyllithium and trideuteromethyl iodide to ortho-phenylenetrithiocarbonate affords the 1,3-benzodithiole (**38a**) showing that the initial reaction with methyllithium involves formation of an *S*-methyl bond and the carboxyl anion equivalent (**38b**);[103] interestingly, use of only half an equivalent of methyllithium leads to formation of the hexathioorthooxalate (**39**), apparently by addition of the anion (**38b**) to the carbon of the neutral trithiocarbonate. Thermodynamic control in the latter addition may account for the difference in behaviour of the two carbanionic reagents. Attempted metallation of selenolo[2,3*b*]selenophane with butyllithium results in formation of the lithiated selenophenes (**40a**) and (**40b**) respectively,[104] their formation requiring attack by the carbanionic butyl group at selenium. Under the same conditions, selenolo[3,2*b*]selenophane reacts without rearrangement to give the expected 2-lithio derivative.

The energy profiles for the reactions of the phosphorus and sulphur ylids, CH_2—PH_3 and CH_2—SH_2, with formaldehyde have been studied theoretically using calculations at the STO-3G level supplemented by 4-31G* calculations at minima and transition states.[105] For Wittig-type reactions leading to alkene formation, both ylids react exothermically with small barriers to yield the four-membered cyclic intermediates (**41**) with the oxygen atom axial on a trigonal bipyramidal P or S atom. In the phosphorus case, pseudo-rotation placing a P—C bond axial, followed by fragmentation to phosphine oxide and alkene has a lower barrier than reversion to starting materials. Analagous fragmentation of the oxathietane has a much higher barrier apparently associated with difficulties in placing a S—C bond in the weaker axial site on the sulphur in the ring. Interestingly, no *cis*-betaines have been found on the Wittig pathways. In Corey–Chaykovsky reactions, leading to epoxides, the open *trans*-betaines are saddle-points on the energy surface. In agreement with experiment this is found on a lower energy pathway for reaction of the sulphur ylid. Wittig condensation of benzaldehyde with phosphorus ylids containing nucleophilic groups such as carboxylate or alkoxide attached near the reaction site are reported to proceed with high E-selectivities.[106] Deprotonation of cyclopropyl-2-propyl phosphonium salts (**42**) by sodium amide in ammonia occurs at the open-chain site to give isopropyl ylids.[107]

The reactivity of a series of cycloimmonium ylids derived from pyridine, pyridazine, pyrimidine, isoquinoline, and phthalazine have been correlated with their frontier orbital properties.[108] The α-lithiation of sterically hindered amides[109] or formamidines of cyclic secondary amines affords a useful route to their α-substituted derivatives.[110] Full papers describing both methods have been published. The modification of the formamidine method which uses amino-alcohols as chiral auxiliaries in enantioselective syntheses of 1,2,3,4-tetrahydroisoquinoline derivatives has also been detailed.[111]

Delocalized Ions

Both α- and β-tricarbonylchromium complexes of 3-*O*-benzyl-17β-*O*-(*tert*-butyldimethylsilyl)estradiol have been prepared, and have been deprotonated at the 6-position by the action of sodium hexamethyldisilazide.[112] The anions react with formaldehyde to introduce a hydroxymethyl group *anti* to the tricarbonylchromium with high selectivity. Addition of 2-lithio-2-methylpropionitrile to the tricarbonylchromium complex of *N*-methyl-1,2,3,4-tetrahydroquinoline in THF at −78° yields a 35:64 mixture of isomeric nitrile (**43a**) and (**43b**) after quenching and cleavage of the chromium with iodine.[113] The isomer ratio shifts to 2:96 if the mixture is allowed to warm up to 0° before work-up, and evidence has been presented that this is the result of equilibration of the initial adducts *via* the starting materials, rather than by 1,5-shifts of the 2-(2-methylpropionitrile) group.

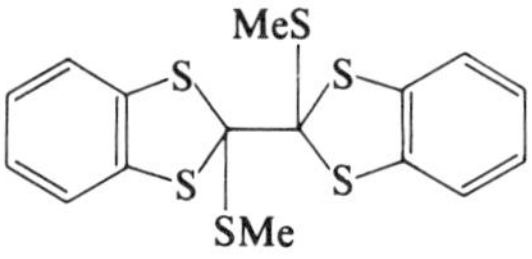

(**39**)

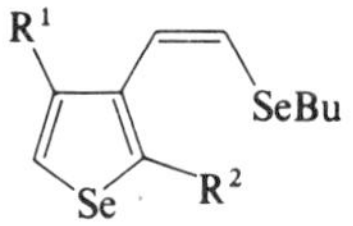

(**40**)

a: R^1 = H; R^2 = Li
b: R^1 = Li; R^2 = H

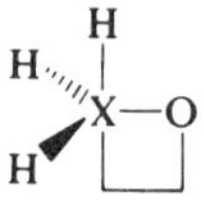

(**41**)

X = S, P

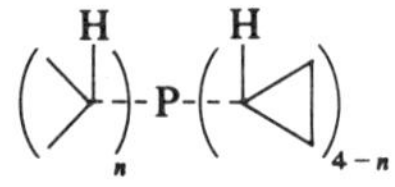

(**42**)

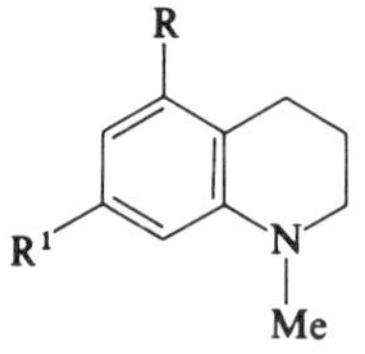

(**43**)

a: R^1 = H; R^2 = $(Me)_2CHCN$
b: R^1 = $(Me)_2CHCN$; R^2 = H

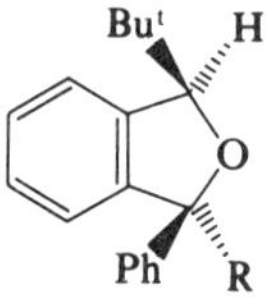

(**44**)

R = H, D

Reduction of substituted isobenzofurans by sodium metal in THF yields an ion which is the result of addition of two electrons and a proton from THF solvent.[114] Quenching of the ion with methanol or D_2O gives the *cis*-dihydroproduct (**44**) which can be equilibrated to its *trans*-isomer by *tert*-butoxide. A pyramidal carbanion is proposed to account for the *cis*-selectivity in kinetic protonation. 7-Alkyl-7,12-dihydropleiadenes have been deprotonated by butyllithium in THF.[115] With a bulky substituent such as isopropyl, alkylation shows appreciable selectivity for formation of the *trans*-7,12-dialkyl-7,12-dihydropleiadenes. The heterocyclic anion (**45**) has been prepared by the action of potassium amide/ammonia or LDA in ethereal solvents on 2-(diethylamino)-5-phenyl-3*H*-azepine.[116] Quenching with D_2O gives 80% D incorporation at C(3) and reaction with alkyl halides and carbonyls affords 3-substituted derivatives. Substituted *gem*-bis(pyrazol-1-yl)alkanes have been prepared by lithiation of the parent alkane and reaction with reactive alkyl halide or carbonyl compound.[117] With LDA as base, under reverse addition conditions, products of substitution at the inter-ring carbon are formed. With other bases, substitution at the ring 5-position can also occur. Adducts of alkylpyridines and dialkylboryl triflates (**46a**) condense with benzaldehyde in presence of triethylamine *via* an "ate" complex to give aldol products (**46b**) with high *erythro*-selectivity.[118] The effect of variation of the boryl alkyl groups and the amine base has been examined.

Substituent effects on the rates of reaction of carbanions derived from 9-substituted fluorenes and substituted diphenylmethanes have been examined.[119] In all cases, free carbanions are more reactive than either Li^+ or Cs^+ ion-pairs towards butyl bromide. In the fluorene series, the 9-substituents include H, CN, COOMe, Ph, CH_2Ph, $SiMe_3$, and Me, and linear correlations of log k with σ_m are reported.[120] The rates of reaction of these carbanions, and others derived from diphenylindene, with benzaldehyde,[121] α,β,β-trifluorostyrene,[122,123] and other electron-deficient alkenes[124] have been determined; for rates of reaction of both ions and ion-pairs there is a log–log correlation with their rates of alkylation by butyl bromide. Variation of solvent increases the rate of addition of 9-phenylfluorenide to trifluorostyrene in the order DMA < HMPA < DMSO < NMP < AN < DME. Again linear log–log relationships are obtained with rates of butylation. The slopes of the correlation lines increase with added ethanol.[125]

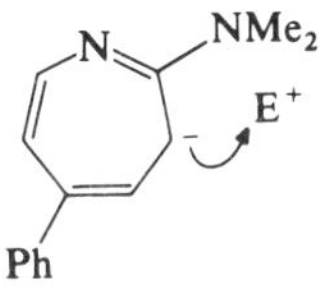

(**45**) 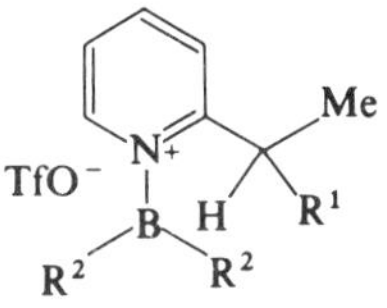 (**46**) (**47**)

a: R^1 = H; R^2 = alkyl
b: R^1 = PhCHOH; R^2 = alkyl

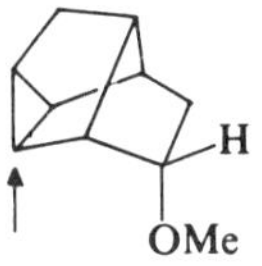

Grignards and Organolithiums

The reaction of phenylmagnesium bromide with 3-chloro-5,6-diphenyl-1,2,4-triazene has been studied.[126] Substituent effects on the reaction of substituted 3-phenylbutan-2-ones with substituted phenylmagnesium bromides have been examined.[127] Stereoselectivity for the formation of the RS, SR-adduct is enhanced by electron-donating substituents in the Grignard reagent. Phenyl substituents in the ketone have little effect.

Intramolecular coordination is important in directing and promoting lithiation of cyclopropanes by alkyllithiums. Treatment of the tricyclic ether (**47**) with isopropyllithium results in lithiation at the indicated site as shown by subsequent reaction with methyl iodide or carbon dioxide.[128] The epimeric ether is inert to the conditions. A range of related bicyclic and tricyclic compounds containing the homocyclopropyl carbinyl ether group have also been studied. Chlorohydrins have been converted to β-oxidoorganolithiums by treatment with butyllithium at $-78°$, followed by lithium naphthalenide.[129] The reagents are stable at low temperature and react with benzyl chloride, benzaldehyde, or diphenyl disulphide. At higher temperatures, decomposition by elimination of Li_2O occurs. One equivalent of lithium bromide stabilizes monohalomethyllithiums.[130]

Treatment of the optically active vinyl halides (**48a**) with two equivalents of butyllithium in ether or THF at low temperature, followed by deuterolysis gives the usual products of metal–halogen exchange and the compound (**48b**) from nucleophilic-type substitution on a carbenoid species (**48c**).[131] Significantly, the product is optically active, showing the indicated inverted configuration. Nucleophilic attack by *tert*-butyllithium on a tight ion-pair, produced by metal-assisted ionization of the C-halogen bond in the carbenoid, is proposed to account for the result.

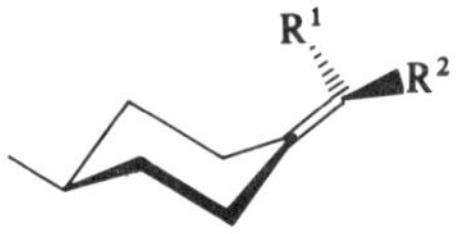

(**48**)

a: R^1 = H; R^2 = Cl, Br
b: R^1 = Bu^t; R^2 = D
c: R^1 = Li; R^2 = Cl, Br

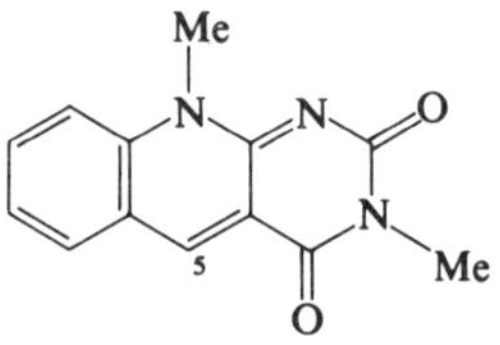

(**49**)

Miscellaneous Reactions

Dicarbanionic oligomers from the reaction of lithium metal with 1,1-diphenylethylene or α-methylstyrene react with elemental sulphur either by electron transfer, yielding hydrocarbons and lithium polysulphides, or by nucleophilic addition, yielding thiaalkanes.[132] The electron-transfer process is favoured by steric congestion at the carbanionic sites, and by their separation by only two carbon atoms. Direct reaction of solid potassium tricyanomethide with chlorine is a convenient preparation of chlorotricyanomethane. Tricyanomethide has been oxidized in aqueous solution by two-electron oxidants such as hypochlorite or permanganate.[133] Sharp end-points are obtained with two equivalents of reagent, but $^+C(CN)_3$ is not observed as a stable species under these conditions.

Esters and amides of 3,4-dihydro-β-carboline-3-carboxylic acid aromatize readily in aqueous base. Carbanions from deprotonation at C(4) and C(3) are in equilibrium, and the aromatization appears to occur by electron transfer from the C(4) carbanion.[134] Use of deuterated substrates has shown that aromatization of *trans*-1,2-dihydrophthalate by 5-carbalumiflavin (**49**) in aqueous carbonate is accompanied by transfer of hydrogen from the dihydrophthalate to the 5-position of the reduced 1,5-carbaflavin.[135] The oxidation of the substrate carbanion thus occurs by hydride transfer to the 5-carbaflavin rather than by one or two electron-transfer pathways. The carbaflavins can replace flavin cofactors in several flavoenzyme dehydrogenases, but it is not clear whether the carbaflavin result should be extrapolated to flavin oxidations. The conversion of azibenzil to benzil azine is catalysed by hydride or carbanions, such as fluorenide, which carry a hydrogen at the carbanionic carbon.[136] The kinetics are consistent with a chain-mechanism initiated by formation of an azibenzil-carbanion adduct (**50**) which transfers hydride, as shown, to azibenzil. Expulsion of nitrogen from the reduced diazo-ketone then forms the benzyl phenyl ketone enolate which continues the chain by adding to an azibenzil molecule. In this unusual process, hydride functions in a manner analagous to a proton in acid-catalysed reactions.

Irradiation of the 2-bromo-1,3-diphenylindenyl anion in the presence of added chloride anion results in halogen exchange.[137] Also formed, in DMSO solution, are 1,3-diphenylindene and the solvent addition product (**51**). Similar results are obtained for the 2-chloro-anion, which requires only about one fifth of the irradiation time for the same product yields. Formation of (**51**) clearly involves H-atom abstraction and radical–radical combination, and the likely intermediate is 1,3-diphenylisoindenylidene, possibly in a triplet state.

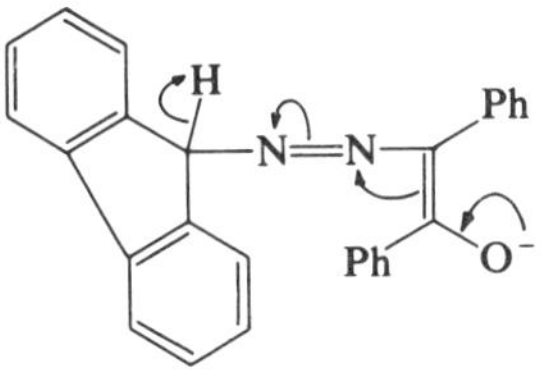

H N=N Ph Ph O⁻

(**50**)

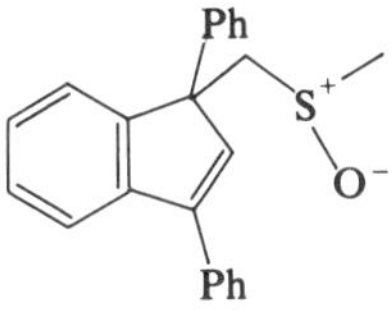

Ph S⁺ O⁻ Ph

(**51**)

Proton-transfer Reactions

Experiments with ^{18}O-labelled OH^- have shown that its major gas-phase reaction with methyl phenyl ether gives phenoxide, mainly by S_N2 displacement, but with smaller amounts of *ipso*-substitution.[138] Deprotonation also occurs, and deuterium labelling indicates that both inter- and intra-molecular hydrogen–deuterium exchanges precede the substitution. Up to six deuterium atoms are incorporated by the $[M-H]^-$ ion in exchange with D_2O, and $PhOCD_3$ gives an $[M-H]^-$ which shows H/D exchange with water. The gas-phase reaction of ethyl phenyl ether is dominated by elimination of ethylene to yield phenoxide. d_5-Ethyl phenyl ether reacts with NH_2^- yielding $C_6H_4DO^-$ solely, showing that aryl proton abstraction precedes elimination in this case. Unlike bare OH^-, mono-solvated hydroxide, $HO^-.H_2O$, does not deprotonate acetonitrile in the gas phase, despite the expectation that the process should also be exothermic.[139] Instead, methyl transfer is observed, and the variation of the pathway is associated with charge development at the potential solvation site in the transition state.

The relative reactivities in proton transfer of diastereotopic hydrogens in CH_3X, with X = SH, SOH, SO_2H, and SH_3, have been examined.[140] The model adopted for the transfer places the hydroxide oxygen 1.5 Å from the relevant hydrogen in a linear O . . . H . . . C array, and MO calculations with a 3-21G* basis set then give results which compare well with those of experimental exchange studies in hydroxylic media. The existence or non-existence of transition states for gas-phase deprotonation of carbon acids has also been examined. Calculations at the 3-21G* level show that barriers for

$$XCH_2^- + CH_3H \longrightarrow XCH_3 + XCH_2^-$$

are 10 kcal mol^{-1} for substituents X = H, F, SH, and Cl. With this intrinsic barrier, the Marcus relationship predicts that transition states should only exist for deprotonation of CH_3X by OH^- when the energy change is less than 20 kcal mol^{-1}. The prediction is confirmed in the cases of X = H or F, where the energy change is less than 20 kcal mol^{-1}, and with X = SH or Cl, where the energy change is more than 20 kcal mol^{-1}.[141]

Tertiary amines, primary and secondary amines, and oxy-anions show decreasing effectiveness as general bases in proton abstraction from nitroethane.[142] The three base types give catalytic constants falling on separate Brønsted lines with increasing β-values of 0.45, 0.60, and 0.72 respectively, in accord with the reactivity–selectivity principle. Quinuclidine has been found to be a more effective catalyst than hexafluoroisopropoxide in base-induced eliminations with transition states near the *E*1*cB*–*E*2 borderline.[143] In both cases the effectiveness of the tertiary amine bases has been attributed to stabilizing electrostatic interaction between developing positive charge on nitrogen and negative charge on the carbanionic carbon. With the primary or secondary amines the effect is attenuated by solvation effects. Rates of deprotonation of 4-(4-nitrophenoxy)-2-butanone by a series of oxy-anion catalysts show severely curved Brønsted plots, but the measured isotope effects give no indication of change in transition-state structure at the curvature point.[144] The curvature is again ascribed to changes in solvation effects. A relationship, based on the Marcus equation, between k_H/k_D values and pK_a and IR stretching frequencies on deuteration of reactants and products, has been derived, and isotope effects in proton abstraction from carbon discussed. This paper also contains a useful compilation of isotope effect data.

Proton transfers between oxygen and carbon have been reviewed.[145] Primary kinetic isotope effects for transfer of proton from methanol to $XC_6H_4CH^-CF_2OMe$, with X = *m*-CF_3, *m*-NO_2, *p*-CN, and *p*-NO_2, have been calculated from the ratios of addition to substitution product in the reaction of methoxide in CH_3OH or CH_3OD with substituted β,β-difluorostyrenes.[146] Extraction of the k_H/k_D value depends on the rates of ejection of F^- from the carbanion not being isotopically sensitive. With X = *p*-NO_2, the isotope effects range from 11.3 at $-70°$ to 6.44 at 25°, and give $A_H/A_D = 1.9$. For *p*-CN, *m*-NO_2, and *m*-CF_3, at 25° they are 2.04, 1.35, and 1.34, respectively, and all decrease with increasing temperature. The results are consistent with a multi-step protonation, with a hydrogen-bonded carbanion as an intermediate whose formation may be rate-limiting. The methyl ether (**52a**) rearranges by 1,3-proton transfer on treatment with diazabicyclooctane, without loss of deuterium or with competing elimination of methanol.[147] The acetate (**52b**) reacts *ca.* 2.3 times faster, partitioning 43:57 between elimination of acetic acid and rearrangement. It has been argued that the

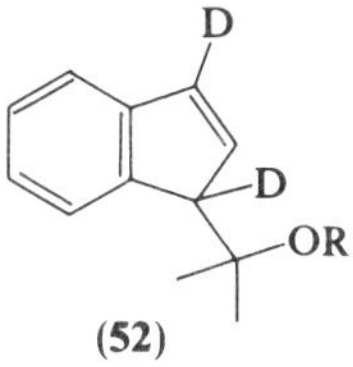

D

D

OR

(52)

a: R = Me
b: R = Ac

ether result, which indicates the intermediacy of a hydrogen-bonded carbanion, may be extended to the acetate and that elimination then occurs from a tightly hydrogen-bonded carbanion.

Rates of C(5) deprotonation and reprotonation of a series of barbituric acid derivatives closely follow substrate acidity.[148] The substitution on the acid varies pK_a predominantly by affecting the deprotonation rates; reprotonation rates change by less than a factor of 50 over a change in pK_a of more than 7 units and never approach the diffusion-controlled limit.

Aqueous solutions of acetophenone enol may be generated by Norrish Type II photocleavage of γ-hydroxybutyrophenone. Rates of ketonization in acid and base have been determined.[149] When combined with enolization rates, the data yields $pK_E = 7.90$ for the keto–enol tautomeric equilibrium. The acidity constant for acetophenone ionizing as a carbon acid is $pK_a = 18.24$. The data also permit an estimate of the rate of proton transfer from H_3O^+ to the β-carbon of the enolate ion, $k = 4.2 \times 10^{10}\ \text{M}^{-1}\,\text{s}^{-1}$, and the possibility has been raised that this very fast reaction involves proton transfer down hydrogen-bonded solvent bridges from acid to substrate.

The kinetics of deuteration of 2-(*p*-substituted benzyl)-4,6-dimethylpyrilium perchlorate in buffered HCOOH have been measured.[150] Exchange at the 4-methyl is *ca.* ten-fold faster than at the 6-methyl, and rates of exchange at the benzylic methylene vary with substituent and give linear log plots against Hammett substituent constants. With 2,3,6-trimethyl-4-phenyl- and 2,6-dimethyl-3,4-dimethyl-pyriliums, exchange at the 2-methyl is three times faster than at the 6-position.[151] Base-catalysed exchange in 3-substituted camphors in aqueous dioxan shows that the kinetic effects of SPh and SePh substituents are only a factor of two or three despite the expected stability of the carbanion.[152]

Yeast enolase is inhibited by nitrophosphonates and by phosphonoacetohydroxamate.[153] These appear to function as analogues of carbanions α to carboxylate groups which are stabilized in the enzyme by coordination to both oxygens of the "aci-carboxylate" contributor to the dianion structure. γ-Glutamyl carboxylation proceeds by initial vitamin K dependent cleavage of the γ-C—H bond followed by carboxylation of the intermediate.[154] At low concentrations, this activated glutamyl residue can incorporate hydrogen. When the reaction is carried out in tritiated or deuterated water, the recovered residues incorporate isotope at the γ-position and a carbanion rather than a radical is favoured as the intermediate.

Treatment of 7-phenyl-1,3,5-cycloheptatriene with aqueous alcoholic base results in initial formation of the 1-, 2-, and 3-phenyl in a constant ratio of 1 : 6 : 3.[155] A much slower process then follows with the mixture tending to the equilibrium composition of 64 : 18 : 18 : 0 for the 1-, 2-, 3-, and 7-phenylcycloheptatrienes, respectively. Deuterium exchange in the 7-phenyl isomer is only slightly faster than isomerization. In contrast, the isomerization of the 7-(methoxycarbonyl)-1,3,5-cycloheptatriene proceeds by sequential formation of the 2-,3-, and 1-substituted compounds, with exchange occurring much faster than isomerization.[156] Curiously, exchange of the 3-(methoxycarbonyl) compound with CH_3OD takes place at vinylic rather than allylic sites giving 1,5-dideutero-3-(methoxycarbonyl)cycloheptatriene. This exchange may in fact occur by intermediates from reversible addition of methoxide to the conjugated ester array. Weak carbon acids such as diphenylmethane or allylbenzene can be deuterated at room temperature under phase-transfer conditions with concentrated $NaOD/D_2O$.[157] Rates of deprotonation of $Cr(CO)_3$-complexed arylmethanes by potassium hydride–crown ether in THF are higher than those of the corresponding uncomplexed hydrocarbons.[158] Exchange kinetics for the reaction of potassium dimsyl in DMSO, lithium cyclohexylamide in CHA, and potassium *tert*-butoxide in DMSO with methyl-, ethyl- and isopropyl-benzene are reported, as are the rates of metallation of the hydrocarbons with butyllithium/TMEDA.[159] Enthalpies have been measured for the reaction of pentylmagnesium bromide with a series of Brønsted acids in diethyl ether. Heats are up to 30 kcal mol^{-1} more negative for oxygen and nitrogen acids compared with carbon acids of the same pK_a.[160] Substituent effects on the phenoxide-catalysed carboxylation of indene have been studied.[161]

The rearrangement of (+)- and (−)-1-methylindene to 3-methylindene induced by the chiral amine base dihydroquinidine proceeds with less than 1% of simple racemization in *o*-dichlorobenzene.[162] The selectivity, $k(+)/k(-)$ is 3.26 in the a protium case. Primary isotope effects for 1-deuterated 1-methylindenes are significantly different, with $k_H/k_D = 5.31$ for the (+)-form and 6.19 for its enantiomer at 30°. The solvent, temperature, and pressure dependence of the primary isotope effects in deprotonation of trinitrotoluene by 1,8-diazabicyclo[5.4.0]undecene have been examined. Isotope effects in acetonitrile, benzonitrile, and 1,2-dichloroethane[163] with k_H/k_D = 19.1, 15.9, and 29.9 at 20° and 1 atm, respectively, are all greater than the semi-classical limit. Values of $E_a^H - E_a^D$ and A_H/A_D, 2.3 and 2.5 kcal mol^{-1} for acetonitrile at 1 atm, provide clear indication of quantum-mechanical tunnelling. Activation volumes are negative showing better solvation of transition state than initial state, but isotope effect decreases with increased pressure.[164] However, the data at 1000 bar, indicates that tunnelling, although reduced, still occurs in the proton-transfer reaction in acetonitrile at high pressure.[165] The deprotonation of 1-(4-nitrophenyl)-1-nitroethane by phenyltetranitromethane shows second-order kinetics.[166] The isotope effects, k_H/k_D, are 8.5, 6.1, and 16 at 25° in acetonitrile, benzonitrile, and chlorobenzene, respectively. The effects on enthalpies of activation are consistent with tunnelling in the proton transfer. Although the chlorobenzene number is notably lower than the $k_H/k_D = 50$ found in earlier work on the reaction of 4-nitrophenylnitromethane with tetramethylguanidine, the general behaviour of the two reacting systems is closely similar.

Secondary kinetic isotope effect, $k_H/k_T = 1.31$ (corresponding to $k_H/k_D = 1.21$) for the $E2$ elimination shown is larger than the estimated equilibrium effect.

$$EtO^- + PhCHTCH_2\overset{+}{N}Me_3 \rightarrow PhCT{=}CH_2 + EtOH + NMe_3$$

Calculations on the model chloride elimination

$$CCHDCH_2Cl + OH^- \rightarrow CCD{=}CH_2 + Cl^- + H_2O$$

show that the experimental result can be reproduced if the H . . . C bond stretch is coupled to the three bending vibrations involving the deuterium.[167] With increasingly strong stretch–bend coupling, and consequently increasing barrier curvature, there are significant tunnel corrections to the semi-classical values of $(k_H/k_D)_{sec}$. There is little relationship between calculated values and the extent of rehybridization in the transition state. In fact, $(k_H/k_D)_{sec}$ better reflects the degree to which the motion of the non-transferred β-hydrogen contributes to reaction coordinate motion. A theoretical model, based on the golden rule of time-dependent perturbation theory, has been developed to calculate rate constants for reactions involving hydrogen transfer by tunnelling.[168] For a simple hydrogen transfer of the type $AH + B \rightarrow A + HB$ the input data is AH and BH stretching potentials derived from spectroscopic data and a distance between equilibrium hydrogen positions estimated from molecular models. The AB potential and the coupling between the AH and BH potentials are regarded as adjustable parameters. The treatment has been applied to a number of experimentally well-characterized reactions including the keto–enol tautomerism of 2-methylacetophenone.[169]

Full details of the kinetic and equilibrium measurements on the acid–base reactions of 1,8-bis(dimethylamino)- and 1,8-bis(diethylamino)-2,7-dimethoxynaphthalene have been reported.[170] The protonated form of the 1,8-diethylamino compound has a $pK_a = 16.3$, making it one of the strongest uncharged bases known, exceeded only by 1,6-diazabicyclo[4.4.4]tetradecane or 1,1,1-cryptand. The deprotonations are general-base-catalysed, and the data is consistent with a mechanism of proton removal from the strongly hydrogen-bonded acid by base attack on a small equilibrium concentration of an open form of the protonated amine.[171]

The equilibria and kinetics of proton transfer between triethylamine and thiocarboxylic acids have been examined.[172]

Revised rates of acid- and base-catalysed proton exchange in N-methylacetamide have been reported.[173] Proton exchange in amides may occur either by N-protonation or by O-protonation and an imidic acid intermediate, and these are, in principle, distinguishable since the N-protonation permits E/Z-isomerization. For most secondary amides, the proportion of E-isomer is too low for accurate measurement, but a study of N-alkylformamides has been undertaken using NMR line-shape analysis and spin saturation transfer methods.[174] For N-formylglycine, formanilide, and N-methylcyanoformamide, the exchange occurs *via* the imidic acids. For N-*tert*-butyl- and N-methyl-formamide, the data are consistent with exchange by both mechanisms. Decreasing solvent polarity shifts the balance in favour of the imidic acid mechanisms. It has been suggested that the previously accepted N-protonation mechanism for NH exchange in protein backbones is not significant, and the pH dependence of rates of exchange of buried protons may be associated with the nature of the transition state for the imidic acid mechanism.[175]

Electrophilic Aliphatic Substitution

The equilibrium in tin–lithium exchange between α-alkoxyorganotin compounds (**53a**) and alkyllithiums to yield α-alkoxyorganolithiums (**53b**) and an alkyltrimethylstannane is rapidly established in DME at $-60°$.[176] The results of a series of competitive equilibrations, monitored by H-NMR, show that an α-alkoxy group has an important stabilizing effect on organolithiums. There is no evidence for stannylate complex formation. Rates are reduced with more hindered trialkystannyl groups, and by change of solvent to THF or ether. The substitution occurs with retention of configuration and further reaction of the α-alkoxyorganolithiums with a range of electrophiles also occurs with retention.

The stereochemistry of S_E' acid cleavage of *cis*- and *trans*-4-*tert*-butylcyclohexenylstannanes (**54a**) has been examined.[177] With CF_3COOD, acting on known mixtures of *cis*- and *trans*-isomers, the product data is consistent with γ-*syn*-cleavage of the *trans*-isomer, but γ-*anti*-cleavage of the *cis*-isomer. With 4-methylcyclohexenyls (**54b**) γ-*anti*-cleavage predominates irrespective of the stereochemistry of the starting material. Favourable σ–π overlap between the C–metal and the double bond is maximized when the C–metal bond is pseudo-axial with respect to a cyclohexene half-chair conformation and, in the absence of strong steric effects, γ-*anti*-protonation of this array should be favoured. In the case of the *trans*-4-*tert*-butyl compound, either this conformation is energetically inaccessible or the necessarily pseudo-axial *tert*-butyl hinders attack by the proton donor. Sulphur dioxide insertion in the same compounds is also γ-specific, but *syn*-stereospecific, in all compounds when the reaction is carried out in chloroform.[178] A cyclic transition

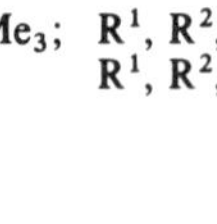

(53)

a: X = $SnMe_3$; R^1, R^2, R^3 = alkyl
b: X = Li; R^1, R^2, R^3 = alkyl

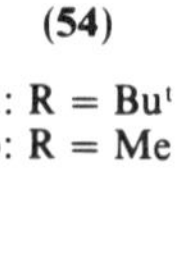

(54)

a: R = Bu^t
b: R = Me

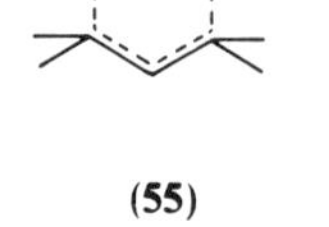

(55)

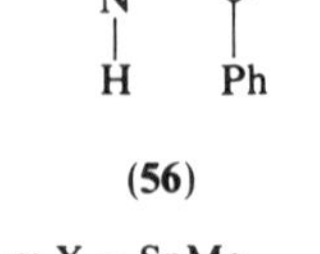

(56)

a: X = $SnMe_3$
b: X = D, $SiMe_3$

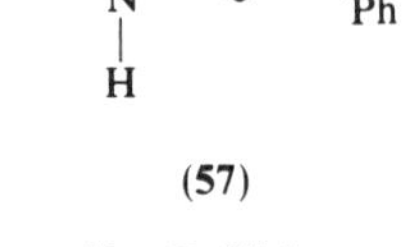

(57)

X = D, $SiMe_3$

state (**55**) is proposed to account for the result. In methanol, competing *syn*- and *anti*-γ-insertion products are observed.

Treatment of the amide (**56a**) with two equivalents of butyllithium at $-70°$ yields solutions of a stable dilithium compound which yields (**56b**) when treated with D_2O or chlorotrimethylsilane.[179] When the solution is warmed to 0° prior to the quench, the rearranged products (**57**) are obtained in good yield. A study of labelled compounds has given evidence for the intermediacy of the doubly lithiated cyclopropanone aminal, but not in concentrations observable by NMR.

Secondary deuterium kinetic isotope effects in the protonolysis of a dialkylmercury and in the brominolysis of a dialkylmercury and a tetraalkyltin have been determined.[180] The α-effects, k_{CH_2}/k_{CD_2}, are normal at *ca.* 1.1 per deuterium and not strongly dependent on solvent, or the nature of entering or leaving group. There is no β-effect. The meaning of these first measurements on S_E2 reactions is not clear, since force constants for C—H at pentacoordinate carbon are not available, but they do indicate a loosening of overall bonding at the reaction site in the transition state.

The rates of reaction of a series of nitrous acid scavengers with aqueous nitrous acid have been measured as a function of acidity.[181] The profiles are consistent with the expected electrophilic nitrosation process but their ordering of reactivity is dependent on acidity. For substrates showing reversible initial nitrosation there is no catalysis by added halide SCN^-, or thiourea. Where nucleophilic catalysis is observed, the data allows estimation of the relative reactivity of the nitrosyl derivatives. Nitrosations of secondary amines by N_2O_3[182] and nitrosyl halides[183,184] have been studied and the effects of added formaldehyde on the kinetics of *N*-nitrosation have been examined; catalysis by formaldehyde is not by formation of new nitrosating agents but by formation of readily nitrosatable intermediates.[185] Mechanisms of nitrosations have been discussed and the limits of the current kinetic treatments have been examined.[186] Nitrosation of *N*-methylacetamide differs from that of secondary amines in that no catalysis by halides is found.[187] General base catalysis by acetate and chloroacetate gives a Brønsted $\beta = 0.49$, and solvent isotope effects are consistent with a slow step involving proton transfer from protonated nitrosamide to the medium. The denitrosation of *N*-methyl-*N'*-cyclohexyl-*N*-nitrosothiourea shows general acid catalysis.[188] With added bromide, the reaction of *N,N*-dimethylethanolamine with hypochlorite gives both *N*-chlorination and -bromination.[189] The decomposition of *N*-bromoalanine in aqueous solution gives acetaldehyde at pH 6 and low bromine concentrations. At higher bromine ratios, nitriles are formed.[190] The nitration of C_{13}–C_{17} alkanes by 58% aqueous nitric acid has been studied and compared with the reaction with $Al(NO_3)_3 \cdot 9\,H_2O$ at high temperatures.[191]

References

1 Lien, M. H., Hopkinson, A. C., and McKinney, M. A., *THEOCHEM*, **14**, 37 (1983); *Chem. Abs.*, **100**, 84919 (1984).

2 Hopkinson, A. C., McKinney, M. A., and Lien, M. H., *J. Comput. Chem.*, **4**, 513 (1983); *Chem. Abs.*, **100**, 120160 (1984).

[3] Hopkinson, A. C., and Lien, M. H., *THEOCHEM*, **13,** 303 (1983); *Chem. Abs.*, **100,** 84909 (1984).
[4] Krespan, C. G., Van-Catledge, F. A., and Smart, B. E., *J. Am. Chem. Soc.*, **106,** 5544 (1983).
[5] Huang, M. B., Goscinski, O., Jonsall, G., and Ahlberg, P., *J. Chem. Soc., Perkin Trans. 2*, **1984,** 1327.
[6] Adams, S. M. and Bank, S., *J. Comput. Chem.*, **4,** 470 (1983); *Chem. Abs.*, **100,** 102418 (1984).
[7] Maercker, A., Bsata, M., Buchmeier, W., and Engelen, B., *Chem. Ber.*, **117,** 2547 (1984).
[8] Amstutz, R., Laube, T., Bernd Schweizer, W., Seebach, D., and Dunitz, J. D., *Helv. Chim. Acta*, **67,** 224 (1984).
[9] Atwood, J. L., Fjeldberg, T., Lappert, M. F., Luong-Thi, N. T., Shakir, R., and Thorne, A. J., *J. Chem. Soc., Chem. Commun.*, **1984,** 1163.
[10] Bushby, J. R., and Tytko, M. P., *J. Organomet. Chem.*, **270,** 265 (1984).
[11] Jackman, L. M., and Scarmoutzos, L. M., *J. Am. Chem. Soc.*, **106,** 4627 (1984).
[12] Wilhelm, D., Clark, T., and Schleyer, P. von R., *J. Chem. Soc., Chem. Commun.*, **1984,** 915.
[13] Goldstein, M. J., and Wenzel, T., *J. Chem. Soc., Perkin Trans. 2*, **1984,** 1655.
[14] Deitrich, H., Mahdi, W., Wolhelm, D., Clark, T., and Schleyer, P. von R., *Angew. Chem. Int. Ed.*, **23,** 621 (1984).
[15] Kos, A. J., Clark, T., and Schleyer, P. von R., *Angew. Chem. Int. Ed.*, **23,** 620 (1984).
[16] Schleyer, P. von R., Kos, A. J., and Kaufmann, E., *J. Am. Chem. Soc.*, **105,** 7617 (1983).
[17] Schleyer, P. von R., Kos, A. J., Wilhelm, D., Clark, T., Boche, G., Decher, G., Etzrodt, H., Deitrich, H., and Mahdi, W., *J. Chem. Soc., Chem. Commun.*, **1984,** 1495.
[18] Boche, G., Decher, G., Etzrodt, H., Deitrich, H., Mahdi, W., Kos, A. J., and Schleyer, P. von R., *J. Chem. Soc., Chem. Commun.*, **1984,** 1493.
[19] Paprott, G., and Seppelt, K., *J. Am. Chem. Soc.*, **106,** 4060 (1984).
[20] Koser, H. G., Renzoni, G. E., and Borden, W. T., *J. Am. Chem. Soc.*, **105,** 6359 (1983).
[21] Andes Hess, B., Schaad, L. J., and Carsky, P., *Tetrahedron Lett.*, **25,** 4721 (1984).
[22] Gerecht, B., Kampchen, T., Hohler, K., Massa, W., Offermann, G., Schmidt, R. E., Seitz, G., and Sutrisno, R., *Chem. Ber.*, **117,** 2714 (1984).
[23] Yoshida, Z-I., Shibata, M., Sakai, A., and Sugimoto, T., *J. Am. Chem. Soc.*, **106,** 6383 (1984).
[24] Hafner, K., and Thiele, G. F., *Tetrahedron Lett.*, **25,** 1445 (1984).
[25] Annastassiou, A. G., Kasmai, H. S., and Naderi, B., *Tetrahedron Lett.*, **25,** 5591 (1984).
[26] Schlosser, M., and Strunk, S., *Tetrahedron Lett.*, **25,** 741 (1984).
[27] Wilhelm, D., Clark, T., Schleyer, P. von R., Courtneidge, J. L., and Davies, A. G., *J. Am. Chem. Soc.*, **106,** 361 (1984).
[28] Verbeek, J., and Brandsma, L., *J. Org. Chem.*, **49,** 3857 (1984).
[29] Wilhelm, D., Clark, T., Schleyer, P. von R., and Davies, A. G., *J. Chem. Soc., Chem. Commun.*, **1984,** 558.
[30] Goldstein, M. J., and Wenzel, T. T., *J. Chem. Soc., Chem. Commun.*, **1984,** 1654.
[31] Norinder, U., Tanner, D., and Wennerstroem, O., *Croat. Chim. Acta*, **56,** 269 (1983); *Chem. Abs.*, **100,** 50849 (1983).
[32] Eliasson, B., and Edlund, U., *J. Chem. Soc., Perkin Trans. 2*, **1983,** 1837.
[33] Eliasson, B., Lejon, T., and Edlund, U., *J. Chem. Soc., Chem. Commun.*, **1984,** 591.
[34] Schneiders, G., Mullen, K., and Huber, W., *Tetrahedron*, **40,** 1701 (1984).
[35] Mohammed, M., *Pak. J. Sci. Ind. Res.*, **26,** 4 (1983); *Chem. Abs.*, **100,** 102570 (1984).
[36] Yuchnovski, I., and Andrew, G., *Izv. Khim.*, **16,** 402 (1983); *Chem. Abs.*, **101,** 6471 (1984).
[37] Yuchnovski, I., and Andrew, G., *Izv. Khim.*, **15,** 339 (1982); *Chem. Abs.*, **99,** 175099 (1983).
[38] Top, S., Jaouen, G., Sayer, B. G., and McGlinchey, M. J., *J. Am. Chem. Soc.*, **105,** 6426 (1984).
[39] Rabideau, P. W., Wetzel, D. M., Husted, C. A., and Lawrence, J. R., *Tetrahedron Lett.*, **25,** 31 (1984).
[40] Okamoto, Y., Takeda, T., and Hatada, K., *Chem. Lett.*, **1984,** 757.
[41] Komatsu, K., Shirai, S., Tomioka, I., and Okamoto, K., *Bull. Chem. Soc. Jpn.*, **57,** 1377 (1984).
[42] Streitweiser, A. J., Bors, D. A., and Kaufmann, M. J., *J. Chem. Soc., Chem. Commun.*, **1983,** 1394.
[43] Washburn, N., *J. Org. Chem.*, **48,** 4287 (1983).
[44] Trimitsis, G. B., and Zimmermen, P., *J. Chem. Soc., Chem. Commun.*, **1984,** 1506.
[45] Fraser, R. R., Bresse, M., and Monsour, T. S., *J. Am. Chem. Soc.*, **105,** 7790 (1984).
[46] Comagnon, P.-L., and Kimny, T., *Bull. Soc. Chim. Belg.*, **93,** 55 (1984).
[47] Rabideau, P. W., and Huser, D. L., *J. Org. Chem.*, **48,** 4266 (1983).
[48] Terekhova, M. I., Petrov, E. S., Vasilevski, S. F., Ivanov, V. F., and Schwartsberg, M. S., *Izv. Akad. Nauk SSR, Ser. Khim.*, **54,** 734 (1984); *Chem. Abs.*, **101,** 54452 (1984).
[49] Tumas, W., Foster, R. F., and Brauman, J. I., *J. Am. Chem. Soc.*, **106,** 4053 (1984).
[50] DePuy, C. H., Bierbaum, V. M., and Danrauer, R., *J. Am. Chem. Soc.*, **106,** 4051 (1984).

[51] Andrist, A. H., Depuy, C. H., and Squires, R. R., *J. Am. Chem. Soc.*, **106**, 845 (1984).
[52] Banfi, L., *Quad. Tec. Sint. Spec. Org.*, **1983**, 1; *Chem. Abs.*, **101**, 22624 (1984).
[53] Zhao, C., and Wen, Z., *Huaxue Tongbao*, **1984**, 8; *Chem. Abs.*, **101**, 54077 (1984).
[54] Heathcock, C. H., *Stud. Org. Chem. (Amsterdam)*, **1984**, 5B (Compr. Carbanion Chem., Pt B), 177; *Chem. Abs.*, **101**, 54081 (1984).
[55] Evans, D. A., *Asymmetric Synth.*, **3**, 1 (1984); *Chem. Abs.*, **101**, 71793 (1984).
[56] Buncel, E., and Durst, T., *Stud. Org. Chem. (Amsterdam)*, **1984**, 5B (Compr. Carbanion Chem., Pt B), 303; *Chem. Abs.*, **101**, 109841 (1984).
[57] Arsenujadis, S., Kyler, K. S., and Watt, D. S., *Org. React. (N. Y.)*, **31**, 1 (1984).
[58] Resnati, G., *Quad. Tec. Sint. Spec. Org.*, **1983**, 107; *Chem. Abs.*, **101**, 22595 (1984).
[59] Seyferth, D., Weinstein, R. M., Wei-Liang, W., Hui, R. C., and Archer, C. M., *Isr. J. Chem.*, **24**, 167 (1984).
[60] Epsztein, R., *Stud. Org. Chem. (Amsterdam)*, **1984**, 5B (Compr. Carbanion Chem., Pt B), 107; *Chem. Abs.*, **101**, 71799 (1984).
[61] Durst, T., *Stud. Org. Chem. (Amsterdam)*, **1984**, 5B (Compr. Carbanion Chem., Pt. B), 239; *Chem. Abs.*, **101**, 71800 (1984).
[62] Barnette, W. E., *J. Am. Chem. Soc.*, **106**, 452 (1984).
[63] Babudri, F., Di Nunno, L., Florio, S., and Valzano, S., *Tetrahedron*, **40**, 1731 (1984).
[64] Schlosser, M., and Lehmann, R., *Tetrahedron Lett.*, **25**, 745 (1984).
[65] Georgiadis, M. P., *J. Heterocycl. Chem.*, **21**, 611 (1984).
[66] Fiandanese, V., Naso, H., and Scilimati, A., *Tetrahedron Lett.*, **25**, 1187 (1984).
[67] Corey, E. J., and Gross, A. W., *Tetrahedron Lett.*, **25**, 495 (1984).
[68] Harris, F. L., and Weiler, L., *Tetrahedron Lett.*, **25**, 1333 (1984).
[69] Reichelt, I., and Reissig, H.-U., *Leibigs Ann. Chem.*, **1984**, 531.
[70] Klemer, A., and Thiemeyer, H., *Leibigs Ann. Chem.*, **1984**, 1094.
[71] Collum, D. B., Kahne, D., Gut, S. A., Depue, R. T., Mohamadi, F., Wanat, R. A., and Van Duyne, G., *J. Am. Chem. Soc.*, **106**, 4865 (1984).
[72] Yasumitsu, T., Uenishi, J., and Ishibashi, H., *Chem. Pharm. Bull.*, **32**, 898 (1984); *Chem. Abs.*, **101**, 89974 (1984).
[73] Tamaru, Y., Uenishi, J., and Ishibashi, H., *Chem. Pharm. Bull.*, **32**, 891 (1984); *Chem. Abs.*, **101**, 71881 (1984).
[74] Yasumitsu, T., Junichi, U., and Hiroyuki, I., *Chem. Pharm. Bull.*, **32**, 945 (1984); *Chem. Abs.*, **101**, 89976 (1984).
[75] Hutchinson, J. H., and Money, T., *Can. J. Chem.*, **62**, 1899 (1984).
[76] Tamaru, Y., Hioki, T., Kawamura, S.-I., Satomi, H., and Yoshida, Z.-I., *J. Am. Chem. Soc.*, **106**, 3876 (1984).
[77] Yamamoto, U., and Maruyama, K., *J. Chem. Soc., Chem. Commun.*, **1984**, 904.
[78] Feit, B.-A., Melamid, U., Speer, H., and Schmidt, R. R., *J. Chem. Soc., Perkin Trans. 1*, **1984**, 775.
[79] Kende, A. S., Fludzinski, P., Hill, J. H., Swenson, W., and Clardy, J., *J. Am. Chem. Soc.*, **106**, 3551 (1984).
[80] Jonczyk, A., and Radwan-Pytlewski, T., *Chem. Lett.*, **1983**, 1557.
[81] Ahlbrecht, H., Farnung, W., and Simon, H., *Chem. Ber.*, **117**, 2622 (1984).
[82] Yamamoto, Y., Yatagai, H., Saito, Y., and Maruyama, K., *J. Org. Chem.*, **49**, 1096 (1984).
[83] Kobahashi, Y., Umeyama, K., and Sato, F., *J. Chem. Soc., Chem. Commun.*, **1984**, 621.
[84] Masson, S., Mothes, U., and Thuillier, A., *Tertrahedron*, **40**, 1573 (1984).
[85] Furuta, K., Misumi, A., Mori, A., Ikeda, N., and Yamamoto, H., *Tetrahedron Lett.*, **25**, 669 (1984).
[86] Furuta, K., Ikeda, N., and Yamamoto, H., *Tetrahedron Lett.*, **25**, 675 (1984).
[87] Misumi, A., Furuta, K., and Yamamoto, H., *Tetrahedron Lett.*, **25**, 671 (1984).
[88] Gawley, R. E., and Nagy, T., *Tetrahedron Lett.*, **25**, 263 (1984).
[89] Narashimhan, N. S., and Ammanamanchi, R., *J. Org. Chem.*, **48**, 3945 (1984).
[90] Grossert, S. J., Hoyle, J., and Hooper, D. L., *Tetrahedron*, **40**, 1135 (1984).
[91] Uno, H., Yoshinori, N., and Maruyama, K., *Tetrahedron*, **40**, 4725 (1984).
[92] Shono, T., Kise, N., and Suzamoto, T., *J. Am. Chem. Soc.*, **106**, 259 (1984).
[93] Atkins, P. J., Gold, V., and Marsh, R., *J. Chem. Soc., Perkin Trans. 2*, **1984**, 1239.
[94] Atkins, P. J., Gold, V., and Wassef, W. N., *J. Chem. Soc., Perkin Trans. 2*, **1984**, 1247.
[95] Buncel, E., Venkatachalam, T. K., and Menon, B. C., *J. Org. Chem.*, **49**, 413 (1984).
[96] Murai, S., Ryu, I., Iniguchi, J., and Sonoda, N., *J. Am. Chem. Soc.*, **106**, 2440 (1984).
[97] Daney, M., Lapouyade, R., and Bouas-Laurent, H., *J. Org. Chem.*, **48**, 5055 (1984).
[98] Paquette, L. A., Uchida, T., and Gallucci, J. C., *J. Am. Chem. Soc.*, **106**, 335 (1984).
[99] Bassindale, A. R., Ellis, R. J., and Taylor, P. G., *Tetrahedron Lett.*, **25**, 2705 (1984).

[100] Ager, D. J., *J. Org. Chem.*, **49**, 168 (1984).
[101] Tamao, K., Kanatani, R., and Kumada, M., *Tetrahedron Lett.*, **25**, 1913 (1984).
[102] Tamao, K., Iwahara, T., Kanatani, R., and Kumada, M., *Tetrahedron Lett.*, **25**, 1909 (1984).
[103] Brown, C. A., Miller, R. D., Lindsay, C. M., and Smith, K., *Tetrahedron Lett.*, **25**, 991 (1984).
[104] Konar, A., *Chem. Scr.*, **22**, 177 (1983).
[105] Volatron, F., and Eisenstein, O., *J. Am. Chem. Soc.*, **106**, 6117 (1984).
[106] Maryanoff, B. E., Duhl-Emswiler, B. A., and Reitz, A. B., *Phosphorus Sulfur*, **18**, 187 (1983); *Chem. Abs.*, **100**, 191021 (1984).
[107] Schier, A., and Schmidbaur, H., *Chem. Ber.*, **117**, 2314 (1984).
[108] Surpateanu, Gh., Constantinescu, M., Luchian, C., Petrovanu, M., Zugravescu, I., and Lablache-Combier, A., *Rev. Roum. Chim.*, **28**, 933 (1983).
[109] Beak, P., and Zajdel, W. J., *J. Am. Chem. Soc.*, **106**, 1010 (1984).
[110] Meyers, A. I., Edwards, P. D., Reiker, W. F., and Bailey, T. R., *J. Am. Chem. Soc.*, **106**, 3270 (1984).
[111] Meyers, A. I., Fuentes, L. M., and Kubota, Y., *Tetrahedron*, **40**, 1361 (1984).
[112] Jaouen, G., Top, S., Laconi, A., Coutourier, D., and Brocard, J., *J. Am. Chem. Soc.*, **106**, 2207 (1984).
[113] Ohlsson, B., and Ullenius, C., *J. Organomet. Chem.*, **267**, C34 (1984).
[114] Smith, G., and Wikman, R. J., *J. Chem. Soc., Chem. Commun.*, **1984**, 1448.
[115] Marcinow, Z., and Rabideau, P. W., *Tetrahedron Lett.*, **25**, 5463 (1984).
[116] Stref, J. W., van der Plas, H. C., and van Veldhuizen, A., *Recl. Trav. Chim. Pays-Bas*, **103**, 225 (1984).
[117] Katritzky, A. R., Abdel-Rahman, A. E., Leahy, D. E., and Schwarz, O. A., *Tetrahedron*, **39**, 4133 (1984).
[118] Hamana, H., and Sugasawa, T., *Chem. Lett.*, **1984**, 1591.
[119] Solov' yanov, A. A., Beletskaya, I. P., and Reutov, O. A., *Zh. Org. Khim.*, **19**, 1846 (1983); *Chem. Abs.*, **100**, 22208 (1984).
[120] Solov' yanov, A. A., Beletskaya, I. P., and Reutov, O. A., *Zh. Org. Khim.*, **19**, 1822 (1983); *Chem. Abs.*, **100**, 22206 (1984).
[121] Solov' yanov, A. A., Beletskaya, I. P., and Reutov, O. A., *Zh. Org. Khim.*, **19**, 2253 (1984); *Chem. Abs.*, **100**, 84992 (1984).
[122] Solov' yanov, A. A., Shtern, M. M., Beletskaya, I. P., and Reutov, O. A., *Zh. Org. Khim.*, **19**, 2233 (1983); *Chem. Abs.*, **100**, 84990 (1984).
[123] Shtern, M. M., *Deposited Doc.*, **1982**, VINITI 3676-186; *Chem. Abs.*, **101**, 190800 (1984).
[124] Solov' yanov, A. A., Shtern, M. M., Beletskaya, I. P., and Reutov, O. A., *Zh. Org. Khim.*, **19**, 1835 (1983); *Chem. Abs.*, **100**, 22207 (1984).
[125] Solov' yanov, A. A., Shtern, M. M., Beletskaya, I. P., and Reutov, O. A., *Vestn. Mosk. Univ., Khim.*, **25**, 182 (1984); *Chem. Abs.*, **101**, 37915 (1984).
[126] Zaher, H. A., Ibrahim, Y. A., Sherif, O., and Mohammady, R., *Indian J. Chem.*, **22B**, 559 (1983); *Chem. Abs.*, **100**, 5495 (1984).
[127] Arjona, O., Gonzales, R. M., Perez-Ossorio, R., Perez-Rubalcaba, A., and Quiroga, M. L., *An. Quim.*, **79C**, 1847 (1983); *Chem. Abs.*, **101**, 190750 (1984).
[128] Klumpp, G. W., Kool, M., Veefkind, A. H., Scakel, M., and Schmitz, R. S., *Recl. Trav. Chim. Pays-Bas*, **102**, 542 (1983).
[129] Barluenga, J., Florez, J., and Yus, M., *J. Chem. Soc., Perkin Trans. 1*, **1983**, 3019.
[130] Tarhouni, R., Kirschleger, B., Rambaud, M., and Villieras, J., *Tetrahedron Lett.*, **25**, 835 (1984).
[131] Duriaswany, M., and Walborsky, H. M., *J. Am. Chem. Soc.*, **106**, 5035 (1984).
[132] Catala, J. M., Boscato, J. F., and Brossas, J., *J. Organomet. Chem.*, **276**, 155 (1984).
[133] Beaumont, R. C., Aspin, K. B., Demas, J. J., Hoggatt, J. H., and Potter, G. E., *Inorg. Chim. Acta*, **84**, 141 (1984).
[134] Previero, A., Barry, L. G., Torreilles, J., Fleury, B., Litellier, S., and Maupas, B., *Tetrahedron*, **40**, 221 (1984).
[135] Bruice, T. C., and Farny, O. L., *J. Chem. Soc., Chem. Commun.*, **1984**, 185.
[136] Bethell, D., and McDowell, L., *J. Chem. Soc., Chem. Commun.*, **1984**, 1408.
[137] Tolbert, L. M., and Siddiqui, S., *J. Am. Chem. Soc.*, **106**, 5538 (1984).
[138] Kleingeld, J. C., and Nibbering, N. M. M., *Tetrahedron*, **39**, 4193 (1983).
[139] Caldwell, G., Rozeboom, M. D., Kiplinger, J. P., and Bartmess, J. E., *J. Am. Chem. Soc.*, **106**, 809 (1984).
[140] Wolfe, S., Stolow, A., and LaJohn, L. A., *Can. J. Chem.*, **62**, 1470 (1984).
[141] Wolfe, S., *Can. J. Chem.*, **62**, 1465 (1984).
[142] Bruice, P. Y., *J. Am. Chem. Soc.*, **106**, 5959 (1984).

[143] Thibblin, A., *J. Am. Chem. Soc.*, **106**, 183 (1984).
[144] Hupe, D. J., and Pohl, E. R., *J. Am. Chem. Soc.*, **106**, 5634 (1984).
[145] Koch, H. F., *Acc. Chem. Res.*, **17**, 137 (1984).
[146] Koch, A. S., and Koch, H. F., *J. Am. Chem. Soc.*, **106**, 4536 (1984).
[147] Thibblin, A., *J. Chem. Soc., Chem. Commun.*, **1984**, 92.
[148] Buckingham, D. A., Clark, C. R., McKeown, R. H., and Ooi, W., *J. Chem. Soc., Chem. Commun.*, **1984**, 1440.
[149] Chiang, Y., Kresge, A. J., and Wirz, J., *J. Am. Chem. Soc.*, **106**, 6392 (1984).
[150] Balaban, A. T., Gheorghiu, M. D., and Balaban, T. S., *J. Labelled Compd. Radiopharm.*, **20**, 1097 (1983); *Chem. Abs.*, **100**, 67587 (1984).
[151] Balaban, A. T., Balaban, T. S., Uncuta, C., Gheorghiu, M. D., and Chiraleu, F., *J. Labelled Compd. Radiopharm.*, **20**, 1105 (1983); *Chem. Abs.*, **100**, 67588 (1984).
[152] Brown, F. C., Morris, D. G., Stephen, D. G., Casadevall, E., Metzger, P., and Gayet, H., *Nouv., J. Chem.*, **7**, 229 (1983).
[153] Anderson, V. E., Weiss, P. M., and Cleland, W. W., *Biochemistry*, **23**, 2779 (1984).
[154] Anton, D. L., and Friedman, P. A., *J. Biol. Chem.*, **258**, 14084 (1983).
[155] Zwaard, A. W., Prins, M. D., and Kloosterziel, H., *Recl. Trav. Chim. Pays-Bas*, **103**, 188 (1984).
[156] Zwaard, A. W., Prins, M. D., and Kloosterziel, H., *Recl. Trav. Chim. Pays-Bas*, **103**, 174 (1984).
[157] Halpern, M., Feldman, D., Sasson, Y., and Rabinovitz, M., *Angew. Chem. Int. Ed. Engl.*, **23**, 54 (1984).
[158] Ceccon, A., Gambaro, A., and Venzo, A., *J. Organomet. Chem.*, **275**, 209 (1984).
[159] Shapiro, I. O., Ranneva, Yu. I., and Shatenstein, A. I., *Zh. Obshch. Khim.*, **53**, 2340 (1983); *Chem. Abs.*, **100**, 34055 (1984).
[160] Holm, T., *Acta Chem. Scand.*, **37B**, 797 (1983).
[161] Mori, H., Minita, K., and Kwan, T., *Chem. Pharm. Bull.*, **31**, 3002 (1983); *Chem. Abs.*, **100**, 138293 (1984).
[162] Matsson, O., Meurling, L., Obenius, U., and Bergson, G., *J. Chem. Soc., Chem. Commun.*, **1984**, 43.
[163] Sugimoto, N., Sasaki, M., and Osugi, J., *Bull. Chem. Soc. Jpn.*, **57**, 366 (1984).
[164] Sugimoto, N., Sasaki, M., and Osugi, J., *J. Chem. Soc., Perkin Trans. 2*, **1984**, 655.
[165] Sugimoto, N., Sasaki, M., and Osugi, J., *J. Am. Chem. Soc.*, **105**, 7676 (1984).
[166] Jarczewski, A., Przemyslaw, P., Mohammed, K., and Leffek, K. T., *Can. J. Chem.*, **62**, 954 (1984).
[167] Saunders, W. H. *J. Am. Chem. Soc.*, **106**, 2223 (1984).
[168] Siebrand, W., Wildman, T. A., and Zgierski, M. Z., *J. Am. Chem. Soc.*, **106**, 4083 (1984).
[169] Siebrand, W., Wildman, T. A., and Zgierski, M. Z., *J. Am. Chem. Soc.*, **106**, 4089 (1984).
[170] Hibbert, F., and Hunter, K. P. P., *J. Chem. Soc., Perkin Trans. 2.*, **1983**, 1895.
[171] Barnett, G. H., and Hibbert, F., *J. Am. Chem. Soc.*, **106**, 2080 (1984).
[172] Pogorelyi, V. K., and Turov, V. V., *Teor. Eksp. Khim.*, **19**, 698 (1983); *Chem Abs.*, **100**, 102653 (1984).
[173] Chrisement, J., Delpeuch, J. J., and Rajierson, W., *J. Chim. Phys. Phys.-Chim. Biol.*, **80**, 747 (1983); *Chem. Abs.*, **101**, 6279 (1984).
[174] Perrin, C. L., Lollo, C. P., and Johnston, E. R., *J. Am. Chem. Soc.*, **106**, 2749 (1984).
[175] Perrin, C. L., and Lollo, C. P., *J. Am. Chem. Soc.*, **106**, 2754 (1984).
[176] Sawyer, J. S., Macdonald, T. L., and McGarvey, G. J., *J. Am. Chem. Soc.*, **106**, 3376 (1984).
[177] Young, D., Kitching, W., and Wickham., G., *Tetrahedron Lett.*, **24**, 5789 (1983).
[178] Young, D., and Kitching, W., *Tetrahedron Lett.*, **24**, 5793 (1984).
[179] Goswami, R., and Corcoran, D. E., *J. Am. Chem. Soc.*, **105**, 7182 (1983).
[180] Bencivengo, D. J., Brownawell, M. L., Li, M.-Y., and San Fillippo, J., *J. Am. Chem. Soc.*, **106**, 3703 (1984).
[181] Fitzpatrick, J., Meyer, T. A., O'Neill, N. E., and Williams, D. L. H., *J. Chem. Soc., Perkin Trans. 2*, **1984**, 927.
[182] Casado, J., Castro, A., Ramon Leis, J., Lopez Quintela, M. A., and Mosquera, M., *Acta Cient. Compostelana*, **19**, 167 (1982); *Chem. Abs.*, **100**, 173970 (1984).
[183] Casado, J., Ramon Leis, J., Mosquera, M., Carlos Paz, L., and Pena, M. E., *Monatsh. Chem.*, **115**, 155 (1984).
[184] Casado, J., Gallastegui, J. R., Losada, M., Paz, L. C., and Vacquez Tato, J., *Acta Cient. Compostelana*, **19**, 209 (1982); *Chem. Abs.*, **100**, 155983 (1984).
[185] Casado, J., Castro, A., Lopez Quintela, M. A., Rodriguez Prieto, F., and Vacquez Tato, J., *Acta Cient. Compostelana*, **19**, 131 (1982); *Chem. Abs.*, **100**, 208743 (1984).
[186] Lopez Quintela, M. A., and Samos, J., *Acta Cient. Compostelana*, **19**, 179 (1982); *Chem. Abs.*, **100**, 173971 (1984).

[187] Casado, J., Castro, A., Leis, J. R., Mosquera, M., and Elena Pena, M., *Monatsh. Chem.*, **115,** 1047 (1984).
[188] Isobe, M., *Bull. Chem. Soc. Jpn.*, **57,** 601 (1984); *Chem. Abs.*, **101,** 22818 (1984).
[189] Antelo, J. M., Armesto, J. L., Casado, J., and Varela, A., *An. Quim.*, **80A,** 308 (1984); *Chem. Abs.*, **101,** 190780 (1984).
[190] Stanbro, W. D., and Lenkevich, M. J., *Int. J. Kinet.*, **15,** 1321 (1983).
[191] Minkov, D., Luvchieva, D., and Kostov, G., *Khim. Ind. (Sofia)*, **1983,** 218; *Chem. Abs.*, **100,** 50772 (1984).

Organic Reaction Mechanisms 1984
Edited by A. C. Knipe and W. E. Watts

CHAPTER 12

Elimination Reactions

R. A. More O'Ferrall

Department of Chemistry, University College, Belfield, Dublin 4, Ireland

E1cB and Borderline *E2* Mechanisms

In the past year a major review[1] of eliminations and several studies of the *E1cB* mechanism and *E1cB–E2* mechanistic borderline have appeared. In general an *E1cB* mechanism is favoured over a concerted mechanism by carbanion-stabilizing β-substituents and a poor leaving group. An interesting change of mechanism has been suggested between elimination of acetate from the β-ketoacetate (**1**) which occurs by an irreversible *E1cB* mechanism and the corresponding by elimination of an intramolecularly bound carboxy group from the keto-lactone (**3**) to give (**4**) which appears to be *E2*.[2] For hydroxide and quinuclidine bases (**3**) is between 50 and 10^4 times more reactive than (**1**). This is inconsistent with rate-determining carbanion formation unless proton loss from (**3**) is unexpectedly more favourable than from (**1**). By comparison proton loss from the cyclic ether (**5**), for which irreversible *E1cB* elimination is indicated by kinetic saturation in quinuclidine buffers, is between 8 and 80 times faster than its non-cyclic analogue (**2**), probably not enough to account for the difference of (**3**) from (**1**).[3] Brønsted exponents (0.69 and 0.42, respectively) for (**3**) and (**1**) also differ whereas those for (**5**) and (**2**) are similar (0.54 and 0.47, respectively). The high reactivity and change of mechanism of (**3**) is ascribed to ring

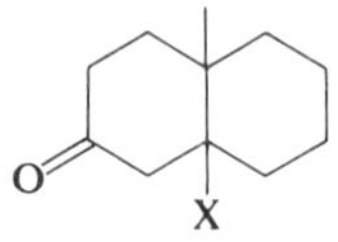

(1) X = OAc
(2) X = OH

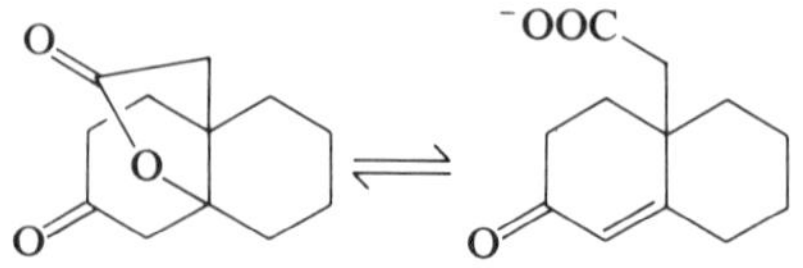

(3)

(4)

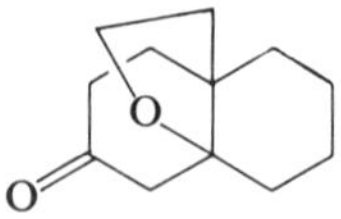

(5)

strain. However the strain is not reflected in the equilibrium because elimination of (3) and (5) is reversible whereas that of (1) and (2) is not; this is consistent with the more favourable entropy change expected of the intermolecular reaction. Presumably release of ring strain develops earlier along the reaction coordinate than the difference in entropy, thereby leading to an imbalance between kinetic and equilibrium behaviour.

The keto-activated eliminations and isomerization of L-glyceraldehyde-3-phosphate and dihydroxyacetone phosphate occur by an "irreversible" *E*1*cB* mechanism *via* a common enediolate intermediate as expected from the lower pK_a of the phosphate compared to the acetate leaving group.[4] This mechanism is indicated by (*i*) similar rate constants for dihydroxyacetone phosphate and ionization of acetone, (*ii*) observation of a buffer-independent reaction implicating the phosphate dianion as an intramolecular base catalyst, and (*iii*) buffer curvature (moderated in D_2O) for reaction in quinuclidine buffers. At high pH values a change in leaving group from phosphate dianion to trianion is inferred.

Reactions of indenyl substrates (6) are characterized by competing elimination and isomerization yielding (7) and (8). Dissection of β-deuterium isotope effects upon rates and product ratios into contributions for the two pathways, assuming that these occur in parallel, leads to characteristically "amplified" and "attenuated" isotope effects for rearrangement and elimination respectively: *e.g.*, the respective values are $k_H/k_D = 26.4$ and 4.5 for reaction of indenyl acetate with DABCO in 65% aqueous DMSO.[5] This implies that the reactions occur not in parallel but via a common (indenyl anion) intermediate. The isotope effects persist for reaction of β-deuterated acetate with DABCO in protic solvents indicating that rearrangement is not accompanied by exchange with solvent and therefore that it occurs intramolecularly within a DABCOH$^+$–indenyl anion ion-pair. Examination

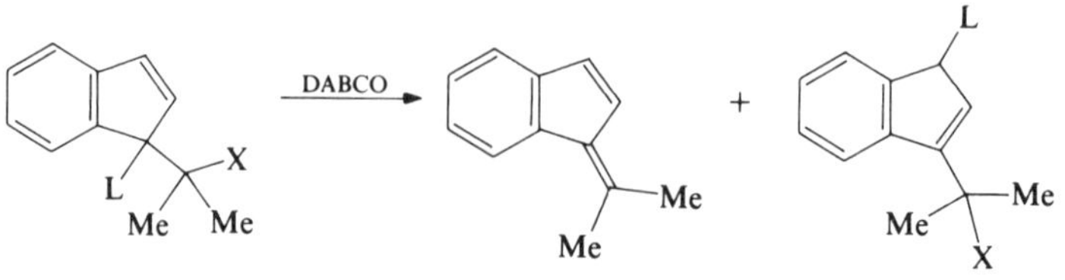

(6) → DABCO → (7) + (8)

L = H, D; X = OMe, OAc

of the solvent and base dependence of this intramolecularity[5] has shown that loss of the acetate leaving group likewise occurs within the ion-pair.[6] The latter reaction may be regarded as a "reverse pre-association", since in the reverse direction attack of acetate on the olefin must follow pre-association of olefin and $DABCOH^+$ catalyst. The influence of charge type of the base upon ionization has also been examined.[7]

Ion-pairs play an important rôle in *E*1*cB* elimination of perfluorinated substrates. These reactions and the reverse additions have been reviewed with emphasis on their proton-transfer steps.[8]

Hydroxide-catalysed dehydration of a β-hydroxycyclopentanone analogue of Prostaglandin E1 in D_2O shows deuterium isotope exchange in competition with elimination consistent with (just) reversible *E*1*cB* elimination;[9] exchange also accompanies the acid reaction. For the methoxychalcones (**9**) reversible *E*1*cB* elimination is indicated by rapid interconversion of *threo*- and *erythro*-reactants and absence of an isotope effect with β-deuterated substrate.[10] Elimination of trimethylamine from (**10**) is assigned an *E*1*cB* mechanism despite reaction in NaOMe–MeOH occurring 10^5 times faster than the expected rate of ionization predicted from a Taft $\rho^*\sigma^*$ correlation.[11] The high reactivity has been ascribed to a field effect of the Me_3N^+ pole and release of non-bonding interactions of the *N*-methyl groups. *E*1*cB* elimination of HCN from a zwitterionic adduct accompanies copolyymerization of *N*-vinylcarbazole and tetracyanoethylene.[12] Irreversible *E*1*cB* elimination of 4-(*p*-nitrophenoxy)-2-butanone has been used to investigate the base dependence of primary hydrogen isotope effects.[13]

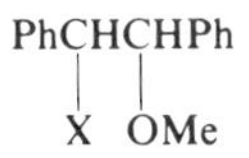

PhCHCHPh (X, OMe)

(9)

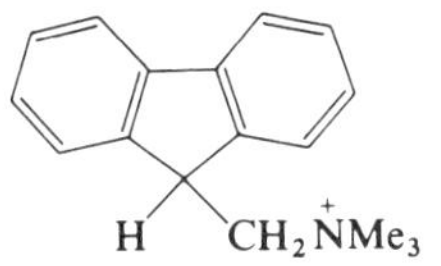

H, $CH_2\overset{+}{N}Me_3$

(10)

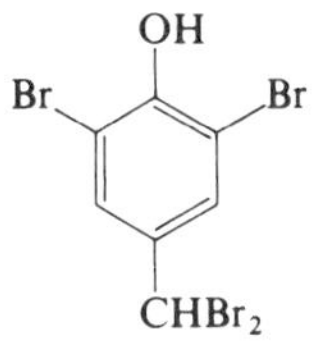

OH, Br, Br, $CHBr_2$

(11)

Hydrolysis of alkyl and aryl allophonate esters by hydroxide ion in aqueous solution features competing *E*1*cB* (PhNHCONRCOOAr $\rightarrow$ PhN=C=O + RNHCOOAr) and acyl-transfer mechanisms of ester hydrolysis with the latter favoured by poor (especially alkoxy) leaving groups.[14] The *E*1*cB* reaction is characterized by a break in the pH profile consistent with ionization of an amide nitrogen ($pK_a = 11$) prior to formation of an isocyanate intermediate, and a sensitive dependence on the aryloxy leaving group.

Chloroalkylphenols are dehydrohalogenated to *o*- or *p*-quinone methides *via* a phenolate ion intermediate.[15] Formation of a quinone methide is also the first step in the conversion of the dibromohydroxybenzal bromide (**11**) to the corresponding aldehyde at pH values above 3 and in substitution of chlorine in *o*-chloromethylphenols by anilines.[16] In less basic media the phenolate ion ceases to be an intermediate and the elimination step is formally *E*1; no concerted reaction is observed. *E*1*cB* elimination with loss of a proton from oxygen is also implicated in the reverse aldol-like reaction of ring-substituted α-trihalomethylbenzyl alcohols to arylaldehydes and haloform.[17] The mechanism is substantiated by a change from first- to zero-order dependence on hydroxide ion consistent with ionization of the reactants with pK_a values in the range 11.4–12.4.

In a study of *E*2 eliminations close to the *E*2–*E*1*cB* mechanistic borderline it has been shown that Brønsted exponents β for elimination of 2,4-dinitrophenylethyl halides to dinitrostyrene in aqueous buffers increase from 0.42–0.54 in the series I, Br, Cl, F (relative rates in H_2O, 14:9:2:1);[18] β also increases between a 2,4-dinitro- and a 4-nitro-aryl group but decreases for phenyl (0.46:0.61:0.51 for the respective bromides). The behaviour indicates an *E*2 mechanism with coupling of C—H and C—halogen bond-breaking in a transition state which changes from a "central" structure, with significant double-bond development and leaving-group bond-breaking, to strongly carbanion-like with accumulation of nitro substituents.

Observation of the halogen reactivity order I > Cl > Br > F has been taken as indicating C—Hal bond-breaking in a transition state and hence an *E*2 rather than irreversible *E*1*cB* mechanism. However, it has been suggested that in strongly carbanionic transition states the order may reflect halogen hyperconjugation.[7] From a measurement of $k_{Cl}/k_F = 0.1$ for *syn*-elimination of stilbene dihalides in Bu^{t}OK/Bu^{t}OH it has been inferred that the mechanism of reaction is probably *E*1*cB* and that, if so, the carbanion intermediate is not stabilized by halogen hyperconjugation.[19] By contrast the predominant, and presumably *E*2, *anti*-mode of elimination shows a "normal" leaving-group effect of $k_{Cl}/k_F = 11$. In a further study of dehydrohalogenation of borderline *E*2 substrates solvent isotope effects have been reported for the reaction of 2,2-di(*p*-chlorophenyl)-1,1-dichloroethane (DDD) and DDT in NaOMe–MeOD.[20]

Concerted reactions may be favoured over step-wise because energy gained in bond-making can be coupled economically with energy required for bond-breaking. A nice distinction is drawn by Dewar between "one-bond" processes, such as proton transfer or S_N2 substitution, in which only one bond is made and one broken in the concerted reaction and multi-bond processes in which several bonds are made and broken.[21] Intuitively, two-bond processes may be expected to have twice the activation energy of a one-bond reaction. Elimination reactions are two-bond processes and it follows that they should be particularly prone to step-wise reactions and to uncoupling of C—H and C—X bond-breaking in a concerted transition state.

*E*1 and *E*2*C* Eliminations: Stereochemistry

Solvolyses of primary alkyl substrates bearing the bulky neutral leaving group 7-phenyltetrahydrodibenzoacridine (**12**) yield both substitution and elimination products.[22] Reactions at 150° in methanol solvent give the usual sterically dominated S_N2 reactivity order, but in CH_3COOH or CF_3COOH α- and β-alkylation increases reactivity suggesting reaction *via* carbonium ion-pairs.

Rate constants for styrene formation from a 1-phenethyl cation intermediate have been measured from ratios of olefin to substitution products in aqueous trifluoro-

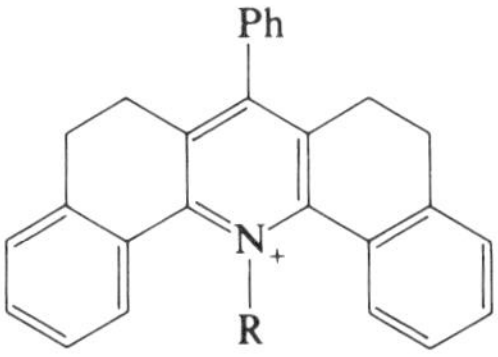

(**12**)

ethanol by assigning a diffusion rate constant 5×10^9 M^{-1} s^{-1} to trapping by azide ion;[23] reactions with carboxylate ions yield a Brønsted correlation with $\beta = 0.14$. Comparison of rate constants for elimination by externally added carboxylate ions and carboxylate ion-pairs is consistent with an ion-pair association constant of 0.04 M^{-1}. Combining rate constants for elimination by H_2O and for protonation of styrene by H_3O^+ gives a pK_a for the 1-phenethyl cation of -11.2.

*E*1 elimination of the stereospecifically deuteriated methyldecahydronapthalenol (**13**) in H_2SO_4–Ac_2O–AcOH yields olefins (**14**) and (**15**) by a *syn*-elimination with $k_H/k_D = 2.2$.[24] The isotope effect implies equilibration between carbocation conformations leading to the two isomeric products and therefore access to a conformation favourable to *anti*-elimination. That *anti*-elimination is nevertheless not seen may imply proton abstraction by an acetoxysulphonate counter-ion within a structured ion-pair. A study of specific acid catalysis of cleavage of alkyltriazenes confirms that initial formation of an amine and alkyldiazonium ion precedes solvolytic reaction of the latter to substitution and elimination products.[25] Solvent effects upon elimination of *tert*-butyl chloride are correlated by a new solvent polarity scale.[26]

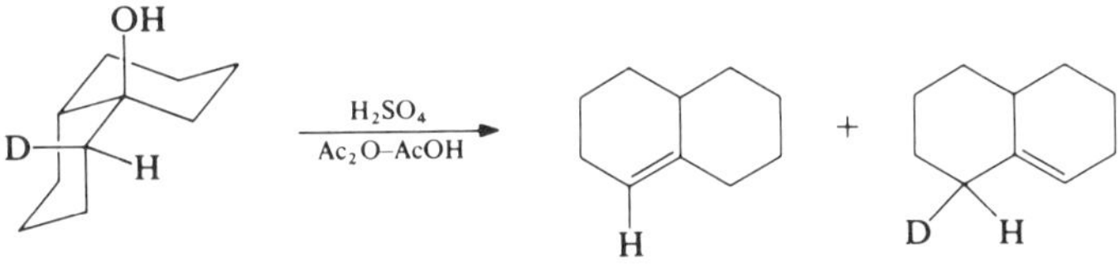

H_2SO_4 / Ac_2O–AcOH

(**13**) (**14**) 1:2.2 (**15**)

Studies of *E2C* elimination have appeared among a group of papers dedicated to the late A. J. Parker.[27–29] *E2C* elimination is characteristic of nucleophilic bases and transition states leaning towards the *E*1 limit of the *E*2 mechanism. The base and structure dependence and anti-periplanar selectivity of the reactions combine to suggest a nucleophilic interaction between base and α-carbon of the substrate at the transition state (**16**). A point at issue has been the extent of α-carbonium ion relative to double-bond character of the transition state. From a reanalysis of alkyl and phenyl α- and β-substituent effects it has been concluded that interaction with a well-developed double bond is an important ingredient in the effects.[27] However a suggestion that α- and β-effects are comparable may be oversimplified. Replacement of a β-H by any bulky substituent has a similarly large influence. Moreover, isopropyl at C_α is surprisingly more effective than methyl, while *tert*-butyl fails to show steric hindrance to the nucleophilic component of the reaction. This behaviour suggests release of non-bonded ground-state interactions (Thorpe–Ingold effect) reflecting the "neighbouring-group" character of the nucleophilic interaction and cyclic structure of the transition state.

Dehydrobromination of substituted stilbene dibromides by AcO^-, CN^-, and Cl^- shows the expected change (between AcO^- and Cl^-) from strong to weak activation by electron-withdrawing β-substituent effects and insensitivity to α-substitution for all bases.[28] The weak effect of electron-donating α-aryl substituents occasions doubt that α-C^+ characteristics are important in the *E2C* transition state. The situation is reminiscent of solvolysis of benzyl substrates where acceleration by both electron-donating and -withdrawing substituents reflects variations in covalent interactions between nucleophile and carbonium ion centre, and of both with the substituent in the aryl ring.

In two papers (sadly also published posthumously) Kwart has further examined the cyclic character of the *E2C* transition state by applying his empirical criterion that non-linear H-transfer transition states show a temperature-independent isotope effect and exalted A_H/A_D. For reaction of the 2-alkyl-2-bromo-3-phenylpropionates (**17**) with lithium bromide in DMF primary isotope effects k_H/k_D (measured by both intra- and inter-molecular methods) have temperature-independent values in the range 2.4–2.7.[30] Remarkably, the intramolecular measurement was based upon an H/D ratio of the product at completion of reaction of a substrate almost certainly consisting of a 50/50 mixture of isotopic diastereomers. Since the product is exclusively *trans*-ethyl cinnamate, stereospecifically *anti*-elimination should lead to an H/D product ratio of 1.0! Equivalence of the bromide leaving group and bromide base in the transition state would formally account for the behaviour, but is hardly likely in practice, and similar behaviour is seen with the amine base proton sponge.[31] Formation of a reaction intermediate having trigonal–bipyramidal structure surrounding C_α and in which the abstractable H and D atoms are equally available to the action of the base has been proposed. Of other possibilities $\alpha'\beta$-elimination is excluded and *syn*-elimination, although reported at a significant level (9 %) for conformationally favourable cyclopentyl derivatives under *E2C* conditions,[29] is normally not expected in *E2C* reactions.

The temperature independence of the primary deuterium isotope effect ($k_H/k_D = 3.43$) observed for *syn*-elimination of *exo*-norbornyl tosylate with sodium alkoxides in triglyme has been used as a criterion for transition-state structure in the reaction.[32] This result is consistent with considerable evidence that *syn*-eliminations occur through a cyclic (and hence non-linear) transition state involving an ion-pair

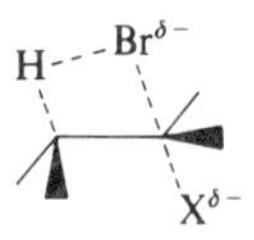

H---$Br^{\delta-}$ $X^{\delta-}$

(**16**)

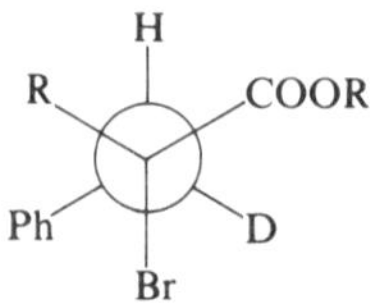

H, R, COOR, Ph, D, Br

(**17**)

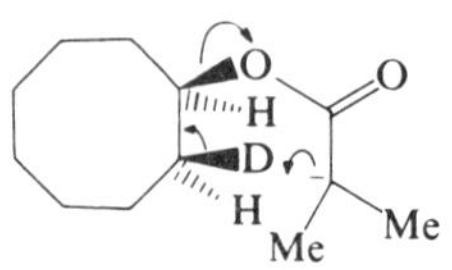

O, H, D, H, O, Me, Me

(**18**)

of the base. Interestingly, a kinetic analysis suggests that for the abnormal *syn*-mode observed for elimination of phenylcyclopentyl brosylate under *E2C* conditions promoted by LiCl in acetone ion-pairs are completely unreactive.[29]

syn-Elimination to *cis*-cyclooctene has been observed for the reaction of stereospecifically deuterated cyclooctyl isobutyrate with magnesium diisopropylamide in ether at $-40°$; an intramolecular β,β' proton abstraction by the ester anion (**18**) is inferred.[33] *syn*-Elimination is also seen in zinc-promoted dehalogenations;[34] for *threo*- and *erythro*-5,6-dihalodecanes, surprisingly, the stereochemistry depends only on the nature of the more electronegative of the two halogens. *syn*-Elimination increases from 4 to 14 to 58% for elimination of IBr to ICl to IF but is unchanged between FBr, FCl, and FI. The behaviour is believed to indicate a variable transition state associated with a small degree of double-bond character. An HOMO method has been used to study the stereochemistry of dehydrohalogenations.[35]

Deslongchamps's suggestions of stereoelectronic control of carbon–oxygen or carbon–nitrogen multiple-bond formation has provoked much interest but few unambiguous experimental tests. However, a large stereoselectivity has now been reported for dehydrochlorination of the E- and Z-hydroximidoyl chloride anions (**19**) and (**20**) to benzonitrile oxide with $k_Z/k_E = 6 \times 10^7$. The E-isomer cannot be

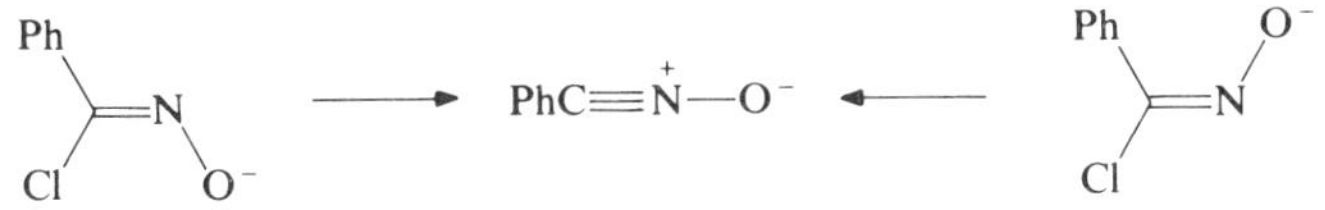

$$\text{Ph(Cl)C=N-O}^- \ \mathbf{(19)} \longrightarrow \text{PhC}\equiv\overset{+}{\text{N}}-\text{O}^- \longleftarrow \text{Ph(Cl)C=N-O}^- \ \mathbf{(20)}$$

(**19**) (**20**)

prepared directly, but preparation and photolysis of the Z-chloroacetate gives a mixture of E- and Z-esters. The E-ester can be hydrolysed *in situ* in aqueous NaOH and the kinetic dependence of the subsequent dehydrochlorination upon hydroxide ion has proved to be consistent with reaction of the hydroximidoyl chloride anion with a pK_a for *O*-protonation of 9.8. For the Z-isomer pH-dependent dehydrochlorination is too fast for measurement above pH 2, but a rate constant for the Z-anion has been inferred by assigning the same pK_a as for the Z-isomer.[36]

The stereoelectronic influence of a keto group upon α-carbanion formation has been investigated for decarboxylation of the *tert*-butylcyclohexanone carboxylic acids (**21**) and (**22**).[37] Remarkably, decarboxylations of both acids and anions are faster for (**21**) than (**22**), by factors of 3 and 15–20, respectively. It has been suggested

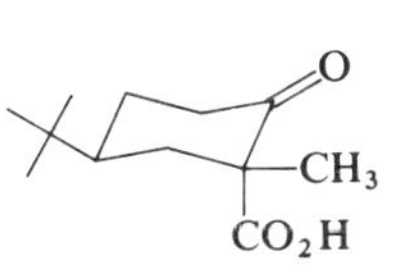

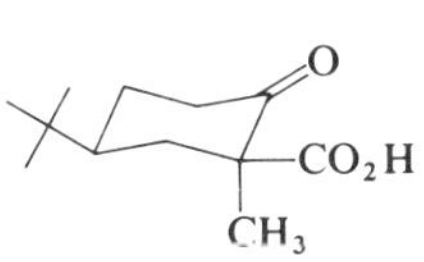

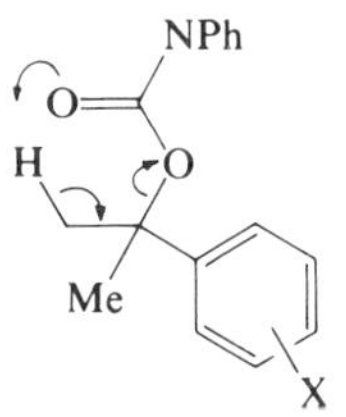

(**21**) (**22**) (**23**)

that the requirement of proton transfer or hydrogen bonding from the carboxyl to the keto group favours (**21**) in the case of the acid and that the anion of (**22**) is destabilized by an unfavourable ion–dipole interaction.

Pyrolytic Eliminations

Carboxylic Acids and Derivatives

Rates of thermal elimination of l-arylethyl carbonates and carbamates, *e.g.* (**23**), correlate with σ^+ to give $\rho^+ = -1.3$ to -2.4 depending on solvent and temperature in the range 130–180°. This is consistent with a concerted transition state, with C—O bond-breaking in advance of proton transfer and carbonium ion character at C_α.[38] Interestingly, for esters, both α-acyl and -cyano substituents increase reactivity;[39] this is probably due mainly to unfavourable dipolar interactions with the acyloxy group in the reactants.

Imbalance in favour of C—O bond-breaking even in simple esters, $XCOOCH_2CH_3$, is implied by a Taft correlation ($\rho^* = 0.32$) consistent with mild acceleration by electron-withdrawing acyl substituents.[40] On the other hand kinetic and product analyses for elimination of a series of secondary alkyl acetates shows the influence of alkyl substituents as mainly steric.[41] Eclipsing of bulky groups in the transition state slows elimination but uneclipsed alkyl substituents increase reactivity.[42] Variation of alkyl substituents at the acyl carbon is without effect.[43]

Rate constants for pyrolysis of 2-alkoxypyridines (**24**) to 2-pyridone and olefins at 220–520° correlate closely with those of acetate esters of corresponding structure, with acceleration by α-alkyl and α-phenyl substituents, reflecting α-C^+ stabilization.[44] Tropolone ethers such as (**25**), which may be regarded as vinylogous esters, undergo thermal elimination to tropolone and olefin.[45] Deuterium labelling indicates that the reaction is stereospecifically *anti*.

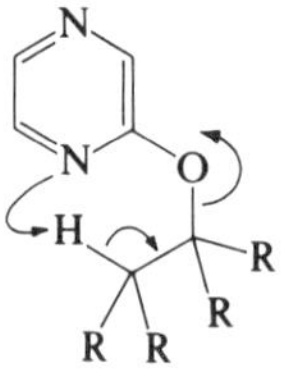

(**24**) 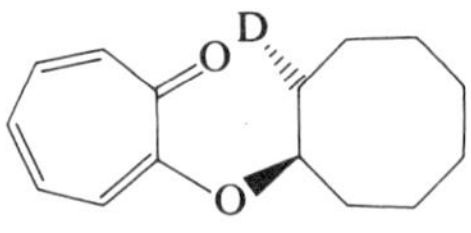 (**25**)

Thermolytic elimination of the spirocyclic amide ester (**26**) to form (**27**) takes place in refluxing toluene and is believed to involve amide participation to give the intermediate ion-pair (**28**).[46] This is supported by trapping (**28**) in ethanol, presumably as (**29**), and hydrolysing to the alcohol epimeric with the original acetate, which on acetylation fails to undergo elimination. The structure of (**26**) has been established by X-ray crystallography.

Comparison of electron-impact-induced elimination of deuterated cholestanyl 3α-, 4α-, and 6α-acetates with the corresponding thermal and pyrolytic reactions with respect to regio- and stereo-chemistry confirms that the reaction is probably

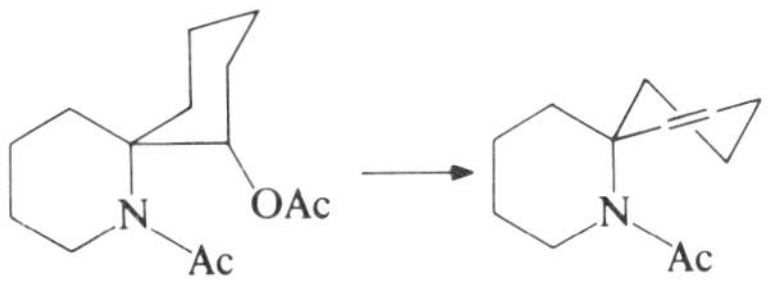

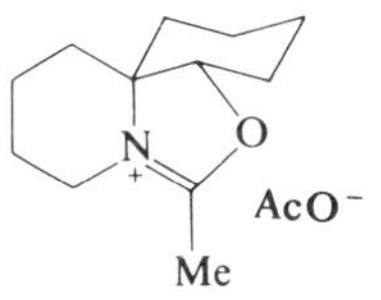

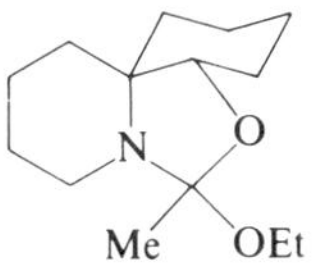

(26) (27) (28) (29)

step-wise and initiated by transfer of a hydrogen radical to the carboxyl group. Differences in products between the electron impact and a photolytic pathway, also a step-wise radical process, have been rationalized in terms of reversibility of the hydrogen abstraction step of the latter.[47]

In other studies of esters, pyrolysis of methyl methacrylate has been investigated in the temperature range 400–800°[48] and CNDO/2 calculations for thermolysis of *tert*-butyl acetate have been found consistent with a concerted mechanism with catalysis by acetic acid.[49] In an unusual thermal reaction, *N*-hydroxypropylenamine esters (**30**) yield *N*-(acetonyl)-β-enamino-aldehydes and 2-acetylpyrroles upon flash pyrolysis at 330–420°; 1,4-elimination of CH_3OH to form an iminoketene has been suggested as the initial step.[50] Low-pressure pyrolysis of phenyl acetate to form ketene and phenol at 1000° is consistent with a concerted rather than a radical chain mechanism, and presumably occurs *via* a four-centre transition state.[51] Pyrolysis of cyclic carbonates in the presence of excess KOH leads to fragmentation with formation of CO_2, CH_2O, and the appropriate alcohol.[52]

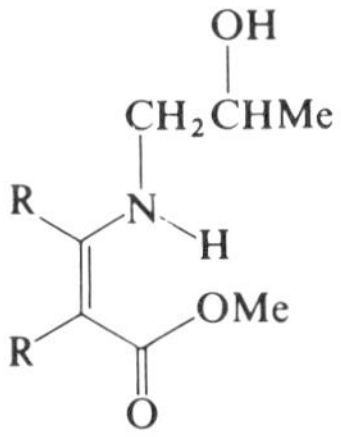

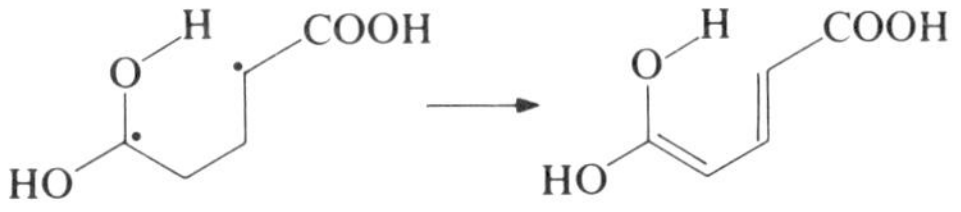

(30) (31) (32)

The scope of dealkoxycarbonylation of β-keto-esters and diesters by heating under reflux with a carboxylic acid for 24–48 h has been investigated and a mechanism for the reaction suggested.[53] A mechanism has also been proposed for the interesting dealkoxycarbonylation of aryl cinnamates to stilbenes;[54] this reaction is synthetically useful with yields of *ca.* 50% from heating in a sealed tube for 1.5 h at 320°. Decarboxylation of phthalide 3-carboxylic acid as a melt in the presence of aromatic aldehydes has been shown to give 3-(arylhydroxymethyl)phthalides.[55]

A shock-tube study of the pyrolysis of acetic acid at 1300–1950 K shows that the principal competing initial reactions give CH_4 and CO_2 or $CH_2{=}C{=}O$ and H_2O by molecular rather than radical mechanisms.[56] Low-pressure pyrolysis of dicarboxylic acids at 610° has been investigated using a tandem mass spectrometer equipped with a curie-point pyrolysis inlet system and m/z product analysis.[57] A principal reaction path is anhydride formation followed by decarboxylation to give a cyclic ketone. Other pathways involve carbonyl-activated 1,5-hydrogen transfer to a (second) carbonyl oxygen coupled with dehydrogenation, *e.g.* in formation of $HOOCCH{=}CHCH_2COOH$ from glutaric acid *via* **(31)** and **(32)**. Minor products, *e.g.* acetone from malonic acid, indicate bimolecular reactions.

Loss of Nitrogen

Effects of single and double substitution at bridgehead positions upon thermolysis of diazabicyclooctenes **(33)** have been interpreted in terms of concerted loss of nitrogen with imbalance of C—N bond-breaking (where $R^1 \neq R^2$) favouring formation of the more strongly stabilized radical centre.[58] However, the possibility of reversible (though not irreversible) formation of a diazenyl radical intermediate **(34)** which subsequently loses N_2 is formally not excluded. Comparing phenyl and vinyl substituent effects with their resonance stabilization energies suggests that the C—N bonds are about half-broken in the transition state. Fluorescence life-times and the temperature dependence of quantum yields for sensitized and unsensitized photolyses of substituted diazabicyclooctenes point to a rough correlation between activation energies for thermolysis and reactions from singlet excited states.[59]

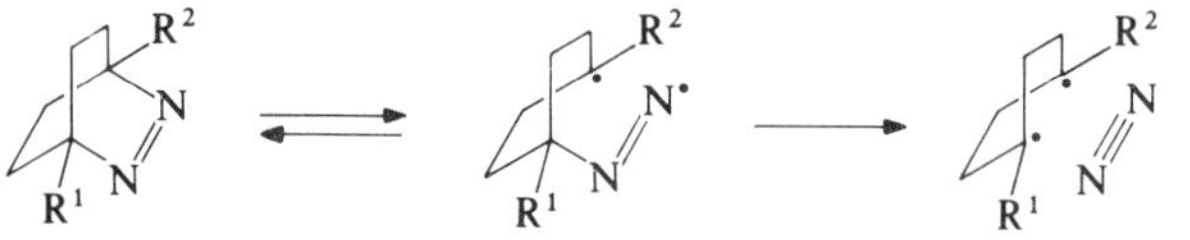

(33) (34)

R^1 = H, Me, Ph, *etc.*
R^2 = H or R^1

On the other hand, in the thermolysis of the diazabicyclohexene **(35)** the pressure dependence of the product proportions of bicyclobutane and butadiene suggests formation of butadiene from a vibrationally excited bicyclobutane in competition with collisional deactivation.[60] It is argued that the implied concentration of energy in the bicyclobutyl rather than nitrogen fragment of bond-cleavage is consistent with step-wise reaction *via* a diazenyl radical but not a concerted process.

(35) 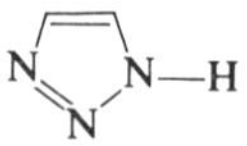 (36)

Flash vacuum pyrolyses of vinyl azide and its isomer ^{1}H-1,2,3-triazole **(36)** yield initially azirine which subsequently rearranges to acetonitrile.[61] On the basis of MNDO calculations and comparison with pyrolysis of methyl azide it has been suggested that formation of azirine occurs by a concerted mechanism and does not involve vinyl nitrene as an intermediate.

Halo and Thio Leaving Groups: Loss of H_2, Silanes

The complex influence of α- and β-halogen substituents upon reactivity in thermal four-centre dehydrohalogenations has been discussed in terms of thermodynamically corrected "intrinsic" activation energies.[62] A Taft correlation of β-substituent effects in gas-phase elimination of ethyl chloride at 440° yields an apparently biphasic plot with $\rho^* = -0.3$ for electron-withdrawing substituents and a more negative ρ^* for alkyl substituents.[63] The behaviour has been ascribed to a change in transition-state structure, but may reflect the complexity of the (small) electronic effects of alkyl groups. At 360° 4-chloro-1-butanone loses HCl 200 times faster than does 1-chloropentane, consistent with neighbouring-group participation by a keto group.[64]

Laser-induced elimination of HCl from $ClCH_2CH_2Cl$ and ClF_2CMe is initiated by C—Cl bond-breaking followed by a hydrogen abstraction chain and a unimolecular termination step that can be made rate-determining.[65] IR multiphonon decomposition of vibrationally excited ClFCHCHClF yields excited HF and HCl in proportions strongly dependent on CO_2 laser fluence.[66] The reaction of trichloroethylene to dichloroacetylene at 450° has been modelled by RRKM theory.[67]

The 1,5-thiazepinone (**37**) undergoes thermal endocyclic elimination of a phenylthio group which recyclizes *via* a vinylic exocyclic displacement of methyl thiolate to yield the benzothiazole (**38**).[68] Thermal decomposition of butanethiol in a low-collision molecular reactor yields mainly ethylene, in contrast to normal formation of 1-butene.[69] Kinetic, product, and isotope measurements for pyrolysis of allyl propargyl sulphide show competing cycloeliminations to propylene and propynthial and allene and propenthial. The thio-aldehyde products react further to a Diels–Alder adduct.[70]

Ph S H H Ph N SMe —200°→ S Ph N Ph

(**37**) (**38**)

The pyrolysis of allyltrimethylsilane (**39**) to Me_4Si, Me_3SiH, the silene dimer (**40**), and vinyltrimethylsilane has been reinvestigated.[71] The pressure dependence of the products and successful trapping of dimethylsilene and Me_3Si radicals point to competing retro-ene elimination and Si—C bond homolysis reactions as primary pyrolytic processes. The ketosilene (**41**) cyclizes spontaneously to the interesting oxasilacyclobutene (**42**) observable in solution by NMR before it decomposes thermally by cycloreversion to trimethylsilyladamantylacetylene.[72] Silene itself has been implicated in the high-temperature decomposition of methylsilane. A shock-tube study of an isotopically substituted reactant (CH_3SiD_3) shows four-centre eliminations of H_2 and CH_4 from $SiCH_2$ (silylene) and $CH_2{=}SiH_2$, respectively.[73] IR multi-phonon elimination of hydrogen from *trans*-CHD=CHD yields isotopic

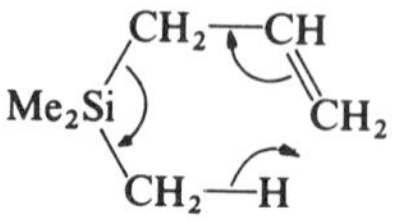

(39) 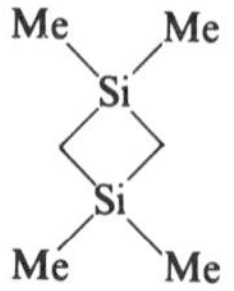 **(40)** 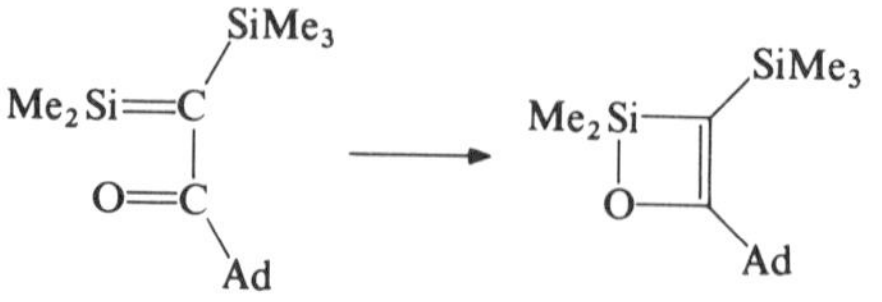 **(41)** **(42)**

hydrogen molecules (H_2, HD, D_2) and acetylenes (C_2H_2, C_2HD, CD_2) in statistical proportions (1:2:1) consistent with competing *cis–trans* isomerization of the reactant.[74]

Cycloeliminations

Cycloreversion of barrelene (**43**) to benzene and acetylene is probably a concerted reaction on the basis of its moderate *A* factor and an activation energy (32.5 kcal mol^{-1}) less than expected for formation of a diradical intermediate.[75] Comparisons with other systems suggest that cycloadditions of acetylene or ethylene to benzene or cyclopentadiene (or their reversal) are concerted, but that their cycloadditions to cyclohexadiene are step-wise. The differences may reflect energies of σ-bond distortion of the reactants required to achieve a geometry consistent with overlap of reacting orbitals in a concerted transition state. This factor has been invoked to explain the slow reversion ($k = 1.67 \times 10^{-4}\ s^{-1}$ at 95°) of the highly unsaturated propellapentaene (**44**) to naphthalene and butadiene.[76] Consistent with the results for barrelene, the corresponding one degree less unsaturated propellapentaene (**45**) shows no conversion to benzocyclohexene (after 90 h at 160°).

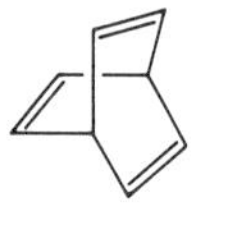

(43) 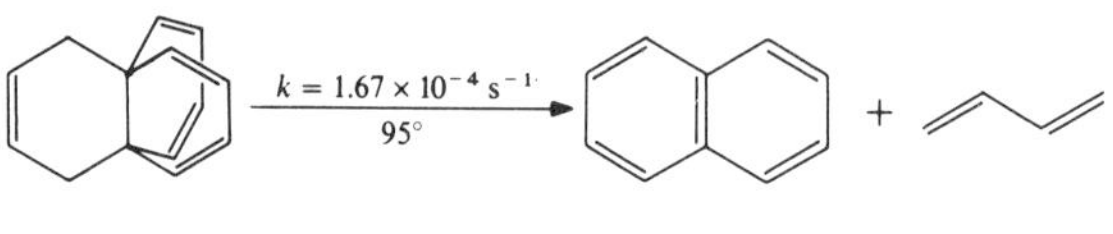 **(44)** 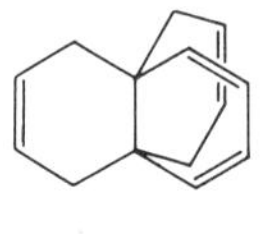 **(45)**

In contrast to the well-known ring-opening of 3,4-dimethylcyclobutenes, "allowed" thermal cycloelimination of the *trans*-perfluoro analogue (**46**) preferentially forms the sterically hindered Z,Z-diene (**47**) with an activation energy 18 kcal mol^{-1} *less* than for the E,E-diene (**48**)![77] A study of thermal isomerization of the tetrafluorobicyclopentane (**49**) to cyclopentene (**50**) shows the reaction to be slower

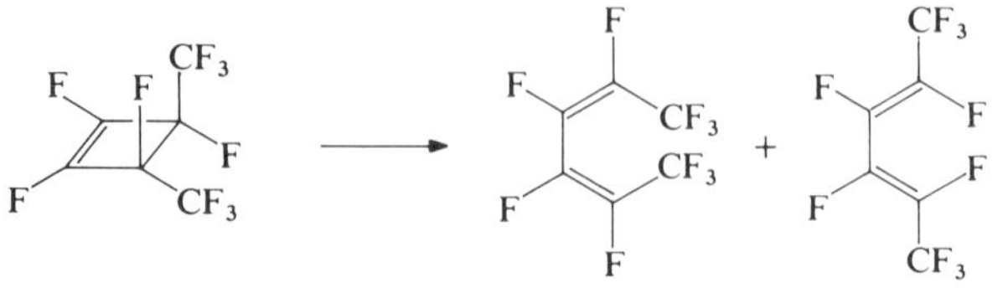

(46) **(47)** **(48)**

than for the corresponding hydrocarbon, consistent with the known inhibiting effect of fluoro substituents on conversion of cyclobutane to ethylene.[78] However thermal interconversion of the *endo*- and *exo*-methyl isomers **(51)** and **(52)** has the same activation energy as the hydrocarbons, suggesting that the slow step for conversion of **(49)** → **(50)** is not the expected bond homolysis but the subsequent rearrangement of hydrogen leading to double-bond formation.

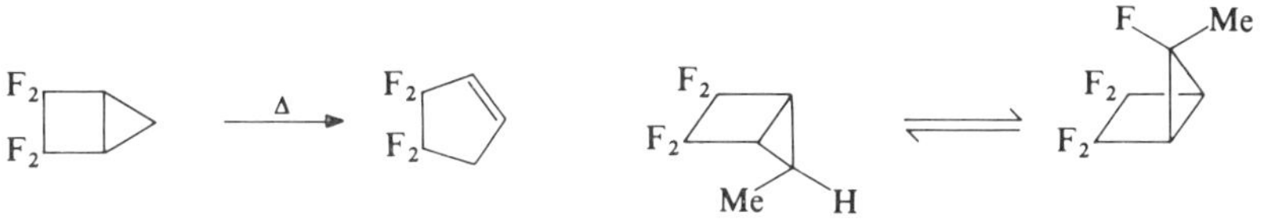

(49) **(50)** **(51)** **(52)**

Flash pyrolysis of the benzotricinnoline **(53)** at 780° gives 11% yield of the tribenzo[12]annulene **(55)**. The product is thought to arise *via* a (symmetry-allowed) $2\pi + 2\pi + 2\pi$ cycloreversion of the intermediate dibenzocyclobutabiphenylene **(54)** despite the large energy barrier expected for such a reaction.[79] A review has appeared of attempts to prepare the as yet inaccessible oxirene (oxacyclopropene) ring system.[80] The corresponding thiirene is observable under matrix isolation and presumably this offers the best prospect for oxirene.[80] Cycloelimination of the saturated thiirane methylthiacyclopropane to propylene and S_2 has been used to monitor production of sulphur atoms accompanying disproportionation of photolytically generated SO to SO_2 and S.[81] There has been a detailed study of products and stereochemistry of thermal and Lewis-acid-catalysed fragmentation of 1,2-dioxolanes.[82]

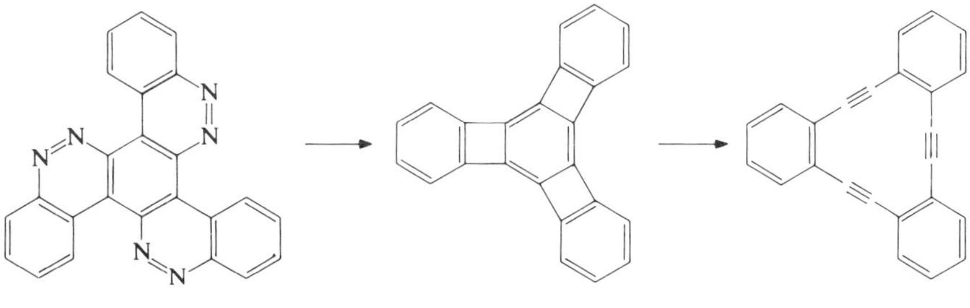

(53) **(54)** **(55)**

Eliminations of Enzymes, Coenzymes and their Models

There have been a number of studies of metabolically important eliminations. By ingenious double-labelling experiments it has been shown that the presumed addition–elimination steps by which phosphoshikimate **(56)** is coupled to phosphoenol pyruvate **(57)** in the shikimate biosynthetic pathway occur with opposite *syn*- and *anti*-stereochemistries (Scheme 1).[83–85] It is suggested that the behaviour corresponds to "least motion" pathways.[84, 85] In a comparison of chemical and

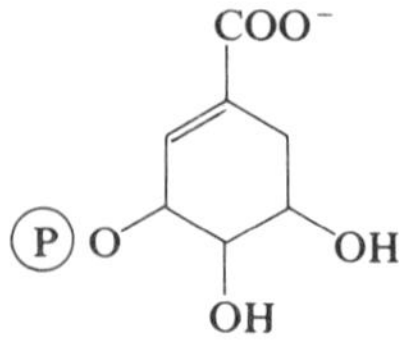

(**56**), SOH

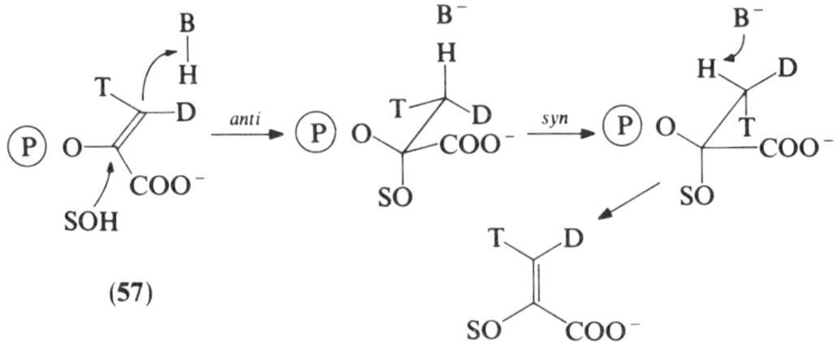

(**57**)

SCHEME 1

enzymatic pathways for elimination and isomerization of L-glyceraldehyde phosphate it has been pointed out[4] that while the reactions involve a common carbanion intermediate, and under chemical catalysis are competitive, the corresponding enzymes triose phosphate isomerase and methylglyoxal synthetase are highly selective. Probably this reflects binding of the substrate at the two enzyme sites in conformations that are respectively favourable and unfavourable to the stereochemical requirements for elimination.

Labelling studies have also shown that dehydration of 3-hydroxybutyl CoA to crotonyl CoA occurs by *syn*-elimination,[86] and that 1,2-hydrogen migration accompanying the complementary "2,1" rearrangement of an amino group in the cobalamin enzyme-catalysed deamination of ethanolamine is stereospecific.[87]

$$HOCHDCH_2NH_2 \rightarrow (CH_3CDO \text{ or } CH_2DCHO) + NH_3$$

In the latter reaction the R-deuteriated substrate yields CH_3CDO and the S-deuteriated substrate gives CH_2DCHO, in contrast to the previously demonstrated non-stereospecific replacement of NH_2 by H at the (H) migration terminus.

Elimination of *p*-nitrothiophenol from a *p*-nitrophenylthiopropanone linked to a peptide fragment has been used to map the active sites of carboxypeptidase and angiotensin-converting enzymes.[88] With suitable modification of the binding group ($R = COO^-$ or phenylalanyl in (**58**)) and choice of diastereomer elimination from both enzymes is observed confirming the structural similarity and availability of a base catalyst at the two active sites.

There have been a number of studies of reactions of pyridoxal. Rates of dehydrochlorination of methyl chlorophenylalaninate and its pyridoxal complex

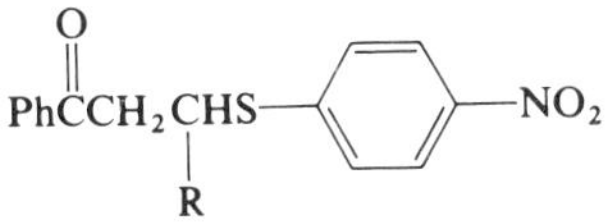

(58)

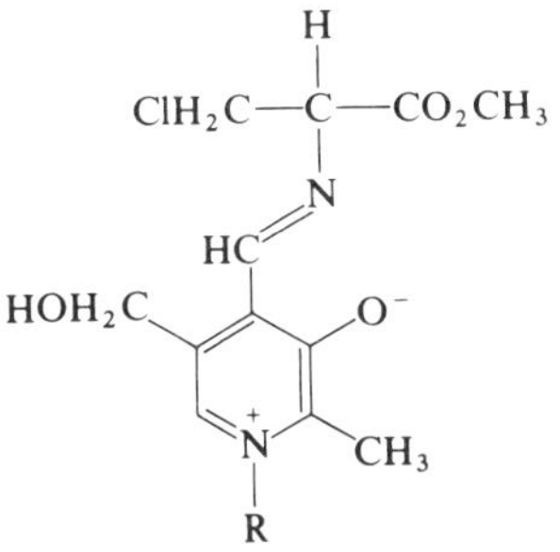

(59)

R = H, Me

(**59**) are reported to be nearly independent of pH in the pH range 6–10. As an explanation of this behaviour hydroxide and water attack on amino- and imino-protonated species would appear most satisfactory, although an alternative possibility has been suggested.[89] The equilibrium for Schiff base formation between phenylalanine and pyridoxal is increased 1000-fold by adding the *N*-substituent, $-CH_2CON([CH_2]_{15}Me)_2$ (R in (**59**)), and incorporating it in bilayer vesicles of peptide lipids; there is a corresponding increase in rate of transamination to pyridoxamine. Addition of Cu(II) perchlorate also accelerates reaction and leads to an observable azomethine anion–Cu(II) complex (λ_{max} = 523 nm).[90]

Azomethine ylids ($R{-}CH{=}\overset{+}{N}H{-}\overset{-}{C}HR$) have been generated by decarboxylation of amino-acid Schiff bases of 4-pyridine aldehyde and trapped (usually stereospecifically) with a variety of dipolarophiles.[91] Pyridine aldehyde may be replaced by less reactive aldehydes such as benzaldehyde, at the expense of more strenuous reaction conditions, *e.g.* benzaldehyde in refluxing DMF for 45 min.[92]

Other Topics

Unstable Unsaturated Species

The "Three-phase Test" has been applied to characterization of benzynes as discrete chemical intermediates.[93] The polymer-bound diaryl iodonium carboxylate (**60**) was heated and stirred at 190° for 24 h in diethylbenzene in the presence of the furoate ester (**61**), also polymer-bound, as trapping agent. After saponification of the polymeric products, 6- and 7-substituted naphthols (**62**) were obtained, consistent with initial formation of the Diels–Alder adducts (**63**). Transfer of a benzyne from one polymer to the other is implied, suggesting a life-time in solution of > *ca.* 0.15 s.

Treatment of the benzobistriazole (**64**) with lead tetraacetate in the presence of dienes leads to dibenzyne adducts under conditions tolerant of functional groups in the diene trap.[94] Treatment of the halobicycloheptenes (**65**) with $KOBu^t$ and KSPh in DMSO at 25° gives the substitution product (**66**), consistent with formation of the cyclobutyne (**67**) as an intermediate.[95] Similarly dehydrohalogenation by $KOBu^t$ followed by trapping with a diene affords evidence of formation of the bicyclic allene (**68**).[96] Trapping experiments also point to formation of the α-methoxy-*o*-

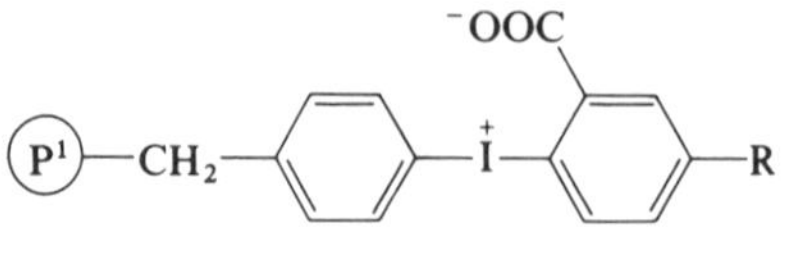

(60)  **(61)**

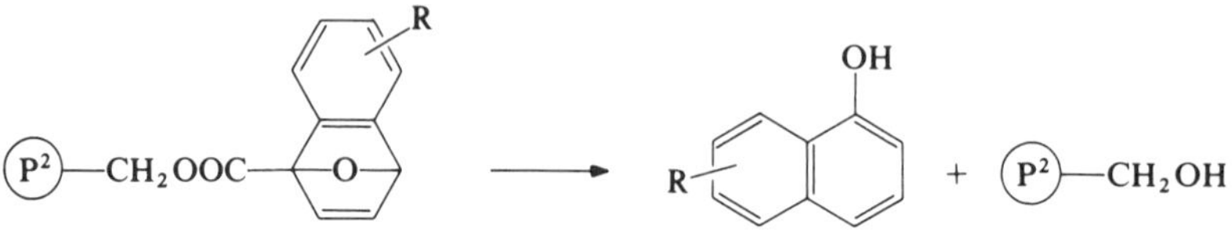

(62) **(63)**

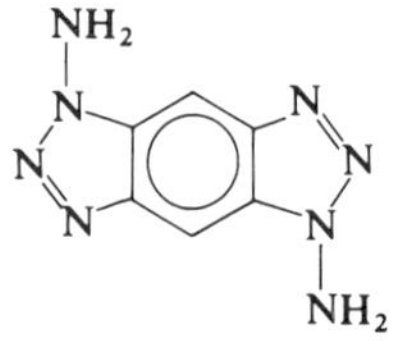

(64)

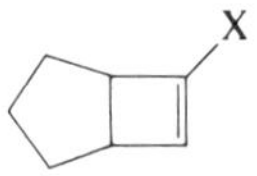

(**65**) X = Hal
(**66**) X = SPh

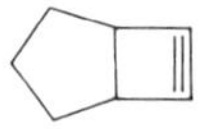

(**67**) 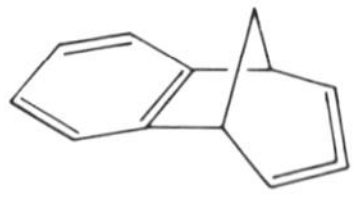 (**68**)

xylylene (**69**) from reaction of *o*-tolualdehyde dimethylacetal with $LiNR_2$. In the absence of a dienophile electrocyclic ring-closure to α-methoxycyclobutene (**70**) competes with dimerization.[97]

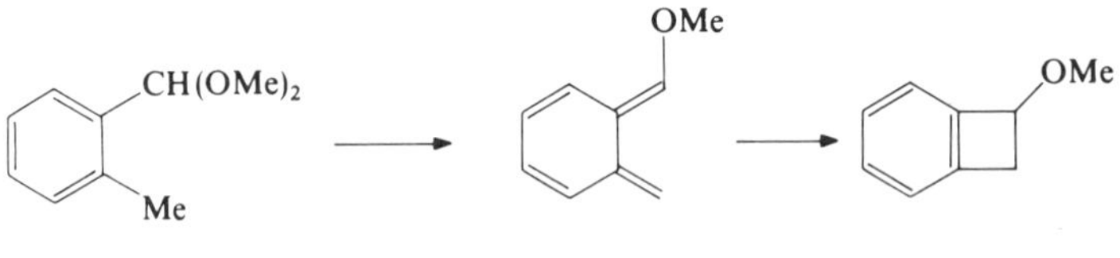

(**69**) (**70**)

Ion–Molecule Reactions

An Ion Cyclotron Resonance study of the reaction of alkoxide ions RO^- with alkyl pentafluorophenyl ethers, C_6F_5OR', shows that displacement of fluoride ion is followed by a second reaction of F^- within an ion–molecule product complex, leading to competitive elimination of aryloxide ion from the new and original alkoxy groups.[98] Deuterium labelling at α- and β-carbon atoms shows that the two alkoxy groups become equivalent, with $k_H/k_D = 1.6$.

Isotope effects have been measured for spontaneous or IR laser-induced elimination of H_2[99] and alkanes[100, 101] from alkoxide ions to yield enolate ions. For elimination of (**71**) primary effects k_H/k_D are 6.0, 2.5, and 1.6 for R = CF_3, Ph, and H, respectively. For the *tert*-butoxide ion an extraordinarily large secondary effect $k_{(CH_3)}/k_{(CD_3)} = 6.9$ indicates near-complete C—C bond-breaking in the transition state and a step-wise rather than a four-centre concerted reaction,[100] as implied also by *ab initio* calculations for elimination of H_2.[99] Substituent effects indicate heterolytic bond-cleavage and a carbanion (or H^-) molecule complex intermediate for leaving alkyl groups with appreciable electron affinity, but otherwise bond homolysis to a radical–molecule complex in which the additional electron is probably not specifically associated with either component of the complex.[101]

$$CH_3C(O^-)(CD_3)CH_2R \longrightarrow [CH_3C(=O)CD_3 + {}^-CH_2R] \longrightarrow CH_3C(O^-){=}CH_2 + RCH_2D$$

(**71**)

Phase-transfer and Micellar Catalysis

In an investigation of the mechanism of phase-transfer catalysis by hydroxide ion, dehydrobromination of 2-bromoethylbenzene in aqueous NaOH in the presence of tetraethylammonium bromide has been shown to be kinetically zero order in substrate and to have a (low) activation energy, consistent with rate-controlling diffusion of HO^- from the aqueous to organic phase.[102] For Hofmann elimination of tetrahexylammonium bromide in a chlorobenzene–aqueous two-phase system, rate constants have been shown to increase by 10^4 between 4M and 20M NaOH, reflecting the large increase in activity (normally described by an acidity function) of the hydroxide ion in the aqueous phase.[103] Ratios of Hofman to Saytseff olefin products have been measured in a facile elimination of 2-bromooctane by solid KOBu^t, NaOEt, NaOMe, and KOH in the presence of tetraalkylammonium salts.[104] Eliminations by solid KOH in the absence of phase-transfer catalyst have also been studied.[105]

An interesting two-phase reductive elimination of stilbene dibromide uses the alkylviologen *N*,*N*′-diacetyl-4,4′-bipyridinium dibromide C_8V^{2+} (**72**), as an electron carrier.[106] Reduction of C_8V^{2+} in an aqeuous phase by dithionate or glucose yields $C_8V^{+\cdot}$ which is extracted into an ethyl acetate organic phase where it disproportionates to C_8V and C_8V^{2+}. The C_8V reduces the stilbene dibromide and re-extraction of C_8V^{2+} completes the cycle. The dithionate may also be replaced by a photosensitizer and electron donor to give a cyclic photoreaction mediated by $C_8V^{+\cdot}$, corresponding to photosynthesis of stilbene through oxidation of the donor [$(NH_4)_3$ EDTA] by stilbene dibromide.

$$C_8H_{17}\text{—}{}^+N(C_5H_4)\text{—}(C_5H_4)N^+\text{—}C_8H_{17}$$

(**72**)

Elimination reactions of DDT and DDM in CTAOH[107] and CTAB[108] micelles at low concentrations of hydroxide ion have been described by a pseudo-phase model with kinetic contributions from water and micellar phases ($k_w[OH^-]_w[S]_w + k_m[OH^-]_m[S_m]$) showing an enhanced concentration and reactivity of substrate in the micellar phase. However, above $[OH^-] = 10^{-2}$ M an extra kinetic term is required, which is interpreted as a phase-transfer contribution arising from reaction of HO^- in the aqueous phase with substrate in the micelles.[107]

Ring-opening Elimination in Solution

In an elegant ^{31}P-NMR study of the stereochemistry of the Wittig reaction the *threo-* and *erythro-β*-hydroxyphosphonium salts (**73**) have been shown to give stereospecifically *trans-* and *cis*-oxaphosphetanes, respectively, when treated with sodium hexamethyldisilazide at −40° in THF.[109] Decomposition of the oxaphosphetanes to Z- and E-olefins is accompanied by *cis–trans* oxaphosphetane isomerization, probably *via* dissociation to ylid and benzaldehyde. Reaction of benzaldehyde and ylid directly gives predominantly *cis*-oxaphosphetane as initial, and presumably kinetically controlled, adduct. The related base-catalysed Peterson olefination of diastereomeric 8-methoxy-7-(trimethylsilyl)-6-tridecanols (**74**) shows stereospecifically *syn*-elimination consistent with reaction *via* a silanolate ion (**75**) analogous to the oxaphosphetane intermediates of the Wittig reaction.[110] However the presence of the vicinal methoxy group leads to competing elimination of methanol with formation of an unsaturated trimethylsilyl ether, which surprisingly appears in the more stable *trans*-configuration irrespective of whether this requires *syn-* or *anti*-elimination.

PhCH(OH)—CH(Ph)$\overset{+}{P}Ph_3$ Br^- ⟶ Ph_3P—O (Ph, Ph) ⟶ PhCH=O + Ph$\overset{-}{C}$H$\overset{+}{P}Ph_3$

(**73**)

AmCH(OH)—CH(SiMe$_3$)—CH(Am)(OMe) ⟶ (**75**)

(**74**)

Am = amyl

In an interesting ring-opening fragmentation, 1-chloro-1-fluoro-2-trimethylsilylmethylcyclopropanes (**76**) and (**78**) react with tetrabutylammonium chloride or fluoride in diglyme at 130° to afford the corresponding fluorodiene (**77**). The reaction conditions appear to preclude carbanion formation and this implies a concerted conrotatory ring-opening and elimination.[111] The notional "backside" displacement of chloride (as in **78**) occurs *ca.* 100-fold faster than "frontside" displacement (as in **76**).

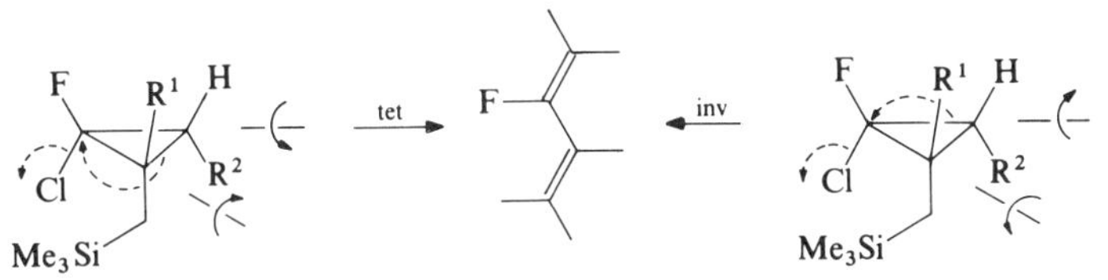

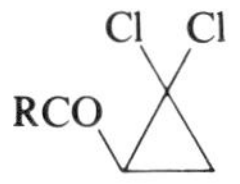

(76) (77) (78) (79)

A mechanism has been proposed for ring-opening of *gem*-dichloroketo-cyclopropanes to keto-esters (**79**) in NaOMe/MeOH.[112] Concerted ring-opening is suggested as the rate-determining step for oxidative conversion of *N*-arylsulphonyloxaziridines (Scheme 2) to arylsulphonylimines by iodide ions on the

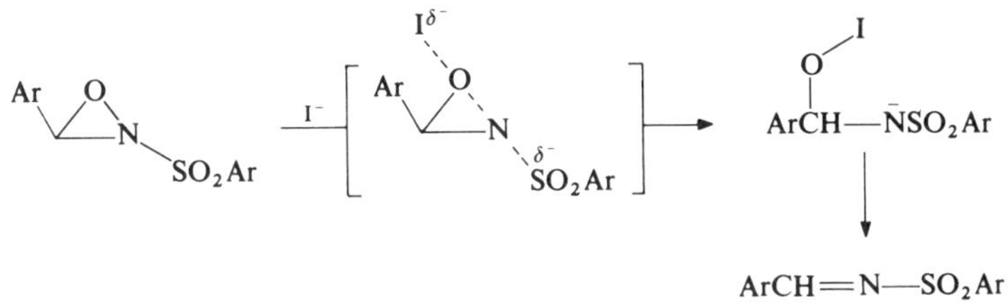

SCHEME 2

basis of ρ values of 0.81 and 0.99 for substitution in the ring and sulphonyl-bound aryl groups, respectively.[113] A ring-opening Hofmann-like elimination to imminium ion is observed as a principal reaction path of diazoniapropellane dications such as (**80**) in aqueous solution at 25° or in the presence of sodium bicarbonate.[114] The ion (**81**) is not hydrolysed, but may be trapped by strong nucleophiles such as CN^- or intramolecularly as in (**82**); related bridgehead ions, however, are hydrolysed.

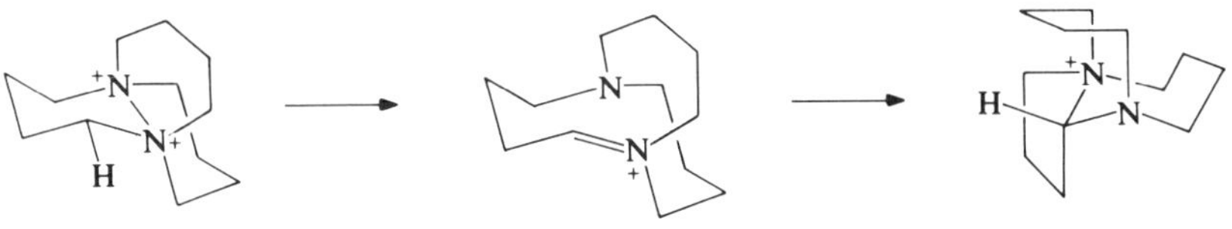

(80) (81) (82)

Other Eliminations

Effective molarities (*EM*) for substitution and elimination have been compared for intramolecular reactions of *o*-(ω-bromoalkoxy)phenoxides as a function of ring size *via* transition states (**83**) and (**84**).[115] For a six-membered transition state substitution is favoured with $EM_E/EM_S = 0.2$, but for medium rings EM_E/EM_S increases from 4 to 50 for $n = 7$–9, falling to 5 for large rings. It has been suggested that the requirement for a linear arrangement of the three atoms involved in proton transfer is responsible for low effective molarities of proton transfer relative to nucleophilic reactions in five- and six-ring transition states; this may be more

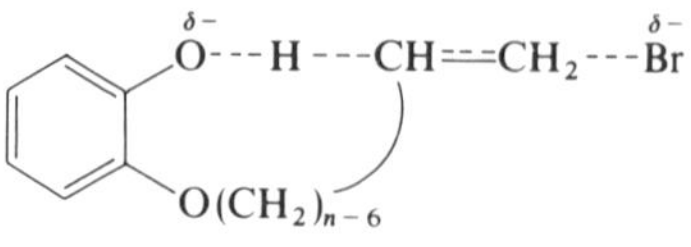

(83)

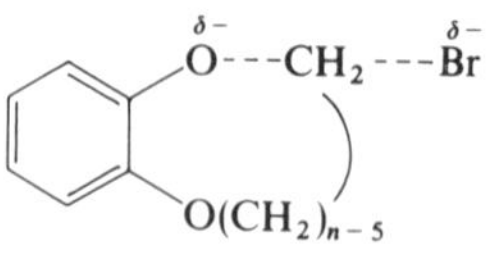

(84)

important than entropic factors associated with "looseness" of proton-transfer transition states.

Stirling has measured relative leaving-group abilities in 1,3-eliminations activated by a phenylsulphonyl group yielding the dimethylcyclopropanes (**85**).[116] He has

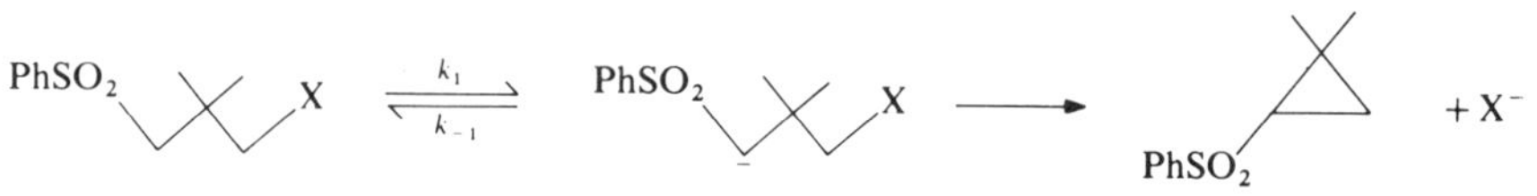

(85)

estimated a "rank" for each leaving group from the expression $\log k_{obs} - \log k_1 + 11$. This is an approximation for $\log k_2$ for expulsion of the leaving group from a carbanion intermediate based on $k_{obs} = k_1 k_2 / k_{-1}$, with $\log k_1$ determined from tritium exchange measurements and $\log k_{-1}$ assigned the value 11, consistent with unactivated protonation of the carbanion by solvent. Comparisons with $E2$ eliminations and S_N2 substitutions show a closer similarity to the latter and, by contrast with 1,2-eliminations, a surprisingly good correlation with leaving-group pK_a, possibly reflecting a product-like transition state with extensive leaving-group bond-breaking.

In a study of $E2$ eliminations the effects of β- and γ-perfluoroalkyl substituents on the rate, orientation, and E/Z-ratio of isomeric olefins for reaction of alkyl iodides in ethanolic NaOH have been investigated.[117] It has been shown that dehydrohalogenation following attack of a halocarbanion on an electrophilic olefin can simulate nucleophilic vinylic displacement of hydrogen (Scheme 3). The behaviour parallels a

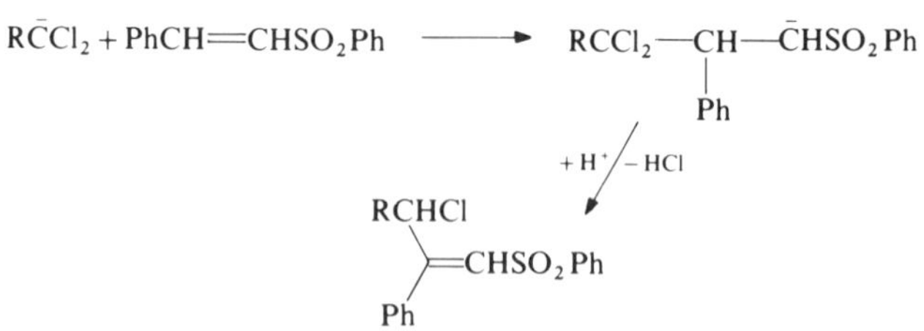

SCHEME 3

similar reaction with nitro- and hetero-aromatic substrates.[118] Dehydrochlorination of trichloroethane has been investigated, by semi-empirical MO calculations, in relation to heterogeneous catalysis by acidic and basic surface sites.[119] There have been a number of experimental studies of olefin-forming dehalogenations,[28, 34, 106] including investigation of an isokinetic relationship for reaction of chalcone dibromides with iodide ion.[120] Kinetics of decomposition of *N*-bromoalanine to acetaldehyde and acetonitrile have been reported.[121]

Eliminations of stannanes represent "electrophilic" eliminations in which the leaving group is a metal and hydrogen is lost as a hydride ion. Reactions of $(RCH_2CH_2)_4Sn$ with the trityltetrafluoroborate $(4\text{-}MeC_6H_4)_3C^+BF_4^-$ as hydride-abstracting agent yield $RCH{=}CH_2$ with rate constants in the order R = Me > Et > H > Pr^i > Bu^t: the mechanism is designated E_E2.[122] With chloroalkylstannanes elimination occurs intramolecularly in the presence of $TiCl_4$ to generate a carbocation centre.[123] For **(86)** electrophilic elimination and substitution compete to give a mixture of cyclic and unsaturated products. Protonation of the metal complexes $Co(\eta\text{-}C_5Me_5)C_2H_4PR_3$ has been suggested as yielding structural analogues of transition states for electrophilic elimination.[124] X-Ray and NMR data show the additional hydrogen as hydride in character, with strong interactions with both the cobalt and an ethylenic carbon atom.

$Me_3Sn\text{–}CH_2CH_2\text{–}(CH_2)_3\text{–}CHCl\text{–}CH_2SPh \xrightarrow{TiCl_4} CH_2{=}CH\text{–}(CH_2)_3\text{–}CH_2CH_2SPh$ + (SPh-substituted cyclopentane)

(86)

$$\begin{matrix} LnM\text{—}R \\ | \\ LnM\text{—}R \end{matrix} \rightleftharpoons \begin{matrix} LnM \\ \| \\ LnM \end{matrix} + \begin{matrix} R \\ | \\ R \end{matrix}$$

(87)

There has been a theoretical discussion of reductive eliminations yielding hydrogen or alkanes and metal–metal double bonds in dinuclear metal complexes **(87)**.[125] The extreme rarity of these reactions compared with the corresponding eliminations from a single metal centre has been ascribed to the stereochemical symmetry constraint for a four-centre concerted reaction.

The base dependence of elimination of bromide and metaphosphate (PO_3^-) ions from 2-bromophosphonate anions to substituted chalcones

$$XC_6H_4CH(PO_3^{2-})CHBrCOC_6H_5 \rightarrow XC_6H_4CH{=}CHCOC_6H_5$$

implicates the phosphonate dianions as reactive species.[126, 127] The reaction appears to occur in a single step with evidence from substituent effects of conjugation of aryl and keto groups through an incipient double bond in the transition state. Interestingly, absence of an isotope effect in CH_3OD and lack of phosphonamide products from added amines implies formation of a "free" metaphosphate ion with a finite life-time. This contrasts with previous studies of metaphosphate transfer, which have shown bonding to, or preassociation of, an external nucleophile.

References

[1] Baciochi, E., in *The Chemistry of Halides, Pseudo-halides and Azides* (Eds. Patai, S., and Rappoport, Z.), Wiley, Chichester, 1983, p. 1173.

[2] Mayer, B. J., Spencer, T. A., and Onan, K. D., *J. Am. Chem. Soc.*, **106**, 6343 (1984).
[3] Mayer, B. J., and Spencer, T. A., *J. Am. Chem. Soc.*, **106**, 6349 (1984).
[4] Richard, J. P., *J. Am. Chem. Soc.*, **106**, 4926 (1984).
[5] Thibblin, A., *Chem. Scr.*, **22**, 182 (1983).
[6] Thibblin, A., *J. Chem. Soc., Chem. Commun.*, **1984**, 92.
[7] Thibblin, A., *J. Am. Chem. Soc.*, **106**, 183 (1984).
[8] Koch, H. F., *Acc. Chem. Res.*, **17**, 137 (1984).
[9] Lee, H.-K., Lambert, H. J., and Schowen, R. L., *J. Pharm. Sci.*, **73**, 306 (1984).
[10] Cabaleiro, M. C., and Garay, R. O., *J. Chem. Soc., Perkin Trans. 2*, **1984**, 179.
[11] Larkin, F., and More O'Ferrall, R. A., *Aust. J. Chem.*, **36**, 1831 (1983).
[12] Kurov, G. N., Svyatkina, L. I., and Pal'chuk, E. G., *Izv. Akad. Nauk SSSR, Ser. Khim.*, **1983**, 2227; *Chem. Abs.*, **100**, 33910 (1984).
[13] Hupe, D. J., *J. Am. Chem. Soc.*, **106**, 1381 (1984).
[14] Al Sabbagh, M. M., Calmon, M., and Calmon, J.-P., *J. Chem. Soc., Perkin Trans. 2*, **1984**, 1233.
[15] Stein, G., Kammerer, H., and Bohmer, V., *J. Chem. Soc., Perkin Trans. 2*, **1984**, 1285.
[16] de la Mare, P. B. D., and Newman, P. A., *J. Chem Soc., Perkin Trans. 2*, **1984**, 231.
[17] Silva e Lins, H., Nome, F., Rezende, M. C., and de Sousa, I., *J. Chem. Soc., Perkin Trans. 2*, **1984**, 1521.
[18] Gandler, J. R., and Yokoyama, T., *J. Am. Chem. Soc.*, **106**, 130 (1984).
[19] Baciocchi, E., and Ruzziconi, R., *J. Org. Chem.*, **49**, 3395 (1984).
[20] Jarczewski, A., Schroeder, G., and Leffek, K. T., *Pol. J. Chem.*, **56**, 521 (1982); *Chem. Abs.*, **100**, 85004 (1984).
[21] Dewar, M. J. S., *J. Am. Chem. Soc.*, **106**, 209 (1984).
[22] Katritzky, A. R., Dega-Szafran, Z., Lopez-Rodriguez, M. L., and King, R. W., *J. Am. Chem. Soc.*, **106**, 5577 (1984).
[23] Richard, J. P., and Jencks, W. P., *J. Am. Chem. Soc.*, **106**, 1373 (1984).
[24] Coxon, J. M., Simpson, G. W., Steel, P. J., and Whiteling, S. C., *Tetrahedron*, **40**, 3503 (1984).
[25] Smith, R. H., Denlinger, C. L., Kupper, R., Koepke, S. R., and Michejda, C. J., *J. Am. Chem. Soc.*, **106**, 1056 (1984).
[26] Stolarova, M., and Bekarek, V., *Acta Univ. Palacki. Olomuc., Fac. Rerum Nat.*, **76**, 131 (1983); *Chem. Abs.*, **100**, 138300 (1984).
[27] Muir, D. M., and Parker, A. J., *Aust. J. Chem.*, **36**, 1667 (1983).
[28] Avraamides, J., and Parker, A. J., *Aust. J. Chem.*, **36**, 1705 (1983).
[29] McLennan, D. J., and Lim, G.-C., *Aust. J. Chem.*, **36**, 1821 (1983).
[30] Kwart, H., and Gaffney, A., *J. Org. Chem.*, **48**, 4502 (1983).
[31] Kwart, H., Gaffney, A., and Wilk, K. A., *J. Org. Chem.*, **48**, 4509 (1983).
[32] Kwart, H., Gaffney, A. H., and Wilk, K. A., *J. Chem. Soc., Perkin Trans. 2*, **1984**, 565.
[33] Aubert, C., Bégué, J.-P., and Biellman, J.-F., *J. Chem. Soc., Chem. Commun.*, **1984**, 351.
[34] Pánková, M., Kocián, O., Kupřička, J., and Závada, J., *Collect. Czech Chem. Commun.*, **48**, 2944 (1983).
[35] Chen, N., and Zhou, B., *Youji Huaxue*, **1983**, 257; *Chem. Abs.*, **99**, 211825 (1983).
[36] Hegarty, A. F., and Mullane, M., *J. Chem. Soc., Chem. Commun.*, **1984**, 229.
[37] Kayser, R. H., Brault, M., Pollack, R. M., Bantia, S., and Sadoff, S. F., *J. Org. Chem.*, **48**, 4497 (1983).
[38] Jordon, E. A., and Thorne, M. P., *J. Chem Soc., Perkin Trans. 2*, **1984**, 647.
[39] Louw, R., Tinkelenberg, A., and Werner, E. S. E., *Recl. Trav. Chim. Pays-Bas*, **102**, 519 (1983).
[40] Rotinov, A., and Chuchani, G., *Int. J. Chem. Kinet.*, **16**, 1267 (1984).
[41] Triana, J. L., and Chuchani, G., *React. Kinet. Catal. Lett.*, **22**, 441 (1983); *Chem. Abs.*, **100**, 67582 (1984).
[42] Arcila, C., Pino, M., Lubinkowski, J., Hernandez, A., and Chuchani, G., *Int. J. Chem. Kinet.*, **16**, 57 (1984).
[43] Cordova, C. M., Martin, L., and Chuchani, G., *React. Kinet. Lett.*, **24**, 243 (1984).
[44] Konakahara, T., Sato, K., Takagi, Y., and Kuwata, K., *J. Chem. Soc., Perkin Trans. 2*, **1984**, 641.
[45] Takeshita, H., and Mametsuka, H., *Can. J. Chem.*, **62**, 2035 (1984).
[46] Holmes, A. B., Raithby, P. R., Russell, K., Stern, E. S., Stubbs, M. E., and Wellard, N. K., *J. Chem. Soc., Chem. Commun.*, **1984**, 1191.
[47] Valente, C., and Eadon. G., *J. Org. Chem.*, **49**, 44 (1984).
[48] Reshetnikov, S. M., Morozova, Z. G., Korvinskaya, L. N., and Skopin, V. V., *Tepl. Protsessy Sroistva Rab. Tel Dvigatelei Letatel'nykh Appar.*, **1982**, 68; *Chem. Abs.*, **99**, 175093 (1983).
[49] Avakayan, V. G., Litmanovich, A. D., and Cherkezgan, V. O., *Izv. Akad. Nauk SSSR, Ser. Khim.*, **1984**, 329; *Chem. Abs.*, **100**, 191192 (1984).

[50] Maujean, A., Grosdemange-Pale, C., Marcy, G., and Chuche, J., *J. Chem. Soc., Chem. Commun.*, **1984,** 1135.
[51] Ghibandi, E., and Colussi, A. J., *J. Chem. Soc., Chem. Commun.*, **1984,** 433.
[52] Bartók, M., Molnár, A., and Bartok-Bozóki, G., *Acta Chim. Acad. Sci. Hung.* **114,** 375 (1983); *Chem. Abs.*, **100,** 138433 (1984).
[53] Brown, R. T., and Jones, M. F., *J. Chem. Res. (S)*, **1984,** 332.
[54] Chamchaang, W., Chantarasiri, N., Chaona, S., Thebtaranonth, C., and Thebtaranonth, Y., *Tetrahedron,* **40,** 1727 (1984).
[55] Prager, R. H., and Schiesser, C. H., *Tetrahedron,* **40,** 1517 (1984).
[56] Mackie, J. C., and Doolan, K. R., *Int. J. Chem. Kinet.*, **16,** 525 (1984).
[57] Dallinga, J. W., Nibbering, N. M. M., and Boerboom, A. J. H., *J. Chem. Soc., Perkin Trans. 2,* **1984,** 1065.
[58] Engel, P. S., Nalepa, C. J., Horsey, D. W., Keys, D. E., and Grow, R. T., *J. Am. Chem. Soc.*, **105,** 7102 (1983).
[59] Engel, P. S., Horsey, D. W. Keys, D. E., Nalepa, C. J., and Soltero, L. R., *J. Am. Chem. Soc.*, **105,** 7108 (1983).
[60] Chang, M. H., Jain, R., and Dougherty, D. A., *J. Am. Chem. Soc.*, **106,** 4211 (1984).
[61] Bock, H., Dammel, R., and Aygan, S., *J. Am. Chem. Soc.*, **105,** 7681 (1983).
[62] Weissman, M., and Benson, S. W., *Int. J. Chem. Kinet.*, **16,** 941 (1984).
[63] Chuchani, G., Martin, I., Rotinov, A., Hernandez, A. J. A., and Reikonnen, N., *J. Phys. Chem.*, **88,** 1563 (1984).
[64] Chuchani, G., and Dominguez, R. M., *Int. J. Chem. Kinet.*, **15,** 1275 (1983).
[65] Wolfrum, J., and Schneider, M., *Proc. SPIE-Int. Soc. Opt. Eng.*, **458,** 46 (1984); *Chem. Abs.*, **101,** 11051 (1984).
[66] Ishikawa, Y.-I., and Arai, S., *Chem. Phys. Lett.*, **103,** 68 (1983).
[67] Kim, H. J., and Choo, K. Y., *Bull. Korean Chem. Soc.*, **4,** 203 (1983); *Chem. Abs.*, **100,** 191180 (1984).
[68] Kaupp, G., *Chem. Ber.*, **117,** 1643 (1984).
[69] Amano, A., Yamada, M., Kamio, T., Takahashi, Y., and Tang, J., *Kenk Hokoku-Asahi Garasu Kogyo Gijutsu Shoreikai,* **43,** 101 (1983); *Chem. Abs.*, **100,** 174112 (1984).
[70] Martin, G., Lugo, N., Ropero, M., and Martinez, H., *Phosphorus Sulfur,* **17,** 47 (1983); *Chem. Abs.*, **100,** 102590 (1984).
[71] Barton, T. J., Burns, S. A., Davidson, I. M. T., Ijadi-Maghsoodi, S., and Wood, I. T., *J. Am. Chem. Soc.*, **106,** 6367 (1984).
[72] Sekiguchi, A., and Ando, W., *J. Am. Chem. Soc.*, **106,** 1486 (1984).
[73] Sawrey, B. A., O'Neal, H. E., Ring, M. A., and Coffey, D., *Int. J. Chem. Kinet.*, **16,** 7 (1984).
[74] Ishikawa, Y., and Shigeyoshi, A., *Chem. Phys. Lett.*, **103,** 68 (1983); *Reza Kagaku Kenkya,* **5,** 121 1983; *Chem. Abs.*, **100,** 138449 (1984).
[75] Martin, H.-D., Urbanek, T., Braun, R., and Walsh, R., *Int. J. Chem. Kinet.*, **16,** 117 (1984).
[76] Paquette, L. A., Jendralla, H., and DeLucca, G., *J. Am. Chem. Soc.*, **106,** 1518 (1984).
[77] Dolbier, W. R., and Koroniak, H., *J. Am. Chem. Soc.*, **106,** 1871 (1984).
[78] Dolbier, W. R., and Dheya, M. A.-F., *J. Am. Chem. Soc.*, **105,** 6349 (1983).
[79] Barton, J. W., and Shepherd, M. K., *Tetrahedron Lett.*, **25,** 4967 (1984).
[80] Lewars, E. G., *Chem. Rev.*, **83,** 519 (1983).
[81] Martinez, R. I., and Herron, J. T., *Int. J. Chem. Kinet.*, **15,** 1127 (1983).
[82] Yoshida, M. Miura, M., Nojima, M., and Kusabayashi, S., *J. Am. Chem. Soc.*, **105,** 6279 (1983).
[83] Knowles, J. R., *Pure Appl. Chem.*, **56,** 1005 (1984).
[84] Grimshaw, C. E., Sogo, S. G., Copley, S. D., and Knowles, J. R., *J. Am. Chem. Soc.*, **106,** 2699 (1983).
[85] Lee, J. J., Asano, Y., Shieh, T.-L., Spreafico, F., Kyungok, L., and Floss, H. G., *J. Am. Chem. Soc.*, **106,** 3367 (1984).
[86] Anderson, V. E., and Hammes, G. G., *Biochemistry,* **23,** 2088 (1984).
[87] Yan, S.-J., McKinnie, B. G., Abacherli, C., Hill, R. K., and Babior, B. M., *J. Am. Chem. Soc.*, **106,** 2961 (1984).
[88] Spratt, T. E., and Kaiser, E. T., *J. Am. Chem. Soc.*, **106,** 6440 (1984).
[89] Kondo, H., Morita, K., Kitamikado, T., and Sunamoto, J., *Bull. Chem. Soc. Jpn.*, **57,** 1031 (1984).
[90] Murakami, Y., Kikuchi, J.-I., Imori, T., and Akiyoshi, K., *J. Chem. Soc., Chem. Commun.*, **1984,** 1434.
[91] Grigg, R., and Thianpatanagul, S., *J. Chem. Soc., Chem. Commun.*, **1984,** 180.
[92] Grigg, R., Aly, M. F., Sridaharan, V., and Thianpatanagul, S., *J. Chem. Soc., Chem. Commun.*, **1984,** 182.

[93] Gavina, F., Luis, S. V., Gil, P., and Costero, A. M., *Tetrahedron Lett.*, **25,** 779 (1984).
[94] Hart, H., and Ok, D., *Tetrahedron Lett.*, **25,** 2073 (1984).
[95] Baumgart, K.-D., and Szeimies, G., *Tetrahedron Lett.*, **25,** 737 (1984).
[96] Balci, M., and Harmandar, M., *Tetrahedron Lett.*, **25,** 237 (1984).
[97] Randall, J. M., and Rickborn, B., *J. Org. Chem.*, **49,** 3694 (1984).
[98] Ingemann, S., and Nibbering, N. M. M., *Nouv. J. Chim.*, **8,** 299 (1984).
[99] Hayes, R. N., Sheldon, J. C., Bowie, J. H., and Lewis, D. E., *J. Chem. Soc., Chem. Commun.*, **1984,** 1431.
[100] Tumas, W., Foster, R. F., Pellerite, M. J., and Brauman, J. I., *J. Am. Chem. Soc.*, **105,** 7464 (1983).
[101] Tumas, W., Foster, R. F., and Brauman, J. I., *J. Am. Chem. Soc.*, **106,** 4053 (1984).
[102] Halpern, M., Sasson, Y., and Rabinovitz, M., *J. Org. Chem.*, **49,** 2011 (1984).
[103] Landini, D., and Maia, A., *J. Chem. Soc., Chem. Commun.*, **1984,** 1041.
[104] Barry, J., Bram, G., Decodts, G., Loupy, A., Pigeon, P., and Sansoulet, J., *J. Org. Chem.*, **49,** 1138 (1984).
[105] D'Incan, E., and Viout, P., *Tetrahedron*, **40,** 3415 (1984).
[106] Goren, Z., and Itamar, W., *J. Am. Chem. Soc.*, **105,** 7764 (1983).
[107] Ionescu, L. G., and Nome, F., *Surfactants Solution (Proc. Int. Symp.)* **2,** 1107 (1984); *Chem. Abs.*, **101,** 129891 (1984).
[108] Stadler, E., Zanette, D., Rezende, M. C., and Nome, F., *J. Phys. Chem.*, **88,** 1892 (1984).
[109] Reitz, A. B., Mutter, M. S., and Maryanoff, B. E., *J. Am. Chem. Soc.*, **106,** 1873 (1984).
[110] Yamamoto, K., Kimura, T., and Tomo, Y., *Tetrahedron Lett.*, **25,** 2155 (1984).
[111] Schlosser, M., Dahan, R., and Cottens, S., *Helv. Chim. Acta*, **67,** 284 (1984).
[112] Banwell, M. G., *J. Chem. Soc., Chem. Commun.*, **1983,** 1453.
[113] Knipe, A. C., McAuley, I. E., Khandelwal, Y., and Kansal, V. K., *Tetrahedron Lett.*, **25,** 1411 (1984).
[114] Alder, R. W., Sessions, R. B., Gmünder, J. O., and Grob, C. A., *J. Chem. Soc., Perkin Trans. 2*, **1984,** 411.
[115] Cort, A. D., Mandolini, L., and Masci, B., *J. Org. Chem.*, **48,** 3979 (1983).
[116] Issari, B., and Stirling, C. J. M., *J. Chem. Soc., Perkin Trans. 2*, **1984,** 1043.
[117] Brace, N. O., Marshall, L. W., Pinson, C. J., and van Wingerden, G., *J. Org. Chem.*, **49,** 2361 (1984).
[118] Mąkosza, M., and Kwast, A., *J. Chem. Soc., Chem. Commun.*, **1984,** 1195.
[119] Bolotin, A. B., Shatkovskaya, D. B., and Gineityte, V. L., *Chem. Phys. Lett.*, **107,** 431 (1984); *Chem. Abs.*, **101,** 89963 (1984).
[120] Raghavan, R. S., Velayutham, K., and Rajendran, V., *Indian J. Chem.*, **23A,** 466 (1984); *Chem. Abs.*, **101,** 129925 (1984).
[121] Stanbro, W. D., and Lenkevich, M. J., *Int. J. Chem. Kinet.*, **15,** 1321 (1983).
[122] Uglova, E. V., Mikhura, I. V., and Reutov, O. A., *Izv. Akad. Nauk SSSR, Ser. Khim.* **1983,** 2632; *Chem. Abs.*, **100,** 120217 (1984).
[123] Murayama, E., Uematsu, M., Nishio, H., and Sato, T., *Tetrahedron Lett.*, **25,** 313 (1984).
[124] Cracknell, R. B., Orpen, A. G., and Spencer, J. L., *J. Chem. Soc., Chem. Commun.*, **1984,** 326.
[125] Trinquier, G., and Hoffman, R., *Organometallics*, **3,** 370 (1984).
[126] Calvo, K. C., and Westheimer, F. H., *J. Am. Chem. Soc.*, **106,** 4205 (1984).
[127] Calvo, K. C., and Berg, J. M., *J. Am. Chem. Soc.*, **106,** 4202 (1984).

Organic Reaction Mechanisms 1984
Edited by A. C. Knipe and W. E. Watts

CHAPTER 13

Addition Reactions: Polar Addition

D. C. Billington

Merck Sharp and Dohme Research Laboratories, Terlings Park, Eastwick Road Harlow, Essex

During the coverage period of this chapter, reviews have appeared on the following topics: polar reactions,[1] reactions of alkenes with electrophiles,[2] palladium-catalysed oxidation of dienes,[3] allene chemistry,[4] non-catalytic polar additions to α,β-unsaturated carbonyl compounds,[5] theoretical studies of alkene addition stereoselectivity,[6] the addition of organometallic reagents to chiral vinyl sulphoxides,[7] the homolytic addition of acetals to alkenes,[8] and stereo-differentiating addition reactions.[9,10] In addition to these articles, two very comprehensive reviews have appeared, covering alkene cyclizations which form new C—C bonds,[11] and the chemistry of halonium ions.[12]

Electrophilic Additions

The hypothesis that remote hetero-atom participation occurs in some classes of polar additions has been re-examined using propellanes as substrates.[13] Substituent–substituent interactions have been quantitatively investigated by using the interactive free-energy relationship (IFER), and this approach is found to give valid results regardless of the reactivity–selectivity dependence observed.[14] The course of ionic additions to *syn*- and *anti*-sesquinorbornene, **(1)** and **(2)**, has been examined.[15] The application of a new stereochemical notation to a number of addition reactions has been described.[16]

Halogen and Related Additions

Evidence that strongly supports the *reversible* formation of bromonium ions during the reaction of Br_2 with alkenes has been presented.[17] The product distribution obtained on bromination of cyclohexene in surfactant solutions is dependent on the order of combination of the reactants.[18] There is evidence for the formation of a discrete carbocation β to silicon being involved in the bromination/rearrangement of (3).[19] The bromination of 5,8-diacetoxy-1,4-dihydro-1,4-ethanonaphthalene (4)

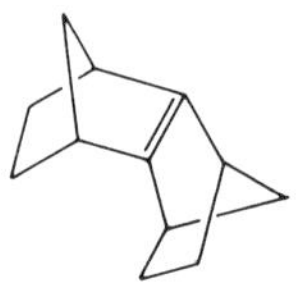

(1) 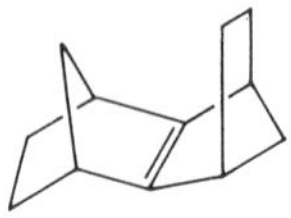 (2) 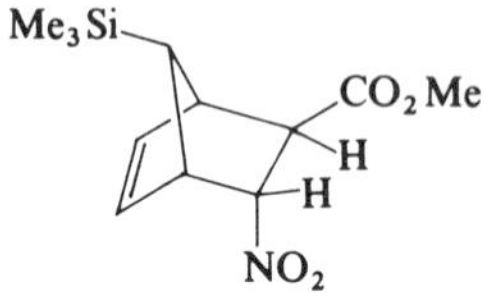 (3)

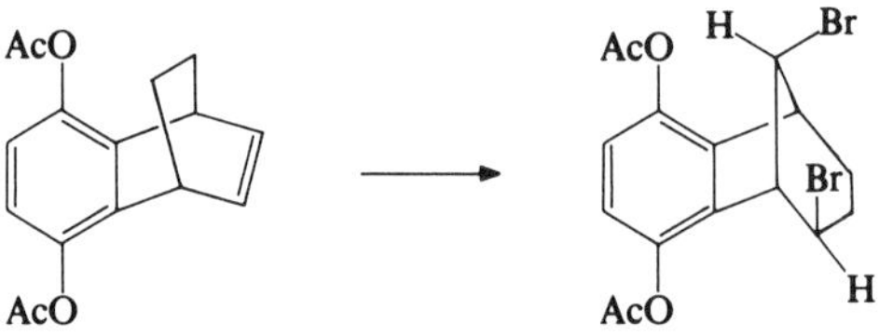

(4)

gives a single product, *via* a Wagner–Meerwein rearrangement with accompanying aryl migration.[20] A non-linear least-squares method for obtaining reliable values for second- and third-order rate constants for the electrophilic bromination of alkenes has been described.[21] Evidence has been presented for the intramolecular rearrangement of a bromonium ion to a thiiranium ion during the bromination of methyl allyl sulphide (5), which results in the formation of (6) (under kinetic control).[22]

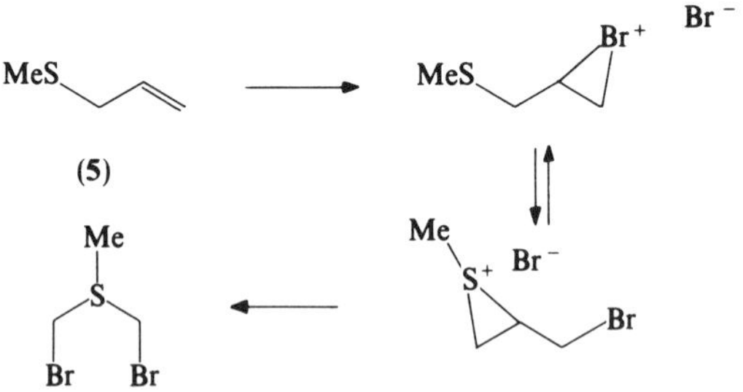

(5)

(6)

The addition of Br_2 to a wide range of *N*-arylimides (7) gives *trans*-diaxial dibromides[23] ($R = H$; $R^1 = Me$; $R^2 =$ 4-OH, 4-OMe, 4-Me, 3-Me, 2-Me, 3-OMe, 4-Br, 4-NO_2; and also $R = Me$, $R^1 = H$, $R^2 =$ as above). The rate constants and activation parameters have been determined for this range of reactions[24] and the effect of ring substituents on the rate of bromination of these compounds in CH_3CN has been reported.[25] The addition of Br_2 and hexachlorocyclopentadiene to a series of compounds (8) shows Br_2 to be more reactive by a factor of *ca.* 10^9, but the two reactions have very similar Hammett ρ constants, suggesting that they proceed *via* rather similar transition states.[26] 2-Chloro-*endo*-benzocyclobutenonorbornene (**9**) gives a *single* product (**10**) on reaction with NBS, and a mechanism has been proposed to account for this, which involves the initial formation of a bromonium ion.[27] The kinetics of bromination and protonation of a range of substituted α-methoxystyrenes has been studied in depth.[28] The contribution of the transition state to the variation in ρ for a series of carbenium ion-forming brominations has been determined.[29] Nucleophilic solvation occurs during the rate-determining step in the bromination of arylalkenes.[30,31] An LFER of σ^* *vs.* the degree of alkoxychlorination of $CH_3CH{=}CH_2OMe$ (R = 13 different groups) has $\rho^* = 0.138$ in CH_3OH.[32] The kinetics of the chlorination (in the dark) of $CH_2{=}CHR$ (R = 12 different groups) relative to the case where R = H, gives a ρ^* of -1.69.[33] Isokinetic relationships have been described and analysed for the chlorination of $RCH_2CH{=}CH_2$ (R = PrO, EtO, CH_3O, AcO, Br, Cl) and 3,3-*bis*(chloromethyl)oxetane.[34]

(**7**)

(**8**)

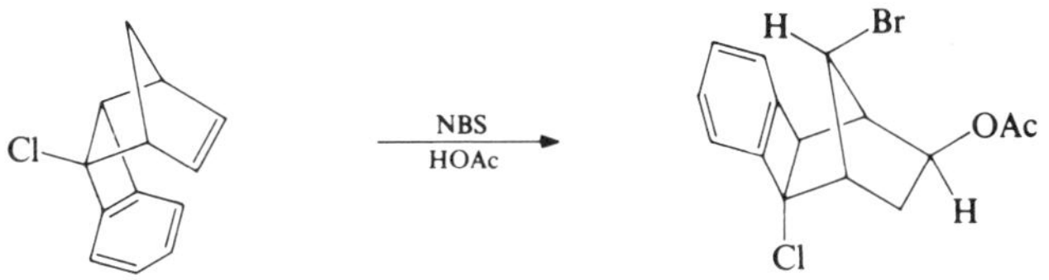

NBS / HOAc

(**9**) (**10**)

Allylic alcohols react with aqueous iodine and acetyl hypoiodite to give iodo-diols and acetoxyiodo-diols with high regio- and stereo-specificity.[35] 1,3-*trans*-Asymmetric induction is observed in the halolactonization of α- and α,β-substituted γ,δ-unsaturated amides (**11**), giving up to 99:1 (**12**):(**13**).[36] *Ab initio* MO calculations with pseudo-potential approximations, for the addition of I_2 to ethylene,

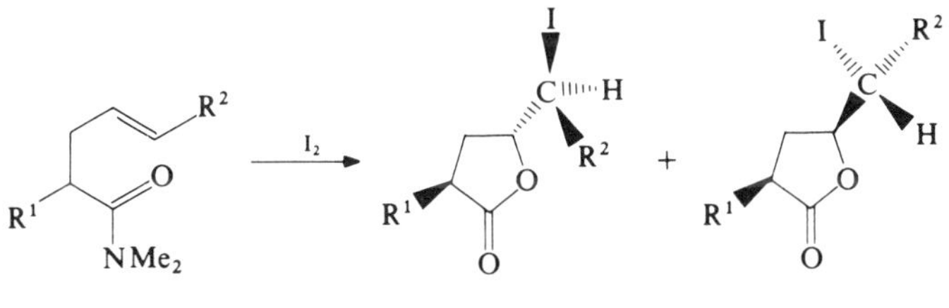

I_2

(**11**) (**12**) (**13**)

indicate that the only possible intermediate is a cyclic iodonium ion,[37] in accord with experimental observations.

Bromo- and chloro-fluorination of 1-phenyl-4-*tert*-butylcyclohexene with NBS or NCS and HF in pyridine gives Markownikoff-type regioselectivity and proceeds in a stereospecifically *anti*-sense.[38] In the addition of ICl and IBr to 1-methyl-4-*tert*-butylcyclohexene, the stereoselectivity is dependent on both the reversible electrophilic first step, and on the nucleophilic second step. The higher stereoselectivity of IBr is rationalized[39] on the basis of Br^- being a softer nucleophile than Cl^-. Bromofluorination of a series of compounds $YC_6H_4CX{=}CHCO_2Et$ where X and Y may be the same or different and consist of H, NO_2, OMe, and Cl, as both E- and Z-isomers, gives Markownikoff-type products.[40] Chlorine monofluoride adds non-regiospecifically and non-stereospecifically to E- and Z-1-bromo-2-chloroethylene and 1,2-dibromoethylene.[41] The addition of ICl to alkenes in nitrobenzene solution has been proposed to involve formation of a π-complex by fast equilibration of the alkene with ICl as the first step.[42] Kinetic analysis of the addition of ICl to vinyl compounds shows that the reaction is second order in ICl, and first order in substrate, and a mechanism has been proposed to account for this, involving formation of a π-complex between ICl and the double bond as a first step, followed by rate-determining reaction of this π-complex with a second molecule of ICl.[43] Addition of HOBr to steroidal dienes proceeds *via* a multi-step mechanism, involving formation of a bromohydrin, rearrangement of this bromohydrin to a bromonium ion, followed by intramolecular cleavage of this ion by the suitably placed hydroxyl function.[44] Nineteen substituted 2,3-unsaturated cholestanes, *e.g.* (**14**), react with HOBr to give cyclic bromo-ethers as sole or major products, *via* cleavage of the 2α, 3α-bromonium ion with $6(O)^n$ participation of the 19*a*-hydroxyl group.[45] A series of papers has appeared analysing the possible $5(O)^n$ *vs.* $6(O)^{\pi,n}$ participation of oxygen as a neighbouring-group in this type of reaction and orders of reactivity for participation are presented.[46–49]

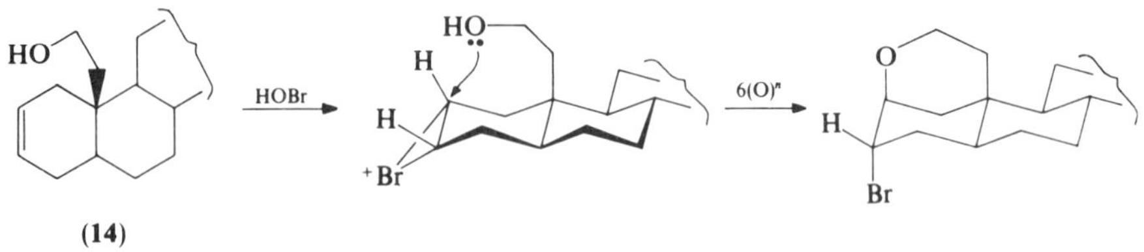

(**14**)

LFER plots for the addition reactions of $RC_6H_4CH{=}CH_2$ with $4\text{-}R^1C_6H_4SCl$ (R = 4-OMe, 4-Me, 3-NO_2, H; R^1 = H, Me, NO_2) contain three straight lines which intersect at $\sigma^{R^+} = 0$.[50] These results may be rationalized in terms of a transition state which varies from zwitterionic to a cyclic sulphonium species.[50] The conclusions of Dalipi and Schmiele[51] concerning the absence of a salt effect in the addition reactions of 2,4-dinitrobenzenesulphenyl chloride with alkenes have been disputed.[52] The carbon isotope effect for α- and β-substitution for ^{13}C *vs.* ^{14}C has been measured for the addition of 2,4-dinitrobenzenesulphenyl chloride to a series of *para*-substituted α- and β-^{14}C-labelled styrenes,[53] and values of 1.0027 to 1.004 for α- and 1.035 to 1.037 for β-labelling were obtained.[53] The addition of benzenesulphenyl chloride to α-methyl- and β-methyl-crotonates and -cinnamates proceeds regiospecifically in the presence of pyridine, and an onium ion has been proposed as an intermediate.[54] A series of arenesulphenyl halides has been shown to add stereospecifically to E- and Z-deuterostyrenes.[55]

Studies on the addition of areneselenyl chlorides to phenylpropenes lead to the conclusion that Se—C bond-making lags behind Se—Cl bond-breaking in the rate-limiting formation of the transition state.[56] Kinetics indicate that the addition of benzenesulphinic acids to *N*-phenylmaleimide proceeds *via* rate-limiting attack of the sulphenic acid sulphur atom on the alkene double bond.[57] 4-Phenyl-1,2,4-triazolinediones, *e.g.* (**15**), add to a range of alkynes giving 1 : 1 and 2 : 1 adducts.[58] These reagents also add to benzonorbornadiene *via* a mechanism similar to that followed by arenesulphenyl halides in their addition reactions with alkenes, and the intermediary of an aziridinium ion has been postulated.[59]

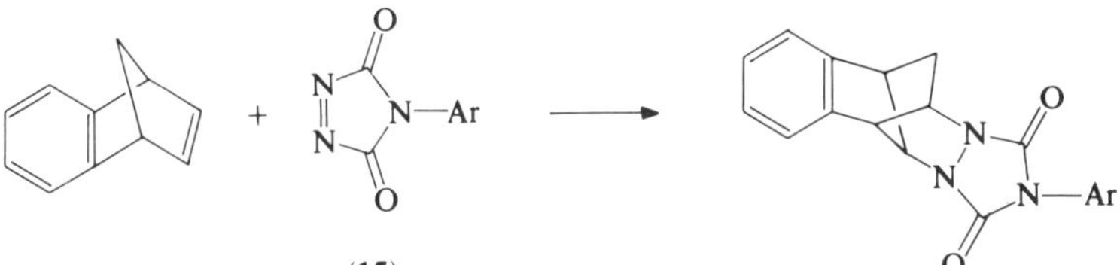

(**15**)

Cyclofunctionalization of hydroxy-alkenes with benzeneselenylchloride, *e.g.* (**16**) → (**17**), occurs with remarkably high regio- and stereo-selectivity.[60] Organoselenium-mediated trapping of a hemiacetal by a suitably disposed alkene is possible, and spiroacetals may be obtained in this way,[61] *e.g.* (**18**) → **19**). Phenylselenyl chloride adds to alkenes in a *trans*-manner, and oxidation followed by treatment with Cl^- ions (Cl_2 or excess PhSeCl is suitable) gives *cis*-1,2-dichlorides.[62]

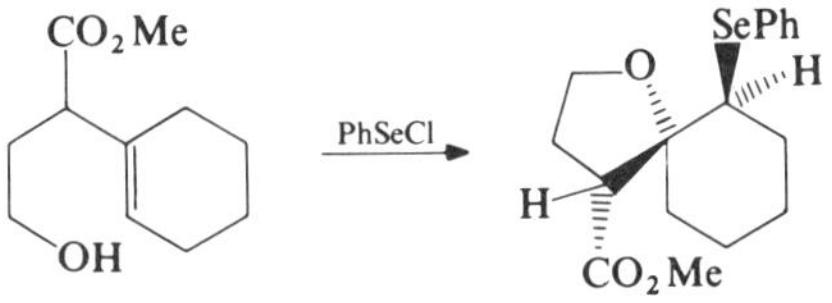

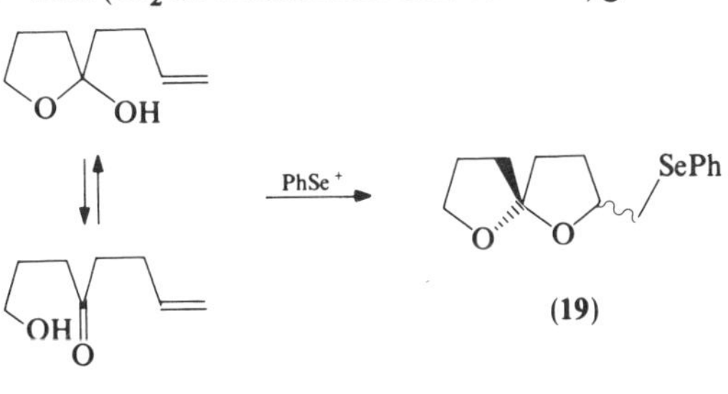

(**16**) (**17**)

(**19**)

(**18**)

Cyanogen bromide adds to $\Delta^{9,10}$-octalin in the presence of $AlCl_3$, to give a mixture of the *trans*-dibromide and *trans*-bromonitrile in low yield.[63]

Addition of Hydrogen Halide

In the addition of HBr to *meta*- and *para*-substituted 3-phenyl-2-propenoic acids, protonation is either the overall rate-determining step, or is comparable in rate to the subsequent nucleophilic attack of bromide ion.[64] The $CuCl_2/KCl/SiO_2$-catalysed reaction of HCl with ethylene and oxygen has been shown to proceed *via* a stepwise oxidation/reduction mechanism.[65] Kinetic studies show that the reaction of HCl with acetylene in the presence of an ($HgCl_2/BiCl_3$) catalyst system follows the rate equation shown below:

$$\text{Rate} = (R + R^1[BiCl_3]/[H^+])[HgCl_2][H^+]^3[C_2H_2]$$

where R = 0.00239 and R^1 = 0.0269 at 25°. Two parallel mechanisms with $\Delta E^{\neq}$ of 13.4 and 14.8 kcal mol^{-1} are proposed to occur in the reaction.[66]

Hydration and Related Reactions

The activation parameters, solvent deuterium isotope effect, and dependence of rates on acid concentration determined for the hydration of *exo*- and *endo*-5-nitro-2-norbornenes and 3-nitronortricyclene, agree with the proposal of rate-determining protonation of the alkene or cyclopropane, and thus support an A–S_E2 mechanism.[67] Similar results are obtained for 5-acetyl-2-norbornenes and 3-acetylnortriclane.[68] The kinetic parameters of the hydration of maleic acid, giving malic acid, have been determined.[69] Micellar catalysts (cationic or non-ionic) accelerate the reversible hydroxylation of the *N*-dodecylacridinium ion by a factor of 10^5–10^6, and this reaction has mechanistic similarities to the oxidation of alcohols by NAD^+ models.[70]

The thermally promoted hydration of phenylallenes in 80% H_2SO_4 has been depicted as proceeding through a vinyl cation intermediate, in contrast to the photochemically promoted hydration of these compounds.[71] Hydroxyalkenes undergo internal hydration of the alkene in the presence of Pd(II) salts, *e.g.* **(20)** → **(21)**, and this stereospecific reaction has been shown to have synthetic potential.[72]

$PdCl_2$ / MeOH

OH H H O H CO_2Me

(20) **(21)**

Kinetic studies of the BF_3-catalysed addition of alcohols to dimethylketene support the previously advanced mechanism, which involves rapid formation of a 1:1 alcohol–BF_3 adduct, which then transfers a proton to the ketene in a subsequent slower step.[73] In the gas-phase protonation of propyne and allene, the kinetically significant step leads predominantly to the 2-propenyl carbocation.[74] The gas-phase basicity of a series of styrenes has been determined, and the results demonstrate the usefulness of the σ^+ constant as a measure of ion stability in solution.[75]

The formation of 2-*exo*-methoxybicyclo[3.2.1]oct-3-ene (**23**) from *endo*-tricyclo [3.2.1.$0^{2,4}$]oct-6-ene (**22**) on treatment with *p*-toluenesulphonic acid in methanol has been ascribed to corner-protonation of the cyclopropane, followed by skeletal rearrangement.[76] Protonation of C(1) of 4-alkyl-4-methoxy-1,4-dihydronaphthalene-1-nitronate anions, *e.g.* (**24**), gives mixtures of *cis*- and *trans*-products, with

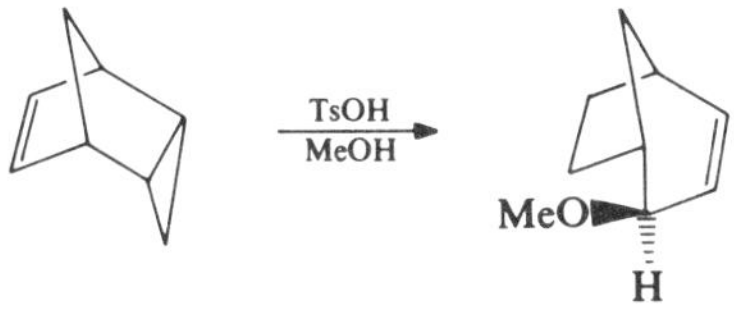

TsOH / MeOH

MeO H

O N O⁻ OMe H R

(**22**) (**23**) 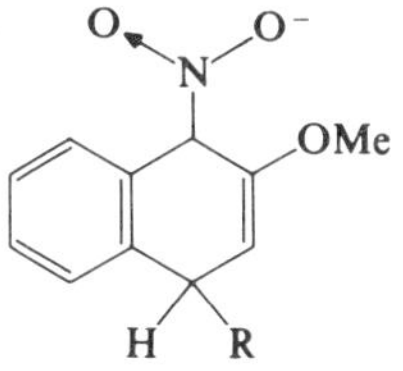 (**24**)

the *cis*-isomers predominating.[77] The Lewis acid-catalysed ene reactions of chloral and bromal with alkenes are completely regiospecific, moderately regioselective, and highly stereoselective.[78] 5-Arylmethylene-3,4-dimethyl-3-pyrrolin-2-ones undergo deuterium substitution at the methylene bridge in ^{2}H-trifluoroacetic acid solution.[79] Cobalt(III) acetate in wet HOAc oxidizes aryl-conjugated alkenes to glycol monoacetates, and a mechanism involving an intermediate coordinated to cobalt has been proposed for this reaction.[80]

Miscellaneous Electrophilic Reactions

Neighbouring-group participation of an η^3-benzyl moiety has been observed during the hydroformylation and hydrosilylation of styrenes.[81] A theoretical study of the stereoselectivity observed in hydroboration reactions has appeared.[82] A simple technique for the preparation of optically active alcohols of "100% optical purity" has been described, which involves the hydroboration of alkenes with monoisocamphenylborane.[83] Evidence has been presented[84] which supports the pre-dissociation of BH_3–Lewis-base complexes prior to attack of boron on the alkene, rather than direct attack of the complexes.[85] The kinetics of the reaction of borinane (**25**) with alkenes show 3/2 rate-order behaviour, suggesting that fast dissociation of the borinane dimer is followed by rate-limiting reaction of the monomeric borane with the alkene.[86] The regioselectivity observed during the hydroboration of a series of ferrocenylalkenes has been rationalized in terms of both steric and electronic factors.[87]

The complex prepared from Cp_2TiCl_2 and $NaBH_4$, in the presence of a crown ether, rapidly and efficiently hydroborates a range of alkenes.[88] $TiCl_3$ reacts with $NaBH_4$ in the presence of the crown ether 18-crown-6 to give a complex which promotes *catalytic* hydroboration of alkenes, dienes, and unsaturated ethers by $NaBH_4$, giving sodium organoborates, which lead to alcohols on treatment with H_2O_2/CH_3ONa.[89]

Studies of the reaction between lithium triethylborohydride and substituted styrenes suggest that the reaction involves *nucleophilic* attack of R_3BH^- on the alkene to form Ar$\bar{C}$HMe which is then trapped by R_3B to give the observed tetraalkyl borate product.[90]

Hydrogen cyanide adds to a wide range of alkynes in the presence of tetrakis(triphenylphosphine)nickel(o) to give stereospecifically *cis*-α,β-unsaturated nitriles.[91] A number of *ortho*-substituted arylacetylenes have been shown to undergo easy intramolecular solvomercuration with $Hg(OAc)_2$ in HOAc, giving benzofurans, benzothiophenes, isocoumarin, and chromone organomercuric chlorides, *e.g.* (**26**) → (**27**).[92] Solvomercuration of alkenes has been shown to be very dependent

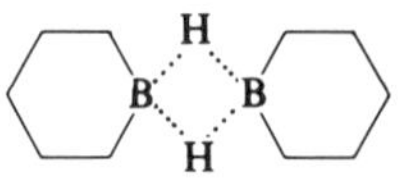

(**25**) 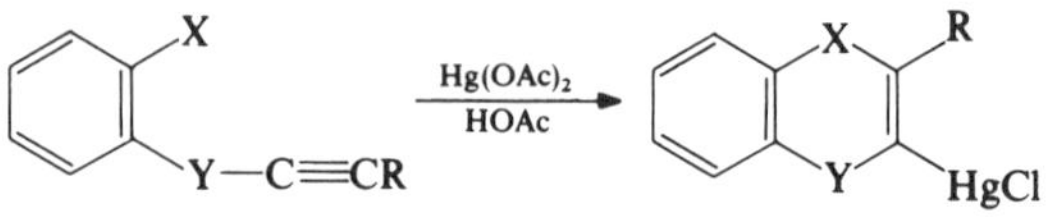 (**26**) (**27**)

X = O, S, CO_2
Y = CO or nothing

on the mercury reagent used and the solvent involved, the most useful reagent being mercury trifluoroacetate in all solvents studied.[93] The mercuric ion-initiated cyclization of δ-alkenylcarbamates occurs stereoselectively *only* under conditions which do not allow equilibration, whereas the E-alkenylcarbamates require equilibrating conditions to give stereoselective results.[94] Cyclization of a series of β, γ-allenic diols and, in one case, a β,δ-allenic diol with silver or mercury salts gives mixtures of dihydropyrans and α-alkenyltetrahydrofurans.[95] Cyclization of δ-allenic alcohols with silver or mercury salts gives 2-(α-alkenyl)tetrahydropyrans with moderate to high stereoselectivity.[96] A mechanism has been proposed to account for the product distribution and stereoselectivity observed in the reaction of alkenes with mercuric oxide/tetrafluoroboric acid and alcohols (or water), giving vicinal diethers (or diols).[97] The 6-*exo*-cyclization of unsaturated hydroperoxides in the presence of mercuric salts, *e.g.* (**28**) → (**29**), gives products whose stereochemistry may be rationalized on the basis of an equatorially substituted chain-like transition state.[98]

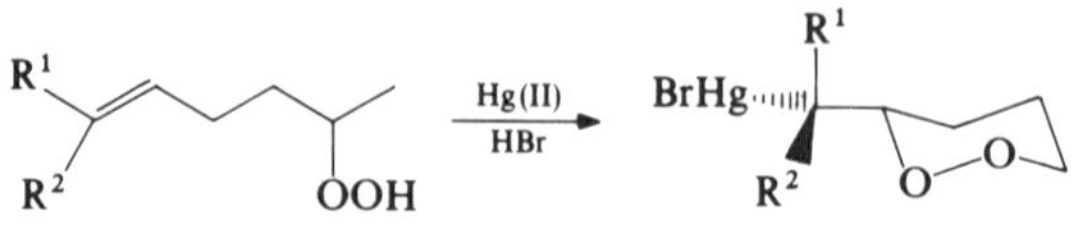

(**28**) (**29**)

In contrast to the accepted mechanism for oxymercuration of alkenes, which involves a reversible reaction, the aminomercuration of alkenes may be reversible or irreversible depending on the mercury salt employed.[99] Stereospecific osmylation of double bonds which are α to ketones in cyclic systems, is possible by prior formation of adducts with *N,S*-dimethyl-*S*-phenylsulphoximine, followed by cleavage of the chiral adjunct.[100] A relatively remote chiral sulphoxide group may induce chirality in the osmium tetraoxide oxidation of alkenes such as (**30**).[101]

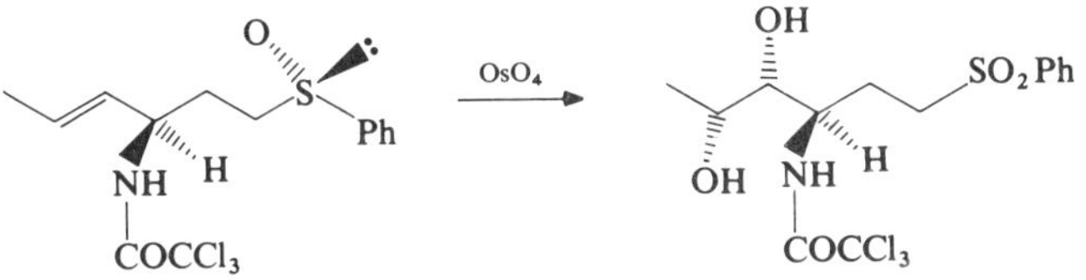

(**30**)

The addition of $SeCl_4$ to a series of maleimides having vinyl or allyl *N*-substituents is second order in substrate, first order in $SeCl_4$, and has an activation entropy of −46 e.u.[102] The addition of $TeCl_4$ to a series of allylic alcohols, esters, acids, and imino-ethers has been reported,[103] and a mechanism involving competing concerted *syn*-addition and radical-chain processes has been proposed.[104]

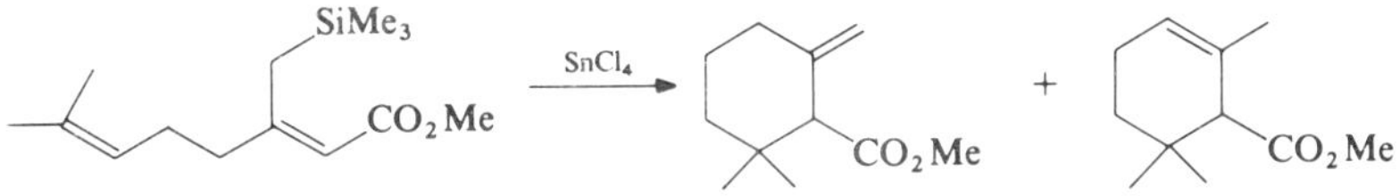

(**31**) 7:1

1,5-Dienes containing an allylsilane, *e.g.* (**31**), may be cyclized in the presence of Lewis acids, with the TMS group exhibiting a strong activating and directing influence on the regioselectivity of the reaction.[105] MINDO/3 calculations have been reported which provide strong support for a stepwise mechanism for biomimetic polyene cyclizations, each step involving the formation of a cyclic π-complex by electrophilic addition to an alkene. This mechanism requires that the observed *trans*-stereospecificity results.[106] The regio- and stereo-chemical outcome of the addition of Brønsted acids to *cis*-cyclooctene *does not* support the involvement of an open-chain carbenium ion in this reaction.[107] The stereoselectivity observed in the electrophilic addition reactions of 6β-methoxy-3α,5-cyclo-5α-cholest-22-ene can be rationalized in terms of a product-like transition-state model.[108] A mechanism has been proposed to account for the observed *syn*, *vic*-ditosylation of alkenes by [hydroxy(tosyloxy)iodo]benzene (**32**).[109]

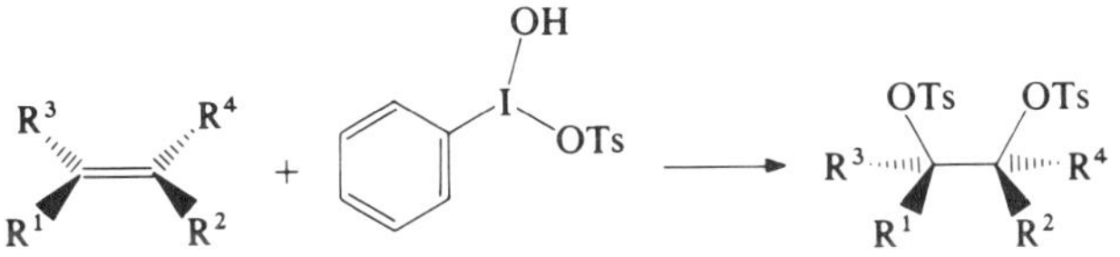

(**32**)

Possible mechanisms for the addition of *p*-$MeC_6H_4SO_2X$ (X = Br, I) to vinylalkynes have been discussed.[110] NMR spectroscopic evidence has been presented[111] that does not endorse the metallacycle mechanism proposed by Green[112] for the Ziegler–Natta polymerization of acetylene.

Nucleophilic Additions

Nucleophiles add regiospecifically 1,4 to the exocyclic enone double bond of α-methylene-cyclopentenones.[113] Enthalpy measurements lead to the conclusion that the transition state for the addition of piperidine to methyl propiolate has a polar character.[114] Silyl enol ethers and ketene silyl acetates add in a Michael fashion to nitroalkenes, giving 1,4-diketones and γ-keto-esters, respectively.[115] Chloroacetylenes may act as Michael acceptors towards enolate ions, giving α-substituted ethynyl ketones and aldehydes as products.[116] The highly stereoselective cyclization of (**33**) giving (**34**) is rationalized as occurring *via* chelation of a

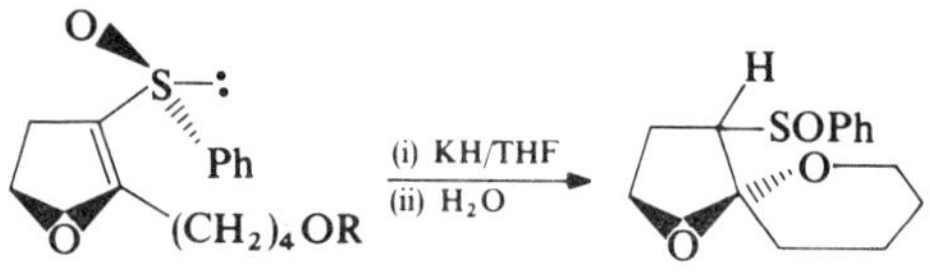

Ph; $(CH_2)_4OR$ — (i) KH/THF (ii) H_2O → H; SOPh

(**33**) (**34**)

potassium ion in the transition state, leading to steric control of the reaction.[117] Readily enolizable ketones add in a 1,4-manner to the acetals of α,β-unsaturated aldehydes, and to ethoxyallene, *via* a mechanism which may involve either direct Michael addition, or *O*-alkylation followed by Claisen rearrangement.[118] The addition of organometallic reagents to compounds containing $C{=}C(CN)_2$ and $C{=}C(CO_2R)_2$ groups has been studied as a model for the addition of these reagents to carbonyl groups.[119] Base-promoted Michael additions of chloromethoxyvinyl phenyl sulphoxides to active methylene compounds have been reported.[120]

Studies of the rate of addition of $\bar{C}H(CN)_2$ to benzylidenemalononitrile show that solvent reorganization contributes significantly to the intrinsic energy barrier of the reaction.[121] Kinetic parameters for the addition of the 9-phenylfluorene carbocation to $PhCF{=}CF_2$ in a series of solvents have been reported.[122]

A series of 1,4-benzoquinones has been shown to undergo attack at the carbonyl group in sodium/liquid ammonia mixtures, rather than the expected 1,4-addition to the enone system.[123] Kinetic analysis of the reaction between ethylene and KOH in liquid ammonia shows that an ordered polarized transition state is involved.[124] Conditions have been delineated which allow attack of weak nucleophiles (*e.g.* OH_2, RNH_2, *etc.*) on unactivated alkenes, *e.g.* (**35**) → (**36**), and these experiments provide a

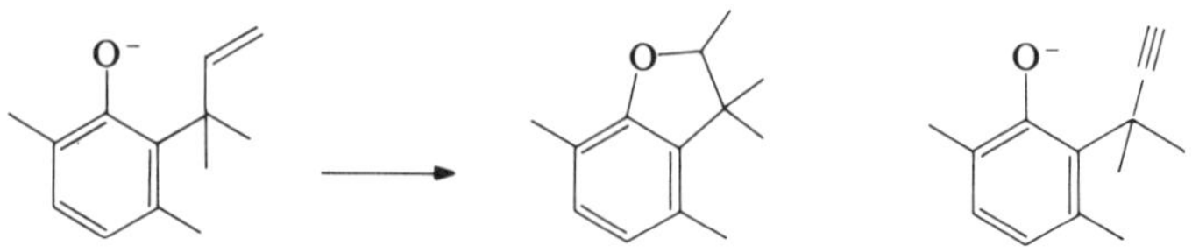

(**35**) (**36**) (**37**)

model for certain enzyme-mediated reactions.[125] In this type of addition, alkynes, *e.g.* (**37**), are more reactive by a factor of *ca.* 10^4, and this may be due to the greater stability of the sp^2-(relative to the sp^3-) hybridized carbanion which is produced.[126]

3-Vinylbicyclobutane-1-carbonitrile reacts with $MeOH/MeO^-$ or $MeSH/MeS^-$ to give 1,3-adducts as E- and Z-mixtures.[127] Thiophenol adds to PhC≡CCOX (X = OH, OMe, NHPh) to give E- and Z-isomers of $PhC(SCH_2Ph)$=CHCOX *via* an ionic mechanism.[128] The reaction of thiophenol with 3-methylbicyclobutane-1-carboxylic acid nitrile proceeds non-regiospecifically, giving a mixture of E- and Z-3-methyl-1-(phenylthio)- and -3-methyl-3-(phenylthio)-cyclobutane-1-carboxylic acid.[129] The outcome of the Lewis acid-catalysed addition of allylic sulphides to methyl propiolate and dimethyl acetylenedicarboxylate is significantly affected by the Lewis acid used.[130] Nucleophilic addition of organo – aluminium species to chiral α,β-unsaturated acetals occurs in a 1,2- and 1,4-sense, with remarkably high asymmetric induction.[131]

The silver ion-mediated cyclization of secondary allenic alcohols is stereoselective, giving *cis*-2,6-disubstituted tetrahydropyrans as major products.[132] Evidence has been presented which indicates that the aminopalladation of alkenes is a reversible reaction.[133]

Acylation of terminal alkynes with $AlCl_3$/acetyl chloride gives β-chloro-α,β-unsaturated ketones with variable amounts of β-chloro-α,γ-unsaturated ketones, and these are proposed to arise *via* the intermediacy of allenic compounds.[134,135] The course of the addition of Grignard reagents and alkyl-lithiums to vinyl(alkoxy)silanes may be rationalized in terms of nucleophile HOMO-vinyl LUMO interactions.[136]

Heteroconjugate addition of various nucleophiles to compounds such as (**38**) provides either *syn*- or *anti*-products diastereoselectively, depending on solvent polarity.[137] Nucleophilic substitution reactions of furans and thiophenes with

R SiMe$_3$ OH SO$_2$Ph —(i) Nu$^-$ (ii) KF→ R Nu H OH SO$_2$Ph and/or R H Nu OH SO$_2$Ph

(**38**)

thiophenoxides proceed *via* an addition–elimination mechanism with a polar transition state.[138] Nucleophilic addition of organolithiums to chiral 1-naphthyloxazoline (**39**) followed by trapping of the intermediate aza-enolate with various electrophiles gives dihydronaphthals, *e.g.* (**40**), with a high degree of diastereofacial selectivity.[139] Rate measurements give a Brønsted α-value of 0.37 for hydride transfer in a series of NAD^+ analogues, and the same value is obtained from Marcus theory, using previously reported symmetrical-exchange rate constants.[140]

The results of a study involving the addition of nucleophiles to diene–iron complexes indicate that (*a*) addition is favoured at an unsubstituted internal position, (*b*) the initial site of addition is not governed by product stability, and (*c*) that equilibration of the initially formed intermediate to a more stable form can

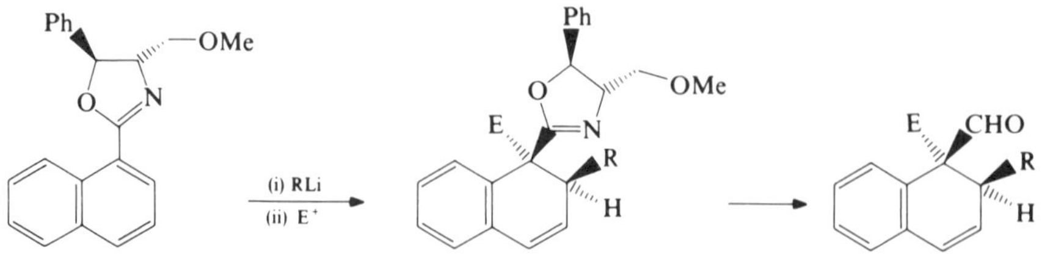

(39)

(40)

occur rapidly at 0°.[141] Maleate dianion coordinated to cobalt has been shown to be susceptible to attack by nucleophiles.[142] The kinetics of cleavage of the adducts formed by nucleophilic addition of morpholine and piperidine to α-cyano-4-nitrostilbene have been analysed.[143] The addition of the lithiospecies (**41**) and (**42**) to α,β-unsaturated ketones gives a range of products which can be rationalized on the basis of an interplay of ion-pairing, steric decompression, and steric hindrance effects.[144]

The orthoquinoid vinyl ketone (**43**) is attacked by methanol in an in-plane *exo*-fashion (*A*), as evidenced by the dramatic rate depression observed when bulky *ortho*-substituents are present.[145] This is in contrast to the reaction of (**43**) with maleic anhydride which appears to approach in an out-of-phase sense (*B*).[145]

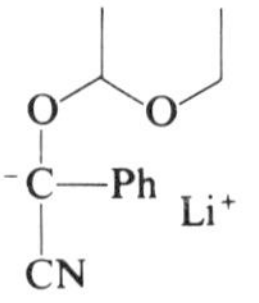

(41) 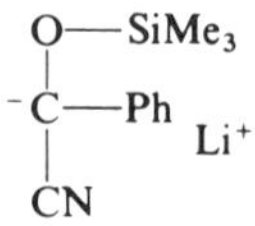 (42) 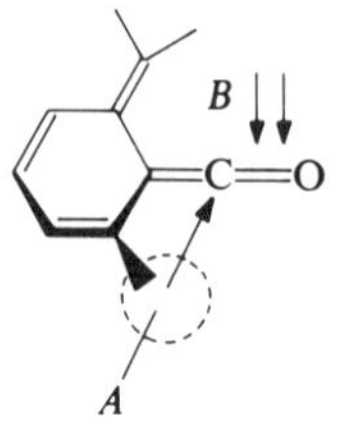 (43)

The stereochemical outcome of the cyclization of homoallylic glycols to 2,3,4-trisubstituted tetrahydrofurans has been studied.[146] The addition of dialkylmagnesium compounds to cyclopentylcyclopropene is first order in each component, and gives *cis*-adducts stereoselectively.[147]

References

1 Morris, D. G., *Ann. Rep. Prog. Chem. Sect. B.*, **79**, 57 (1983).
2 Greene, F. D., *Methods Stereochem. Anal.*, **3**, 197 (1983); *Chem. Abs.*, **101**, 129827 (1984).
3 Bäckvall, J.-E., *Acc. Chem. Res.*, **16**, 335 (1983).
4 Pasto, D. J., *Tetrahedron*, **40**, 2805 (1984).
5 Tomioka, K., and Koga, K., *Asymmetric Synth.*, **2**, 201 (1983); *Chem. Abs.*, **101**, 37797 (1984).
6 Houk, K. N., *Methods Stereochem. Anal.*, **3**, 1 (1983); *Chem. Abs.*, **101**, 129823 (1984).
7 Posner, G. H., *Asymmetric Synth.*, **2**, 225 (1983); *Chem. Abs.*, **101**, 37798 (1984).
8 Rakhmankulov, D. L., Zorin, V. V., Sapiev, O. G., Zlotskii, S. S., and Bartok, M., *Acta Phys. Chem.*, **29**, 165 (1983); *Chem. Abs.*, **101**, 89903 (1984).

[9] Morrison, J. D. (Ed.), "Stereodifferentiating addition reactions: Part A," *Asymmetric Syntheses*, Vol. 2, Academic Press, New York, 1983; *Chem. Abs.*, **101**, 54055 (1984).
[10] Morrison, J. D. (Ed.), "Stereodifferentiating addition reactions: Part B", *Asymmetric Syntheses*, Vol. 3, Academic Press, New York, 1984; *Chem. Abs.*, **101**, 54054 (1984).
[11] Bartlett, P. A., *Asymmetric Synth.*, **3**, 341 (1984); *Chem. Abs.*, **101**, 37792 (1984).
[12] Koser, G. F., *Chem. Halides, Pseudo-Halides, Azides*, **2**, 1265 (1983); *Chem. Abs.*, **100**, 138199 (1984).
[13] Mundy, B. P., and Wilkening, D., *J. Org. Chem.*, **49**, 3779 (1984).
[14] Dubois, J.-E., Ruasse, M. F., and Argile, A., *J. Am. Chem. Soc.*, **106**, 4840 (1984).
[15] Bartlett, P. D., Roof, A. A. M., Subramanyam, R., and Winter, W. J., *J. Org. Chem.*, **49**, 1875 (1984).
[16] Toromanoff, E., *Actual. Chim.*, **1984**, 13; *Chem. Abs.*, **101**, 129872 (1984).
[17] Brown, R. S., Gedge, R., Slebocka-Tilk, H., Buschek, J. M., and Kopecky, K. R., *J. Am. Chem. Soc.*, **106**, 4515 (1984).
[18] Bianchi, M. T., Cerichelli, G., Mancini, G., and Marinelli, F., *Tetrahedron Lett.*, **25**, 5205 (1984).
[19] Michael, J. P., Blom, N. F., and Boeyens, J. C. A., *J. Chem. Soc., Perkin Trans. 1*, **1984**, 1739.
[20] Smith, W. B., Saint, G., and Johnson, L., *J. Org. Chem.*, **49**, 3771 (1984).
[21] Schmid, G. H., and Toyonaga, B., *J. Org. Chem.*, **49**, 761 (1984).
[22] Bland, J. M., and Stammer, C. H., *J. Org. Chem.*, **48**, 4393 (1983).
[23] Salakhov, M. S., Musaera, N. F., Nagiev, V. A., and Akhundora, F. G., *Deposited Doc.*, **1983**, VINITI 1317; *Chem. Abs.*, **100**, 208719 (1984).
[24] Musaera, N. F., Salakhov, M. S., Nagiev, V. A., Casanova, A. A., and Polador, P. M., *Deposited Doc.*, **1983**, VINITI 2277; *Chem. Abs.*, **101**, 109970 (1984).
[25] Musaera, N. F., Salakhov, M. S., Nagiev, V. A., Pulatova, Sh. M., and Zul'falter, Sh. R., *Zh. Org. Khim.*, **20**, 629 (1984); *Chem. Abs.*, **101**, 37898 (1984).
[26] Musaera, N. F., and Salakhov, M. S., *React. Kinet. Catal. Lett.*, **22**, 305 (1983); *Chem. Abs.*, **100**, 120211 (1984).
[27] Ghenciulescu, A., Gheorghiu, M. D., Enescu, L., Dinulescu, I. G., and Airam, M., *Rev. Roum. Chim.*, **29**, 49 (1984).
[28] Ruasse, M.-F., and Dubois, J.-E., *J. Am. Chem. Soc.*, **106**, 3230 (1984).
[29] Ruasse, M.-F., Argile, A., and Dubois, J.-E., *J. Am. Chem. Soc.*, **106**, 4846 (1984).
[30] Ruasse, M.-F., and Lefebure, E., *J. Org. Chem.*, **49**, 3210 (1984).
[31] Ruasse, M.-F., and Zhang, B.-L., *J. Org. Chem.*, **49**, 3207 (1984).
[32] Shinoda, K., and Yasuda, K., *Toyama Kogyo Koto Senmon Gakko Kiyo.*, **18**, 5 (1984); *Chem. Abs.*, **101**, 151129 (1984).
[33] Shinoda, K., and Yasuda, K., *Toyama Kogo Koto Senmon Gakko Kiyo.*, **18**, 1 (1984); *Chem. Abs.*, **101**, 151128 (1984).
[34] Garabadzhiu, A. V., and V'yunov, K. A., *Teor. Eksp. Khim.*, **20**, 75 (1984); *Chem. Abs.*, **101**, 6282 (1984).
[35] Chamberlin, A. R., and Mulholland, R. L., *Tetrahedron*, **40**, 2297 (1984).
[36] Tamaru, Y., Mizutain, M., Furakana, Y., Kawamura, S., Yoshida, Z., Yanagi, K., and Minobe, M., *J. Am. Chem. Soc.*, **106**, 1079 (1984).
[37] Jin, S., and Liu, R., *Int. J. Quantum Chem.*, **25**, 699 (1984); *Chem. Abs.*, **100**, 191084 (1984).
[38] Gregorcic, A., and Zupan, M., *J. Org. Chem.*, **49**, 333 (1984).
[39] Adinolfi, M., Barone, G., Lanzetta, R., Laonigro, G., Mangoni, L., and Parrilli, M., *Tetrahedron*, **40**, 2183 (1984).
[40] Hamman, S., and Beguin, C. G., *J. Fluorine Chem.*, **23**, 515 (1983); *Chem. Abs.*, **100**, 155978 (1984).
[41] Bogaslavskaya, L. S., and Temovskoi, L. A., *Zh. Org. Khim.*, **19**, 1881 (1983); *Chem. Abs.*, **100**, 22209 (1984).
[42] Amirtha, N., Viswanathan, S., and Ganesan, R., *Monatsh. Chem.*, **115**, 53 (1984); *Chem. Abs.*, **100**, 102506 (1984).
[43] Amirtha, N., Viswanathan, S., and Ganesan, R., *Monatsh. Chem.*, **115**, 35 (1984).
[44] Kocovsky, P., Stary, I., Turacek, F., and Hanus, V., *Collect. Czech. Chem. Commun.*, **48**, 2994 (1983).
[45] Kocovsky, P., *Collect. Czech. Chem. Commun.*, **48**, 3606 (1983).
[46] Kocovsky, P., *Collect. Czech. Chem. Commun.*, **48**, 3660 (1983).
[47] Kocovsky, P., *Collect. Czech. Chem. Commun.*, **48**, 3618 (1983).
[48] Kocovsky, P., *Collect. Czech. Chem. Commun.*, **48**, 3629 (1983).
[49] Kocovsky, P., *Collect. Czech. Chem. Commun.*, **48**, 3643 (1983).
[50] Chumakov, L., V., Borisov, A. V., and Bodrikov, I. V., *Izv. Vyssh. Uchebn. Zaved. Khim. Khim. Technol.*, **26**, 1033 (1983); *Chem. Abs.*, **100**, 33894 (1984).

[51] Dalipi, S., and Schmide, G. H., *J. Org. Chem.*, **47,** 5027 (1982).
[52] Zefirov, N. S., and Bodrikov, I. V., *Zh. Org. Khim.*, **19,** 2225 (1983); *Chem. Abs.*, **100,** 50798 (1984).
[53] Kanska, M., and Fry, A., *J. Am. Chem. Soc.*, **105,** 7666 (1983).
[54] Bhongle, N. N., and Gote, V. N., *Indian J. Chem.*, **22B,** 632 (1983); *Chem. Abs.*, **100,** 34061 (1984).
[55] Bodrikov, I. V., Borisov, A. V., Smit, W. A., and Lutsenko, A. I., *Tetrahedron Lett.*, **25,** 4983 (1984).
[56] Schmid, G. H., and Garratt, D. G., *J. Org. Chem.*, **48,** 4169 (1983).
[57] Matsuda, I., Akiyama, K. Furuta, H., and Mizuta, M., *Bull. Chem. Soc. Jpn.*, **57,** 219 (1984).
[58] Cheng, C.-C., and Greene, F. D., *J. Org. Chem.*, 49, 2917 (1984).
[59] Adam, W., Carballeira, N., Scheutzow, D., Peters, K., Peters, E.-M., and von Schnering, H. G., *Chem. Ber.*, **117,** 1139 (1984).
[60] Schultz, A. G., and Sundararaman, P., *J. Org. Chem.*, **49,** 2455 (1984).
[61] Doherty, A. M., Ley, S. V., Lygo, B., and Williams, D. J., *J. Chem. Soc., Perkin Trans. 1,* **1984,** 1371.
[62] Morella, A. M., and Ward, A. D., *Tetrahedron Lett.*, **25,** 1197 (1984).
[63] Weber, W., and Ginsburg, D., *Chem. Scr.*, **22,** 166 (1983).
[64] Cervinka, O., and Kriz, O., *Collect. Czech. Chem. Commun.*, **48,** 2952 (1983).
[65] Bakshi, Y. M., Dmitrieva, M. P., and Gel'bshtein, A. I., *Kinet. Katal.*, **25,** 136 (1984); *Chem. Abs.*, **100,** 191080 (1984).
[66] Hao, J., Zhang, Q., and Chem. R., *Gaodeng Xuexiao Huaxue Xuehaco*, **5,** 221 (1984); *Chem. Abs.*, **101,** 37900 (1984).
[67] Lajunen, M., and Kukkonen, H., *Acta. Chem. Scand.*, **A37,** 447 (1983); *Chem. Abs.*, **100,** 50806 (1984).
[68] Lajunen, M., and Hintsanen, A., *Acta. Chem. Scand.*, **A37,** 545 (1983); *Chem. Abs.*, **100,** 138301 (1984).
[69] Kasa, Z., and Deak, G., *Hung. J. Ind. Chem.*, **11,** 313 (1983); *Chem. Abs.*, **101,** 6267 (1984).
[70] Shinkai, S., Tsuno, T., and Manahe, O., *J. Chem. Soc., Perkin Trans. 2,* **1984,** 661.
[71] Rafizadeh, K., and Yates, K., *J. Org. Chem.*, **49,** 1500 (1984).
[72] Semmelhack, M. F., and Bodurow, C., *J. Am. Chem. Soc.*, **106,** 1496 (1984).
[73] Poon, N. L., and Satchell, D. P. N., *J. Chem. Soc., Perkin Trans. 2,* **1984,** 1083.
[74] Fornarini, S., Speranza, M., Allina, M., Cacace, F., and Giacomello, P., *J. Am. Chem. Soc.*, **106,** 2498 (1984).
[75] Harrison, A. G., Houriet, R., and Tidwell, T. T., *J. Org. Chem.*, **49,** 1302 (1984).
[76] Battiste, M. A., Coxon, J. M., Simpson, G. W., Steel, P. J., and Jones, A. J., *Tetrahedron,* **40,** 3137 (1984).
[77] Baccolini, G., Bartoli, G., Bosco, M., and Dalpozzo, R., *J. Chem. Soc., Perkin Trans. 2,* **1984,** 363.
[78] Benner, J. P., Gill, G. B., Parrott, S. J., Wallace, B., and Begley, M. J., *J. Chem. Soc., Perkin Trans. 1,* **1984,** 315.
[79] Ribo, J. M., and Trull, F. R., *Monatsh. Chem.*, **114,** 1087 (1983).
[80] Hirano, M., and Morimoto, T., *J. Chem. Soc., Perkin Trans. 2,* **1984,** 1033.
[81] Crabtree, R. H., Mella, M. F., and Quirk, J. M., *J. Am. Chem. Soc.*, **106,** 2913 (1984).
[82] Houk, K. N., Rondan, N. G., Wu, Y.-D., Metz., J. T., and Paddon-Row, M. N., *Tetrahedron,* **40,** 2257 (1984).
[83] Brown, H. C., and Singaram, B., *J. Am. Chem. Soc.*, **106,** 1797 (1984).
[84] Brown, H. C., and Chandrasekharan, J., *J. Am. Chem. Soc.*, **106,** 1863 (1984).
[85] Clark, T., Wilhelm, D., and Schleyer, P. von R., *J. Chem. Soc. Chem. Commun.*, **1983,** 606.
[86] Brown, H. C., Chandrasekharan, J., and Nelson, D. J., *J. Am. Chem. Soc.*, **106,** 3768 (1984).
[87] Sterzo, C., and Ortaggi, G., *J. Chem. Soc., Perkin Trans. 2,* **1984,** 345.
[88] Lee, H. S., Isagawa, K., and Otsuji, Y., *Chem. Lett.*, **1984,** 363.
[89] Lee, H. S., Isagawa, K., Togda, H., and Otsuji, Y., *Chem. Lett.*, **1984,** 673.
[90] Brown, H. C., and Kim, S.-C., *J. Org. Chem.*, **49,** 1064 (1984).
[91] Jackson, W. R., and Lovel, C. G., *Aust. J. Chem.*, **36,** 1975 (1983).
[92] Larock, R. C., and Harrison, L. W., *J. Am. Chem. Soc.*, **106,** 4218 (1984).
[93] Brown, H. C., Kurek, J. T., Rei, M.-H., and Thompson, K. L., *J. Org. Chem.*, **49,** 2551 (1984).
[94] Harding, K. E., and Marman, T. H., *J. Org. Chem.*, **49,** 2838 (1984).
[95] Chilot, J.-J., Doutheau, A., and Gore, J., *Bull. Soc. Chim. Fr. II,* **1984,** 307.
[96] Audin, P., Doutheau, A., and Gore, J., *Bull. Soc. Chim. Fr. II,* **1984,** 297.
[97] Bartuenga, J., Alonso-Cires, L., Campos, P. J., and Asensio, G., *Tetrahedron,* **40,** 2563 (1984).
[98] Porter, N. A., and Zuraw, P. J., *J. Org. Chem.*, **49,** 1345 (1984).
[99] Barluenga, J., Perez-Prieto, J., Bayon, A. M., and Asensio, G., *Tetrahedron,* **40,** 1199 (1984).
[100] Johnson, C. R., and Barbachyn, M. R., *J. Am. Chem. Soc.*, **106,** 2459 (1984).
[101] Hauser, F. M., and Ellenberger, S. R., *J. Am. Chem. Soc.*, **106,** 2458 (1984).

[102] Salakhova, R. S., Musaeva, N. F., Salakhov, M. S., Mamedor, E. Sh., and Shakhtakhtinskii, T. N., *Dokl. Akad. Nauk. Az. SSR,* **39,** 37 (1983); *Chem. Abs.,* **100,** 138315 (1984).
[103] Engman, L., *J. Am. Chem. Soc.,* **106,** 3977 (1984).
[104] Bäckvall, J.-E., Bergman, J., and Engman, L., *J. Org. Chem.,* **48,** 3918 (1983).
[105] Armstrong, R. J., and Weiler, L., *Can. J. Chem.,* **61,** 2530 (1983).
[106] Dewar, M. J. S., and Reynolds, C. H., *J. Am. Chem. Soc.,* **106,** 1744 (1984).
[107] Nordlander, J. E., Kotian, K. D., Raff, D. E., and Njoroge, G. F., *Prepr.-Am. Chem. Soc. Div. Pet. Chem.,* **28,** 285 (1983); *Chem. Abs.,* **101,** 22660 (1984).
[108] Hirano, T., Takatsuto, S., and Ikekana, N., *J. Chem. Soc., Perkin Trans. 1,* **1984,** 1775.
[109] Rebrovic, L., and Koser, G. F., *J. Org. Chem.,* **49,** 2462 (1984).
[110] Kolosov, E. Y., Sadnichuk, M. D., and Nadezhina, L. S., *Zh. Org. Khim.,* **19,** 2298 (1983); *Chem. Abs.,* **100,** 84993 (1984).
[111] Clarke, T. C., Yannoni, C. S., and Katz, T. J., *J. Am. Chem. Soc.,* **105,** 7787 (1983).
[112] Green, M. L. H., *Pure Appl. Chem.,* **50,** 27, (1978).
[113] Siwapinyoyos, T., and Thebtaranonth, Y., *Tetrahedron Lett.,* **25,** 353 (1984).
[114] Breus, V. A., Neklyudov, S. A., Solomonov, B. N., and Konovalov, A. I., *Zh. Org. Khim.,* **19,** 1613 (1983); *Chem. Abs.,* **99,** 211879 (1983).
[115] Miyashita, M., Yanami, T., Kumazawa, T., and Yoshikoshi, A., *J. Am. Chem. Soc.,* **106,** 2149 (1984).
[116] Kende, A. S., Fludzinski, P., Hill, J. H., Swenson, W., and Clardy, J., *J. Am. Chem. Soc.,* **106,** 3551 (1984).
[117] Iwata, G., Hattori, K., Uchida, S., and Imanishi, T., *Tetrahedron Lett.,* **25,** 2995 (1984).
[118] Coates, R. M., and Hobbs, S. J., *J. Org. Chem.,* **49,** 140 (1984).
[119] Kruger, D., Sopchik, A. E., and Kingsbury, C. A., *J. Org. Chem.,* **49,** 778 (1984).
[120] Kinoshita, H., Hori, I., Oishi, T., and Ban, Y., *Chem. Lett.,* **1984,** 1517.
[121] Bernasconi, C. F., Zitomer, J. L., Fox, J. P., and Howard, K. A., *J. Org. Chem.,* **49,** 482 (1984).
[122] Solov'yanov, A. A., Shtern, M. M., Beletskaya, I. P., and Reutov, O. A., *Vestn. Mosk. Univ., Ser. Z: Khim.,* **25,** 182 (1984); *Chem. Abs.,* **101,** 37915 (1984).
[123] Chudek, J. A., Foster, R., and Reid, F. J., *J. Chem. Soc., Perkin Trans. 2,* **1984,** 287.
[124] Esikova, I. A., Zuz'micher, A. I., and Yufit, S. S., *Kinet. Katal.,* **25,** 50 (1984); *Chem. Abs.,* **101,** 6288 (1984).
[125] Evans, C. M., and Kirby, A. J., *J. Chem. Soc., Perkin Trans. 2,* **1984,** 1259.
[126] Evans, C. M., and Kirby, A. J., *J. Chem. Soc., Perkin Trans. 2,* **1984,** 1269.
[127] Razin, V. V., and Vasin, V. A., *Zh. Org. Khim.,* **20,** 736 (1984); *Chem. Abs.,* **101,** 109921 (1984).
[128] Rasterkiene, L., Vidagiriene, V., Talaikyte, Z., and Lukoseviciute, N., *Zh. Org. Khim.,* **20,** 713 (1984); *Chem. Abs.,* **101,** 109920 (1984).
[129] Razin, V. V., and Pokorskaya, O. V., *Zh. Org. Khim.,* **19,** 1771 (1983); *Chem. Abs.,* **99,** 211834 (1983).
[130] Hayakawa, K., Kamikawaju, Y., Wakita, A., and Kanematsu, K., *J. Org. Chem.,* **49,** 1985 (1984).
[131] Fujiwara, J., Fukutani, Y. Hasegawa, M., Maruoka, K., and Yamoto, H., *J. Am. Chem. Soc.,* **106,** 5004 (1984).
[132] Gallagher, T., *J. Chem. Soc., Chem. Commun.,* **1984,** 1554.
[133] Bäckvall, J.-E., and Bjorkman, E. E., *Acta Chem. Scand.,* **B38,** 91 (1984).
[134] Manoiu, D., Manoiu, M., Dinulescu, I. G., and Avram, M., *Rev. Roum. Chim.,* **29,** 193 (1984).
[135] Manoiu, D., Manoiu, M., Dinulescu, G. I., and Avram, M., *Rev. Roum. Chim.,* **29,** 201 (1984).
[136] Tamao, K., Kanatani, R., and Kumada, M., *Tetrahedron Lett.,* **25,** 1905 (1984).
[137] Isobe, M., Funabashi, Y., Ichikawa, Y., Mio, S., and Goto, T., *Tetrahedron Lett.,* **25,** 2021 (1984).
[138] Knoppova, V. Vegh, D., and Kovac, J., *Top. Furan Chem. Proc. Symp. Furan Chem., 4th,* **1983,** 146; *Chem. Abs.,* **101,** 110004 (1984).
[139] Barner, B. A., and Meyers, A. I., *J. Am. Chem. Soc.,* **106,** 1865 (1984).
[140] Kreevoy, M. M., and Lee, I.-N. S., *J. Am. Chem. Soc.,* **106,** 2550 (1984).
[141] Semmelhack, M. F., and Le Hanh, T. M., *J. Am. Chem. Soc.,* **106,** 2715 (1984).
[142] Hammershoi, A., Sargeson, A. M., and Steffen, W. L., *J. Am. Chem. Soc.,* **106,** 2819 (1984).
[143] Bernasconi, C. F., and Murray, C. J., *J. Am. Chem. Soc.,* **106,** 3257 (1984).
[144] Seuron, N., and Seyden-Penne, J., *Tetrahedron,* **40,** 635 (1984).
[145] Schiess, P., Eberle, M., Huys-Francotte, M., and Wirz, J., *Tetrahedron Lett.,* **25,** 2201 (1984).
[146] Williams, D. R., Grote, J., and Harigaya, Y., *Tetrahedron Lett.,* **25,** 5231 (1984).
[147] Richleg, H. G., and Watkins, E. K., *J. Chem. Soc., Chem. Commun.,* **1984,** 772.

Organic Reaction Mechanisms 1984
Edited by A. C. Knipe and W. E. Watts

CHAPTER 14

Addition Reactions: Cycloaddition

R. M. PATON

Department of Chemistry, University of Edinburgh

Introduction

Although the volume of literature devoted to mechanistic aspects of cycloaddition reactions remains fairly constant, the number of papers in which synthetic applications are described has increased greatly. The latter have been excluded from the present survey.

The extent to which cycloaddition reactions are synchronous continues to be a subject for debate. Dewar has reviewed[1] recent evidence concerning the mechanisms of pericyclic reactions and finds no clear support for synchronicity. Indeed he concludes that "synchronous multi-bond mechanisms are normally prohibited". Brown and Houk,[2] on the other hand, have performed *ab initio* calculations (STO-3G basis set) for the Diels–Alder reaction of butadiene and ethylene which indicate that there is a concerted synchronous transition state. The value of activation volume measurements in assessing the relative position of the transition states in various cycloadditions (2 + 4, 4 + 6, 2 + 2 + 2) has been stressed.[3] The computer program CAMEO, which is designed to predict reaction products given starting materials and conditions, has been further developed[4] to tackle the problem of periselectivity. The rôle of σ/π interaction in controlling stereoselectivity has been critically assessed.[5,6] The feasibilities of various cycloaddition reactions have been examined by topological analysis.[7] A two volume work on 1,3-dipolar cycloadditions has recently been published.[8] In addition to detailed coverage of all the major classes of 1,3-dipoles, there are also chapters devoted to cycloreversions, theoretical aspects, and synthetic applications. The use of acetylene equivalents in cycloaddition reactions has been reviewed.[9] Cycloadditions involving allenes,[10] cycloimmonium ylides,[11] and aza analogues of sulphur dioxide[12] have been summarized as parts of wider surveys of these compounds. Reviews dealing with a single class of cycloaddition are covered in the appropriate section of this chapter.

2 + 2-Cycloaddition

The MINDO/3 method has been used[13] to study 2 + 2-cycloadditions between mono-substituted alkenes; depending on the substituents the mechanism is predicted to be one-step with a tricentric transition state or to involve intermediates with high charge transfer. 2-Cyanonaphthalene on irradiation not only forms photodimers but can also give a cycloadduct with naphthalene itself.[14] A mechanism involving a singlet exciplex is favoured[15] for the photoreactions of 8-methoxypsoralen with both electron-rich and electron-poor olefins. The regio- and stereo-chemistry of the reaction between 1-benzoylindoles and alkenes has been investigated;[16] 1,3-disubstituted cyclobutanes predominate. The first examples of asymmetric induction involving a 2C + 2C-photocycloaddition have been described.[17] *exo*-Adducts predominate in the photosensitized reaction of dichlorovinylidene carbonate with dimethyl *endo*-bicyclo[2.2.1]hept-5-ene-2,3-dicarboxylate.[18] Intramolecular 2 + 2-cycloadditions of *N*-allyl vinylogous imides have been studied[19] in greater detail and new adducts isolated. Photolysis of 2-alkenoyloxycyclohex-2-enones affords head-to-head products[20] similar to those found for the 2-pentenyl analogue. The photodimerization of cinnamate esters catalysed by BF_3 and $SnCl_4$ shows greater regio- and stereo-selectivity in the solid state than in solution.[21] Both unsymmetrical and symmetrical cyclobutanes result from the solid-state topochemical 2 + 2-cycloaddition of 2-chloro-substituted cinnamic acids cocrystallized as a molecular complex;[22] each product is formed at a different rate. The double photodimerization of 1,4-dicinnamoylbenzene crystals is also regarded[23] as a topochemical process. A study[24] of the dimerization of coumarin in micellar media and micro-emulsions has suggested that the orientation and concentration of molecules at the interfacial film favours reaction *via* the singlet excimer. The photoreactions of cinnamate esters in micro-emulsions are similar to those in the solid state.[25] The 2 + 2-cycloadditions of α-enones have been reviewed.[26,27]

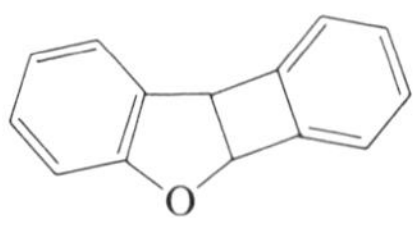

(1) 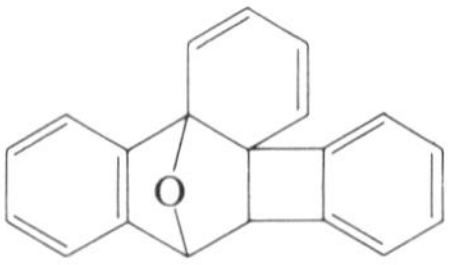 (2) 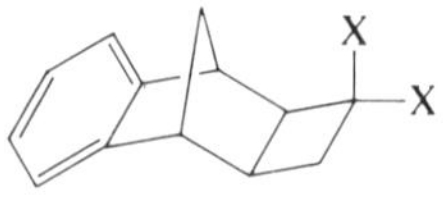 (3) X = $PhSO_2$

The stereochemistry of cycloadduct formation from cyclopentyne and 2,3-dihydrofuran has been determined;[28] a mechanism involving antarafacial participation of the alkyne in a concerted process is thought to be most likely. Cyclobutano[*a*]indene is also a reactive strained olefin which readily undergoes cycloaddition reactions including $_2\pi + _2\pi$-dimerization.[29] A non-concerted pathway involving diradical intermediates is believed[30] to operate for the formation of (**1**) and (**2**) from benzyne and benzofuran. The use of 1,1-di(benzenesulphonyl)ethylene as an acetylene equivalent is being developed. Benzonorbornadiene with $(PhSO_2)_2C{=}CH_2$ yields (**3**), probably *via* a zwitterionic mechanism.[31]

MINDO/3 calculations have been performed[32] for the Lewis-acid-catalysed reaction between hydroxyethene and acrolein; the main effect of the catalyst appears to be to advance the first transition state and to increase the zwitterionic character of the intermediate. Very low activation energies are predicted[33] for hole-catalysed cyclodimerization of olefins. A two-step mechanism *via* a zwitterion explains the formation of Dewar benzene and benzvalene derivatives, (**4**) and (**5**), from the reaction of tetracyanoethylene (TCNE) with the cyclobutadiene (**6**) (Scheme 1).[34]

(**6**) (**4**) (**5**)

SCHEME 1

Cyclobutane formation in competition with Diels–Alder addition has been observed[35] in cation radical cycloadditions. A kinetic study has been made[36] of the reaction of TCNE with 9-vinylcarbazole and 10-vinylphenothiazine. Activation parameters have been determined[37] for the $SnCl_2$- and $PdCl_2$-promoted cycloreversion of quadricyclane to norbornadiene; the results have been interpreted in terms of the Dewar–Chatt–Duncanson model of metal-ion bonding to olefins. The effect of Lewis acids on the 2 + 2-cycloreversion of Cookson's cage ketones has been studied.[38] The length of the bridge X appears to be significant in determining the ease of thermal reaction for (**7**).[39] Spiro-fusion of a fluorene unit activates cyclopropanes for $_\sigma2 + _\pi2$-cycloadditions to TCNE; a mechanism involving electron transfer has been proposed.[40]

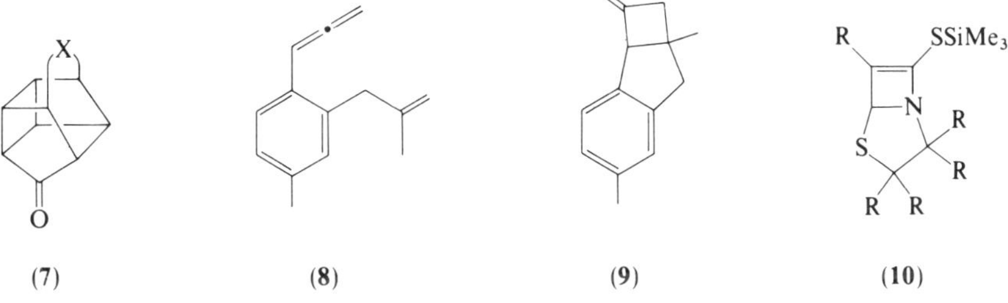

(**7**) (**8**) (**9**) (**10**)

Comparison of chemo- and stereo-selectivities for the cycloadditions of mono- and di-alkylallenes to fumarate and maleate esters with those found for the same allenes with tetrachloroethylene indicate that the former also proceed *via* two-step diradical mechanisms.[41] (Dimethylamino)allene is reactive towards electrophilic alkenes forming 2 + 2-adducts under mild conditions.[42] A diradical mechanism is also considered[43] likely for the conversion of (**8**) to (**9**). The first examples of thermal 2 + 2-cycloadditions to butatrienes at the central double bond have been reported.[44] The triphenylallenyl cation undergoes step-wise reaction with alkenes yielding, after deprotonation, 3-(diphenylmethylene)-2-phenylcyclobutanes.[45]

Photocycloaddition of alkenes to quinoxalin-2(1*H*)-ones occurs regiospecifically at the C=N double bond *via* the triplet state of the heterocycle.[46] 2 + 2-Cycloaddition of alkynyl silyl sulphides to 4,5-dihydrothiazoles leads to thiopenam derivatives,[47] presumably through the adduct (**10**). The stereochemical course of *β*-lactam formation from ester enolates and *N*-(trimethylsilyl)imines depends partly on the geometry of the former;[48] E-enolates give mainly *cis*-*β*-lactams, while Z-enolates yield nearly equal mixtures of *cis*- and *trans*-products. Dinucleotide analogues in which 6-azauracil is connected by a trimethylene chain to uracil or thymine undergo internal "photodimerization" forming azacyclobutanes, a process similar to that found for the corresponding bis-pyrimidine system.[49] The involvement of diradicals and carbenes in the formation of oxetanes and vinyldihydrofurans from tetramethylethylene and ynones has been discussed.[50,51] The photocycloaddition of benzophenone to some rigid dicyclopropylethenes has also been examined.[52] Tetramethylethylene and stilbene undergo photocycloaddition to *N*-methylthiophthalimide forming spirothietanes.[53] 2,4,6-Tri-(*tert*-butyl)thiobenzaldehyde with alkoxy-, alkylthio-, and phenyl-allenes yields mainly 2-methylene-4-arylthietanes in accord with the results of MNDO calculations.[54] There is further support for the involvement of 2 + 2-cycloadditions and cycloreversions involving M=C and C=C species in olefin metathesis and related reactions.[55] For the reaction of acetyl nitrate with cyclic olefins a mechanism has been proposed[56] which involves 2 + 2-cycloaddition of the linear nitryl cation to the carbon–carbon double bond forming an isoxazetidinonyl cation (**11**). The reaction of triazolinediones with olefins has been examined[57] further; in view of the ease of the process for hindered alkenes, such as adamantylideneadamantane, a perpendicular planes approach of the reactants is considered most likely. The parallel alkene and azo units in the rigid molecule (**12**) photocyclize in near quantitative yield forming 1,2-diazetidine (**13**);[58] radiation-induced denitrogenation, the normal reaction for similar compounds without a nearby alkene group, is suppressed.

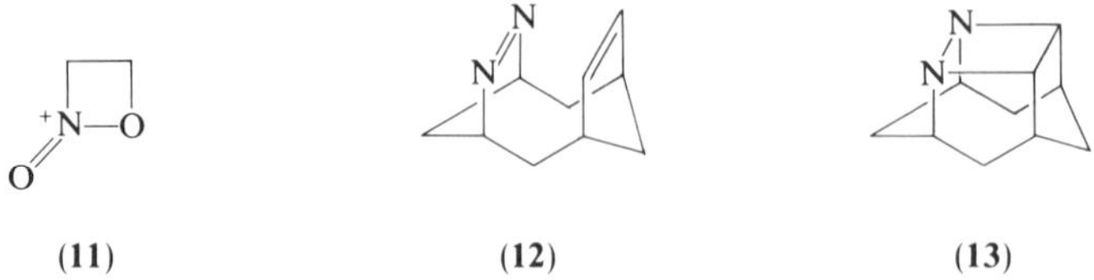

(11) (12) (13)

The ketene radical cation reacts readily with ethene forming the cyclobutanone radical cation, the ease of the process compared with that for neutral ketene being explained[59] by a lowering of its LUMO energy bringing it closer to the HOMO of the alkene. A concerted $2_s + 2_a$-mechanism has been proposed[60] for the regio- and stereo-specific cycloadditions of ketenes to tricarbonyl(η^4-polyene)iron complexes. With silyl enol ethers the formation of a rearranged adduct in addition to the expected 2 + 2-product has been taken[61] as evidence for an ionic pathway. The additions of dichloroketene to several bridgehead bicyclic olefins have been studied;[62] steric factors appear to dictate facial preference while remote π-systems such as etheno or benzo bridges have a strong effect on regioselectivity. The reactivity of chlorocyanoketene towards olefins is similar to that of $Cl_2C{=}C{=}O$.[63] Unlike some *cis*-azo compounds, triazolinediones form 1:1 2 + 2-cycloadducts with diphenylketene.[64] An alternative mechanism has been proposed[65] for the photocycloaddition of ketenimines to ketones involving triplet exciplex formation with charge and possibly electron transfer.

In view of the increasing use[66–68] of intramolecular 2 + 2-photocycloadditions for constructing polycyclic ring-systems of natural products, the scope of the process has been the subject of further investigation. The first example involving an alkyne linked to an α-enone has been described.[66]

2 + 3-Cycloaddition

1,3-Dipolar cycloaddition reactions have been comprehensively surveyed in a very useful two-volume book.[8] The influence of donor–acceptor and localization interactions on the reactivity of the addends has been the subject of review.[69]

The debate surrounding the Huisgen (concerted) and Firestone (diradical) mechanisms continues. The latest of several theoretical treatments involves qualitative valence bond theory to examine the rôle of extended diradicals; it was concluded[70] that the primary electronic structure of the initial transition state determines whether or not the cycloaddition is concerted or proceeds in two steps *via* a second transition state. In his aforementioned survey of pericyclic reactions Dewar suggests[1] that a third mechanism be considered alongside those of Huisgen and Firestone: *viz*., a two-stage process in which the second step (the formation of the second bond) is rate-determining.

A stepwise mechanism involving zwitterionic intermediates is favoured[71] for the cycloaddition of cyclopropene ketals and electron-deficient olefins. The gas-phase reaction of the 2-formylallyl anion with fluoroalkenes has been studied[72] by Fourier transform ion cyclotron resonance mass spectrometry. Diphenylcyclopropenone reacts readily with carboximidates and carboximidamides to give 2-pyrrolin-4-ones.[73] Further examples of Ni(0)-catalysed 3 + 2-addition of methylenecyclopropanes to electron-poor alkenes have been described.[74] Thermal ring-opening cycloadditions of cyclopropyl derivatives with activated olefins have been reviewed.[75]

The *C*-adamantylylide (**14**) is sufficiently stable for X-ray examination;[76] although the central ylidic system is essentially linear, MINDO/3 calculations on model compounds suggest that the arrangement is quite flexible. Kinetic studies[77] of the reactions of electron-deficient olefins with nitrile ylides, generated by pulsed laser photolysis of 2*H*-azirines, provide evidence for reversible formation of an ylide–olefin complex.

Cycloaddition of benzonitrile *N*-phenylimide to phosphindole oxides occurs stereo- and regio-specifically;[78] as expected the dipole approaches from the less hindered side. The same dipole adds to 2,3,4,5-tetraarylfulvenes exclusively at the exocyclic C=C, but 6,6-diphenylfulvene reacts at one of the ring double bonds.[79] Further examples of the generation of nitrile imines by thermolysis of Δ^4-1,3,4,2-oxadiazaphospholes have been reported.[80] The reactivity of *C*-ethoxycarbonyl-*N*-phenylnitrilimine towards olefins is similar to that of diphenylnitrilimine.[81] Diarylnitrilimines react with benzotriazole to form a tetrazole by cycloaddition at the N(2)—N(3) double bond, and an acyclic 1,3-dipolar product by nucleophilic addition at N(1).[82] Cycloaddition to 2*H*-1,3-benzothiazines has also been described.[83] *C*-Acetyl- and *C*-ethoxycarbonyl-nitrilimines react preferentially at the carbon–nitrogen double bond of conjugated imines.[84] With $PhSO_2C{\equiv}CMe$ in the presence of base, pyrazoles are formed from both the alkyne and its allene tautomer.[85] The reactions of benzonitrile *N*-methylimide with alkynes, alkenes, nitriles, and CS_2 have been the subject of further investigation.[86]

Several new dipolarophiles have been examined. The recently confirmed ability of the arenediazonium group to act as an electrophilic dienophile has suggested its possible use as a dipolarophile;[87] diazomethane, azomethine, and thiocarbonyl ylides are all found to react. Benzonitrile oxide, phenyl azide, and diphenyldiazomethane cycloadd to the π-bond of bicyclo[2.1.0]pent-2-ene,[88] whereas dienophiles tend to react at the strained σ-bond. Steric and electronic effects have been invoked[89] to explain the site-, regio-, and stereo- selectivities observed for cycloaddition to the dienes (**15**; X = O, H_2, HOH; $n = 0,1$). Site selectivity for the reactions of nitrile imines and oxides, azides, and nitrones to 7-oxabicyclo[2.2.1]heptadienes is found to be critically dependent on substituents in the 1,3-dipole;[90] thermolysis of the adducts results in the alternative Diels–Alder cycloreversion. Triarylphosphaalkenes undergo 1,3-dipolar cycloaddition to nitrile oxides and azides, but Diels–Alder reactions have not been observed.[91] The stable phosphaalkyne $P{\equiv}CBu^t$ and ClP=CPhSiMe form a variety of 1,3-dipolar cycloadducts.[92,93] The first examples of cycloaddition of nitrile imines and oxides to mono-substituted thioketenes have been described.[94] The exocyclic thione of (**16**) is an efficient dipolarophile towards nitrilium betaines.[95] Thiirene dioxides react with nitrile ylides and imines, the isolated products being formed by extrusion of SO_2 from the initial adducts.[96] The sulphoxide (**17**) also reacts at the alkene double bond.[97] 1,3-Dipolar cycloadditions to adamantanethione are in accord with FMO predictions when allowance is made for steric factors.[98,99]

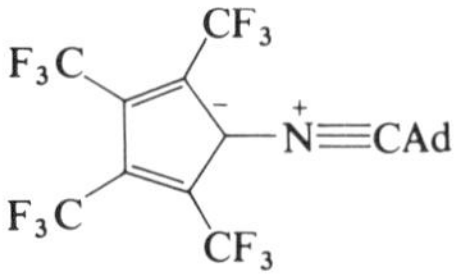

F_3C, CF_3, $\overset{+}{N}{\equiv}CAd$, F_3C, CF_3

(**14**)

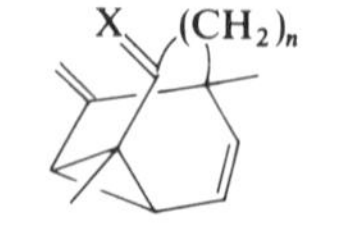

X, $(CH_2)_n$

(**15**)

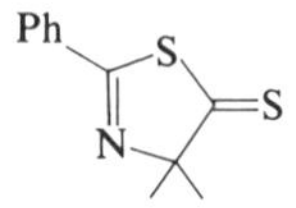

Ph, S, S, N

(**16**)

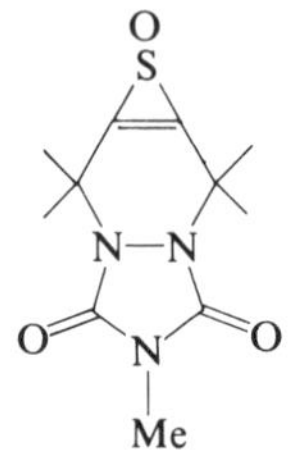

O, S, N—N, O, N, O, Me

(**17**)

The cycloaddition reactions of nitrile oxides continue to attract attention. A detailed study[100] has been made of the addition of aromatic nitrile oxides to alkyl- and halo-substituted benzoquinones; the frontier orbital approach can be used to rationalize[101,102] the site- and regio-selectivities. Evidence for the involvement of nitrile oxides as intermediates in the reactions of primary nitroalkanes with *p*-toluenesulphonic acid[103] or Ac_2O/NaOAc[104] and of arylbromonitromethanes with triphenylphosphine[105] is provided by trapping experiments. The effect of substituents on the reaction of aromatic nitrile oxides with aliphatic nitriles has been examined[106] in more detail; V-shaped Hammett plots were obtained, similar to those observed for other 1,3-dipolar cycloadditions. The dipolarophilicity of aromatic nitriles is enhanced by *ortho*-acetylamino groups, a solvent dependent effect which can be attributed to hydrogen bonding.[107] The influence of substituents at N(1) of indoles has been studied;[108] electron-withdrawing groups reduce reactivity. 2,3-Dihydrothiophene is less regioselective than 2,3-dihydrofuran due to differences in the energies and coefficients of the frontier orbitals.[109] With E-cinnamaldehyde, 4-formylisoxazolines are the major product, the opposite regiochemistry to that found for acrolein.[110] The reaction of $PhSO_2CNO$ with *N*-substituted 1,2-dihydropyridines yields only pyrido[3,4-*d*]isoxazolines.[111] With PhCOCH=CHSMe, 4-benzoylisoxazoles are formed in accord with FMO predictions.[112] Contrary to earlier reports, trifluoroacetonitrile oxide does dimerize.[113] Diastereoface selection in the reactions of nitrile oxides with alkenes has been the subject of several investigations,[114–118] mainly associated with their use in natural product synthesis. Both chiral alkenes and chiral nitrile oxides have been studied. The selectivity depends, as anticipated, on the proximity of the developing asymmetric centre to the existing chiral centre. Benzonitrile oxide adds preferentially *anti* to the less crowded face of *cis*-3,5-disubstituted cyclopentenes.[119] Nitrile sulphide-derived products are reported[120] to be formed on irradiation of thiocarboxamides under aerobic conditions.

By studying their reactions with diazomethane the order of dipolarophilic reactivity of perfluoroalkenes has been established[121] as $(R_F)_2C{=}C(R_F)_2 > (R_F)_2C{=}CFR_F \gg (R_F)_2C{=}CF_2$, $R_FCF{=}CFR_F$ (R_F = perfluoroalkyl). Highly regioselective addition occurs with the carbon of the dipole becoming attached to the site in the alkene most susceptible to nucleophilic attack; the results were interpreted in terms of frontier orbital interactions. Fluoroallene and 1,1-difluoroallene add to diazoalkanes using exclusively their non-fluorinated double bond consistent with a dipole–HOMO-controlled process;[122] both steric and FO considerations need to be invoked to explain the regiochemistry of the products. 2-Diazopropane readily cycloadds to imidazo[1,2-*b*]pyridazine at C(7)—C(8), the partially localized double bond, in spite of the reported low reactivity of the pyridazine moiety in such systems.[123,124] The regiochemistry of the reaction between Me_2CN_2 and polyenic esters and nitriles has been examined;[125] whether Δ^2- or Δ^1-pyrazolines are formed depends critically on the substituents. As for diazocyclopentadienes the preferred mode of cycloaddition of diazoindene to electron-deficient alkynes and alkenes is 1,3 rather than 1,7.[126]

The reaction of phenyl azide and azomethine ylides with 3-methoxycarbonyl-1,2-

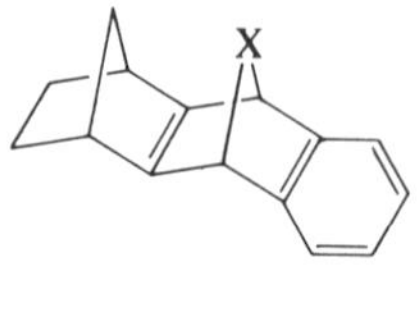

(**18**) X = O
(**19**) X = CH_2

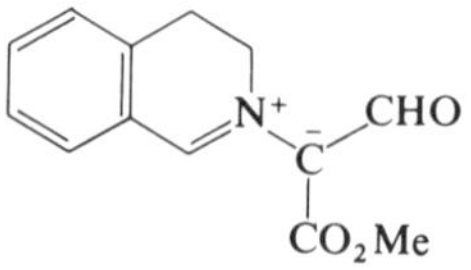

(**20**)

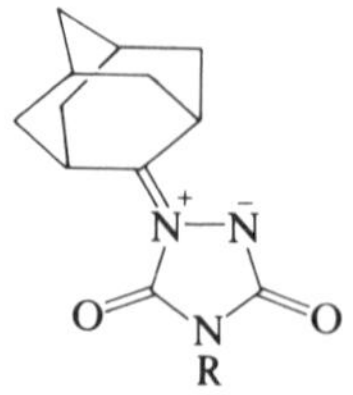

(**21**)

dihydronaphthalenes and indenes is diastereoselective.[127,128] The double bond in (**18**) is more reactive towards phenyl azide than that in (**19**).[129] CNDO/2 and CNDO/S calculations have been used[130] to explain the regioselectivities observed for the reactions of 6,7-dialkoxy-3,4-dihydroisoquinolinium methylides. A variety of cycloaddition reactions has been reported[131] for (**20**), the first isolable azomethine ylide in which the 1,3-dipole system is not part of an aromatic nucleus. On thermolysis, 2,2-dichloro-1,3-diphenylaziridine undergoes ring-opening with C—C rather than the usual C—N bond fission to afford the isomeric azomethine ylide, which can be trapped by DMAD.[132] Non-stabilized azomethine ylides have been generated by CsF-induced desilylation of (trimethylsilyl)methylimmonium salts and their cycloaddition reactions studied.[133–136] Treatment of *N*-(trimethylsilyl)imines with water generates non-stabilized azomethine ylides (*e.g.* PhCH—$\overset{+}{N}CH_2^-$), which cycloadd to olefinic dipolarophiles in a regioselective and stereospecific manner.[137] The reactivities of various heterocyclic immonium ylides have been assessed using CNDO/2 calculations.[138] The variety of addition and cycloaddition products formed from DMAD with pyrazolo- and triazolo-[1,2-*a*]benzotriazoles can be explained in terms of diradical or zwitterionic intermediates.[139] Adamantanone azine reacts with triazolinediones to form a 1,3-dipole (**21**) which can be trapped by DMAD or phenylisocyanate; the closely related adamantanone *O*-oxide forms a 1,3,4-trioxolane adduct with *p*-nitrobenzaldehyde.[140]

The reaction of nitrones with electron-deficient alkenes closely follows frontier orbital predictions;[141] 5-substituted isoxazolidines are usually the major products as expected for nitrone LUMO–dipolarophile HOMO control, but a greater proportion of 4-substituted adducts is formed as the electron affinity of the dipolarophile increases. Oxazoline *N*-oxides are more reactive than the corresponding pyrroline oxides in accord with FMO predictions.[142] The formation of isoxazolidines from *N*,α-diphenyl nitrone with benzo[*b*]thiophene *S*-oxides and *S*,*S*-dioxides has been studied;[143] the observed regiochemistry was interpreted in terms of FO interactions on the basis of CNDO/S calculations and *PE* data. A fuller discussion of X—Y=ZH systems as potential 1,3-dipoles has been published.[144–146] Various α-amino-acids ($R_2C(NH_2)CO_2H$) react with carbonyl compounds (R'_2CO) to generate, after decarboxylation, 1,3-dipoles of the form $R_2C{=}\overset{+}{N}H{-}\overset{-}{C}R'_2$, which can be trapped inter- or intra-molecularly by olefinic dipolarophiles.[147,148] Some examples of cycloaddition of oximes to alkenes have been described,[149] which are believed to proceed *via* the nitrone tautomer $R_2C=$

$\overset{+}{N}H—O^-$. 5-Methyl-2,4-diphenyl-3*H*-pyrrol-3-one dimerizes to give isoxazolidine (**22**) rather than a head-to-tail dioxadiazine.[150] A more detailed account has been presented[151] of the high diastereofacial selectivity in the cycloaddition of chiral nitrones to electron-rich dipolarophiles. Further examples illustrating the 1,3-dipolar character of nitronic esters have been reported.[152]

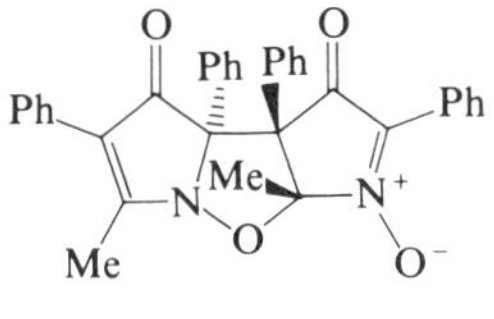

(**22**)

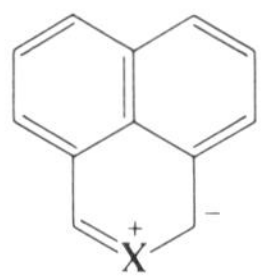

(**23**) X = S
(**24**) X = NMe

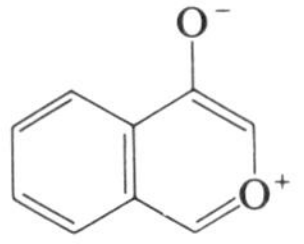

(**25**)

The differing behaviour of the naphthothiocarbonyl ylide (**23**) and azomethine ylide (**24**) towards maleimides has been rationalized in terms of FO theory;[153] the former gives mainly *exo*- and the latter *endo*-adducts. The variations in regioselectivity observed for the cycloaddition of 2-substituted 3-oxidopyrylium betaines to styrene, ethyl vinyl ether, and acrylonitrile have been interpreted in terms of FO theory.[154] The benzopyrylium oxide (**25**) yields 3 + 2-adducts with a variety of dipolarophiles including electron-rich, electron-poor, and electronically unbiased olefins.[155] Only $4\pi + 2\pi$-adducts are formed from 2,4,6-triphenylpyrylium-3-olate with 6,6-diphenylfulvene.[156] Cellulose-derived levoglucosenone (**26**) undergoes thermal 3 + 2-cycloreversion to 3-oxidopyrylium, which subsequently reacts with further (**26**) yielding (**27**) in addition to a series of dimers (Scheme 2).[157] Thiobenzophenone *S*-ethylide and *S*-benzylide, generated from $Ph_2C{=}S$ and the appropriate diazoalkane *via* Δ^3-1,3,4-thiadiazolines, have been trapped by *N*-phenylmaleimide, TCNE, and DMAD.[158] Both non-radical and diradical character for carbonyl ylides is indicated by MO treatment.[159] For the cycloaddition of aryl isocyanates to 1-oxa-4-azobuta-1,3-dienes forming 3-arylimidazoline-2,4-diones, the rate of reaction is greatly enhanced by electron-withdrawing groups in the isocyanate; the key rôle of the isocyanate nitrogen lone-pair electrons in this unusual cycloaddition is emphasized.[160] The formation of 1-aminopyrroles from 1,2-diazabutadienes and ynamines represents a rare case of 1,3-cycloaddition to a heterodiene.[161]

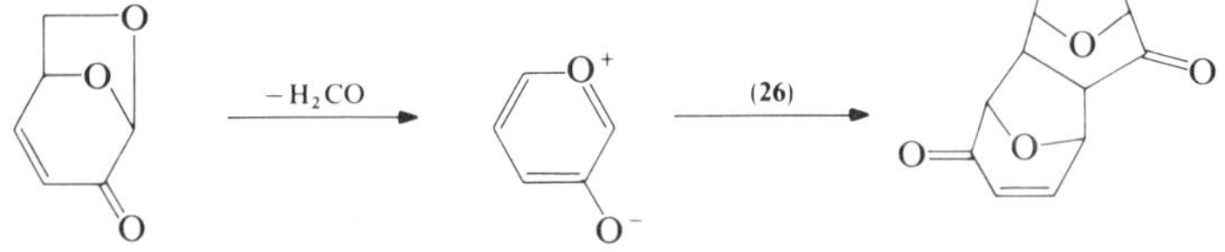

(**26**) $\xrightarrow{-H_2CO}$ $\xrightarrow{(\mathbf{26})}$ (**27**)

SCHEME 2

There has been a further increase in the volume of literature dealing with intramolecular 1,3-dipolar cycloadditions and the synthetic exploitation of these reactions. Further examples of intramolecular trapping of carbonyl ylides by alkenes have been described.[162,163] A systematic study has been made[163] of (**28**), generated *in situ* by thermal ring-opening of the corresponding oxiranes. For non-activated alkenes ($X = CH_2$) cycloaddition only takes place when the dipole and dipolarophile are in close promimity ($n = 3$), but when the alkene is activated (X = OCO) the process is much more facile and larger rings can be prepared (*e.g.* $n = 11$). The first examples of intramolecular 1,3-dipolar cycloaddition to a furan ring have been described.[164] Further details have been published[165] of the intramolecular trapping of arene nitrile *N*-(*o*-alkenylthio)- and (*o*-alkynylthio)-imines. Several examples of 3 + 2-cycloaddition between parallel alkene and 1,3-dipolar groups have been reported.[166–168] The azomethine imine (**29**), generated by deprotonation of the corresponding *N*-methyl compound, forms the pyrazolidine (**30**), a reaction previously unknown for non-activated components.[166] Similarly the azoxy compound (**31**) affords the oxadiazoline (**32**).[167] The Δ^4-isoxazoline (**33**), formed from DMAD and the nitrone (**34**), acts as a precursor for the azomethine ylide (**35**), which readily undergoes intramolecular cycloaddition to suitably placed alkenes (Scheme 3).[169] Intramolecular ($3^+ + 2$)-cycloaddition of aldehyde and ketone

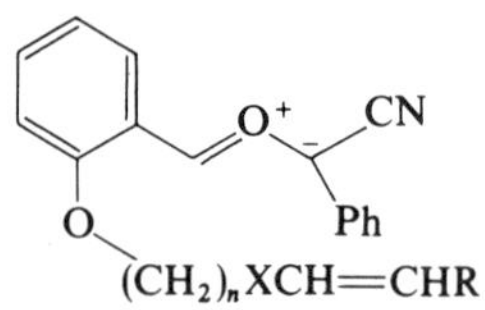

(**28**)

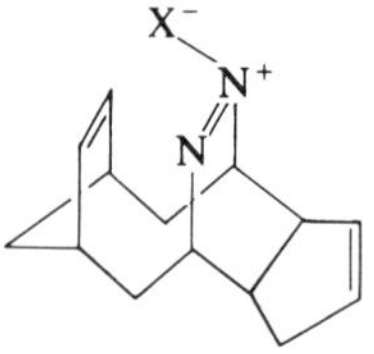

(**29**) $X = CH_2$
(**31**) $X = O$

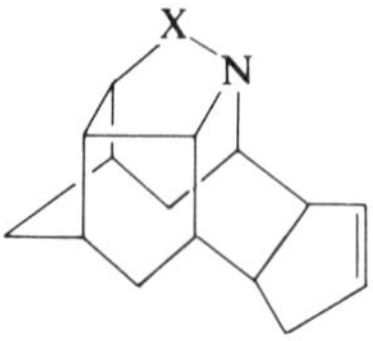

(**30**) $X = CH_2$
(**32**) $X = O$

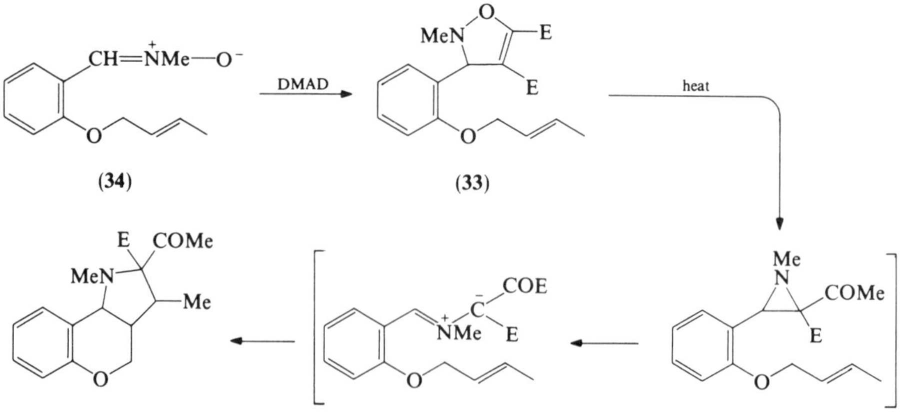

(**34**) (**33**) (**35**)

$E = CO_2Me$

SCHEME 3

hydrazones to alkenes forming chromeno[4,3-*c*]pyrazoles have been described.[170] Studies of intramolecular 2 + 3-cycloadditions with mainly synthetic objectives include reactions of nitrile oxides,[171-174] azides,[175] azomethine ylides,[176,177] azomethine imines,[178] and nitrones.[179]

2 + 4-Cycloaddition

Diels–Alder reactions have been the subject of several reviews.[5,6,180-187] Topics covered include control of regio- and stereo-chemistry,[5,6,180] and synthetic applications of asymmetric cycloadditions.[181,182] There have also been reviews in the Chinese[186] and Polish[187] languages.

From his assessment of the recent literature Dewar[1] has suggested that the process is likely to be non-synchronous in *all* cases, even when both diene and dienophile are symmetrical. He proposes a mechanism similar to that described earlier for 1,3-dipolar cycloadditions as an alternative to current two-step diradical (or zwitterionic) and one-step hypotheses: *viz.* a two-stage process where a diradical or zwitterionic-like intermediate is formed without specific activation from diene or dienophile, and where the rate-determining step is its activated conversion into product. Support for these conclusions was provided by kinetic studies of the maleic anhydride–furan system.[188] In contrast, Brown and Houk,[2] on the basis of *ab initio* calculations, favour a concerted synchronous transition state for the reaction of butadiene and ethylene. Bertran *et al.*[189] have examined the effect of correlation energy for the same reaction using Moeller–Plesset perturbation theory and conclude that its introduction stabilizes preferentially asynchronous structures. The value of activation volume measurements in assessing the relative position of transition states has also been discussed.[3]

The influence of alkyl, halogen, and sulphur substituents on Diels–Alder reactivity and selectivity has been the subject of further examination.[190,191] Regioselectivity for cycloadditions involving thio-substituted butadienes under thermal and catalysed conditions have been explained[192] by using primary and secondary interactions of the frontier orbitals. Further kinetic studies[193] have been made of the gas-phase reaction of cyclohexa-1,3-diene and alkenes; a diradical calculation model was used to interpret the results. 9-Phenylethynylanthracene readily forms 4 + 2-photodimers, but attempted reactions of anthracene itself with acetylene or phenylacetylene only yielded dianthracene.[194] Pyrido[3,2-*g*]quinoline[195] and anthra[2,3-*b*]thiophene[196] both react with dienophiles across the 5,10-positions in accord with theoretical predictions. Cyclopentadienone can act as either diene or dienophile.[197,198] Attack of *N*-phenyltriazolinedione (PTAD) on napadiene (**36**) occurs at the less hindered face.[199]

The reaction of 6,6-dimethylfulvene with 2-acetoxyacrylonitrile gives, in addition to the expected 4 + 2-product, an adduct formally derived from 2 + 4-addition to 1-isopropylidenecyclopentadiene.[200] The 3,7-dehydrotropone (**37**) behaves more like a fulvene than a tropone undergoing reversible site-, regio-, and stereo-specific dimerization to (**38**).[201] The origins of the high *endo*/*exo* product ratio in the Diels–Alder reaction of cyclopentadiene and methyl vinyl ketone in water have been

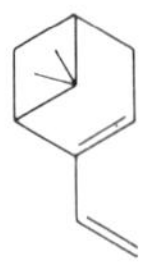

(36) 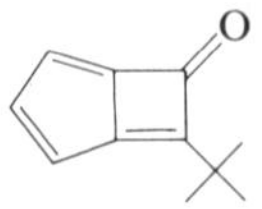 (37) 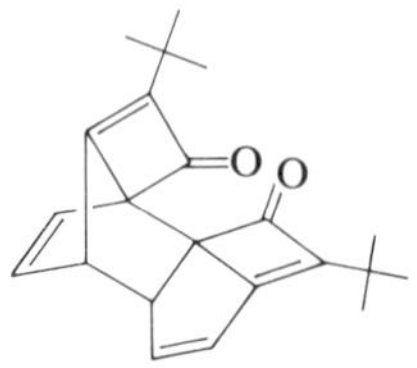 (38) 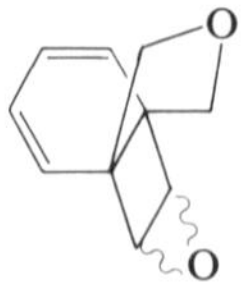 (39)

the subject of more detailed investigation;[202] the effect is *not* due to aggregation as previously suggested, but the ratio is influenced by salts (*e.g.* lithium or guanidinium chlorides) which change the hydrophobicity of the medium. X-Ray crystal-structure analysis[203] has confirmed that the attack of PTAD on the propellanes (**39**) occurs *syn* to the five-membered ether ring whose α-hydrogens exert less repulsion than the α-epoxy hydrogens or the epoxide oxygen attached to the cyclobutane. 2+4-Addition of PTAD to 1,2-benzocycloheptatriene, a cycloheptatriene *not* in equilibrium with its norcaradiene isomer, has been observed.[204] A concerted mechanism is favoured[205] for the cycloaddition of homophthalic anhydrides to alkenes, but a step-wise process involving Michael addition cannot be ruled out. The Diels–Alder reactions of 3-carbomethoxy-2-pyrone have been examined in greater detail.[206] *N*-Phenylmaleimide adds to phenol across the 2- and 5-positions yielding a mixture of *exo*- and *endo*-bicyclooctenones.[207] The structures of the adducts between *N*-methylisoindole and maleimides have been determined.[208]

The *ortho*-quinoid vinylketene (**40**), generated by flash photolysis of benzocyclobutenone, behaves as an electron-rich diene system rather than as a simple ketene and gives a Diels–Alder adduct with maleic anhydride.[209] Its diphenylmethylene analogue (**41**) undergoes intramolecular cycloaddition forming norcaradiene (**42**).[210] 9-Aryl- and 9-vinyl-barbaralenes (**43**; R = Ar, $CH{=}CH_2$) readily undergo a retro-Diels–Alder reaction yielding the corresponding cyclohepta-

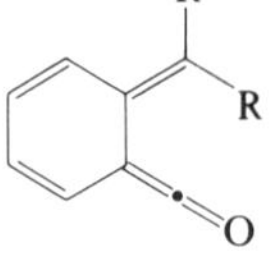

(**40**) R = H
(**41**) R = Ph

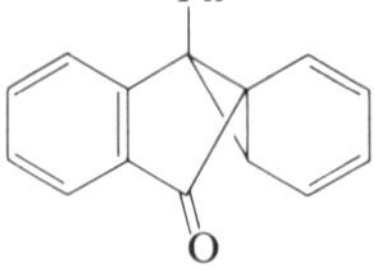

(42) 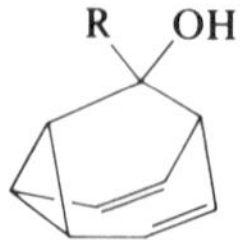 (43)

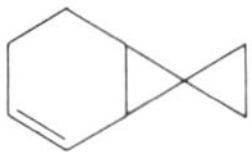

(44) 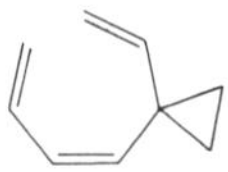 (45)  (46)

trienes,[211] thus demonstrating that it is not necessary for the fragmented 4π-component to be incorporated in an aromatic nucleus as earlier suggested.[212] The reversible cycloreversion of the bicyclo[4.1.0]hept-2-ene system has been studied;[213] the interconversion of (**44**) and (**45**) is believed to be a pericyclic process, while the competitive conversion of (**44**) into (**46**) is likely to involve a diradical intermediate.

Further work has been carried out on the cycloaddition of cyclic and acyclic dienophiles to hexachlorocyclopentadiene.[214–219] The effect of ring size on activation volume for its reaction with cycloalkenes has been studied;[214] a trend toward an early transition state with low $\Delta V^{\neq}/\overline{\Delta V}$ values was noted as the alkene varied from cyclopentene to cyclodecene.

The thiiranoradialene system (**47**) has been studied further;[220,221] (**47**; R = Me) undergoes 4 + 2-cycloaddition to singlet oxygen without any detectable *S*-oxidation, ene reaction, or 2 + 2-addition, to give the endoperoxide (**48**), which itself acts as a dienophile forming (**49**) with furan.[220] Further examples of diene-transmissive Diels–Alder reactions have been described.[222–226]

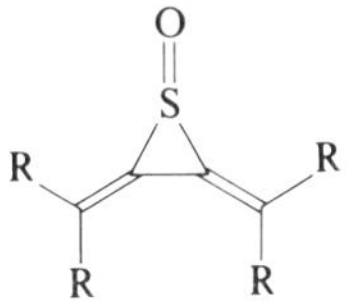

(47) 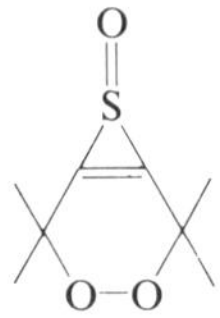 (48) 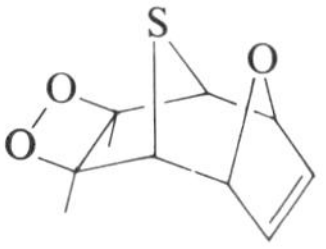 (49)

The cycloaddition reactions of isodicyclopentadienes continue to attract attention. Each of the three isomers, (**50**)–(**52**), forms two adducts with TCNE;[227] (**50**) gives, unusually, mainly a 4 + 2-*exo*-product, (**51**) both *exo*- and *endo*-adducts, while (**52**) forms a 2 + 2- as well as an *exo*-4 + 2-product. (**50**) with 2-methyltropone yields (**53**) *via* the bond-shift isomer (**52**) even though the latter is the least reactive of the three isomeric dienes.[228] Divergent π-facial stereoselection is found[229] for the

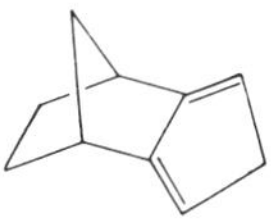

(50) 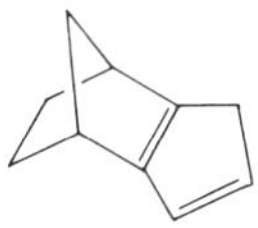 (51) 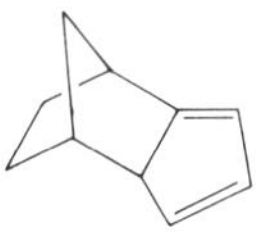 (52)

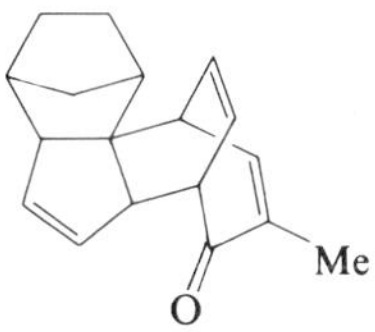

(53) 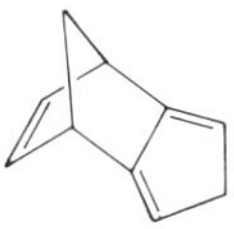 (54) 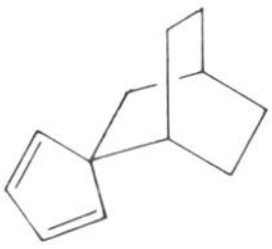 (55)

reaction of (**50**) and its dehydro analogue (**54**); (**50**) gives predominantly *exo*-products, whereas *endo*-adducts are formed from (**54**). Trimethylsilyl derivatives of (**50**) have also been examined.[230] Little stereoselection is observed for the spiro compound (**55**).[231]

In view of the synthetic potential of exocyclic dienes and tetraenes such as (**56**) and (**57**), the mechanisms of their reactions are being studied in detail.[184, 232–236] Complexation of the endocyclic double bond of (**56**) with d^8- and d^6-metal carbonyls increases the reactivity of the *exo-s-cis*-butadiene unit;[232] 4 + 2-dimerization to products derived from adduct (**58**) is also favoured whereas Fe atoms promote formation of the 2 + 2-dimer (**59**).[233] Substituents at the bridgehead have been shown to influence the regiochemistry of such cycloadditions.[234] The field has been reviewed.[184]

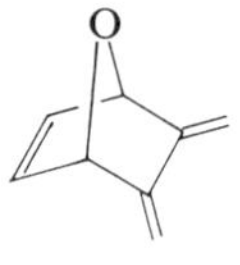

(**56**) 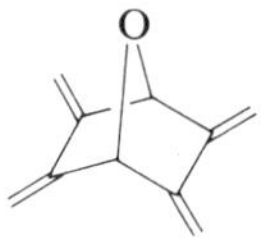 (**57**) 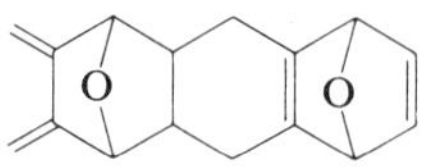 (**58**) 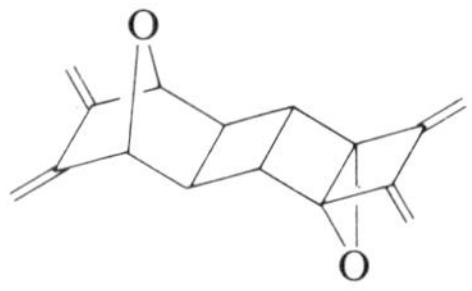 (**59**)

The effect of substituents on the Diels–Alder reaction of cycloalkenones and isoprene has been the subject of MINDO/3 calculations.[237] The same method has been used to study the acid-catalysed reaction of acrolein with 1-hydroxybutadiene, and it was concluded[238] that the effect of the catalyst is to increase the two-step character of the process. CNDO/2 calculations have been applied to the α-allenic ketone–diene system.[239]

The high reactivity of the alkene units in (**60**) can be attributed to strain;[240] with cyclopentadiene it readily forms a bis-2 + 4-adduct. Cyclooctatetraene reacts with diphenylisobenzofuran to give three 1 : 1 and one 2 : 1 adducts;[241] the isomerization of monoadduct (**61**) to the cage compound (**62**) is noteworthy in that it appears to involve an unprecedented 4 + 2-cycloaddition in which a dialkyl-substituted benzene ring acts as the 2π-component. Methylenecyclopropene, the simplest cross-conjugated cyclic hydrocarbon, undergoes 2 + 4-cycloaddition to cyclopentadiene at the endocyclic double bond.[242] The strained bicyclic allene, 6,7-benzobicyclo[3.2.1]octa-2,3-diene, has been trapped with diphenylisobenzofuran.[243] Complexation with silver ion promotes 2 + 4-cycloadditions of *trans*-cycloheptenes.[244] Benzyne reacts with 9-(*tert*-butylthio)anthracene yielding trip-

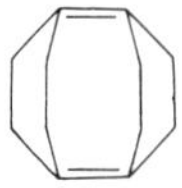

(**60**) 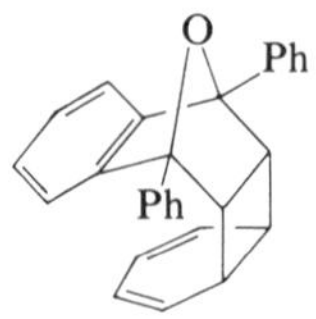 (**61**) 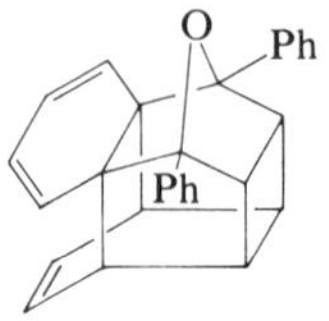 (**62**) 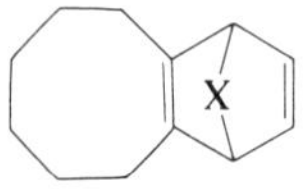 (**63**)

tycene and naphthacene adducts, rather than the usual products derived from attack at sulphur.[245] The adducts (**63**) formed from cyclooctyne and dienes show varying thermal stabilities;[246] cycloreversion to benzocyclooctene is rapid when the expelled group X is CO_2, SO_2, CO, and N_2, but more forcing conditions are required for X = CH_2CH_2.

The effect of solvent on the kinetics of reaction between CH_2=CMeCMe=CH_2 and 1,4-naphthoquinone has been determined;[247] a hyperbolic correlation was obtained with the acceptor number of the solvent (which itself correlates with the LUMO of the solvent), suggesting that the solvent acts as an acceptor on a donor reactant. Chloroprene adds to anthracene-1,4,9,10-tetraone exclusively at the external C(2)—C(3) double bond.[248] The novel 1,4- and 1,6-quinones of azulene have been trapped as their 2 + 4-adducts with cyclopentadiene, reaction taking place at the C(2)—C(3) double bond.[249] In contrast, the 1,5- and 1,7-isomers are much less reactive and do not add to cyclopentadiene; with diphenylisobenzofuran 2:1 products are formed by a combination of 4 + 2- and 4 + 6-reactions.[250] Further examples of cycloaddition reactions of sulphonylalkenes have been reported.[251-254] The dienophilic character of 1,4-benzodithiin tetroxide appears to be similar to that of naphthoquinone.[252] *S*-Vinyl-*S*-aryl sulphoximes are more reactive than phenyl vinyl sulphone.[253] A fuller account has been published[254] of the use of 1,2-bis(phenylsulphonyl)ethenes as acetylene equivalents; their high reactivity allows a variety of adducts to be prepared, including (**64**) from indene, presumably *via* isoindene. The Dewar furan (**65**) has been trapped as its 2 + 4-adduct with furan.[255]

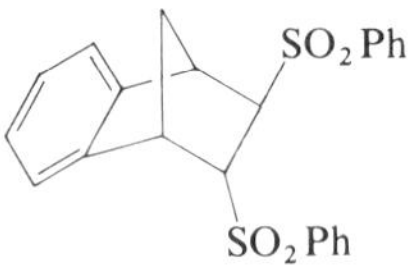

SO₂Ph SO₂Ph (**64**) Me—N O O Me Me O (**65**) AcO O AcO OEt (**66**) AcO O AcO OEt (**67**)

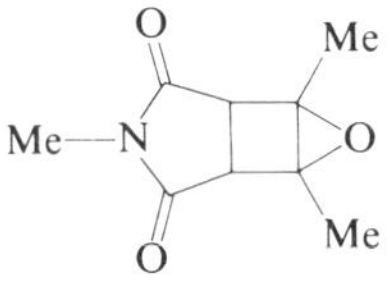

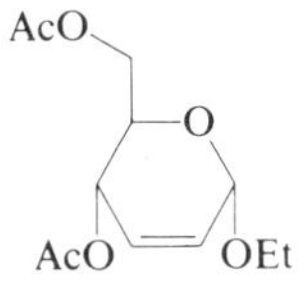

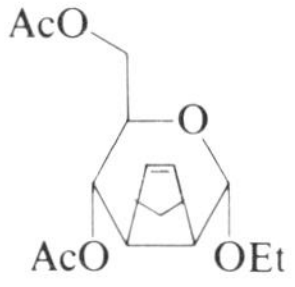

The reactivities of a series of ketene hemiacetals in inverse-electron-demand Diels–Alder reactions with monoaryltetrazines have been measured;[256] $(MeS)_2C$=CH_2 proves to be 5–6 orders of magnitude more reactive than $(Me_2N)_2C$=CH_2. α,β-Acetylenic chromium–carbene complexes, like their olefinic counterparts, readily cycloadd to a variety of alkenes.[257] More detailed investigation of the 2 + 4- and 2 + 2-reactions of fluorinated allenes has reinforced the hypothesis that the former process is concerted and the latter step-wise.[258]

There is an increasing amount of work on cycloaddition to chiral dienophiles.[259-265] Diastereoselectivity has been observed for the reactions of dienes with 2-oxazolidinones,[259] ethyl arylsulphinylmethylenepropionate,[260] acrylate esters of lactic acid,[261] and camphor-derived dienophiles.[262] Some enantiomeric selectivity was obtained[263] for the formation of 4-vinylcyclohexene from butadiene catalysed by chiral diazadieneiron(0). Face selectivity is observed for the reaction of the unsaturated pyranoside (**66**) with cyclopentadiene, only the α-D-*manno*-adduct (**67**) being isolated.[266] The degree of diastereoselectivity for the cycloaddition of bis-ether-1,3-dienes with TCNE has been presented[267] as evidence for cooperativity in these reactions. The presence of Lewis acid catalysts can change the transition state from synchronous to non-synchronous; their rôle in enhancing asymmetric induction is attributed[268] to increased steric interaction at one end of the dienophile.

Interest in hetero-Diels–Alder reactions has been maintained. Aryl 2-furyl and 2-thienyl thioketones react as a heterodiene rather than a cyclic diene with dienophiles such as styrene, norbornene, and maleic anhydride.[269] The formation of pyrans from α-enones and phosphacumulene ylides has been studied.[270] The C═C—C═S unit of α,β-unsaturated thio-esters can act as a heterodiene yielding, for example, thiapyrans with alkenes.[271,272] Triethyl azomethinetricarboxylate reacts with electron-rich alkenes using C═N—C═O as the 4π-component yielding 2-ethoxy-5,6-dihydro-1,3-oxazines.[273]

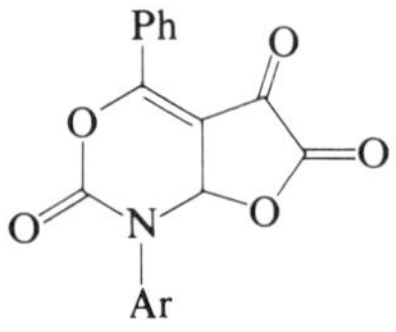

(68) 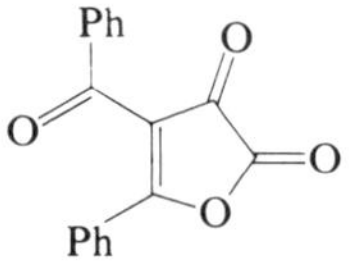 **(69)** 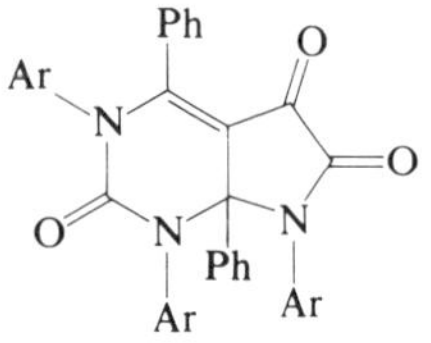 **(70)**

On the basis of isotopic labelling experiments a mechanism involving initial formation of the 4 + 2-adduct (**68**) has been suggested[274] for the conversion of furandiones (**69**) into pyrrolo[2,3-*b*]pyrimidines (**70**) in the presence of aryl isocyanates.

Phenylglyoxal anils can act as amphoteric Diels–Alder reagents reacting as a dienophile at C═N or as a heterodiene using the C═N and the conjugated Kekulé double bond;[275] its reactions with norbornene show low selectivity.[276] The electron-rich C═N bonds of oxazoline, thiazoline, and 1-methylpyrazolines are effective dienophiles towards the *s-cis*-azine system of tetrazines bearing electron-withdrawing substituents.[277] Hexafluoroacetone reacts with cyanamides to form 4*H*- and not 2*H*-1,3,5-oxadiazines as previously reported; a mechanism (Scheme 4) involving 2 + 2- and 4 + 2-cycloadditions was proposed.[278] Hexafluorothioacetone behaves similarly.

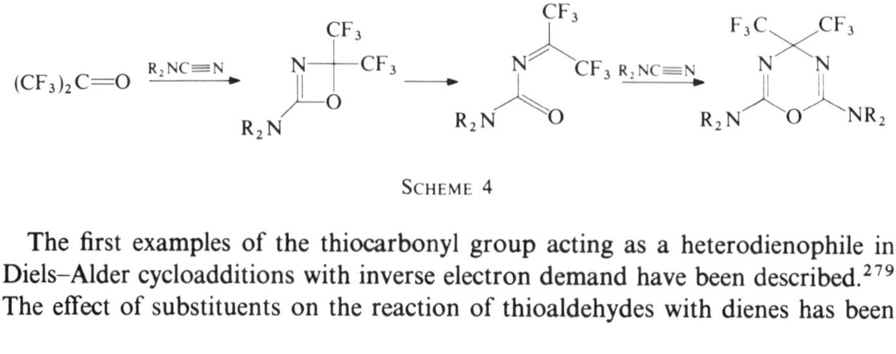

$(CF_3)_2C{=}O \xrightarrow{R_2NC\equiv N}$ [4-membered ring: N, C(CF_3)$_2$, O, C–R_2N] $\longrightarrow$ R_2N–C(=O)–N=C(CF_3)$_2$ $\xrightarrow{R_2NC\equiv N}$ [oxadiazine: F_3C, CF_3, N, N, O, R_2N, NR_2]

SCHEME 4

The first examples of the thiocarbonyl group acting as a heterodienophile in Diels–Alder cycloadditions with inverse electron demand have been described.[279] The effect of substituents on the reaction of thioaldehydes with dienes has been studied[280] and the results compared with predictions from *ab initio* MO calculations. Adamantanethione reacts regioselectively with α,β-unsaturated carbonyl compounds to yield 1,3-oxathiacyclohex-5-enes,[281, 282] apparently the first examples of a thione reacting with a heterodiene; the results were interpreted in terms of frontier MO theory.

The phosphine *P*-sulphide (**71**), generated as a transient intermediate by treatment of the corresponding phosphine with sulphur, reacts as a dienophile at its 1,6-positions with 2,3-dimethylbutadiene and with DMAD as a diene across the 1,4-positions (Scheme 5).[283] Further details of 4 + 2-dimerization of 2*H*-phospholes have been published.[284] Ph-P(S)=CH_2, the first example of a methylenephosphine free from steric crowding, has been trapped by $H_2C{=}CMeCMe{=}CH_2$.[285] A further case of 2 + 4-cycloaddition to a stable silaethene has been reported.[286]

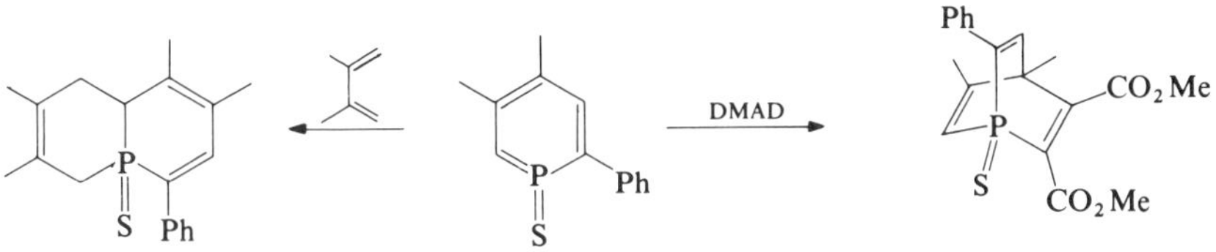

S Ph

P S Ph

DMAD

Ph P S CO_2Me CO_2Me

(**71**)

SCHEME 5

The scope and regiochemistry of the reaction between 1,3-dienes and *p*-nitrobenzenediazonium salts yielding 1,6-dihydropyridazines have been examined in more detail;[287, 288] all the evidence points to a concerted cycloaddition involving the diazonium-HOMO and the diene-LUMO. The same diazonium salt with cyclopentadiene yields 1 : 2 and 1 : 3 adducts, (**72**) and (**73**), and not solely (**74**) as previously reported; the proposed mechanism[288] involves initial azo coupling followed by successive Diels–Alder additions. Naphthalene forms a 4 + 2-adduct (**75**) on photolysis with *N*-methyltriazolinedione;[289] no such reaction occurs with benzene. Protonation has been shown[290] to increase the dienophilicity of the azo linkage.

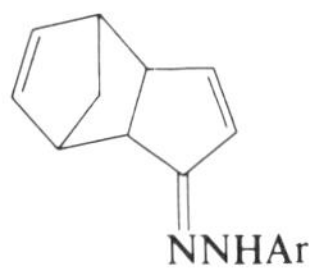

NNHAr

(**72**)

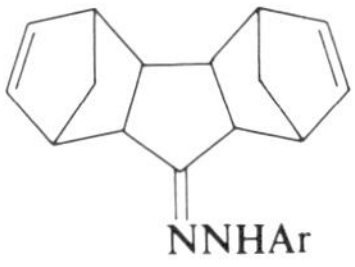

NNHAr

(**73**)

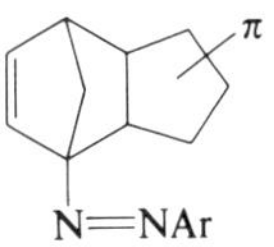

π N=NAr

(**74**)

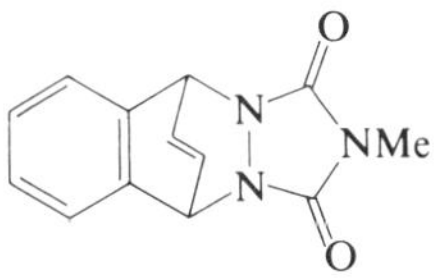

O N NMe N O

(**75**)

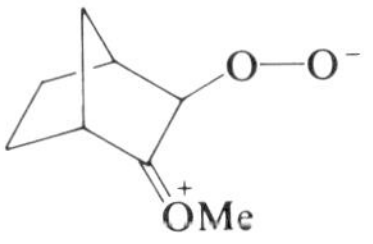

O—O^- $\overset{+}{O}$Me

(**76**)

O O O OMe Me

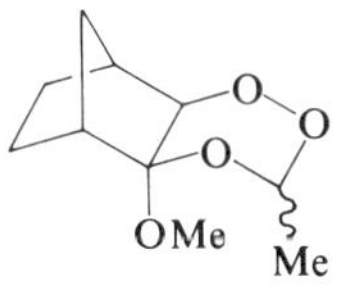

(**77**)

The transient zwitterionic peroxide (**76**), formed from 1-methoxynorbornene and singlet oxygen, can be captured by acetaldehyde giving the 1,2,4-trioxane (**77**).[291] Solvent dependence of the rates of 2 + 4- and 2 + 2-addition of $^1O_2$ to simple mono- and di-enes indicates weak charge separation in the transition state for the former in accord with a concerted mechanism, but for the latter the estimated dipole moment (6.7 D) is much greater, consistent with a two-step pathway *via* a zwitterionic intermediate.[292] Treatment of silyl- and germanium-protected trisulphides with Ph_3PBr_2 generates S_2, which forms Diels–Alder adducts with dienes but, unlike $^1O_2$, does not appear to undergo 2 + 2-cycloadditions or the ene reaction.[293] The regioselectivity of the 2 + 4-reactions of acylnitroso compounds has been examined,[294] and is found to be consistent with either a normal diene-HOMO-controlled or an inverse-electron-demand diene-LUMO-controlled process. Acyl- and sulphonyl-thionitroso compounds have been generated[295] by extrusion from the 4 + 2-adducts of thiophene-*S,N*-ylides with alkenes such as norbornene, and may be trapped by Diels–Alder or ene reactions with dienes.

The origins of diastereofacial control in the Lewis-acid-catalysed formation of 2,3-dihydro-4-pyrones from aldehydes and activated dienes have been investigated;[296, 297] the stereochemistry of the products depends on the metal used. Interaction between chiral lanthanide complex catalysts, *e.g.* $Eu(hfc)_3$, and the chiral auxiliary results in substantial enantioselectivity for the reaction of aldehydes with 1-menthyloxy-3-trimethylsilyloxybutadienes.[298] Slectivity is also observed for the reactions of sulphines derived from proline[299] and α-D-galactopyranose-6-ulose.[300]

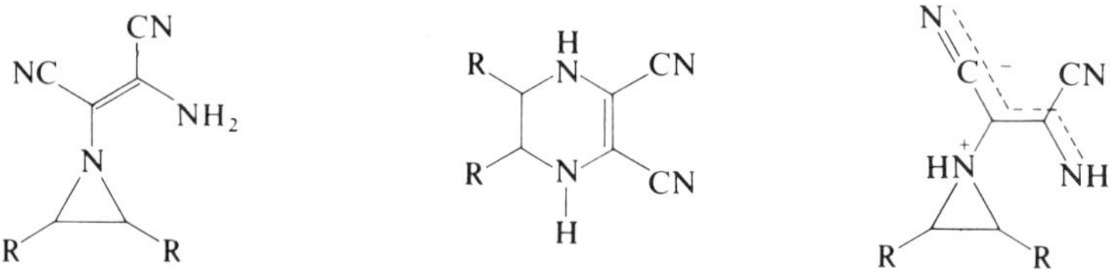

(**78**)

(**79**)

(**80**)

Both aziridines (**78**) and tetrahydropyrazines (**79**) are formed from the reaction of diiminosuccinonitrile with nucleophilic olefins; the proposed mechanism[301] involves initial interaction between the alkene-HOMO and the nitrogen terminus of the heterodiene-LUMO, forming the zwitterion (**80**), followed by protonation or ring-carbon migration.

The mechanism of the cation radical Diels–Alder dimerization of butadiene has been investigated by mass spectrometry;[302] the results are consistent with a two-step pathway involving the cations of an acyclic C_8H_{12} species and 4-vinylcyclohexene. A similar mechanism *via* an acyclic intermediate is also indicated[303] for the reaction of the methyl vinyl ether radical cation and butadiene. The mechanism of the photochemical dimerization of cyclohexa-1,3-diene is solvent-dependent;[304] in benzene the reaction probably proceeds by way of a triplex, but in a polar solvent such as acetonitrile radical cations are involved. Isopyrazoles and 4,5-dihydropyridazines have been studied[305, 306] as electron-deficient heterodienes in $(4^+ + 2)$-

cycloadditions with inverse electron demand. The use of phenol-clay catalysts for Diels–Alder reactions has been described.[307–309] α-Acyl- and α-cyano-α-chlorosulphides undergo Lewis-acid-catalysed ($2^+ + 4$)-cycloaddition to conjugated dienes.[310] A new reaction mode has been discovered[311] for benzvalene in which it acts as the four-electron component in concerted six-electron cycloadditions.

The increased interest shown over recent years in intramolecular Diels–Alder cycloadditions has been maintained. While there have been a number of studies dealing with mechanistic aspects of these reactions, the preponderance of work has been aimed at synthetic applications. Intramolecular photocycloadditoon[312] of two acylbenzo-1,4-quinone moieties separated by a polymethylene bridge parallels that of the previously described intermolecular process. Phenanthrenetrione derivatives (**81**) are formed by reaction of one benzoquinone with the dienol tautomer of the other (Scheme 6); as expected different regio- and stereo-selectivities are observed in the two cases.

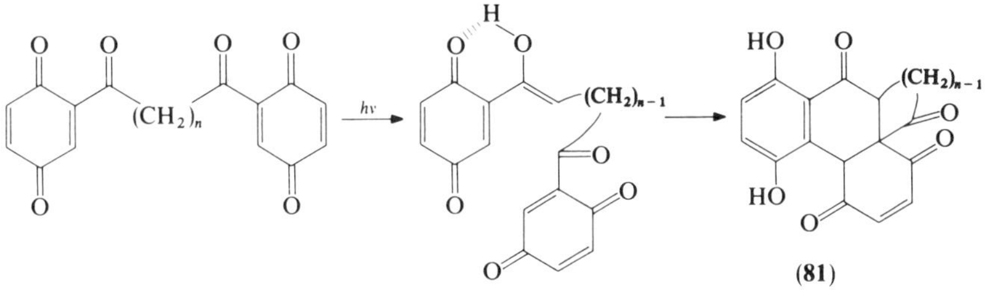

$(CH_2)_n$ $h\nu$ $(CH_2)_{n-1}$ HO HO $(CH_2)_{n-1}$

(**81**)

SCHEME 6

The factors influencing the conversion of (**82**) into (**83**) have been examined;[313] the reaction is facilitated when R is an oxygen substituent. Side-chain keto-ester functionality appears to have an adverse effect on intramolecular cycloadditions

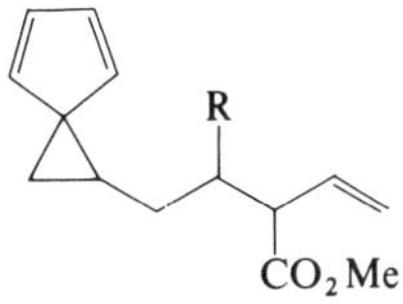

R CO_2Me (**82**)

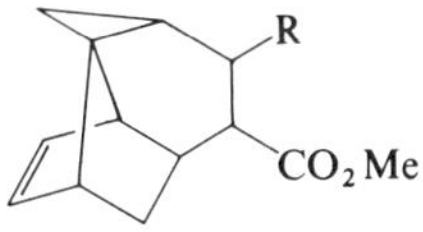

R CO_2Me (**83**)

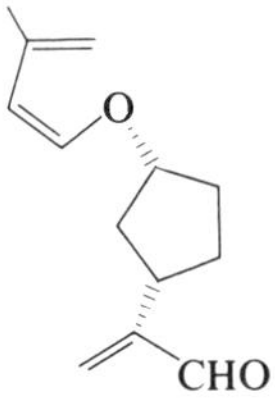

CHO (**84**)

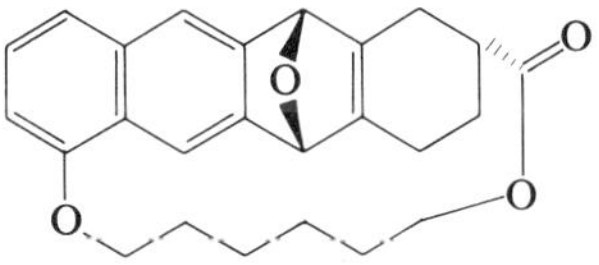

(**85**)

$O(CH_2)_nOCOCH{=}CH_2$

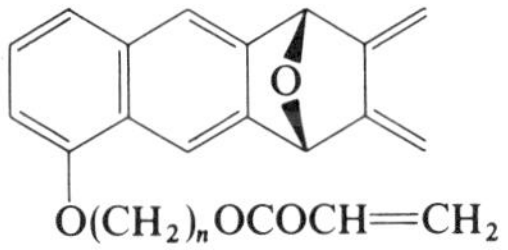

(**86**) $n = 6$
(**87**) $n = 3$

involving tri-substituted cyclopentadienes.[314] The ease of reaction of the *cis*-dienol ether **(84)** is attributed[315] to an unusual boat transition state; its *trans*-isomer fails to react. Intramolecular Diels–Alder reactions are being developed as a regioselective route to anthracycline precursors. For the formation of **(85)** from **(86)** *endo*-face attack is favoured in accord with Alder's rule; only polymerization was observed for the shorter chain analogue **(87)**.[316]

Complete stereoselectivity is observed for the conversion of **(88)** into **(89)**.[317] Further examples of benzene rings acting as the 4π-component have been reported.[318,319] The ease of the process for the formation of **(90)** from **(91)** can be attributed[318] to hydrogen bonding providing favourable geometry. Intramolecular Diels–Alder reactions of unactivated polyene systems can be accomplished using protic acid or amminium cation radical catalysis.[320] Increase in rate brought about by increasing solvent density and viscosity is said to be in accord with predictions based on vibrational activation theory.[321]

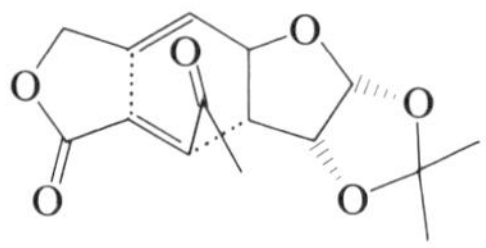

(88)

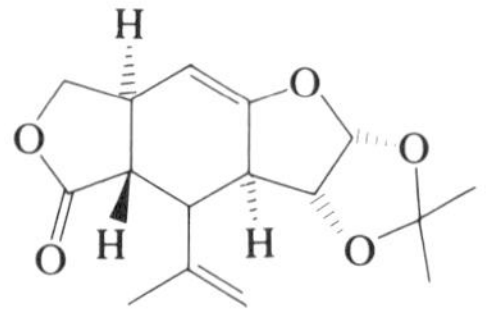

(89)

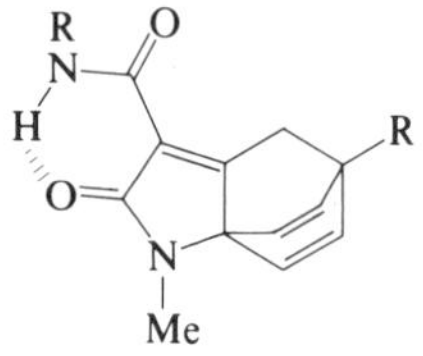

(90)

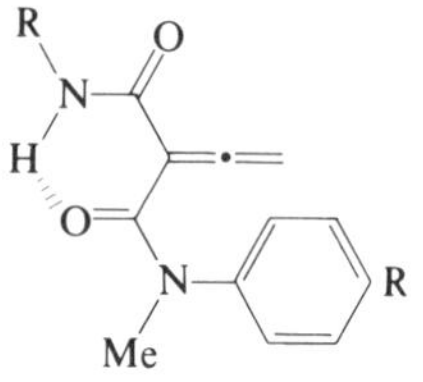

(91)

The number of papers dealing with intramolecular hetero-Diels–Alder reactions has increased. Unactivated 1,2-diazines do not normally undergo intermolecular cycloadditions with unactivated dienophiles, but entropic assistance allows the corresponding intramolecular reaction to proceed in reasonable yield.[322] Quinolizidines and indolizidines are formed from 1-azadienes.[323–325] For *N*-acyl derivatives, generated by thermal elimination of acetic acid from *O*-acetylhydroxylamines, a predominantly *exo*-stereochemical pathway is followed.[323] Intramolecular cycloadditions to thiazoles,[326] triazines,[327] and pyrroles[328,329] have also been described. Stereochemical control of intramolecular Diels–Alder reactions continues to be a focus of attention.[330–338] 1,3,9-Trienes incorporating amide, amine, or ester groups in the chain linking diene and dienophile have been studied;[339] a preference for *cis*-adducts was noted. Pyrano[3,4-*b*]indol-3-ones provide stable equivalents of indole-2,3-quinodimethanes.[340] The synthetic applications of these reactions[341] and of intramolecular Diels–Alder reactions in general[183] have been reviewed.

Miscellaneous Cycloadditions

The first example of a 1 + 2-cycloaddition involving carbon monosulphide has been described;[342] with $R_2NC{\equiv}CNR_2$, 2,3-bis(dialkylamino)cyclopropenethiones are formed. The presence of electron-withdrawing substituents accelerates the reaction between aryl isocyanides and ynediamines.[343] A 1 + 2 nitrene–olefin cycloaddition is a key step in the proposed mechanisms[344] for the reactions of $CH_2{=}C(OMe)_2$ and $CH_2{=}CMeCMe{=}CH_2$ with 2-diazo-4,5-dicyanoimidazole. Phosphenium ions react readily with dienes forming 3-phospholenium salts;[345, 346] both cheletropic and step-wise pathways can be considered equally viable. Further examples of cheletropic reactions between germylenes and 1,3-dienes have been reported;[347] Me_2Ge behaves as a donor, reacting faster with more electron-deficient dienes, thus exhibiting the "inverse" type of addition rather than the "normal" one. Cycloheptatriene in the presence of $TiCl_4$–Et_2AlCl undergoes 6 + 2-cycloaddition to butadiene forming (**92**), and thence (**93**) by an intramolecular Diels–Alder reaction;[348] the first stage is facilitated by complexion of both the trienophile and the triene to the catalyst which enhances the reactivity of the former and changes the symmetry of the frontier orbitals of the latter.

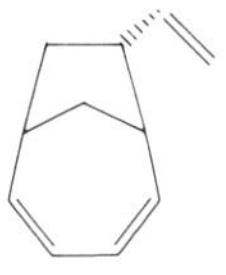

(**92**) 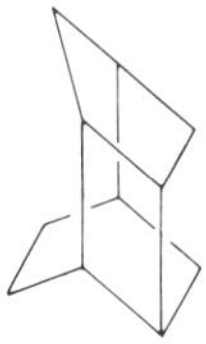 (**93**) 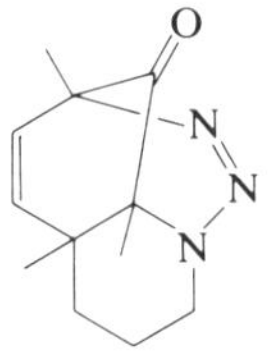 (**94**) 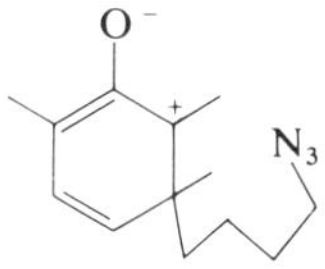 (**95**)

The formation of seven-membered carbocycles by cycloaddition of allyl cations to 1,3-dienes has been reviewed;[349] the rôle of concerted and step-wise pathways was critically assessed. Some regioselectivity is found in the 4 + 3-reaction of 2-methylfuran with 1,1- and 1,3-dichloroalkan-2-ones in the presence of $LiClO_4/Et_3N$;[350] intramolecular examples have also been described.[351] The formation of (**94**) from the photochemically generated zwitterion (**95**) represents a new reaction for azides.[352]

The nature of the interactions between the 4π-components of pentalene and *s*-indacene dimers have been studied[353] using MNDO/1 and MINDO/3 calculations for distances between 2 Å and 5 Å; contrary to qualitative predictions of frontier orbital analysis these interactions are shown to be *de*stabilizing for distances exceeding 2.5 Å. The structure of the 4 + 4-adduct resulting from dimerization of 1-methyl-5,6-diphenylpyrazin-2-one in the solid state has been established[354] as (**96**) by X-ray crystallography. Comparison of the thermal and photochemical decomposition of (**97**) and (**98**) to benzene and anthracene provides a clear demonstration of orbital symmetry control of both the ground and excited state reactions.[355] 6 + 4-Cycloadditions of unsymmetrical tropones with unsymmetrical dienes proceed with high regioselectivity; it is considered[356] that primary rather than secondary orbital interactions are the controlling factor. Isodicyclopentadiene (**50**) reacts predominantly at its *exo*-face with tropone and oxyallyl cations forming, respectively, 4 + 6- and 4 + 3-adducts;[357] this contrasts with its 4 + 2-reactions which usually give *endo*-products. Further examples of intramolecular 6 + 4-cycloadditions involving fulvenes have been reported.[358]

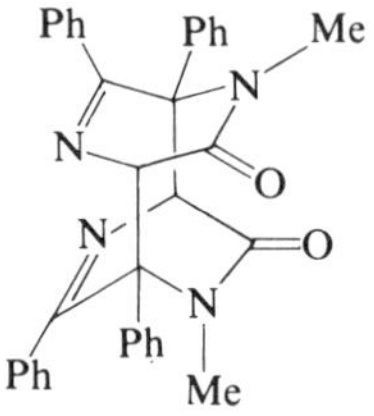

(96) 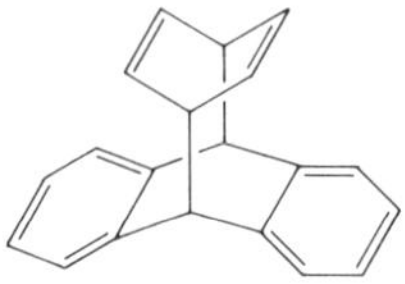 (97) 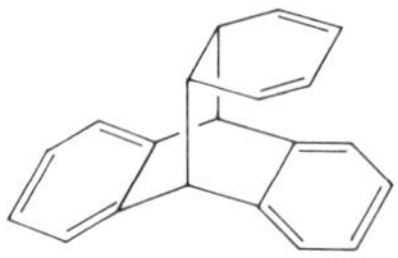 (98)

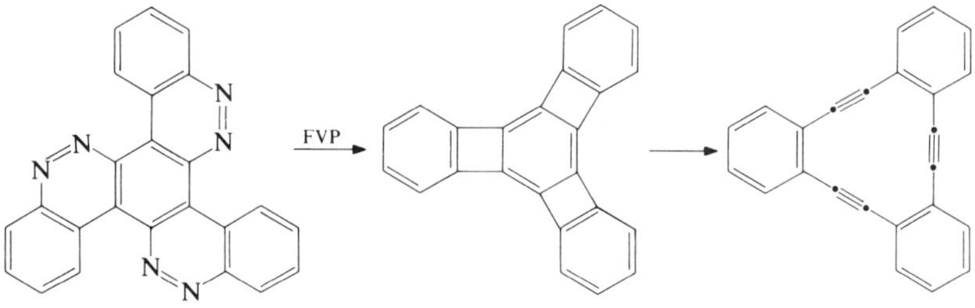

(100) (101) (99)

The value of activation volume measurements in assessing the position of the transition state in 4 + 6- and 2 + 2 + 2-cycloadditions has been stressed.[3] For the titanium-catalysed cyclotrimerization of acetylenes to arenes a mechanism has been put forward[359] which involves concerted trimerization of the three alkenes coordinated to a divalent centre. Nickel-catalysed cyclotrimerization of hex-3-yne has also been investigated.[360] The formation of **(99)** on flash vacuum pyrolysis of **(100)** might possibly involve a novel symmetry-allowed 2 + 2 + 2-cycloreversion of the benzene ring of postulated intermediate **(101)**.[361] Cobalt-mediated 2 + 2 + 2-cycloadditions have been reviewed.[362,363] The stereochemical outcome of 2 + 2 + 2-annulations appears to be critically dependent on substituents.[364] The possible mechanisms for the cobalt-mediated 2 + 2 + 2-reactions of enediynes have been discussed;[365,366] the stereochemistry of the products is explicable in terms of an intermediate cobaltacyclopentadiene. The formation of **(102)** from the cyclobu-

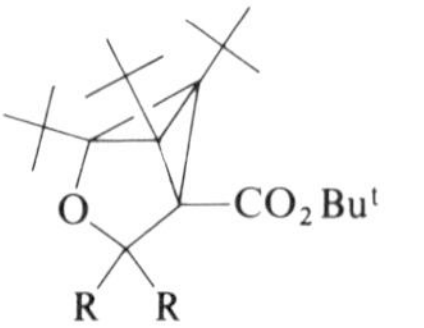

(102) 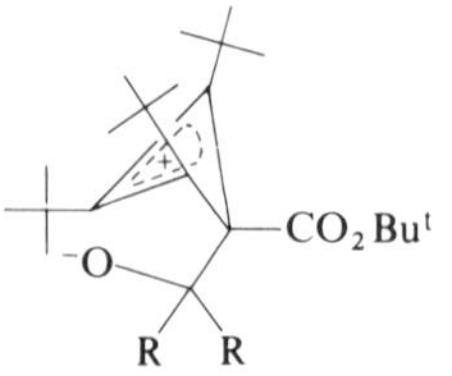 (103) 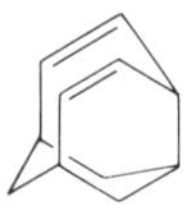 (104)

tadiene derivative (**6**) with acceptor-substituted ketones such as acetyl cyanide can be formally interpreted as a *cis, trans*-2 + 2 + 2-cycloaddition; a step-wise pathway *via* zwitterion (**103**), similar to that reported[34] for the addition of TCNE, is considered most likely.[367]

Homo-Diels–Alder reactions have been reported[368] for (**104**), the first case involving a 1,4-cyclooctadiene system; its high reactivity is attributed to through-space interactions. The rôle of exciplexes in photocycloadditions of benzene and 1,3-dioxoles has been examined in greater detail.[369] A step-wise mechanism involving an intermediate aziridinium ion is favoured[370] for the formation of rearranged urazoles from the reaction of triazolinediones with bicycloalkenes.

References

1 Dewar, M. J. S., *J. Am. Chem. Soc.*, **106,** 209 (1984).
2 Brown, F. K., and Houk, K. N., *Tetrahedron Lett.*, **25,** 4609 (1984).
3 Jenner, G., *Bull. Soc. Chim. Fr. II,* **1984,** 275.
4 Burnier, J. S., and Jorgensen, W. L., *J. Org. Chem.*, **49,** 3001 (1984).
5 Gleiter, R., and Paquette, L. A., *Acc. Chem. Res.*, **16,** 328 (1983).
6 Paquette, L. A., *Methods Stereochem. Anal.*, **3,** 41 (1984); *Chem. Abs.*, **101,** 129824 (1984).
7 Ponec, R., *Collect. Czech. Chem. Commun.*, **49,** 455 (1984).
8 Padwa, A. (Ed.), *1,3-Dipolar Cycloaddition Chemistry,* Wiley, New York, 1984.
9 De Lucchi, O., and Modena, G., *Tetrahedron,* **40,** 2585 (1984).
10 Pasto, D. J., *Tetrahedron,* **40,** 2805 (1984).
11 Surpateanu, G., and Lablache-Gombier, A., *Heterocycles,* **22,** 2079 (1984).
12 Bussas, R., Kresze, G., Münsterer, H., and Schwöbel, A., *Sulfur Reports,* **2,** 215 (1983).
13 Duran, M., and Bertran, J., *THEOCHEM,* **16,** 239 (1984); *Chem. Abs.*, **100,** 208756 (1984).
14 Albini, A., and Giannantonio, L., *J. Org. Chem.*, **49,** 3862 (1984).
15 Shim, S. C., and Kim, Y. Z., *J. Photochem.*, **23,** 83 (1983).
16 Ikeda, M., Ohno, K., Mohri, S.-i., Takahashi, M., and Tamura, Y., *J. Chem. Soc., Perkin Trans. 1,* **1984,** 405.
17 Bruneel, K., De Keukeleire, D., and Vandewalle, M., *J. Chem. Soc., Perkin Trans. 1,* **1984,** 1697.
18 Ried, W., Bellinger, O., and Bats, J. W., *Chem. Ber.*, **116,** 3794 (1983).
19 Swindell, C. S., deSolms, S. J., and Springer, J. P., *Tetrahedron Lett.*, **25,** 3797 (1984).
20 Ikeda, M., Takahashi, M., Uchino, T., and Tamura, Y., *Chem. Pharm. Bull.*, **32,** 538 (1984).
21 Lewis, F. D., Oxman, J. D., and Huffman, J. C., *J. Am. Chem. Soc.*, **106,** 466 (1984).
22 Sarma, J. A. R. P., and Desiraju, G. R., *J. Chem. Soc., Chem. Commun.*, **1984,** 145.
23 Hasegawa, M., Nohara, M., Saigo, K., Mori, T., and Nakanishi, H., Tetrahedron Lett., **25,** 561 (1984).
24 Sakellariou-Fargues, R., Maurette, M.-T., Oliveros, E., Riviere, M., and Lattes, A., *Tetrahedron,* **40,** 2381 (1984).
25 Amarouche, H., de Bourayne, C., Riviere, M., and Lattes, A., *C. R. Hebd. Seances Acad. Sci., Ser. 2,* **298,** 121 (1984).
26 Horspool, W. M., *Photochemistry,* **14,** 207 (1983).
27 Serebryakov, E. P., *Izv. Akad. Nauk SSSR, Ser. Khim.*, **1984,** 97; *Chem. Abs.*, **100,** 155887 (1984).
28 Gilbert, J. C., and Baze, M. E., *J. Am. Chem. Soc.*, **106,** 1885 (1984).
29 Warrener, R. N., Harrison, P. A., Sterns, M., and Russell, R. A., *J. Chem. Soc., Chem. Commun.*, **1984,** 546.
30 Anthony, I. J., and Wege, D., *Aust. J. Chem.*, **37,** 1283 (1984).
31 De Lucchi, O., Pasquato, L., and Modena, G., *Tetrahedron Lett.*, **25,** 3643 (1984).
32 Duran, M., and Bertran, J., *J. Chem. Soc., Perkin Trans. 2,* **1984,** 197.
33 Pabon, R. A., and Bauld, N. L., *J. Am. Chem. Soc.*, **106,** 1145 (1984).
34 Regitz, M., and Eisenbach, P., *Chem. Ber.*, **117,** 1991 (1984).
35 Pabon, R. A., Bellville, D. J., and Bauld, N. L., *J. Am. Chem. Soc.*, **106,** 2730 (1984).
36 Kurov, G. N., Svyatkina, L. I., and Palchuk, E. G., *Izv. Akad. Nauk SSSR, Ser. Khim.*, **1983,** 2227; *Chem. Abs.*, **100,** 33910 (1984).

[37] Patrick, T. B., and Bechtold, D. S., *J. Org. Chem.*, **49,** 1935 (1984).
[38] Mehta, G., Reddy, D. S., and Reddy, A. V., *Tetrahedron Lett.*, **25,** 2275 (1984).
[39] Yamashita, Y., and Mukai, T., *Chem. Lett.*, **1984,** 1741.
[40] Nishida, S., Murakami, M., Mizuno, T., Tsuji, T., Oda, H., and Shimizu, N., *J. Org. Chem.*, **49,** 3428 (1984).
[41] Pasto, D. J., and Yang, S. -H., *J. Am. Chem. Soc.*, **106,** 152 (1984).
[42] Klop, W., Klusener, P. A. A., and Brandsma, L., *Recl. Trav. Chim. Pays-Bas*, **103,** 85 (1984).
[43] Summermatter, W., and Heimgartner, H., *Helv. Chim. Acta*, **67,** 1298 (1984).
[44] Gotthardt, H., and Jung, R., *Tetrahedron Lett.*, **25,** 4217 (1984).
[45] Mayr, H., and Bäuml, E., *Tetrahedron Lett.*, **25,** 1127 (1984).
[46] Nishio, T., *J. Org. Chem.*, **49,** 827 (1984).
[47] Schaumann, E., Förster, W. -R., and Adiwidjaja, G., *Angew. Chem. Int. Ed.*, **23,** 439 (1984).
[48] Ha, D. -C., Hart, D. J., and Yang, T. -K., *J. Am. Chem. Soc.*, **106,** 4819 (1984).
[49] Golankiewicz, K., Jankowska, J., and Koroniak, H., *Heterocycles*, **22,** 67 (1984).
[50] Wolff, S., and Agosta, W. C., *J. Am. Chem. Soc.*, **106,** 2363 (1984).
[51] Saba, S., Wolff, S., Schröder, C., Margaretha, P., and Agosta, W. C., *J. Am. Chem. Soc.*, **105,** 6902 (1983).
[52] Nitta, M., Kuroki, T., and Sugiyama, H., *Rikogaku Kenkyusho Hokoku, Waseda Daigaku*, **1983,** 39; *Chem. Abs.*, **100,** 5521 (1984).
[53] Coyle, J. D., and Rapley, P. A., *Tetrahedron Lett.*, **25,** 2247 (1984).
[54] Hofstra, G., Kamphuis, J., and Bos, H. J. T., *Tetrahedron Lett.*, **25,** 873 (1984).
[55] Hamilton, J. G., and Rooney, J. J., *J. Chem. Soc., Faraday Trans. 1*, **80,** 129 (1984).
[56] Borisenko, A. A., Nikulin, A. V., Wolfe, S., Zefirov, N. S., and Zyk, N. V., *J. Am. Chem. Soc.*, **106,** 1074 (1984).
[57] Cheng, C. -C., Seymour, C. A., Petti, M. A., Greene, F. D., and Blount, J. F., *J. Org. Chem.*, **49,** 2910 (1984).
[58] Albert, B., Berning, W., Burschka, C., Hünig, S., and Prokschy, F., *Chem. Ber.*, **117,** 1465 (1984).
[59] Dass, C., and Gross, M. L., *J. Am. Chem. Soc.*, **106,** 5775 (1984).
[60] Goldschmidt, Z., Antebi, S., Cohen, D., and Goldberg, I., *J. Organomet. Chem.*, **273,** 347 (1984).
[61] Raynolds, P. W., and DeLoach, J. A., *J. Am. Chem. Soc.*, **106,** 4566 (1984).
[62] Erden, I., *Tetrahedron Lett.*, **25,** 1535 (1984).
[63] Fishbein, P. L. and Moore, H. W., *J. Org. Chem.*, **49,** 2190 (1984).
[64] Hall, J. H., and Krishnan, G., *J. Org. Chem.*, **49,** 2498 (1984).
[65] Perumal, S. I., *Indian J. Chem.*, **23B,** 293 (1984); *Chem. Abs.*, **101,** 90071 (1984).
[66] Koft, E. R., and Smith, A. B., *J. Am. Chem. Soc.*, **106,** 2115 (1984).
[67] Koft, E. R., and Smith, A. B., *J. Org. Chem.*, **49,** 832 (1984).
[68] Crimmins, M. T., and DeLoach, J. A., *J. Org. Chem.*, **49,** 2077 (1984).
[69] Samuilov, Y. D., and Konovalov, A. I., *Russ. Chem. Rev.*, **53,** 332 (1984).
[70] Harcourt, R. D., and Little, R. D., *J. Am. Chem. Soc.*, **106,** 41 (1984).
[71] Boger, D. L., and Brotherton, C. E., *J. Am. Chem. Soc.*, **106,** 805 (1984).
[72] Kleingeld, J. C., and Nibbering, N. M. M., *Recl. Trav. Chim. Pays-Bas*, **103,** 87 (1984).
[73] Yoshida, H., Sogame, S., Bando, S., Nakajima, S., Ogata, T., and Matsumoto, K., *Bull. Chem. Soc. Jpn.*, **56,** 2849 (1983).
[74] Binger, P., and Wedemann, P., *Tetrahedron Lett.*, **24,** 5847 (1983).
[75] Tsuji, T., and Nishida, S., *Acc. Chem. Res.*, **17,** 56 (1984).
[76] Janulis, E. P., Wilson, S. R., and Arduengo, A. J., *Tetrahedron Lett.*, **25,** 405 (1984).
[77] Padwa, A., Rosenthal, R. J., Dent, W., Filho, P., Turro, N. J., Hrovat, D. A., and Gould, I. R., *J. Org. Chem.*, **49,** 3174 (1984).
[78] Sedqui, A., Vebrel, J., and Laude, B., *Chem. Lett.*, **1984,** 965.
[79] Dhar, D. N., and Ragunathan, R., *Tetrahedron*, **40,** 1585 (1984).
[80] Tanaka, K., Igarashi, T. -r., Maeno, S., and Mitsuhashi, K., *Bull. Chem. Soc. Jpn.*, **57,** 2689 (1984).
[81] Shimizu, T., Hayashi, Y., Nishio, T., and Teremura, K., *Bull. Chem. Soc. Jpn.*, **57,** 787 (1984).
[82] Sayanna, E., Venkataratnam, R. V., and Thyagarajan, G., *Heterocycles*, **22,** 1561 (1984).
[83] Fodor, L., El-Gharib, M.-S., Szabo, J., Bernath, G., and Sohar, P., *Heterocycles*, **22,** 537 (1984).
[84] Prajapati, D., Sandhu, J. S., and Baruah, J. N., *J. Chem. Res. (S)*, **1984,** 56.
[85] Bruche, L., and Zecchi, G., *J. Heterocycl. Chem.*, **20,** 1705 (1983).
[86] Fliege, W., Grashey, R., and Huisgen, R., *Chem. Ber.*, **117,** 1194 (1984).
[87] Bronberger, F., and Huisgen, R., *Tetrahedron Lett.*, **25,** 65 (1984).
[88] Adam, W., Beinhauer, A., De Lucchi, O., and Rosenthal, R. J., *Tetrahedron Lett.*, **24,** 5727 (1983).
[89] Nitta, M., Omata, A., and Okada, S., *Bull. Chem. Soc. Jpn.*, **57,** 1505 (1984).

[90] Fisera, L., Povazanec, F., Zalupsky, P., Kovac, J., and Pavolvic, D., *Collect. Czech. Chem. Commun.*, **48,** 3144 (1983).
[91] van der Knaap, T. A., Klebach, T. C., Visser, F., Lourens, R., and Bickelhaupt, F., *Tetrahedron*, **40,** 991 (1984).
[92] Rösch, W., and Regitz, M., *Angew. Chem. Int. Ed.*, **23,** 900 (1984).
[93] Märkl, G., and Trötsch, I., *Angew. Chem. Int. Ed.*, **23,** 901 (1984).
[94] Corsaro, A., Chiacchio, U., Alberghina, G., and Purrello, G., *J. Chem. Res. (S)*, **1984,** 370.
[95] Büchel, T., Prewo, R., Bieri, J. H., and Heimgartner, H., *Helv. Chim. Acta*, **67,** 534 (1984).
[96] Komatsu, M., Yoshida, Y., Uesaka, M., Oshiro, Y., and Agawa, T., *J. Org. Chem.*, **49,** 1300 (1984).
[97] Ando, W., Hanyu, Y., Takata, T., and Ueno, K., *J. Am. Chem. Soc.*, **106,** 2216 (1984).
[98] Katada, T., Eguchi, S., and Sasaki, T., *J. Chem. Soc., Perkin Trans. 1*, **1984,** 2641.
[99] Sasaki, T., Eguchi, S., and Katada, T., *Heterocycles*, **21,** 682 (1984).
[100] Shiraishi, S., Holla, B. S., and Imamura, K., *Bull. Chem. Soc. Jpn.*, **56,** 3457 (1983).
[101] Hayakawa, T., Araki, K., and Shiraishi, S., *Bull. Chem. Soc. Jpn.*, **57,** 1643 (1984).
[102] Hayakawa, T., Araki, K., and Shiraishi, S., *Bull. Chem. Soc. Jpn.*, **57,** 2216 (1984).
[103] Shimizu, T., Hayashi, Y., and Teramura, K., *Bull. Chem Soc. Jpn.*, **57,** 2531 (1984).
[104] Rahman, A., Younas, M., and Kahn, N. A., *J. Chem. Soc., Pak.*, **5,** 243 (1984); *Chem. Abs.*, **101,** 54969 (1984).
[105] Coutouli-Argyropoulou, E., *Tetrahedron Lett.*, **25,** 2029 (1984).
[106] Beltrame, P., Gelli, G., and Loi, A., *J. Heterocycl. Chem.*, **20,** 1609 (1983).
[107] Corsaro, A., Chiacchio, U., Caramella, P., and Purrello, G., *J. Heterocycl. Chem.*, **21,** 949 (1984).
[108] Albini, F. M., Albini, E., Bandiera, T., and Caramella, P., *J. Chem. Res. (S)*, **1984,** 36.
[109] Caramella, P., Bandiera, T., Grünanger, P., and Marione Albini, F., *Tetrahedron*, **40,** 441 (1984).
[110] De Sarlo, F., Guarna, A., and Brandi, A., *J. Heterocycl. Chem.*, **20,** 1505 (1983).
[111] Dubey, S. K., and Knaus, E. E., *J. Org. Chem.*, **49,** 123 (1984).
[112] Argyropoulos, N. G., Coutouli-Argyropoulou, E., and Pistikopoulos, P., *J. Chem. Res. (S)*, **1984,** 362.
[113] Middleton, W. J., *J. Org. Chem.*, **49,** 919 (1984).
[114] Kozikowski, A. P., Kitagawa, Y., and Springer, J. P., *J. Chem. Soc., Chem. Commun.*, **1983,** 1460.
[115] Jones, R. H., Robinson, G. C., and Thomas, E. J., *Tetrahedron*, **40,** 177 (1984).
[116] Wade, P. A., Singh, S. M., and Pillay, M. K., *Tetrahedron*, **40,** 601 (1984).
[117] Houk, K. N., Moses, S. R., Wu, Y.-D., Rondan, N. G., Jäger, V., Schohe, R., and Fronczek, F. R., *J. Am. Chem. Soc.*, **106,** 3880 (1984).
[118] Kozikowski, A. P., and Ghosh, A. K., *J. Org. Chem.*, **49,** 2762 (1984).
[119] Caramella, P., Albini, F. M., Vitali, D., Rondan, N. G., Wu, Y.-D., Schwartz, T. R., and Houk, K. N., *Tetrahedron Lett.*, **25,** 1875 (1984).
[120] Machida, M., Oda, K., and Kanaoka, Y., *Tetrahedron Lett.*, **25,** 409 (1984).
[121] Bryce, M. D., Chambers, R. D., and Taylor, G., *J. Chem. Soc., Perkin Trans. 1*, **1984,** 509.
[122] Dolbier, W. R., Burkholder, C. R., and Winchester, W. R., *J. Org. Chem.*, **49,** 1518 (1984).
[123] Stanovnik, B., Kupper, M., Tisler, M., Leban, I., and Golic, L., *J. Chem. Soc., Chem. Commun.*, **1984,** 268.
[124] Stanovnik, B., Furlan, B., Sarka, A., Tisler, M., and Zlicar, M., *Heterocycles*, **22,** 2479 (1984).
[125] Franck-Neumann, M., and Miesch, M., *Bull. Soc. Chim. Fr. II*, **1984,** 362.
[126] Padwa, A., and Goldstein, S. I., *Can. J. Chem.*, **62,** 2506 (1984).
[127] Vebrel, J., and Carrié, R., *Tetrahedron*, **39,** 4163 (1983).
[128] Vebrel, J., Greé, D., and Carrié, R., *Can. J. Chem.*, **62,** 939 (1984).
[129] Bartlett, P. D., and Combs, G. L., *J. Org. Chem.*, **49,** 625 (1984).
[130] Bende, Z., Toke, L., Weber, L., Toth, G., Janke, F., and Csonka, G., *Tetrahedron*, **40,** 369 (1984).
[131] Huisgen, R., and Niklas, K., *Heterocycles*, **22,** 21 (1984).
[132] De Kimpe, N., Sulmon, P., De Buyck, L., Verhé, R., Schamp, N., Declercq, J.-P., and Van Meerssche, M., *J. Chem. Res. (S)*, **1984,** 82.
[133] Padwa, A., Haffmanns, G., and Tomas, M., *J. Org. Chem.*, **49,** 3314 (1984).
[134] Tsuge, O., Kanemasa, S., Kuraoka, S., and Takenaka, S., *Chem. Lett.*, **1984,** 279.
[135] Tsuge, O., Kanemasa, S., Kuraoka, S., and Takenaka, S., *Chem. Lett.*, **1984,** 281.
[136] Tsuge, O., Kanemasa, S., Takenaka, S., and Kuraoka, S., *Chem. Lett.*, **1984,** 465.
[137] Tsuge, O., Kanemasa, S., Hatada, A., and Matsuda, K., *Chem. Lett.*, **1984,** 801.
[138] Surpateanu, G., Constantinescu, M., Luchian, C., Petrovanu, M., Zugravescu, I., and Lablache-Combier, A., *Rev. Roum. Chim.*, **28,** 933 (1983).
[139] Albini, A., Bettinetti, G., and Minoli, G., *J. Org. Chem.*, **49,** 2670 (1984).
[140] Akasaka, T., Sonobe, H., Sato, R., and Ando, W., *Tetrahedron Lett.*, **25,** 4757 (1984).

141 Padwa, A., Fisera, L., Koehler, K. F., Rodriguez, A., and Wong, G. S. K., *J. Org. Chem.*, **49**, 276 (1984).
142 Ashburn, S. P., and Coates, R. M., *J. Org. Chem.*, **49**, 3127 (1984).
143 Bened, A., Durand, R., Pioch, D., Geneste, P., Guimon, C., Guillouzo, G. P., Declercq, J. -P., Germain, G., Briard, P., Rambaud, J., and Roques, R., *J. Chem. Soc., Perkin Trans. 2*, **1984,** 1.
144 Grigg, R., Gunaratne, N., and Kemp, J., *J. Chem. Soc., Perkin Trans. 1*, **1984,** 41.
145 Grigg, R., Jordan, M., Tangthongkum, A., Einstein, F. W. B., and Jones, T., *J. Chem. Soc., Perkin Trans. 1*, **1984,** 47.
146 Grigg, R., and Thianpantanagul, S., *J. Chem. Soc., Perkin Trans. 1*, **1984,** 653.
147 Grigg, R., and Thianpantanagul, S., *J. Chem. Soc., Chem. Commun.*, **1984,** 180.
148 Grigg, R., Aly, M. F., Sridharan, V., and Thianpatanagul, S. *J. Chem. Soc., Chem. Commun.*, **1984,** 182.
149 Dalgard, N. K. A., Larsen, K. E., and Torssell, K. B. G., *Acta Chem. Scand.*, **38B,** 423 (1984).
150 Haddadin, M. J., Amalu, S. J., and Freeman, J. P., *J. Org. Chem.*, **49,** 2824 (1984).
151 DeShong, P., Dicken, C. M., Leginus, J. M., and Whittle, R. R., *J. Am. Chem. Soc.*, **106,** 5598 (1984).
152 Bellandi, C., De Amici, M., De Micheli, C., and Gandolfi, R., *Heterocycles*, **22,** 2187 (1984).
153 Ikeda, M., Tamura, Y., Yamagishi, M., Goto, Y., Niiya, T., and Sumoto, K., *Heterocycles*, **22,** 981 (1984).
154 Sammes, P. G., and Street, L. J., *J. Chem. Res. (S)*, **1984,** 196.
155 Sammes, P. G., and Whitby, R. J., *J. Chem. Soc., Chem. Commun.*, **1984,** 702.
156 Friedrichsen, W., Seidel, W., and Debaerdemaeker, T., *J. Heterocycl. Chem.*, **20,** 1621 (1983).
157 Furneaux, R. H., Mason, J. M., and Miller, I. J., *J. Chem. Soc., Perkin Trans. 1*, **1984,** 1923.
158 Huisgen, R., and Xingya, L., *Heterocycles*, **20,** 2363 (1983).
159 Yamaguchi, K., *THEOCHEM*, **12,** 101 (1983); *Chem. Abs.*, **99,** 174931 (1983).
160 Moskal, J., Moskal, A., and Milart, P., *Monatsh. Chem.*, **115,** 187 (1984).
161 Burger, K., and Rottegger, S., *Tetrahedron Lett.*, **25,** 4091 (1984).
162 Brokatzky-Geiger, J., and Eberbach, W., *Tetrahedron Lett.*, **25,** 1137 (1984).
163 Brokatzky-Geiger, J., and Eberbach, W., *Chem. Ber.*, **117,** 2157 (1984).
164 Tsuge, O., Ueno, K., and Kanemasa, S., *Chem. Lett.*, **1984,** 285.
165 Bruché, L., Garanti, L., and Zecchi, G., *J. Chem. Soc., Perkin Trans. 1*, **1984,** 2535.
166 Hünig, S., and Prokschy, F., *Chem. Ber.*, **117,** 2099 (1984).
167 Hünig, S., and Schmitt, M., *Tetrahedron Lett.*, **25,** 1725 (1984).
168 Fischer, G., Hunkler, D., and Prinzbach, H., *Tetrahedron Lett.*, **25,** 2459 (1984).
169 Tsuge, O., Ueno, K., and Kanemasa, S., *Chem. Lett.*, **1984,** 797.
170 Shimizu, T., Hayashi, Y., Nakano, M., and Teramura, K., *Bull. Chem. Soc. Jpn.*, **57,** 134 (1984).
171 Confalone, P. N., and Ko, S. S., *Tetrahedron Lett.*, **25,** 947 (1984).
172 Kozikowski, A. P. and Stein, P. D., *J. Org. Chem.*, **49,** 2301 (1984).
173 Kozikowski, A. P., Hiraga, K., Springer, J. P., Wang, B. C., and Xu, Z. -B., *J. Am. Chem. Soc.*, **106,** 1845 (1984).
174 Kozikowski, A. P., Chen, Y. -Y., Wang, B. C., and Xu, Z. -B., *Tetrahedron*, **40,** 2345 (1984).
175 Kozikowski, A. P., and Greco, M. N., *J. Org. Chem.*, **49,** 2310 (1984).
176 Wang, C. -L. J., Ripka, W. C., and Confalone, P. N., *Tetrahedron Lett.*, **25,** 4613 (1984).
177 Confalone, P. N., and Huie, E. M., *J. Am. Chem. Soc.*, **106,** 7175 (1984).
178 Jacobi, P. A., Martinelli, M. J., and Polanc, S., *J. Am. Chem. Soc.*, **106,** 5594 (1984).
179 Toy, A., and Thompson, W. J., *Tetrahedron Lett.*, **25,** 3533 (1984).
180 Gleiter, R., and Boehm, M. C., *Methods Stereochem. Anal.*, **3,** 105 (1983); *Chem. Abs.*, **101,** 129825 (1984).
181 Oppolzer, W., *Angew. Chem. Int. Ed.*, **23,** 876 (1984).
182 Welzel, P., *Nachr. Chem., Tech. Lab.*, **31,** 979 (1983); *Chem. Abs.*, **100,** 84903 (1984).
183 Fallis, A. G., *Can. J. Chem.*, **62,** 183 (1984).
184 Vogel, P., *Methods Stereochem. Anal.*, **3,** 147 (1983); *Chem. Abs.*, **101,** 129826 (1984).
185 Konovalov, A. I., *Usp. Khim.*, **52,** 1852 (1983); *Chem. Abs.*, **100,** 50681 (1984).
186 Xu, X., and Xie, Z., *Huaxue Tongbao*, **1984,** 11; *Chem. Abs.*, **101,** 54076 (1984).
187 Koszuk, J., and Bodalski, R., *Wiad. Chem.*, **37,** 549 (1983); *Chem. Abs.*, **101,** 71830 (1984).
188 Dewar, M. J. S., and Pierini, A. B., *J. Am. Chem. Soc.*, **106,** 203 (1984).
189 Ortega, M., Oliva, A., Lluch, J. M., and Bertran, J., *Chem. Phys. Lett.*, **102,** 317 (1983).
190 Shakhova, S. K., Rar, L. F., Polkovnikov, B. D. Vainberg, N. N., and Elyanov, B. S., *Izv. Akad. Nauk SSSR, Ser. Khim.*, **1984,** 334; *Chem. Abs.*, **100,** 208809 (1984).
191 Bridges, A. J., and Fischer, J. W., *J. Org. Chem.*, **49,** 2954 (1984).

192 Alston, P. V., Gordon, M. D., Ottenbrite, R. M., and Cohen, T., *J. Org. Chem.*, **48**, 5051 (1983).
193 Huybrechts, G., Poppelsdorf, H., Maesschalck, L., and Van Mele, B., *Int. J. Chem. Kinet.*, **16**, 93 (1984).
194 Becker, H. D., and Andersson, K., *J. Photochem.*, **26**, 75 (1984).
195 Quast, H., and Schön, N., *Liebigs Ann. Chem.*, **1984**, 877.
196 Lindley, W. A., MacDowell, D. W. H., and Petersen, J. L., *J. Org. Chem.*, **48**, 4419 (1983).
197 Gavina, F., Costero, A. M., Gil, P., and Luis, S. V., *J. Am. Chem. Soc.*, **106**, 2077 (1984).
198 Baraldi, P. G., Barco, A., Benetti, S., Pollini, G. P., Polo, E., and Simoni, D., *J. Chem. Soc., Chem. Commun.*, **1984**, 1049.
199 Ashkenazi, P., Benn, R., and Ginsburg, D., *Helv. Chim. Acta*, **67**, 583 (1984).
200 Oku, A., Nozaki, Y., Hasegawa, H., Nishimura, J., and Harada, T., *J. Org. Chem.*, **48**, 4374 (1983).
201 Szechner, B., Rey, M., Dreiding, A. S., and Grieb, R., *Helv. Chim. Acta*, **67**, 1386 (1984).
202 Breslow, R., and Maitra, U., *Tetrahedron Lett.*, **25**, 1239 (1984).
203 Ashkenazi, P., Kaftory, M., Sayrac, T., Maier, G., and Ginsberg, D., *Helv. Chim. Acta*, **66**, 2709 (1983).
204 Balci, M., and Atasoy, B., *Tetrahedron Lett.*, **25**, 4033 (1984).
205 Tamura, Y., Sasho, M., Nakagawa, K., Tsugoshi, T., and Kita, Y., *J. Org. Chem.*, **49**, 473 (1984).
206 Boger, D. L., and Mullican, M. D., *J. Org. Chem.*, **49**, 4033 (1984).
207 Bryce-Smith, D., Gilbert, A., McColl, I. S., and Yianni, P., *J. Chem. Soc., Chem. Commun.*, **1984**, 951.
208 Kovtunenka, V. A., Voitenko, Z. V., Tyltin, A. K., Turov, A. V., and Babichev, F. S., *Ukr. Khim. Zh. (Russ. Ed.)*, **49**, 1287 (1983); *Chem. Abs.*, **100**, 173943 (1984).
209 Schiess, P., Eberle, M., Huys-Francotte, M., and Wirz, J., *Tetrahedron Lett.*, **25**, 2201 (1984).
210 Kuzuya, M., Miyake, F., and Okuda, T., *J. Chem. Soc., Perkin Trans. 2*, **1984**, 1471.
211 Miyashi, T., Ahmed, A., and Mukai, T., *J. Chem. Soc., Chem. Commun.*, **1984**, 179.
212 RajanBabu, T. V., Eaton, D. F., and Fukunaga, T., *J. Org. Chem.*, **48**, 652 (1983).
213 Sellers, S. F., Dolbier, W. R., Koroniak, H., and Al-Fekri, D. M., *J. Org. Chem.*, **49**, 1033 (1984).
214 Jenner, G., Papadopoulos, M., and le Noble, W. J., *Nouv. J. Chim.*, **7**, 687 (1983).
215 Alekperov, N. A., Mishiev, R. D., Salakhov, M. S., Kurbanova, R. A., and Shukyurova, M. B., *Azerb. Khim. Zh.*, **1983**, 51; *Chem. Abs.*, **100**, 208807 (1984).
216 Salakhov, M. S., Musaeva, N. F., Suleimanov, S. N., Salakhova, R. S., and Kopylova, T. A., *Monomery Polim.*, **1983**, 8; *Chem. Abs.*, **101**, 110023 (1984).
217 Salakhov, M. S., Orudzhev, K. D., Musaeva, N. F., Gasanov, G. M., and Gasamova, A. A., *Deposited Doc.*, **1982** VINITI 4249; *Chem. Abs.*, **100**, 50824 (1984).
218 Salakhov, M. S., Musaeva, N. F., Tsulatova, S. M., Zulfaliev, S. R., and Nagiev, V. A., *Deposited Doc.*, **1982** VINITI 770; *Chem. Abs.*, **100**, 208806 (1984).
219 Musaeva, N. F., Salakhov, P. S. M., Nagiev, V. A., and Zulfaliev, S. R., *Deposited Doc.*, **1982** VINITI 771; *Chem. Abs.*, **101**, 22741 (1984).
220 Ando, W., Hanyu, Y., Takata, T., Sakurai, T., and Kobayashi, K., *Tetrahedron Lett.*, **25**, 1483 (1984).
221 Takata, T., Hanyu, Y., and Ando, W., *Heterocycles*, **21**, 497 (1984).
222 Tsuge, O., Kanemasa, S., Sakoh, H., and Wada, E., *Chem. Lett.*, **1984**, 273.
223 Tsuge, O., Kanemasa, S., Sakoh, H., and Wada, E., *Chem. Lett.*, **1984**, 277.
224 Tsuge, O., Wada, E., Kanemasa, S., and Sakoh, H., *Chem. Lett.*, **1984**, 469.
225 Tsuge, O., Wada, E., Kanemasa, S., and Sakoh, H., *Bull. Chem. Soc. Jpn.*, **57**, 3221 (1984).
226 Tsuge, O., Kanemasa, S., Sakoh, H., and Wada, E., *Bull. Chem. Soc. Jpn.*, **57**, 3234 (1984).
227 Bartlett, P. D., and Wu, C., *J. Org. Chem.*, **49**, 1880 (1984).
228 Paquette, L. A., Hathaway, S. J., and Gallucci, J. C., *Tetrahedron Lett.*, **25**, 2659 (1984).
229 Paquette, L. A., Green, K. E., and Hsu, L.-Y., *J. Org. Chem.*, **49**, 3650 (1984).
230 Paquette, L. A., Charumilind, P., and Gallucci, J. C., *J. Am. Chem. Soc.*, **105**, 7364 (1983).
231 Burnell, D. J., Goodbrand, H. B., Kaiser, S. M., and Valenta, Z., *Can. J. Chem.*, **62**, 2398 (1984).
232 Voiget, P., Bonivento, M., Roulet, R., and Vogel, P., *Helv. Chim. Acta*, **67**, 1630 (1984).
233 Voiget, P., Bonivento, M., Roulet, R., and Vogel, P., *Helv. Chim. Acta*, **67**, 1638 (1984).
234 Métral, J.-L., and Vogel, P., *Tetrahedron Lett.*, **25**, 5387 (1984).
235 Tamariz, J., and Vogel, P., *Tetrahedron*, **40**, 4549 (1984).
236 Tornare, J.-M., and Vogel, P., *J. Org. Chem.*, **49**, 2510 (1984).
237 Lee, I., and Han, E. S., *Haksurwon Nonmunjip, Chayon Kwahak Pyon*, **22**, 43 (1983); *Chem. Abs.*, **101**, 109891 (1984).
238 Bertran, J., Branchadell, V., Duran, M., and Oliva, A., *THEOCHEM*, **16**, 191 (1984); *Chem. Abs.*, **101**, 6210 (1984).

[239] Han, E. S., Lee, I., and Chang, B. D., *Bull. Korean Chem. Soc.*, **4,** 197 (1983); *Chem. Abs.*, **100,** 208796 (1984).
[240] Wiberg, K. B., Matturro, M. G., Okarma, P. J., and Jason, M. E., *J. Am. Chem. Soc.*, **106,** 2194 (1984).
[241] Saito, K., Omura, Y., Maekawa, E., and Gassman, P. G., *Tetrahedron Lett.*, **25,** 2573 (1984).
[242] Billups, W. E., Lin, L.-J., and Casserly, E. W., *J. Am. Chem. Soc.*, **106,** 3698 (1984).
[243] Balci, M., and Harmandar, M., *Tetrahedron Lett.*, **25,** 237 (1984).
[244] Jendralla, H., and Spur, B., *J. Chem. Soc., Chem. Commun.*, **1984,** 887.
[245] Nakamura, N., *Chem. Lett.*, **1983,** 1795.
[246] Molz, T., König, P., Goes, R., Gauglitz, G., and Meier, H., *Chem. Ber.*, **117,** 833 (1984).
[247] Corsico Corda, A., Desimoni, G., Ferrari, E., Righetti, P. P., and Tacconi, G., *Tetrahedron*, **40,** 1611 (1984).
[248] Kimura, Y., Suzuki, M., Matsumoto, T., Abe, R., and Terashima, S., *Chem. Lett.*, **1984,** 473.
[249] Scott, L. T., Grütter, P., and Chamberlain, R. E., *J. Am. Chem. Soc.*, **116,** 4852 (1984).
[250] Scott, L. T., and Adams, C. M., *J. Am. Chem. Soc.*, **106,** 4857 (1984).
[251] Rao, Y. K., and Nagarajan, M., *Synthesis*, **1984,** 757.
[252] Nakayama, J., Nakamura, Y., and Hoshino, M., *Heterocycles*, **21,** 752 (1984).
[253] Glass, R. S., Reineke, K., and Shanklin, M., *J. Org. Chem.*, **49,** 1527 (1984).
[254] De Lucchi, O., Lucchini, V., Pasquato, L., and Modena, G. *J. Org. Chem.*, **49,** 596 (1984).
[255] Pitt, I. G., Russell, R. A., and Warrener, R. N. *J. Chem. Soc., Chem. Commun.*, **1984,** 1466.
[256] Müller, K., and Sauer, J., *Tetrahedron Lett.*, **25,** 2541 (1984).
[257] Wulff, W. D., and Yang, D. C., *J. Am. Chem. Soc.*, **106,** 7565 (1984).
[258] Dolbier, W. R., and Burkholder, C. R., *J. Org. Chem.*, **49,** 2381 (1984).
[259] Evans, D. A., Chapman, K. T., and Bisaha, J., *J. Am. Chem. Soc.*, **106,** 4261 (1984).
[260] Koizumi, T., Hakamada, I., and Yoshii, E., *Tetrahedron Lett.*, **25,** 87 (1984).
[261] Poll, T., Heimchen, G., and Bauer, B., *Tetrahedron Lett.*, **25,** 2191 (1984).
[262] Oppolzer, W., Chapuis, C., and Bernardinelli, G., *Helv. Chim. Acta*, **67,** 1397 (1984).
[263] tom Dieck, H., and Dietrich, J. *Chem. Ber.*, **117,** 694 (1984).
[264] Oppolzer, W., and Chapuis, C., *Tetrahedron Lett.*, **25,** 5383 (1984).
[265] Masamune, S., Reed, L. A., Davis, J. T., and Choy, W., *J. Org. Chem.*, **48,** 4441 (1983).
[266] Gnichtel, H., Gumprecht, C., and Luger, P., *Liebigs Ann. Chem.*, **1984,** 1531.
[267] Grée, R., Kessabi, J., Mosset, P., Martelli, J., and Carrié, R., *Tetrahedron Lett.*, **25,** 3697 (1984).
[268] Tolbert, L. M., and Ali, M. B., *J. Am. Chem. Soc.*, **106,** 3806 (1984).
[269] Ohmura, H., and Motoki, S., *Bull. Chem. Soc. Jpn.*, **57,** 1131 (1984).
[270] Bestmann, H. J., and Schmid, G., *Tetrahedron Lett.*, **25,** 1441 (1984).
[271] Lawson, K. R., Singleton, A., and Whitham, G. H., *J. Chem. Soc., Perkin Trans. 1*, **1984,** 865.
[272] Lawson, K. R., Singleton, A., and Whitham, G. H., *J. Chem. Soc., Perkin Trans. 1*, **1984,** 859.
[273] Hall, H. K., and Miniutti, D. L., *Tetrahedron Lett.*, **25,** 943 (1984).
[274] Kollenz, G., Penn, G., Dolenz, G., Akcamur, Y., Peters, K., Peters, E.-M., and von Schnering, H. G., *Chem. Ber.*, **117,** 1299 (1984).
[275] Lucchini, V., Prato, M., Quintily, U., and Scorrano, G., *J. Chem. Soc., Chem. Commun.*, **1984,** 48.
[276] Destro, F., Lucchini, V., and Prato, M., *Tetrahedron Lett.*, **25,** 5573 (1984).
[277] Seitz, G., Dhar, R., Mohr, R., and Overheu, W., *Arch. Pharm. (Weinheim, Ger.)*, **317,** 237 (1984).
[278] Burger, K., and Simmerl, R., *Liebigs Ann. Chem.*, **1984,** 982.
[279] Seitz, G., Mohr, R., Overheu, W., Allmann, R., and Nagel, M., *Angew. Chem. Int. Ed.*, **23,** 890 (1984).
[280] Vedejs, E., Perry, D. A., Houk, K. N., and Rondan, N. G., *J. Am. Chem. Soc.*, **105,** 6999 (1983).
[281] Katada, T., Eguchi, S., Esaki, T., and Sasaki, T., *J. Chem. Soc., Perkin Trans. 1*, **1984,** 1869.
[282] Katada, T., Eguchi, S., Esaki, T., and Sasaki, T., *J. Chem. Soc., Perkin Trans. 1*, **1984,** 2649.
[283] Alcaraz, J.-M., and Mathey, F., *J. Chem. Soc., Chem. Commun.*, **1984,** 508.
[284] Charrier, C., Bonnard, H., de Lauzon, G., and Mathey, F., *J. Am. Chem. Soc.*, **105,** 6871 (1983).
[285] Deschamps, E., and Mathey, F., *J. Chem. Soc., Chem. Commun.*, **1984,** 1214.
[286] Wiberg, N., and Wagner, G., *Angew. Chem. Int. Ed.*, **22,** 1005 (1983).
[287] Bronberger, F., and Huisgen, R., *Tetrahedron Lett.*, **25,** 57 (1984).
[288] Huisgen, R., and Bronberger, F., *Tetrahedron Lett.*, **25,** 61 (1984).
[289] Kjell, D. P., and Sheridan, R. S., *J. Am. Chem. Soc.*, **106,** 5368 (1984).
[290] Nelsen, S. F., Blackstock, S. C., and Frigo, T. B., *J. Am. Chem. Soc.*, **106,** 3366 (1984).
[291] Jefford, C. W., Kohmoto, S., Boukouvalas, J., and Burger, U., *J. Am. Chem. Soc.*, **105,** 6498 (1983).
[292] Gollnick, K., and Griesbeck, A., *Tetrahedron Lett.*, **25,** 725 (1984).
[293] Steliou, K., Gareau, Y., and Harpp, D. N., *J. Am. Chem. Soc.*, **106,** 799 (1984).

294 Boger, D. L., and Patel, M., *J. Org. Chem.*, **49**, 4098 (1984).
295 Meth-Cohn, O., and van Vuuren, G., *J. Chem. Soc., Chem. Commun.*, **1984**, 1144.
296 Danishefsky, S. J., Pearson, W. H., and Harvey, D. F., *J. Am. Chem. Soc.*, **106**, 2455 (1984).
297 Danishefsky, S. J., Pearson, W. H., and Harvey, D. F. *J. Am. Chem. Soc.*, **106**, 2456 (1984).
298 Bednarski, M., and Danishefsky, S., *J. Am. Chem. Soc.*, **105**, 6968 (1983).
299 van den Broek, L. A. G. M., Porskamp, P. A. T. W., Hattiwanger, R. C., and Zwanenburg, B., *J. Org. Chem.*, **49**, 1691 (1984).
300 Jurczak, J., Bauer, T., and Jarosz, S., *Tetrahedron Lett.*, **25**, 4809 (1984).
301 Fukunaga, T., and Begland, R. W., *J. Org. Chem.*, **49**, 813 (1984).
302 Groenewold, G. S., and Gross, M. L., *J. Am. Chem. Soc.*, **106**, 6569 (1984).
303 Groenewold, G. S., and Gross, M. L., *J. Am. Chem. Soc.*, **106**, 6575 (1984).
304 Calhoun, G. C., and Schuster, G. B., *J. Am. Chem. Soc.*, **106**, 6870 (1984).
305 Beck, K., Höhn, A., Hünig, S., and Protschy, F., *Chem. Ber.*, **117**, 517 (1984).
306 Hünig, S., and Prokschy, F., *Chem. Ber.*, **117**, 534 (1984).
307 Laszlo, P., and Lucchetti, J., *Tetrahedron Lett.*, **25**, 1567 (1984).
308 Laszlo, P., and Lucchetti, J., *Tetrahedron Lett.*, **25**, 2147 (1984).
309 Laszlo, P., and Lucchetti, J., *Tetrahedron Lett.*, **25**, 4387 (1984).
310 Ishibashi, H., Kitano, Y., Nakatani, H., Okada, M., Ikeda, M., Okura, M., and Tamura, Y., *Tetrahedron Lett.*, **25**, 4231 (1984).
311 Christl, M., Brunn, E., and Lanzendörfer, F., *J. Am. Chem. Soc.*, **106**, 373 (1984).
312 Miyagi, Y., Muruyama, K., Tanaka, N., Sato, M., Tomizu, T., Isogawa, Y., and Kashiwano, H., *Bull. Chem. Soc. Jpn.*, **57**, 791 (1984).
313 Gallacher, G., Ng, A. S., Attah-Poku, K., Antczak, K., Alward, S. J., Kingston, J. F., and Fallis, A. G., *Can. J. Chem.*, **62**, 1709 (1984).
314 Alward, S. J., and Fallis, A. G., *Can. J. Chem.*, **62**, 121 (1984).
315 Koreeda, M., and Luengo, J. I., *J. Org. Chem.*, **49**, 2079 (1984).
316 Tamariz, J., and Vogel, P., *Angew. Chem. Int. Ed.*, **23**, 74 (1984).
317 Fraser-Reid, B., Benko, Z., Guiliano, R., Sun, K. M., and Taylor, N., *J. Chem. Soc., Chem. Commun.*, **1984**, 1029.
318 Himbert, G., Diehl, K., and Maas, G., *J. Chem. Soc., Chem. Commun.*, **1984**, 900.
319 Fukazawa, Y., Kikuchi, M., and Ito, S., *Tetrahedron Lett.*, **25**, 1505 (1984).
320 Gassman, P. G., and Singleton, D. A., *J. Am. Chem. Soc.*, **106**, 6085 (1984).
321 Firestone, R. A., and Saffar, S. G., *J. Org. Chem.*, **48**, 4783 (1983).
322 Boger, D. L., and Coleman, R. S., *J. Org. Chem.*, **49**, 2240 (1984).
323 Cheng, Y-S., Lupo, A. T., and Fowler, F. W., *J. Am. Chem. Soc.*, **105**, 7696 (1983).
324 Whitesell, M. A., and Kyba, E. P., *Tetrahedron Lett.*, **25**, 2119 (1984).
325 Ihara, M., Kirihara, T., Kawaguchi, A., Fukumoto, K., and Kametani, T., *Tetrahedron Lett.*, **25**, 4541 (1984).
326 Jacobi, P. A., Weiss, K. T., and Egbertson, M., *Heterocycles*, **22**, 281 (1984).
327 Seitz, G., and Dietrich, S., *Arch. Pharm. (Weinheim, Ger.)*, **317**, 379 (1984).
328 Jung, M. E., and Rohloff, J. C., *J. Chem. Soc., Chem. Commun.*, **1984**, 630.
329 Eddaif, A., Laurent, A., Mison, P., and Pellissier, N., *Tetrahedron Lett.*, **25**, 2779 (1984).
330 Yoshioka, M., Nakai, H., and Ohno, M., *J. Am. Chem. Soc.*, **106**, 1133 (1984).
331 Remiszewski, S. W., Whittle, R. R., and Weinreb, S. M., *J. Org. Chem.*, **49**, 3243 (1984).
332 Sternbach, D. D., Rossana, D. M., and Onan, K. D., *J. Org. Chem.*, **49**, 3427 (1984).
333 Edwards, M. P., Ley, S. V., Lister, S. G., Palmer, B. D., and Williams, D. J., *J. Org. Chem.*, **49**, 3503 (1984).
334 Shishido, K., Shimada, S., Fukumoto, K., and Kametani, T., *Chem. Pharm. Bull.*, **32**, 922 (1984).
335 Tamaru, Y., Ishige, O., Kawamura, S.-i., and Yoshida, Z.-i., *Tetrahedron Lett.*, **25**, 3583 (1984).
336 Brun, P., Teneglia, A., Zahra, J. P., and Waegell, B., *Tetrahedron Lett.*, **25**, 3827 (1984).
337 Brun, P., Teneglia, A., Zahra, J. P., and Waegell, B., *Tetrahedron Lett.*, **25**, 3831 (1984).
338 Evans, D. A., Chapman, K. T., and Bisaha, J., *Tetrahedron Lett.*, **25**, 4071 (1984).
339 Martin, S. F., Williamson, S. A., Gist, R. P., and Smith, K. M., *J. Org. Chem.*, **48**, 5170 (1983).
340 Moody, C. J., *J. Chem. Soc., Chem. Commun.*, **1984**, 925.
341 Magnus, P., Gallacher, T., Brown, P., and Pappalardo, P., *Acc. Chem. Res.*, **17**, 35 (1984).
342 Krebs, A., Güntner, A., Senning, A., Moltzen, E. K., Klabunde, K. J., and Kramer, M. P., *Angew. Chem. Int. Ed.*, **23**, 729 (1984).
343 Krebs, A., Güntner, A., Versteylen, S., and Schulz, S., *Tetrahedron Lett.*, **25**, 2333 (1984).
344 Padwa, A., and Tohidi, M., *J. Chem. Soc., Chem. Commun.*, **1984**, 295.
345 SooHoo, C. K., and Baxter, S. G., *J. Am. Chem. Soc.*, **105**, 7443 (1983).

[346] Cowley, A. H., Kemp, R. A., Lasch, J. G., Norman, N. C., and Stewart, C. A., *J. Am. Chem. Soc.*, **105,** 7444 (1983).
[347] Köcher, J., and Neumann, W. P., *J. Am. Chem. Soc.*, **106,** 3861 (1984).
[348] Mach, K. M., Antropiusova H., Petrusova, L., Hanus, V., Turecek, F., and Sedmera, P., *Tetrahedron*, **40,** 3295 (1984).
[349] Hoffmann, H. M. R., *Angew. Chem. Int. Ed.*, **23,** 1 (1984).
[350] Föhlisch, B., Flogaus, R., Oexle, J., and Schädel, A., *Tetrahedron Lett.*, **25,** 1773 (1984).
[351] Föhlisch, B., and Herter, R., *Chem. Ber.*, **117,** 2580 (1984).
[352] Schultz, A. G., Myong, S. O., and Puig, S., *Tetrahedron Lett.*, **25,** 1011 (1984).
[353] Böhm, M. C., Bickert, P., Hafner, K., and Boekelheide, V., *Proc. Natl. Acad. Sci. U. S. A.*, **81,** 2589 (1984).
[354] Kaftory, M., *J. Chem. Soc., Perkin Trans. 2*, **1984,** 757.
[355] Yang, N. C., Chen, M.-J., and Chen., P., *J. Am. Chem. Soc.*, **106,** 7310 (1984).
[356] Garst, M. E., Roberts, V. A., Houk, K. N., and Rondan, N. G., *J. Am. Chem. Soc.*, **106,** 3882 (1984).
[357] Paquette, L. A., Hathaway, S. J., Kravetz, T. M., and Hsu, L.-Y., *J. Am. Chem. Soc.*, **106,** 5741 (1984).
[358] Wu, T.-C., Mareda, J., Gupta, Y. N., and Houk, K. N., *J. Am. Chem. Soc.*, **105,** 6996 (1983).
[359] Meijer-Veldman, M. E. E., and de Liefde Meijer, H. J., *J. Organomet. Chem.*, **260,** 199 (1984).
[360] Mauret, P., and Alphonse, P., *J. Organomet. Chem.*, **276,** 249 (1984).
[361] Barton, J. W., and Shepherd, M. K., *Tetrahedron Lett.*, **25,** 4967 (1984).
[362] Vollhardt, K. P. C., *Chem. Future, Proc. IUPAC Congr., 29th*, **1983,** 181; *Chem. Abs.*, **100,** 208621 (1984).
[363] Vollhardt, K. P. C., *Angew. Chem. Int. Ed.*, **23,** 539 (1984).
[364] Danishefsky, S., Harrison, P., Silvestri, M., and Segmuller, B., *J. Org. Chem.*, **49,** 1319 (1984).
[365] Sternberg, E. D., and Vollhardt, K. P. C., *J. Org. Chem.*, **49,** 1564 (1984).
[366] Sternberg, E. D., and Vollhardt, K. P. C., *J. Org. Chem.*, **49,** 1574 (1984).
[367] Fink, J., and Regitz, M., *Tetrahedron Lett.*, **25,** 1711 (1984).
[368] Yamaguchi, R., Ban, M., Kawanisi, M., Osawa, E., Jaime, C., Buda, A. B., and Katsumata, S., *J. Am. Chem. Soc.*, **106,** 1512 (1984).
[369] Leismann, H., Mattay, J., and Scharf, H.-D., *J. Am. Chem. Soc.*, **106,** 3985 (1984).
[370] Adam, W., and Carballeira, N., *J. Am. Chem. Soc.*, **106,** 2874 (1984).

Organic Reaction Mechanisms 1984
Edited by A. C. Knipe and W. E. Watts

CHAPTER 15

Molecular Rearrangements

A. W. MURRAY

Department of Chemistry, University of Dundee

Aromatic Rearrangements

Benzene Derivatives

The potential energies of the 1,2-hydrogen shift in protonated benzene and its derivatives have been estimated by MCNDO/3,[1] and the relative rates and Arrhenius parameters for selected 1,2-aryl migrations have been calculated.[2] An isokinetic point has been found[3] for the degenerate rearrangement of heptamethylbenzenium tetrachloroaluminate on passing from the solution to the crystalline state, and a D-labelling method has been used to study the effect of crystal structure on the degenerate 1,2-phenyl shift in the 1-phenyl-1,2,3,4,5,6-hexamethylbenzenonium ion.[4] Although the automerization of phenylium ions is in contradiction with theoretical predictions of a high energy barrier for the process, evidence has been obtained,[5] from phenylation studies of methyl halides, to suggest that the above process is a degenerate rearrangement *via* consecutive 1,2-hydrogen shifts in the phenylium ion. The distribution of products in the degenerate rearrangement of triarylvinyl cations has been investigated,[6,7] and the cyclization of 1,2-diarylethanes (**1**) in SbF_5–HF at 0°, to yield a variety of tricyclic phenanthrenones, has been observed.[8] A Lewis-acid-catalysed 1,2-aryl shift in α-haloalkylaryl acetals has provided a convenient route to α-arylalkanoic acids,[9] while 2-arylpropanoic acids have been prepared[10] by oxidative aryl migration in aryl α-seleno- and aryl α-telluro-ethyl ketones; see (**2**) → (**3**).

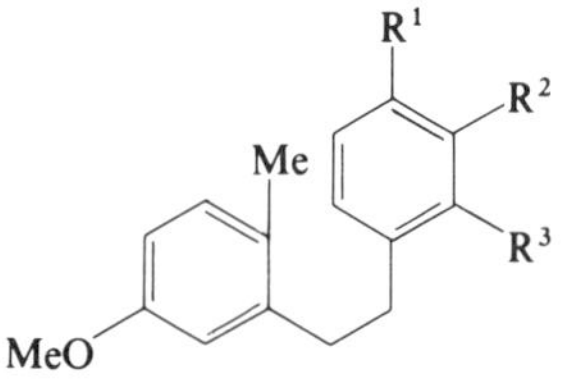

R[1] R[2] R[3] Me MeO

(**1**)

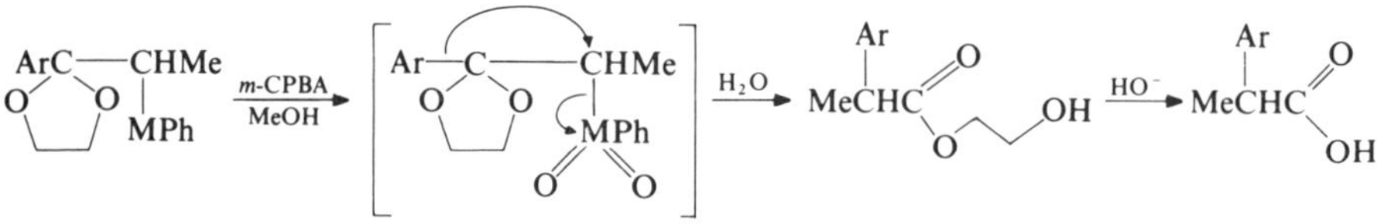

ArC—CHMe, O O, MPh —(m-CPBA, MeOH)→ [Ar—C—CHMe, O O, MPh, O O] —(H_2O)→ MeCHC(Ar)(=O)O–CH₂CH₂–OH —(HO^-)→ MeCHC(Ar)(=O)OH

(**2**) (**3**)

M = Se, Te

In the presence of trifluoroacetic acid, *N*-phenyl-*N'*-phenoxyurea (**4**) is found to undergo a novel rearrangement to *N*-(4'-hydroxy-2-biphenylyl)urea (**6**) and *N*-carbamoyl-2-hydroxydiphenylamine (**7**) (see Scheme 1).[11] By analogy with the benzidine rearrangement one plausible mechanism for the formation of (**6**) consists of a [5,5]-sigmatropic rearrangement of the enolized phenoxyurea (**5**). The kinetics of the Wallach rearrangement of *para*-monosubstituted azoxybenzenes have been determined[12] by use of UV spectroscopy, and an extension of the Wallach rearrangement studies has been applied to the phenylazoxypyridine series.[13] A study of the kinetics of the Fischer–Hepp rearrangement of *N*-nitrosomonoarylamines at elevated acidities has been undertaken,[14] while various 4-nitrosodiarylamines have been synthesized by the above rearrangement.[15] A substantial nitrogen kinetic isotope effect, but no carbon kinetic isotope effect, has been observed[16] in the nitramine rearrangement of *N*-nitro-*N*-methylaniline. Although the absence of the latter is not a conclusive proof, it is consistent with, and suggestive of, non-

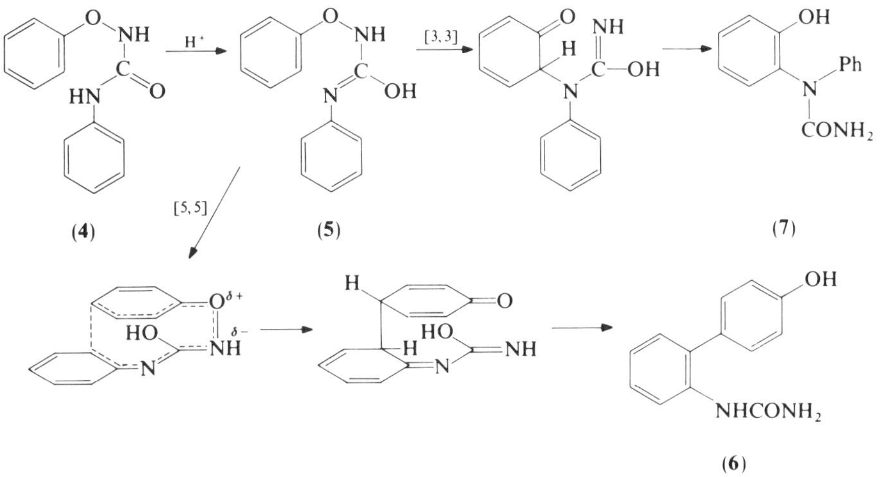

SCHEME 1

concertedness. So a two-step process, scission followed by recombination, is supported for the first time by direct probes of the bond-breaking and bond-making steps. The novel rearrangement of aniline to 2-methylpyridine has been catalysed by zeolite.[17]

A study of substituent effects[18] appears to argue against the involvement of a nitrenium ion intermediate in the acid-catalysed rearrangement of *N*-phenylhydroxylamines to aminophenols; the results rather suggest that the intermediate is better represented by an imine structure with the positive charge at the 4-position in the aromatic ring. However, more recently, evidence has been obtained[19] for the intermediacy of a nitrenium ion in this rearrangement. A Hammett study of a series of methanesulphonate esters of monosubstituted *N*-arylhydroxamic acids has demonstrated[20] that their rearrangement proceeds *via* cleavage of the N—O bond of the ester in a heterolytic manner to produce an acylarylnitrenium ion. In much the same way, the reaction of similarly activated derivatives of *N*-acetoxy-2-acetamidofluorene (**8**) with *in vitro* nucleophiles has been reported[21] to produce products resulting from formal addition of the nucleophile to the 4-position of the substrate. Thus initial attack *para* to the 2-acetamido moiety, *via* the intermediate nitrenium ion (**9**), is postulated to produce the dienone imine (**10**) which is converted into (**11**; X = OH), and then into (**12**) by loss of water. Sulphinanilides (**13**) have been converted into anilinosulphoxides (**14**) and (**15**) in a hydrochloric acid-catalysed rearrangement which appears to be an example of an electrophilic aromatic substitution with the acid aiding in the creation of the electrophile.[22]

The kinetics of the Smiles rearrangement of 2-(acetylamino)ethyl 2,6-dinitrophenyl ether and of *N*-acetyl-*N*-(2-hydroxyethyl)-2,6-dinitroaniline with potassium methoxide in DMSO–MeOH have been investigated,[23] while the reversible formation of a transient spiro-Meisenheimer complex of a mononitrobenzene derivative has been established[24] in the Smiles rearrangement of 4-nitrophenyl

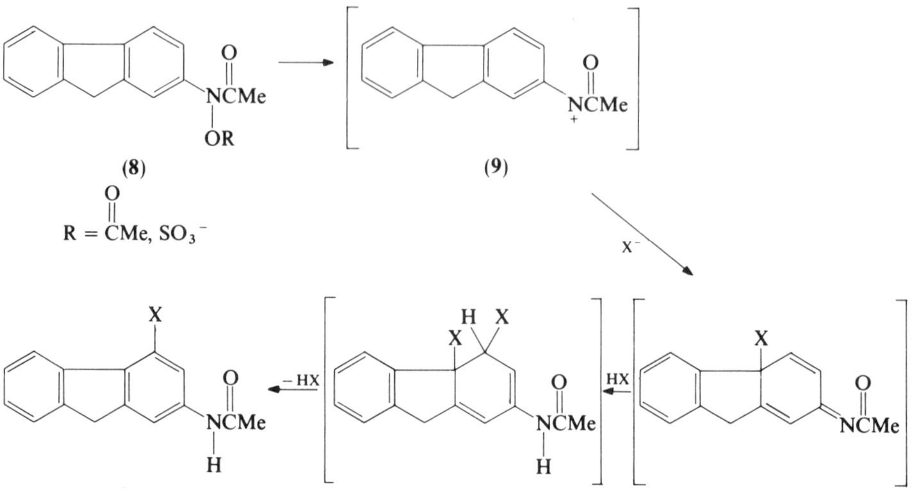

(8) (9)

R = CMe, SO_3^-

(12) (11) (10)

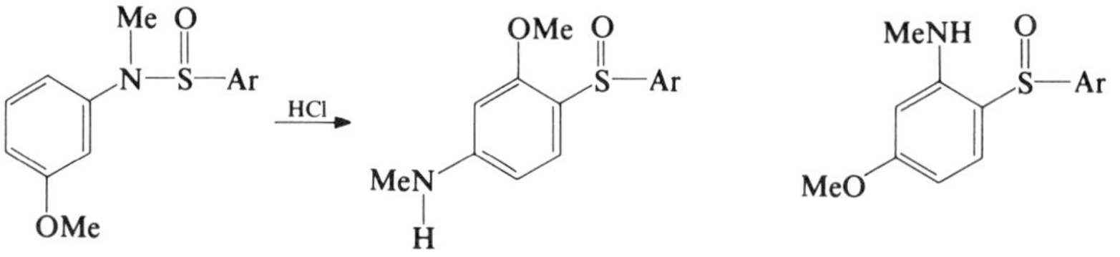

(13) (14) (15)

Ar = *p*-tolyl

N-hydroxycarbamate. The first example of an alkoxide-accelerated Smiles rearrangement has been reported.[25] Thus the rearrangement of *N*-(2-hydroxyethyl)aryloxyacetamides in THF–18-crown-6 and potassium hydroxide produces useful yields of *N*-(2-hydroxyethyl)anilines. An interesting halogen-activated Smiles rearrangement of (*o*-halophenoxy)aniline derivatives (**16**) has produced ring-substituted 10-[3-(dimethylamino)propyl]phenoxazines (**17**).[26] 3-Aryl-1,2-cyclohexanediones (**19**) have been conveniently prepared by the photolysis of the corresponding 2-[(arylsulphonyl)oxy]cyclohex-2-enones (**18**).[27] A possible mechanism for this conversion is presented in Scheme 2. Treatment of dialkyl aryl phosphates with Bu^nLi or LDA has been shown to result in a rearrangement that involves fission of an O—P bond and formation of a C—P bond, yielding dialkyl (2-hydroxyaryl) phosphonates.[28]

Sulphonate ester rearrangements have been induced in solution by the ordered structure of a smectic B solvent.[29] Rate constants have been determined for the conversion of *p*-alkylbenzenesulphonic acids to their *meta* isomers,[30] and for the isomerization of *o*- and *p*-aminobenzenesulphonic acids.[31] The latter process has also been studied in the presence of mercury(II) sulphate.[32]

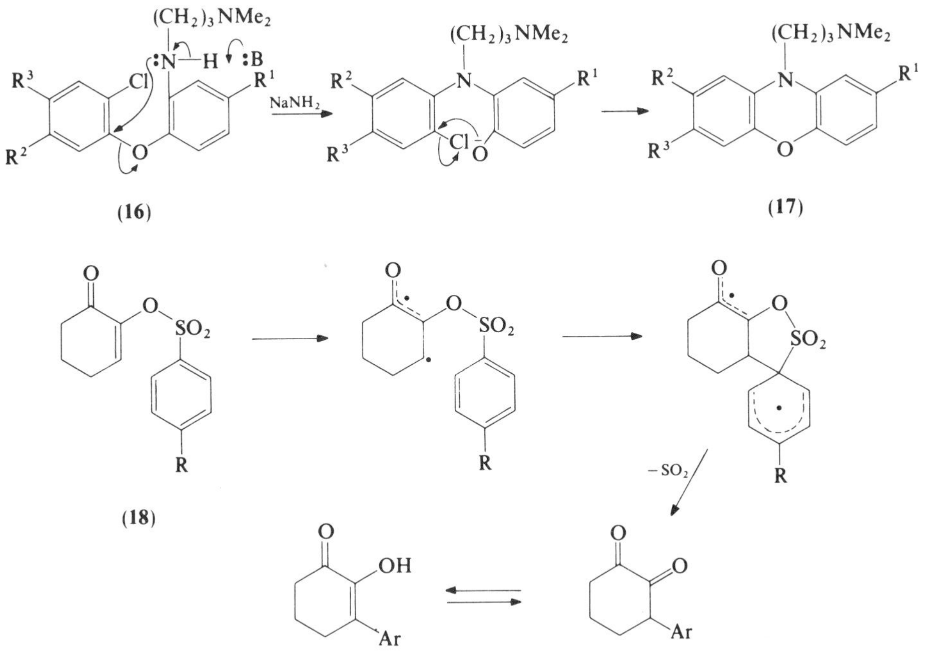

(16)

(17)

(18)

(19)

SCHEME 2

The photo-Fries rearrangement of a number of anilides on the surface of dry silica gel has been examined.[33] It has been shown[34] that the photorearrangement of phenyl benzoate in aqueous medium changes in the presence of cyclodextrins or amylose in that the *ortho*/*para* migration ratio is altered in favour of the *para*-position. The photo-Fries rearrangement of 2-naphthyl benzoate has yielded the 6- and 8-benzoyl derivatives,[35] and an unusual Fries rearrangement of 1-acetoxy-4-hydroxy-5-methoxynaphthalenes to the corresponding 3-acetylquinols involves *meta*-migration of the acetyl groups.[36] The Fries rearrangement of mixed esters of polyhydroxy-fluoroanthenes has been reported,[37] and the photo-Fries rearrangement has been applied to the construction of xantho[2,3-*g*]tetralin and xantho[2,3-*f*]tetralin systems.[38]

Acid-catalysed rearrangement of [6]paracyclophane to the *meta*- and *ortho*-isomers has been reported,[39] while under the influence of hydrogen chloride/aluminium trichloride, [8]paracyclophane appears to isomerize to [8]metacyclophane which rearranges further to benzocyclodecene.[40] An anomalous rearrangement that occurs during the hydrolysis of diazotized *syn*-4-amino[2.2](1,4)naphthalenoparacyclophane to the corresponding *syn*-4-hydroxy- and isomeric 17-hydroxyphane has been postulated to proceed *via* loss of nitrogen, formation of a bond between C(4) and C(17) to form a cationic intermediate, followed by hydrogen migration.[41]

A recent paper[42] has drawn attention to the fact that the first-order kinetics reported for the thermal rearrangement of azulene to naphthalene in the gas phase do not demand a monomolecular mechanism; in fact, the kinetics are equally compatible with a radical-chain mechanism initiated by bimolecular coupling of two azulene molecules. New mechanistic insights into automerization in benzenoid hydrocarbons has been obtained[43] from the thermal rearrangement of benz[*a*]anthracene-5-^{13}C, while examination of ($^{13}C_2$)biphenylene formed by the gas-phase pyrolysis of doubly labelled benzyne precursors [(1,6-$^{13}C_2$)phthalic anhydride and (2a,3-$^{13}C_2$)benzocyclobutenedione] has shown that the principal pyrolytic process leads to overall 1,2 → 1,3 rearrangement of the C_6H_4 carbon skeleton, either in an intermediate before decarbonylation, or in benzyne itself.[44]

Heterocyclic Derivatives

New rearrangements of nitrogenous heteroaromatic compounds have been reviewed,[45] and molecular rearrangements of five-membered ring heterocycles have been conveniently classified into five categories based on the number of participating side-chain atoms.[46] These include the rearrangements which operate by a ring-contraction–ring-expansion mechanism: the Dimroth rearrangement, the Cornforth rearrangement, the Boulton–Katritzky scheme, the bond-switch rearrangements with sulphur as the pivot atom, and many other isolated cases.

3-Nitro-4-alkyl(or aryl)aminopyridines (**20**) have been converted into 3-acetamido-4-alkyl(or aryl)amino(1*H*)pyridine-2-ones (**21**) by a catalytic hydrogenation–rearrangement in acetic acid (see Scheme 3).[47] A new base-catalysed rearrangement of *o*-carbomethoxy(diazoacetyl)pyridines has been reported.[48] 1-Allyl-2,4,6-trimethyl- and 1-allyl-2,4,6-triphenyl-pyridinium cations have been isomerized by mild alkali into the corresponding 1-propenylpyridinium cations and strong base converts the latter into oxazatricyclononene isomers of the quaternary

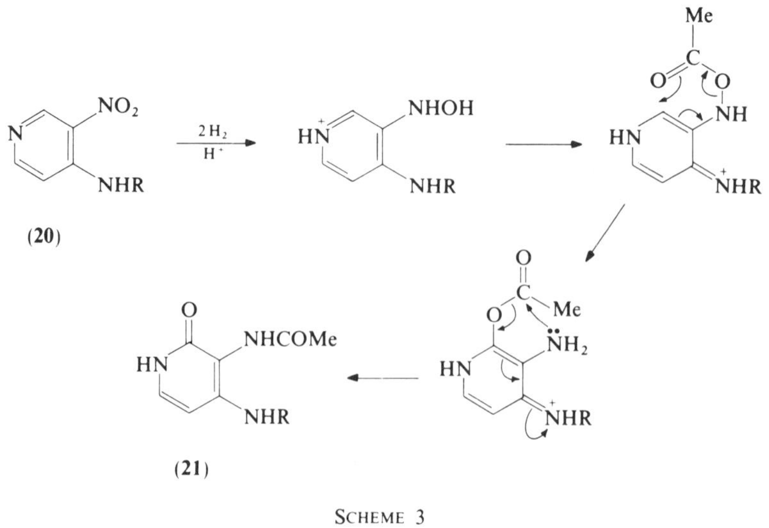

(**20**) $\xrightarrow[H^+]{2H_2}$ … → (**21**)

SCHEME 3

hydroxides. One possible sequence that would explain this complex conversion is outlined in Scheme 4,[49] namely, isomerization of the 1-allyl group to propenyl, addition of base to the α-position of the pyridinium ring to give a dihydropyridine (**22**) which ring-opens to form the enolic form (**23**) of the vinylogous amide. Internal Diels–Alder cycloaddition (**23a**) to give (**24**) followed by internal cyclization *via* nucleophilic attack of the hydroxyl group on the imino group would then afford (**25**). The authors considered two further mechanisms for the rearrangement. The

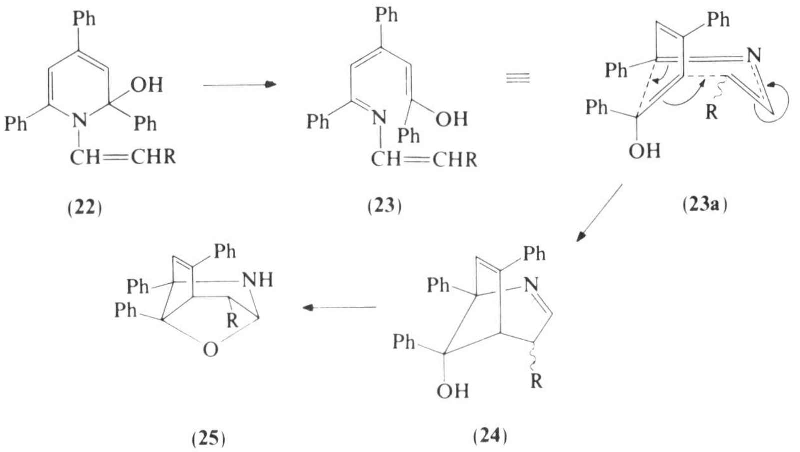

(**22**) (**23**) (**23a**) (**25**) (**24**)

SCHEME 4

pyrylium cation (**26**) has been shown to coexist in water with the enedione pseudobase (**26a**) and the anion (**26b**).[50] The kinetic and equilibrium interrelationships of these species and the corresponding 2*H*-pyran and oxodienol over a wide pH range have been studied and elucidated. The ion (**26**) can be converted into the pyridinium ion (**29**) by the addition of acidic solutions of the species to amine buffers.[51] The conversion is considered to proceed *via* tautomerization of the divinylogous amide (**27**) to the iminoenol (**28**), and electrocyclization of the latter. The potential of pyrylium salts for the selective transformation of primary amines into the corresponding pyridinium cations and hence into other functionalities has been discussed in some detail,[52,53] and the kinetics of the reaction of 4-(4-methoxy-3-sulphophenyl)-2,6-bis(4-sulphophenyl)pyrylium perchlorate with gelatin and chymotrypsin has been studied.[54] The Dimroth rearrangement of 2,4-diaminothiopyranylium halides has been used to synthesize 1-substituted 4-amino-2(1*H*)-pyridinethiones.[55]

4-Aroyl-1*H*-pyrazolo[3,4-*d*]pyrimidines undergo aryl migration to yield 4-aryl-4,5-dihydro-1*H*-pyrazolo[3,4-*d*]pyrimidine-4-carboxylic acids[56] and a series of dihydropyrimidines has been converted into 2-aminopyridines on base treatment.[57] A heterocyclic ring-contraction from six-membered (**30**) to five-membered, *e.g.* (**32**), with the loss of a nitrogen atom, has been discovered by Brown *et al.*[58] who speculate

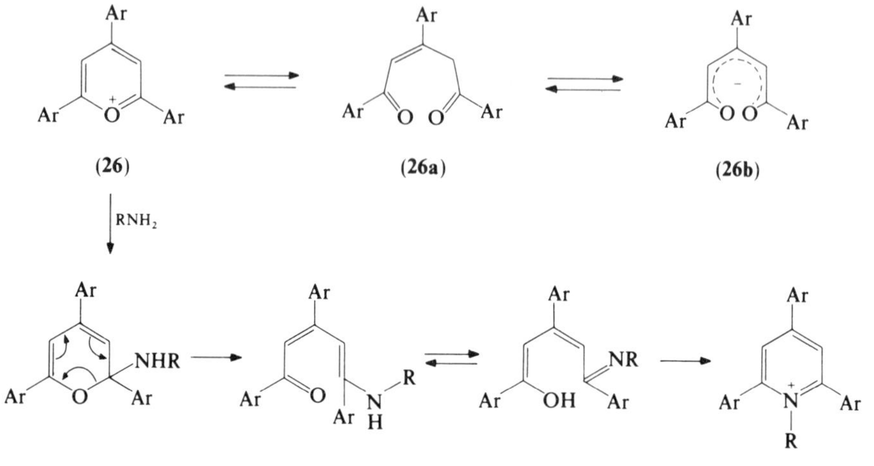

Ar Ar Ar
Ar O Ar Ar O O Ar Ar O O Ar

(26) (26a) (26b)

RNH_2

Ar Ar Ar Ar
NHR R NR
Ar O Ar Ar O N Ar OH Ar Ar N Ar
Ar H R

(27) (28) (29)

that the mechanism involves N—N cleavage to give the intermediate azomethine (**31**); two distinct ring-closure pathways could thereby lead to the five-membered ring products (see Scheme 5).

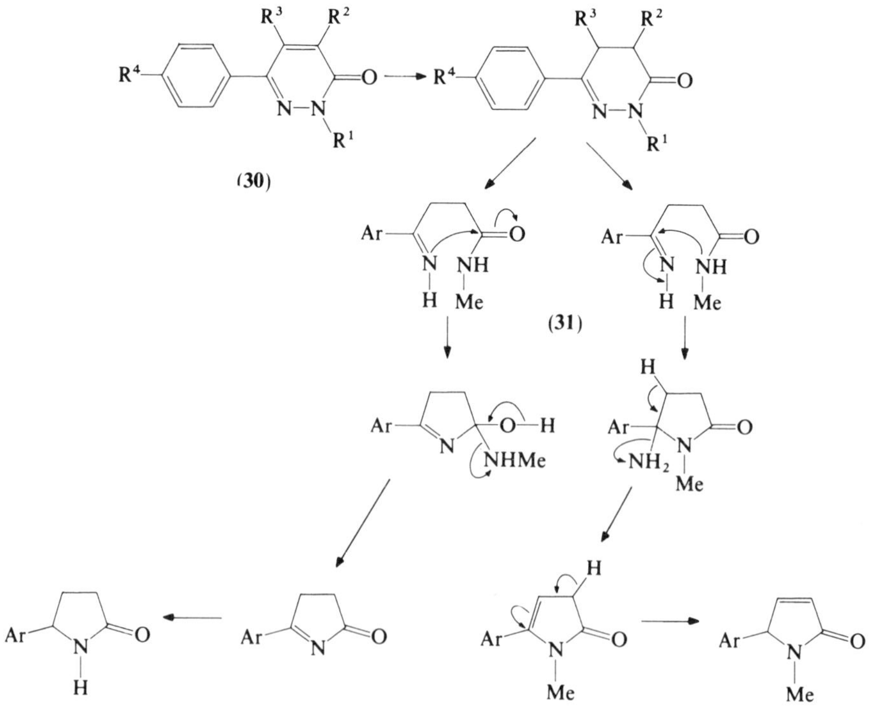

R^3 R^2 R^3 R^2
R^4 O R^4 O
N—N N—N
R^1 R^1

(30)

Ar O Ar O
N NH N NH
H Me H Me

(31)

H
Ar O—H Ar O
N NHMe NH_2 N
Me

H
Ar O Ar O Ar O Ar O
N N N N
H Me Me

(32)

SCHEME 5

A novel ring transformation of 1,3-oxazines (**33**) into pyrroles (**34**) by the reaction with the soft cyanide anion has been reported,[59] while nucleophilic attack on 4-acetoxy-2*H*-1,4-benzoxazin-3(4*H*)-one has been found to yield 2-, 5-, 6-, and 7-substituted derivatives of the benzoxazinone depending on the reaction conditions.[60] A mechanism involving an oxonium ion intermediate has been presented for the formation of the 2-substituted product. Heating dipyridazino[4,5-*b*:4′,5′-*e*][1,4]thiazines with K_2CO_3 in DMF was shown to afford dipyridazino[4,5-*b*:4′,5′-*d*]pyrroles.[61] The mechanism of this ring-contraction is thought to involve the intermediacy of an anion containing a thiirane ring. On the other hand, the acid- or base-catalysed ring-contraction of 1,4-benzthiazin-2,3-4*H*-dione (**35**) to benzthiazole (**36**) has been shown by a ^{14}C-labelling study[62] (outlined in Scheme 6) to proceed *via* hydrolytic opening of the thiolactone bond in (**35**).

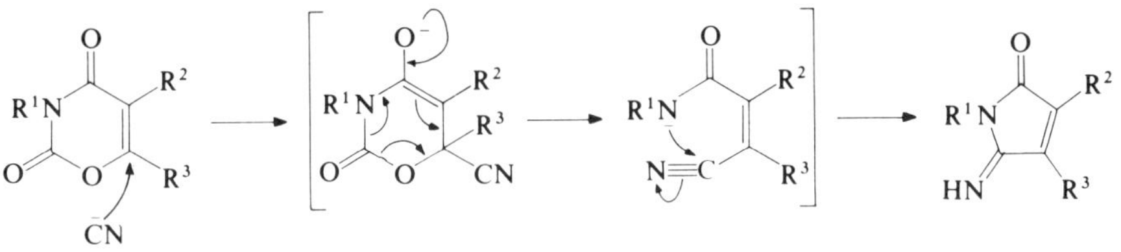

(33) (34)

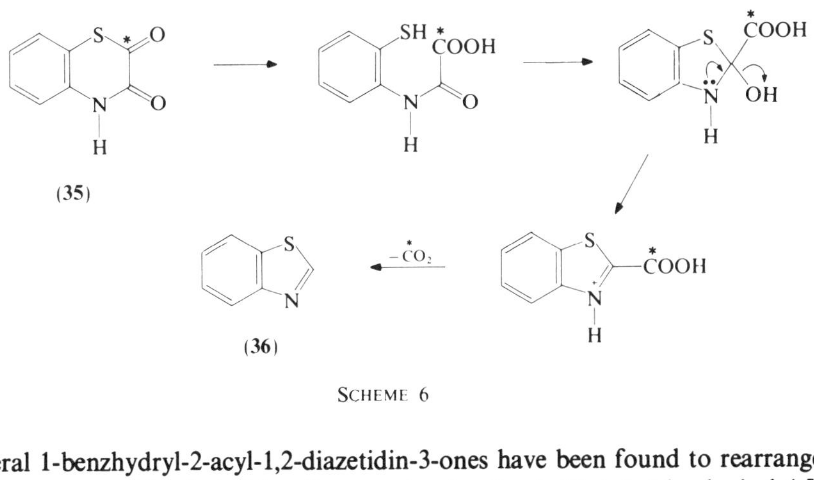

(35) (36)

SCHEME 6

Several 1-benzhydryl-2-acyl-1,2-diazetidin-3-ones have been found to rearrange upon treatment with ethyl chloroformate to give 2-substituted 4-benhydryl-4,5-dihydro-1,3,4-oxadiazin-6-ones.[63] A novel heterocyclic rearrangement of benzolypyrazolines into phenylpyridazinones has been reported.[64] Surprisingly, it has been found that *N*-acyl-α-aminonitriles and oxalyl derivatives do not produce 2(1*H*)-pyrazinones, as do α-aminonitriles. Instead, the reaction proceeds *via* 5(4*H*)-iminooxazole intermediates (**37**), which can undergo aromatization to yield oxazolones, or rearrange by a Dimroth-type rearrangement to afford 4(5*H*)-imidazolones (**38**).[65] New 1,3-oxazine derivatives (**40**) have been prepared[66] by an extraordinarily

(37) (38)

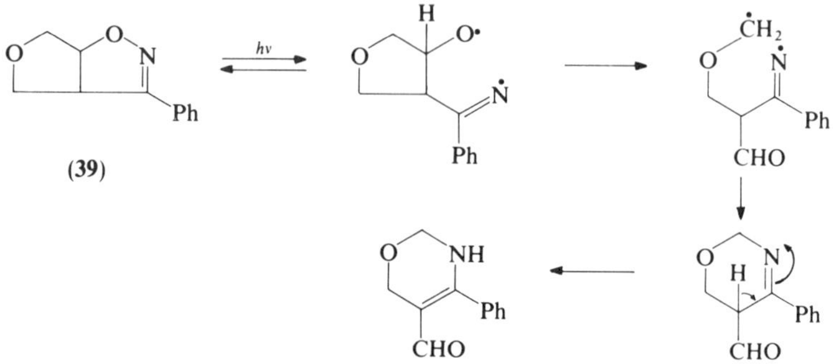

(39) (40)

unambiguous photoinduced rearrangement of furoisoxazolines (**39**). A variety of rearrangement products has been obtained by heating 3-alkoxybenzisothiazole 1,1-dioxides,[67] an unusual S → N methyl migration has been observed[68] on pyrolysis of *N*,*N*-dimethyl-*N*′- (4-methyl-2-thiazolyl)-*S*-methylisothiourea, and the first observation of a rearrangement occurring in the cycloaddition of anhydrohydroxythiazolium hydroxide mesoinic systems with olefinic diapolarophiles has been reported.[69] Non stereospecific ring-expansion reactions of five-membered heterocyclic sulphoxides bearing a heteroatom at the β-position to the sulphinyl group have been described.[70] The reactions, which are induced by Ac_2O or *p*-toluenesulphonic acid, are considered to proceed *via* sulphonium ion intermediates. The same authors have described[71] the thermal ring transformations of 1,4- and 1,5-benzothiazepine 1-oxides in the presence of catalytic amounts of *p*-toluenesulphonic acid. *trans*-Sulphoxides, having a methyl group *cis* to the sulphoxide oxygen, undergo ring-expansion to benzothiazocine derivatives, whereas the *cis* isomers afford ring-contraction products. The *m*-CPBA-catalysed rearrangement of a 3*H*-1,2-diazepine 2-oxide to a 3-alkanyl-3*H*-pyrazole 2-oxide has been reported.[72] The activation parameters for the rearrangement of 1*H*-1,2-diazepines to the corresponding 1-iminopyridinium ylides have been derived from ^{1}H-NMR-determined kinetics,[73] and 2,3-dihydro-1*H*-benz[*b*]azepines have been obtained by alkoxide-induced rearrangement of 2,5-dihydro-1*H*-benz[*b*]azepines.[74]

5-(2-Aminophenyl)-1,3,4-oxadiazole-2-(3*H*)-one has been found to rearrange above its melting point to 3-amino-2,4(1*H*,3*H*)-quinazolinedione.[75] Chemical and kinetic evidence has been presented[76] to support the intermediacy of cyclic compound (**42**) and open-chain anion (**43**) in the base-catalysed rearrangement of 2-phenylamino-1,3,4-oxadiazole (**41**) into 4-phenyl-1,2,4-triazolin-5-one (**44**). Detailed kinetic studies of substituent effects in the rearrangement of the (z)-phenylhydrazones of some 5-alkyl-[77] and 5-aryl-3-benzoyl-1,2,4-oxadiazoles[78] into 4-acyl- and 4-aroylamino-2,5-diphenyl-1,2,3-triazoles have been undertaken. The first kinetic results on the ring-transformation equilibrium of a 5-[(1-aminoethylidene)amino]-3-methyl-1,2,4-thiadiazole system have been reported,[79] while interesting transformations in thiadiazole[80] and isothiazole[81] ring systems have been explained in terms of bond switch at π-hypervalent sulphur. Nucleophile-induced transformations of 3-substituted 5-trifluoromethyl-1,3,4-thiadiazol-2(3*H*)-ones into 3-acylated thiadiazoles have been discovered,[82] as has a new ring-expansion reaction of 4,5-dihydro-1,3,4-thiadiazoles (**45**) into dihydro-1,3,4-thiadiazines (**46**) (see Scheme 7).[83] 1,4,2-Benzodithiazines, themselves, have been converted into 2-imino-1,3-benzodithioles.[84] An unusual feature of this rearrangement is that it occurs smoothly only when the benzodithiazine is heated without solvent; it does not take place in non-polar solvents.

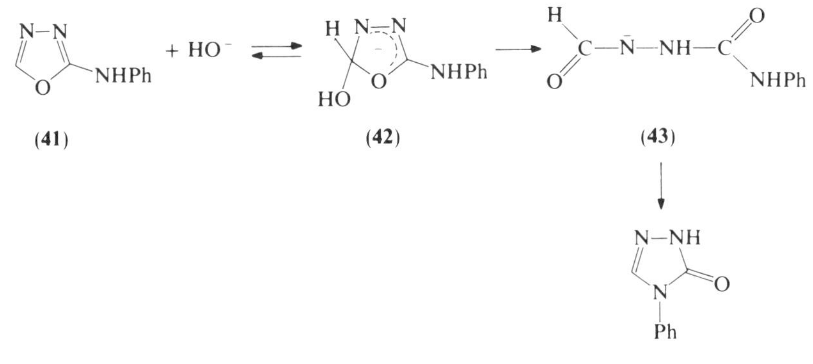

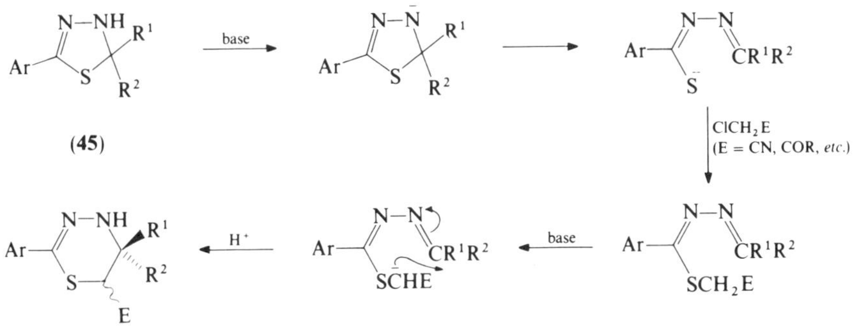

SCHEME 7

Triazinoindazolones (**48**) have been synthesized by the rearrangement of oxadiazoles (**47**) in base,[85] while interesting N to O and O to O alkyl group

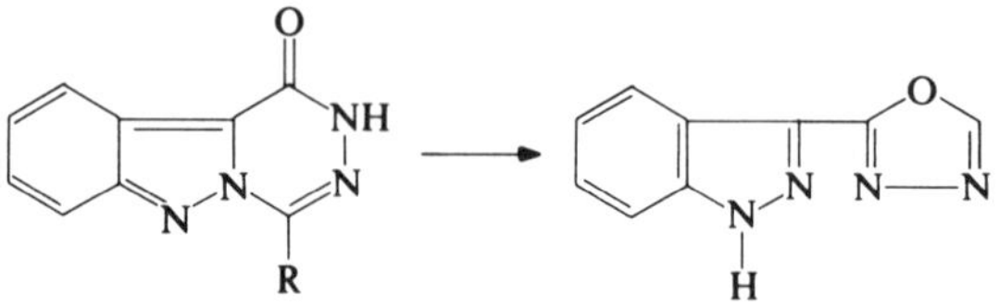

(47) (48)

migrations have been observed[86] during selective akylations of 3-alkoxy- and 3-akylamino-1,2,4-triazine *N*-oxides.

Cyclohexadiene Derivatives

Photorearrangements of dienones and quinones have been reviewed,[87] while a report of the overall photochemical conversion of 4-azidoalkyl-2,5-cyclohexadienones [**49**; R = $(CH_2)_3N_3$, $O(CH_2)_2N_3$] to bicyclic triazenes (**52**; X = CH_2, O)[88] has provided direct evidence for the intermediacy of zwitterions of the type (**51**) in the photochemistry of bicyclo[3.1.0]hex-3-en-2-ones (**50**), the initial photoproducts of 2,5-cyclohexadienones. The dienone–phenol rearrangement of spirodienones (**53**) to (**54**) has been studied[89] and the further transformation into the phenol (**55**) has been noted in one case (**54**; $n = 2$), while 6-methyl-6-nitrocyclohexa-2,4-dienones containing no substituent in the 2-position of the dienone have been observed to undergo regiospecific rearrangement to 6-nitro-*o*-cresols.[90] The structure of the novel product obtained from the oxidation of ketone (**56**), and similar

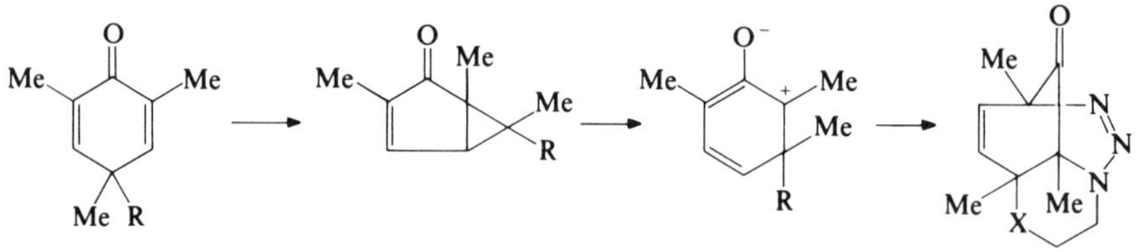

(49) (50) (51) (52)

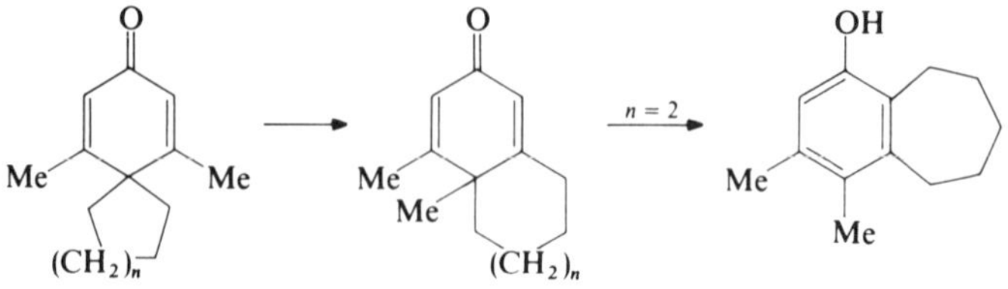

(53) (54) (55)

spiro-ketones, with 2,3-dichloro-5,6-dicyano-1,4-benzoquinone (DDQ) has been determined as (**57**).[91] Mechanistically, the formation of (**57**) can occur by the initial abstraction of a hydride ion at a benzylic position by DDQ resulting in a carbocation, followed by migration of bond '*a*' and subsequent loss of a proton. Spironaphthalenones of type (**58**) have been shown to undergo isomerization to (**59**) upon UV irradiation,[92] while benzazepines (**62**) substituted at C(2) have been prepared by treatment of 6-hydroxylated tetrahydroisoquinolinium salts (**60**) with

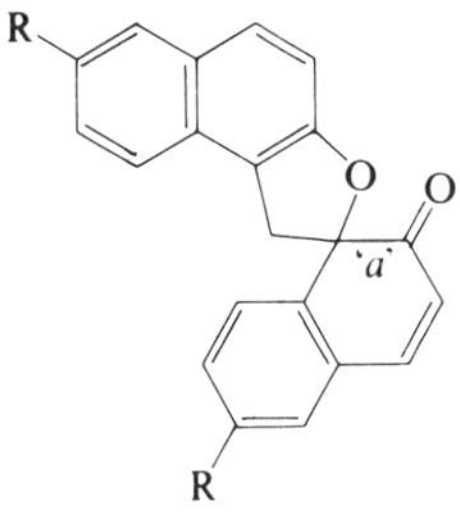

(**56**) 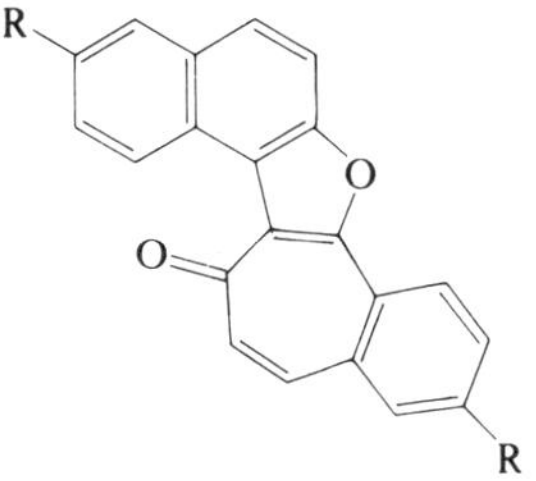 (**57**)

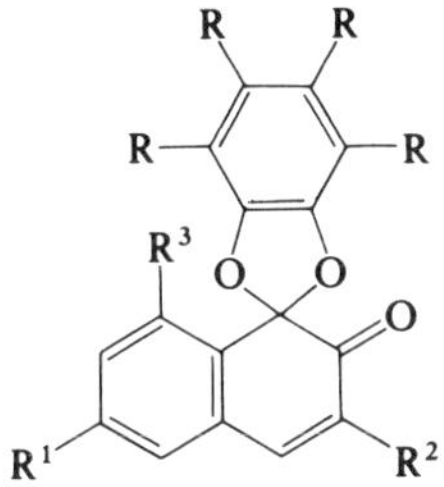

(**58**) 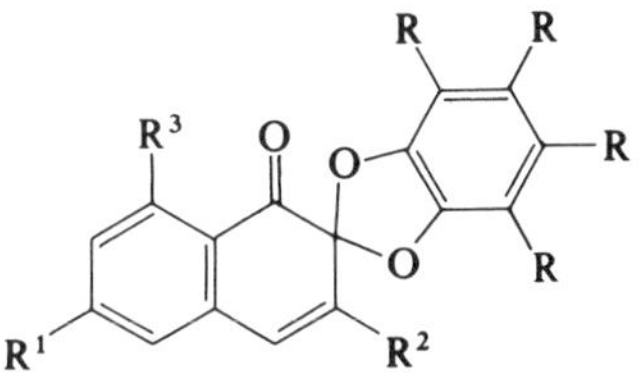 (**59**)

potassium *tert*-butoxide.[93] The general sequence depicted in Scheme 8, which proceeds through the intermediacy of a quinone methide (**61**), seems particularly attractive.

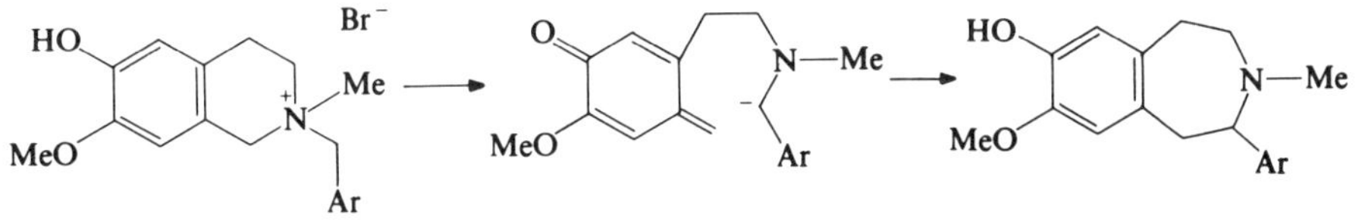

(**60**) (**61**) (**62**)

SCHEME 8

Allenylbenzenes (**64**) have been prepared[94] by the treatment of cyclohexadienols (**63**) with *p*-toluenesulphonic acid in ether at $-15°$, in a process which can be formally described as a [3,4]-sigmatropic rearrangement.

Electrocyclic Reactions

Pericyclic reactions have been reviewed,[95] and procedures to treat the competitions between most thermal pericyclic reactions have been developed and implemented in

HO H R Me → R Me

(63) **(64)**

a computer program.[96] Recent work by Dewar[97] has suggested that the current picture of pericyclic reactions is not correct, multibond processes of this type not being in general synchronous. The paper presents in detail arguments which lead to a new and apparently very powerful rule limiting the possible mechanisms of chemical reactions. A simple method of analysis of nodal properties of wave-functions has been introduced[98] to help characterize the nature of chemical reactions.

A theoretical study of the electronic influence of substituents on the thermally allowed electrocyclic interconversion of cyclobutene and butadiene has been undertaken,[99] and kinetic and thermodynamic studies of the electrocyclic ring-opening of the *cis*- and *trans*-perfluoro-3,4-dimethylcyclobutene systems have been presented.[100] This latter study shows that the *trans*-isomer ring-opens to the (Z, Z)-diene with an activation energy of *ca.* 18 kcal mol^{-1} less than it ring-opens to the (E, E)-diene. A paper has appeared[101] which provides a discussion of both conrotatory and disrotatory processes and, specifically, a tentative characterization of both stationary points, for the electrocyclic isomerization of cyclobutene to *cis*-butadiene. Recently, the facile electrocyclic ring-opening of gas-phase 1-methyl- and 3-methyl-cyclobutene radical cations has been achieved[102] by using collisional activation decomposition spectroscopy and Fourier transform mass spectroscopy. Thermal electrocyclic ring-opening of 1,2,3,6-tetrahydrobenzocyclobutene-3,6-dione has been found to take place readily at 140° to generate 5,6-dimethylene-2-cyclohexene-1,4-dione. However, when alkyl groups are present in the four-membered ring, intramolecular rearrangements take place. Thus, thermolysis of 1,4,4b,5,6,7,8,8a-octahydrobiphenylene-1,4-dione to give 1,4,7,8,9,10-hexahydro-benzocyclooctene-1,4-dione has been accounted for in terms of conrotatory ring-opening to an (E, Z)-cyclooctadiene followed by a [1, 5]-sigmatropic hydrogen shift.[103]

The electrocyclization rates of some donor–acceptor substituted trienes have been determined[104] and found to be virtually independent of the donor–acceptor properties of the substituents. It has been noted[105] that, although photolysis of 2-(carbomethoxy)spiro[5.5]undeca-1,3-dien-7-one (**65**) at 300 nm yields the oxa-di-π-methane rearrangement product, 11-carbomethoxy-*trans*-tricyclo-[5.4.$0^{7,11}$]undeca-9-en-2-one (**66**), the main primary process on photolysis at 254 nm is electrocyclic opening of the cyclohexadiene system to afford (**67**) and (**68**). It has also been demonstrated[106] that the photochemical formation of 6-phenylbenzobicyclo[3.1.0]hex-2-ene (**70**) from 2-phenyl-1,2-dihydronaphthalene

(69) occurs simultaneously *via* two distinct pathways, *viz. via* an electrocyclic ring-opening, single-bond rotation, and a final cycloaddition step, as formulated in Scheme 9, and through a di-π-methane rearrangement.

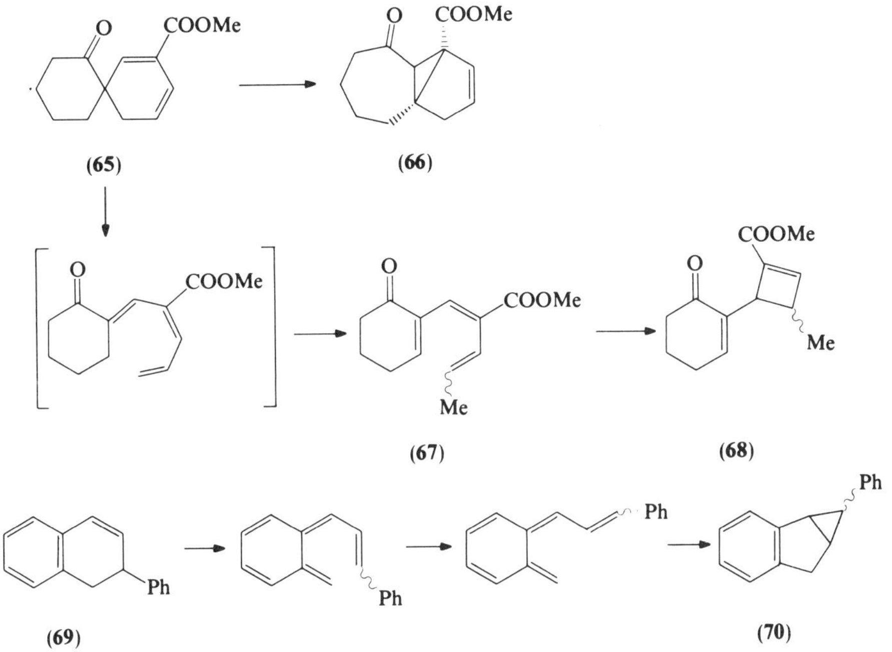

(65) (66) (67) (68) (69) (70)

SCHEME 9

A recent study[107] has shown that seven-membered ring annulation onto a polyene framework undergoing electrocyclization results in accelerated rates compared to that of similar six-membered ring systems, and an unprecedented, preparatively useful route to benzo-fused oxa[10]annulene **(73)** from cyclobuta[*d*]naphthalene **(71)** has been uncovered.[108] Of complementary mechanistic significance is the notable involvement of the carbonyl group of **(72)** in a "symmetry-disallowed" bond-reorganization scheme which evidently proceeds at the expense of a Cope rearrangement.

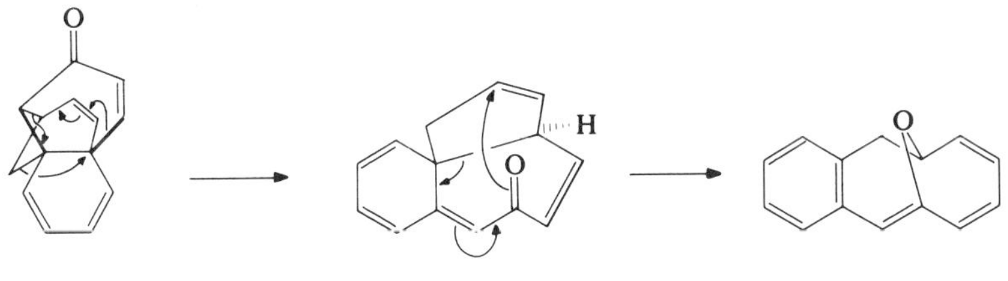

(71) (72) (73)

An internal conrotatory electrocyclization followed by a cascade of $10\pi/6\pi$-pericyclic processes has been postulated[109] to explain the observed pentacyclic hydrocarbons produced on thermal activation of heptahendecafulvadiene (**74**). The mechanism for the formation of 8b,9a-dihydro-9*H*-cyclopropa[*e*]pyrene from photolysis of 8,16-methano[2.2]metacyclophane-1,9-diene has been analysed in some detail.[110] This process represents a model case for adiabatic electrocyclic ring-closure in the excited singlet state.

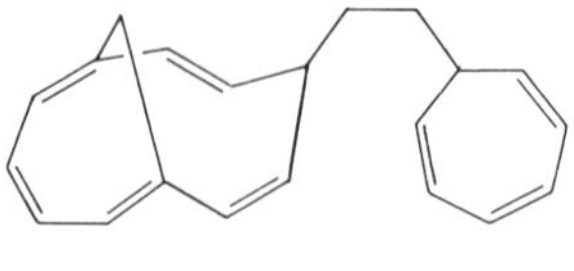

(**74**)

The first definitive example of the intramolecular cyclization of an acetylene to a benzene derivative in a [2 + 2 + 2] reaction has been observed[111] for a 15-membered trioxahexasilatriyne, while the trimethylsilyl group has been shown to have a profound effect on the regiochemistry of the ene reaction of (E)-1-methyl-2-trimethylsilyl olefins during the formation of *cis*-fused [3.3.0]bicyclooctanes.[112] During the transformation of (3-cyclohexenyl)diallylcarbinols of the type (**75**) into crotonylcyclohexenes (**76**), the bicyclo[3.3.1]nonenone (**78**) was also formed,[113] presumably by an ene reaction (**77**). From the kinetics of the gas-phase pyrolysis of a series of *cis*-1,1-dimethyl-2-alkenylcyclopropanes, the preference for the *endo*-transition state in the homodienyl hydrogen shift (retro-ene reaction) has been estimated to be at least 12 kcal mol^{-1}.[114] Presumably the large *endo*-preference is not an orbital-symmetry effect, since both *endo*- and *exo*-reactions can occur by formally "allowed" pathways, but rather it is associated with orbital overlap factors. Interestingly, the concerted nature of the *endo*-reaction pathway in this case implies

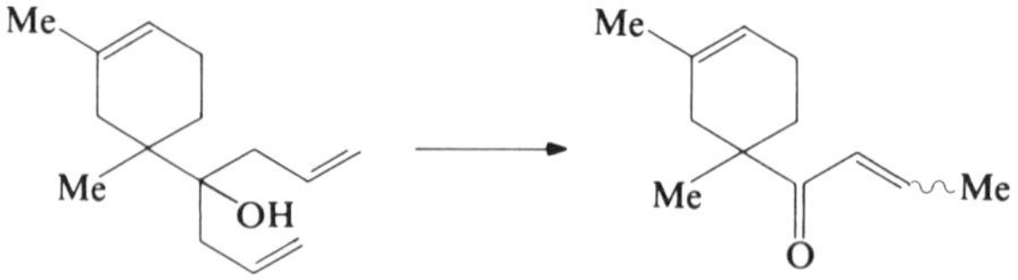

Me Me OH (**75**) → Me Me O Me (**76**)

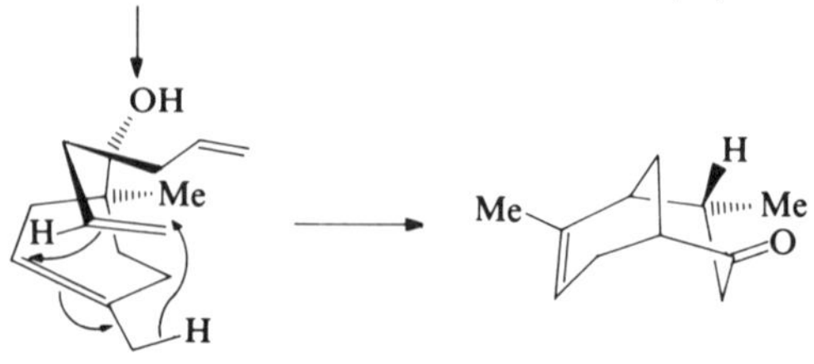

OH Me H H (**77**) → Me H Me O (**78**)

that a high degree of chirality transfer should prevail in the homodienyl hydrogen shift from an appropriately substituted alkyl group.

4,9-Dihydropyrazolo[1,5-*b*]isoquinolines have been produced in high yields, with relatively few side-products, from the corresponding allenyl azines,[115] and the electrocyclic reactions of azines, α,β-unsaturated keto-azines, and allenylic keto-azines have been reviewed by the same group.[116,117] The 8π-electron 1,7-electrocyclization of 1-aryl-3-diazoalkenes (**79**) to benzo-1,2-diazepines (**81**) (see Scheme 10) has been extended to 1-thienyl-3-diazoalkenes,[118] where it has been found that the thienyl ring is more reactive to electrocyclization than the analogous arene ring. A mechanistic study[119] of the reaction of (**79**) to give (**81**) has shown that the cyclization step is an equilibrium and that the second step is a wholly intramolecular migration. On the other hand, the electrocyclization of the related *o*-alkenylaryldiazoalkanes (**82**; X = H) to give 1*H*-2,3--benzodiazepines (**83**) was found to take place readily for substrates with a *cis*-hydrogen at the cyclization site, but was blocked by phenyl or methyl groups at that position.[120] Such compounds react *via* loss of nitrogen to produce products such as naphtho[*a*]cycloheptenes by a new intramolecular carbene reaction. The blocking effect of *cis*-methyl and -phenyl groups has been attributed to steric hindrance in a helical transition state (**84**) for the

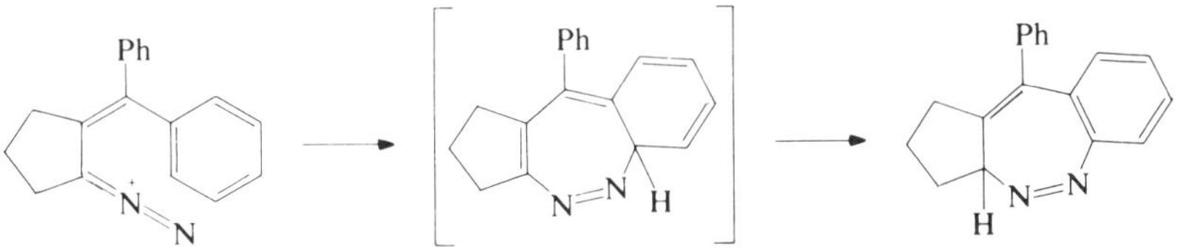

(**79**) (**80**) (**81**)

SCHEME 10

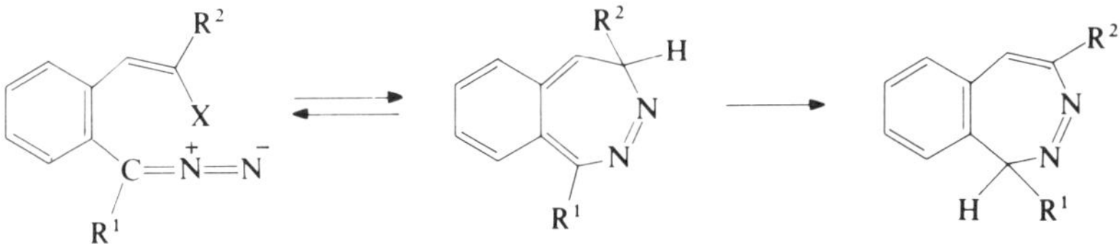

(**82**) (**83**)

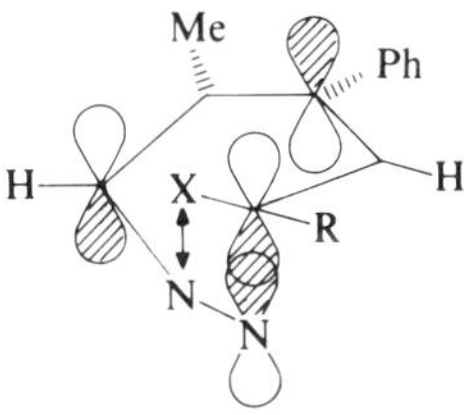

(**84**) 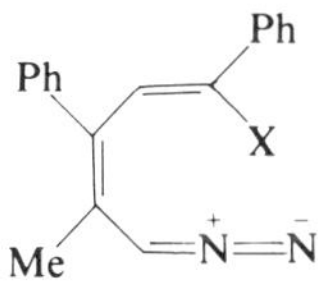 (**85**) 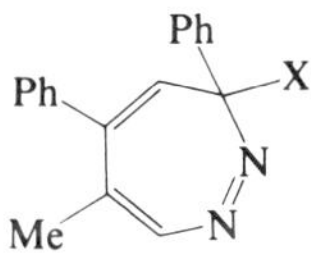 (**86**)

1,7-electrocyclization. A further extension of this work[121] to a study of the periselectivity of cyclization in a range of α,β: γ,δ-unsaturated diazo compounds containing only olefinic unsaturation has been described. Compounds (**85**; X = Me) having a *cis*-methyl group were found to react only *via* 1,5-ring-closure and subsequent vinyl-group and hydrogen migrations to give pyrazoles. Photolysis of these α,β:γ,δ-unsaturated diazo compounds (**85**; X = H) at low temperatures has established that the process proceeds by direct 1,7-electrocyclization to afford the intermediate diazepine (**86**). Pyrroles (**88**) were isolated when arylidene phenylglycine esters (**87**) were heated with 2 moles of dimethyl acetylenedicarboxylate. A mechanism involving an 8π-electrocyclization has been presented for this reaction (see Scheme 11).[122] 2-Pyrazolines (**89**) have been found to rearrange to pyrazoles (**90**) in an intramoleculr concerted double group transfer of hydrogen atoms in a non-catalysed thermal reaction,[123] while the cyclization of imines of (2-

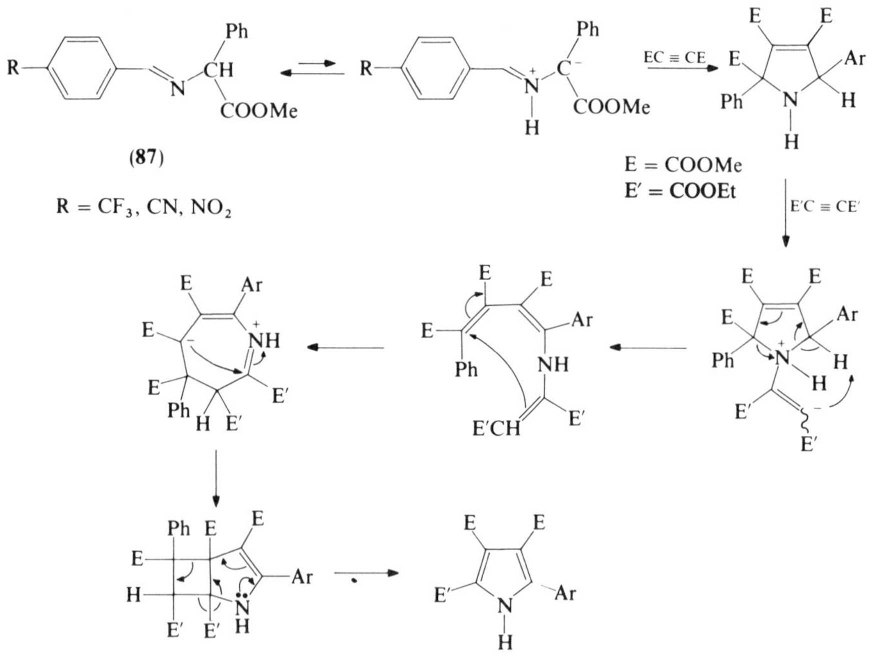

(**87**)

R = CF_3, CN, NO_2

E = COOMe
E′ = COOEt

(**88**)

SCHEME 11

(**89**) (**90**)

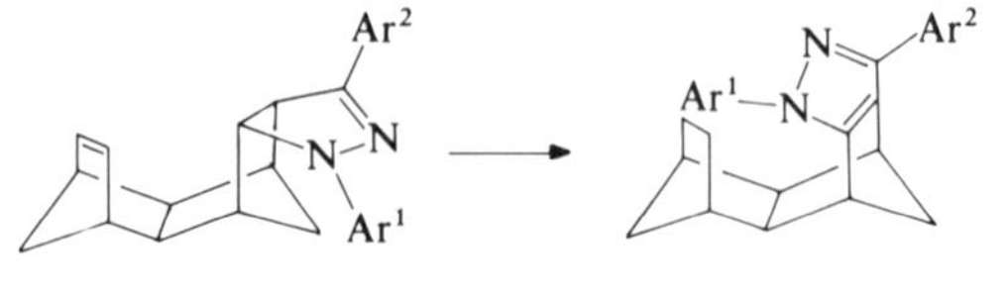

aminophenyl)acetic acid (**91**) to yield 2-substituted indole-3-carboxylic acids (**92**) has been viewed[124] as a 1,5-dipolar 6π-electrocyclization. β-Aminoacetylenes with a terminal acetylenic bond have been shown to undergo a thermal retro-ene reaction in which nitrogen acts as the hydrogen donor. On the other hand, alkyl substitution

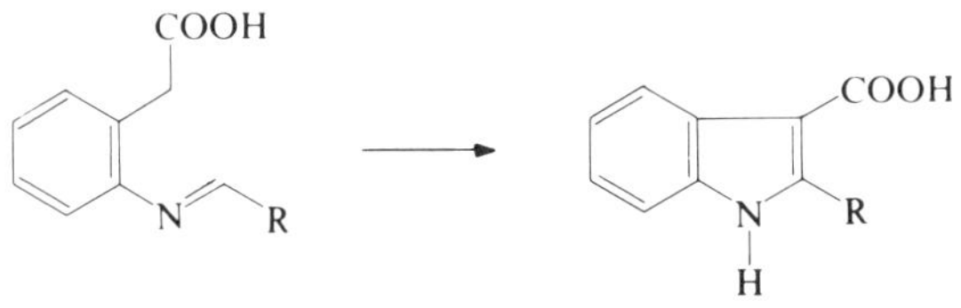

(**91**) (**92**)

on the terminus of the acetylenic bond as in (**93**) results in a novel facile cyclization to a 2-alkylidenepyrrolidine (**94**) in a reaction which can be represented as an 8-electron $[_{\sigma}2_s + _{\pi}2_a + _{\pi}2_s + _{\sigma}2_s]$ process.[125] A concerted ene process has been established[126] for the Et_2AlCl-catalysed cyclization of the 1,6-dienoate, ethyl (Z)-4,4-bis(ethoxycarbonyl)-4-{*N*-[(E)-3-methyl-2-butenyl]trifluoroacetamido}-2-butenoate to ethyl *trans*-[2,2-bis(ethoxycarbonyl)-4-(1-methylvinyl)-1-(trifluoroacetyl)pyrrolidin-3-yl]acetate.

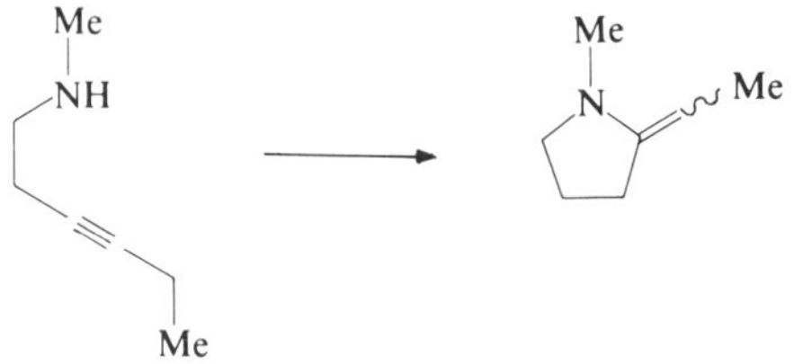

(**93**) (**94**)

Sigmatropic Rearrangements

[3,3]-*Migrations*

Reviews have appeared on chirality transfer *via* sigmatropic rearrangements[127] and on the catalysis of Claisen and Cope rearrangements,[128] and a detailed account of Hg(II)- and Pd(II)-catalysed [3,3]-sigmatropic rearrangements has been presented.[129] A report on recent developments in allene chemistry[130] has included the [3,3]-sigmatropic rearrangements of such molecules. The $[i,j]$-sigmatropic (i,j = odd numbers) reactions of conjugated polyenes have been studied by the MO method,[131] and a general MNDO method has been described[132] for the location of transition states in the Claisen and Cope reactions. An empirical approach to substituent effects in the Claisen rearrangement, utilizing the thermochemistry of coupled non-concerted alternative paths, has been presented.[133] It has been shown that an electron-donating substituent (*e.g.* oxygen) in the λ-allylic position (*i.e.* 6-

position) can accelerate the Claisen rearrangement relative to its unsubstituted counterpart, and this effect has been rationalized in terms of the "vinylogously anomeric" nature of the cleaving C—O bond. Curran and Suh[134] have reported this effect in the clean reaction of bis-alkyl ether (**95**) to the product (**96**) by two Claisen rearrangements. They show that the first of these is accelerated by the ring oxygen while the second goes at a normal rate, and they have applied this substituent-controlled Claisen rearrangement to the synthesis of pseudomonic acid intermediates.[135] Differential thermal analysis of the rearrangement of alkyl aryl ethers

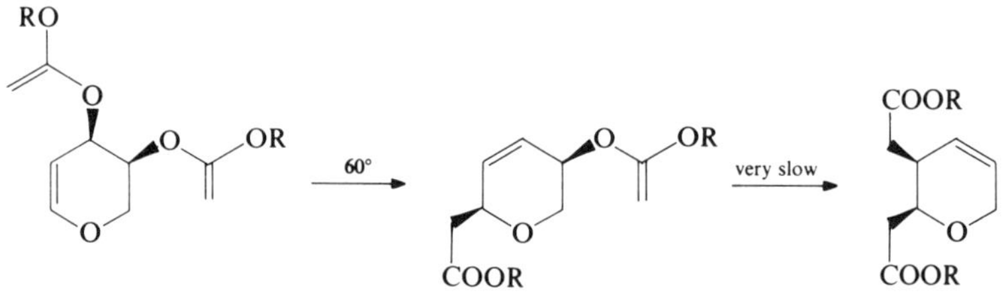

(**95**) (**96**)

has indicated[136] an intramolecular mechanism *via* a pseudo-cyclic complex for these species. Regioselective trifluoroacetic acid-catalysed rearrangement of allyl 5-allyloxy-2-hydroxybenzoate, with concomitant cyclization of the initial product, has permitted the preparation of a number of phenolic fungal metabolites.[137] A Mitsunobu alkylation–Claisen sequence and an unprecedented aromatic substitution of an anthraquinone have been employed[138] as key steps in the synthesis of a useful aklavinone precursor from 1,8-dihydroxyanthraquinone, and a Claisen rearrangement has been used in a new route to 3-methylenecoumarins *via* Lewis-acid-catalysed rearrangement of methyl α-aryloxymethylacrylates.[139] The recent preparation of a stereoselectively labelled [9-^{2}H,^{3}H]-chorismate (**97**) has led to the conclusion that rearrangement to prephenate (**98**) by chorismate mutase involves a chair transition state.[140,141] It has been reported[142] that the treatment of simple

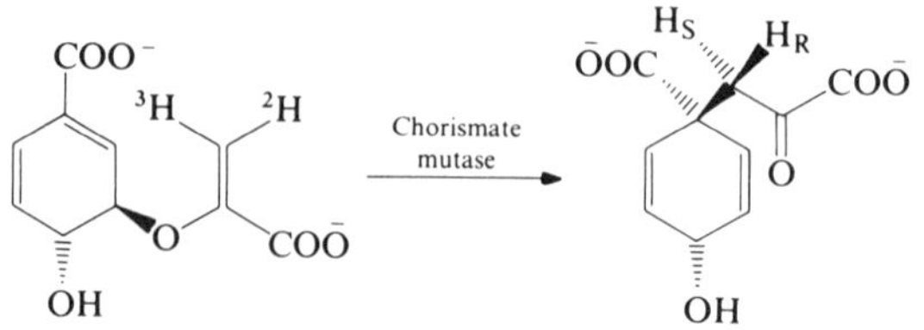

(**97**) (**98**)

allyl vinyl ether substrates with organoaluminium amphoteric reagents results in [3,3]-sigmatropic rearrangement under mild conditions, while Büchi and Vogel[143] have developed a modification of the Claisen rearrangement for primary and secondary allylic alcohols that does not require catalysis by mercury salts or mineral acids, and gives aldehydes directly (see Scheme 12). Quaternary centres are hard to

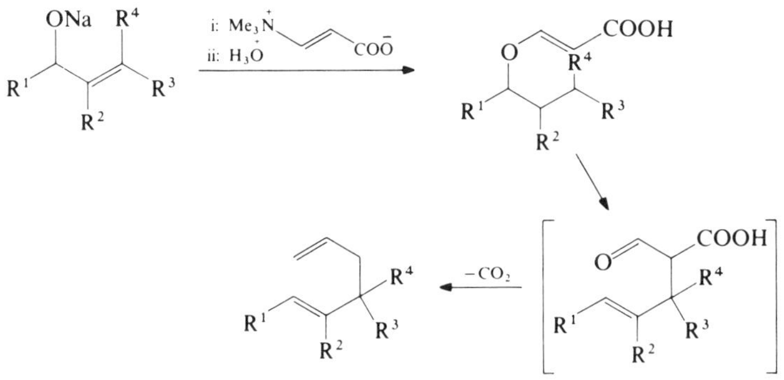

SCHEME 12

generate, and neighbouring quarternary centres even more so. Denmark and Harmata[144] have developed a version of the Claisen rearrangement which accomplishes this. Addition of an allylic alcohol to an allenic sulphone **(99)** gives **(100)** whose anion is found to undergo a [3,3]-sigmatropic shift to **(101)** at 50°. The extra substituents appear to accelerate the Claisen step and do not alter the expected regiochemistry. Application of the Claisen rearrangement to 1,5-anhydro-4,6-*O*-benzylidene-1,2-dideoxy-D-ribo-hex-1-enitol[145] supports previous studies in showing that α-D-*C*-glycopyranosides can be prepared in excellent yields by this method. The Claisen rearrangement has been evaluated as a means for the stereoselective creation of functionalized geminal substituents at C(2) and C(3) of hexapyrano systems,[146] while the Claisen rearrangement of an allyl vinyl ether residue elaborated at C(4) of a pyranoside ring has been found to occur with *β*-stereoselectivity leading predominantly to the isomer having the ethylene and acetaldehyde groups in equatorial and axial orientations, respectively. These studies have been directed at the synthesis of verrucarol from D-glucose.[147] Structurally equivalent derivatives of pyranosides and cyclohexanes have been found to undergo

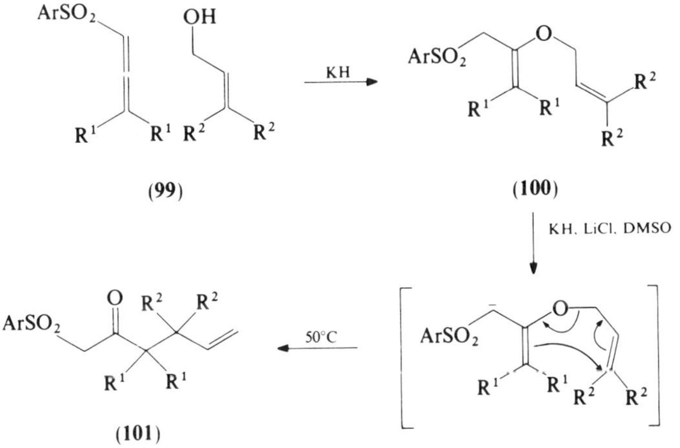

(99) **(100)** **(101)**

the spiro-Claisen rearrangement with different stereochemical results. A rationalization for the course observed with the pyranosides is suggested which invokes interaction of the oxygen lone pair with a developing electron-deficient centre at the spiro carbon.[148] A Claisen rearrangement has been used in the synthesis of 1-vinylmenth-4(8)-ene from geraniol,[149] and a Claisen rearrangement-based methodology for acyclic diastereoselection has been employed in the synthesis of tirandamycic acid.[150] Claisen rearrangements of 3,6-bis(aryloxy)cyclohexenes,[151] of 3-allyloxy- and 3-propargyloxy-thiophenes,[152] and of 3-allyloxy-4-bromo(and iodo)xanthones[153] have been studied. 1-Allyl-1,8-dihydropyrrolo[2,3-*b*]indoles have been photochemically rearranged to their 2*H*-isomers,[154] and the regioselective Claisen rearrangement of several ethyl allyloxyindole-2-carboxylates has been described.[155] It has been demonstrated that the Claisen rearrangement of *meta*-substituted arylpropargyl ethers in poly(ethyleneglycol) is not regiospecific.[156] A ^{1}H-NMR study[157] has provided conclusive evidence for the involvement of two sequential Claisen rearrangements in the thermal conversion of 1,4-diaryloxybut-2-ynes (**102**) into 11a-methylpterocarpans (**103**).

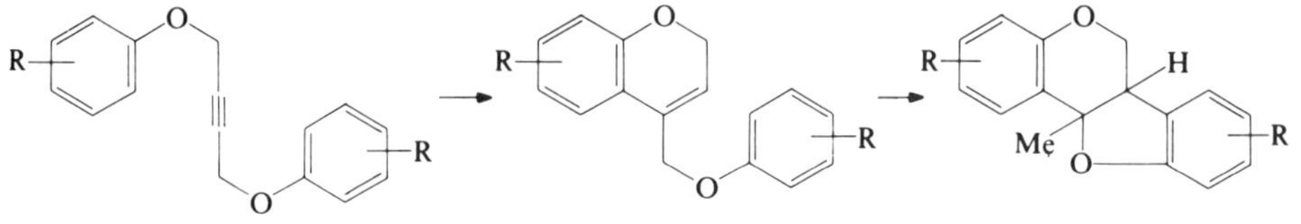

(**102**) (**103**)

A new method for the stereoselective synthesis of dihydropyrans of a variety of substitution patterns, which involves [3,3]-sigmatropic reorganization of 6-alkenyl-4-oxapyran-2-ones, has been described,[158] while a central feature of a recent($\pm$)-aphidicolin synthesis[159] is the Claisen rearrangement of the vinyldihydropyran derivative (**104**) to the spiroketone (**105**). A combination of a methylenation of the

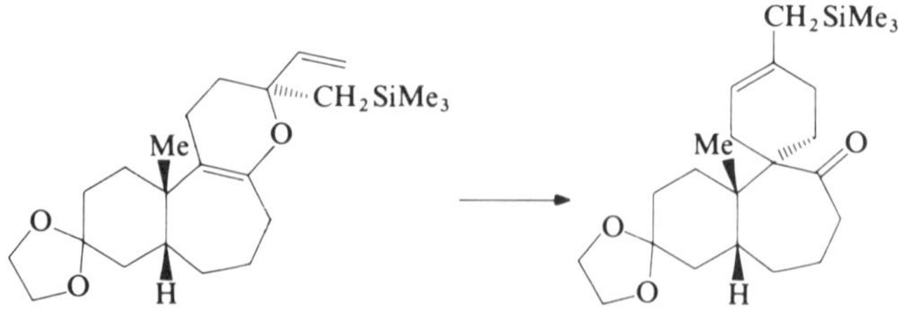

(**104**) (**105**)

lactone (**106**) using the Tebbe reagent, followed by a Claisen rearrangement of the resulting enol ether, lies at the heart of Paquette's synthesis[160] of precapnelladiene (**107**). The ester enolate Carroll rearrangement, (**108**) → (**109**), has been used[161] as a mild procedure for the preparation of γ,δ-unsaturated ketones and allyl-substituted

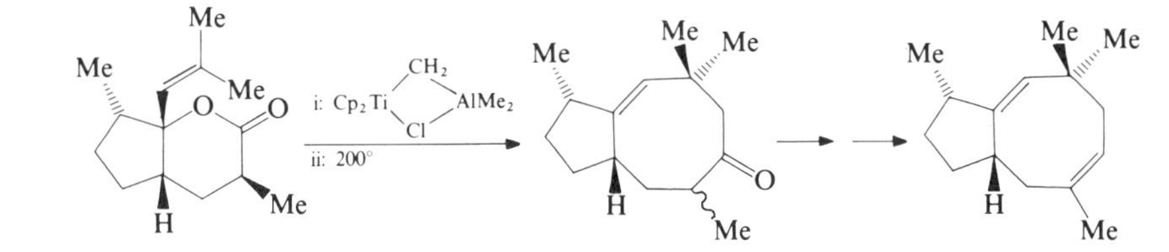

(106) (107)

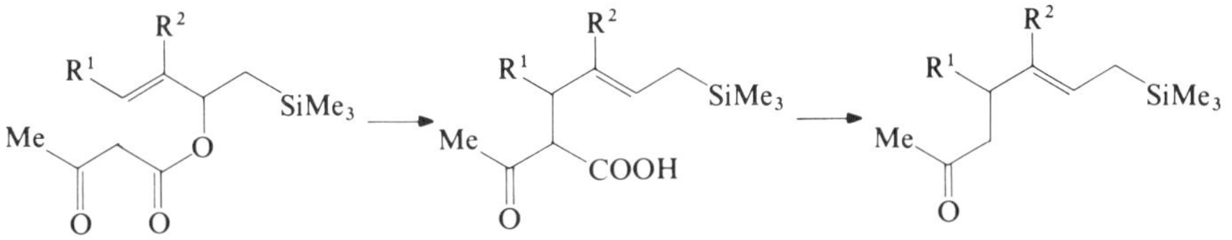

(108) (109)

acetoacetic acids, and the Claisen rearrangement of β-hydroxyallylic ester dianions has been utilized to provide stereocontrolled access to three contiguous chiral centres on an acyclic framework.[162] A diastereoselective synthesis of four isomers of 3-methyl-4-methoxycarbonyl-5-hydroxyhex-1-ene was achieved[163] from easily available (E)- or (Z)-2-butenyl 3-hydroxybutanoate by the ester enolate Claisen rearrangement by selecting appropriate reaction conditions, while the ester enolate Claisen rearrangement of the optically active allyl glycolate, *viz.* (R)-1-methyl-(E)-2-butenyl hydroxyacetate, was found to proceed with high diastereoselectivity and with high chirality transmission to give (2R, 3S)-2-hydroxy-3-methyl-(E)-4-hexenoic acid.[164] The Ireland–Claisen rearrangement of *O*-protected allylic glycolates has been shown to proceed in good yields with moderate to high diastereoselectivities.[165] Chelation control of enolate geometry is apparently responsible for the direction and magnitude of the stereoselection observed. A "chelation-controlled" Ireland–Claisen rearrangement of allylic glycolate esters has also been described[166] in which the stereocontrol of the prochiral sp^2 sites is achieved by the allylic oxygen substituent. The general utility of chiral α-silylallylic alcohols in the ester enolate Claisen rearrangement has been demonstrated in a study[167] which establishes the α-silyl unit as an extremely important "chirality inducing grouping", and the synthesis of an ionophore synthon having four asymmetric carbons has been achieved[168] by sequential aldol addition, Claisen rearrangement, and hydroboration. An enantiospecific synthesis of (1S, 2S)-(+)-2-methyl-3-cyclopenten-1-ol (**112**) has been described which involves the $TiCl_4$-promoted cyclization of the chiral allylic silane (**111**) which itself was obtained by Claisen rearrangement of (R, E)-1-trimethylsilyl-1-buten-3-ol (**110**).[169]

Kinetic isotope effects in the Ireland–Claisen rearrangement have indicated[170] that the transition state of the rearrangement closely resembles an oxaallyl–allyl radical pair and that the remarkable facility of this rearrangement is not due to "greater concert" but to the enhanced stability of the 2-(trimethylsiloxy)-1-oxaallyl

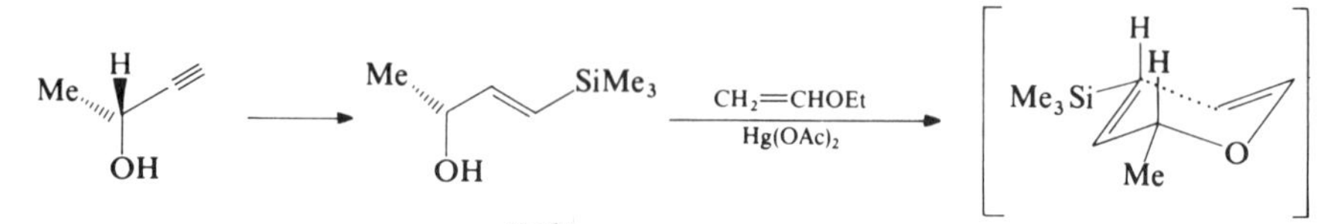

(110)

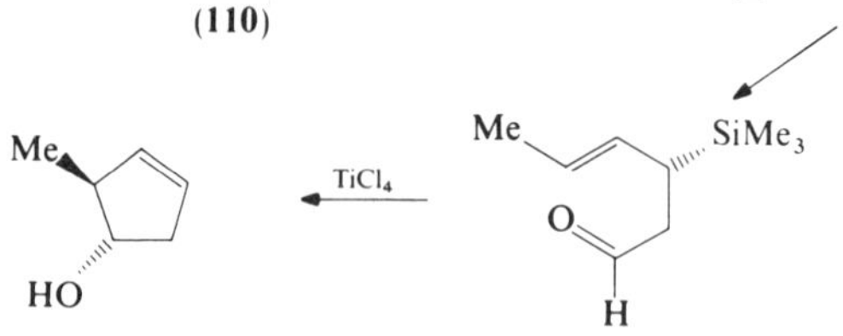

(112) (111)

moiety resulting in alteration in transition-state structure and energy to resemble this species. The Claisen rearrangement of (S)-1-buten-3-yl propanoate *via* the (E)- and (Z)-silyl ketene acetal has been shown to proceed with 96–100% of 1,4-chirality transfer to afford (E, R)- and (E, S)-2-methyl-4-hexenoic acid, respectively.[171] Similar methodology, using (trifluoromethyl)ketene silyl acetals has been successfully used[172] to prepare a broad variety of α-CF_3 acid derivatives with different additional functionalities. 12β-Hydroxypicrasan-3-one, a compound having the correct relative stereochemistry of all the six ring juncture chiral centres of the picrasane skeleton, has been synthesized from a known tricyclic compound, using the orthoester Claisen rearrangement and lead tetraacetate oxidation as key reactions,[173] while 4-penten-4-olides have been synthesized from allylic alcohols *via* an orthoester Claisen reaction.[174] A facile synthesis of γ,δ-unsaturated esters has been achieved by making use of a Pd(II)-catalysed orthoester Claisen rearrangement,[175] and the ketal Claisen rearrangement has been studied with a number of simple unsymmetrical ketals in order to establish the regioselectivity associated with the transformation. The results demonstrate[176] that the rearrangement can exhibit substantial selectivity and proceed in fair-to-good yield. Application of the enolate-Claisen rearrangement to unsaturated lactones, a method for the stereospecific synthesis of substituted carbocycles (see Scheme 13) has been found useful to provide hydroazulenes suitable for the synthesis of Dictyol diterpenes. Thus, perhydroazulenes (**114**) have been obtained[177] by the stereospecific alicyclic Claisen rearrangement of silyl enolates derived from the appropriate lactones (**113**), and (±)-muscone, has been prepared by the Ireland–Claisen rearrangement of *O*-silylated lactone enolates.[178]

The presence of an amino substituent at the 2-vinyl position in substituted allyl vinyl ethers has been shown to facilitate low-temperature rearrangements through [3,3]- or [1,3]-sigmatropic shifts, leading to aminoaldehydes; see (**115**) and (**116**), respectively.[179] A practical approach to the synthesis of (±)-acivicin (**119**; R = Cl) and (±)-bromoacivicin (**119**; R = Br), which centres on the synthesis of alkene (**118**) by a thermal [3,3]-sigmatropic rearrangement of trichloroimidate (**117**), has been developed,[180] and a new efficient synthesis of α-amino-acids from allyl alcohol

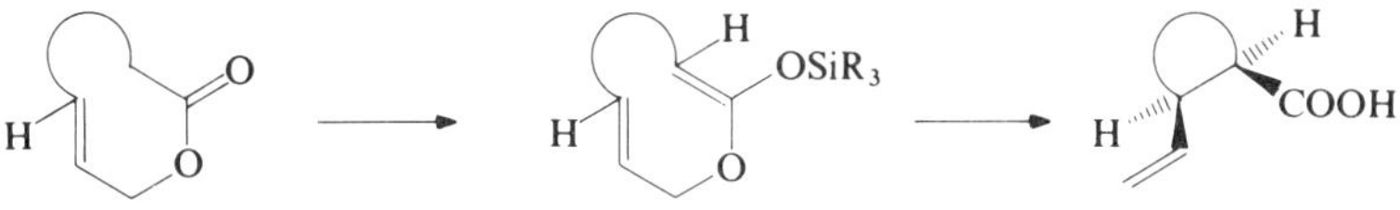

H O O → H H OSiR$_3$ O → H H COOH

SCHEME 13

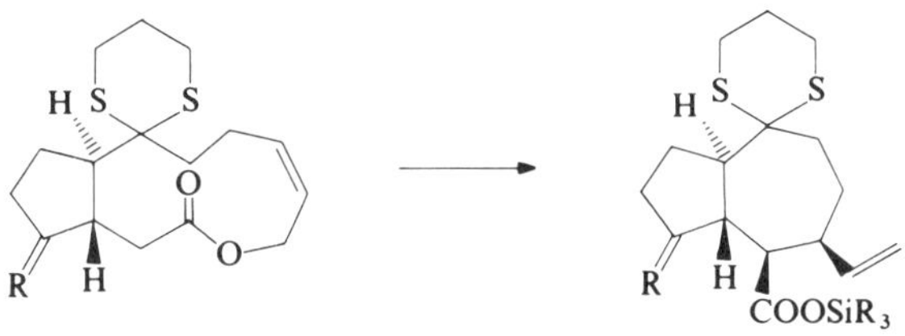

(113) → (114)

$R = NNMe_2, O(CH_2)_2O$

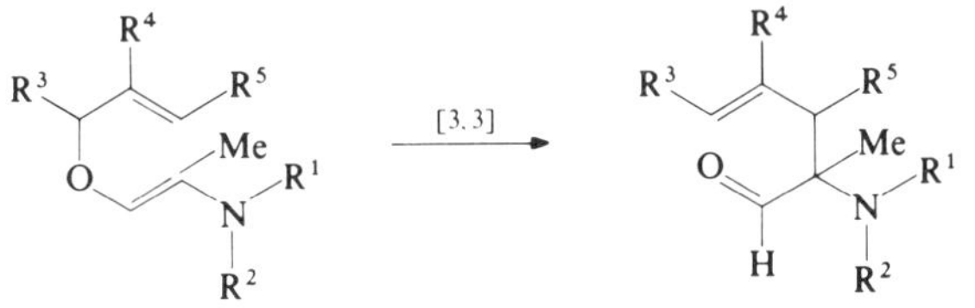

(115)

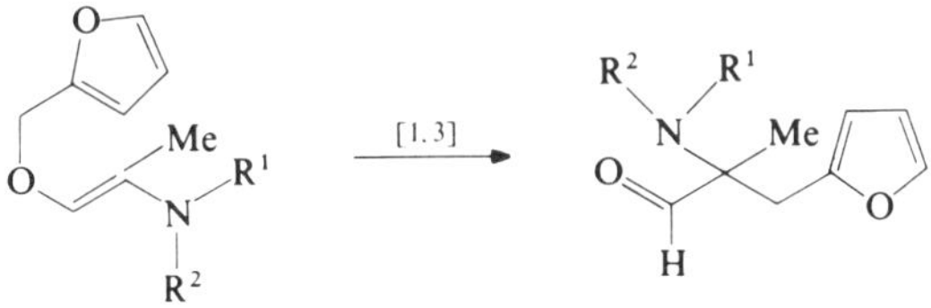

(116)

derivatives, *via* an acetimidate rearrangement, has been described (see Scheme 14).[181] The effect of alkyl substituents on the amino-Claisen rearrangement has been studied.[182] The introduction of a γ-methyl group on the allyl function was found to accelerate the rearrangement, whereas the introduction of two γ-methyl groups produced a complex reaction in which both sigmatropic and dissociative processes were observed. A recently reported[183] new route to 4-substituted indoles, involves the aza-Claisen rearrangement of *meta*-substituted *N*-allylanilines, and studies have been made of the Claisen rearrangements of a series of *N*-allylanilines[184] and of *N*-(1-methyl-2-butenyl)-2,5-xylidine.[185] The scope of the thermal [3,3]-sigmatropic rearrangement of a number of 2-allyloxy-substituted 4,5-diphenyloxazoles has been examined.[186] This system was found to undergo a facile aza-Claisen rearrangement

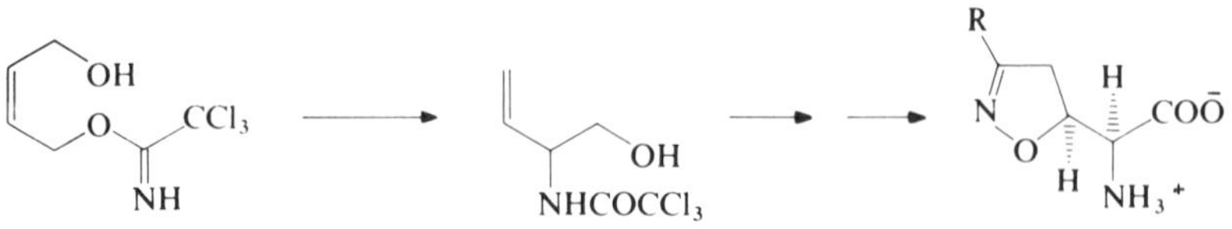

(117) (118) (119)

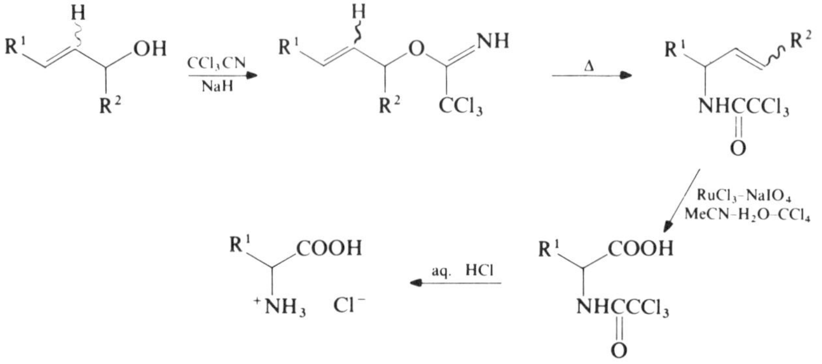

SCHEME 14

to give 3-allyl-substituted 4,5-diphenyl-4-oxazolin-2-ones. In contrast to the thermal results, irradiation of the 2-allyloxy- or benzyloxy-substituted oxazole gave rise to a mixture of 3- and 5-substituted oxazolinones. The photolysis is considered to proceed *via* C—O bond scission to generate a radical pair which subsequently recombines to produce the mixture of oxazolinones. Allylic enamines have been observed to undergo a modified aza-Claisen rearrangement to yield iminium salts upon alkylation and heating at 80°. The salts can be hydrolysed to aldehydes in an overall procedure which allows generation of quaternary carbon centres under mild reaction conditions.[187] A new version of the amino-Claisen rearrangement of the *N*-vinylisoquinuclidene system has been described[188] and the methodology has been successfully developed into inducing the amino-Claisen rearrangement of *N*-tryptophylisoquinuclidene into *N*-tryptophylisoquinoline with *tert*-butylpropiolate.[189]

The formation of uncondensed binuclear heterocyclic compounds in the thio-Claisen rearrangement of heteroaromatic sulphides has been reported,[190] the thio-Claisen rearrangement of indol-2-yl propargyl sulphide has been utilized[191] as a key step in a novel synthesis of 4-(aminomethyl)tetrahydrothiopyrano[2,3-*b*]indoles, while several 2-(allylthio)tropones and 2-(propargylthio)tropones have been found to yield 8(*H*)-cyclohept[*b*]thiophen-8-one derivatives *via* a thio-Claisen rearrangement.[192] The S $\rightarrow$ N allylic transposition of 3-(allylthio)-1,2,4-triazin-5(4*H*)-ones[193] has been performed successfully by catalysis of a Pd(II) salt, where it has been observed that the regioselectivities of the rearrangement [N(2) *vs.* N(4)] are highly dependent on the substitution pattern of the allylic moiety. Acid-catalysed isopropenyl acetate treatment of vinyl sulphoxides has been shown to result in a formal oxygen transposition from sulphur to the β-olefinic carbon in a presumed [3,3]-sigmatropic rearrangement of an acetoxysulphonium salt. A possible rationale for the reaction is shown in Scheme 15.[194] The reaction of allyl sulphides with

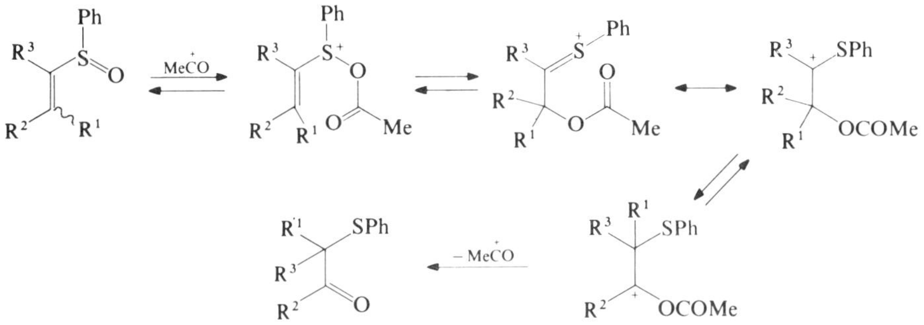

SCHEME 15

dichloroketene, followed by a new "ketene" Claisen rearrangement has been adapted[195] for the efficient ring-expansion of α-alkenyl cyclic sulphides **(120)** (see Scheme 16).

The tandem Claisen–ene reaction has been described[196] as a method for the stereocontrolled synthesis of functionalized cyclopentanoids, and Ziegler and Lim[197] have further developed a tandem Claisen–Cope rearrangement as a method

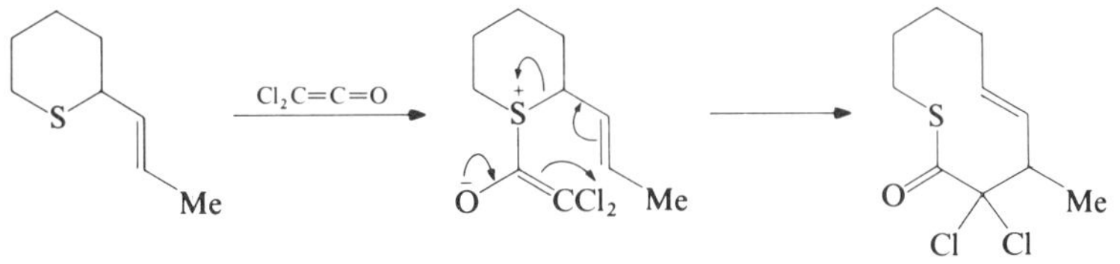

SCHEME 16

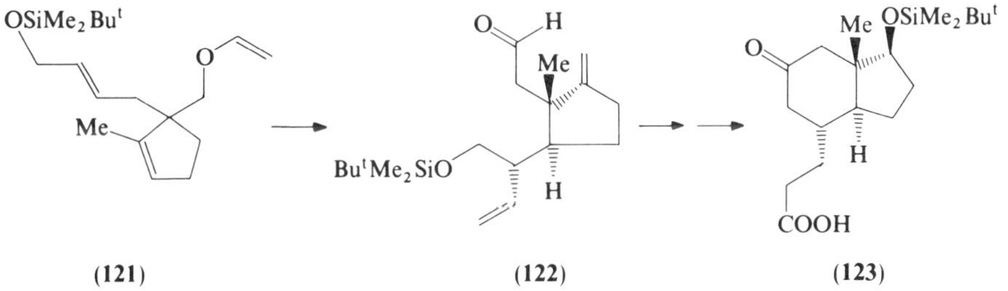

of stereocontrol in the synthesis of Stork's corticosteroid intermediate **(123)** from the vinyl ether **(121)**; the latter rearranges to establish the *trans*, *trans*-stereochemistry of the potential six-membered ring of **(122)**. New data on the stereochemistry of the tandem Claisen–aza-Cope rearrangements in the cyclization of *C*-phenylglycine derivatives have been presented.[198]

Recent *ab initio* studies[199] have indicated that the Cope rearrangement proceeds *via* a concerted pericyclic transition state of C_{2h} symmetry, in disagreement with previous MINDO results which found the lowest energy C_{2h} species to be a metastable diradical intermediate. Moreover, the present study indicates that in the Cope rearrangement, bond-making and bond-breaking occur in unison, thus also contradicting the earlier general assertion that multibond reactions cannot normally be synchronous. A perturbational treatment[200] has suggested that, irrespective of its donor or acceptor character, a substituent always enhances the Cope rearrangement rate. A variety of 4,8-substituted bicyclo[5.1.0]octa-2,5-dienes has been prepared and their ability to undergo Cope rearrangement has been found to depend on the pattern of substitution.[201] The influence of methyl substituents on the 1,7-octadien-3-yne to methylene-vinylcyclopentene rearrangement has been investigated.[202] Whereas methyl groups in the 6-position appear to induce the formation of 1,4-cycloheptadienes, methyl substituents in the 1-position lead to aromatic compounds. The distinct differences observed[203] in the gas-phase kinetic parameters for (**124**) and (**125**) have indicated a change of mechanism of the Cope rearrangement from concerted [for (**124**)] to step-wise [for (**125**)], apparently due to the cyclopropyl substituent effect. A detailed comparison of the theoretical and experimental data for the kinetics of the Cope rearrangement of 1,5-hexadiene has been made.[204]

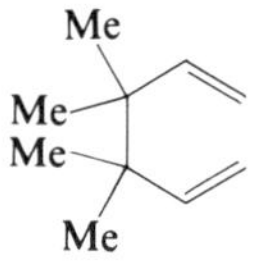

Me
Me
Me
Me

(**124**)

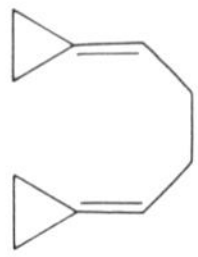

(**125**)

Reaction paths for the degenerate Cope rearrangement of semibullvalene and several of its derivatives have been calculated[205] by the MNDO MO method. NMR data for a variety of substituted semibullvalenes have been reported and the Cope rearrangement of these compounds has been analysed kinetically,[206] while the Cope rearrangement in bullvalene has been studied by dynamic deuterium NMR spectroscopy in liquid crystalline solvents.[207] ^{1}H- and ^{13}C-NMR spectroscopy have also been used to study the degenerate Cope rearrangement in 2,6-barbaralane dicarbonitrile,[208] and the Cope rearrangement of tricyclo[$4.2.2.2^{2,5}$]dodeca-1,5-diene to 2,5-dimethylenetricyclo[$4.2.2.0^{1,6}$]decane has been described.[209] When the 4,4-diarylphthalazin-1(4*H*)-one (**126**) was warmed to room temperature extrusion of nitrogen occurred to result in the formation of a quinonoid ketene (**127**). This species exhibited quite remarkable behaviour in that it underwent a novel intramolecular cycloaddition with an aromatic ring bearing a *p*-dialkylamino substituent, followed by a Cope-type rearrangement, to yield a benzazulenone derivative (**128**).[210]

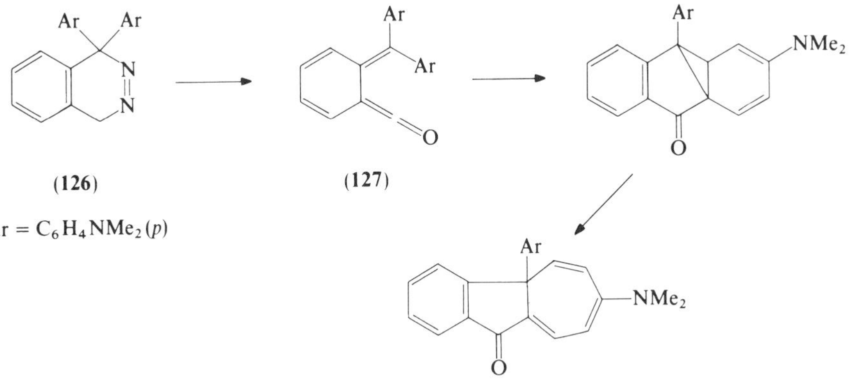

(126)

$Ar = C_6H_4NMe_2(p)$

(127)

(128)

Base-catalysed rearrangements of oxy-Cope systems have been reviewed,[211] and it has been observed[212] that the acceleration of the Cope rearrangement due to the presence of an anionic group on C(3) occurs in the gas phase as well as in solution. Alkyl substitution at C(3) results in an appreciable rate difference between the prototypical anionic Cope rearrangement systems (**129**; $R^1 = R^2 = R^3 = H$) and (129; $R^1 = Me, R^2 = R^3 = H$) in both the gas and condensed phases, ascribable to an intrinsic structural effect and not to relative ion-pairing or solvation effects. Hex-5-en-1-yn-3-ols (**130**) have been smoothly converted into α,δ-diethylenic ketones (**131**) by treatment with one molar equivalent of silver trifluorosulphonate,[213] and

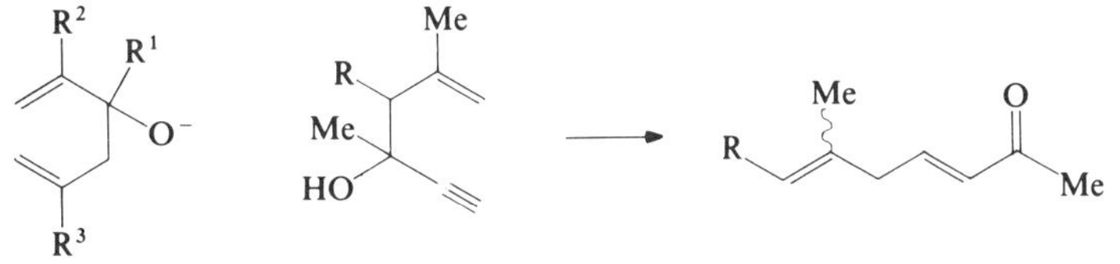

(129) (130) (131)

tertiary 1,5-hexadien-3-ols have been converted into δ-ethylenic ketones by treatment with one molar equivalent of mercuric trifluoroacetate followed by demercuration of the intermediate α-mercuro-ketone with sodium borohydride.[214] The rearrangement of optically active vinyl carbinol (**132**) to (**133**) has been found[215] to be concerted when carried out with KH–THF, but non-concerted in KOH–MeOH, and an anionic oxy-Cope rearrangement has been used as the key step in a total synthesis of (±)-poitediol and (±)-4-epipoitediol.[216]

The reaction of *N*-phenylhydroxylamine derivatives with electron-withdrawing allenes has been shown to yield *O*-alkylated anilines in a hetero-Cope rearrangement, (**134**) → (**135**);[217] these react further to form 1,2-disubstituted indoles (**136**). A similar rearrangement of the intermediate *N*-phenyl-*O*-vinylhydroxylamine derivat-

(132)

[3,3]

fragmentation

racemic (133)

(133)

ives (**137**), formed by treatment of *N*-phenylhydroxamic acids with vinyl acetate in the presence of Li_2PdCl_4, has yielded 2,3-unsubstituted *N*-acylindoles (**138**).[218]

(134) (135) (136)

X = SO_2Ph, $PO(OEt)_2$, *etc.*

(137) (138)

Thermal isomerization of 1-methyl-2-(3-butenylidene)pyrrolidine has been found to yield 1,2-dimethyl-3-allyl-2-pyrroline by [3,3]-sigmatropic rearrangement[219] although attempted photoisomerization of the reactant afforded only starting material. The tandem cationic aza-Cope–Mannich cyclization reaction (see Scheme 17) has allowed a variety of polysubstituted pyrrolidines to be assembled in excellent yields from the reaction of aldehydes (or ketones) and allylically oxygenated homoallylic amines.[220] An extension of this chemistry to the construction of more complex ring systems, *e.g.* for the efficient stereocontrolled assembly of the octahydroindole system, has been detailed.[221] A useful new method for the regiocontrolled construction of unsaturated azacyclic systems has been reported;[222] thus, cationic aza-Cope rearrangement of the iminium ion (**139**) to the allylsilane iminium ion isomer (**140**), followed by cyclization to (**141**), has the advantage of

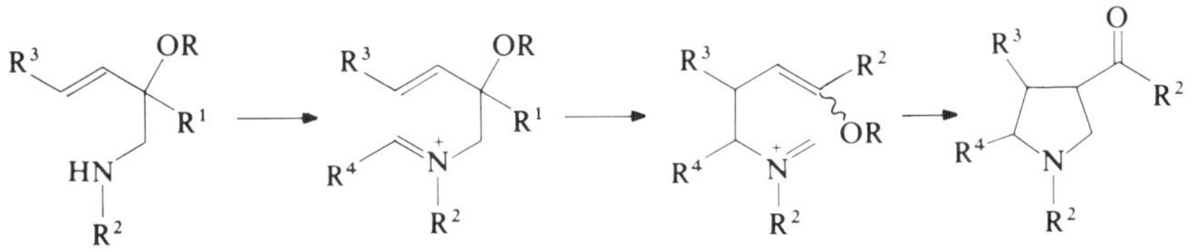

SCHEME 17

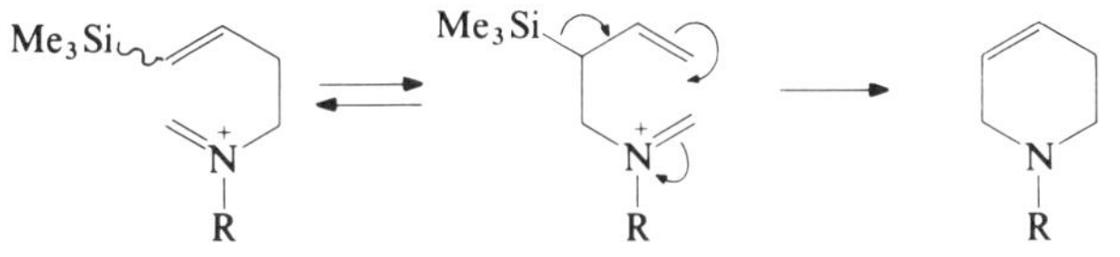

(139) (140) (141)

complete regiocontrol of the double-bond position. The synthesis of the γ-allenic amino-acid, 4-amino-5,6-heptadienoic acid (**143**; $R^1 = R^2 = H$), has been achieved[223] by a novel aza-Cope rearrangement of the acyliminium ion (**142**).

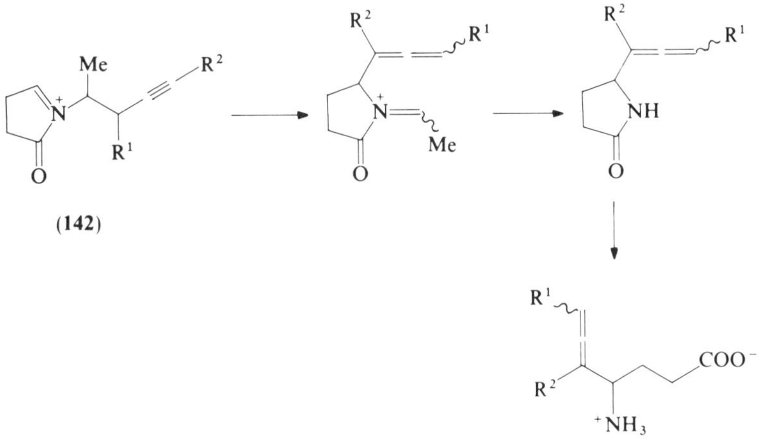

(142)

(143)

Finally, a concerted [3,3]-sigmatropic rearrangement [see (**145**)] has been postulated[224] to account for the thermal rearrangement of diazabicyclooctadienone (**144**) to the pyrazolopyran (**146**). Interestingly, this reaction involves conversion of a carbonyl group in to an enol ether, the reverse of the usual oxa-Cope reaction.

[2,3]-Migrations

The transition-state structure and activation barriers for [1,2]-, [2,3]-, and [3,5]-sigmatropic hydrogen rearrangements involving central carbon atoms have been determined by MINDO/3 calculations.[225] According to the authors,[226] the chirality transfer observed in the conversion of (s)-1-methyl-2-methylenecyclododecyl allyl

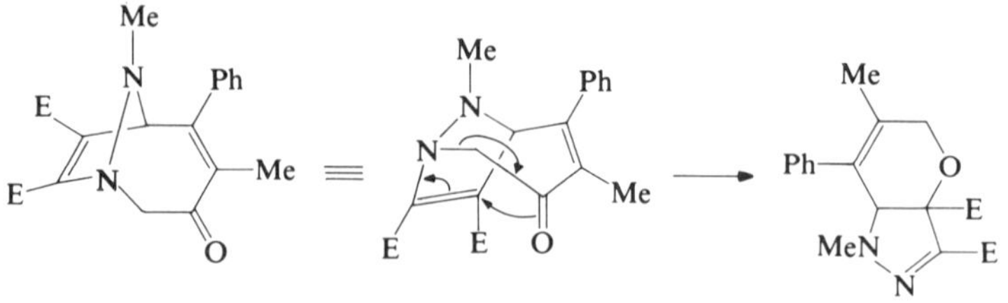

(144) (145) (146)

E = COOMe

ether (**147**) to alcohol (**148**) is in accord with a chair-like envelope transition state for the rearrangement. High diastereoselection and complete transfer of chirality has been observed[227] in the [2,3]-Wittig rearrangement of optically active allyl (Z)-2-methylhex-4-en-3-yl ether, and 1,4-chirality transfer has been achieved[228] *via* [2,3]-Wittig rearrangement of the chiral allylic propargyl ether system.

(147) (148)

The synthetic potential of the asymmetric [2,3]-Wittig rearrangement of chiral allylic benzyl ethers for the concurrent control of diastereo- and enantio-selection has been developed,[229] and conformational preferences in an intramolecular reaction have been exploited in an asymmetric variant of the [2,3]-Wittig rearrangement.[230] Thus, the readily available chiral acetylene (S)-(−)-(**149**) afforded ether (**150**) which rearranged to produce the elm bark beetle pheromone (**151**) after hydrogenation. The [2,3]-Wittig rearrangement of properly designated (E)- and (Z)-crotyl propargyl ether has been shown to exhibit a remarkably high degree of *threo*- and *erythro*-selection, respectively.[231] Some useful transformations of the rearrangement products are also described within the context of the formal synthesis of (±)-oudemansin. The high level of diastereoselection is maintained in the reaction of the α-methylcrotyl counterparts, together with the exclusive formation of the (E)-olefinic bond. A simple synthesis of (+)-verrucarinolactone, in which the key reaction is an asymmetric [2,3]-Wittig sigmatropic rearrangement involving a chiral

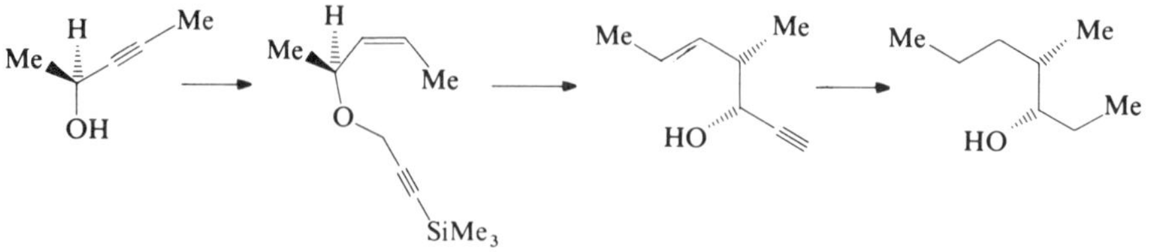

(149) (150) (151)

azaenolate as the migrating terminus, has been reported.[232] (+)-Polyzonimine (**154**), a terpenoid insect repellant produced by a millipede, has been synthesized by a reaction sequence which utilizes the asymmetric [2,3]-sigmatropic rearrangement of the ammonium ylide, formed from the salt (**152**), to generate the chiral intermediate (**153**).[233]

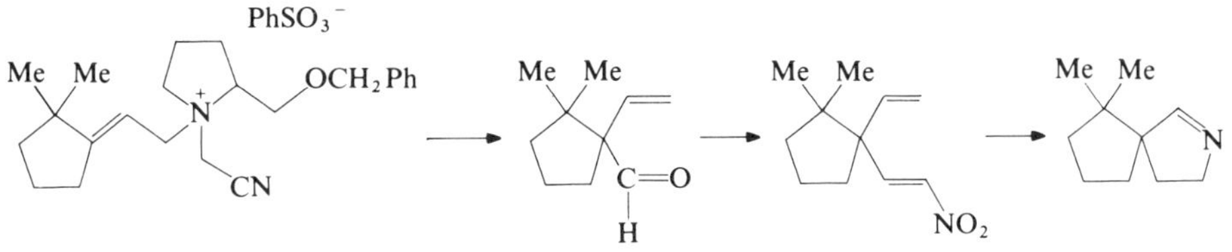

Me Me PhSO$_3^-$ OCH$_2$Ph CN (**152**) → Me Me C=O H (**153**) → Me Me NO$_2$ → Me Me N (**154**)

The results of rhodium (II)-acetate-catalysed reactions of diazo-esters with allyl acetals and thioketals have been reported.[234] These results demonstrate the effectiveness of heteroatom substitution at the allylic carbon in orienting product formation through the formal [2,3]-sigmatropic rearrangement pathway. Functionalization at C(5) of the triacetic acid lactone methyl ether (**157**) has been achieved by a transfer from the 6-position.[235] Thus, treatment of sulphide (**155**) with an excess of ethyl diazoacetate has provided pyrone (**156**) through [2,3]-sigmatropic

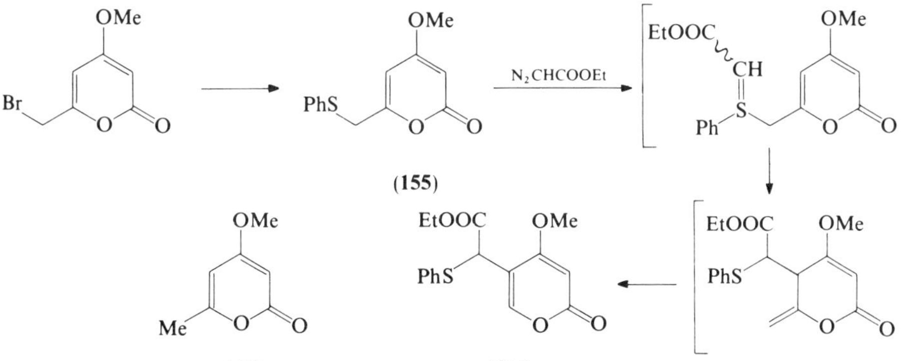

OMe Br O O → OMe PhS O O (**155**) $\xrightarrow{N_2CHCOOEt}$ [EtOOC CH S Ph OMe O O] ↓ [EtOOC PhS OMe O O] → EtOOC PhS OMe O O (**156**); OMe Me O O (**157**)

rearrangement, and the [2,3]-sigmatropic rearrangement of alkoxycarbonyl-stabilized ylides from *gem*-disulphides has provided the first entry to betweenanenes with a sulphur atom directly attached to the encapsulated olefinic unit.[236] In the presence of $SnCl_4$, α-acyl- and α-cyano-α-chlorosulphides (**158**) react with conjugated dienes in a $[2^+ + 4]$ fashion to afford cycloadducts of the type (**159**) which are converted with base *in situ* into the vinylcyclopropane derivatives (**161**) *via* the ylide intermediates (**160**).[237] An allyl sulphoxide–sulphenate rearrangement has been used[238] as a key reaction in the stereocontrolled construction of the 4β-hydroxy-2,5-dien-1-one system of withanolide D and related compounds. On

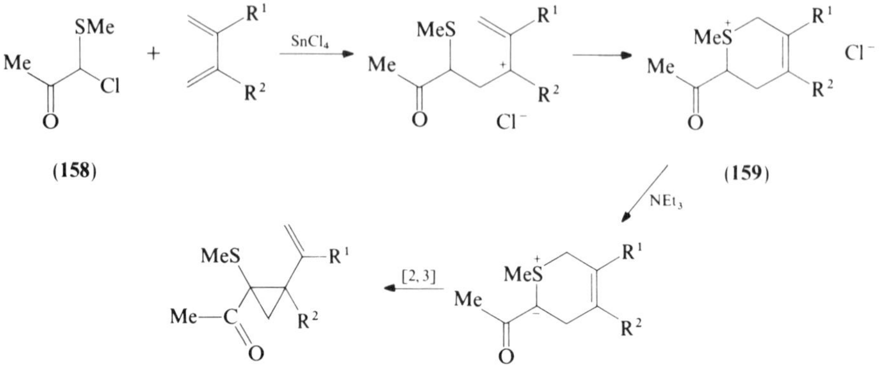

(158) (159) (160) (161)

pyrolysis, 2-sulphinylated 2-alkenoate esters (**162**) have been found to undergo migration of their carbon–carbon double bond and then [2,3]-sigmatropic rearrangement to benzenesulphenate esters (**163**) followed by thermal extrusion of benzenesulphenic acid, probably *via* ethyl (2E)-4-phenylsulphinyl-2-alkenoates (**164**) to afford ethyl (2E, 4E)-2,4-alkadienoates (**165**) (see Scheme 18).[239] Dramatic solvent effects have been observed in the thermolysis of allyl sulphinates. Moreover, heating chiral *trans*- and *cis*-allyl sulphinates, *i.e.* (s)-(−)-(**166**), in DMF at 90–120° provides chiral sulphones (**167**) in good yields with very high stereospecificity. The authors conclude[240] that these reactions occur by a [2,3]-sigmatropic rearrangement and that the transfer of asymmetry from sulphur to carbon occurs because the steric

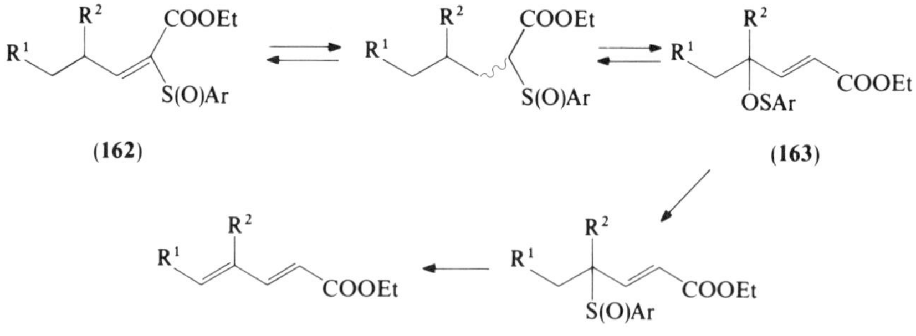

(162) (163) (164) (165)

SCHEME 18

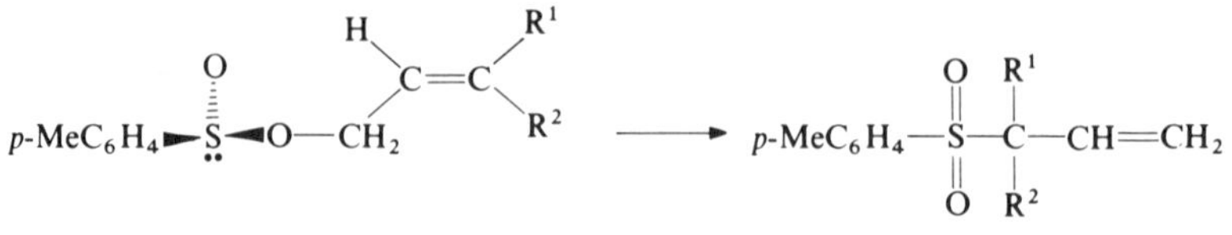

(166) (167)

interference between the tolyl group and the substituents R^1 and R^2 is much more severe in the cyclic intramolecular transition state **(168)** than in **(169)**.

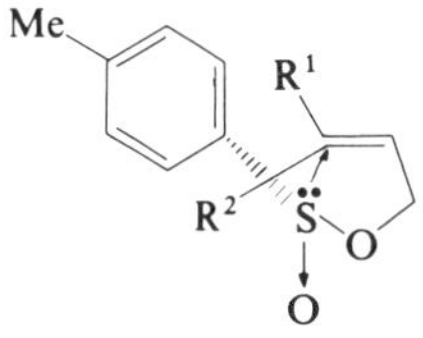

(168) 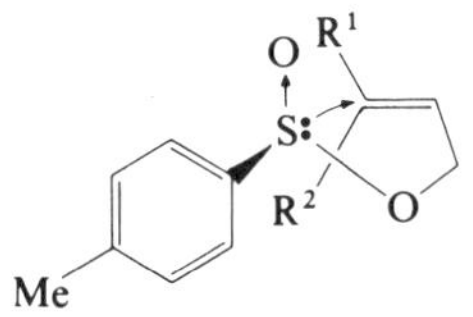 **(169)**

Gassman *et al.*[241] have demonstrated that the use of [2,3]-sigmatropic rearrangements of ylides generated from tetrahydroquinoline-derived azasulphonium salts provides a one-pot procedure for the synthesis of derivatives of the 5,6-dihydro-4*H*-pyrrolo[3.2.1-*ij*]quinoline system. The base-catalysed [2,3]-sigmatropic rearrangement of a series of *N*-aryl-substituted azasulphonium chlorides has been studied,[242] and the thermal ring-expansion of 2-vinylthiacycloalkane *N*-imides has been discussed in terms of a [2,3]-sigmatropic rearrangement.[243]

Anhydrous chloramine T in methanol has been established[244] as a highly effective reagent for converting allylic phenyl selenides **(170)** into the corresponding rearranged *N*-allylic-*p*-toluenesulphonamides **(172)**. This reaction presumably proceeds *via* an allylic selenimide intermediate **(171)** which undergoes a [2,3]-sigmatropic rearrangement. Improved, one-step procedures have been developed[245, 246] for the direct conversion of allylic phenyl selenides into *tert*-butoxycarbonyl-protected and carbobenzyloxy-protected allylic primary amines using more stable chlorosodiocarbamate reagents.

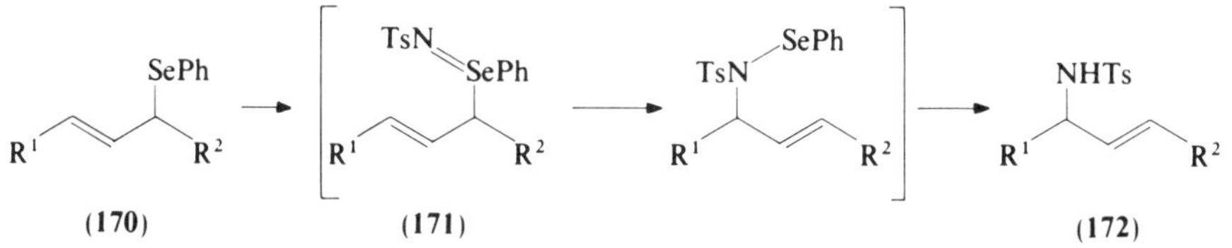

(170) **(171)** **(172)**

Wittig and Wittig–Horner reactions with α-phenylseleno-aldehydes **(173)** have converted them into allylic selenides **(174)** whose corresponding selenoxides are found to undergo [2,3]-sigmatropic rearrangement to provide a route to α-hydroxy-β,γ-ethylenic esters and ketones **(175)**,[247] while oxidation and subsequent [2,3]-

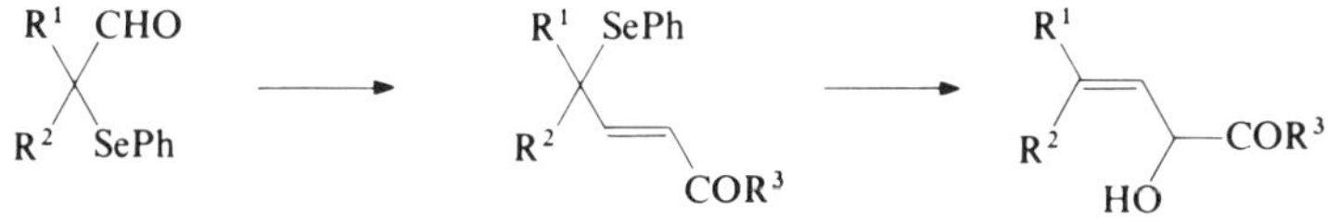

(173) **(174)** **(175)**

sigmatropic rearrangement of 1-substituted 3-phenylseleno-1,3-butadienes has led to α-functionalized α-hydroxyallenes.[248] An interesting methodology involving phenylselenolactonization coupled with an allylic selenoxide rearrangement has been used for functionalizing a range of dienylacetic acids.[249]

Finally, an interesting effect of a perfluorobutyl substituent has been observed during the rearrangement of prop-2-ynyl diethylphosphites to allenic phosphonates.[250] Thus, although (**176**; R^3 = H,Me) undergoes ready rearrangement to (**177**), (**176**; R^3 = n-C_4F_9) only undergoes [2,3]-sigmatropic rearrangement when R^1 and R^2 are fairly large. On the other hand, unexpected HF elimination [see (**178**)] occurs from (**176**; R^1 = H, R^2 = Me, R^3 = n-C_4H_9).

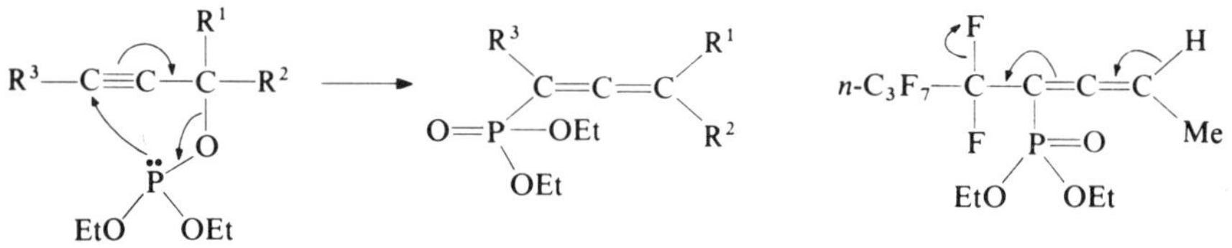

(**176**) (**177**) (**178**)

[1,3]-Migrations

Two groups, $\{E,\sigma_v\}$ and $\{E, C_2\}$, have been proposed to describe the orbital symmetry of various reaction paths with respect to suprafacial and antarafacial [1,3]-sigmatropic rearrangements.[251] The [1,3]- and [1,2]-sigmatropic shifts of hydrogen in propene have been investigated[252] at the MC-SCF level, and sigmatropic hydrogen rearrangements in the gas-phase decarboxylation and isomerization of but-3-enoic acid have been studied[253] by the MINDO/3RHF method. A quantum-chemical study[254] of the photochemical [1,3]-OH shift in 2-propen-1-ol has offered an explanation for the stereochemical outcome of the reaction, while the steric course of the deprotonation at C(4) in the enzyme-catalysed allylic rearrangement of the N-acetylcysteinamine thio-ester of (E)-dec-2-enoic acid to the corresponding thio-ester of (Z)-dec-3-enoic acid has been defined,[255] as has the overall steric course of the reaction.[256] The degenerate [1,3]-shift of the phenyl group in benzoylketene has been demonstrated by ^{13}C-labelling,[257] and examples of intermolecular borotropy have been observed in N-borylated derivatives of phosphoric amides.[258]

A striking reversal of the stereochemical course of potassium-alkoxide-accelerated [1,3]-sigmatropic rearrangements has been induced by the use of complexing agents (*e.g.* crown ether) for the potassium ion.[259] Norbornadiene acetals (**179**; R^1 = COOMe, CONMe$_2$, CHO) have been found to undergo facile rearrangement to the cycloheptatrienes (**182**). The transitory appearance of (**183**) during this transformation was established, although continued heating led to complete conversion into (**182**).[260] This presumably involves valence tautomerism (**183**) ⇌ (**181**), walk rearrangement of (**181**) to (**180**), and ring-opening to (**182**) (see Scheme 19). The remarkable ease of this walk rearrangement and the strong

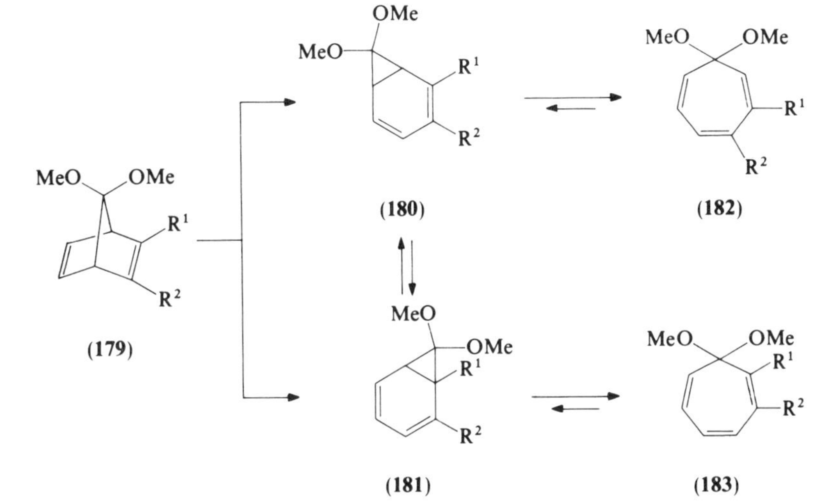

(179) (180) (181) (182) (183)

SCHEME 19

acceleration of the formal [1,3]-shift involved in this transformation is in agreement with predictions for a concerted forbidden process aided by donor and acceptor substituents. Irradiation of mortonin c (**184**) produced the photo-product (**186**) and its C(8) epimer. The authors[261] envisage the formation of the tetrahydrofuran derivative (**185**) as arising from a photochemically induced ring-contraction of the tetrahydrooxepine nucleus by way of a concerted [1,3]-sigmatropic rearrangement.

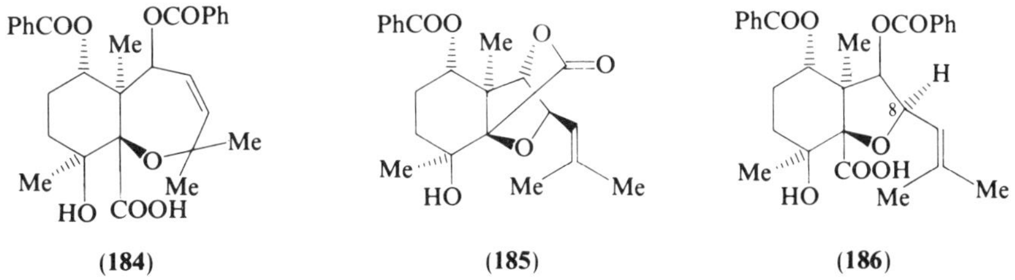

(184) (185) (186)

The primary photochemical step in the highly stereoselective photorearrangement of benzodihydrofuran-2-carboxylic acid ester (**187**) to phenol (**189**) is best described[262] as a concerted [1,3]-migration of C(2) to give the spiro(cyclopropane-2′,4′-cyclohexadien-1′-one) (**188**), which then undergoes a thermal hydrogen-atom

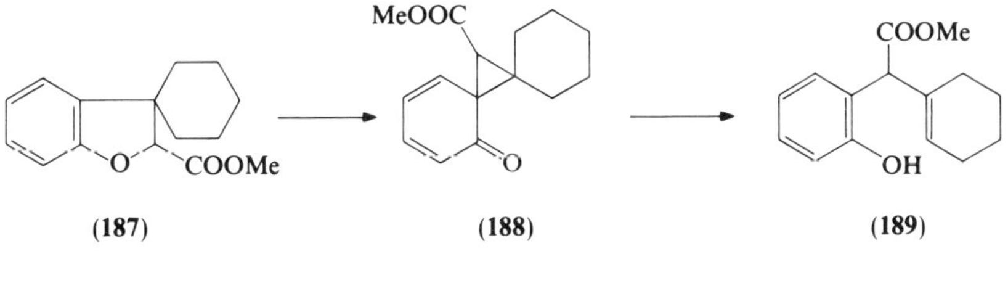

(187) (188) (189)

transfer to generate the phenol. The unique property of semiconductors has been applied[263] to the photochemical [1,3]-sigmatropic rearrangement of 2,2-bis(4-methoxyphenyl)-1-(dideuteromethylene)cyclopropane to 2,2-bis(4-methoxyphenyl)-3,3-dideutero-1-methylenecyclopropane.

A novel synthetic method for α-methylene-γ-butyrolactones involving a thioallylic rearrangement as a key step has been developed.[264] The formation of 2-(*tert*-butoxydimethylsilyl)-5-methyl-3-trimethylsilyl-1,3,5-hexatriene (**192**) on UV irradiation of 4-(*tert*-butoxydimethylsilyl)-1-methyl-5-trimethylsilyl-1,3-cyclohexadiene (**190**) has been explained[265] in terms of successive [1,3]-sigmatropic hydrogen shifts in (**190**) followed by rapid ring-opening and rearrangement of the resulting product (**191**) (see Scheme 20). The [1,3]-sigmatropic rearrangement of silaimine intermediates has been reported,[266] while a recent study[267] has demonstrated that, although (silylamino)phosphines containing P—H bonds undergo the same types of oxidation and methylation as their *P*-alkyl analogues, occasionally the P—H bonds as well as the Si—N bonds come into play, leading to rearranged N—H phosphine products. For example, oxidation of these phosphines by the silyl peroxide $Bu^tO_2SiMe_3$ is accompanied by a [1,3]-silyl shift from nitrogen to oxygen.

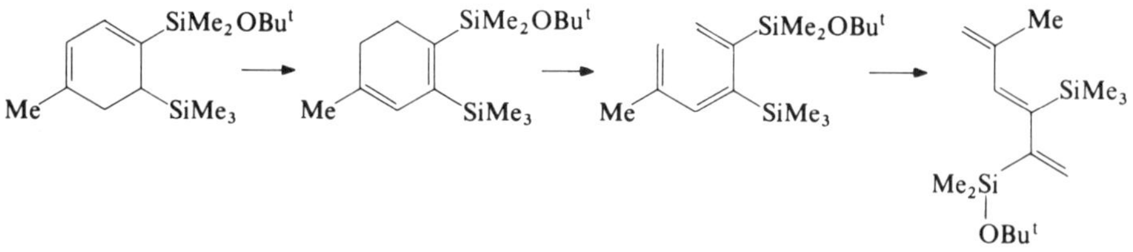

(**190**) (**191**) (**192**)

SCHEME 20

The intramolecular [1,3]-hydrogen shift of 1,6-dihydropyridines (**193**; R = CN, $CONH_2$) to the corresponding 1,4-dihydropyridines (**194**) in the solid state is considered to be non-concerted due to the constraints of the system.[268] 2,6-Diazabicyclo[3.2.0]hepta-2,6-dienes (**196**), presumably arising from a [1,3]-sigmatropic rearrangement, have been isolated[269] as intermediates in the phototransformation of 4*H*-1,2-diazepines (**195**) to 6*H*-1,4-diazepines (**197**).

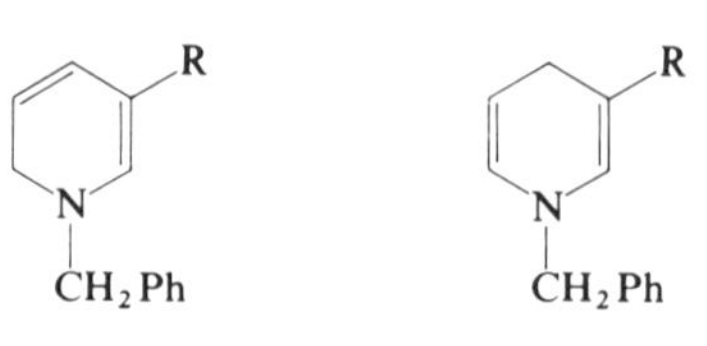

(**193**) (**194**)

(**195**) (**196**) (**197**)

[1,5]-Migrations

It has been reported that *ab initio* transition structures for [1,5]-sigmatropic hydrogen shifts of 1,3-pentadiene[270] and cyclopentadiene[271] show that H-transfer occurs in a significantly non-linear fashion; although the transfer angle and imaginary frequency for cyclopentadiene are smaller than those for 1,3-pentadiene isomerization. A MO model supported by energy-optimized MNDO calculations has been presented[272] for the substituent effects on the rates of the [1,5]-sigmatropic reaction over cyclopentadienylborane compounds. The propensity of simple silyl-substituted isodicyclopentadienes for [1,5]-silatropic migrations has been examined,[273] and a labelling study of the [1,5]-sigmatropic shift transforming the isodicyclopentadiene (**198**) to the more stable isomer (**199**) has confirmed the existence of a theoretically predicted interaction of σ- and π-orbitals.[274] Further evidence in support of a polar transition state in cyclopentadiene [1,5]-shifts has come from a study[275] of the thermolysis of phenyl- and cyano-substituted spiro[4.4]nona-1,3-dienes. The alkoxide accelerated [1,5]-sigmatropy of alkyl, vinyl, aryl, and cyclopropyl groups in cyclopentadiene systems has been observed[276] and evidence presented for the concerted nature of these processes, while the reversible [1,5]-sigmatropic migration of a nitro group in a cyclopentadiene ring has been observed.[277] A thorough study of the [1,5]-proton shift of 1H-phospholes to give 2H-phospholes has been undertaken,[278,279] and pentamethylcyclopentadienylphosphanes have been shown to possess a fluxional structure as a result of [1,5]-sigmatropic rearrangements of the phosphorus group.[280]

(**198**) (**199**)

A general mechanistic scheme has been proposed[281] to account for the products formed in the vapour-phase thermolyses of *gem*-dimethylated 2-cyclohexenones. Such a mechanism includes a [1,5]-sigmatropic methyl shift from the enol form. A study[282] of the [1,5]-migratory aptitudes for double- and triple-bonded groups in a series of benzocycloheptatrienes has established that the migratory aptitudes of the latter groups depend strongly on the migratory framework and, in particular, on variations of strain in bridged transition states. [1,5]-Hydrogen shifts from activated positions to a carbonyl group are considered to play a crucial rôle in the pyrolysis of saturated α,ω-dicarboxylic acids.[283] Thermal [1,5]-sigmatropic rearrangement of vitamin-D-type vinylallenes has been used as a means of preparing 3-deoxy-1α-hydroxy-14-epiprevitamin D_3.[284] Several indole alkaloids with an "*aspidosperma*" framework have been converted into "*vinca*" derivatives under flow thermolysis conditions by a rearrangement best interpreted in terms of two [1,5]-sigmatropic shifts and by the cleavage of a *gem*-hydroxy methoxycarbonyl function to a carbonyl group.[285] 2-Methoxyphenylmagnesium bromide on reaction with 2-nitropyridiene N-oxide, followed by demethylation, was found to afford 2-(2′-hydroxyphenyl)pyridine N-oxide which underwent thermal deoxidation to the free base. The process has been interpreted[286] as a [1,5]-sigmatropic shift from nitrogen to oxygen in the hydroxyl moiety in the second ring to give a hydroperoxide which undergoes decomposition. A recent study[287] has demonstrated that 1-(1-pyrrolidinyl)benzenes with an imino or a thiocarbonyl substituent at the 2-position are useful precursors for the synthesis of quinazoline and benzothiazine derivatives, respectively. For example, the formation of 5*H*-pyrrolo[1,2-*a*][3,1]benzothiazines (**201**) can be illustrated by the pathway depicted in Scheme 21, which involves a thermal superfacial [1,5]-hydrogen shift of the zwitterionic intermediate (**200**).

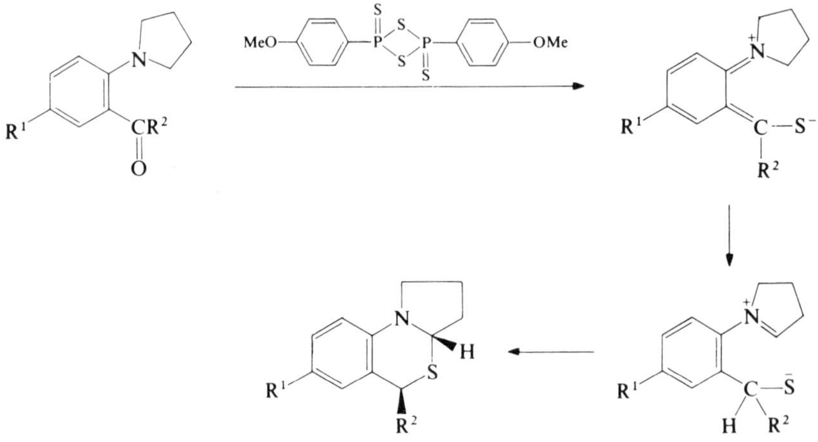

(**201**) (**200**)

SCHEME 21

Miscellaneous

"Walk" rearrangements in [n.1.0]-bicyclic compounds have been reviewed,[288] and a detailed research account of the effect of strain on the rearrangement of allylic hydroperoxides has appeared.[289] A model which, instead of formulating hydrogen transfer in terms of barrier penetration, describes it as a radiationless transition between potential-energy surfaces has been applied[290] to a number of hydrogen-transfer reactions such as the [1,4]-sigmatropic hydrogen transfer in hexahydrocarbazole. New examples of [1,3]- and [1,5]-migration of chlorine *via* radical intermediates have been observed during the reaction of 1,1,1-trichloro-2-hydroxy-4-methylpent-3-ene with inorganic acid chlorides.[291] The interesting overall rearrangement of pentaspirohexadecanes (**202**) to (**203**) has been described in terms of six-fold [1,2]-shifts with inversion at both the migration origins and termini.[292]

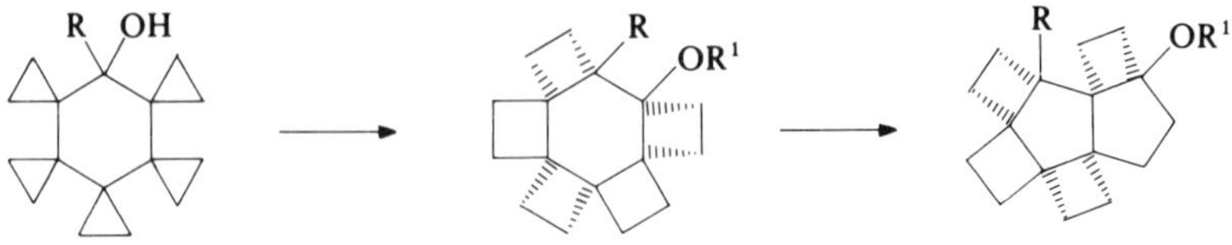

(**202**) (**203**)

The "walk" rearrangement of cyclopentadienylphosphines and related compounds has been analysed on the basis of a quantum-chemical study.[293] Although according to orbital symmetry considerations the reaction takes place with retention of configuration at the migrating atom, it appears that inversion of configuration is also feasible. The recent observation[294] that the *cis*-configuration is preserved during the "walk" rearrangement in the *cis*-bicyclo[6.1.0]nonatriene system has ruled out the participation of a diradical intermediate in this process.

Anionic Rearrangements

Ab initio methods have indicated that the anionic vinylidene–acetylene rearrangement proceeds with high activation energy possibly *via* a perpendicular transition state.[295] Evidence has been provided from a study[296] of the reactions of cyclopropylcarbinyl halides with (trimethylstannyl)alkalis, to show that the cyclopropylcarbinyl anion can undergo rearrangement to the 3-butenyl anion in competition with other processes. The bicyclo[3.3.0]octadienediyl dianion (**204**) has been shown to ring-open to the cyclooctatetraene dianion (**205**),[297] and a surprising carbanion-induced ring-contraction of dibenzo[*a,e*]cyclooctene to indeno[2,1-*a*]indene has been reported.[298]

Another illustration of a ketonic carbonyl group activating proton-abstraction from carbon more than one bond removed has been exemplified by the preparation of 3,3-dimethyltricyclo[6.3.0.0^{1,5}]undecan-4-one (**208**) from the tricyclodecene (**206**) *via* β-enolate rearrangement of the tricycloundecanone (**207**).[299] This sequence provides a new entry into the tricyclo[6.3.0.0^{1,5}]undecane ring system

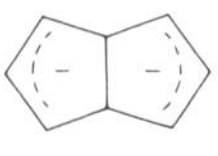

(**204**) 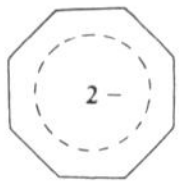 (**205**)

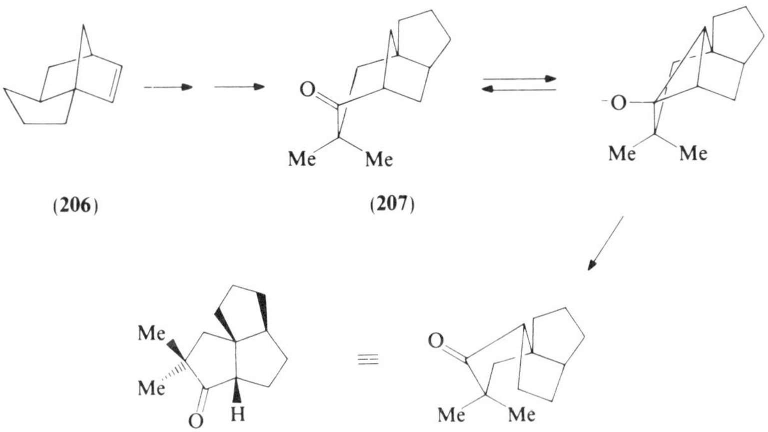

(**206**) (**207**)

(**208**)

which is characteristic of a number of recently encountered naturally occurring terpenes. The effect of strongly basic conditions on the *exo*- and *endo*-isomers of 7,7-dimethyltricyclo[3.2.1.0^{2,4}]octan-6-one has been examined.[300] While the *exo*-isomer is stable, the *endo*-isomer readily rearranges to 4,4-dimethyltricyclo-[3.3.0.0^{2,8}]octan-3-one which subsequently undergoes reduction of the three-membered ring. The initial transformation is one of the very few reported examples of an γ-enolate rearrangement. A new general ring-expansion of cyclic ketones by rearrangement of β-hydroxyselenides of type (**209**) has been employed[301] to introduce the dimethyl groups required in, for example, β-cuperenone (**210**; R = Ph) (see Scheme 22).

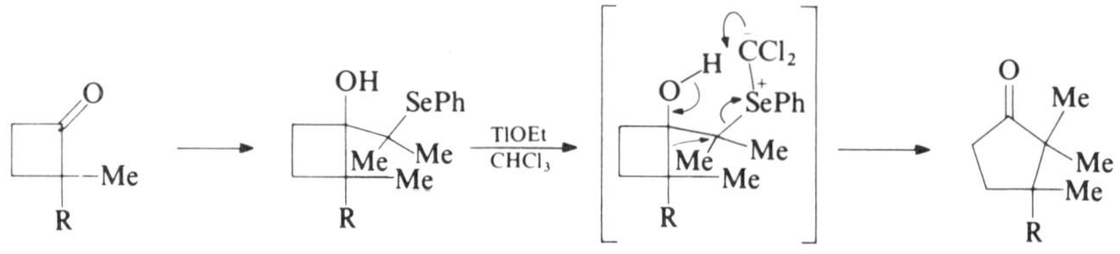

(**209**) (**210**)

SCHEME 22

The influence of deuterium substitution on the rate of the base-catalysed rearrangement of 2,3-dioxabicyclo[2.2.1]heptane has been examined.[302] New 1,3-dioxanes have been obtained from the rearrangement of glycidic esters with base,[303] and the mechanism shown in Scheme 23 has been postulated[304] to account for the base-catalysed conversion of α-acyloxyacetates into 2-hydroxy-3-keto-esters. The overall result of the process is equivalent to the acylation of the carbanion of an α-hydroxy-ester by a carboxylic acid, a transformation normally difficult to achieve. Tertiary α-chloroketimines on reaction with cyanide in methanol have been found to undergo a competitive reaction between α-cyanoaziridine formation and production of 1-(alkylamino)cyclopropanecarbonitriles.[305] A malonate anion induced Favorskii-type rearrangement has been observed in the reaction of methyl-substituted α-chlorocyclohexanones with sodiomalonates,[306] while iodosobenzene treatment of 17β-hydroxy-5α-androstan-3-one was found to give an unexpectedly good yield of the Favorskii acid.[307]

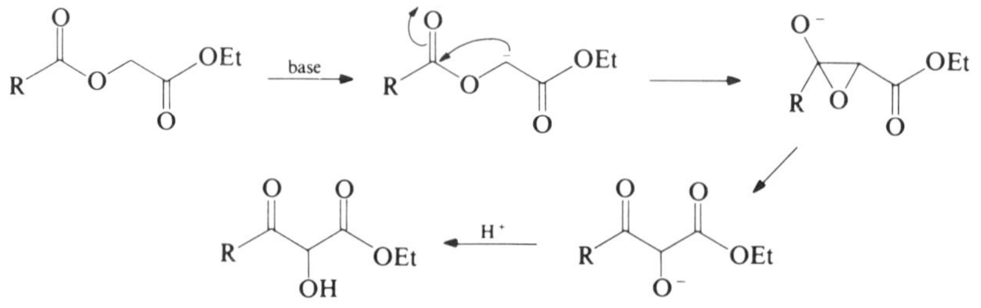

SCHEME 23

The Stevens rearrangements of dialkyldiallylammonium salts,[308] of alkynyl-(oxocycloalkyl)ammonium salts,[309] of various dialkylallyl(3-chlorophenyl-propargyl)ammonium halides,[310] of ammonium salts containing 2-oxocyclopentyl groups,[311] and of sulphur sulphylides[312] have been studied, and instances of participation of α-naphthylacyl groups in the above rearrangement have been reported.[313] The Stevens rearrangement has also been used to synthesize unsaturated α-dialkylamino-acids[314] and their esters,[315] and α-siloxyamines have been prepared by a Stevens-type base-promoted rearrangement of siloxyammonium salts.[316] 10-Methoxycarbonyl-10*H*-benzo[3,4]cyclopenta[1,2-*b*]thiopyran-9-one (**213**) has been obtained by rhodium(II)-catalysed decomposition and cyclization of methoxycarbonyl[2-(2′-thienyl)benzoyl]diazomethane (**211**) followed by spontaneous Stevens rearrangement of the initially formed sulphur ylide (**212**).[317] The formation of methyl ethyl ether from the treatment of trimethyloxonium hexachloroantimonate with the strongly hindered base, 2,2,6,6-tetramethylpiperidyllithium, has been attributed to the formation of an oxygen ylide followed by a

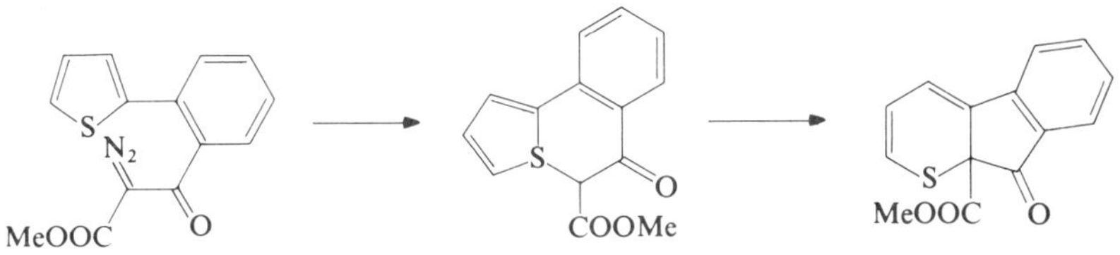

(**211**) (**212**) (**213**)

Stevens-type rearrangement.[318] The reported thermal rearrangement of 1-fluoren-9-ylidene-1,2,5-triphenyl-λ^5-phosphole (**214**) provides the first authentic example of a 1,2-Stevens rearrangement of a phosphorus ylide to a spiro-1,2-dihydrophosphorinine (**215**),[319] while the base-induced rearrangement of the *O*-methanesulphonyl derivatives of *N*-(alkylphenylphosphinoyl)hydroxylamines[320] has yielded products resulting from selective migration of the phenyl group. When

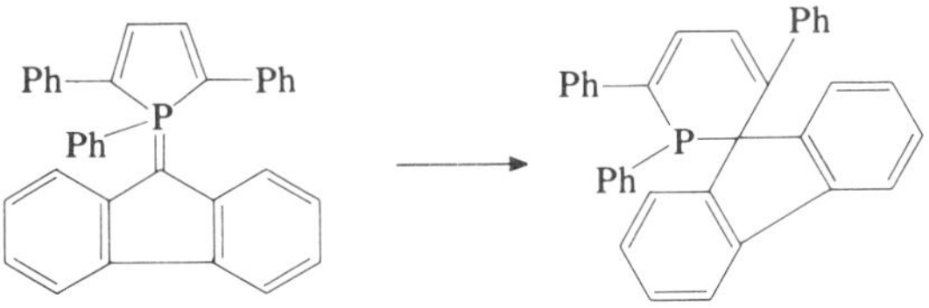

(**214**) (**215**)

treated with *n*-butyllithium, 2,3-disubstituted 1,2-thiazetidine 1,1-dioxides (**216**) were found to undergo dimerization and rearrangement to yield a new type of six-membered heterocycle (**217**); (see Scheme 24).[321] It has been shown[322] that in the presence of sulphides the thermal decomposition of aryl azides generally affords 2-substituted anilines such as (**219**) by Sommelet–Hauser rearrangement of the intermediate *N*-arylsulphimides (**218**), which themselves result from arylnitrene attack at the sulphur atom of the sulphide (see Scheme 25).

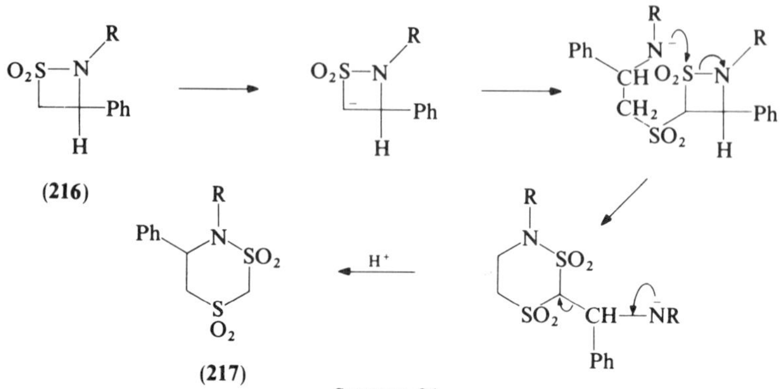

(**216**)

(**217**)

SCHEME 24

(**218**)

(**219**)

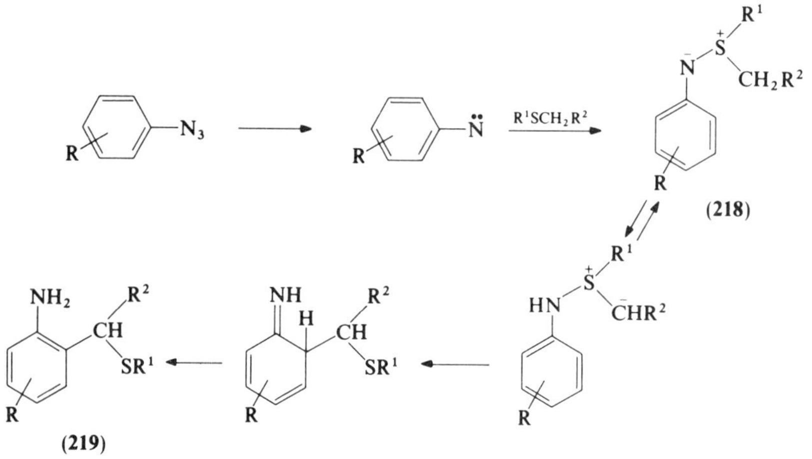

SCHEME 25

α-Oxoketenedithioacetals (**220**), derived from propiophenones, have been observed to undergo facile base-catalysed rearrangement to give the corresponding 3-alkylthio-2-alkylthiomethylacrylophenones (**221**),[323] but when these studies were extended to ketene dithioacetals derived from indan-1-one, 1,2,3,4-tetrahydronaphthalen-1-one, and 2,3-dihydro-1-benzothiopyran-4-one, respectively, the reaction did not follow the same course of rearrangement, except in the case of the thiopyranone. The mechanism governing the formation of the products from these substrates has been discussed.[324] Both [1,2]- and [1,4]-migrations of alkyl and aryl groups on sulphur have been observed during the thermolyses of 2-alkyl(or aryl)-1-benzoyl-1*H*-2-thianaphthaleniumides and 3,4-dihydro-1*H*-2-thianaphthaleniumide derivatives,[325] and a novel rearrangement of 1,4-ylidic thiaanthracenes has been described; thus, the 1,4-ylide from 9-mesityl-10-methyl-10-

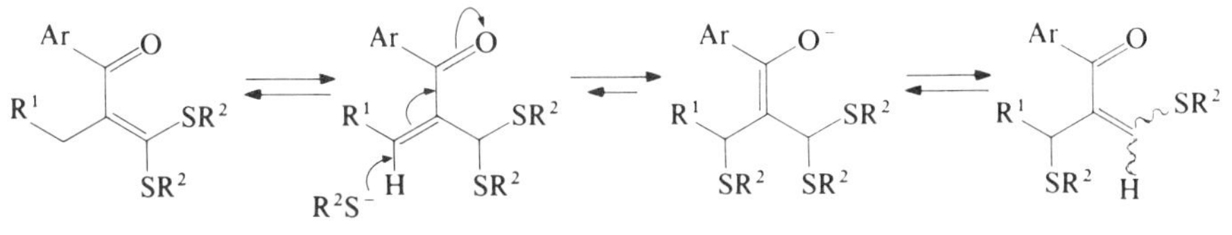

(220) (221)

thiaanthracene affords 3-methyl-9-mesitylthioxanthene. The authors propose[326] that such abnormal alkyl-group shifts are caused by the steric effect of the bulky substituents at the 9-position of the thioanthracenes which prevents the normal [1,4]-sigmatropic rearrangement of the 10-alkyl group to the 9-position. A [1,4]-trimethylsilyl migration has likewise been shown to occur in 9,9-bis (trimethylsilyl)-10-lithio-9,10-dihydroanthracene; the resulting carbanion, on protonation with water, gives rise exclusively to the *cis*-9,10-bis(trimethylsilyl)-9,10-dihydroanthracene. The same migration was also observed in a series of 9-(trimethylsilyl)-9-alkyl-10-lithio-9,10-dihydroanthracenes.[327]

A new intramolecular rearrangement that appears to involve oxy-anion to silanion isomerization has been described.[328] The base-induced rearrangements of 1-(trimethylsilyl)propargyl alcohols[329] (see Scheme 26) and of 1-(trimethylsilyl)allylic alcohols[330] have been studied. The latter study has opened up another indirect methodology for determining both regiochemistry and stereochemistry of silyl enol ethers. Treatment of alcohol (**222**) with KH–THF resulted in rapid migration of the trimethylsilyl group from carbon to oxygen and formation of (**225**), an outcome that has been explained[331] by a homo-Brook rearrangement, namely deprotonation of (**222**) to the highly basic alkoxide (**223**), which possibly exists as the pentavalent silicon anion (**224**), and subsequent protonation at carbon. Silyl enol ethers with sterically hindered silyl groups have been transformed into the corresponding α-silyl ketones by treatment with *n*-butyllithium–potassium *tert*-butylate.[332] This 1,3-oxygen-to-carbon silyl rearrangement seems to be triggered by allylic rather than vinylic metallation since enol ethers without an allylic α-proton

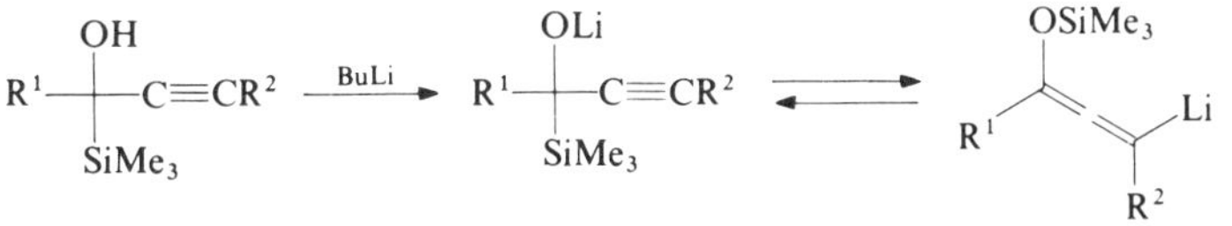

SCHEME 26

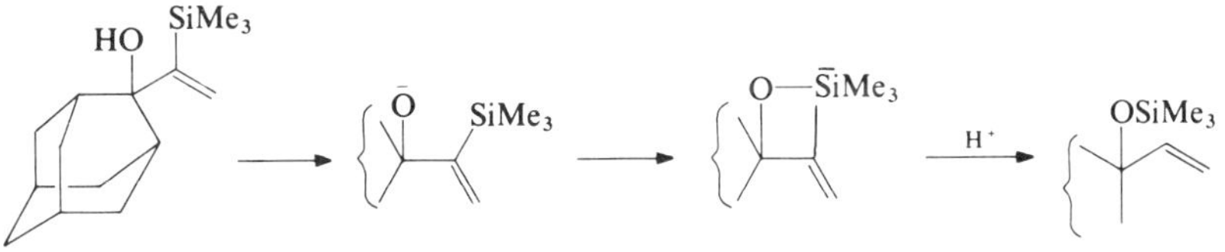

(222) (223) (224) (225)

are inert under the reaction conditions. A corresponding 1,4-oxygen to carbon silyl rearrangement has been unearthed[333] in the conversion of γ-silyloxyalkyl phenyl thio-ethers into γ-silyl alcohols on their treatment with lithium di-*tert*-butylbiphenyl, while anions generated from the trialkylsilyl ethers of methyl ketoximes (**226**) have been found to undergo reversible anionic rearrangement with 1,4-migration of the silyl group; see (**226**) ⇄ (**227**).[334] An acyllithium-to-lithium enolate conversion, achieved by utilizing the well-known 1,2-silicon shift, has provided a useful shortcut to acylsilane enolates.[335] Finally, an anionic shift of iron-coordinated dialkylaminosilyl ligands to the η^5-cyclopentadienyl ligand has been observed.[336]

$R^1C(Me)=N-OSiMe_2R^2$ (**226**) ⇄ [i: LDA; −78°C; ii: H_2O / heat] $R^1C(CH_2SiMe_2R^2)=N-OH$ (**227**)

Cationic Rearrangements

A detailed account of non-classical carbocations has appeared[337] and reviews have been presented on 1,2-shifts of carbocations,[338] on rearrangements of alkyl cations in solution,[339] and on carbon skeletal rearrangements *via* pyrimidal carbocations.[340] Rearrangements in chemical ionization mass spectra have been reviewed[341] and the rate constants for the isomerization of carbocations in collision complexes have been determined.[342] The activation barriers for 1,2-shifts in carbocations have been calculated[343] for different groups by MINDO/3 and they have been correlated to the eigenvalues and to the degree of localization of the LUMO. Cross-polarization, magic-angle-spinning ^{13}C-NMR spectroscopy of carbocations at very low temperatures and static methods at near liquid helium temperature have been used[344] to establish barriers to some rearrangements that have so far eluded solution-state studies. A theoretical study[345] of the Meyer–Schuster reaction has provided strong support for an intramolecular solvolytic mechanism for the rearrangement. The electronic structure of the reaction intermediate corresponds to an alkynyl cation interacting electrostatically with a water molecule, while for non-aqueous solvents the result agrees with the formation of a stable alkynyl carbocation. A quantum-chemical study of protonated intermediates in the Rupe and Meyer–Schuster rearrangement mechanisms has been undertaken,[346] while the reported appearance of the hexamethylbutadiene radical cation when di-*tert*-butylacetylene was treated with aluminium trichloride in dichloromethane has implied a novel molecular rearrangement of the acetylene.[347] One possible mechanism for this transformation is illustrated in Scheme 27. The effect of structural factors and solvent on the 1,2-bromine migration of 1(2)-bromo-2(1)-acetoxypropanes has been investigated.[348] A mechanistic investigation of the acid-catalysed rearrangements of 1- and 2-methyl-1-phenylallyl alcohols has been undertaken,[349] and a study has been carried out [350] to optimize the acid-catalysed rearrangement of diols (**228**) to diols (**229**). Stereoelectronic requirements for the

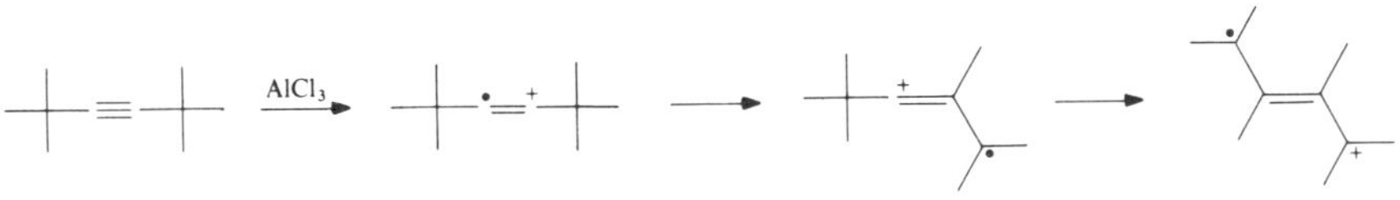

SCHEME 27

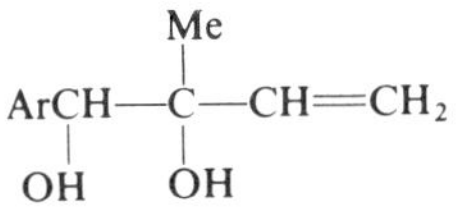

(228)

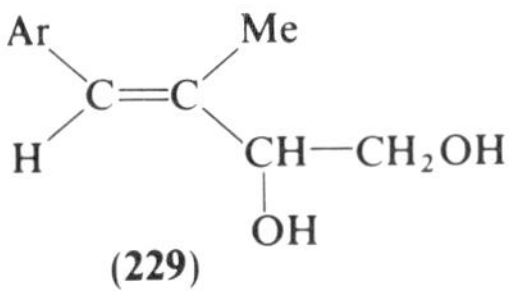

(229)

1,2-rearrangements of vinyl cations have been examined,[351,352] a route has been discovered to the novel $(C_7H_{11})^+$ cation,[353] and the $(C_{11}H_{11})^+$ armilenyl cation structure has been resolved by contemporary ^{13}C-NMR techniques.[354] Interestingly, this study has excluded the existence of the entirely organic sandwich form whose cyclopentadienide ring is spanned in C_2 symmetry by two mutually perpendicular 1,3-dehydroallyl ligands. 9-Vinyl- and 9-aryl-9-barbaralyl cations have been studied by ^{1}H- and ^{13}C-NMR spectroscopy and found to be in the charge range where the change from the Cope to the divinylcyclopropyl methyl cation mechanism takes place from degenerate rearrangements in the barbaralyl system.[355]

An attempt[356] to clarify the mechanism of formation and decay of the reactive intermediates in the decomposition of 3-chlorodiazirines has resulted in the proposal of two competitive pathways for olefin formation, one of which involves proton abstraction from a carbocation, itself formed by protonation of a carbene. A facile route to amino-substituted cyclopentadiene anions has been opened up[357] by the reaction of triaminocyclopropenyl cations with vinyllithium. The preparation of the previously unknown *cis*-1-methylcyclopropylcarbinyl cation (**230**) has been reported.[358] At -100°C it rearranges rapidly to the stable *trans*-isomer (**232**), but instead of a direct rotation of the C(α)—C(1) bond, (**230**) finds a lower energy route involving at least seven intermediates, including the observable 1-ethylallyl cation (**231**). This facile rearrangement of (**230**) to (**232**) only serves to emphasize the very facile nature of carbocation rearrangements. The silver-trifluoroacetate-assisted reaction of methyl 12-bromooleate with H_2O_2 has been found to afford cyclopropane hydroperoxides, presumably *via* a homoallylic cation rearrangement.[359] A novel photoreaction of styrylcyclopropanes *via* photoinduced electron transfer and subsequent 1,2-migration of the cyclopropyl group has been reported.[360] The major products, α-cyclopropylbenzyl ketones, are accounted for by the path shown in

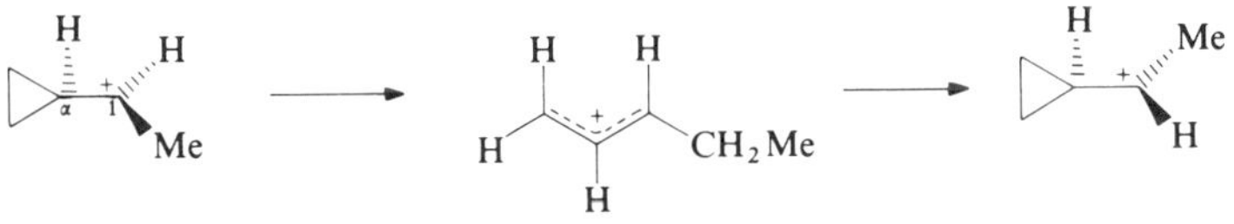

(230) (231) (232)

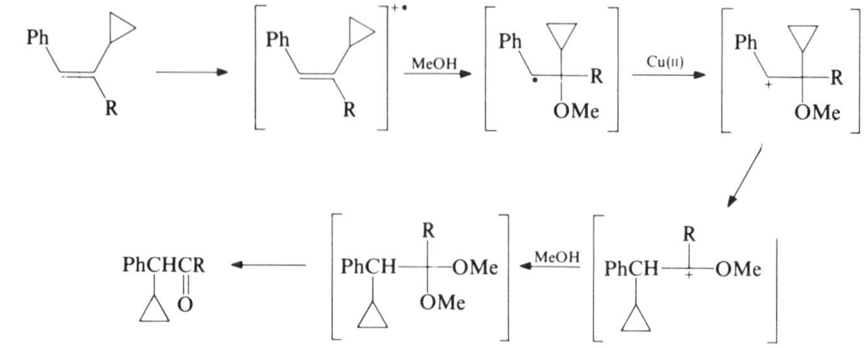

SCHEME 28

Scheme 28. 2-(2,3-Dialkyl-2-cyclopropenyl)-2-propanols have been found to undergo a highly selective ring-expansion in protic or strongly ionizing media, leading exclusively to cyclobutenes.[361] A homoaromatic cyclobutenyl cation has been identified as an intermediate in this reaction. Interestingly it has been observed[362] that 1-bromo-1-cyclopropyl-2-methyl-prop-1-ene (**233**) does not yield the expected cycloadduct on reaction with cyclohexene and $AgSbF_6$. Instead, the first-generated cyclopropylvinyl cation (**234**) rearranges to the allyl cation (**235**) which, after addition to cyclohexene, affords the diene (**236**) after a number of unusual rearrangements (see Scheme 29). The formation of spiro-diene (**238**) on pyrolysis of 7-bromo-7-(trimethyltin)norcar-2-ene (**237**) has been accounted for by proposing[363] a Skattebol-style rearrangement *via* a cationic mechanism as illustrated in Scheme 30. A photochemical rearrangement of the 1,2-benzotropilidene cation to the benzonorcaradiene cation and thence to the 1-hydro-2-methylenenaphthalene cation has been carried out in solid argon,[364] while acid treatment was observed to

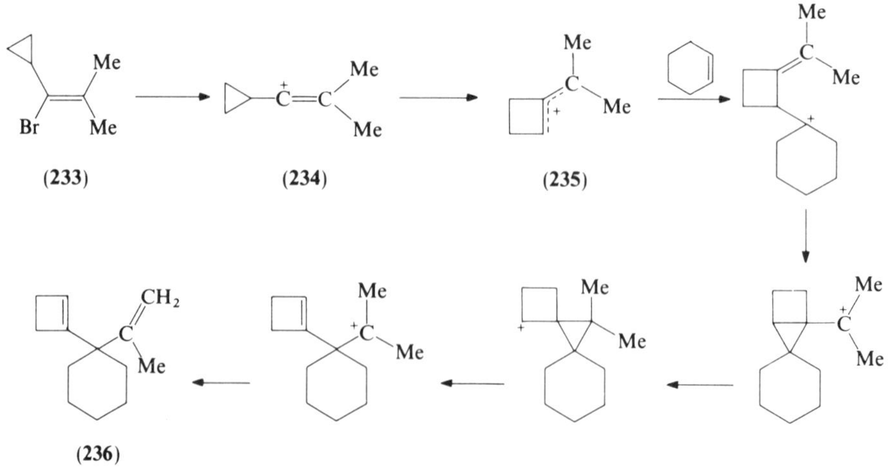

SCHEME 29

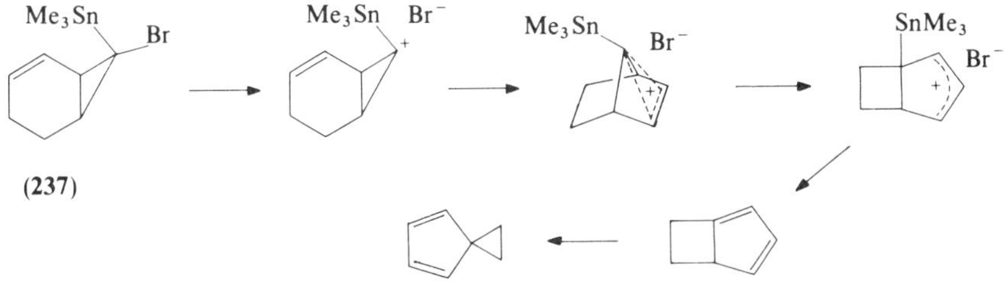

(237)

(238)

SCHEME 30

transform (7-cycloheptatrienyl)diphenylcarbinol into triphenylethene, presumably by way of the norcaradiene.[365] Ionizations of 2-substituted 1,1-dimethyl-3,4-methano-1,2,3,4-tetrahydronaphthalen-2-ols **(239)** do not appear to yield the corresponding 2-cations **(240)** but instead the benzylic cations **(241)** as the result of a cyclopropylcarbinyl–cyclopropylcarbinyl rearrangement.[366] A Wagner–Meerwein-type rearrangement has been reported during the thermally induced conversion of 7-chloro-6b,7a-dihydro-7*H*-cycloprop[*a*]acenaphthylene into phenalenyl.[367] The highly plausible mechanism outlined in Scheme 31 has been proposed[368] for the novel rearrangement of dihydroazulene **(242)** to the indene **(243)**.

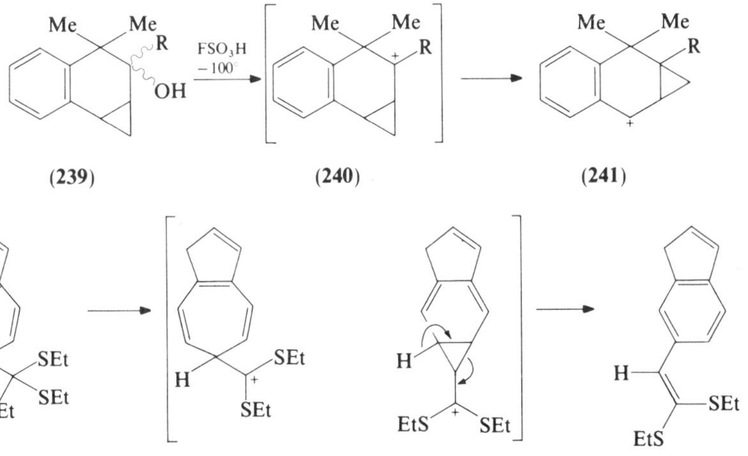

(239) (240) (241)

(242) (243)

SCHEME 31

The possible formation of a bridged carbocation intermediate in the lactonization of (E)-β-tert-butylcinnamic acid has been discussed on the basis of deuteration and ^{13}C-NMR experiments,[369] and the first direct experimental evidence has been obtained[370] for a degenerate carbocation rearrangement during the reaction of dimethyltitanium dichloride with a ketone. [*m.n.*3]Propella-γ-lactones **(245)** have been conveniently prepared[371] by the acid-catalysed rearrangement of 3-hydroxy-acids (**244**), while as a result of the heteroatom dipole effect, (**247**) was obtained as the unexpected product in the pinacol rearrangement of (**246**).[372] Ring-expansion reactions of succinoin derivatives (**248**), themselves produced by the Lewis-acid-catalysed aldol-type reaction of 1,2-bis(trimethylsiloxy)cyclobutene and a ketal or an acetal, have proved useful in the synthesis of a variety of compounds such as 1,3-cyclopentanediones, 4-keto-acids and -esters, 3-alkylidenecyclopentan-1-ones, and 2- and 3-cyclopenten-1-ones.[373] Optically pure α-methyl-β,γ-unsaturated ketones

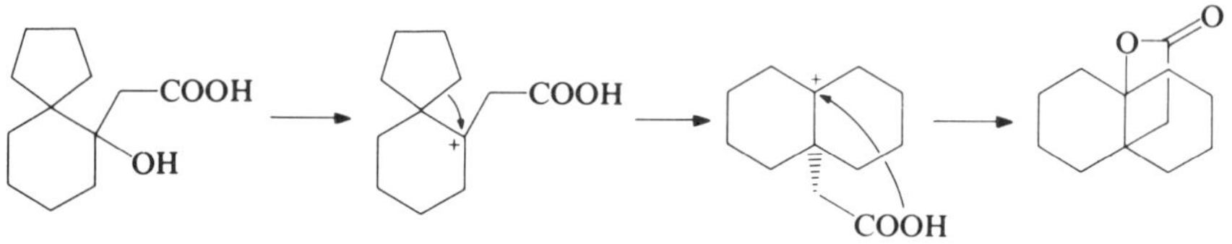

COOH COOH OH COOH O O

(**244**) (**245**)

HO OH O O O RO OH R^2 R^3 O

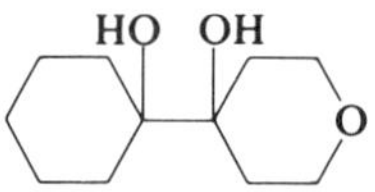

(**246**) 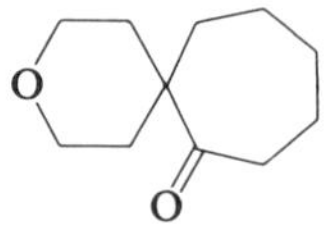 (**247**) 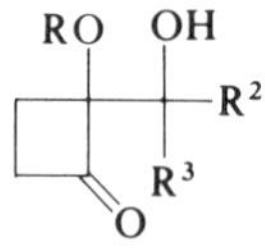 (**248**)

have been synthesized by an Et_3Al-mediated pinacol-type rearrangement in which alkenyl groups migrate stereospecifically with complete retention of the olefin geometry.[374] A reductive pinacol-type rearrangement of chiral α-mesyloxyketones leading to enantiomerically pure 2-aryl- or 2-alkenyl-propan-1-ols has been effected by organoaluminiums,[375] and the same reagents have been used[376] to promote the migration of an alkyl group to an optically active migration terminus in the synthesis of optically pure α-alkyl ketones such as the ant alarm pheromone (**250**). This process probably proceeds *via* (**249**). A deoxygenative 1,2-hydride shift rearrangement converting cyclic *cis*-diol monotosylates into inverted secondary alcohols has been achieved by using an excess of lithium triethylborohydride in THF. This new conversion provides convenient access to inverted deoxypentofuranosyl compounds that are difficult to obtain by conventional methods.[377] A practical two-step procedure for the ring-expansion of ketones to 1,2-keto-thioketals has been

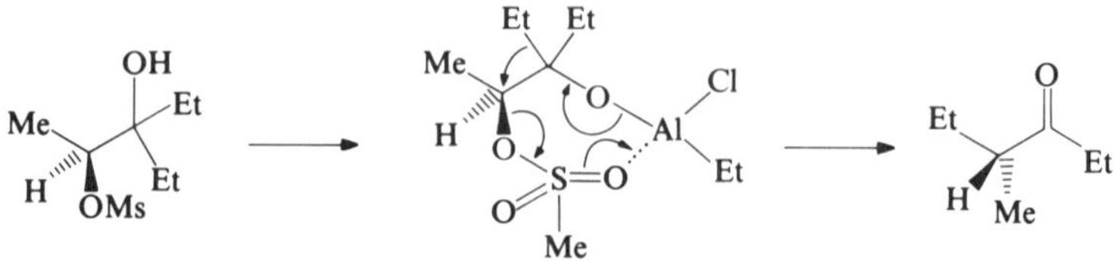

OH Me Et H Et OMs Et Et Me Cl H O Al O S=O Et O Me O Et Et H Me

(**249**) (**250**)

developed,[378] and butadienylphosphine sulphide has been readily transformed into β-keto-phosphine sulphide *via* a pinacol rearrangement.[379] An efficient ring-contraction of stereoisomeric 1,2-benzo-4-bromo-3-hydroxycyclohept-1-enes and their derivatives, (**251**) to (**253**), has been reported and a likely transition state, *viz.* (**252**), has been suggested for the process.[380] Treatment of (E)-1-trimethylsilylcyclooctene with bromine afforded (Z)-1-bromocyclooct-4-ene.[381] A remarkable cleavage of a Si—Me bond by trifluoroacetic acid has been attributed[382] to the large anchimeric assistance by a γ-OMe group [and thus the stability of the bridged cation (**254**)], as depicted in Scheme 32. The reactions of ketenes with

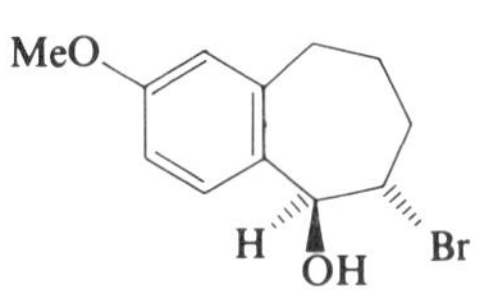

(**251**) 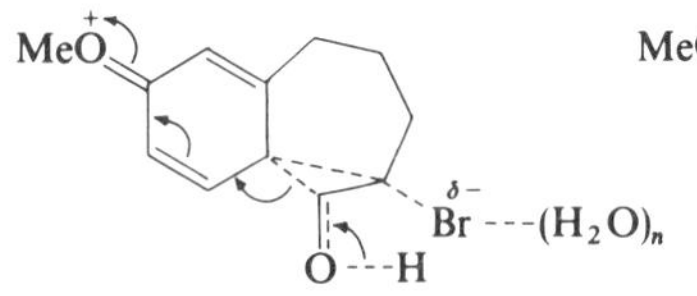 (**252**) 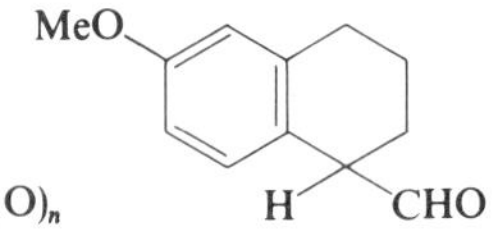 (**253**)

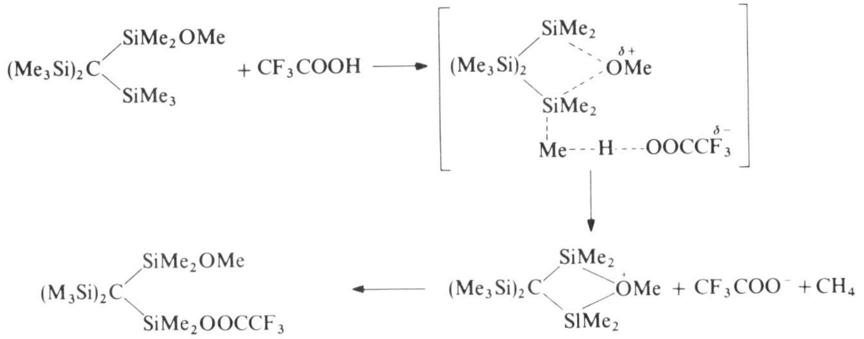

(**254**)

SCHEME 32

various silyl enol ethers and the chemistry of some of the reaction products have been investigated.[383] The formation of a rearranged adduct (**256**) in addition to the normally expected cyclobutanone, has been taken as an indication that the reaction proceeds *via* the carbocationic intermediate (**255**). One of the most intriguing features of the reported cyclization of (ω-haloalkynyl)silanes (**257**) with organoaluminium reagents is that the reaction can proceed with rearrangement of the carbon skeleton as represented by the conversion of (**258**) into (**259**).[384] A Lewis-acid-mediated rearrangement of 1-(trimethylsilyl)prop-2-ynyl trimethylsilyl ethers (**260**) has been reported.[385] The reaction is considered to proceed through initial coordination of aluminium to the acetylenic linkage of (**260**) which facilitates the 1,2-rearrangement of the silyl group to the neighbouring carbon atom followed by removal of another silyl group from oxygen.

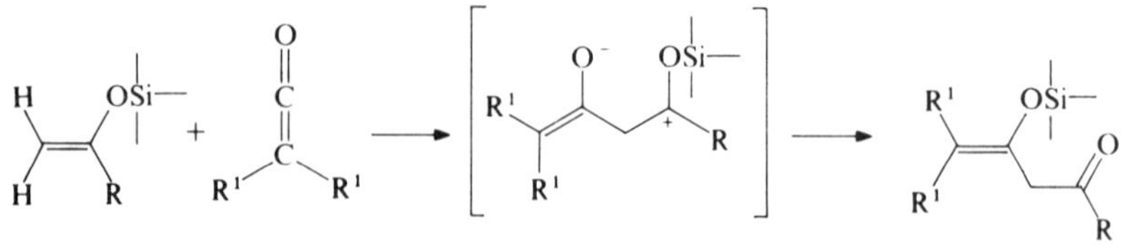

(255) (256)

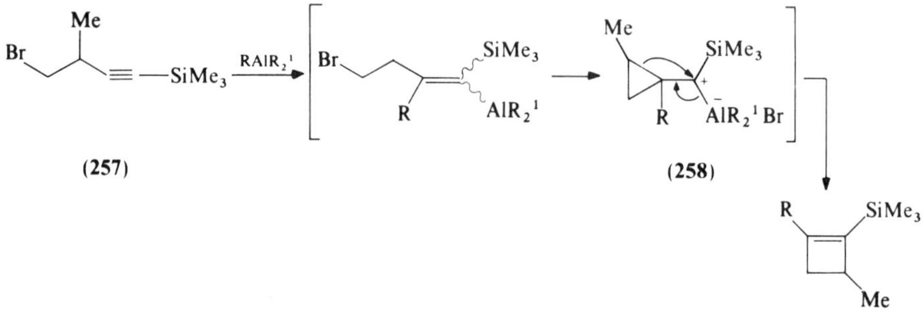

(257) (258)

(259)

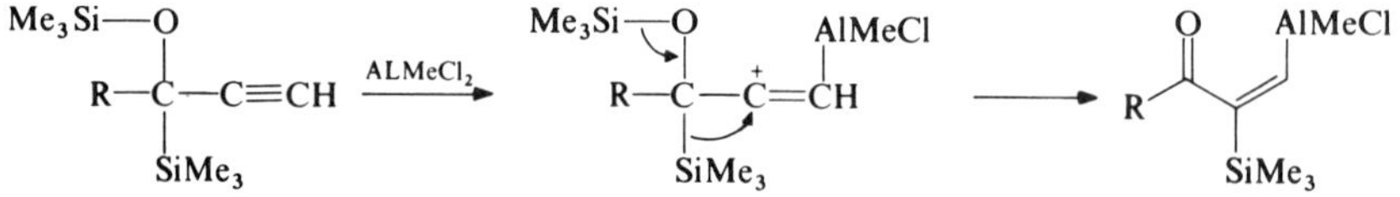

(260)

2,2-Disubstituted vinylsilane epoxides (**261**) have been shown to rearrange stereospecifically in the presence of $BF_3 \cdot OEt_2$ to the silyl enol ethers of 2,2-disubstituted aldehydes (**262**)[386] (see Scheme 33). A highly remarkable boron-trifluoride-catalysed rearrangement of 2,2-di-*tert*-butyloxirane into 2,2,3,3,4,4-

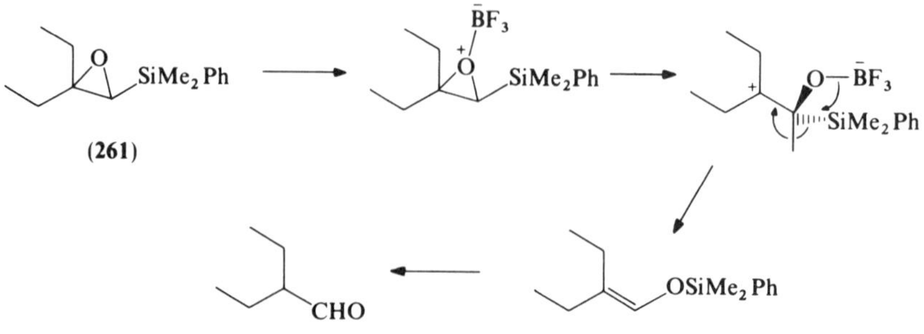

(261)

(262)

SCHEME 33

hexamethyltetrahydrofuran has been reported.[387] The reaction of substituted 2-chloro- and 2,3-dichloro-oxiranes with silver tetrafluoroborate has been observed[388] to give rise to products obtained by 1,2-methyl and 1,2-hydride shifts of the corresponding α-keto-carbocations. Rigid, bisected oxiranylcarbinyl mesylates, *e.g.* (**263**), appear to be solvolysed *via* an oxiranylcarbinyl cation which rearranges to a 2-oxahomoallyl cation, while the perpendicular isomers, *e.g.* (**264**), give displacement with inversion.[389] The acid-catalysed aromatization of [1,4-$^2H_2$]-1,4-epoxy-1,4-dihydronaphthalene and [4-^{2}H]-1,4-epoxy-1-methyl-1,4-dihydronaphthalene has been shown to result in partial retention of deuterium at the 2-position of 1-naphthol and 4-methyl-1-naphthol, respectively,[390] while *syn*-4a, 10:9,9a-diepoxy-4a,9,9a,10-tetrahydroanthracene has been isomerized by $ZnCl_2$ into the bicyclic acetal, 6,11-epoxy-6,11-dihydrodibenzo[*b*,*e*]oxepine.[391]

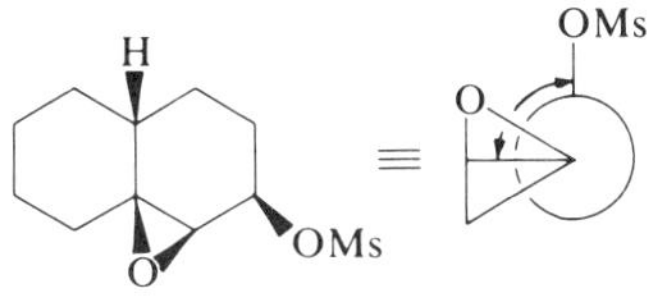

H OMs O OMs
(263)

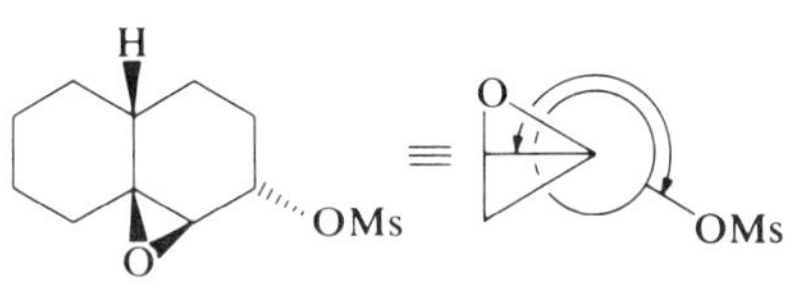

H O OMs O OMs
(264)

An approach to the synthesis of hydroxy-substituted spiro-ketals has been described,[392] based on the stereoselective formation of 4-hydroxytetrahydropyrans formed *via* a cation–olefin cyclization step; this methodology has been used in the synthesis of (±)-talaromycin B.[393] The stereocontrolled transformation of 1,3-dioxolane precursors bearing a neighbouring hydroxyl group [see (**265**)] has been reported.[394] A process which involves initial carbocation capture by the neighbouring hydroxyl at C(9) of (**265**) leading to orthoester (**266**), opening of which affords the dioxolenium cation (**267**), has been proposed. This cation then suffers internal nucleophilic attack at C(9) to provide the observed tetrahydrofuran (**268**). Although each of the oxygens of orthoester (**266**) are suitable leaving groups, the observed result demonstrates that ring-closure is feasible only in cases where two or three bridging methylenes separate intramolecular backside collinear displacement. NMR studies have indicated[395] that tetrahydrofurfural acetals in super-acids are cleaved to yield carbocations which readily undergo rearrangement.

A study of the effects of the *trans*-trimethylene bridge on the behaviour of norbornyl cations in the *trans*-trimethylenenorbornene system has shown[396] that carbocationic rearrangements in this system are dominated by the great relief of internal strain associated with the bridge. A 6,2-methyl migration, analogous to the well-known 6,2-hydride shift, has been reported in the norbornane system,[397] and it has been shown that the rearrangement of chloro-olefin (**269**) to both isomers of ketone (**270**) occurs through well-precedented carbocation pathways, without the intrusion of an *endo*-3,2-methyl migration in a substituted norbornyl cation.[398] Acid solvolysis of 4,5-*exo*-epoxybrendane has provided the noradamantane system functionalized at C(2) and C(8).[399] The principal features of the mechanism of

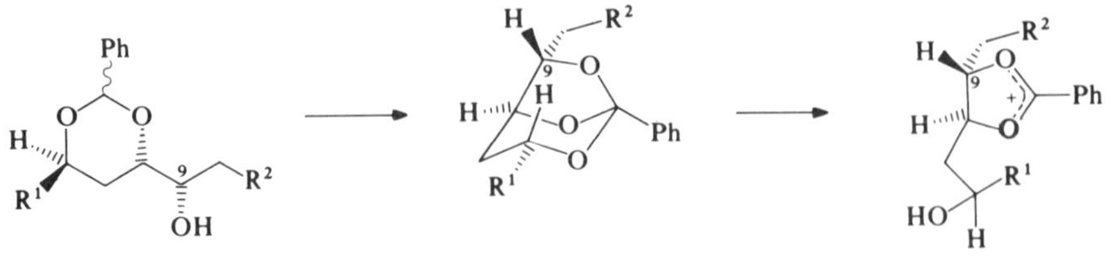

(265) (266) (267)

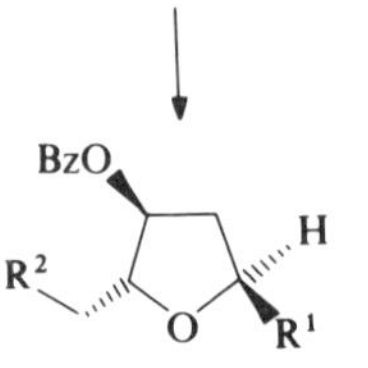

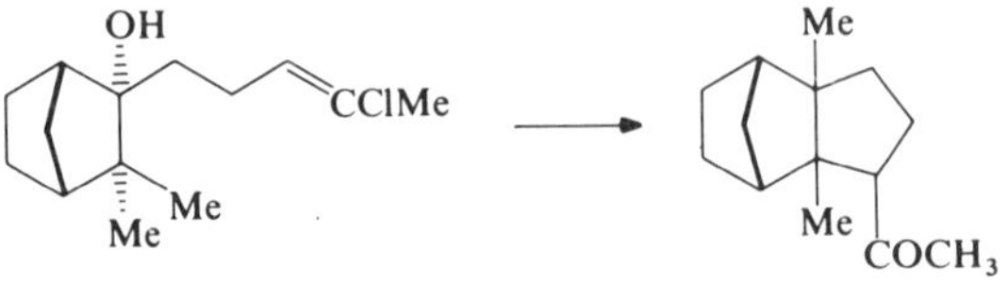

(269) (270) (268)

deamination of *endo*- and *exo*-bicyclo[3.2.1]octan-3-ylamines and their derivatives have been compared with those proposed earlier for the deamination of 4-*tert*-butylcyclohexylamines.[400] The main difference appears to be the non-stereospecific formation of a common non-classical carbocation (**271**) for both *endo*- and *exo*-compounds of the bicyclic series *via* classical precursors. Evidence has been presented[401] for the involvement of a discrete carbocation β to silicon during the bromination-induced rearrangement of 7-trimethylsilylnitronorbornenes, while a search for long-range aryl migration in the solvolysis of suitably positioned aryl derivatives of tricyclo[3.2.1.0^{2,4}]octane and bicyclo[3.2.1]octane has established[402] that a second aryl group is not necessary in order to observe long-range aryl migration coupled with electrocyclic ring-opening. Photosolvolysis of 2-D-*exo*-benzonorbornenyl-2-chloride to *exo*-2-benzonorbornenyl methyl ether has been observed to proceed with scrambling of the deuterium between C(1) and C(2), indicative of migration of the *anti*-aryl group.[403] The ring-expansion of several 1-substituted bicyclo[2.1.1]hexan-2-ones has been investigated[404] in order to compare the effect of substitution on the migratory aptitude of the bridgehead carbon in relation to the methylene carbon. In all the cases examined it was observed that there is a marked preference for migration of the methylene carbon, with the substituted substrates displaying higher regioselectivity than the parent ketone. In the rearrangement of both 1-fluoro- and 1-chloro-bicyclo[2.1.1]hexan-2-one, migration of the methylene carbon occurs exclusively. Direct irradiation of dichlorodibenzobicyclo[2.2.2]octadienes (**272**; X = H, Cl, OMe) has yielded Wagner—Meerwein-rearranged compounds[405] in which the rearrangement stereochemistry is consistent with the idea that, in the principal photochemical process, migration with retention

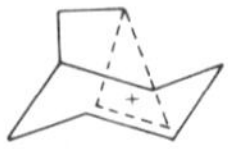

(271)

of configuration is concerted with the loss of chloride ion. On the other hand, cycano, aceto, or nitro substituents in the 10-position of 7,8-dichlorodibenzobicyclo[2.2.2]octadienes completely inhibit photorearrangements and photosolvolyses seen in these compounds when such substituents are not present. The authors[406] rationalize their results in terms of the inability of the excited states of the light-absorbing chromophores in these compounds to transfer electrons to the carbon–chlorine bonds remote from these chromophores. A novel bicyclo[2.2.2]- to bicyclo[3.2.1]-alkadiene rearrangement has been noted[407] on heating tricyclic lactams **(273)** in xylene. The conversion into novel tricyclic products **(275)** probably proceeds *via* the well-stabilized zwitterion **(274)** which results from anionotropic migration of the $C^{(c)}{=}C^{(d)}$ vinyl group in **(273)** from position (*a*) to position (*b*). A Wagner–Meerwein shift of the $C^{(c)}H_2$ group from $C^{(b)}$ to $C^{(a)}$ finally completes the rearrangement. The bromonium-ion-induced rearrangement of the

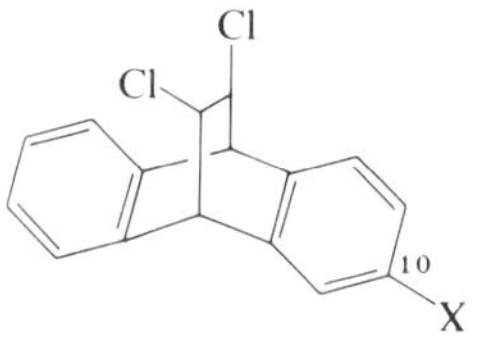

(272)

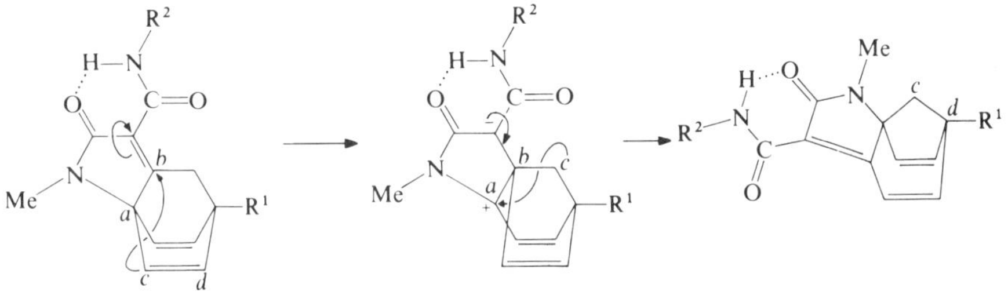

(273) (274) (275)

Diels–Alder adduct **(276)** to the rearranged bromohydrin **(278)** has been postulated[408] to result from attack of Br^+ at the less-hindered face of the double bond with participation of the electron-deficient sulphonamide nitrogen to form a species such as **(277)** which collapses to **(278)** under nucleophilic attack. The unusual

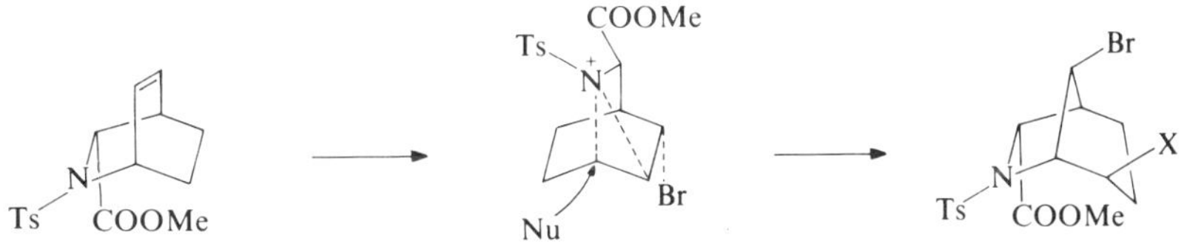

(276) (277) (278)

X = OH, Br

cascade of carbocation rearrangements characteristic of the *anti*-6-tetracyclo[4.4.0.$1^{1,4}$.$1^{7,10}$]dodecyl cation has been described,[409] while a decisive electronic effect of a bridgehead methoxy substituent on the electrophilic cage-opening of 1,3-bishomocubanones has been observed,[410] as illustrated by the acid-catalysed fragmentation of 4-methoxypentacyclo[5.3.0.$0^{2,5}$.$0^{3,9}$.$0^{4,8}$]decan-6-ones **(279)** to tricyclodecenediones **(280)**.

The acid-catalysed rearrangements of the pentaspirane (**281**) have been shown to afford either the propellane (**282**) or the hexacyclic compound (**283**), depending on the reaction conditions. The rearrangements proceed diastereospecifically and are rationalized as five- and nine-fold 1,2-shifts, respectively.[411] It has been established[412] that the solvolysis of [4.2.2]propellan-7-yl derivatives is critically dominated by the *exo*/*endo* stereochemistry of the leaving group. The acid-catalysed rearrangement of [*m.n.*2]propellanones (m = 3–5, n = 3,4) has been studied[413] to ascertain the effect of ring size on the mode of rearrangement, while a similar

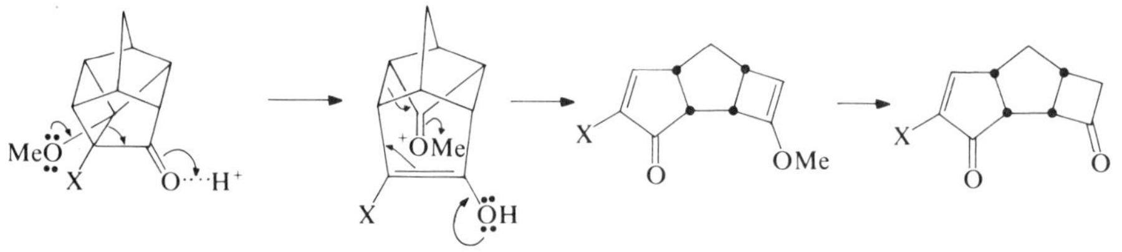

(279)
x = Br, H

(280)

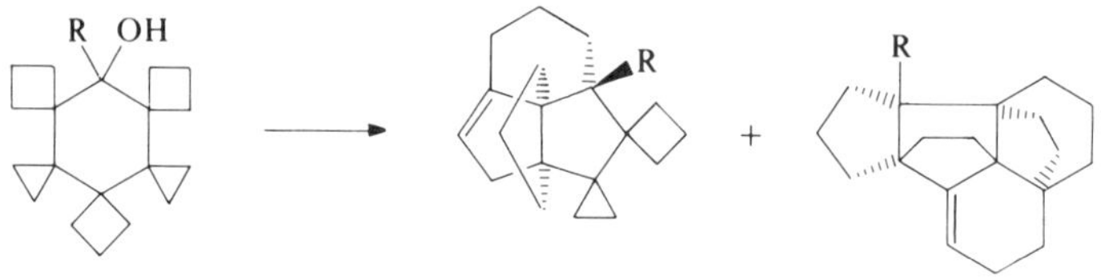

(281) (282) (283)

study[414] of model tricyclo[4.3.2]propellane derivatives (**284**) has indicated that the observed products arise *via* a concerted 1,2-shift of the cyclobutane bond having an antiperiplanar alignment with the leaving group. Descarboxyquadrone model compounds have actually been synthesized by utilizing the acid-catalysed rearrangement of tricyclo[4.3.2]propellanones.[415] An investigation of the reactivity of 2,4-methano-2,4-didehydroadamantane, a prototype small propellane, has concluded[416] that the central bond in small propellanes is actually a limiting form of the carbon–carbon single bond, the other limiting form being the unusual single bond between two carbon atoms in aliphatic compounds. A theoretical study[417] of the trimantane rearrangement has thrown up several reasonable rearrangement paths together with the structures of the isomers in these paths. The two triepoxide stereoisomers of 2,8,9-trimethylene[3.3.3]propellane have been shown to rearrange

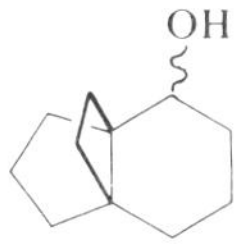

(284) 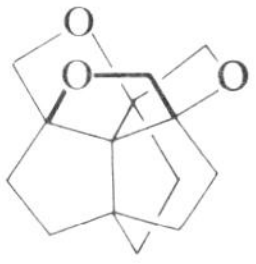 (285)

under the influence of Lewis acids or heat to the topologically nonplanar trioxahexaquinane (**285**).[418]

A new route to the tricyclo[6.3.0.0^{1,5}]undecane system has been described[419] involving the carbocation-mediated transannular cyclization of a suitably functionalized bicyclo[6.3.0]undecane derivative as the pivotal step. Brief thermolysis of the hexachlorinated bridged tricyclotriadecatetraenes (**286**; R = Bu^t, Ph) in air unexpectedly gave 2,3-dichlorobenz[*e*]indenones (**289**). Although currently available evidence precludes a fully secure rationalization of the transformation, one pathway that has been suggested[420] involves ring-expansion of the hexachloronorbornene element (*via* a chlorinated norcarene) and hydrolysis to give the bicyclic cation (**287**) from which a number of routes to the tricyclic ketone (**288**) can be envisaged. Loss of HCl from (**288**) and successive cross-cyclization – ring-opening steps, then provides a rational route to (**289**).

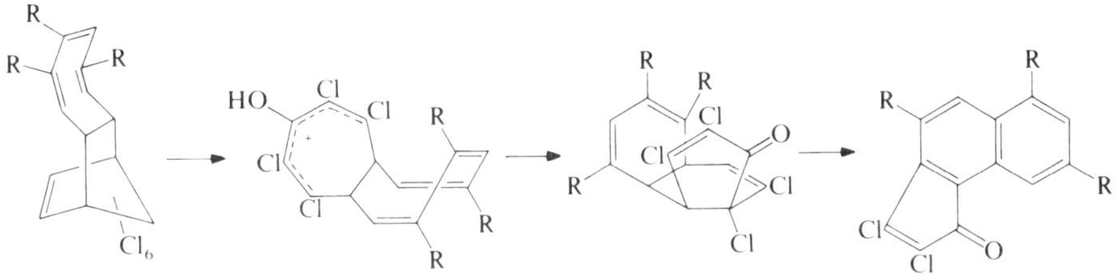

(286) (287) (288) (289)

The solvolysis of *N*-alkylacridinium salts in phenol and in benzoic acid has been found to give products which are interpreted[421] as being formed *via* primary carbocations which undergo partial rearrangement before being trapped. The Lewis-acid-induced reaction of *N*-carboethoxy-4-hydroxy-1,2,3,4-tetrahydropyridine with various carbon nucleophiles, a reaction which bears a close resemblance to the well-known Ferrier rearrangement of glycals, has been studied as a route to 2-substituted Δ^3-piperidines.[422] A carbocation species has been invoked[423] in the rearrangement of *cis*- and *trans*-1-(2-aryl-1,3-dithian-5-yl)-2-thioureas to 5-aryl-3-imino-7,7a-dihydro-1*H*,3*H*,5*H*-thiazolo[3,4-*c*]thiazole, and a sulphur-stabilized cationic intermediate has been postulated[424] in an interesting phenylselenenyl-chloride-induced ring-expansion of 1,3-dithiolans and 1,3-dithians (scc Scheme 34). An unprecedented reaction of an ethylene dithioketal has been reported.[425] Thus 4-homoadamantanone ethylene dithioketal (**290**) produced on bromination 4,5-[(thioethano)thio]-4-homoadamantene (**291**) in a reaction in

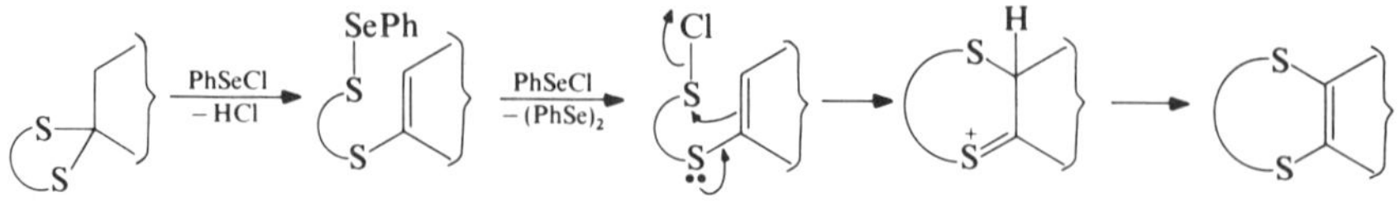

SCHEME 34

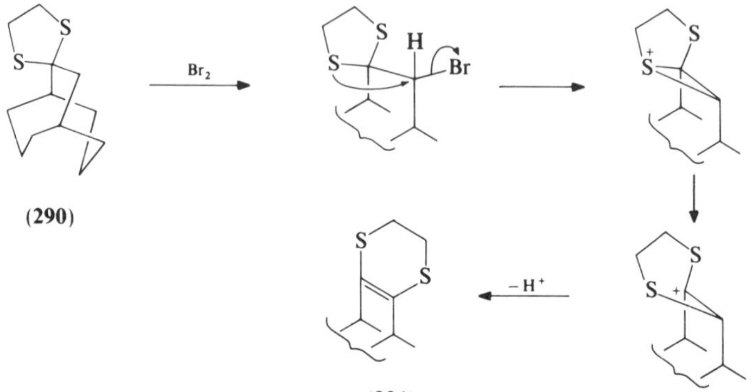

(290)

(291)

which the authors suggest α-bromination of **(290)** followed by sulphur participation.

Stannous triflate has been found to convert β-keto-sulphoxides into α-thiocarbocations, which in turn react with silyl enol ethers to yield 2-arylsulphenyl-1,4-diketones,[426] while treatment of sulphoxides with ketene silyl acetal in the presence of zinc iodide induces a Pummerer rearrangement and the formation of α-siloxysulphides.[427] Several chlorotrimethylsilane-induced Pummerer rearrangements have been reported,[428] and unexpected products arising from the Pummerer rearrangement of 3-methoxycarbonylthian-4-one *S*-oxide have been rationalized[429] by invoking the intermediacy of a common thiiranium ion in the process, while the successful synthesis and chlorinolysis of α-mercaptodimethyl sulphone have provided additional support for the contention that Pummerer rearrangements may occur during the chlorinolysis of α-sulphonylsulphinyl chlorides.[430]

The photoinduced Wolff rearrangement of α-diazo-β-ketophosphonates has provided a novel entry to substituted phosphonoacetates.[431] The vinylogous Wolff rearrangement of a number of acyclic and monocyclic β,γ-unsaturated α′-diazoketones to give γ,δ-unsaturated esters has been described.[432, 433] The rearrangement was promoted by copper sulphate or copper triflate in the presence of various alcohols and is postulated to involve insertion of the diazo carbon into the β,γ-olefinic bond to form a bicyclo[2.1.0]pentan-2-one which subsequently undergoes fragmentation to an unsaturated ketene and then capture by the alcohol.

Rearrangements in Natural-product Systems

These studies are classified thus because they involve rearrangements of natural-product systems that may well proceed *via* carbocations.

Backbone rearrangements of triterpenes have been studied,[434] and a new example of a backbone rearrangement under novel conditions has been reported whereby the silyl ether of methyl lithocholate has been transformed into D-homoandrostane derivatives.[435] Other backbone rearrangements that have been reported include Wagner–Meerwein rearrangements in taraxasterol and pseudotaraxasterol,[436] and the acid-catalysed rearrangement of hederagenin.[437] Isomultiflorenol has been partially synthesized from dendropanoxide by a backbone rearrangement[438] and the synthesis of A-homo-B-norsteroids by way of the Lewis-acid-catalysed rearrangement of steroidal epoxides has been described.[439] Cyclosadol and other isomers including lanostane- and curcurbitane-type isomers with (23E)- and (23Z)-, Δ^{23}- and Δ^{24}-unsaturated side-chains have been obtained from the acid-catalysed isomerizations of cycloartanol[440] and 24-methylenecycloartanol,[441] and a new regioselective rearrangement of Δ^{16}- steroids into D-homoazaandrostanes has been noted.[442] 5-Bromo-6β-chloro-5α-cholestan-3β-ol has been converted into epoxyhomonorcholestene by a silver-perchlorate-catalysed Westphalen rearrangement,[443] and an intermolecular mechanism involving a quinonoid zwitterionic intermediate has been suggested[444] to account for the acid-catalysed rearrangement of steroidal 14α-hydroxyketones with 17-alkoxy-16-en-15-ones, in the presence of aqueous or alcoholic hydrochloric acid. A $BF_3 \cdot OEt_2$-induced rearrangement of methyl isopimarate-8,9-epoxide has produced compounds with a new diterpene skeleton[445] in which the nature of the bicyclo[3.2.1]octane moiety constituting the C/D ring system means that these compounds can be related to stemodane and stemorane diterpenes. A study has been made of the rearrangement of the eudesmane sesquiterpene, (+)-rosifoliol, to valencane structures.[446] This biogenetically significant methyl migration has been accomplished *via* Lewis-acid-catalysed epoxide opening. The acid-catalysed rearrangement of hinesol epoxides and related compounds to eudesmane derivatives has been reported,[447] while in a recent approach to the taxane family of diterpenes, the optically active starting material (**292**) has been used to furnish the rearranged epoxide (**293**) which underwent fragmentation to (**294**) when treated with $Ti(OPr^i)_4$.[448] The acid-catalysed rearrangements of α-

Me Me Me Me O (292) — $BF_3 \cdot OEt_2$ → Me Me Me Me OH — Bu^tOOH → Me Me Me Me OH O (293)

↓ $Ti(OPr^i)_4$

Me HO Me Me Me H O

(294)

santonin and 6-epi-α-santonin have been reinvestigated,[449] and the major product from the formic-acid-induced rearrangement of lauren-1-ene has been confirmed as (1S,4S,8R,9S,12R)-4,8,12,14,15-pentamethyltetracyclo[10.3.0.$0^{1,9}$.$0^{4,9}$]pentadec-14-ene.[450] Evidence has been presented[451] to suggest that erythrolide A (**296**) is produced in nature from diterpenoid (**295**) by a unique naturally occurring di-π-methane rearrangement. A ring-expansion of a 6,5- to a 6,6-system has provided a stereocontrolled entry into the polyhydrochromanones. The steric and electronic requirements of such a process have been probed and, for the synthesis of verrucarol, a procedure involving the synthesis of a norapotrichothecene, followed by rearrangement to the trichothecane skeleton [see (**297**) → (**298**)] has been developed.[452]

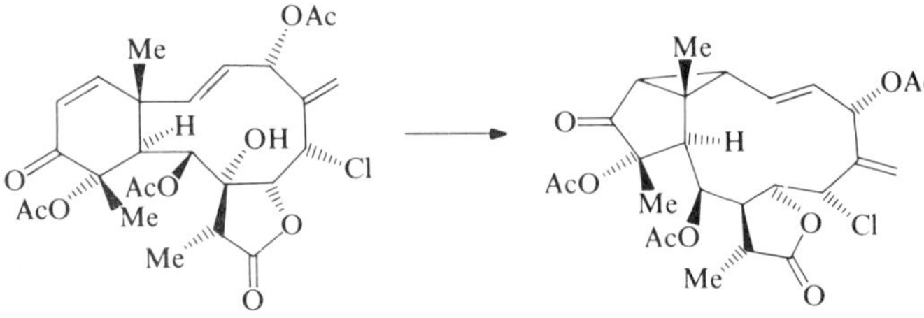

(**295**) (**296**)

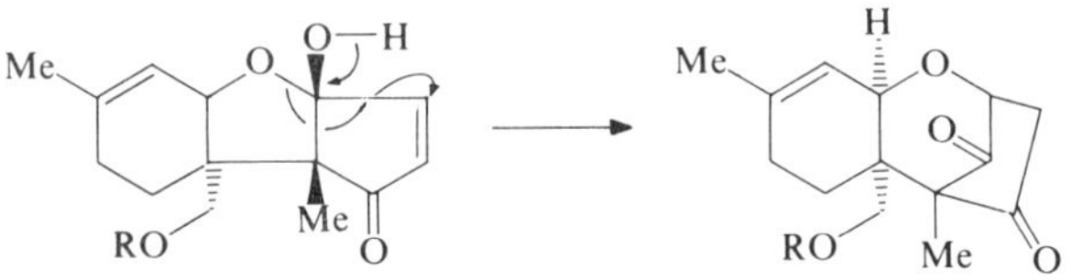

(**297**) (**298**)

Solvolysis of methyl 4,6-*O*-cyclohexylidene-2-deoxy-2-trifluoroacetamido-3-*O*-trifluoromethylsulphonyl-α-D-glucopyranoside (**299**) and its derivatives in hot methanol has been shown to yield ring-contraction products, *e.g.* (**300**), having a bicyclic D-xylofuranoside structure (see Scheme 35).[453] The formation of 4,6-acetoxonium and benzoxonium ion derivatives of methyl gluco-, manno-, galacto-, and ido-pyranosides from the 6-azido-6-deoxy compounds on treatment with nitrosonium ion has been described.[454]

A convenient conversion of codeine into 6-demethoxythebaine has been outlined,[455] and a novel rearrangement product, resulting from a non-classical D ring migration, has been obtained by the action of acetic anhydride on thebaine.[456] Rearrangements involving rupture of the Ph—N—C—N chromophore and formation of a carbonyl group at C(3) have been observed[457] on treating the indole alkaloids, desformocorymine and dihydrocorymine, with trifluoroacetic acid, while the observed facile conversion of (−)-vincadifformine into *N*-methyltetrahydromeloscine, represents the first *in vitro* skeletal rearrangement of an *Aspidosperma* to a *Melodinus* alkaloid.[458]

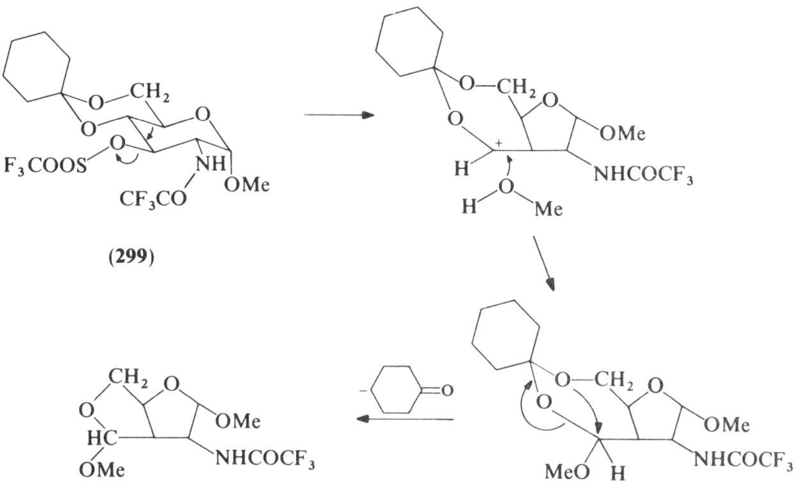

(299)

(300)

SCHEME 35

Rearrangements Involving Electron-deficient Heteroatoms

The Beckmann rearrangement has been reviewed,[459] and the effect of trimethylsilyl groups on that rearrangement has been discussed.[460] The Beckmann rearrangement of cyclododecanone oxime to dodecalactam in the presence of hydrochloric acid has been reported,[461] and the Beckmann rearrangement of cyclohexanone oxime has been achieved in the vapour phase over boria-hydroxyapatite catalyst,[462] the basic property of which appears to play an important rôle in effecting the rearrangement selectively. 2-Chloro-2-nitrosofenchane has been observed[463] to rearrange in a Beckmann-like manner under the influence of $AlCl_3$. The carbamoylation of oximes by isocyanates has been shown to involve initial formation of an *N*-acyl nitrone, which can then rearrange to the *O*-carbamoylated oxime or decompose to an *N*-hydroxyurea,[464] and an investigation[465] of the ring-expansions of *N*-methyl nitrones formed from cyclic conjugated ketones, often referred to as the Barton rearrangement, has surprisingly shown that the reaction does not proceed at all well for many simple systems. Functionalized alkenyl oxines have been found to cyclize to Δ^1-pyrrolines when treated with trimethylsilyl polyphosphate in refluxing carbon tetrachloride,[466] and nitrogen-containing heterocycles have been derived from a variety of carbocyclic frameworks *via* the successive rearrangement–alkylation sequence of hydroxylamine carbonates with trialkylaluminiums; see **(301)** → **(302)**.[467] In contrast to the previously reported Beckmann rearrangement of oxime sulphonates, this method is characterized by the ultimate potentiality of introducing a new carbon–carbon bond in a diastereoselective manner at the α-position of amines.

Sodium bromite in the presence of a catalytic amount of sodium bromide in aqueous sodium hydroxide has been developed[468] as a new reagent for the Hofmann

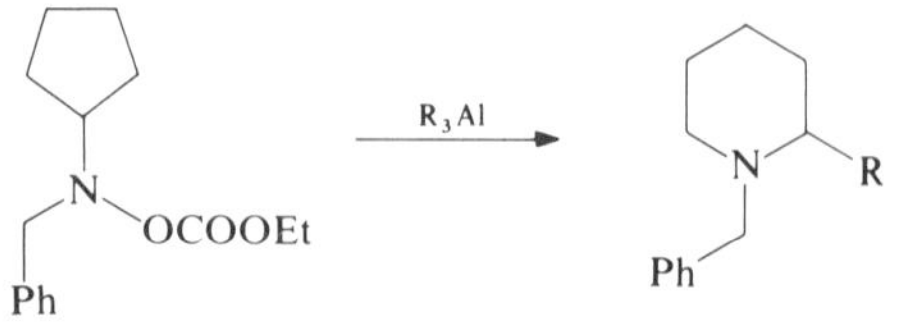

(301) $\xrightarrow{R_3Al}$ (302)

rearrangement. Studies on phase-transfer-catalysed Hofmann rearrangements have appeared,[469, 470] while the observed conversion of aliphatic amides into amines with [*I,I*-bis(trifluoroacetoxy)iodo]benzene, in effect an "acidic Hofmann rearrangement", has been shown to occur with complete retention of configuration in the migrating group.[471] Curtius rearrangements of ethyl 3-carboethoxyamino-2-benzofuran carboxylate,[472] of cyclic silicon azides,[473] and of adamantyl azides,[474] have been investigated, and phase-transfer catalysis by quaternary ammonium salts in the Curtius rearrangement has been described.[475] 2-Trimethylsilylethanol has been used to trap isocyanates produced by the Curtius rearrangement of acyl azides and the resulting trimethylsilylethyl carbamates have been cleaved with tetra-*n*-butylammonium fluoride to liberate the primary amines.[476] The reported synthesis[477] of an unusually hybridized phosphorus cation from an azidophosphonium salt represents the first example of a Curtius-type rearrangement involving a charged atom, while the photolysis of diphenyl- and *tert*-butyl(phenyl)-phosphinic acids has been shown to yield the corresponding phosphonamidate in high yield by a Curtius-like rearrangement.[478] 4-Aminoquinoline has been obtained in good yield by an amide modification of the Lossen rearrangement of 4-quinolinecarbohydroxamic acid,[479] and the acid-catalysed breakdown of 11b-azido-5,6,7a,11b-tetrahydrobenzocyclobuta[1,2-*f*]7*H*-benzocycloheptene has been observed to lead to the unexpected formation of benzazocine.[480] Certain conclusions concerning the general mechanism of the Schmidt reaction as applied to tertiary amides have been drawn from this study.

A convenient method has been reported for the Baeyer–Villiger oxidation of carbonyl compounds using a 1-alkoxycarbonyl-1,2,4-triazole–H_2O_2 system.[481] Regioselectivity in Baeyer–Villiger oxidation of tetracyclic ketones of the type (**303**)has been interpreted[482] in terms of torsional effects (see Scheme 36), while the Baeyer–Villiger oxidation of a number of β,β,γ-trisubstituted cyclopentanones and

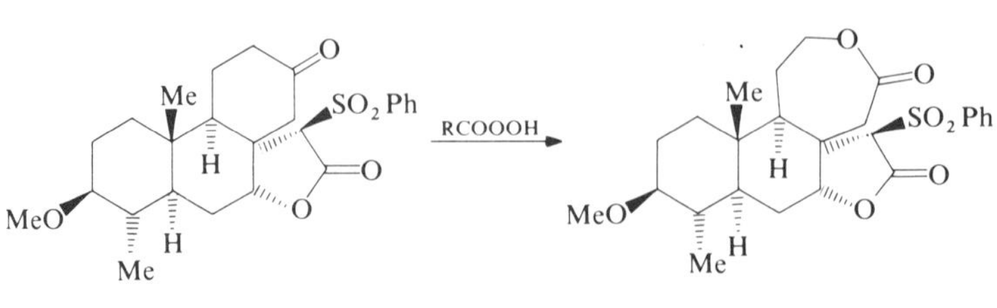

(303) $\xrightarrow{RCOOOH}$

SCHEME 36

cyclohexanones has been shown to lead to preferential migration of the α-carbon relative to the α′-carbon by a considerable factor. The origin of the preference is suggested to be steric and is ascribed to the greater relief of non-bonded interactions as the α–β bond lengthens in going to the transition state. Thus, in the carbonyl-addition intermediate, the 1,3-diaxial-like interaction between the hydroxyl and the closest γ-carbon is diminished to a greater extent when the α-carbon migrates relative to the α′-carbon.[483]

Metal-catalysed Rearrangements

Reviews have appeared on the rearrangements and isomerizations of organometallic compounds[484] and on the carbon-skeleton rearrangements of aldoses by molybdate,[485] while the effect of calcium ion on the rearrangement of aldos-2-uloses to aldonic acids has been examined.[486]

The rearrangement of dichlorotitanacyclobutane **(304)** to the (olefin)–titanium–methylidene complex **(305)** has been studied,[487] and the rearrangement of a titanocene derivative involving migration of the organic fragment from metal to sulphur [see **(306)** → **(307)**] has been reported.[488] A new synthetic method involving the σ–π rearrangement of a hydrido-2,4-pentadienylzirconium

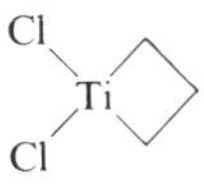

(304)

$(H_2C{=}CH_2)Ti({=}CH_2)Cl_2$

(305)

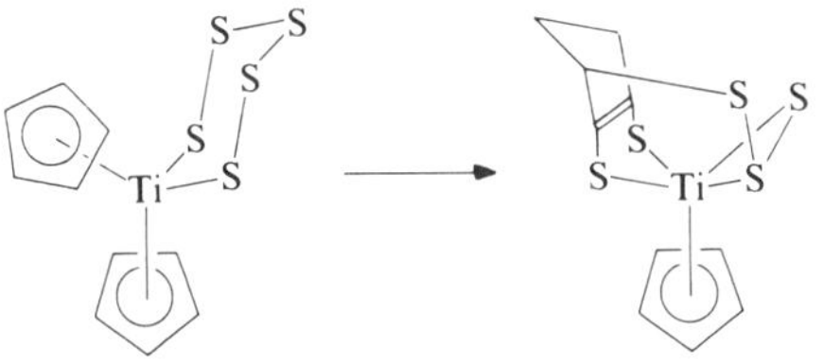

(306) → **(307)**

species has been noted.[489] The salt $[Mo(\eta\text{-}C_7H_8)_2]BF_4$ has been shown to rearrange thermally to yield $[Mo(\eta\text{-}C_7H_7)\ (\eta\text{-}C_7H_9)]BF_4$,[490] the kinetics and mechanism of the transfer hydrogenation and double-bond migration of hex-1-ene catalysed by molybdenum complexes have been investigated,[491] and molybdenum(II)acetate dimer has been used to induce the regiospecific ring-cleavage of α,β-epoxysilanes to form enolate intermediates which react with aldehydes to give the corresponding α,β-unsaturated carbonyl compounds.[492] A novel and stereospecific synthesis of imides which utilizes the molybdenum-carbonyl-induced reaction of azirines with carbanions derived from β-dicarbonyl compounds, has been described.[493] A possible mechanism for this conversion, illustrated for 2-phenylazirine and diethyl malonate anion, is outlined in Scheme 37. An unusual transform-

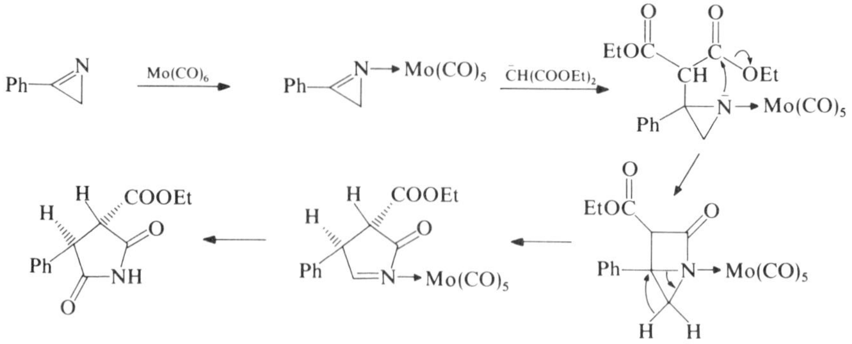

SCHEME 37

ation (**308**) to (**312**), involving the apparent insertion of a carbonyl ligand into a carbon–sulphur bond of a vinyl ligand, has been reported to proceed *via* a σ-vinyl complex (**309**), and the authors have proposed[494] that this tungsten-promoted 1,3-sulphur shift can be accommodated by the mechanism in Scheme 38, in which nucleophilic attack on coordinated CO in (**309**) is followed by ring-closure to give (**310**) and subsequent ring-opening to yield the η^2-vinyl complex (**311**).

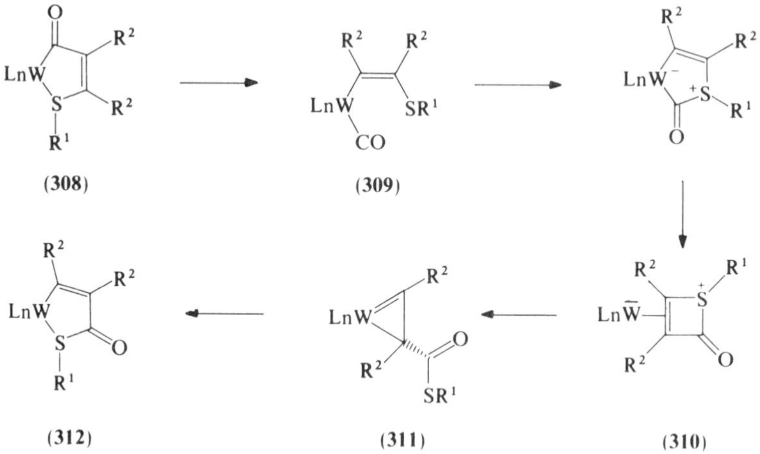

(**308**) (**309**) (**312**) (**311**) (**310**)

SCHEME 38

The unprecedented rearrangement and migration of a phosphorus-bound allyl group in an iron phosphoranide into an iron-bound vinyl group has been observed.[495] The mechanism of this process is considered to involve insertion of iron into an allylic or a terminal vinylic carbon–hydrogen bond followed by a 1,3-proton shift to the terminal olefinic or allylic carbon atom, respectively, with concomitant phosphorus–carbon bond-cleavage.

1,2-Bis(σ-methylene)benzenetetracarbonyliron has been rearranged to *o*-xylylenetricarbonyliron under both photochemical and thermal conditions,[496] and the rearrangement of bridging alkylidyneiron complexes to bridging alkenyliron complexes has been reported.[497] Bridge rearrangements of multi-bridged ferrocenophanes[498] in the presence of aluminium trichloride, and reverse methyl migration from a methyl iron complex to trimethylaluminium,[499] have been observed, while the electron-transfer catalysis of a ligand rearrangement around an iron–iron bond has been described.[500] A systematic study of the addition of nucleophiles to (η^4-1,3-diene)tricarbonyliron complexes has been undertaken.[501] The study has demonstrated that kinetically controlled addition at $-78°$ is strongly preferred at an unsubstituted internal position, but that reversal of the initial addition can be rapid below 0°. Formation of the more stable η^3-allyl complex then occurs *via* nucleophilic addition at the terminal position and, in special cases, *via* hydride migration. The rôle of ferric salts on the transformation of phenylated bishomocubane to the [2 + 2]-cycloreversion product has been compared[502] with the effect of other metal catalysts on this substrate. Hexamethyl(Dewar benzene) has been isomerized to hexamethylbenzene under visible light irradiation in the presence of stoichiometric or catalytic amounts of cationic arene–iron(II) complexes,[503] while the effect of complexation of diiron hexacarbonyl complexes with 3*H*-1,2-diazepines on the rate of [1,5]-sigmatropic hydrogen migration in the diazepine system has been examined.[504] Evidence for reversible aryl migration from iron to nitrogen of five-coordinate σ-bonded iron aryl porphyrins has been presented.[505] It has been proposed[506] that the most likely mechanism for the hematin-catalysed rearrangement of hydroperoxylinoleic acid to epoxy-alcohols is a unique oxygen-rebound process in which the metal complex reduces the hydroperoxide group to an alkoxyl radical and transfers the hydroperoxy oxygen to intermediate carbon-centred radicals generated by alkoxyl radical cyclization. Although the reactive species in the cytochrome P-450 cycle has not been observed, oxidation of several olefinic substrates with purified and microsomal cytochrome P-450, and synthetic porphyrin–iodosylarene model systems, has indicated[507] that the oxygen-rebound mechanism already described for aliphatic hydroxylation by simple iron peroxide systems suffices to explain the cytochrome P-450 cycle as well. The enantioface-discriminating isomerization of olefins by chiral ruthenium complexes has been described,[508] while ruthenium(III) tris(acetylacetonate) has proved to be an efficient catalyst for the isomerization of *O*-alkylvinyl carbinols, thus allowing a facile route to alkyl ethyl ketones.[509] The intramolecular rearrangement of the σ–π-acetylide ligand in $Os_3H(CO)_{10}(C{\equiv}CC_6H_5)$ has been studied by ^{13}C-NMR spectroscopy.[510] Model studies[511] of the reaction of dimethyl bromomethylmalonate with vitamin B_{12} have provided support for the concept that, in the coenzyme-B_{12}-dependent enzymic carbon-skeleton rearrangement of methylmalonyl-SCoA to succinyl-SCoA, the substrate attaches itself to the central cobalt atom of vitamin B_{12}, while the first non-enzymic rearrangement of a derivative of a *bona fide* substrate belonging to the coenzyme-B_{12}-dependent carbon skeleton rearrangements has been described.[512] Rearrangements characterized as diretentive diatropic migrations have been observed[513] during Co(I)alamin-catalysed reduction of 4β-(*tert*-butyl)-1α-(1-methylvinyl)cyclohexanecarbaldehyde, and a novel cobalt-carbonyl-catalysed ring-enlargement of cyclobutanones with a hydrosilane and carbon monoxide represents the first example of the catalytic incorporation of CO into a ketonic carbon (see Scheme 39).[514] The rhodium-catalysed isomerization of α-, β-, and γ-silylated allyl alcohols has been successfully applied to the selective synthesis of acylsilanes, α-silyl ketones, and β-silyl ketones, respectively.[515] The direct, regiospecific, rhodium-catalysed carbonylation of aziridines to β-lactams has been reported,[516] as has the rhodium(I)-catalysed isomerization of *N*-allylimines to 2-aza-1,3-dienes,[517] and chiral diphosphine rhodium(I) complexes, which exhibit high catalytic activity and excellent enantioselectivity for the isomerization of prochiral allylamines, have been discovered.[518] The authors believe that the isomerization provides a new practical method for the synthesis of optically pure aliphatic enamines and aldehydes.

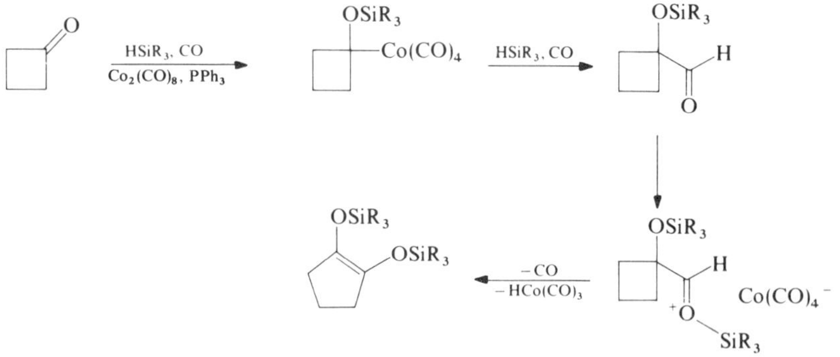

HSiR₃, CO / $Co_2(CO)_8$, PPh_3 → $OSiR_3$, $Co(CO)_4$ → HSiR₃, CO → $OSiR_3$, H, O → $OSiR_3$, H, ^+O, SiR_3, $Co(CO)_4^-$ → −CO / −$HCo(CO)_3$ → $OSiR_3$, $OSiR_3$

SCHEME 39

The rearrangement of dienols $R_2C(OH)CH{=}CHC(X){=}CR_2$ possessing a non-terminal arylthio or alkylthio substituent, to the vicinally substituted dienols $R_2C{=}CHCH{=}C(X)CR_2(OH)$, has been effected by nickel chloride in aqueous *tert*-butyl alcohol.[519] The palladium (0)-catalysed rearrangement of a vinylic epoxide intermediate has been utilized[520] as a key step in a recent synthesis of (R)-4-hydroxy-2-benzyloxymethylcyclopent-2-en-1-one from D-glucose. Diastereomeric mixtures of 2-substituted 1-vinylcyclohexyl acetates (or benzoates) have been rearranged stereoselectively to 2-substituted (E)-β-acetoxy(or benzoyloxy)ethylidenecyclohexanes by catalysis with bis(acetonitrile)palladium(II)chloride, [521] and the same catalyst has been used in a stereospecific 1,3-diene synthesis from bis(E,Z)-allylic acetates.[522] A mechanistic study of the Pd(0) and Pd(II)-catalysed rearrangements of (E,Z)-4-acetyl[^{17}O]oxyhepta-2,5-diene has been carried out using ^{17}O-NMR spectroscopy.[523] The rearrangement of allylic sulphinates to sulphones has been facilitated by palladium catalysts.[524] The observed stereospecificity of this catalysis suggests that the transition state for the transformation is somewhat ionic, involving a palladium chelate, unlike the intramolecular cyclic intermediates in [2,3]-sigmatropic rearrangements. A palladium-complex-catalysed rearrangement of allylic sulphinates to sulphones has been reported,[525] while palladium(II) catalysts have been shown to be effective in the [3,3]-sigmatropic rearrangement of propargyl thionophosphates to allenyl thiolophosphates.[526] An unusual rearrangement of a phosphine ligand in the coordination sphere of palladium has been reported,[527] and the reverse migration of allylic groups from palladium to phosphorus, and the formation of diaryl(allyl)phosphine in the (π-allyl)palladium(II)-induced cleavage of triarylphosphines has been demonstrated.[528] Palladium dodecatungstophosphate supported on silica has proved to be an active and selective catalyst for the isomerization of pentane and hexane in the presence of hydrogen.[529] The reactivity of di-η-chlorobis[3-(nitrosooxy)bicyclo[2.2.1]hept-2-yl-*C*,*N*]dipalladium (**313**) appears to be modified by added reagents. Thus, in the presence of $CuCl_2$, the thermal decomposition leading to epoxynorbornane is completely suppressed and an unequivocal rearrangement of the norbornane framework is observed[530] (see Scheme 40). An attempt to delineate the rôle played by (η^4-norbornadiene)$PdCl_2$ in

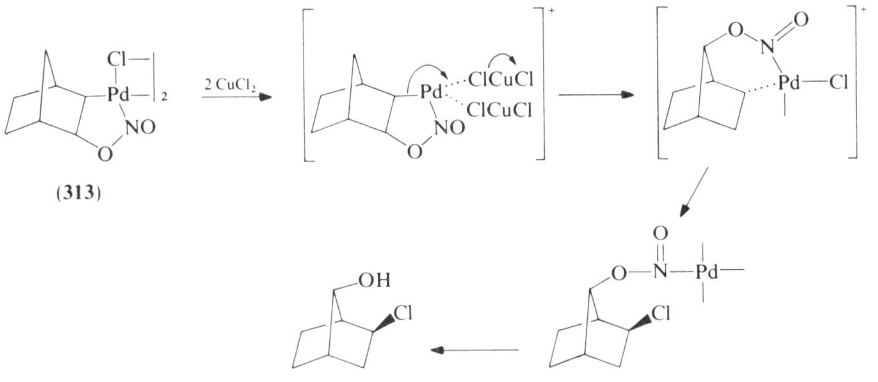

(313)

SCHEME 40

a recently reported photo-generated valence isomerization of quadricyclene to norbornadiene has been made,[531] and the conversion of quadricyclane into norbornadiene has been promoted by $PdCl_2$ itself.[532] The palladium-catalysed oxidation of aminoalkenes (**314**) to cyclic imines (**315**) under "Wacker" conditions has been reported,[533] while the phenylation of allylic butenols has been achieved using $PdCl_2$–$NaHCO_3$, in a reaction which appears to proceed by a highly regioselective 1,2-hydrogen shift *via* a Wacker-type intermediate.[534] The skeletal rearrangement of ^{13}C-labelled hexanes on platinum–nickel bulk alloys has been examined,[535] surface structure and temperature dependence of *n*-hexane skeletal rearrangements catalysed over platinum single-crystal surfaces have been investigated,[536] and the ease of rearrangement of (bicyclo[2.2.1]hept-2-ene)-bis(triphenylphosphine)platinum(0) complexes, which proceeds *via* the insertion of the platinum atom into the vinyl carbon–halogen bond, has been found to depend upon the identity of the halogen and the stability of the precursor haloolefin–platinum(0)complex.[537] A study considering the theoretical aspects of the shift of an alkyl (or aryl) group from a coordinated phosphine to the 16-electron d^8 transition metal to which it is coordinated, has been undertaken.[538] This

Pd(II) ···· OH_2 $^+NH_3$ (314) → Me C(=O) ... $\overset{+}{N}H_3$ → (315) Me N H

rearrangement at first seems to resemble an anionic [1,2]-sigmatropic shift, but a detailed analysis of the relatively low barrier calculated for the reaction shows how all traces of the forbiddeness of the reaction have vanished.

The dynamic structural behaviour of 2-alkenylzinc compounds has been examined.[539] The regio- and stereo-chemical control exerted in hitherto unknown mercury(II)-mediated lactonization of cyclopropane acid derivatives has been examined.[540] For the most part the products derive from apparent backside entry of the internal nucleophile with good to excellent stereocontrol at the electrophilic cyclopropane carbon. Overall, the cyclization process represents an operational equivalent of a carbon electrophilic based cyclofunctionalization of unsaturated acids. Quadricyclanes have been rapidly isomerized to norbornadienes by mercuric halides under very mild conditions. Corner-mercuration is assumed to be the essential step in this transformation.[541] Copper(II) and tin(II) salts have also been reported to show exceptional heterogeneous catalytic behaviour in achieving this conversion.[542] An interesting quasi-concerted allylic rearrangement has been reported in the reaction of allylic chlorides with methyltin(methanethiolate),[543] and diorganoarsenic(III) diphenyldithiophosphinates have been prepared by an unexpected sulphotropic rearrangement of diphenylphosphinyl diorganodithioarsinates.[544]

The skeletal isomerization of cyclohexene on alumina and related catalysts has been studied,[545] various paraffins have been isomerized at temperatures as low as 0° over alumina treated with CF_3Cl,[546] and the stereoselective rearrangement of β-allenic amides to (2E),(4Z)-dienamides has been promoted by alumina.[547]

Rearrangements Involving Ring-opening and Ring-closure

Three-membered Rings

New rearrangements implicating small rings have been reviewed.[548] A complete kinetic analysis of thermal stereomutations among the eight 2,3-dideutero-2-(methoxymethyl)spiro[cyclopropane-1,1′-indenes] has been carried out.[549] The study has revealed that two trimethylenes produced through two distinct cyclopropane carbon–carbon bond cleavages, are implicated, since all three possible one-centre epimerizations are seen. One-centre epimerizations have also been found to be decisively dominant in the thermal stereomutations of phenylcyclopropane.[550] On thermolysis 1-alkoxycarbonyl-2-arylcyclopropanes have been found to yield 2-carbonyl-1-naphthol derivatives in an unprecedented rearrangement of the cyclopropane ring which involves the formation of a carbon–carbon bond between arene and carbonyl carbon atoms situated on vicinal atoms of the cyclopropane ring.[551] The acid-catalysed rearrangements of methyl 1R-*cis*-2,2-dimethyl-3-(2-oxopropyl)cyclopropane esters and related compounds have been examined,[552] and the thermal rearrangement of a number of *gem*-dihalodiphenylcyclopropanes has been found to afford the corresponding halophenylindenes.[553] The product distribution observed from the isomerization of the four possible 22,23-methylenecholesterol acetates has indicated[554] that, in the absence of ring constraints, proximate chiral centres influence the acid-promoted isomerization of cyclopropanes. In the reactions studied, the most favoured mode of reaction was the loss of a tertiary proton, *trans*-antiperiplanar to the breaking (or broken) cyclopropane bond. Moreover, the recent synthesis of all four *trans*-isomers of the marine cyclopropyl sterols, petrosterol and its diastereomers, along with the unambiguous assignment of their absolute stereochemistry, has allowed the examination in unprecedented detail of the influence of relative stereochemistry around the cyclopropane ring on the acid-catalysed ring-opening reaction.[555] Products of 1,2-, 1,3-, and 1,5- hydride shifts and of a 1,2-methyl migration were seen to result from the acid-catalysed isomerization reaction, and a correlation was noted between some of the reaction types and the relative stereochemistry around the cyclopropane ring. Investigation of the thermal rearrangement of cycloprop[*c*]isoquinolines (**316**) has revealed two distinct reaction paths.[556] In cases where either R^1 or R^2 = H, heating causes the interconversion of the *endo*-and *exo*-isomers, (**316**; R^1 = H) and (**317**; R^1 = H) and the eventual formation of the 1*H*-2-benzazepine system (**318**) in high yield. When neither R^1 nor R^2 was hydrogen, the rearrangement followed an alternative path to give 5*H*-2-benzazepines (**319**). Interestingly, cycloheptatriene and its derivatives (**320**), in sulphur dioxide as solvent, are quantitatively converted into the corresponding benzenemethanesulphinic acids (**321**).[557] The mechanisms depicted in Scheme 41, although not proven, are both consistent with the experimental data.

Tricyclic compounds such as (**322**), which incorporate cyclopropylsilyl ether moieties, have been observed[558] to rearrange to *cis*-decalones (**323**) on acid hydrolysis, while the rearrangement of α-phenyl-1,4-dilithio-*N*-phenylamide (**324**) to the more stable β-phenyl-1,4-dilithio derivative (**326**) is considered to proceed through a transient dilithiocyclopropane intermediate (**325**).[559]

Chemiselective addition of an electrophile to the double bond of vinylcyclopropanols (**327**) is considered to initiate ring-expansion to the cyclobutanone (**328**) in a new α-substitution–spiroannulation procedure,[560] while a recent account[561] of the thermal rearrangement of vinylcyclopropane derivatives has emphasized the importance of this rearrangement for the annellation of functionalized cyclopentenes. The vinylcyclopropane-to-cyclopentene rearrangement has been used by Salaün *et al.* on silyl ethers (**329**) to give masked ketones (**330**).[562] The reaction appears to work well when spiro-compounds (**331**) are the products, and spirovetivanes have been made this way.[563] A new synthesis of (±)-grandisol from (+)-Δ^2-carene, involving the photoinduced vinylcyclopropane rearrangement as the key step, has been described,[564] and the potential of the divinylcyclopropane rearrangement of substances such as (**332**), for the preparation of C(8)-functionalized bicyclo[3.2.1]octa-2,6-dienes (**333**) has been outlined.[565] Evidence has been pro-

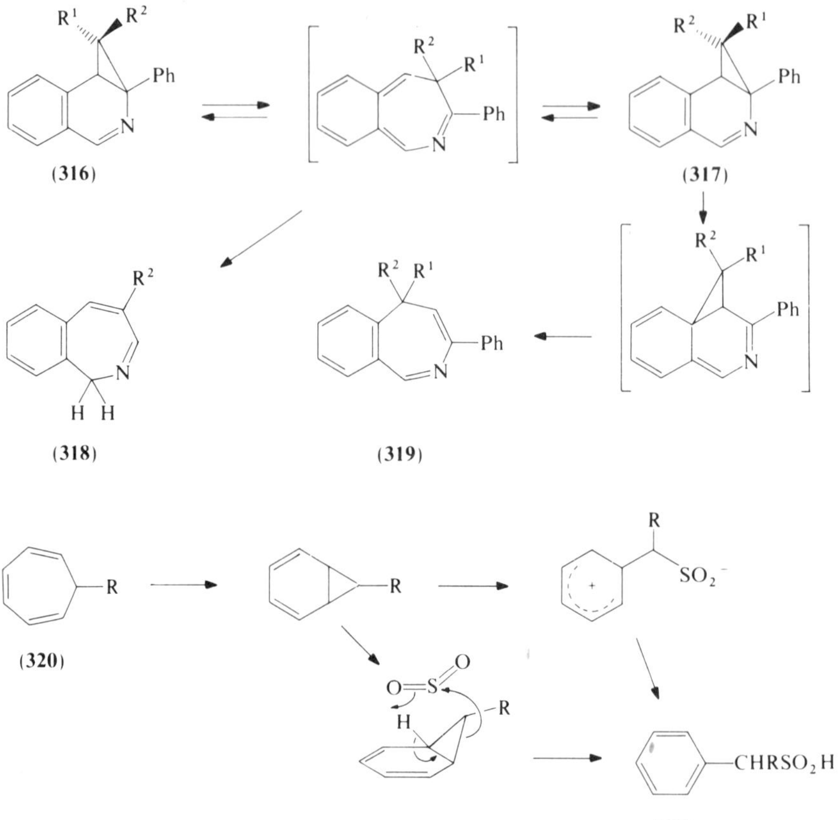

(316) (317) (318) (319) (320) (321)

SCHEME 41

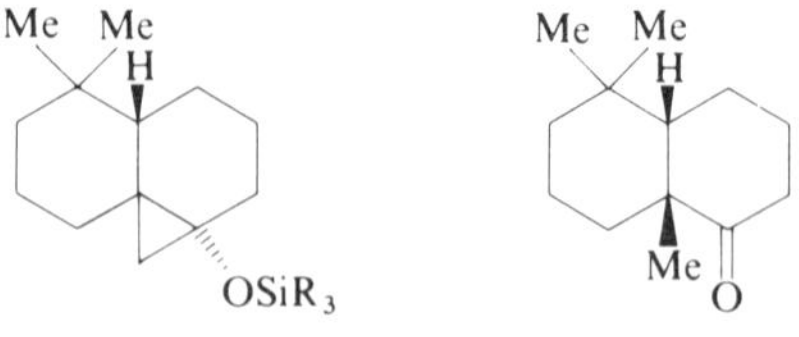

(322) (323)

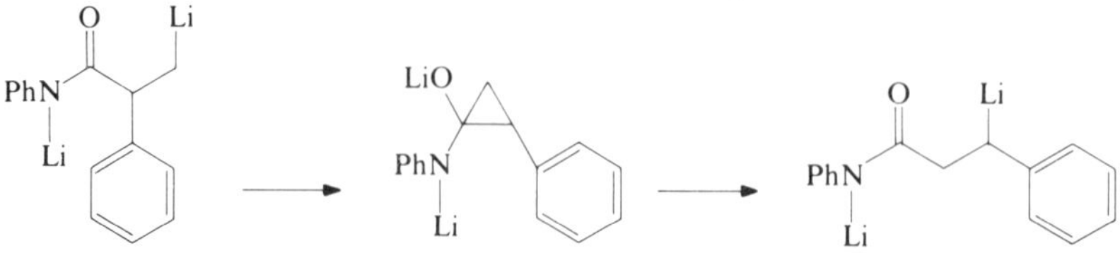

(324) (325) (326)

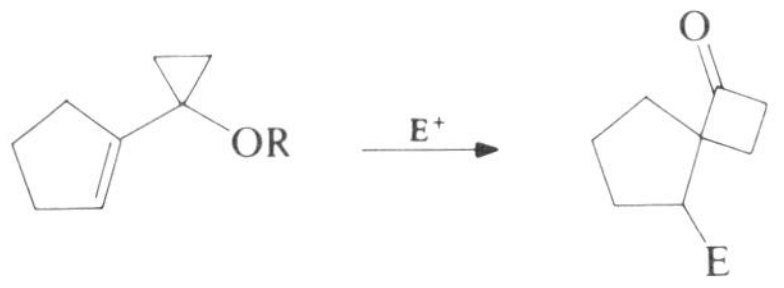

(327) (328)

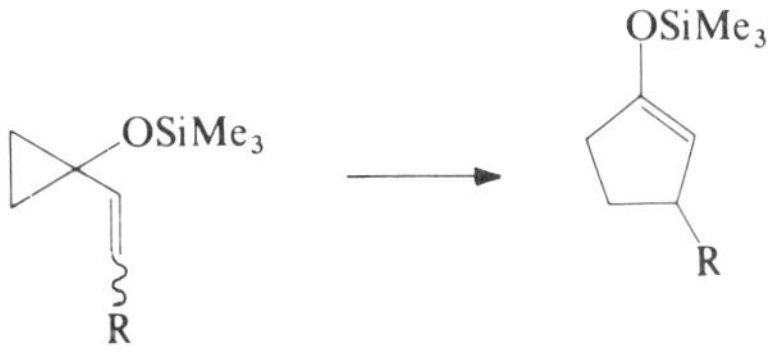

(329) (330) 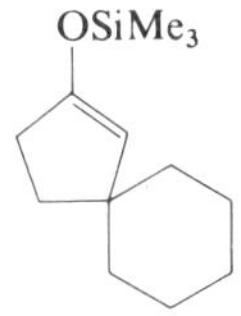 (331)

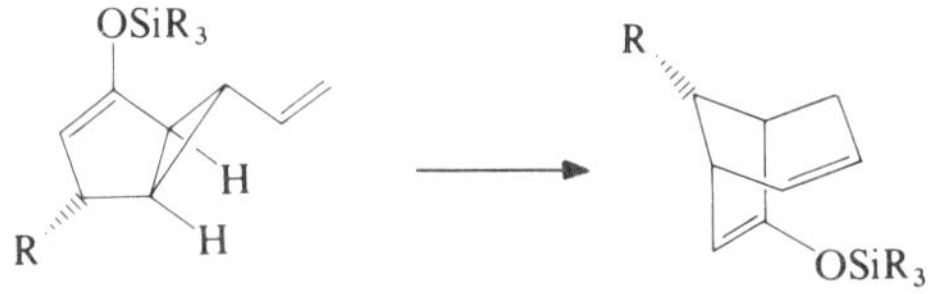

(332) (333)

duced[566] from stereochemistry and kinetic deuterium isotope effects to indicate that the thermal [1,3]-rearrangement of (−)-(R,R)-*trans*-2-methyl-1-(1-*tert*-butylvinyl)cyclopropane is not a concerted process. A face-to-face 1,3-biradical resulting from stretching of the C(1)—C(2)bond, and its closure with least-motion control, has been invoked to rationalize the stereochemistry of the process. Tetraaryl-[567] and alkylaryldihydro-semibullvalene[568] derivatives have been shown to exhibit vinylcyclopropane-type rearrangements. The thermolysis of 4-(methyldiphenylcyclopropenyl)but-1-enes has been found to yield substituted tricyclo-[3.2.0.0^{2,7}]heptanes in good yield. The facility of this reaction has been associated[569] with the considerable relief in bond-angle strain of the cyclopropene ring. A novel cyclization reaction of alkylthiodiphenylcyclopropenium ions with acyclic 1,3-diketones to give cyclopentadienols has been discovered.[570] The mechanism for the formation of (**335**) is not clear, but most probably involves the formation of intermediate corresponding to (**334**), ring-expansion, and proton movement as shown in Scheme 42. The thermal conversion of a 3-cyclopropenyl-substituted 2*H*-azirine into a pyridine in high yield has been rationalized[571] by a mechanism involving formation of a nitrile ylide intermediate, followed by intramolecular dipolar cycloaddition to give an azabenzvalene, which subsequently rearranges to the pyridine(see Scheme 43). The thermal chemistry of a series of cyclopropenyl-substituted oxazolinones has also been investigated. These substances also yield a nitrile ylide intermediate adjacent to the cyclopropene ring, and similarly afford pyridines. Flash thermolysis of a potential 1*H*-azirene precursor has

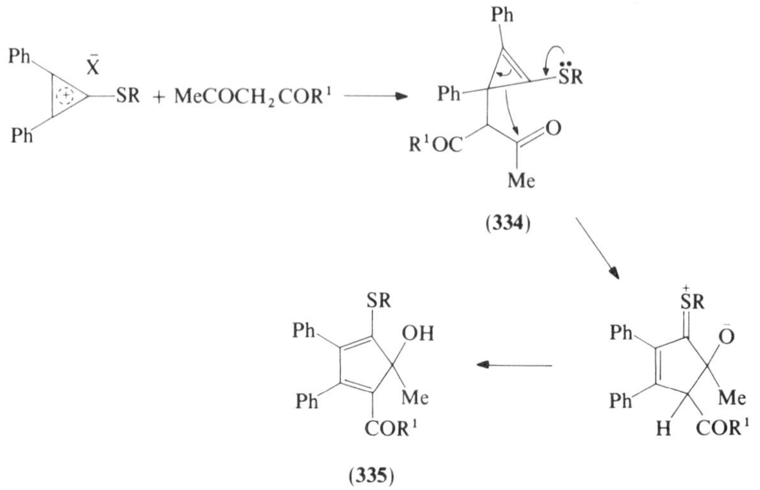

SCHEME 42

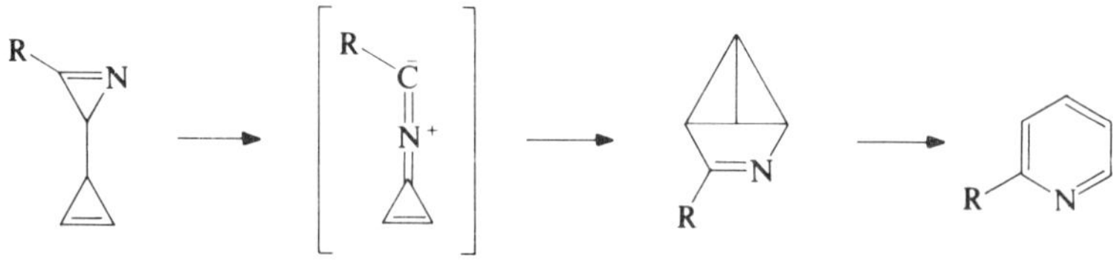

SCHEME 43

(336)

been reported;[572] thus, (**336**) has been observed to isomerize to 2-carboethoxy-2-azabicyclo[3.2.1]octa-3,6-diene which itself decomposes to cyclopentadiene, cyclohexa-1,3-diene, and other products at temperatures of *ca.* 700°. The details of a novel rearrangement involving the formation of a five-membered lactam system on photocyclization of *N*-chloroacetylamines, have been described,[573] and a dual mechanism involving an α-lactam intermediate has been postulated in order to account for the products of the reaction. The formation of thioketones (**339**) from the photolysis of *N*-acylthiobenzamides (**337**) has been reasonably explained[574] in terms of ring-opening of an aziridine (**338**) which is produced by cyclization of (**337**). On the basis of product and kinetic data, the intervention of an aziridinium ion has been postulated[575] in the formation of rearranged urazoles during the cycloaddition of triazolinediones with benzonorbornadienes, norbornenes, and related bicycloalkenes.

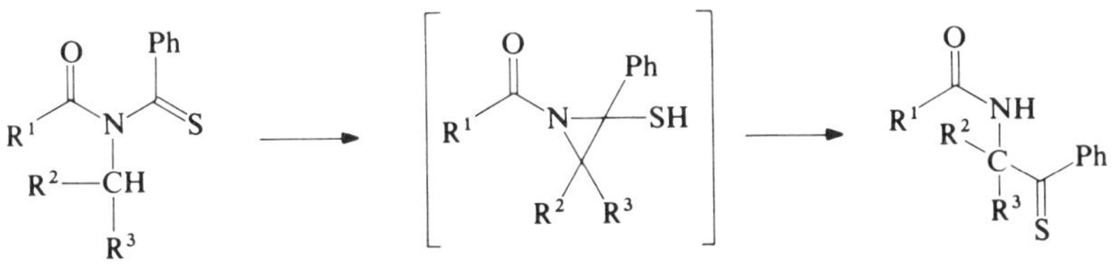

(337) (338) (339)

The base-promoted isomerizations of epoxides have been reviewed[576], an account of the rearrangements of epoxides to carbonyl compounds and allylic alcohols has appeared[577], and molecular-orbital calculations have been performed on the alkylidene–cyclopropanone rearrangement.[578] The photolysis of methano-epoxyenone (**340**) and its diastereomer (**341**) has been examined[579] in an attempt to obtain information about the influence of the cyclopropane ring on the formation and reactivity of ylide intermediates. Both systems show the typical behaviour of ^1n, π^*-excited epoxyenones, with no participation of the additional three-membered ring. Upon thermal activation, 1,2-epoxy-(z)-3-hexen-5-ynes have been transformed predominantly into 2-vinylfurans,[580] and the thermal rearrangement of *trans*-2,3-epoxy-1,3-diphenylbutan-1-one has been found to proceed without loss of optical activity and with inversion of configuration at the migration terminus.[581] A new formal total synthesis of (±)-modhephene, employing a chelation-controlled regioselective epoxide–carbonyl rearrangement [see (**342**) → (**343**)] as the key step, has been described.[582] Epoxyisocyanates and hence 3- and 4-oxazolin-2-ones have been prepared by the thermal rearrangement of α,β-epoxyacyl azides,[583] and an unusual acid-mediated rearrangement of *o*-nitrostyrene oxide to 1-(hydroxy-methyl)-2,1-benzisoxazol-3(1*H*)-one has been uncovered[584] during a recent attempt at synthesizing the known photolabile protecting group, (*o*-nitrophenyl)-ethylene glycol. A transient cyclic epoxyallene has been invoked[585] to account for

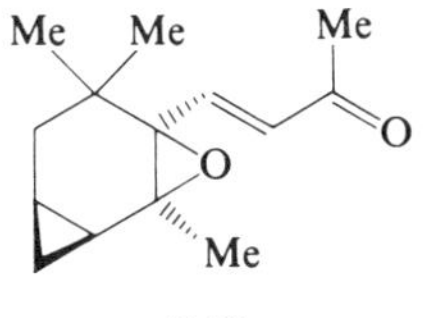

(340) 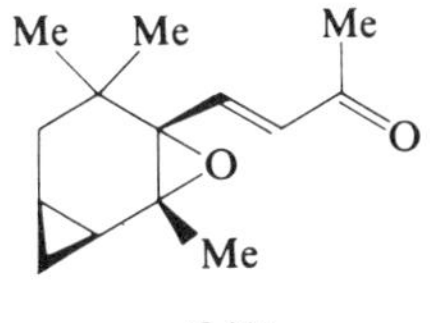 (341)

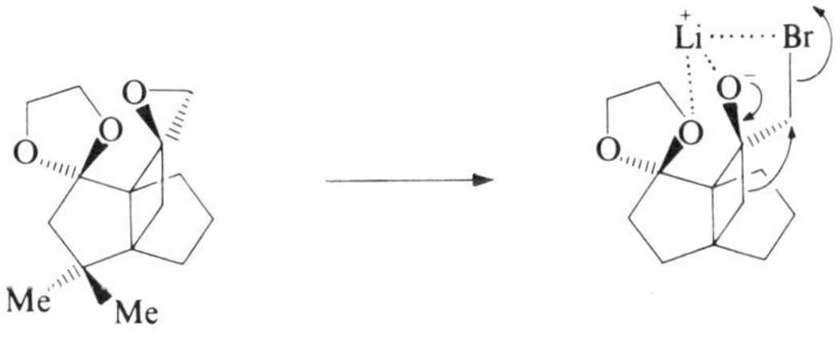

(342) (343)

the products of isomerization of diepoxycyclooctanes by *N*-lithioethylenediamine. The photochemical isomerizations of tetrahydronaphthalenic and anthracenic 1,4-endoperoxides into diepoxides have been further investigated,[586] and carbonyl oxides in solution have been shown to rearrange to esters, probably *via* diradical intermediates.[587]

The major products of photolysis of sterodial nitronate salts have been identified as hydroxamic acids, ketones, and alkenes, which are presumed to arise from a common *N*-hydroxyoxaziridine intermediate. [588] A kinetic study of the thermal isomerization of oxaziridines derived from *para*-substituted *C*-arylaldimines and *C*-diarylketimines to the isomeric nitrones has shown that the rearrangement is facilitated by electron-donating substituents on the phenyl ring.[589] *N*-acyl-substituted 2-azaadamantanes have been prepared by the photoinduced ring-contraction of epoxy-4-azahomoadamantanes. The conversion has been rationalized[590] in terms of stereoelectronic control theory; in other words, the geometrical arrangement of the nitrogen lone pair and the skeletal C(5)—C(6) bond in an *anti*-relation in the 4-azahomoadamantane is favourable for the rearrangement.

Finally, an intramolecular bromonium ion to thiiranium ion rearrangement has been described.[591]

Four-membered Rings

Selective cyclobutane formation has been observed in competition with the Diels–Alder mode in the cation radical cycloaddition of electron-rich alkenes to certain conjugated dienes. Moreover, efficient transformation of the resulting vinylcyclobutanes by an "anionic oxy" rearrangement has provided a further option of convenient, indirect access to the formal Diels–Alder adducts.[592] A novel acid-catalysed skeletal rearrangement of 7-oxy-7-vinyl derivatives of 2-azabicyclo[3.2.0]heptane-3,4-dione to 2-azatricyclo[4.3.0.0^{4,9}]nonane-3,7-diones has been described. The formation of the products has been explained in terms of intramolecular Prins-type cyclization with concomitant 1,2-shift.[593] Irradiation of 7-cyano-2,3,-benzobicyclo[4.2.0]octa-2,4-dien-6-ol in the presence of HgO–I_2 has been shown to yield 7-cyano-2,3-benzobicyclo[5.1.0]octa-2,4-dien-6-one,[594] and the base-catalysed ring-expansion of 7-chloro-2-oxabicyclo[4.2.0]oct-4-en-3-ones to 2*H*-oxocin-2-ones has been reported.[595] The formation of this new oxacyclooctatrienone system can be illustrated as in Scheme 44. At low temperatures, benzvalene (**344**) and sulphur dioxide have been found to react to give the four-

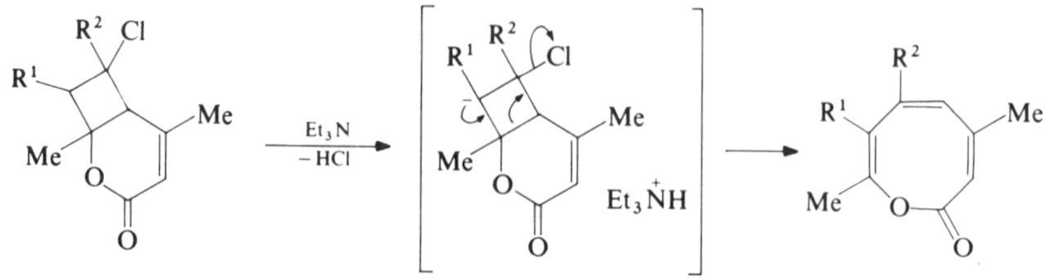

R^2 Cl R^1 Me Me O O —Et_3N, −HCl→ [R^1 R^2 Cl Me Me O O $Et_3\overset{+}{N}H$] → R^2 R^1 Me Me O O

SCHEME 44

membered sulphone (**345**) which rearranges by way of the zwitterion (**346**) to yield the sultine (**347**),[596] while cyclobutadiene (**348**; R = Bu^t) has been observed to react with tetracyanoethylene to yield the novel equilibrating system, 5,5,6,6-tetracyano-5,6-dihydro(Dewar)benzene (**349**)/3,3,4,4-tetracyano-3,4-dihydrobenzvalene (**350**).[597]

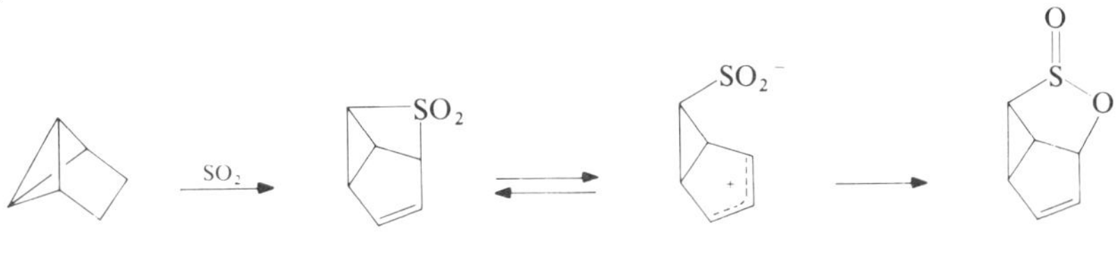

(344) (345) (346) (347)

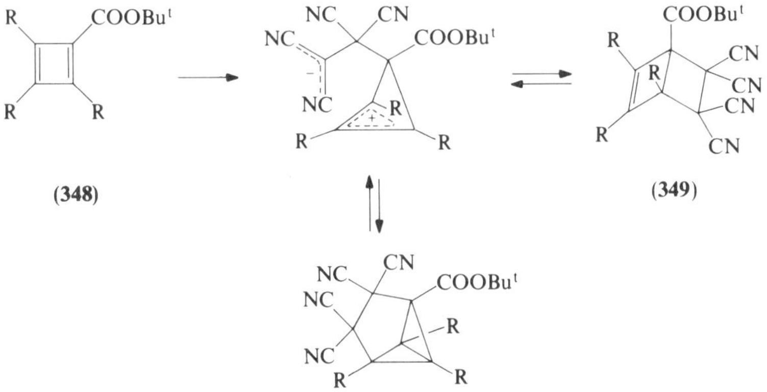

(348) (349) (350)

The photochemical conversion of 7,7-diphenyl-2,3-epoxybicyclo[3.2.0]heptan-6-one (**351**) into 2-oxabicyclo[2.1.1]hexan-3-one (**353**) has been reported.[598] The process is considered to proceed through a retro-[2 + 2]-reaction leading to the relatively stable alkenylketene (**352**). Scission of the allylic C—O bond followed by rapid addition across the cumulene moiety would then lead to the observed product. An interesting intramolecular transfer of an acetal alkoxy group has been observed[599] during the alkaline hydrolysis, and in the $LiAlH_4$ reduction, of α-methyl-α-[(1-*tert*-butoxy-2-methyl-2,3-epoxypropyl)oxy]-β-propiolactone. The mechanism outlined[600] in Scheme 45 reasonably accounts for the addition–rearrangement of 3,4-dihydro-2-methoxy-2*H*-pyran (**354**) with (4-methylphenyl)-sulphonyl isocyanate. A number of $BF_3 \cdot OEt_2$-catalyzed rearrangements of diazopenicillanate derivatives which appear to proceed *via* penam C(5)—C(6) bond-cleavage have been described,[601,602] while 2-vinyl-1,3-thiazetidines such as (**355**) have been found to undergo a novel and high-yielding rearrangement on hydrogenation with heterogeneous catalysts to give thiazolidines (**356**). Reaction with the homogeneous catalyst, $Rh(Ph_3P)_3Cl$, results in an alternative high-yielding

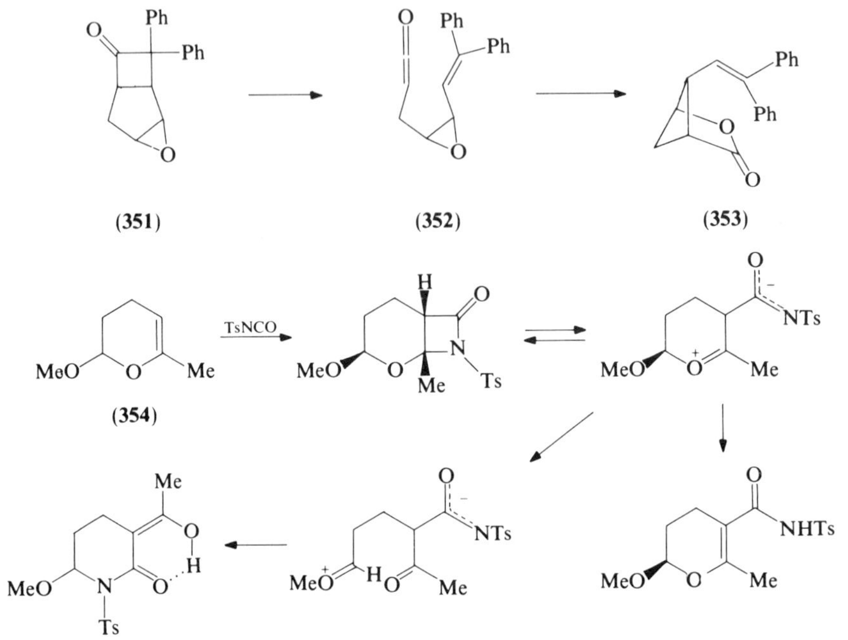

(351) (352) (353)

(354)

SCHEME 45

rearrangement to afford the thiazine (**357**).[603] Finally, ^{13}C-labelling studies and cross-over experiments have indicated[604] that the cyclization of 2-cyano-3-mercapto-3-(methylthio)acrylamide (**358**) with benzoic acid in the presence of polyphosphate ester to give (**360**) probably proceeds by way of a thiallylic rearrangement of the initial ring-closure product (**359**) (see Scheme 46).

Five-membered Rings

A surprisingly high degree of retention of chirality has been observed[605] during the thermal automerization of methyl 1,2-diphenylcyclopentane-1-carboxylate. Derivatives of dihydromethylenomycin have been found to undergo a facile cleavage of the five-membered ring ketone and subsequent reclosure to give 3(2*H*)-furanone derivatives under weakly basic conditions,[606] while novel rearrangements of 5,5-disubstituted 4-methylenefuran-2,3-diones and indeed, the principle of the furandione–furandione isomerization, have been discussed in some detail.[607] Chloroindolenines of tetrahydro-*β*-carbolines which have a 4′-(2′-carbomethoxy)-butyrate chain at position 1 have been rearranged under basic conditions into *α*-methyleneindolines (**363**). Although several pathways may explain this transformation, the authors have favoured[608] a mechanism in which the first-formed malonate anion (**361**) intramolecularly attacks C(2) of the indole. *C*-Ring contraction, with chloride expulsion, is followed by decarbalkoxylation of the intermediate (**362**), which occurs under extremely mild conditions because of the triactivated nature of the ester-bearing carbon.

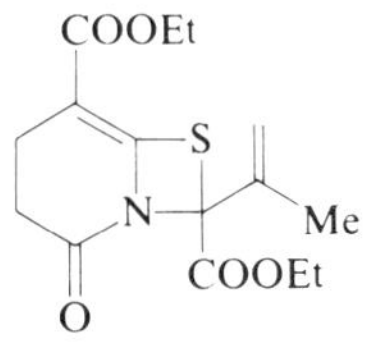

(355) 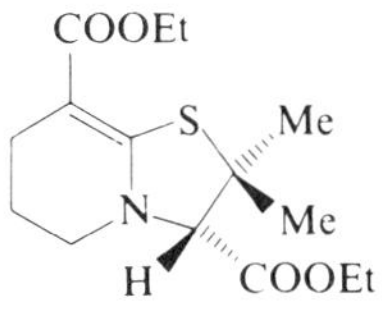 (356) 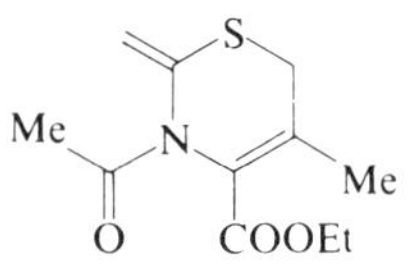 (357)

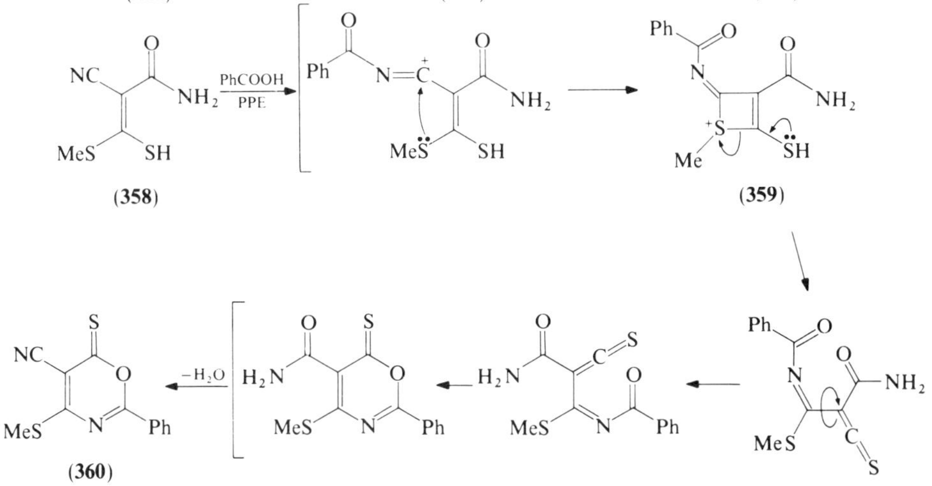

(358) (359)

(360)

SCHEME 46

(361)

E = COOEt

(363) (362)

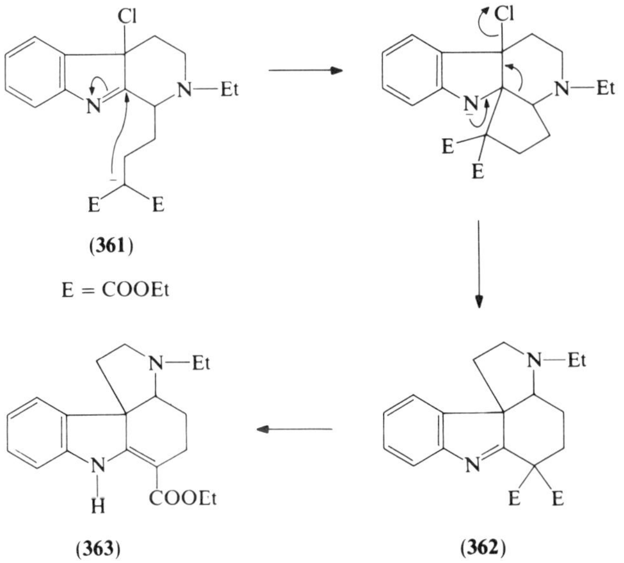

Several novel rearrangements of the reactive intermediates produced on intramolecular 1,3-dipolar cycloaddition of alkylazide-enones have been encountered,[609] and cyclic 2-(acylmethyl)-1,3-diketones (**364**) have been converted into 1-acyl-1*H*-pyrroles (**366**) by ammonium acetate in acetic acid, *via* rearrangement of a hydroxypyrroline intermediate (**365**). It is suspected[610] that a key reaction in the process is a transannular interaction between an amide nitrogen atom and a carbonyl group in a medium ring (see Scheme 47). The facile conversion of 3-

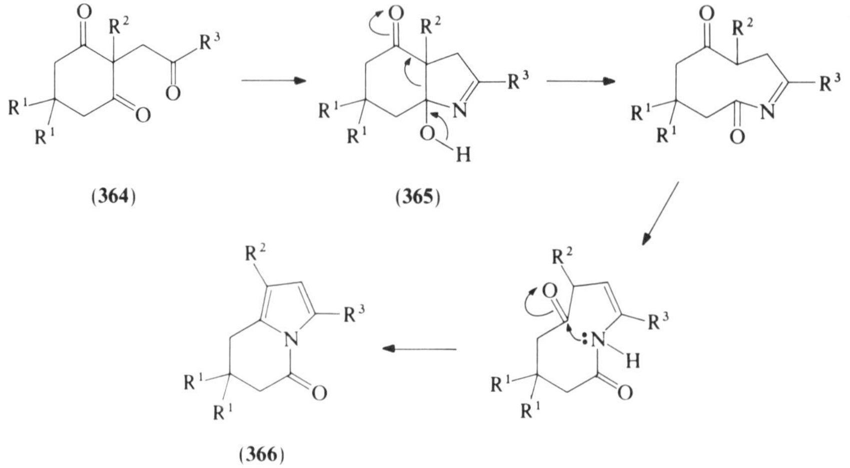

(**364**) (**365**) (**366**)

SCHEME 47

unsubstituted anthranilium salts (**367**) into *N*-alkylbenzoazetinones (**368**) on treatment with triethylamine, has been described,[611] and indirect evidence has been presented[612] for the existence of the ring-chain tautomer equilibrium (**367**) $\rightleftharpoons$ (**368**). Ionized *N*-aryl-3,4-diphenylisoxazol-5(2*H*)-one isomers have been found to react unimolecularly in the gas phase mainly by CO_2 elimination.[613] The ring-contraction appears to resemble the reported behaviour of the corresponding neutral compounds in solution.

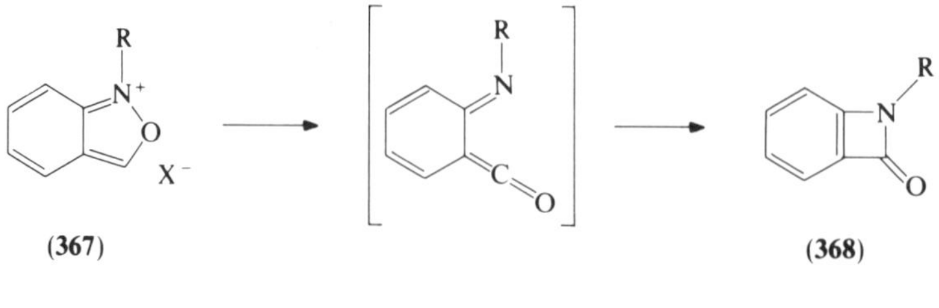

(**367**) (**368**)

An interesting ring-expansion of 2-(tetrahydrofuryl)dimethylhydrosilane to an oxasilacyclohexane has been observed,[614] and a thorough understanding of the mechanism of the solvent-induced rearrangement of 2,3-dioxabicyclo-[2.2.1]heptane has been gleaned from kinetic and product studies with the monodeuterated peroxide. These model studies have inspired and guided parallel studies on prostaglandin endoperoxides, and as a result evidence has been presented to explain the formation of levulinaldehyde derivatives for the solvent-induced fragmention of these intermediates.[615]

Six-membered and Larger Rings

The photoisomerization of 5-hydroxy-6-cholestanones to lactones is considered to involve ketene intermediates formed by migration of the 7α-hydrogen; the stereospecificity of the migration is independent of hydrogen bonding and has been attributed to slowing of the rotation about the C(9)—C(10) bond in the alkyl acyl diradicals that are the ketene precursors.[616] Substituted cyclooctadienones (**370**) have been obtained[617] from the reaction of β-keto-esters (**369**) with dimethyl acetylenedicarboxylate (see Scheme 48), and 2-(2-Δ'-pyrrolinyl)-4-cyclopentene-1,3-diones and pyrrolidino[2,1-*b*]azepine-1,5-diones have been obtained from the

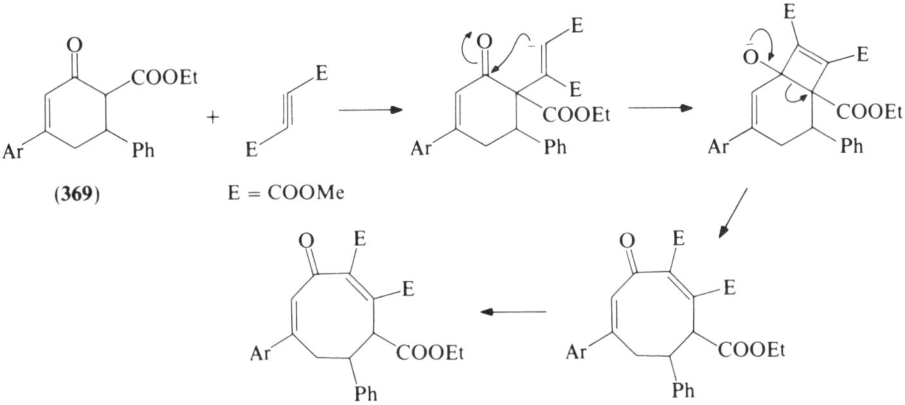

(**369**) E = COOMe

(**370**)

SCHEME 48

rearrangement of 2-(azidopropyl)-1,4-benzo- and-1,4-naphtho-quinones *via* intermediate triazolines.[618] As a consequence of a new structural assignment for pseudoanisatin (**371**), it has now been suggested[619] that its base-catalysed isomerization product (**372**) results from an unusual steric inversion at C(3) and C(4) in

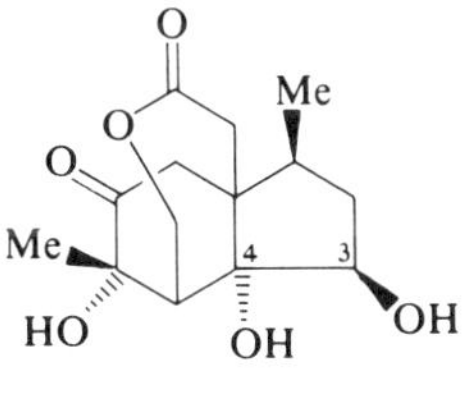

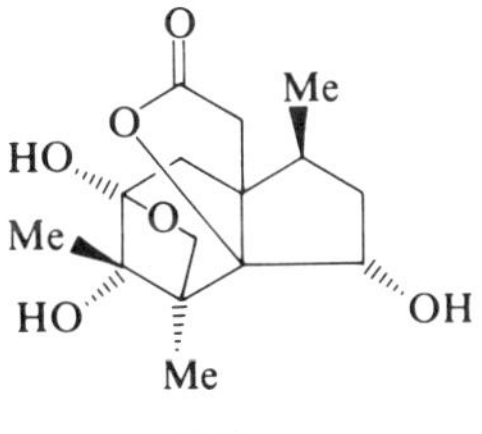

(**371**) (**372**)

addition to trans-lactonization into a stable γ-lactone ring, and acetal formation between C(7) and C(4). The steric inversion at C(4) is of special interest since it involves conversion of a *cis*-function into a *trans*-function. The same authors have also described[620] a photochemical rearrangement of pseudoanisatin to an α-hydroxyketone. A critical re-examination of the acid-catalysed rearrangement product of aucubigenin (**373**) has led to the reassignment of this product as 1,10-anhydro-3,4-dihydro-4-decarbomethoxy-3α-hydroxygardenogenin (**374**). Its formation is rationalized[621] by the pathway in Scheme 49. The stereospecific rearrangement of the ionophore antibiotic X-14547A (indanomycin) and related derivatives has been found to occur readily in acetonitrile solution containing lithium tetrafluoroborate. The product of the reaction has been identified[622] as a novel tetracyclic compound derived by ring-cleavage of the tetrahydropyranyl ring with concomitant double-bond migration and intramolecular cyclization involving a pyrrole group. Extensive rearrangement of 2-bromomercurio-9-oxabicyclo[3.3.1]nonane has been described.[623] A novel Wessely–Moser rearrangement of a 5-mercaptochromone to the corresponding 5-hydroxythiochromone has been reported,[624] and 2-alkyl-3-cyclopentenones have been prepared from tetrahydrothiopyran-4-one by a "one-pot" Ramberg–Bäcklund reaction of 6-alkyl-1,4-dioxathiaspiro[4,5]decane-8,8-dioxides, followed by acid-catalysed dedioxolanation.[625]

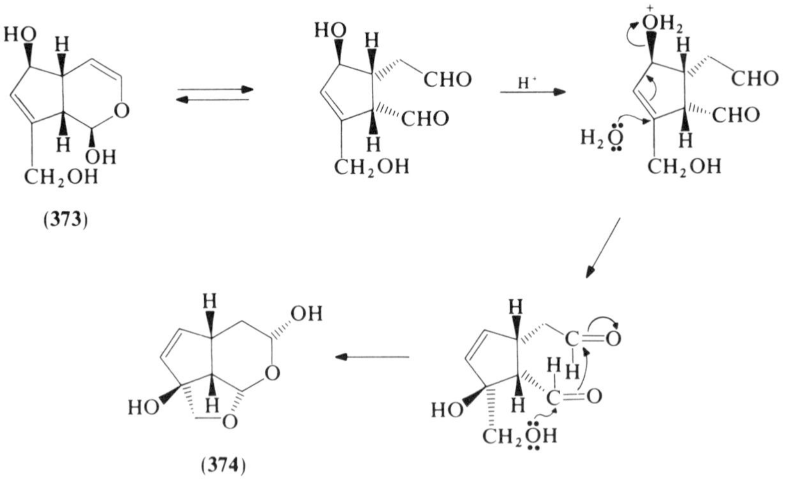

SCHEME 49

The photolysis of 1,4-epidioxy-1,4-dimethyl-9,10-diphenyl-1,2,3,4-tetrahydroanthracene to 2,5-epoxy-2,5-dimethyl-6,11-diphenyl-2,3,4,5-tetrahydronaphtho[2,3-*b*]oxepine and 5*a*,8-dimethyl-9-phenyl-5*a*,6,7,8-tetrahydronaphtho[3,2,1-*k*,1]-xanthene-8-ol has been explained[626] by the various paths of evolution of the diradical arising from the homolyte cleavage of the peroxidic bond and, in particular, by the preferred attack of a *peri*-phenyl substituent which leads to the latter product. Polyfunctional hexahydro-1(2*H*)-pentalenones (**376**) have been prepared[627] by the rearrangement of bicyclic 1,2-oxazine *N*-oxide derivatives (**375**), and several unusual bicyclic systems containing a bridgehead nitrogen have been prepared[628] by the photolysis of 5-(methoxycarbonyl)-2-phenyl-1,2,4-triazolo[1,2-*a*]pyridazine-1,3-dione in various solvents.

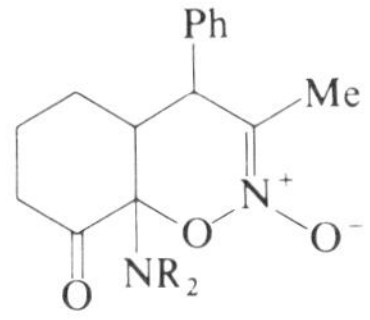

(375) 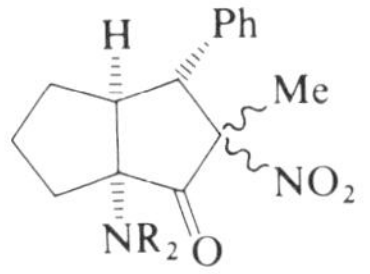 (376)

Isomerizations

A review has appeared on the reactive intermediates involved in the interconversion of C_3H_4 hydrocarbons,[629] and a critical summary of the oxirene literature[630] has included the isomerization reactions of that species. The energetics of the rearrangement of thioformaldehyde to thiohydroxycarbene have been calculated,[631] *ab initio* calculations have been made on the isomerization of diphosphene 1-sulphide to thiadiphosphene,[632] and the first example of dispiro-spiro ansa isomerism in a phosphazene structure has been reported.[633] *N*-Nitroso compounds have been transformed into the previously unknown *C*-nitroso compounds *via* an HNO complex,[634] and nitriles have been obtained from isonitriles in almost quantitative yield, and with almost complete retention of stereochemistry, by a convenient flash-pyrolytic procedure.[635]

An enediolate intermediate has been invoked[636] to explain the acid- and base-catalysed isomerizations of triose phosphates, and it has been noted[637] that protonated 2,5-dimethylfuran and 2-ethylfuran appear to undergo isomerization in the gas phase through common intermediates before cleavage. The reaction pathway is considered to require ring-enlargement and ring-contraction steps. The gas-phase thermal isomerization of hexachlorocyclopropane to hexachloropropene has been interpreted[638] as a unimolecular process which occurs with chlorine-atom migration, and for the first time a reaction has been identified[639] that enables a study of electronic effects on C—Cl bonds in an equilibrium situation. The reaction in question, the $\alpha \rightleftharpoons \alpha'$ rearrangement of α-chlorodibenzyl ketones, has been found to correlate well with the difference in Taft σ° values with $\rho + 0.75$.

A novel method for the introduction of deuterium into long-chain carbon compounds has been described.[640] The procedure is based upon isomerization of a triple bond of an $(n+1)$-alkyn-n-ol to an ω-alkyn-n-ol using *N*-deuterated reagents and solvents. It has been observed[641] that electron-withdrawing substituents in the 3-position of medium-ring cycloalkenones shift the equilibrium towards the 3-cycloalkenone and away from the 2-cycloalkenone, relative to the parent compound. *N*-Benzyl-2-cyano-Δ^3-piperideines have been readily isomerized to 2-cyano-Δ^4-

piperideines on alumina.[642] A quantum-mechanical study of the equilibrium between 1,4- and 1,6-dialkylcyclooctatetraenes has been undertaken,[643] and the isomerizations of 6-cyanotricyclo [5.5.0.0^{2,5}]dodeca-3,6,8,10,12-pentaene have been reported for the first time.[644] The minimum-energy paths of valence isomerization of aza-substituted cyclobutadienes have been calculated,[645] while the cycloaddition of diazoalkanes to diazobicyclo[2.2.0]hexenes and subsequent extrusion of nitrogen has been found to afford diazatricyclo[3.2.0.0^{2,4}]heptanes that are easily valence-isomerized to novel dihydrodiazepines.[646] The laser-induced infrared multiphoton isomerization reactions of 2,4- and 1,3-hexadienes have been described.[647] The stereoselective transformation of 2,4-alkadienoic esters to their 3,5-dienoic isomers has been achieved[648] using LDA, and a unique one-step, double isomerization (2E,4Z $\rightleftarrows$ 2Z,4E) of 6-oxo-2,4-heptadienoic acid has been catalysed by maleylacetone *cis–trans* isomerase.[649] Although simple Schiff-base formation between vitamin A aldehydes and aliphatic amines was found to have only a small effect on the *cis–trans* isomerization of these compounds,[650] protonation of the Schiff bases enhanced the rate of isomerization considerably.[651] The relative thermodynamic stabilities for geometrical isomers of several silyl ketene acetals derived from propanoate esters have been determined.[652] In each case, the more stable isomer is that in which the ester alkoxy group is adjacent to the unsubstituted vinylic site. The observation that increases in the size of the ester alkoxy group lead to diminished equilibrium ratios is of interest from both a theoretical and a practical viewpoint. A study of solvent and substituent effects on the thermal *cis–trans* isomerization of some 4-diethylaminoazobenzenes has supported an inversion mechanism rather than a rotational one for the process,[653] while experimental results and *ab initio* calculations have indicated[654] that *N*-arylketenimines and *N*-arylimines invert their configuration by a coupled mechanism in which a rotation around the *N*-aryl bond is coupled to inversion at nitrogen. The photochemical geometrical isomerization of a $>C{=}\overset{|}{C}{-}\overset{|}{C}{=}N{-}$ system has been studied in 1-methoxyimino-3-phenylprop-2-ene.[655] Spectroscopic evidence has been produced to support the occurrence of the first reported *trans–cis* isomerization of a diphosphene.[656]

Electrochemical oxidation of quadricyclene has resulted in its isomerization to norbornadiene by a redox chain mechanism,[657] while the valence isomerization between coloured acylnorbornadienes and quadricyclanes has been presented as a promising model for visible (solar) light energy conversion.[658] The valence isomerization of phane-bridged *cis*-trioxatris-*o*-homobenzenes has been examined[659] in an attempt to find how much the deformation, forced upon the trioxide six-membered ring by phane-bridging, influences its kinetic stability. No significant effect was noted. It has been suggested[660] that the thermal naphthalene automerization proceeds by a reversible valence isomerization *via* a laterally bridged benzvalene (**377**) (see Scheme 50). It has been shown[661] that the presence of two trifluoromethyl groups in the thiophene ring, provided they are placed in the 2,3-positions, is sufficient to produce Dewar thiophenes on photolysis. Activation parameters for the isomerization of 2,4-di-*tert*-butyl-and 2,3,4,5-tetrachloronorcaradienes to cycloheptatrienes have been determined.[662] The tricyclic annulene 7b-

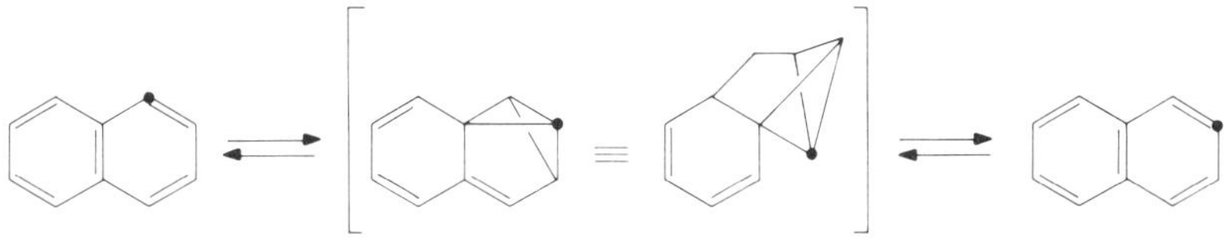

• = ^{13}C label

(377)

SCHEME 50

methyl-7b*H*-cyclopent[*cd*]indene (**378**) has been rearranged to the 2a*H*-isomer (**379**) on heating.[663] The isomerization is considered to arise because of the extra driving force caused by the formation of a benzene ring, despite the fact that the starting material has some aromatic stabilization. The existence of a thermal equilibrium between aromatic *syn*-1,6:8,13-bisoxido[14]annulene and its olefinic *anti*-isomer has been deduced from stereochemical and theoretical studies,[664] while variable temperature ^{13}C-NMR and electronic spectra have indicated that 2,7-methanothia[9]annulene is in equilibrium with its valence isomer 9-thiatricyclo[4.3.1.0^{1,6}]deca-2,4,7-triene.[665]

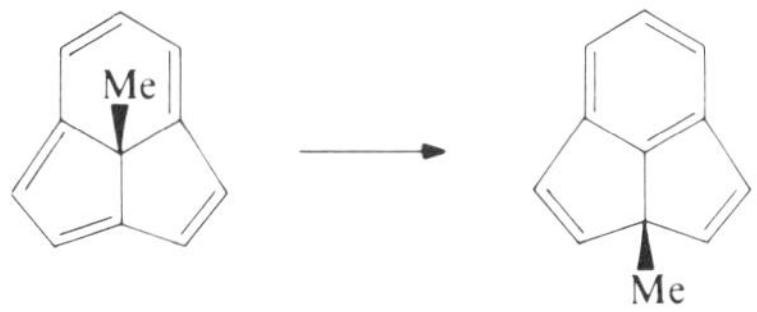

Me

Me

(378) (379)

Tautomerism

The light-induced tautomerism of *β*-dicarbonyl compounds has been reviewed,[666] and an insight into the valence tautomerization (**380**) ⇄ (**381**) has been obtained using dibenzylketene and benzoyl(thiobenzoyl)ketene as examples.[667] The former was found to dimerize, whereas the latter cyclized to the thietone, and an equilibrium existed in the gas phase.

X=Y—ZH systems have been considered in general terms and divided into four classes according to the number of constituent atoms that possess lone-pair electrons.[668]

High-resolution solid-state NMR spectroscopy has been applied to the determination of the ring-chain tautomerism of 2-[2-(methylamino)ethyl]-3-nitrobenzaldoxime,[669] the ring-chain tautomerization of 2-cyano-substituted benzamides and benzenesulphonamides has been studied[670] using IR and NMR spectroscopy, and an examination of structural effects on the ring-chain tautomerism of 1-hydroxyphthalans has been undertaken.[671] The acid–base tautomeric partitioning ratios of *p*- and *m*-aminobenzoic acids have been determined[672] by a novel ^{13}C-NMR method, the benzenoid–quinoid tautomerism of imines of 4-hydroxy-5,6-benzocinnamaldehyde has been examined in various solvents,[673] and

(380) (381)

X = O, S

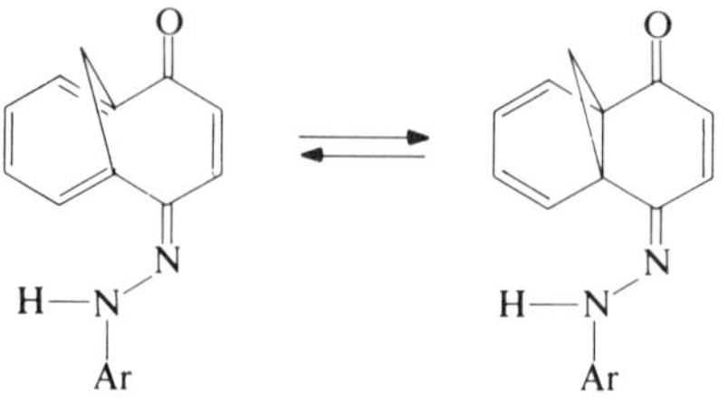

(382)

the tautomerism of thiobenzoylhydrazones of aroylacetones and aroylacetaldehydes has been investigated.[674] Bridged quinone hydrazones of type (**382**) have been observed to exist as tautomeric mixtures of their norcaradiene and cycloheptatriene forms,[675] while a ^{13}C-NMR study of the effect of the π-acceptor ability of aryl substituents on the valence tautomerism of 7-aryl-2,5-di-*tert*-butyl-cyclohepta-1,3,5-trienes has indicated[676] that conjugative interaction between the cyclopropane ring in the norcaradiene structure and the aryl group, increases with increase in the π-acceptor ability of the aryl group. The influence of substituents on the valence tautomeric equilibrium of 4,8-substituted homotropilidenes has been investigated,[677] while MNDO calculations for (3)-, (4)-, and (5)-pericyclynes have indicated[678] that interactions between triple bonds are hyperconjugative in the π-system and homoconjugative in the σ-system. The geometry for (3)-pericyclyne, with linear acetylene units and internal angles of 60°, was also found to be a minimum on the MNDO potential surface, corresponding to tetracyclo[6.1.0.0^{2,4}.0^{5,7}]nona-1,4,7-triene (tricyclopropabenzene), a valence tautomer which has yet to be synthesized.

An *ab initio* study[679] of the tautomerism of uracil, thymine, 5-fluorouracil, and cytosine has shown that the tautomeric equilibria of both uracil and cytosine are sensitive to phase change, and it has been suggested that at least two, and possibly three, tautomers of cytosine may be observed in the gas phase. The amino $\rightleftharpoons$ imino tautomerism of 2-aminothiazoles and their salts has been studied,[680] as have the effects of substituents and solvent on the annular tautomerism of substituted 5-aryltetrazoles.[681]

Addendum

Photochemical Studies

The di-π-methane rearrangement has been reviewed,[682] and the reactivity of aryl vinyl di-π-methane systems has been studied.[683] The photolyses of spirocyclobutanones incorporating a cyclohexa-, cyclohepta-, or cycloocta-diene moiety have been described.[684] On triplet excitation these compounds have been found to undergo isomerization *via* a 1,2-acyl shift involving one or more double bonds of the diene system (see Scheme 51). A SINDO study of the photoisomerization of cyclopentanone has been carried out.[685] The topological course of the photochemical 1,3-acetyl shift in (R)-(+)-cyclopent-2-enyl methyl ketones and the kinetics of the

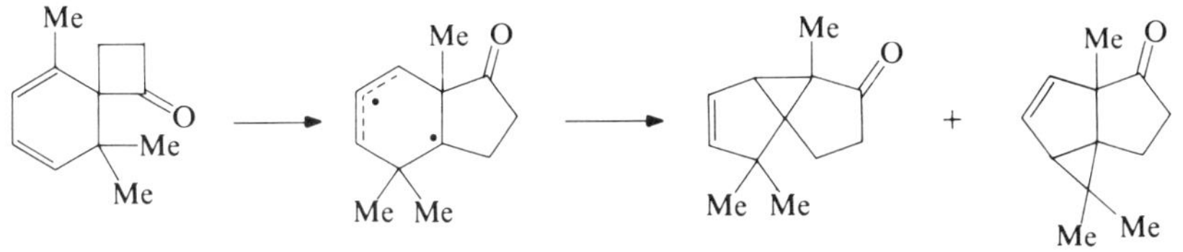

SCHEME 51

process have been studied,[686] and a detailed study of the photolysis of 2,2,7,7-tetramethylcyclohepta-3,5-dien-1-one has indicated[687] that, in such seven-membered 3,5-dien-1-one systems, both cyclization and 1,2-acyl shifts are competitive photoprocesses of the triplet excited states. The adduct (383) of bicyclohept-1-en-1-yl and 2,5-dimethyl-*p*-benzoquinone, on exposure to light, has been found to isomerize to the pentacyclic product (**384**)[688] by a long-range hydrogen migration

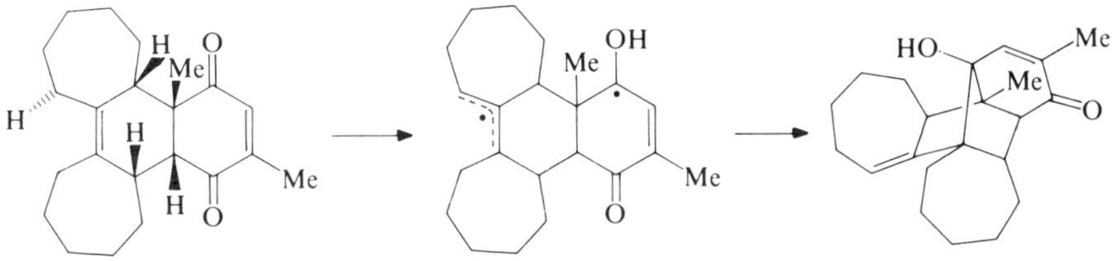

(**383**) (**384**)

similar to that previously observed in analogous systems under electron impact. Finally, the divergent photobehaviour of *exo*- and *endo*- 7-methylbicyclo[4.1.0]hept-2-ene upon direct and toluene-sensitized photolysis has been investigated.[689]

References

1. Ichikawa, H., Nakajima, Y., and Harrison, A. G., *Shitsuryo Bunseki*, **31**, 189 (1983); *Chem. Abs.*, **100**, 208834 (1984).
2. Franz, J. A., Barrows, R. D., and Camaioni, D. H., *Energy Res. Abstr.*, **9**, 14361 (1984); *Chem. Abs.*, **101**, 151154 (1984).
3. Borodkin, G. I., Nagy, S. M., Mamatyuk, V. I., Shakirov, M. M., and Shubin, V. G., *J. Chem. Soc., Chem. Commun.*, **1983**, 1533.
4. Borodkin, G. I., Nagy, S. M., Mamatyuk, V. I., Shakirov, M. M., and Shubin, V. G., *Zh. Org. Khim.*, **20**, 552 (1984); *Chem. Abs.*, **101**, 54236 (1984).
5. Speranza, M., Keheyan, Y., and Angelini, G., *J. Am. Chem. Soc.*, **105**, 6377 (1983).
6. Rappoport, Z., Fiakpui, C. Y., Yu, X. -D., and Lee, C. C., *J. Org. Chem.*, **49**, 570 (1984).

[7] Kitamura, T., Kobayashi, S., Taniguchi, H., Faikpui, C. Y., Lee, C. C., and Rappoport, Z., *J. Org. Chem.*, **49**, 3167 (1984).
[8] Berrier, C., Jacquesy, J. C., Gesson, J. P., and Renoux, A., *Tetrahedron*, **40**, 1983 (1984).
[9] Castaldi, G., Belli, A., Uggeri, F., and Giordano, C., *J. Org. Chem.*, **48**, 4658 (1983).
[10] Uemura, S., Fukuzawa, S. -i., Yamauchi, T., Hattori, K., Mizutaki, S., and Tamaki, K., *J. Chem. Soc.,Chem. Commun.*, **1984**, 426.
[11] Endo, Y., Terashima, T., and Shudo, K., *Tetrahedron Lett.*, **25**, 5537 (1984).
[12] Yamamoto, J., Masuda, Y., Sumida, T., Aimi, H., and Umezu, M., *Tottori Daigaku Kogakubu Kenkyu Hokoku*, **14**, 166 (1983); *Chem. Abs.*, **100**, 191110 (1984).
[13] Buncel, E., Keum, S. R., Cygler, M., Varughese, K. I., and Birnbaum, G. I., *Can. J. Chem.*, **62**, 1628 (1984).
[14] Nikitenkova, L. P., Karakotov, S. D., Strepikheev, Y. A., and Naumova, I. I., *Zh. Obshch. Khim.*, **54**, 1375 (1984); *Chem. Abs.*, **101**, 110052 (1984).
[15] Shevchuk, A. S., Pavelko, N. V., Shein, V. D., and Kurapin, A. V., *Osnov. Organ. Sintez i Neftekhimiya, Yaroslavl*, **1983**, 53; *Chem. Abs.*, **101**, 151508 (1984).
[16] Shine, H. J., Zygmunt, J., Brownawell, M. L., and San Filippo, J., *J. Am. Chem. Soc.*, **106**, 3610 (1984).
[17] Chang, C. D., and Perkins, P. D., *Zeolites*, **3**, 298 (1983).
[18] Kohnstam, G., Petch, W. A., and Williams, D. L. H., *J. Chem. Soc., Perkin Trans. 2*, **1984**, 423.
[19] Sternson, L. A., and Chandrasakar, R., *J. Org. Chem.*, **49**, 4295 (1984).
[20] Gassman, P. G., and Granrud, J. E., *J. Am. Chem. Soc.*, **106**, 1498 (1984).
[21] Gassman, P. G., and Granrud, J. E., *J. Am. Chem. Soc.*, **106**, 2448 (1984).
[22] Anderson, K. K., and Malver, O., *J. Org. Chem.*, **48**, 4803 (1983).
[23] Sekiguchi, S., Hirai, M., and Tomoto, N., *J. Org. Chem.*, **49**, 2378 (1984).
[24] Fitzgerald, L. R., Blakeley, R. L., and Zerner, B., *Chem. Lett.*, **1984**, 29.
[25] Baker, W. R., *J. Org. Chem.*, **48**, 5140 (1983).
[26] Schmidt, D. M., and Bonvicino, G. E., *J. Org. Chem.*, **49**, 1664 (1984).
[27] Feigenbaum, A., Pete, J. -P., and Scholler, D., *J. Org. Chem.*, **49**, 2355 (1984).
[28] Dhawan, B., and Redmore, D., *J. Org. Chem.*, **49**, 4018 (1984).
[29] De Maria, P., Lodi, A., Samori, B., Rustichelli, F., and Torquati, G., *J. Am. Chem. Soc.*, **106**, 653 (1984).
[30] Popkova, I. A., and Kozlov, V. A., *Izv. Vyssh. Uchebn. Zaved., Khim. Khim. Tekhnol.*, **27**, 35 (1984); *Chem. Abs.*, **100**, 174028 (1984).
[31] Khelevin, R. N., *Zh. Obshch. Khim.*, **54**, 747 (1984); *Chem. Abs.*, **101**, 54242 (1984).
[32] Khelevin, R. N., *Zh. Org. Khim.*, **20**, 791 (1984); *Chem. Abs.*, **101**, 110038 (1984).
[33] Abdel-Malik, M. M., and De Mayo, P., *Can. J. Chem.*, **62**, 1275 (1984).
[34] Chênevert, R., and Voyer, N., *Tetrahedron Lett.*, **25**, 5007 (1984).
[35] Pathak, V. P., Saini, T. R., and Khanna, R. N., *Monatsh. Chem.*, **114**, 1269 (1983).
[36] Chorn, T. A., Giles, R. G. F., Green, I. R., Hugo, V. I., Mitchell, P. R. K., and Yorke, S. C., *J. Chem. Soc., Perkin Trans. 1*, **1984**, 1339.
[37] Azarov, A. S., Shenbor, M. I., Zaichenko, N. L., Cherkashin, M. I., and Epshtein, S. A., *Zh. Org. Khim.*, **19**, 1717 (1983); *Chem. Abs.*, **100**, 6026 (1984).
[38] Lown, J. W., and Sondhi, S. M., *J. Org. Chem.*, **49**, 2844 (1984).
[39] Tobe, Y., Ueda, K. -i., Kakiuchi, K., and Odaira, Y., *Chem. Lett.*, **1983**, 1645.
[40] Noble, K. -L., Hopf, H., and Ernst, L., *Chem. Ber.*, **117**, 455 (1984).
[41] Mori, N., and Tachibana, T., *J. Am. Chem. Soc.*, **106**, 6115 (1984).
[42] Scott, L. T., *J. Org. Chem.*, **49**, 3021 (1984).
[43] Scott, L. T., Tsang, T. -H., and Levy, L. A., *Tetrahedron Lett.*, **25**, 1661 (1984).
[44] Barry, M., Brown, R. F. C., Eastwood, F. W., Gunawardana, D. A., and Vogel, C., *Aust. J. Chem.*, **37**, 1643 (1984).
[45] Sagitullin, R. S., *Khim. Geterotsikl. Soedin.*, **1984**, 563; *Chem. Abs.*, **101**, 38280 (1984).
[46] L'abbé, G., *J. Heterocycl. Chem.*, **21**, 627 (1984).
[47] Rivalle, C., Bisagni, E., Rousseau, C., and Chardon, A. L., *Tetrahedron*, **40**, 2457 (1984).
[48] Chapyshev, S. V., and Kartsev, V. G., *Dokl. Akad. Nauk SSSR*, **272**, 125 (1983); *Chem. Abs.*, **100**, 138353 (1984).
[49] Katritzky, A. R., and Rubio, O., *J. Org. Chem.*, **49**, 448 (1984).
[50] Katritzky, A. R., De Rosa, M., and Grzeskowiak, N. E., *J. Chem. Soc., Perkin Trans. 2*, **1984**, 841.
[51] Katritzky, A. R., Mokrosz, J. L., and De Rosa, M., *J. Chem. Soc., Perkin Trans. 2*, **1984**, 849.
[52] Katritzky, A. R., Yang, Y. -K., Gabrielsen, B., and Marquet, J., *J. Chem. Soc., Perkin Trans. 2*, **1984**, 857.

[53] Katritzky, A. R., and Leahy, D. E., *J. Chem. Soc., Perkin Trans. 2,* **1984,** 867.
[54] Katritzky, A. R., Mokrosz, J. L., and Lopez-Rodriguez, M. L., *J. Chem. Soc., Perkin Trans. 2,* **1984,** 875.
[55] Schweiger, K., *Monatsh. Chem.,* **114,** 581 (1983).
[56] Higashino, T., Matsushita, Y., Takemoto, M., and Hayashi, E., *Chem. Pharm. Bull.,* **31,** 3951 (1983).
[57] Barluenga, J., Tomas, M., and Gotor, V., *J. Chem. Res. (S),* **1984,** 154.
[58] Brown, G. R., Foubister, A. J., and Wright, B., *J. Chem. Soc., Chem. Commun.,* **1984,** 1373.
[59] Yogo, M., Hirota, K., and Maki, Y., *J. Chem. Soc., Chem. Commun.,* **1984,** 332.
[60] Hashimoto, Y., Ishizaki, T., Shudo, K., and Okamoto, T., *Chem. Pharm. Bull.,* **31,** 3891 (1983).
[61] Kaji, K., Nagashima, H., and Oda, H., *Chem. Pharm. Bull.,* **32,** 1423 (1984).
[62] Kollenz, G., and Seidler, P., *Monatsh. Chem.,* **115,** 623 (1984).
[63] Taylor, E. C., Davies, H. M. L., and Lavell, W. T., *J. Org. Chem.,* **49,** 2204 (1984).
[64] Rao, C. B., Raju, P. V. N., Flammany, R., Maquestian, A., and Elguero, J., *Heterocycles,* **20,** 2385 (1983).
[65] Verschave, P., Vekemans, J., and Hoornaert, G., *Tetrahedron,* **40,** 2395 (1984).
[66] Fišera, L., L'Audár, S., Timpe, H. -J., Zálupský, P., and Štibrányi, L., *Collect. Czech. Chem. Commun.,* **49,** 1193 (1984).
[67] Unterhalt, B., and Brunisch, F., *Arch. Pharm. (Weinheim, Ger.),* **317,** 807 (1984).
[68] Yoda, R., Yamamoto, Y., and Matsushima, Y., *Chem. Pharm. Bull.,* **32,** 2224 (1984).
[69] Potts, K. T., Bordeaux, K., Kuehnling, W., and Salsbury, R., *J. Chem. Soc., Chem.Commun.,***1984,** 213.
[70] Ueda, N. Shimizu, H., Kataoka, T., and Hori, M., *Tetrahedron Lett.,* **25,** 757 (1984).
[71] Shimizu, H., Ueda, N., Kataoka, T., and Hori, M., *Heterocycles,* **22,** 1025 (1984).
[72] Argo, C. B., Robertson, I. R., and Sharp, J. T., *J. Chem. Soc., Perkin Trans. 1,* **1984,** 2611.
[73] Streith, J., Frost, J. R., Strub, H., and Goursot, P., *Nouv. J. Chim.,* **8,** 223 (1984).
[74] Uriac, P., Bonnic, J., and Huet, J., *J. Chem. Res. (S),* **1984,** 142.
[75] Davidson, J. S., *Monatsh. Chem.,* **115,** 565 (1984).
[76] Noto, R., Buccheri, F., and Werber, G., *J. Chem. Soc., Perkin Trans. 2,* **1984,** 537.
[77] Frenna, V., Vivona, N., Consiglio, G., and Spinelli, D., *J. Chem. Soc., Perkin Trans. 2,* **1984,** 541.
[78] Frenna, V., Vivona, N., Caronia, A., Consiglio, G., and Spinelli, D., *J. Chem. Soc., Perkin Trans. 2,* **1984,** 785.
[79] Yamamoto, Y., and Akiba, K. -y., *J. Am. Chem. Soc.,* **106,** 2713 (1984).
[80] Yamamoto, Y., and Akiba, K. -y., *J. Am. Chem. Soc.,* **106,** 2713 (1984).
[81] Ohkata, K., Ohyama, Y., Watanabe, Y., and Akiba, K. -y., *Tetrahedron Lett.,* **25,** 4561 (1984).
[82] Kristinsson, H., Winkler, T., and Mollenkopf, M., *Helv. Chim. Acta,* **66,** 2714 (1983).
[83] Evans, D. M., Taylor, D. R., and Myers, M., *J. Chem. Soc., Chem. Commun.,* **1984,** 1444.
[84] Nakayama, J., Sakai, A., Tokita, S., and Hoshino, M., *Heterocycles,* **22,** 27 (1984).
[85] Robba, M., and Lancelot, J. C., *J. Heterocycl. Chem.,* **21,** 91 (1984).
[86] Jovanovic, M. V., *Tetrahedron Lett.,* **25,** 1677 (1984).
[87] Horspool, W. M., *Photochemistry,* **14,** 207 (1983); *Chem. Abs.,* **100,** 50671 (1984).
[88] Schultz, A. G., Myong, S. O., and Puig, S., *Tetrahedron Lett.,* **25,** 1011 (1984).
[89] Hagenbruch, B., and Hünig, S., *Chem. Ber.,* **116,** 3884 (1983).
[90] Cross, G. G., Fischer, A., Henderson, G. N., and Smyth, T. A., *Can. J. Chem.,* **62,** 1446 (1984).
[91] Kasturi, T. R., Reddy, G. M., Raju, G. J., and Sivaramakrishnan, R., *J. Chem. Soc., Chem. Commun.,* **1984,** 677.
[92] Kasturi, T. R., Prasad, K. B. G., Raju, G. J., and Ramamurthy, V., *Tetrahedron Lett.,* **25,** 2253 (1984).
[93] Smith, S., Elango, V., and Shamma, M., *J. Org. Chem.,* **49,** 581 (1984).
[94] Summermatter, W., and Heimgartner, H., *Helv. Chim. Acta,* **67,** 1298 (1984).
[95] Gill, G. B., *Annu. Rep. Prog. Chem., Sect. B,* **79,** 35 (1983); *Chem. Abs.,* **100,** 173902 (1984).
[96] Burnier, J. S., and Jorgensen, W. L., *J. Org. Chem.,* **49,** 3001 (1984).
[97] Dewar, M. J. S., *J. Am. Chem. Soc.,* **106,** 209 (1984).
[98] Ponec, R., *Collect. Czech. Chem. Commun.,* **49,** 455 (1984).
[99] Jensen, A., and Kunz, H., *Theor. Chim. Acta,* **65,** 33 (1984); *Chem. Abs.,* **101,** 109877 (1984).
[100] Dolbier, W. R., Koroniak, H., Burton, D. J., Bailey, A. R., Shaw, G. S., and Hansen, S. W., *J. Am. Chem. Soc.,* **106,** 1871 (1984).
[101] Breulet, J., and Schaefer, H. F., *J. Am. Chem. Soc.,* **106,** 1221 (1984).
[102] Dass, C., Sack, T. M., and Gross, M. L., *J. Am. Chem. Soc.,* **106,** 5780 (1984).
[103] Kanao, Y., and Oda, M., *Bull. Chem. Soc. Jpn.,* **57,** 615 (1984).

[104] Marvell, E. N., Hilton, C., and Cleary, M., *J. Org. Chem.*, **48,** 4272 (1983).
[105] Oren, J., Schleifer, L., Shmueli, U., and Fuchs, B., *Tetrahedron Lett.*, **25,** 981 (1984).
[106] Lamberts, J. J. M., and Laarhoven, W. H., *J. Am. Chem. Soc.*, **106,** 1736 (1984).
[107] Gerdes, J. M., and Okamura, W. H., *J. Org. Chem.*, **48,** 4030 (1983).
[108] Paquette, L. A., Jendralla, H., Jelich, K., Korp, J. D., and Bernal, I., *J. Am. Chem. Soc.*, **106,** 433 (1984).
[109] Beck, A., Knothe, L., Hunkler, D., and Prinzbach, H., *Tetrahedron Lett.*, **25,** 1785 (1984).
[110] Wirz, J., Persy, G., Rommel, E., Murata, I., and Nakasuji, K., *Helv. Chim. Acta,* **67,** 305 (1984).
[111] Sakurai, H., Eriyama, Y., Hosomi, A., Nakadaira, Y., and Kabuto, C., *Chem. Lett.*, **1984,** 595.
[112] Ziegler, F. E., and Mikami, K., *Tetrahedron Lett.*, **25,** 127 (1984).
[113] Thomas, A. F., and Lander-Schouwey, M., *Helv. Chim. Acta,* **67,** 191 (1984).
[114] Daub, J. P., and Berson, J. A., *Tetrahedron Lett.*, **25,** 4463 (1984).
[115] Schweizer, E. E., Hsueh, W., Rheingold, A. L., and Durney, R. L., *J. Org. Chem.*, **48,** 3889 (1983).
[116] Schweizer, E. E., and Lee, K. -J., *J. Org. Chem.*, **49,** 1959 (1984).
[117] Schweizer, E. E., and Lee, K. -J., *J. Org. Chem.*, **49,** 1964 (1984).
[118] Miller, T. K., and Sharp, J. T., *J. Chem. Soc., Perkin Trans. 1,* **1984,** 223.
[119] Miller, T. K., Sharp, J. T., Sood, H. R., and Stefaniuk, E., *J. Chem. Soc., Perkin Trans. 2,* **1984,** 823.
[120] Munro, D. P., and Sharp, J. T., *J. Chem. Soc., Perkin Trans. 1,* **1984,** 849.
[121] Robertson, I. R., and Sharp, J. T., *Tetrahedron,* **40,** 3113 (1984).
[122] Grigg, R., Gunaratne, H. Q. N., and Kemp, J., *Tetrahedron Lett.*, **25,** 99 (1984).
[123] Mackenzie, K., Proctor, G. J., and Woodnutt, D. J., *Tetrahedron Lett.*, **25,** 977 (1984).
[124] Grigg, R., and Gunaratne, H. Q. N., *J. Chem. Soc., Chem. Commun.*, **1984,** 661.
[125] Viola, A., and Locke, J. S., *J. Chem. Soc.,Chem.Commun.*, **1984,** 1429.
[126] Oppolzer, W., and Mirza, S., *Helv. Chim. Acta,* **67,** 730 (1984).
[127] Hill, R. K., *Asymmetric Synth.*, **3,** 503 (1984); *Chem. Abs.*, **101,** 37834 (1984).
[128] Lutz, R. P., *Chem. Rev.*, **84,** 205 (1984).
[129] Overman, L. E., *Angew. Chem. Int. Ed.*, **23,** 579 (1984).
[130] Pasto, D. J., *Tetrahedron,* **40,** 2805 (1984).
[131] Zhao, X., and Yuan, M., *Fenzi Kexue Yu Huaxue Yanjiu,* **4,** 53 (1984);*Chem. Abs.*, **101,** 54119 (1984).
[132] Dewar, M. J. S., Healy, E. F., and Stewart, J. J. P., *J. Chem. Soc., Faraday Trans. 2,* **80,** 227 (1984).
[133] Gajewski, J. J., and Gilbert, K. E., *J. Org. Chem.*, **49,** 11 (1984).
[134] Curran, D. P., and Suh, Y. -G., *J. Am. Chem. Soc.*, **106,** 5002 (1984).
[135] Curran, D. P., and Suh, Y. -G., *Tetrahedron Lett.*, **25,** 4179 (1984).
[136] Kuliev, K. I., and Rzaeo, A. S., *Izv. Vyssh. Uchebn, Zaved., Khim. Khim. Technol.*, **26,** 920 (1983); *Chem. Abs.*, **99,** 211914 (1983).
[137] Harwood, L. M., *J. Chem. Soc., Perkin Trans. 1,* **1984,** 2577.
[138] Murphy, R. A., and Cava, M. P., *Tetrahedron Lett.*, **25,** 803 (1984).
[139] Sunitha, K., Balasubramanian, K. K., and Rajagopalan, K., *Tetrahedron Lett.*, **25,** 3125 (1984).
[140] Knowles, J. R., *Pure Appl. Chem.*, **56,** 1005 (1984).
[141] Sogo, S. G., Widlanski, T. S., Hoare, J. H., Grimshaw, C. E., Berchtold, G. A., and Knowles, J. R., *J. Am. Chem. Soc.*, **106,** 2701 (1984).
[142] Takai, K., Mori, I., Oshima, K., and Nozaki, H., *Bull. Chem. Soc. Jpn.*, **57,** 446 (1984).
[143] Büchi, G., and Vogel, D. E., *J. Org. Chem.*, **48,** 5406 (1983).
[144] Denmark, S. E., and Harmata, M. A., *Tetrahedron Lett.*, **25,** 1543 (1984).
[145] Tulshian, D. B., and Fraser-Reid, B., *J. Org. Chem.*, **49,** 518 (1984).
[146] Tulshian, D. B., Tsang, R., and Fraser-Reid, B., *J. Org. Chem.*, **49,** 2347 (1984).
[147] Tulshian, D. B., and Fraser-Reid, B., *Tetrahedron,* **40,** 2083 (1984).
[148] Fraser-Reid, B., Tulshian, D. B., Tsang, R., Lowe, D., and Box, V. G. S., *Tetrahedron Lett.*, **25,** 4579 (1984).
[149] Ho, T. -L., and Liu, S. -H., *J. Chem. Soc., Perkin Trans. 1,* **1984,** 615.
[150] Ziegler, F. E., and Wester, R. T., *Tetrahedron Lett.*, **25,** 617 (1984).
[151] Sen, B. K., and Majumdar, K. C., *Indian J. Chem.*, **22B,** 1184 (1983).
[152] Banks, M. R., Barker, J. M. and Huddleston, P. R., *J. Chem. Res. (S),* **1984,** 27.
[153] Patel, G. N., and Trivedi, K. N., *Indian J. Chem.*, **22B,** 755 (1983).
[154] Moody, C. J., and Ward, J. G., *J. Chem. Soc., Chem. Commun.*, **1984,** 646.
[155] Moody, C. J., *J. Chem. Soc., Perkin Trans. 1,* **1984,** 1333.
[156] Rao, U., and Balasubramanian, K. K., *Heterocycles,* **22,** 1351 (1984).
[157] Ramakanth, S., Narayanan, K., and Balasubramanian, K. K., *Tetrahedron,* **40,** 4473 (1984).
[158] Burke, S. D., Armistead, D. M., and Schoenen, F. J., *J. Org. Chem.*, **49,** 4320 (1984).
[159] Ireland R. E., Dow, W. C., Godfrey, J. D., and Thaisrivongs, S., *J. Org. Chem.*, **49,** 1001 (1984).

[160] Kinney, W. A., Coghlan, M. J., and Paquette, L. A., *J. Am. Chem. Soc.*, **106**, 6868 (1984).
[161] Wilson, S. R., and Price, M. F., *J. Org. Chem.*, **49**, 722 (1984).
[162] Kurth, M. J., and Yu, C. -M., *Tetrahedron Lett.*, **25**, 5003 (1984).
[163] Fujisawa, T., Tajima, K., Ito, M., and Sato, T., *Chem. Lett.*, **1984**, 1169.
[164] Fujisawa, T., Tajima, K., and Sato, T., *Chem. Lett.*, **1984**, 1669.
[165] Burke, S. D., Fobare, W. F., and Pacofsky, G. J., *J. Org. Chem.*, **48**, 5221 (1983).
[166] Cha, J. K., and Lewis, S. C., *Tetrahedron Lett.*, **25**, 5263 (1984).
[167] Ireland, R. E., and Varney, M. D., *J. Am. Chem. Soc.*, **106**, 3668 (1984).
[168] Heathcock, C. H., Jarvi, E. T., and Rosen, T., *Tetrahedron Lett.*, **25**, 243 (1984).
[169] Mikami, K., Maeda, T., Kishi, N., and Nakai, T., *Tetrahedron Lett.*, **25**, 5151 (1984).
[170] Gajewski, J. J., and Emrani, J., *J. Am. Chem. Soc.*, **106**, 5733 (1984).
[171] Nagatsuma, M., Shirai, F., Sayo, N., and Nakai, T., *Chem. Lett.*, **1984**, 1393.
[172] Yokozawa, T., Nakai, T., and Ishikawa, N., *Tetrahedron Lett.*, **25**, 3991 (1984).
[173] Miyaji, K., Nakamura, T., Hirota, H., Igarashi, M., and Takahashi, T., *Tetrahedron Lett.*, **25**, 5299 (1984).
[174] Guenther, H. J., Guntrum, E., and Jaeger, V., *Liebigs Ann. Chem.*, **1984**, 15.
[175] Ohshima, M., Murakami, M., and Mukaiyama, T., *Chem. Lett.*, **1984**, 1535.
[176] Daub, G. W., McCoy, M. A., Sanchez, M. G., and Carter, J. S., *J. Org. Chem.*, **48**, 3876 (1983).
[177] Begley, M. J., Cameron, A. G., and Knight, D. W., *J. Chem. Soc., Chem. Commun.*, **1984**, 827.
[178] Brunner, R. K., and Borschberg, H. J., *Helv. Chim. Acta*, **66**, 2608 (1983).
[179] Barluenga, J., Aznar, F., Liz, R., and Bayod, M., *J. Chem. Soc., Chem. Commun.*, **1984**, 1427.
[180] Vyas, D. M., Chiang, Y., and Doyle, T. W., *Tetrahedron Lett.*, **25**, 487 (1984).
[181] Takano, S., Akiyama, M., and Ogasawara, K., *J. Chem. Soc., Chem. Commun.*, **1984**, 770.
[182] Katayama, H., Tachikawa, Y., Takatsu, N., and Kato, A., *Chem. Pharm. Bull.*, **31**, 2220 (1983).
[183] Danishefsky, S. J., and Phillips, G. B., *Tetrahedron Lett.*, **25**, 3159 (1984).
[184] Abdrakhmanov, I. B., Sharafutdinov, V. M., Nigmatullin, N. G., Mustafin, A. G., Saraeva, Z. N., and Tolstikov, G. A., *Izv, Akad. Nauk SSSR, Ser. Khim.*, **1983**, 1273; *Chem. Abs.*, **100**, 5529 (1984).
[185] Abdrakhmanov, I. B., Shabaeva, G. B., Mustafin, A. G., and Tolstikov, G. A., *Zh. Org. Khim.*, **20**, 663 (1984); *Chem. Abs.*, **101**, 130311 (1984).
[186] Padwa, A., and Cohen, L. A., *J. Org. Chem.*, **49**, 399 (1984).
[187] Gilbert, J. C., and Senaratne, K. P. A., *Tetrahedron Lett.*, **25**, 2303 (1984).
[188] Kunng, F. -A., Gu, J. -M., Chao, S., Chen, Y., and Mariano, P. S., *J. Org. Chem.*, **48**, 4262 (1983).
[189] Chao, S., Kunng, F. -A., Gu, J. -M., Ammon, H. L., and Mariano, P. S., *J. Org. Chem.*, **49**, 2708 (1984).
[190] Anisimova, A. V., Panov, S. M., Sizoi, V. F., Nepogod'ev, S. A., and Viktorova, E. A., *Khim. Geterotsikl. Soedin.*, **1983**, 1348 *Chem. Abs.*, **100**, 68110 (1984).
[191] Takada, S., and Makisumi, Y., *Chem. Pharm. Bull.*, **32**, 872 (1984).
[192] Takeshita, H., Uchida, K., and Mametsuka, H., *Heterocycles*, **20**, 1709 (1983).
[193] Mizutani, M., Sanemitsu, Y., Tamaru, Y., and Yoshida, Z. -i., *J. Org. Chem.*, **48**, 4585 (1983).
[194] De Lucchi, O., Marchioro, G., and Modena, G., *J. Chem. Soc., Chem. Commun.*, **1984**, 513.
[195] Vedejs, E., and Buchanan, R. A., *J. Org. Chem.*, **49**, 1840 (1984).
[196] Ziegler, F. E., and Mencel, J. J., *Tetrahedron Lett.*, **25**, 123 (1984).
[197] Ziegler, F. E., and Lim, H., *J. Org. Chem.*, **49**, 3278 (1984).
[198] Steglich, W., *Chem. Future, Proc. IUPAC Congr., 29th*, 211 (1983); *Chem. Abs.*, **101**, 55493 (1984).
[199] Osamura, Y., Kato, S., Morokuma, K., Feller, D., Davidson, E. R., and Borden, W. T., *J. Am. Chem. Soc.*, **106**, 3362 (1984).
[200] Delbecq, F., and Nguyen, T. A., *Nouv. J. Chim.*, **7**, 505 (1983).
[201] Kettenring, J. K., and Maas, G., *Tetrahedron*, **40**, 391 (1984).
[202] Kirsch, R., Priebe, H., and Hopf, H., *Tetrahedron Lett.*, **25**, 53 (1984).
[203] Kaufmann, D., and de Meijere, A., *Chem. Ber.*, **117**, 1128 (1984).
[204] Slanina, Z., *Collect. Czech. Chem. Commun.*, **48**, 3027 (1983).
[205] Miller, L. S., Grohmann, K., and Dannenberg, J. J., *J. Am. Chem. Soc.*, **105**, 6862 (1983).
[206] Schnieders, C., Müllen, K., Braig, C., Schuster, H., and Sauer, J., *Tetrahedron Lett.*, **25**, 749 (1984).
[207] Poupko, R., Zimmerman, H., and Luz, Z., *J. Am. Chem. Soc.*, **106**, 5391 (1984).
[208] Jackman, L. M., Ibar, G., Freyer, A. J., Goerlach, Y., and Quast, H., *Chem. Ber.*, **117**, 1671 (1984).
[209] Wiberg, K. B., Matturro, M. G., Okarma, P. J., and Jason, M. E., *J. Am. Chem. Soc.*, **106**, 2194 (1984).
[210] Kuzuya, M., Miyake, F., and Okuda, T., *J. Chem. Soc., Perkin Trans. 2*, **1984**, 1471.
[211] Swaminathan, S., *J. Indian Chem. Soc.*, **61**, 99 (1984); *Chem. Abs.*, **101**, 109855 (1984).
[212] Rozeboom, M. D., Kiplinger, J. P., and Bartmess, J. E., *J. Am. Chem. Soc.*, **106**, 1025 (1984).

[213] Bluthe, N., Malacria, M., and Gore, J., *Tetrahedron Lett.*, **25,** 2873 (1984).
[214] Bluthe, N., Malacria, M., and Gore, J., *Tetrahedron,* **40,** 3277 (1984).
[215] Uma, R., Swaminathan, S., and Rajagopalan, K., *Tetrahedron Lett.*, **25,** 5825 (1984).
[216] Gadwood, R. C., Lett, R. M., and Wissinger, J. E., *J. Am. Chem. Soc.*, **106,** 3869 (1984).
[217] Blechert, S., *Tetrahedron Lett.*, **25,** 1547 (1984).
[218] Martin, P., *Helv. Chim. Acta,* **67,** 1647 (1984).
[219] Červinka, O., Fábryová, A., Josef, J., Sermek, V., and Smrčková, S., *Collect. Czech. Chem. Commun.*, **48,** 3407 (1983).
[220] Overman, L. E., Kakimoto, M. -a., Okazaki, M. E., and Meier, G. P., *J. Am. Chem. Soc.*, **105,** 6622 (1983).
[221] Overman, L. E., Mendelson, L. T., and Jacobsen, E. J., *J. Am. Chem. Soc.*, **105,** 6629 (1983).
[222] Overman, L. E., Malone, T. C., and Meier, G. P., *J. Am. Chem. Soc.*, **105,** 6993 (1983).
[223] Castelhano, A. L., and Krantz, A., *J. Am. Chem. Soc.*, **106,** 1877 (1984).
[224] Moore, J. A., Rothenberger, O. S., Fultz, W. C., and Rheingold, A. L., *J. Org. Chem.*, **49,** 1261 (1984).
[225] Cho, J. K., Lee, I., Oh, H. K., and Cho, I. H., *Taehan Hwahakhoe Chi,* **28,** 217 (1984).
[226] Marshall, J. A., and Jenson, T. M., *J. Org. Chem.*, **49,** 1707 (1984).
[227] Tsai, D. J. -S., and Midland, M. M., *J. Org. Chem.*, **49,** 1842 (1984).
[228] Sayo, N., Shirai, F., and Nakai, T., *Chem. Lett.*, **1984,** 255.
[229] Sayo, N., Kitahara, E. -i., and Nakai, T., *Chem. Lett.*, **1984,** 259.
[230] Sayo, N., Azuma, K. -I., Mikami, K., and Nakai, T., *Tetrahedron Lett.*, **25,** 565 (1984).
[231] Mikami, K., Azuma, K. -I., and Nakai, T., *Tetrahedron,* **40,** 2303 (1984).
[232] Mikami, K., Fujimoto, K., Kasuga, T., and Nakai, T., *Tetrahedron Lett.*, **25,** 6011 (1984).
[233] Sugahara, T., Komatsu, Y., and Takano, S., *J. Chem. Soc., Chem. Commun.*, **1984,** 214.
[234] Doyle, M. P., Griffin, J. H., Chinn, M. S., and van Leusen, D., *J. Org. Chem.*, **49,** 1917 (1984).
[235] de March, P., Moreno-Mañas, M., and Ripoll, I., *Synth. Commun.*, **14,** 521 (1984).
[236] Nickon, A., Rodriguez, A., Shirhattl, V., and Ganguly, R., *Tetrahedron Lett.*, **25,** 3555 (1984).
[237] Ishibashi, H., Kitano, Y., Nakatani, H., Okada, M., Ikeda, M., Okura, M., and Tamura, Y., *Tetrahedron Lett.*, **25,** 4231 (1984).
[238] Gamoh, K., Hirayama, M., and Ikekawa, N., *J. Chem. Soc., Perkin Trans. 1,* **1984,** 449.
[239] Tanikaga, R., Nozaki, Y., Nishida, M., and Kaji, A., *Bull. Chem. Soc. Jpn.*, **57,** 729 (1984).
[240] Hiroi, K., Kitayama, R., and Sato, S., *J. Chem. Soc., Chem. Commun.*, **1983,** 1470.
[241] Gassman, P. G., Roos, J. J., and Lee, S. J., *J. Org. Chem.*, **49,** 717 (1984).
[242] Sato, S., Tomita, K., Fujita, H., and Sato, Y., *Heterocycles,* **22,** 1045 (1984).
[243] Sashida, H., and Tsuchiya, T., *Heterocycles,* **22,** 1303 (1984).
[244] Fankhauser, J. E., Peevey, R. M., and Hopkins, P. B., *Tetrahedron Lett.*, **25,** 15 (1984).
[245] Fitzner, J. N., Shea, R. G., Fankhauser, J. E., and Hopkins, P. B., *Synth. Commun.*, **14,** 605 (1984).
[246] Shea, R. G., Fitzner, J. N., Fankhauser, J. E., and Hopkins, P. B., *J. Org. Chem.*, **49,** 3647 (1984).
[247] Lerouge, P., and Paulmier, C., *Tetrahedron Lett.*, **25,** 1983 (1984).
[248] Lerouge, P., and Paulmier, C., *Tetrahedron Lett.*, **25,** 1987 (1984).
[249] Pearson, A. J., Ray, T., Richards, I. C., Clardy, J., and Silveira, L., *Tetrahedron Lett.*, **24,** 5827 (1983).
[250] Collet, H., Calas, P., and Commeyras, A., *J. Chem. Soc., Chem. Commun.*, **1984**, 1152.
[251] Feng, J., Li, Z., Ba, Y., Zhang, Q., and Sun, J., *Jilin Daxue Ziran Kexue Xuebao,* **1984,** 109; *Chem. Abs.*, **101,** 22755 (1984).
[252] Bernardi, F., Robb, M. A., Schlegel, H. B., and Tonachini, G., *J. Am. Chem. Soc.*, **106,** 1198 (1984).
[253] Lee, I., and Cho, J. K., *Bull. Korean Chem. Soc.*, **5,** 51 (1984); *Chem. Abs.*, **101,** 6346 (1984).
[254] Dormans, G. J. M., Fransen, H. R., and Buck, H. M., *J. Am. Chem. Soc.*, **106,** 1213 (1984).
[255] Schwab, J. M., and Klassen, J. B., *J. Chem. Soc., Chem. Commun.*, **1984,** 296.
[256] Schwab, J. M., and Klassen, J. B., *J. Chem. Soc., Chem. Commun.*, **1984,** 298.
[257] Wentrup, C., and Netsch, K. -P., *Angew. Chem. Int. Ed.*, **23,** 802 (1984).
[258] Nöth, H., and Storch, W., *Chem. Ber.*, **117,** 2140 (1984).
[259] Bhupathy, M., and Cohen, T., *J. Am. Chem. Soc.*, **105,** 6978 (1983).
[260] Bleasdale, C., and Jones, D. W., *J. Chem. Soc., Chem. Commun.*, **1984,** 1200.
[261] Sánchez, A., Martínez, M., and Rodríguez-Hahn, L., *Tetrahedron,* **40,** 1005 (1984).
[262] Schultz, A. G., Napier, J. J., and Sundararaman, P., *J. Am. Chem. Soc.*, **106,** 3590 (1984).
[263] Okada, K., Hisamitsu, K., Takahashi, Y., Hanaoka, T., Miyashi, T., and Mukai, T., *Tetrahedron Lett.*, **25,** 5311 (1984).
[264] Mandai, T., Mori, K., Hasegawa, K., Kawada, M., and Otera, J., *Tetrahedron Lett.*, **25,** 5225 (1984).

[265] Ishikawa, M., Oda, M., Nishimura, K., and Kumada, M., *Bull. Chem. Soc. Jpn.*, **56**, 2795 (1983).
[266] Kazoura, S. A., and Weber, W. P., *J. Organomet. Chem.*, **268**, 19 (1984).
[267] O'Neal, H. R., and Neilson, R. H., *Inorg. Chem.*, **23**, 1372 (1984).
[268] Carelli, V., Liberatore, F., Tortorella, S., and Moracci, F. M., *Gazz. Chim. Ital.*, **113**, 569 (1983); *Chem. Abs.*, **101**, 110036 (1984).
[269] Demlehner, U., Sauer, J., Nöth, H., and Lindner, H. J., *Tetrahedron Lett.*, **25**, 5627 (1984).
[270] Hess, B. A., and Schaad, L. J., *J. Am. Chem. Soc.*, **105**, 7185 (1983).
[271] Rondan, N. G., and Houk, K. N., *Tetrahedron Lett.*, **25**, 2519 (1984).
[272] Schoeller, W. W., *J. Chem. Soc., Dalton Trans.*, **1984**, 2233.
[273] Paquette, L. A., Charumilind, P., and Gallucci, J. C., *J. Am. Chem. Soc.*, **105**, 7364 (1983).
[274] Washburn, W. N., and Hillson, R. A., *J. Am. Chem. Soc.*, **106**, 4575 (1984).
[275] Replogle, K. S., and Carpenter, B. K., *J. Am. Chem. Soc.*, **106**, 5751 (1984).
[276] Battye, P. J., and Jones, D. W., *J. Chem. Soc., Chem. Commun.*, **1984**, 990.
[277] Mikhailov, I. E., Dushenko, G. A., Zhdanov, Y. A., Olekhnovich, L. P., and Minkin, V. I., *Dokl. Akad. Nauk SSSR*, **275**, 1431 (1984); *Chem. Abs.*, **101**, 110048 (1984).
[278] Charrier, C., Bonnard, H., de Lauzon, G., Holand, S., and Mathey, F., *Phosphorus Sulphur*, **18**, 51 (1983).
[279] Charrier, C., Bonnard, H., de Lauzon, G., and Mathey, F., *J. Am. Chem. Soc.*, **105**, 6871 (1983).
[280] Jutzi, P., and Saleske, H., *Chem. Ber.*, **117**, 222 (1984).
[281] Lange, G. L., Nye, M. J., Pereira, V. A., Stratton, V., and Yurkevich, T., *Can. J. Chem.*, **62**, 1903 (1984).
[282] Battye, P. J., and Jones, D. W., *J. Chem. Soc., Chem. Commun.*, **1984**, 1458.
[283] Dallinga, J. W., Nibbering, N. M. M., and Boerboom, A. J. H., *J. Chem. Soc., Perkin Trans. 2*, **1984**, 1065.
[284] Jeganathan, S., Johnston, A. D., Kuenzel, E. A., Norman A. W., and Okamura, W. H., *J. Org. Chem.*, **49**, 2152 (1984).
[285] Hugel, G., and Lévy, J., *Tetrahedron*, **40**, 1067 (1984).
[286] Antkowiak, W. Z., and Gessner, W. P., *Tetrahedron Lett.*, **25**, 4045 (1984).
[287] Verboom, W., Hamzink, M. R. J., Reinhoudt, D. N., and Visser, R., *Tetrahedron Lett.*, **25**, 4309 (1984).
[288] Klaerner, F. G., *Top. Stereochem.*, **15**, 1 (1984).
[289] Frimer, A. A., *J. Photochem.*, **25**, 211 (1984); *Chem. Abs.*, **101**, 190660 (1984).
[290] Siebrand, W., Wildman, T. A., and Zgierski, M. Z., *J. Am. Chem. Soc.*, **106**, 4089 (1984).
[291] Töke, L., Bende, Z., Bitter, I., Tóth, G., Simon, P., and Soós, R., *Tetrahedron*, **40**, 4507 (1984).
[292] Fitjer, L., Wehle, D., Noltemeyer, M., Egert, E., and Sheldrick, G. M., *Chem. Ber.*, **117**, 203 (1984).
[293] Schoeller, W. W., *Z. Naturforsch.*, **38B**, 1635 (1983).
[294] Klaerner, F. G., and Glock, V., *Angew. Chem. Int. Ed.*, **23**, 73 (1984).
[295] Frenking, G., *Chem. Phys. Lett.*, **100**, 484 (1983); *Chem. Abs.*, **100**, 33921 (1984).
[296] Alnajjar, M. S., Smith, G. F., and Kuivila, H. G., *J. Org. Chem.*, **49**, 1271 (1984).
[297] Wilhelm, D., Clark, T., Schleyer, P. von R., and Davies, A. G., *J. Chem. Soc., Chem. Commun.*, **1984**, 558.
[298] Hellwinkel, D., Hasselbach, H. -J., and Laemmerzahl, F., *Angew. Chem. Int. Ed.*, **23**, 705 (1984).
[299] Jurlina, J. L., Patel, H. A., and Stothers, J. B., *Can. J. Chem.*, **62**, 1159 (1984).
[300] Cheng, A. K., Ghosh, A. K., and Stothers, J. B., *Can. J. Chem.*, **62**, 1385 (1984).
[301] Laboureur, J. L., and Krief, A., *Tetrahedron Lett.*, **25**, 2713 (1984).
[302] Zagorski, M. G., and Salomon, R. G., *J. Am. Chem. Soc.*, **106**, 1750 (1984).
[303] Marsura, A., Luu Duc, C., and Gellon, G., *Tetrahedron Lett.*, **25**, 4509 (1984).
[304] Lee, S. D., Chan, T. H., and Kwon, K. S., *Tetrahedron Lett.*, **25**, 3399 (1984).
[305] De Kimpe, N., Sulmon, P., Verhé, R., De Buyck, L., and Schamp, N., *J. Org. Chem.*, **48**, 4320 (1983).
[306] Sakai, T., Tabata, H., and Takeda, A., *J. Org. Chem.*, **48**, 4618 (1983).
[307] Daum, S. J., *Tetrahedron Lett.*, **25**, 4725 (1984).
[308] Kocharyan, S. T., Grigoryan, V. V., and Babayan, A. T., *Arm. Khim. Zh.*, **36**, 523 (1983); *Chem. Abs.*, **100**, 34094 (1984).
[309] Kocharyan, S. T., Karapetyan, V. E., and Babayan, A. T., *Arm. Khim. Zh.*, **36**, 586 (1983); *Chem. Abs.*, **100**, 67868 (1984).
[310] Manasyan, L. A., Chukhadzhyan, E. O., and Babayan, A. T., *Arm. Khim. Zh.*, **36**, 591 (1983); *Chem. Abs.*, **100**, 138665 (1984).
[311] Karapetyan, V. E., Kocharyan, S. T., and Babayan, A. T., *Arm. Khim. Zh.*, **36**, 702 (1983); *Chem. Abs.*, **101**, 23013 (1984).

[312] Tronchet, J. M. J., and Eder, H., *J. Carbohydr. Chem.*, **2,** 139 (1983).
[313] Kocharyan, S. T., Karapetyan, V. E., and Babayan, A. T., *Arm. Khim. Zh.*, **37,** 58 (1984); *Chem. Abs.*, **101,** 54670 (1984).
[314] Kocharyan, S. T., Razina, T. L., and Babayan, A. T., *Arm. Khim. Zh.*, **36,** 581 (1983); *Chem. Abs.*, **100,** 51068 (1984).
[315] Kocharyan, S. T., Grigoryan, V. V., and Babayan, A. T., *Arm. Khim. Zh.*, **36,** 576 (1983); *Chem. Abs.*, **100,** 51069 (1984).
[316] Okazaki, R., and Tokitoh, N., *J. Chem. Soc., Chem. Commun.*, **1984,** 192.
[317] Storflor, H., Skramstad, J., and Nordenson, S., *J. Chem. Soc., Chem. Commun.*, **1984,** 208.
[318] Rimmelin, P., Taghavi, H., and Sommer, J., *J. Chem. Soc., Chem. Commun.*, **1984,** 1210.
[319] Gilheany, D. G., Kennedy, D. A., Malone, J. F., and Walker, B. J., *J. Chem. Soc., Chem. Commun.*, **1984,** 1217.
[320] Harger, M. J. P., and Smith A., *J. Chem. Soc., Chem. Commun.*, **1984,** 1140.
[321] Meyle, E., and Otto, H. -H., *J. Chem. Soc., Chem. Commun.*, **1984,** 1084.
[322] Benati, L., Montevecchi, P. C., and Spagnolo, P., *J. Chem. Soc., Perkin Trans. 1*, **1984,** 625.
[323] Apparao, S., Ila, H., and Junjappa, H., *J. Chem. Soc., Perkin Trans. 1*, **1983,** 2837.
[324] Apparao, S., Datta, A., Ila, H., and Junjappa, H., *J. Chem. Soc., Perkin Trans. 1*, **1984,** 921.
[325] Kataoka, T., Tomoto, A., Shimizu, H., and Hori, M., *J. Chem. Soc., Perkin Trans. 1*, **1983,** 2913.
[326] Ohno, S., Shimizu, H., Kataoka, T., and Hori, M., *J. Org. Chem.*, **49,** 2472 (1984).
[327] Daney, M., Lapouyade, R., and Bouas-Laurent, H., *J. Org. Chem.*, **48,** 5055 (1983).
[328] Brook, A. G., and Chrusciel, J. J., *Organometallics*, **3,** 1317 (1984).
[329] Kato, M., and Kuwajima, I., *Bull. Chem. Soc. Jpn.*, **57,** 827 (1984).
[330] Kato, M., Mori, A., Oshino, H., Enda, J., Kobayashi, K., and Kuwajima, I., *J. Am. Chem. Soc.*, **106,** 1773 (1984).
[331] Wilson, S. R., and Georgiadis, G. M., *J. Org. Chem.*, **48,** 4143 (1983).
[332] Corey, E. J., and Rücker, C., *Tetrahedron Lett.*, **25,** 4345 (1984).
[333] Rücker, C., *Tetrahedron Lett.*, **25,** 4349 (1984).
[334] Mora, J., and Costa, A., *Tetrahedron Lett.*, **25,** 3493 (1984).
[335] Murai, S., Ryn, I., Iriguchi, J., and Sonoda, N., *J. Am. Chem. Soc.*, **106,** 2440 (1984).
[336] Thum, G., Riés, W., Greissinger, D., and Malisch, W., *J. Organomet. Chem.*, **252,** C67 (1983).
[337] Barkash, V. A., *Top. Curr. Chem.*, **116,** 1 (1984).
[338] Shubin, V. G., *Top. Curr. Chem.*, **117,** 267 (1984).
[339] Reutov, O. A., *Usp. Khim.*, **53,** 462 (1984); *Chem. Abs.*, **101,** 71818 (1984).
[340] Schwarz, H., Thies, H., and Franke, W., *NATO ASI Ser., Ser. C*, **1984,** 118; *Chem. Abs.*, **100,** 173910 (1984).
[341] Kingston, E. E., Shannon, J. S., and Lacey, M. J., *Org. Mass Spectrom.*, **18,** 183 (1983).
[342] Ausloos, P., and Lias, S. G., *Int. J. Mass Spectrom. Ion Proc.*, **58,** 165 (1984); *Chem. Abs.*, **101,** 130039 (1984).
[343] Frenking, G., *Tetrahedron*, **40,** 377 (1984).
[344] Myhre, P. C., Yannoni, C. S., and Macho, V., *Prepr. -Am. Chem. Soc., Div. Pet. Chem.*, **28,** 271 (1983).
[345] Andres, J., Arnan, A., Silla, E., Bertran, J., and Tapia, O., *Theochem.*, **14,** 49 (1983); *Chem. Abs.*, **100,** 50727 (1984).
[346] Andres, J., Silla, E., and Tapia, O., *Theochem*, **14,** 307 (1983); *Chem. Abs.*, **100,** 120155 (1984).
[347] Courtneidge, J. L., and Davies, A. G., *J. Chem. Soc., Chem. Commun.*, **1984,** 136.
[348] Brusova, G. P., Permin, A. B., and Reutov, O. A., *Izv. Akad. Nauk SSSR, Ser. Khim.*, **1984,** 431: *Chem. Abs.*, **100,** 191118 (1984).
[349] Shandala, M. Y., Salih, R. G., and Jamel, A. M., *Iraqi J. Sci.*, **23,** 68 (1982); *Chem. Abs.*, **100,** 50831 (1984).
[350] Gharbi-Benarous, J., Morales-Rios, M. S., and Dana, G., *J. Org. Chem.*, **49,** 2039 (1984).
[351] Collins, C. J., Fuchs, K. A., and Hanack, M., *Prepr. - Am. Chem. Soc., Div. Pet. Chem.*, **28,** 352 (1983); *Chem. Abs.*, **101,** 22661 (1984).
[352] Collins, C. J., Martínez, A. G., Alvarez, R. M., and Aguirre, J. A., *Chem. Ber.*, **117,** 2815 (1984).
[353] Kirmse, W., Siegfried, R., and Streu, J., *J. Am. Chem. Soc.*, **106,** 2465 (1984).
[354] Goldstein, M. J., and Dinnocenzo, J. P., *J. Am. Chem. Soc.*, **106,** 2473 (1984).
[355] Jonsäll, G., and Ahlberg, P., *J. Chem. Soc., Chem. Commun.*, **1984,** 1125.
[356] Liu, M. T. H., Chishti, N. H., Tencer, M., Tomioka, H., and Izawa, Y., *Tetrahedron*, **40,** 887 (1984).
[357] Bartmann, E., and Mengel, T., *Angew. Chem. Int. Ed.*, **23,** 225 (1984).
[358] Falkenberg-Andersen, C., Ranganayakulu, K., Schmitz, L. R., and Sorensen, T. S., *J. Am. Chem. Soc.*, **106,** 178 (1984).

[359] Bascetta, E., and Gunstone, F. D., *J. Chem. Soc., Perkin Trans. 1*, **1984,** 2217.
[360] Mizuno, K., Yoshioka, K., and Otsuji, Y., *J. Chem. Soc., Chem. Commun.*, **1984,** 1665.
[361] Vincens, M., Dumont, C., and Vidal, M., *Bull. Soc. Chim. Fr. II*, **1984,** 59.
[362] Brennenstuhl, W., and Hanack, M., *Tetrahedron Lett.*, **25,** 3437 (1984).
[363] Warner, P. M., and Herold, R. D., *Tetrahedron Lett.*, **25,** 4897 (1984).
[364] Kelsall, B. J., Andrews, L., and Trindle, C., *J. Phys. Chem.*, **87,** 4898 (1983).
[365] Mizumoto, K., Okada, K., and Oda, M., *Tetrahedron Lett.*, **25,** 2999 (1984).
[366] Kelly, D. P., Leslie, D. R., and Smith, B. D., *J. Am. Chem. Soc.*, **106,** 687 (1984).
[367] Hass, C., Kirste, B., Kurreck, H., and Schlömp, G., *J. Am. Chem. Soc.*, **105,** 7375 (1983).
[368] Hünig, S., and Ort, B., *Angew. Chem. Int. Ed.*, **23,** 237 (1984).
[369] Jalander, L., *Tetrahedron Lett.*, **25,** 457 (1984).
[370] Posner, G. H., and Kogan, T. P., *J. Chem. Soc., Chem. Commun.*, **1983,** 1481.
[371] Fujita, T., Watanabe, S., Suga, K., Higuchi, Y., and Sotoguchi, T., *J. Org. Chem.*, **49,** 1975 (1984).
[372] Mundy, B. P., Kim, Y., and Warnet, R. J., *Heterocycles*, **20,** 1727 (1983).
[373] Shimada, J. -i., Hashimoto, K., Kim, B. H., Nakamura, E., and Kuwajima, I., *J. Am. Chem. Soc.*, **106,** 1759 (1984).
[374] Suzuki, K., Katayama, E., and Tsuchihashi, G. -i., *Tetrahedron Lett.*, **25,** 1817 (1984).
[375] Suzuki, K., Katayama, E., Matsumoto, T., and Tsuchihashi, G. -i., *Tetrahedron Lett.*, **25,** 3715 (1984).
[376] Tsuchihashi, G. -i., Tomooka, K., and Suzuki, K., *Tetrahedron Lett.*, **25,** 4253 (1984).
[377] Hannsske, F., and Robins, M. J., *J. Am. Chem. Soc.*, **105,** 6736 (1983).
[378] Knapp, S., Trope, A. F., Theodore, M. S., Hirata, N., and Barchi, J. J., *J. Org. Chem.*, **49,** 608 (1984).
[379] Pietrusiewicz, K. M., Monkiewicz, J., and Bodalski, R., *Pol. J. Chem.*, **57,** 637 (1983); *Chem. Abs.*, **101,** 72848 (1984).
[380] Bhattacharya, S., Mandal, A. N., Chaudhuri, S. R. R., and Chatterjee, A., *Tetrahedron Lett.*, **25,** 3007 (1984).
[381] Dhanak, D., Reese, C. B., and Williams, D. E., *J. Chem. Soc., Chem. Commun.*, **1984,** 988.
[382] Eaborn, C., Lickiss, P. D., and Ramadan, N. A., *J. Chem. Soc., Perkin Trans. 2*, **1984,** 267.
[383] Raynolds, P. W., and DeLoach, J. A., *J. Am. Chem. Soc.*, **106,** 4566 (1984).
[384] Boardman, L. D., Bagheri, V., Sawada, H., and Negishi, E. -i., *J. Am. Chem. Soc.*, **106,** 6105 (1984).
[385] Enda, J., and Kuwajima, I., *J. Chem. Soc., Chem. Commun.*, **1984,** 1589.
[386] Fleming, I., and Newton, T. W., *J. Chem. Soc., Perkin Trans. 1*, **1984,** 119.
[387] De Kimpe, N., De Buyck, L., Verhé, R., and Schamp, N., *Tetrahedron*, **40,** 3291 (1984).
[388] Keul, H., Pfeffer, B., and Griesbaum, K., *Chem. Ber.*, **117,** 2193 (1984).
[389] Clark, G. R., *Tetrahedron Lett.*, **25,** 2839 (1984).
[390] Cooke, M. D., Dransfield, T. A., and Vernon, J. M., *J. Chem. Soc., Perkin Trans. 2*, **1984,** 1377.
[391] Defoin, A., Baranne-Lafont, J., and Rigaudy, J., *Bull. Soc. Chim. Fr. II*, **1984,** 145.
[392] Kay, I. T., and Williams, E. G., *Tetrahedron Lett.*, **24,** 5915 (1983).
[393] Kay, I. T., and Bartholomew, D., *Tetrahedron Lett.*, **25,** 2035 (1984).
[394] Williams, D. R., Harigaya, Y., Moore, J. L., and D'sa, A., *J. Am. Chem. Soc.*, **106,** 2641 (1984).
[395] Karakhanov, R. A., Skurko, M. R., Ramazanov, O. M., Kantor, E. A., Bartok, M., and Bucsi, I., *Acta Phys. Chem.*, **29,** 181 (1983); *Chem. Abs.*, **101,** 22826 (1984).
[396] Clemans, G. B., Samaritoni, J. G., Holloway, R. J., and Edinger, W., *J. Org. Chem.*, **49,** 3457 (1984).
[397] Vaughan, W. R., Gross, B. A., Burkle, S. E., Langell, M. A., Caple, R., and Oakes, D. B., *J. Org. Chem.*, **48,** 4792 (1983).
[398] Baldwin, J. E., and Barden, T. C., *J. Am. Chem. Soc.*, **105,** 6656 (1983).
[399] Majerski, Z., and Hamersak, Z., *J. Org. Chem.*, **49,** 1182 (1984).
[400] Maskill, H., and Wilson, A. A., *J. Chem. Soc., Perkin Trans. 2*, **1984,** 1369.
[401] Michael, J. P., Blom, N. F., and Boeyens, J. C. A., *J. Chem. Soc., Perkin Trans. 1*, **1984,** 1739.
[402] Wilt, J. W., Curtis, V. A., Congson, L. N., and Palmer, R., *J. Org. Chem.*, **49,** 2937 (1984).
[403] Morrison, H., and De Cárdenas, L. M., *Tetrahedron Lett.*, **25,** 2527 (1984).
[404] Della, E. W., and Pigou, P. E., *Aust. J. Chem.*, **36,** 2261 (1983).
[405] Cristol, S. J., Seapy, D. G., and Aeling, E. O., *J. Am. Chem. Soc.*, **105,** 7337 (1983).
[406] Cristol, S. J., Bindel, T. H., Hoffmann, D., and Aeling, E. O., *J. Org. Chem.*, **49,** 2368 (1984).
[407] Himbert, G., Diehl, K., and Maas, G., *J. Chem. Soc., Chem. Commun.*, **1984,** 900.
[408] Holmes, A. B., Raithly, P. R., Thompson, J., Baxter, A. J. G., and Dixon, J., *J. Chem. Soc., Chem. Commun.*, **1983,** 1490.
[409] Paquette, L. A., DeLucca, G., Korp, J. D., Bernal, I., Swartzendruber, J. K., and Jones, N. D., *J. Am. Chem. Soc.*, **106,** 1122 (1984).
[410] Klunder, A. J. H., Ariaans, G. J. A., and Zwanenburg, B., *Tetrahedron Lett.*, **25,** 5457 (1984).

411 Fitjer, L., Kühn, W., Klages, U., Egert, E., Clegg, W., Schormann, N., and Sheldrick, G. M., *Chem. Ber.*, **117**, 3075 (1984).
412 Tobe, Y., Ohtani, M., Kakiuchi, K., and Odaira, Y., *J. Org. Chem.*, **48**, 5114 (1983).
413 Kakiuchi, K., Itoga, K., Tsugaru, T., Hato, Y., Tobe, Y., and Odaira, Y., *J. Org. Chem.*, **49**, 659 (1984).
414 Smith, A. B., and Wexler, B. A., *Tetrahedron Lett.*, **25**, 2317 (1984).
415 Kakiuchi, K., Nakao, T., Takeda, M., Tobe, Y., and Odaira, Y., *Tetrahedron Lett.*, **25**, 557 (1984).
416 Mlinarić-Majerski, K., and Majerski, Z., *J. Am. Chem. Soc.*, **105**, 7389 (1983).
417 Tanaka, N., Kan, T., and Iizuka, T., *J. Chem. Inf. Comput. Sci.*, **23**, 177 (1983); *Chem. Abs.*, **99**, 211899 (1983).
418 Paquette, L. A., Williams, R. V., Vazeux, M., and Browne, A. R., *J. Org. Chem.*, **49**, 2194 (1984).
419 Mehta, G., and Rao, K. S., *Tetrahedron Lett.*, **25**, 3481 (1984).
420 Greenfield, S., Mackenzie, K., and Muir, K. W., *Tetrahedron Lett.*, **25**, 3133 (1984).
421 Katritzky, A. R., Dega-Szafran, Z., Lopez-Rodriguez, M. L., and King, R. W., *J. Am. Chem. Soc.*, **106**, 5577 (1984).
422 Kozikarski, A. P., and Park, P. -u., *J. Org. Chem.*, **49**, 1674 (1984).
423 Borgulya, J., Daly, J. J., Schönholzer, P., and Bernauer, K., *Helv. Chim. Acta*, **67**, 1827 (1984).
424 Francisco, C. G., Freire, T., Hernández, R., Salazar, J. A., and Suárez, E., *Tetrahedron Lett.*, **25**, 1621 (1984).
425 Majerski, Z., Marinić, Z., and Šarac-Arneri, R., *J. Org. Chem.*, **48**, 5109 (1983).
426 Shimizu, M., Akiyama, T., and Mukaiyama, T., *Chem. Lett.*, **1984**, 1531.
427 Kita, Y., Yasuda, H., Tamura, O., Itoh, F., and Tamura, Y., *Tetrahedron Lett.*, **25**, 4681 (1984).
428 Lane, S., Quick, S. J., and Taylor, R. J. K., *Tetrahedron Lett.*, **25**, 1039 (1984).
429 Lane, S., Quick, S. J., and Taylor, R. J. K., *J. Chem. Soc., Perkin Trans. 1*, **1984**, 2549.
430 Ahern, T. P., Haley, M. F., Langler, R. F., and Trenholm, J. E., *Can. J. Chem.*, **62**, 610 (1984).
431 Callant, P., D'Haenens, L., Van der Eycken, E., and Vandewalle, M., *Synth. Commun.*, **14**, 163 (1984).
432 Smith, A. B., Toder, B. H., and Branca, S. J., *J. Am. Chem. Soc.*, **106**, 3995 (1984).
433 Smith, A. B., Toder, B. H., Richmond, R. E., and Branca, S. J., *J. Am. Chem. Soc.*, **106**, 4001 (1984).
434 Ohta, S., Takai, M., Yokoyama, Y., Tsuyuki, T., Takahashi, T., and Tori, M., *Tennen Yuki Kagobutsu Toronkai Koen Yoshishu*, **26**, 375 (1983); *Chem. Abs.*, **100**, 192108 (1984).
435 Bégué, J. -P., and Bonnet-Delpon, D., *J. Chem. Soc., Chem. Commun.*, **1984**, 402.
436 Anjaneyulu, A. S. R., Anjaneyulu, V., Prasad, K. H., Rao, G. S., and Suryanarayana, P., *Indian J. Chem.*, **22B**, 1179 (1983).
437 Singh, A. K., and Dhar, D. N., *Bull. Soc. Chim. Belg.*, **93**, 129 (1984).
438 Takai, M., Tori, M., Tsuyuki, T., Takahashi, T., Itai, A., and Iitaka, Y., *Chem. Pharm. Bull.*, **32**, 2464 (1984).
439 Bridge, A. W., and Morrison, G. A., *J. Chem. Soc., Perkin Trans. 1*, **1983**, 2933.
440 Shimizu, N., Itoh, T., Saito, M., and Matsumoto, T., *J. Org. Chem.*, **49**, 709 (1984).
441 Shimizu, N., Itoh, T., Ichinohe, Y., and Matsumoto, T., *Bull. Chem. Soc. Jpn.*, **57**, 1425 (1984).
442 Kamernitskii, A. V., Turuta, A. M., Fadeeva, T. M., Bogdanov, V. S., Cherepanova, E. G., and Korobov, A. A., *Izv. Akad. Nauk SSSR, Ser. Khim.*, **1983**, 2376; *Chem. Abs.*, **100**, 121436 (1984).
443 Kasal, A., *Collect. Czech. Chem. Commun.*, **49**, 892 (1984).
444 Turner, A. B., *J. Chem. Soc., Chem. Commun.*, **1984**, 711.
445 Taran, M., and Delmond, B., *J. Chem. Soc., Chem. Commun.*, **1984**, 716.
446 Ramage, R., and Southwell, I. A., *J. Chem. Soc., Perkin Trans. 1*, **1984**, 1323.
447 Itokawa, H., Nakanishi, H., and Mihashi, S., *Chem. Pharm. Bull.*, **31**, 1991 (1983).
448 Holton, R. A., *J. Am. Chem. Soc.*, **106**, 5731 (1984).
449 Waring, A. J., *J. Chem. Soc., Perkin Trans. 2*, **1984**, 373.
450 Hanton, L. R., Simpson, J., and Weavers, R. T., *Aust. J. Chem.*, **36**, 2581 (1983).
451 Van Engen, D., and Clardy, J., *J. Am. Chem. Soc.*, **106**, 5026 (1984).
452 Trost, B. M., McDougal, P. G., and Haller, K. J., *J. Am. Chem. Soc.*, **106**, 383 (1984).
453 Tsuchiya, T., Ajito, K., Umezawa, S., and Ikeda, A., *Carbohydr. Res.*, **126**, 45 (1984).
454 Jacobsen, S., *Acta Chem. Scand.*, **38B**, 157 (1984).
455 Beyerman, H. C., Crabbendam, P. R., Lie, T. S., and Maat, L., *Recl.: J. R. Neth. Chem. Soc.*, **103**, 112 (1984).
456 Allen, A. C., Cooper, D. A., Moore, J. M., and Teer, C. B., *J. Org. Chem.*, **49** 3462 (1984).
457 Massiot, G., Lavaud, C., Vercauteren, J., Le Men-Olivier, L., Levy, J., Giulhem, J., and Pascard, C., *Helv. Chim. Acta*, **66**, 2414 (1983).

[458] Danieli, B., Lesma, G., Palmisano, G., Riva, R., Riva, S., Demartin, F., and Masciocchi, N., *J. Org. Chem.*, **49,** 4138 (1984).
[459] Brossi, M., and Kaenel, H., *SLZ, Schweiz. Lab. -Z.* **41,** 23 (1984); *Chem. Abs.*, **100,** 138180 (1984).
[460] Hudrlik, P. F., Waugh, M. A., and Hudrlik, A. M., *J. Organomet. Chem.*, **271,** 69 (1984).
[461] Levashova, L. A., Lazareva, M. P., Gromoglasov, Y. A., and Korot'ko, L. V., *Deposited Doc., 1982; Chem. Abs.*, **101,** 71990 (1984).
[462] Izumi, Y., Sato, S., and Urabe, K., *Chem. Lett.*, **1983,** 1649.
[463] Lub, J., and de Boer, T. J., *Recl. Trav. Chim. Pays-Bas*, **103,** 328 (1984).
[464] Tashchi, V. P., Rukasov, A. F., Orlova, T. I., Ivanov, A. P., Tashchi, O. A., Baskahov, Y. A., and Putsykin, Y. G., *Zh. Org. Khim.*, **20,** 988 (1984); *Chem. Abs.*, **101,** 109981 (1984).
[465] Prager, R. H., Raner, K. D., and Ward, A. D., *Aust. J. Chem.*, **37,** 381 (1984).
[466] Gawley, R. E., and Termine, E. J., *J. Org. Chem.*, **49,** 1946 (1984).
[467] Fujiwara, J., Sano, H., Maruoka, K., and Yamamoto, H., *Tetrahedron Lett.*, **25,** 2367 (1984).
[468] Kajigaeshi, S., Nakagawa, T., Fujisaki, S., Nishida, A., and Noguchi, M., *Chem. Lett.*, **1984,** 713.
[469] Wang, Z., Wang, S., Xu, Z., Wang, Y., and Wu, S., *Huaxue Tongbao*, **1983,** 11; *Chem. Abs.*, **100,** 120465 (1984).
[470] Wang, S., Wang, Z., Xu, Z., Wang, Y., Wu, S., and Huang, X., *Jilin Daxue Ziran Kexue Xuebao*, **1984,** 89; *Chem. Abs.*, **101,** 90361 (1984).
[471] Loudon, G. M., Radhakrishna, A. S., Almond, M. R., Blodgett, J. K., and Boutin, R. H., *J. Org. Chem.*, **49,** 4272 (1984).
[472] Patil, V. M., Sangapure, S. S., and Agasimundin, Y. S., *Indian J. Chem.*, **23B,** 132 (1984).
[473] Baceiredo, A., Bertrand, G., Mazerolles, P., and Majoral, J. P., *Nouv. J. Chim.*, **7,** 645 (1983).
[474] Zlobin, V. A., Kosolapov, V. T., Moiseev, I. K., and Tarasov, A. K., *Izv. Vyssh. Uchebn. Zaved., Khim. Khim. Tekhnol.*, **27,** 401 (1984); *Chem. Abs.*, **101,** 90317 (1984).
[475] Zlobin, V. A., Kosolapov, V. T., and Tarasov, A. K., *Izv. Vyssh. Uchebn. Zaved., Khim. Khim. Tekhnol.*, **27,** 169 (1984); *Chem. Abs.*, **100,** 209083 (1984).
[476] Capson, T. L., and Poulter, C. D., *Tetrahedron Lett.*, **25,** 3515 (1984).
[477] Mulliez, M., Majoral, J. -P., and Bertrand, G., *J. Chem. Soc., Chem. Commun.*, **1984,** 284.
[478] Harger, M. J. P., and Westlake, S., *J. Chem. Soc., Perkin Trans. 1*, **1984,** 2351.
[479] Eckstein, Z., Lipczynska-Kochany, E., and Leszczynska, E., *Liebigs Ann. Chem.*, **1984,** 395.
[480] Adam, G., Andrieux, J., Plat, M., Viossat, B., and Rodier, N., *Bull. Soc. Chim. Fr. II*, **1984,** 101.
[481] Tsunokawa, Y., Iwasaki, S., and Okuda, S., *Chem. Pharm. Bull.*, **31,** 4578 (1983); *Chem. Abs.*, **100,** 208840 (1984).
[482] Suryawanshi, S. N., Swenson, C. J., Jorgensen, W. L., and Fuchs, P. L., *Tetrahedron Lett.*, **25,** 1859 (1984).
[483] Dave, V., Stothers, J. B., and Warnhoff, E. W., *Can. J. Chem.*, **62,** 1965 (1984).
[484] Deeming, A. J., *Mech. Inorg. Organomet. React.*, **2,** 319 (1984); *Chem. Abs.*, **100,** 209924 (1984).
[485] Yano, S., and Yoshikawa, S., *Kagaku (Kyoto)*, **38,** 890 (1983); *Chem. Abs.*, **100,** 139488 (1984).
[486] Vuorinen, T., *Carbohydr. Res.*, **133,** 329 (1984).
[487] Rappe, A. K., and Upton, T. H., *Organometallics*, **3,** 1440 (1984).
[488] Giolando, D. M., Rauchfuss, T. B., and Wilson, S. R., *J. Am. Chem. Soc.*, **106,** 6455 (1984).
[489] Yasuda, H., Nagasuna, K., Akita, M., Lee, K., and Nakamura, A., *Organometallics*, **3,** 1470 (1984).
[490] Green, M. L. H., and Newman, P. A., *J. Chem. Soc., Chem. Commun.*, **1984,** 816.
[491] Tatsumi, T., Shibagaki, M., and Tominaga, H., *J. Mol. Catal.*, **24,** 19 (1984); *Chem. Abs.*, **101,** 110055 (1984).
[492] Hirao, T., Fujihara, Y., Tsuno, S., Ohshiro, Y., and Agawa, T., *Chem. Lett.*, **1984,** 367.
[493] Alper, H., and Mahatantila, C. P., *J. Am. Chem. Soc.*, **106,** 2708 (1984).
[494] Davidson, J. L., and Carlton, L., *J. Chem. Soc., Chem. Commun.*, **1984,** 964.
[495] Vierling, P., and Riess, J. G., *J. Am. Chem. Soc.*, **106,** 2432 (1984).
[496] Ullah, S. S., Kabir, S. E., Rahman, A. K. F., and Karim, M., *Indian J. Chem.*, **23B,** 103 (1984).
[497] Casey, C. P., Marder, S. R., and Fagan, P. J., *J. Am. Chem. Soc.*, **105,** 7197 (1983).
[498] Hisatome, M., Kawajiri, Y., Watanabe, J., Yoshioka, M., and Yamakawa, K., *J. Organomet. Chem.*, **266,** 147 (1984).
[499] Komiya, S., Katoh, M., Ikaruja, T., Grubbs, R. H., Yamamoto, T., and Yamamoto, A., *J. Organomet. Chem.*, **260,** 115 (1984).
[500] Lhadi, E. K., Patin, H., and Darchen, A., *Organometallics*, **3,** 1128 (1984).
[501] Semmelhack, M. F., and Le, H. T. M., *J. Am. Chem. Soc.*, **106,** 2715 (1984).
[502] Yokoyama, K., Saegusa, Y., Mujashi, T., Kabuto, C., and Mukai, T., *Chem. Lett.*, **1984,** 89.
[503] Romań, E., Hernández, S., and Barrera, M., *J. Chem. Soc., Chem. Commun.*, **1984,** 1067.
[504] Argo, C. B., and Sharp, J. T., *J. Chem. Soc., Perkin Trans. 1*, **1984,** 1581.

[505] Lancon, D., Cocolios, P., Guilard, R., and Kadish, K. M., *J. Am. Chem. Soc.*, **106,** 4472 (1984).
[506] Dix, T. A., and Marnett, L. J., *J. Am. Chem. Soc.*, **105,** 7001 (1983).
[507] Groves, J. T., and Subramanian, D. V., *J. Am. Chem. Soc.*, **106,** 2177 (1984).
[508] Matteoli, U., Bianchi, M., Frediani, P., Menchi, G., Botteghi, C., and Marchetti, M., *J. Organomet. Chem.*, **263,** 243 (1984).
[509] Georgoulis, C., Valéry, J. M., and Ville, G., *Synth. Commun.*, **14,** 1043 (1984).
[510] Koridze, A. A., Kizas, O. A., Kolobova, N. E., and Petrovskii, P. V., *Izv. Akad. Nauk SSSR, Ser. Khim.*, **1984,** 472; *Chem. Abs.*, **101,** 55303 (1984).
[511] Dowd, P., and Shapiro, M., *Tetrahedron*, **40,** 3063 (1984).
[512] Dowd, P., Shapiro, M., and Kang, J., *Tetrahedron*, **40,** 3069 (1984).
[513] Wan, T. S., and Fischili, A., *Helv. Chim. Acta*, **67,** 1883 (1984).
[514] Chatani, N., Furukawa, H., Kato, T., Murai, S., and Sonoda, N., *J. Am. Chem. Soc.*, **106,** 430 (1984).
[515] Sato, S., Okada, H., Matsuda, I., and Izumi, Y., *Tetrahedron Lett.*, **25,** 769 (1984).
[516] Alper, H., and Urso, F., *J. Am. Chem. Soc.*, **105,** 6737 (1983).
[517] Grigg, R., and Stevenson, P. J., *Synthesis*, 1009 (1983).
[518] Tani, K., Yamagata, T., Akutagawa, S., Kumobayashi, H., Taketomi, T., Takaya, H., Miyashita, A., Noyori, R., and Otsuka, S., *J. Am. Chem. Soc.*, **106,** 5208 (1984).
[519] Kyler, K. S., Bashir-Hashemi, A., and Watt, D. S., *J. Org. Chem.*, **49,** 1084 (1984).
[520] Achab, S., Cosson, J. -P., and Das, B. C., *J. Chem. Soc., Chem. Commun.*, **1984,** 1040.
[521] Tamaru, Y., Yamada, Y., Ochiai, H., Nakajo, E., and Yoshida, Z., *Tetrahedron*, **40,** 1791 (1984).
[522] Oehlschlager, A. C., Mishra, P., and Dhami, S., *Can. J. Chem.*, **62,** 791 (1984).
[523] Curzon, E., Golding, B. T., Pierpoint, C., and Waters, B. W., *J. Organomet. Chem.*, **262,** 263 (1984).
[524] Hiroi, K., Kitayama, R., and Sato, S., *J. Chem. Soc., Chem. Commun.*, **1984,** 303.
[525] Lu, X., and Huang, Y., *Huaxue Xuebao*, **42,** 835 (1984); *Chem. Abs.*, **101,** 191243 (1984).
[526] Yamada, Y., Suzukamo, G., Yoshioka, H., Tamaru, Y., and Yoshida, Z. -i., *Tetrahedron Lett.*, **25,** 3599 (1984).
[527] Fenske, D., and Christidis, A., *Z. Naturforsch.*, **38b,** 1295 (1983); *Chem. Abs.*, **100,** 121324 (1984).
[528] Goel, A. B., *Tetrahedron Lett.*, **25,** 4599 (1984).
[529] Suzuki, S., Kogai, K., and Ono, Y., *Chem. Lett.*, **1984,** 699.
[530] Chauvet, F., Heumann, A., and Waegell, B., *Tetrahedron Lett.*, **25,** 4393 (1984).
[531] Borsub, N., and Kutal, C., *J. Am. Chem. Soc.*, **106,** 4826 (1984).
[532] Patrick, T. B., and Bechtold, D. S., *J. Org. Chem.*, **49,** 1935 (1984).
[533] Pugin, B., and Venanzi, L. M., *J. Am. Chem. Soc.*, **105,** 6877 (1983).
[534] Smadja, W., Ville, G., and Cahiez, G., *Tetrahedron Lett.*, **25,** 1793 (1984).
[535] Aeiyach, S., Garin, F., Hilaire, F., Legare, P., and Maire, G., *J. Mol. Catal.*, **25,** 183 (1984).
[536] Davis, S. M., Zaera, F., and Somorjai, G. A., *J. Catal.*, **85,** 206 (1984).
[537] Gassman, P. G., and Cesa, I. G., *Organometallics*, **3,** 119 (1984).
[538] Ortiz, J. V., Havlas, Z., and Hoffmann, R., *Helv. Chim. Acta*, **67,** 1 (1984).
[539] Hoffmann, E. G., Nehl, H., Lekmkuhl, H., Seevogel, K., and Stempfle, W., *Chem. Ber.*, **117,** 1364 (1984).
[540] Collum, D. B., Mohamadi, F., and Hallock, J. S., *J. Am. Chem. Soc.*, **105,** 6882 (1983).
[541] Rood, I. D. C., and Klumpp, G. W., *Recl. Trav. Chim. Pays-Bas*, **103,** 303 (1984).
[542] Fife, D. J., Morse, K. W., and Moore, W. M., *J. Am. Chem. Soc.*, **105,** 7404 (1983).
[543] Katz, H. E., and Starnes, W. H., *J. Org. Chem.*, **49,** 2758 (1984).
[544] Silaghi-Dumitrescu, L., and Haiduc, I., *J. Organomet. Chem.*, **252,** 295 (1983).
[545] Campelo, J. M., Garcia, A., Gutierrez, J. M., Luna, D., and Marinas, J. M., *Can. J. Chem.*, **62,** 1455 (1984).
[546] Kurosaki, A., and Okazaki, S., *Chem. Lett.*, **1983,** 1741.
[547] Tsuboi, S., Nooda, Y., and Takeda, A., *J. Org. Chem.*, **49,** 1204 (1984).
[548] Jefford, C. W., and Zuber, J. A., *Rev. Port. Quim.*, **24,** 1 (1982); *Chem. Abs.*, **101,** 130263 (1984).
[549] Baldwin, J. E., and Black, K. A., *J. Am. Chem. Soc.*, **106,** 1029 (1984).
[550] Baldwin, J. E., and Barden, T. C., *J. Am. Chem. Soc.*, **106,** 6364 (1984).
[551] Alonso, M. E., Chitty, A. W., Pekerar, S., and Borgo, De L. M., *J. Chem. Soc., Chem. Commun.*, **1984,** 1542.
[552] Mahamulkar, B. G., Kulkarni, G. H., and Mitra, R. B., *Indian J. Chem.*, **22B,** 910 (1983).
[553] Kostikov, R. R., Varakin, G. S., and Ogloblin, K. A., *Zh. Org. Khim.*, **19,** 1625 (1983); *Chem. Abs.*, **100,** 51201 (1984).
[554] Zimmerman, M. P., Li, H. -t., Duax, W. L., Weeks, C. M., and Djerassi, C., *J. Am. Chem. Soc.*, **106,** 5602 (1984).

[555] Proudfoot, J. R., and Djerassi, C., *J. Am. Chem. Soc.*, **106**, 5613 (1984).
[556] Motion, K. R., Robertson, I. R., and Sharp, J. T., *J. Chem. Soc., Chem. Commun.*, **1984**, 1531.
[557] De Lucchi, O., Filipuzzi, F., and Lucchini, V., *Tetrahedron Lett.*, **25**, 1407 (1984).
[558] Scheller, M. E., Mathies, P., Petter, W., and Frei, B., *Helv. Chim. Acta*, **67**, 1748 (1984).
[559] Goswami, R., and Corcoran, D. E., *J. Am. Chem. Soc.*, **105**, 7182 (1983).
[560] Trost, B. M., and Mao, M. K. -T., *J. Am. Chem. Soc.*, **105**, 6753 (1983).
[561] Ramaiah, M., *Synthesis*, **1984**, 529.
[562] Ollivier, J., and Salaün, J., *Tetrahedron Lett.*, **25**, 1269 (1984).
[563] Barnier, J. P., and Salaün, J., *Tetrahedron Lett.*, **25**, 1273 (1984).
[564] Sonawane, H. R., Nanjundiah, B. S., and Kumar, M. U., *Tetrahedron Lett.*, **25**, 2245 (1984).
[565] Piers, E., Jung, G. L., and Moss, N., *Tetrahedron Lett.*, **25**, 3959 (1984).
[566] Gajewski, J. J., and Warner, J. M., *J. Am. Chem. Soc.*, **106**, 802 (1984).
[567] Greenfield, S., and Mackenzie, K., *Tetrahedron Lett.*, **25**, 2255 (1984).
[568] Greenfield, S., and Mackenzie, K., *Tetrahedron Lett.*, **25**, 2259 (1984).
[569] Padwa, A., Rieker, W. F., and Rosenthal, R. J., *J. Org. Chem.*, **49**, 1353 (1984).
[570] Yoshida, H., Nakajima, M., Ogata, T., Matsumoto, K., Acheson, R. M., and Wallis, J. D., *Bull. Chem. Soc. Jpn.*, **56**, 3015 (1983).
[571] Padwa, A., Cohen, L. A., and Gingrich, H. L., *J. Am. Chem. Soc.*, **106**, 1065 (1984).
[572] Lewars, E., and Young, A., *Chem. Ind. (London)*, **1984**, 585.
[573] Kumar, B., Mehta, R. M., Kaira, S. C., and Kaur, N., *J. Chem. Soc., Perkin Trans. 1*, **1984**, 1387.
[574] Sakamoto, M., Aoyama, H., and Omote, Y., *J. Org. Chem.*, **49**, 1837 (1984).
[575] Adam, W., and Carballeira, N., *J. Am. Chem. Soc.*, **106**, 2874 (1984).
[576] Crandall, J. K., and Apparu, M., *Org. React. (N. Y.)*, **29**, 345 (1983); *Chem. Abs.*, **99**, 193955 (1983).
[577] Smith, J. G., *Synthesis*, **1984**, 629.
[578] Ortiz, J. V., *J. Org. Chem.*, **48**, 4744 (1983).
[579] Bischofberger, N., Frei, B., and Jeger, O., *Helv. Chim. Acta*, **67**, 136 (1984).
[580] Roser, J., and Eberbach, W., *Tetrahedron Lett.*, **25**, 2455 (1984).
[581] Bach, R. D., and Domagala, J. M., *J. Chem. Soc., Chem. Commun.*, **1984**, 1472.
[582] Tobe, Y., Yamashita, S., Yamashita, T., Kakiuchi, K., and Odaira, Y., *J. Chem. Soc., Chem. Commun.*, **1984**, 1259.
[583] Lemmens, J. M., Blommerde, W. W. J. M., Thijs, L., and Zwanenburg, B., *J. Org. Chem.*, **49**, 2231 (1984).
[584] Wierenga, W., Harrison, A. W., Evans, B. R., and Chidester, C. G., *J. Org. Chem.*, **49**, 438 (1984).
[585] Apparu, M., and Barrelle, M., *Bull. Soc. Chim. Fr. II*, **1984**, 156.
[586] Rigaudy, J., Barnne-Lafont, J., Ranjon, A., and Casper, A., *Bull. Soc. Chim. Fr. II*, **1984**, 187.
[587] Sawaki, Y., and Ishiguro, K., *Tetrahedron Lett.*, **25**, 1487 (1984).
[588] Edge, G. J., Imam, S. H., and Marples, B. A., *J. Chem. Soc., Perkin Trans. 1*, **1984**, 2319.
[589] Boyd, D. R., Coulter, P. B., Hamilton, W. J., Jennings, W. B., and Wilson, V. E., *Tetrahedron Lett.*, **25**, 2287 (1984).
[590] Eguchi, S., Asai, K., and Sasaki, T., *J. Chem. Soc., Chem. Commun.*, **1984**, 1147.
[591] Bland, J. M., and Stammer, C. H., *J. Org. Chem.*, **48**, 4393 (1983).
[592] Pabon, R. A., Bellville, D. J., and Bauld, N. L., *J. Am. Chem. Soc.*, **106**, 2730 (1984).
[593] Sano, T., Toda, J., and Tsuda, Y., *Heterocycles*, **22**, 53 (1984).
[594] Suginome, H., Liu, C. F., and Tokuda, M., *J. Chem. Soc., Chem. Commun.*, **1984**, 334.
[595] Shimo, T., Somekawa, K., Kuwakino, J., Uemura, H., Kumamoto, S., Tsuge, O., and Kanemasa, S., *Chem. Lett.*, **1984**, 1503.
[596] Christl, M., Brunn, E., and Lanzendörfer, F., *J. Am. Chem. Soc.*, **106**, 373 (1984).
[597] Regitz, M., and Eisenbarth, P., *Chem. Ber.*, **117**, 1991 (1984).
[598] Davies, H. G., Roberts, S. M., Wakefield, B. J., Winders, J. A., and Williams, D. J., *J. Chem. Soc., Chem. Commun.*, **1984**, 640.
[599] Jedlinski, Z., Klimek-Slezak, R., and Kowalczuk, M., *J. Org. Chem.*, **49**, 2427 (1984).
[600] Chan, J. H., and Hall, S. S., *J. Org. Chem.*, **49**, 195 (1984).
[601] Jephcote, V. J., John, D. I., Edwards, P. D., Luk, K., and Williams, D. J., *Tetrahedron Lett.*, **25**, 2915 (1984).
[602] Jephcote, V. J., and John, D. I., *Tetrahedron Lett.*, **25**, 2919 (1984).
[603] Capps, N. K., Davies, G. M., and Young, D. W., *Tetrahedron Lett.*, **25**, 4157 (1984).
[604] Yokoyama, M., Kodera, M., and Imamoto, T., *J. Org. Chem.*, **49**, 74 (1984).
[605] Doering, W. von E., and Birladeanu, L., *J. Am. Chem. Soc.*, **105**, 7461 (1983).
[606] Amemiya, S., Kojima, K., and Sakai, K., *Chem. Pharm. Bull.*, **32**, 457 (1984).

[607] Kollenz, G., Ott, W., Ziegler, E., Peters, E. M., Peters, K., Von Schnering, H. G., Formacek, V., and Quast, H., *Leibigs Ann. Chem.*, **1984,** 1137.
[608] Vercauteren, J., Massiot, G., and Levy, J., *J. Org. Chem.*, **49,** 3230 (1984).
[609] Sha, C. -K., Ouyang, S. -L., Hsieh, D. -Y., and Hseu, T. -H., *J. Chem. Soc., Chem. Commun.*, **1984,** 492.
[610] Sammes, M. P., Maini, P. N., and Katritzky, A. R., *J. Chem. Soc., Chem. Commun.*, **1984,** 354.
[611] Meer, R. K. V., and Olofson, R. A., *J. Org. Chem.*, **49,** 3373 (1984).
[612] Olofson, R. A., and Meer, R. K. V., *J. Org. Chem.*, **49,** 3377 (1984).
[613] Liguori, A., Sindona, G., and Uccella, N., *Tetrahedron,* **40,** 925 (1984).
[614] Lukevics, E., Gevorgyan, V., Goldberg, Y., Popelia, J., Gavars, M., Gaukhman, A., and Shimanska, M., *Heterocycles,* **22,** 987 (1984).
[615] Salomon, R. G., Miller, D. B., Zagorski, M. G., and Coughlin, D. J., *J. Am. Chem. Soc.*, **106,** 6049 (1984).
[616] Stiver, S., and Yates, P., *Tetrahedron Lett.*, **25,** 3289 (1984).
[617] Afsah, E. M., Abou-Elzahab, M. M., Zimaity, M. T., and Proctor, G. R., *Monatsh. Chem.*, **115,** 1065 (1984).
[618] Schultz, A. G., and McMahon, W. G., *J. Org. Chem.*, **49,** 1676 (1984).
[619] Kouno, I., Irie, H., and Kawano, N., *J. Chem. Soc., Perkin Trans. 1,* **1984,** 2511.
[620] Kouno, I., and Kawano, N., *Chem. Pharm. Bull.*, **31,** 4193 (1983).
[621] Bianco, A., Guiso, M., Iavarone, C., Passacantilli, P., and Trogolo, C., *Tetrahedron,* **40,** 1191 (1984).
[622] Edwards, M. P., and Ley, S. V., *J. Chem. Soc., Perkin Trans. 1,* **1984,** 1761.
[623] Bloodworth, A. J., and Eggelte, H. J., *J. Chem. Soc., Perkin Trans. 1,* **1984,** 1009.
[624] Suschitzky, J. L., *J. Chem. Soc., Chem. Commun.*, **1984,** 275.
[625] Matsuyama, H., Miyazawa, Y., Takei, Y., and Kobayashi, M., *Chem. Lett.*, **1984,** 833.
[626] Rigaudy, J., Caspar, A., Baranne-Lafont, J., and Chassagnard, C., *Bull. Soc. Chim. Fr. II,* **1984,** 195.
[627] Barbarella, G., Bruckner, S., Pitacco, G., and Valentin, E., *Tetrahedron,* **40,** 2441 (1984).
[628] Sheradsky, T., and Moshenberg, R., *J. Org. Chem.*, **49,** 587 (1984).
[629] Steinmetz, M. G., Srinivasan, R., and Leigh, W. J., *Rev. Chem. Intermed.*, **5,** 57 (1984); *Chem. Abs.*, **100,** 138218 (1984).
[630] Lewars, E. G., *Chem. Rev.*, **83,** 519 (1983).
[631] Pope, S. A., Hillier, I. H., and Guest, M. F., *J. Chem. Soc., Chem. Commun.*, **1984,** 623.
[632] Matsushita, T., Hirotsu, K., Higuchi, T., Nishimoto, K., Yoshifuji, M., Shibayama, K., and Inamoto, N., *Tetrahedron Lett.*, **25,** 3321 (1984).
[633] Contractor, S. R., Hursthouse, M. B., Parkes, H. G., Shaw, L. S., Shaw, R. A., and Yilmaz, H., *J. Chem. Soc., Chem. Commun.*, **1984,** 675.
[634] Müller, R. -P., Murata, S., Nonella, M., and Huber, J. R., *Helv. Chim. Acta,* **67,** 953 (1984).
[635] Meier, M., and Rüchardt, C., *Tetrahedron Lett.*, **25,** 3441 (1984).
[636] Richard, J. P., *J. Am. Chem. Soc.*, **106,** 4926 (1984).
[637] Robin, D., Sarraf, M., Audier, H. E., and Fétizon, M., *Tetrahedron Lett.*, **25,** 3815 (1984).
[638] Cosa, J. J., and Hamity, M., *Int. J. Chem. Kinet.*, **15,** 1179 (1983).
[639] Warnhoff, E. W., and Nakamura, A., *Tetrahedron Lett.*, **25,** 503 (1984).
[640] Abrams, S. R., *J. Org. Chem.*, **49,** 3587 (1984).
[641] Mease, R. C., and Hirsch, J. A., *J. Org. Chem.*, **49,** 2925 (1984).
[642] Bonin, M., Romero, J. R., Grierson, D. S., and Husson, H. -P., *J. Org. Chem.*, **49,** 2392 (1984).
[643] Tosi, C., *J. Comput. Chem.*, **5,** 248 (1984); *Chem. Abs.*, **101,** 89952 (1984).
[644] Sugihara, Y., Wakabayashi, S., and Murata, I., *J. Am. Chem. Soc.*, **105,** 6718 (1983).
[645] Glukhovtsev, M. N., Simkin, B. Y., and Minkin, V. I., *Zh. Org. Khim.*, **19,** 1353 (1983); *Chem. Abs.*, **99,** 211784 (1983).
[646] Martin, H. -D., Höchstetter, H., and Steigel, A., *Tetrahedron Lett.*, **25,** 297 (1984).
[647] Buechele, J. L., Weitz, E., and Lewis, F. D., *J. Phys. Chem.*, **88,** 868 (1984).
[648] Tsuboi, S., Kuroda, A., Masuda, T., and Takeda, A., *Chem. Lett.*, **1984,** 1541.
[649] Feliu, A. L., Smith, K. J., and Seltzer, S., *J. Am. Chem. Soc.*, **106,** 3046 (1984).
[650] Lukton, D., and Rando, R. R., *J. Am. Chem. Soc.*, **106,** 258 (1984).
[651] Lukton, D., and Rando, R. R., *J. Am. Chem. Soc.*, **106,** 4525 (1984).
[652] Wilcox, C. S., and Babston, R. E., *J. Org. Chem.*, **49,** 1451 (1984).
[653] Marcandalli, B., Liddo, L. P. -D., Di Fede, C., and Bellobono, I. R., *J. Chem. Soc., Perkin Trans. 2,* **1984,** 589.
[654] Jochims, J. C., Lambrecht, J., Burkert, U., Zsolnai, L., and Huttner, G., *Tetrahedron,* **40,** 893 (1984).
[655] Okami, A., Arai, T., Sakuragi, H., and Tokumaru, K., *Chem. Lett.*, **1984,** 289.

[656] Caminade, A. -M., Verrier, M., Ades, C., Paillous, N., and Koenig, M., *J. Chem. Soc., Chem. Commun.*, **1984**, 875.
[657] Yasufuku, K., Takahashi, K., and Kutal, C., *Tetrahedron Lett.*, **25**, 4893 (1984).
[658] Hirao, K. -i., Ando, A., Hamada, T., and Yonemitsu, O., *J. Chem. Soc., Chem. Commun.*, **1984**, 300.
[659] Stöbbe, M., Behrens, U., Adiwidjaja, G., Gölitz, P., and de Meijere, A., *Angew. Chem. Int. Ed.*, **22**, 867 (1983).
[660] Burger, U., and Mareda, J., *Tetrahedron Lett.*, **25**, 177 (1984).
[661] Kobayashi, Y., Kawada, K., Ando, A., and Kumadaki, I., *Tetrahedron Lett.*, **25**, 1917 (1984).
[662] Rubin, M. B., *Tetrahedron Lett.*, **25**, 4697 (1984).
[663] McCague, R., Moody, C. J., and Rees, C. W., *J. Chem. Soc., Perkin Trans. 1*, **1984**, 165.
[664] Vogel, E., Tückmantel, W., Schlög, K., Widhalm, M., Kraka, E., and Cremer, D., *Tetrahedron Lett.*, **25**, 4925 (1984).
[665] Okazaki, R., Hasegawa, T., and Shishido, Y., *J. Am. Chem. Soc.*, **106**, 5271 (1984).
[666] Markov, P., *Chem. Soc. Rev.*, **13**, 69 (1984).
[667] Wentrup, C., Winter, H. -W., Gross, G., Netsch, K. -P., Kollenz, G., Ott, W., and Biedermann, A. G., *Angew. Chem. Int. Ed.*, **23**, 800 (1984).
[668] Grigg, R., Gunaratne, H. Q. N., and Kemp, J., *J. Chem. Soc., Perkin Trans. 1*, **1984**, 41.
[669] Kessler, H., Oschkinat, H., Zimmermann, G., Möhrle, H., Biegholdt, M., Arz, W., and Förster, H., *Chem. Ber.*, **117**, 702 (1984).
[670] Valters, R., Kampare, R., Valtere, S., Balode, D., and Bace, A., *Khim. Geterotsikl. Soedin.*, **1983**, 1635; *Chem. Abs.*, **100**, 120239 (1984).
[671] Smith, J. G., and Dibble, P. W., *Tetrahedron*, **40**, 1667 (1984).
[672] Laufer, D. A., Gelb, R. I., and Schwartz, L. M., *J. Org. Chem.*, **49**, 691 (1984).
[673] Andreeva, I. M., Babeshko, O. M., Bondarenko, E. M., Medyantseva, E. A., and Minkin, V. I., *Zh. Org. Khim.*, **19**, 2146 (1983); *Chem. Abs.*, **100**, 138352 (1984).
[674] Yakimovich, S. I., Zelenin, K. N., Nikolaev, V. N., Koshmina, N. V., Alekseev, V. V., and Khrustalev, V. A., *Zh. Org. Khim.*, **19**, 1875 (1983); *Chem. Abs.*, **100**, 22225 (1984).
[675] Neidlein, R., and Radke, C. M., *Helv. Chim. Acta*, **66**, 2626 (1983).
[676] Takeuchi, K., Fujimoto, H., Kitagawa, T., Fujii, H., and Okamoto, K., *J. Chem. Soc., Perkin Trans. 2*, **1984**, 461.
[677] Maas, G., and Kettenring, J. K., *Chem. Ber.*, **117**, 575 (1984).
[678] Dewar, M. J. S., and Holloway, M. K., *J. Chem. Soc., Chem. Commun.*, **1984**, 1188.
[679] Scanlan, M. J., and Hillier, I. H., *J. Am. Chem. Soc.*, **106**, 3737 (1984).
[680] Tóth, G., and Podányi, B., *J. Chem. Soc., Perkin Trans. 2*, **1984**, 91.
[681] Butler, R. N., Garvin, V. C., Lumbroso, H., and Liégeois, C., *J. Chem. Soc., Perkin Trans. 2*, **1984**, 721.
[682] Adam, W., Carballeira, N., De Lucchi, O., and Hill, K., *Methods Stereochem. Anal.*, **3**, 241 (1983); *Chem. Abs.*, **101**, 129828 (1984).
[683] Zimmerman, H. E., and Swafford, R. L., *J. Org. Chem.*, **49**, 3069 (1984).
[684] Lyle, T. A., Mereyala, H. B., Pascual, A., and Frei, B., *Helv. Chim. Acta*, **67**, 774 (1984).
[685] Müller-Remmers, P. L., Mishra, P. C., and Jug, K., *J. Am. Chem. Soc.*, **106**, 2538 (1984).
[686] Sadler, D. E., Wendler, J., Olbrich, G., and Schaffner, K., *J. Am. Chem. Soc.*, **106**, 2064 (1984).
[687] Hoshi, N., Uda, H., Sato, K., and Hagiwara, H., *J. Chem. Soc., Perkin Trans. 1*, **1984**, 769.
[688] Weisz, A., Kaftory, M., Vidavsky, I., and Mandelbaum, A., *J. Chem. Soc., Chem. Commun.*, **1984**, 18.
[689] Leigh, W. J., and Srinivasan, R., *J. Org. Chem.*, **48**, 3970 (1983).

Author Index

In this index bold figures relate to chapter numbers, roman figures are reference numbers

Cumulative Subject Index, 1980–84